Integrity of Scientific Research

Joel Faintuch • Salomão Faintuch
Editors

Integrity of Scientific Research

Fraud, Misconduct and Fake News in the Academic, Medical and Social Environment

Springer

Editors
Joel Faintuch
Hospital das Clinicas
University of Sao Paulo
Sao Paulo, São Paulo, Brazil

Salomão Faintuch
Beth Israel Deaconess Medical Center
& Harvard Medical School
Boston, MA, USA

ISBN 978-3-030-99679-6 ISBN 978-3-030-99680-2 (eBook)
https://doi.org/10.1007/978-3-030-99680-2

This Springer imprint is published by the registered company Springer Nature Switzerland AG
The registered company address is: Gewerbestrasse 11, 6330 Cham, Switzerland

Foreword

Research integrity concerns the consequences of the behaviour of researchers on the validity of their findings and on the trust in their work among peers and in society at large. This does not mean that the drivers of that behaviour solely depend on individual virtuousness. Increasingly, it has become clear that the research climate in the labs and research groups plus the—sometimes perverse—incentives within the system of science strongly influence research integrity. Consequently, a diversity of perspectives and stakeholders is needed to optimize research integrity. This book contains a multitude of chapters, authors, and topics that reflect the emerging state of the art—with its occasional confusion, inconsistency, and contradiction—quite well.

For most of us, the first association of research integrity is with research misconduct, which is often framed as the three "mortal sins" of fabrication, falsification, and plagiarism (FFP). Spectacular cases of research misconduct typically get a lot of media attention and often serve as wake-up calls. Increasingly, it has been recognized that a lack of research integrity can also consist of minor misbehaviours which are often labelled rather euphemistically as questionable research practices (QRPs). Examples are selective reporting, p-hacking, and HARK-ing (hypothesizing after results are known). Arguably QRPs on the aggregate level do much more harm because, while FFP seems to be relatively rare, QRPs turn out to be alarmingly common.

It might be argued that we should look more at the positive side of the coin and focus on the norms for good research. The behaviours at issue are usually labelled as responsible conduct of research (RCR) or responsible research practices (RRPs). The idea is that it is better to stimulate the good end of the spectrum of researcher conduct rather than to focus solely on detecting and sanctioning of research misconduct. Carrots are usually more effective than sticks. This attitude is also reflected in recent codes of conduct for research integrity, like the one for the Netherlands that specifies 61 norms for good research. An allegation of research misconduct is only eligible for not following one third of these norms. Recent codes of conduct try to be not only positive and aspirational but also practical in the sense of being clear about the do's and don'ts of research.

Next to the researchers, important stakeholders in research integrity are research institutes, scholarly journals, and funding agencies. When these stakeholders act in concert, interventions aimed at promoting responsible research practices arguably will be more effective. To me the most important drivers of improvement are the engagement in Open Science practices and focussing the assessment of researchers on behaviours that strengthen research integrity. Open Methods, Open Codes, and Open Data increase transparency and enable checks on selective reporting, re-analyses of data sets, and replication studies. Research and researchers need to be trusted for their findings to have any impact on health, society, and nature. Trust needs to be deserved by being trustworthy. Being open and transparent provides the fundament for that.

Careers in research are determined by assessments for grants, promotion, and tenure. It is important to use criteria for these assessments that reflect the behaviours that matter for research integrity. Sadly, that is not always the case and then assessment criteria can become perverse incentives. Such as having a dominant focus on numbers of publications and citations, including its derivatives like Impact Factors and the Hirsch Index. Recently, mutually reinforcing initiatives like the San Francisco Declaration on research assessment, the Leiden Manifesto for research metrics, and the Hong Kong principles for assessing researchers have been launched, which urge for using better assessment criteria like engaging in Open Science practices and being a good peer reviewer and mentor.

The chapters of this book together form a rich source for exploring the broad and expanding field of research integrity. Because the topic interests you—why else would you read the foreword of the book—you should consider participation in the next World Conference on Research Integrity. When you want to learn more about ways to foster research integrity, I can also strongly recommend the websites of the Standard Operating Procedures for Research Integrity (SOPs4RI) project and the Embassy of Good Science.

Lex Bouter

Department of Epidemiology and Data Science
Amsterdam University Medical Centers
Amsterdam, The Netherlands

Department of Philosophy
Faculty of Humanities
Vrije Universiteit,
Amsterdam, The Netherlands

Preface

From the thousands to the trillions, it´s the same slippery, treacherous road

The above sentence was not plagiarized from a "noir" novel or from the press release of a B-rated horror movie. Even though the words look familiar, the authors themselves concocted them, maybe inspired by an occasional novel or movie. What do they actually mean?

Typical money involved in "contract cheating", or students paying for third parties to fake their thesis or dissertation, does not exceed a couple of thousand dollars, and could be as little as a few hundreds. Predator journals, books, and meetings charge in the same range, and gifts or trips offered by pharmaceutical laboratories or medical device manufactures to professionals often operate in this bandwidth, even though substantially higher values are possible.

At the other extreme, the market valuation of the four digital multinationals Google, Amazon, Facebook, and Apple (GAFA) comes close to the trillion-dollar benchmark, if not beyond it. Are their disbursements with questionable engagements also in the astronomical range? Technically, these organizations are cleaner and more ethical than most, and they vow for a single cent spent in extraneous or illegal activities. Yet it cannot be denied the colossal power they are endowed with, in the form of detailed personal information accumulated with millions or billions of customers all over the world. Even though wrongdoing has rarely if ever been demonstrated, none of the GAFA companies is transparent with regard to the degree of security awarded to these datasets, the nature and purpose of the algorithms attached to them, and particularly the organizations they share them with, both public and private. Actually, nearly all large corporations nowadays use cookies and apps to collect Olympic amounts of personal informations, almost never fully disclosing the ethics and the governance behind so much sensitive data storage.

Academic misconduct has not been linked to breaches of commercial privacy, only political meddling as demonstrated in the paradigmatic case of Facebook–Cambridge Analytica, most prominently in the period of 2016–2018. Nevertheless, anti-vaccine campaigns and other direct or indirect public health threats are more difficult to rule out.

Article publication is perhaps more meticulously scrutinized nowadays than researcher integrity, laboratory records, or protocol funding and manipulation. Indeed the ICMJE (International Committee of Medical Journal

Editors) started requesting research registration, a procedure that substantially enhances article reliability and transparency in 2005, before the Food and Drug Administration/USA (2007) considered it mandatory.

One reason is that publication universally exposes the name of the journal and the editor-in-chief, as indirect accomplices of the misdeeds in case of fraudulent procedures. Even though they typically have no involvement with authors or laboratories, and the worst accusation could be of a sloppy journal review job, not of actual responsibility for cheating or deception.

Of course, the subject can become more shady depending on the business context. In the last two decades, the emerging field of open access books, congress proceedings, and notably journals has earned a bad name for the so-called predator organizations. As discussed in several chapters of the book, scores of international entrepreneurs offer prompt and seamless electronic publication of any scientific material for a fee, in peer-reviewed indexed journals. The catch is that both review and indexation are falsified and not trustworthy. Those journals, books, and conferences lack endorsement from reputable scientific and professional organizations and often promise publication in as short a period as weeks or even days, which is incompatible with decent peer review, provided payment is received. They are also the culprit of spam messages overflowing scientists' mailboxes.

The worst risk is for the unaware reader in the healthcare professions, and particularly for public health authorities who eventually trust those articles and reports. Unsubstantiated and fallacious diagnostic and therapeutic recommendations could be adopted, with unforeseen consequences. For the naïve (or not so naïve) author who believes he or she is divulging the research in a serious publication, financial waste and the wrong type of recognition could arise. Also fooled (or careless) universities and research laboratories which accept curricula built on predator publications could discover they hired or promoted the wrong scientist, to the detriment of more serious candidates.

Incidentally, traditional and respected publishers are under fire as well because of the regular fees they charge for open access, as these companies also offer this alternative, nor just dishonest competitors. There is no fraud or misconduct implied, just criticism for their commercial practices. The standard bearer of the war is the Coalition S, established by a group of European national research funders as well as other organizations, with the support of the European Commission and the European Research Council (ERC) (www.coalition-s.org). According to the official statement, all European research results funded by any organization should be published in open access journals, platforms, or repositories, without embargo (promptly and freely available to all interested parties).

The original plan should become effective in 2021, and bypassed payment considerations. Copyright barriers should be abolished in such circumstances. Given the resistance of many publishers, who alleged major losses or even risk of bankruptcy, a model for low-cost reimbursement was created, and the deadline was postponed. At this moment, a number of publishers announced full or partial compliance with the plan, typically with some degree of copyright retainment, whereas others are reluctant.

Is money the link between this large and nominally incoherent spectrum of facts and a book on integrity, fraud, and fake news? Financial profits and personal benefits do not underlie the entire universe of academic or professional misconduct and lack of integrity, which is diverse and grows by the day. Yet for many, it remains the quintessence in the fight for improved human morals, behaviours, and deeds, especially in a twenty-first century largely moved by material rewards. Such assumption notwithstanding, this book does not adopt a market approach for assessing or valuating current status of integrity or lack thereof in the sciences and professions.

Anti-vaccine activism is a major threat to responsible research and ethical management of public health, even though the economic undertones are not easily distinguishable. Science denialism as a whole, from "scholarly" controversy till outright quackery, has similarly engaged the rich and the poor, the gold diggers and the misguided idealists. Harassment, bullying, and sexual misconduct are other examples of money blind immoral and dishonest behaviours, even though those in the upper echelons of the social and economic ladder seem more prone to them.

Economists are comparatively scarce in the pages of the book, and more of them would unquestionably enhance the publication. Nevertheless, it is debatable whether a reductionist monetary focus would faithfully mirror the myriads and often elusive shades of academic and professional integrity. Respected ethicists and experienced professors with extensive medical and scientific background were summoned as well, to complement the editorial team and contribute with solid and deep-rooted expertise in their respective fields.

It was not the purpose of the book to become encyclopaedic. The field is constantly evolving, and it would be risky for any publication to advertise itself as all-inclusive. Still this is arguably the most complete and authoritative text to appear in recent times, delving with a variety of questionable practices in healthcare and social sciences, professions, publications, the Internet, and the academic environment. Plagiarism, misconduct, harassment, misrepresentation, conflict of interest, informed consent, ethnic and sexual prejudice, fraud, cheating, and fake news are addressed, of course. However, the text did not shy away from less established or borderline topics such as data sharing and biobanking, cyberbullying, image forgeries, body donation, human rights, animal rights, telehealth, curbside consultations, artificial intelligence, and other emerging technologies.

Corrective and preventive initiatives were especially sought after, as millennia of legal experience have demonstrated that sanctions alone are rarely sufficient. These are complemented by guidelines and recommendations from several parts of the world, as well as useful Internet sites and a glossary. In synthesis, no stone was left unturned in the effort to bring the latest and most practical information.

Sao Paulo, Brazil — Joel Faintuch
Boston, MA, USA — Salomão Faintuch

Additional Reading

https://www.coalition-s.org/about/

https://www.sciencemag.org/news/2019/05/meet-plan-s-open-access-mandate-journals-mull-setting-papers-free-publication

https://www.researchprofessionalnews.com/rr-news-europe-infrastructure-2021-1-batch-of-elsevier-journals-given-transformative-plan-s-status/

https://www.nytimes.com/2018/04/04/us/politics/cambridge-analytica-scandal-fallout.html

www.icmje.org

Contents

Part I

Introduction

Past and Current Status of Scientific, Academic, and Research Fraud

1

Joel Faintuch and Salomao Faintuch

Abstract

In the quest to advance knowledge, researchers publish approximately 2 million scientific articles each year, in close to 30,000 peer-reviewed journals. Even though there was hardly a time when science was more questioned, distorted, and defamed, this is still the closest thing to the established body of knowledge for universities, academic laboratories, and other government and private organizations and agencies worldwide. Are plagiarism, fraud, and misconduct indeed infiltrating and rotting the revered shrines of wisdom? Or is alleged scientific decay a consequence of relatively few overambitious, incompetent, or truly rogue researchers, who sooner or later are identified and banned from the community? In the distant past, fraud was often in the eye of the beholder. As experimental design was rarely adequate, measuring instruments were primitive or nonexistent, and modern statistics were lacking, conflicts were often decided by the strongest side, namely, the one backed by the authorities, not necessarily the correct one. Fortunately, much institutional advance has occurred, and current fraudsters exceptionally get away with it. This chapter briefly reviews the state of the art in academic research.

1.1 Introduction

It is tempting to state that fraud is as old as science. However, in order to be coherent, one has first to establish the initial milestone of science. Was it introduced by Pythagoras of Samos (571–490 BCE), the author of the classic Pythagorean theorem ($a^2 + b^2 = c^2$), about the sides of a right triangle? This theorem is so highly regarded that it is not uncommonly included in time capsules, notably space-time capsules aimed at remote civilizations (?), based on the principle that it is an easily recognizable universal truth.

However, there is evidence that around 1900 BCE, well over one millennium before Pythagoras, the same theorem was familiar to Mesopotamians and to Egyptians [1, 2]. Did modern science actually begin with plagiarism? Or are we insufficiently familiar with historical trajectories, so that we are unable to agree even about the authorship of that theorem?

This leads us to another possibility, namely, that fraud, like beauty, is in the eye of the beholder. If a given behavior is deemed

J. Faintuch (✉)
Hospital das Clinicas, Sao Paulo University Medical School, Sao Paulo, Brazil
e-mail: j.faintuch@hc.fm.usp.br

S. Faintuch
Department of Radiology, Beth Israel Deaconess Medical Center, Harvard Medical School, Boston, MA, USA
e-mail: sfaintuc@bidmc.harvard.edu

J. Faintuch, S. Faintuch (eds.), *Integrity of Scientific Research*,
https://doi.org/10.1007/978-3-030-99680-2_1

fraudulent by contemporary academic and regulatory consensus, then it should be accepted as such. Although too pragmatic and utilitarian for some tastes, that's how the concept has actually evolved in recent decades. As recently as 1981, Patricia Woolf, a professor at Princeton University, wrote with awe and astonishment about four episodes of scientific fraud and misconduct in the USA, asking the community whether that was the rule or just a cluster of outliers [3].

Sadly, it was not as exceptional as estimated. In a more recent survey of 2155 academic psychologists at major US universities involved with research, nearly one in ten had introduced false data into a study. Most admitted engaging in selective reporting, omitting dependent measures, collecting additional data after statistical analysis was conducted, converting unexpected findings into predicted ones, as well as excluding data if they did not fit the hypothesis [4].

Currently, scientific integrity and professional ethics are not being taken for granted any more. They are being embedded in the bylaws and ostensibly enforced by scores of academic, government, and private institutions worldwide.

1.2 Variants of Fraud and Misconduct

Brain adaptability and plasticity are miracles of nature that occur all the time. This means that man and woman are endless creators, inventors, and redevelopers, often for the good of humanity. No less frequently are these abilities employed for less noble purposes, some of them downright unacceptable and dishonest. If this happens in the medical, scientific, and general academic environment, repercussions can be exponential, as honorable conduct is taken for granted within these ecosystems.

Classically, fraud is classified as fabrication, falsification, and plagiarism. However, forms of deceit widely vary, including ghost writing, other authorship improprieties, misrepresentation of data, statistical concealing or "spin," and fake news. Moreover, harassment and prejudice, be it sexual, professional, or ethnical, along with even indirect collaboration with abuses such as coercion and bullying, are similarly being enrolled among academic misconducts.

It is alleged that deception is so widespread in nature that some species would not survive without a number of tricks. Indeed, predator animals not uncommonly display bright colors to attract and distract victims, whereas prey might similarly appeal to such resources as color, environmental camouflage, and even bad smell to evade capture. Preverbal children, or those who at age three still resent a limited vocabulary, can already be fully proficient in deception and manipulative behavior [5].

Every scientific investigation involving humans should in principle adhere to informed consent, particularly when some type of harm or damage can emerge (physical, emotional, professional, social, financial). Informed consent forms are evolving all the time, often generating heated conflicts as each country and organization adopts new rules and legislations, and authorities seem unable to agree with a perfect and universal model. Yet it is obvious that some are openly fallacious and misleading.

By the same token, nearly all protocols involve expenses and require a budget and its corresponding sponsor. When grants originate from government agencies and other public sources, suspicions of objectionable intentions by the funder are understandably rare. The same may not be true for private financing, notably when the study addresses such controversial and highly profitable topics as smoking, alcohol, sugar-rich soft drinks, ultra-processed industrialized foods, other consumer products and services, and even standard new drugs, vaccines, and devices to be licensed by the pharmaceutical or medical supply industries.

Within the realm of conflict of interest, a whole list of modalities can be found in the literature, both financial and nonfinancial. Cheating is another title that encompasses multiple categories, be it during research (participant cheating, investigator dishonesty) or academic career (student reports, theses, books, Internet releases) [6–9].

1.3 Integrity in Scientific Publication

For most purposes, science only starts existing from the moment it is publicly reported in full and acknowledged by the community, either positively or with restrictions. Secret research is not a novelty of course, within such domains as military activities and sensitive industrial processes. However, at least in the biomedical field, without publication in peer-reviewed journals, new drugs and invasive medical devices will hardly be approved for commercialization by regulatory agencies.

Scientific publication is big business and allegedly a particularly profitable one. An estimated 30,000 journals publish 2 million scientific articles annually [10], generating over 19 billion pounds ($27 billion US dollars) in global publishing revenues and employing over 100,000 workers. The range is comparable to the worldwide music and movie industries, however, with a much more enviable profit of 30–40% [11].

In the show business, authors (musicians, artists, scriptwriters, directors) have to be rewarded with contracts, royalties, and other benefits, which reach princely sums in the case of first-class professionals. General production and marketing costs including labor, travel, and other budgetary items can also be staggering. Worst of all, productions can simply flop and leave behind enormous losses. Scientists of the highest caliber freely transfer full and irreversible copyright of their texts and images to the publishers or else have to pay in order to retain such rights, in the so-called open-access models. Publication and marketing costs are considered moderate, and despite growing competition, few established publishers are driven out of the market.

Given the vast money involved, it is not surprising that the industry has attracted a huge following of less reputable competitors, best known as predator journals (including books, congress proceedings, and technical and scientific conferences). Their "modus operandi," almost invariably open access, looks uncannily familiar and trustworthy, except that it is viscerally fraudulent and unreliable. Open access is actually a perfectly legitimate approach, conceived by traditional publishers and transparently conducted, as discussed in other chapters. Sadly, the concept was also hijacked by rogue organizations, giving it the bad image it undeservedly acquired.

All of these obscure organizations do need to invest in advertisement, mostly in the form of millions of daily spam messages, yet quite a few authors are more than willing to comply. Why would they do it?

At the scientist's end, as well as that of the pharmaceutical sponsor sometimes backing his research, there are no less robust and often conflictual drivers at play. Publication is not an option; it is a must in many circumstances. Should a standard route and conventional journal be selected, sometimes crowned with rejection after multiple submissions and precious wasted time? Or maybe a shortcut will do as well?

Given the critical and antagonistic forces from all sides, including not negligible doses of cheating and deception, it is not difficult to understand that imputations of conflict of interest, poor stewardship, dishonesty, and misconduct can be sprinkled on most players, not just openly predator organizations and their preys. On occasions, not even "bona fide" authors and journals escape unscathed from the blanket malevolence against the entire scientific publishing system, fueled by not very rare scandals.

Of course professional journal editors notably in the medical field have long adopted stringent ethical standards, and universities as well as multiple funding agencies have not wasted time either. Yet fraudsters are ubiquitous, creative, strongly motivated, and at the same time naive, believing they will never be caught and will somehow get away. Consequently, publication retractions are becoming more prominent and troublesome, not less so [12, 13] (Table 1.1).

1.4 Retractions in the Past

Calculation mistakes and overlooked details during formal studies have occurred at all times and

Table 1.1 Common examples of academic or research impropriety [4–9]

Fraud	Context
Fabrication	Nonexistent data, images, or study participants
Falsification	Purposeful selection, misinterpretation, distortion, or adulteration of findings and conclusions
Plagiarism	Undisclosed copying of ideas, methods, results, or images from self or from others, text recycling
Misconduct	
Data storage and handling	Irresponsible or careless data manipulation, lack of data protection, confidentiality violation, clandestine or illegal data access
Study participants	Animal abuse, inadequate human consent regarding aims and benefits, concealment of risks and harms of the study, coercion or exploitation of participants
Authorship	Ghost writers, authorship bargaining or kidnapping, questionable personal credits
Study protocol	Failure of registration and ethical approval, breach of good clinical practices (GCP), and offensive, immoral, or unscientific projects
Conflict of interest	Lack of transparency regarding relationships with funders, undisclosed commercial or personal interests, nepotism
Publication	Noncompliance with journal or publisher ethics, nonprofessional language (sensational science), salami slicing (multiple articles from the same study)
Data sharing	Lack of sharing with legitimate stakeholders (study participants, co-investigators, other authorized parties), breach of data confidentiality, clandestine data disclosure or commercialization
Personal misconduct	Harassment, prejudice, bullying, disrespectful behavior or publications
Academic mentoring	Malicious or dishonest mentoring of students, fellows, and junior staff, nepotism
Financial misconduct	Misappropriation, misuse of research funds
Academic career	False or exaggerated academic titles, qualifications, publications, professional experience
Miscellaneous	Cheating, deceit, fake news, scams, pseudoscience, malicious misconduct allegations, retaliation against whistleblowers

eventually were brought to public attention by the authors themselves or by others. Two examples from past centuries were relatively recently unveiled by library archeologists. One involves the renowned French mathematician Henri Poincaré (1854–1912), and the other has tangential relationship with the electricity experiments of Benjamin Franklin (1706–1790). In the first case, the author won a prize from *Acta Mathematica* in 1989 for a complex calculation which was actually flawed. In the second case, the investigator Benjamin Wilson wrote a letter to *Philosophical Transactions of the Royal Society* in 1756, aiming to clarify some possibly wrong details of his previous communication, which was related to Franklin's studies as well [14, 15]. It is important to emphasize that in none of these cases was intentional fraud suspected, and the retractions were spontaneous and transparent.

1.5 Do "Honest Mistakes" Exist?

There is evidence that no less eminent and revered names than Galileo Galilei (1564–1642), Isaac Newton (1643–1727), and Gregor Mendel (1822–1884) faked their data or at least were not accurate in their descriptions [16–18]. The landmark postulates by Robert Koch (1843–1910), the discoverer of the tuberculosis bacillus [19], are not entirely original, as Friedrich Gustav Jakob Henle (1809–1885) had previously expressed three of the four postulates [20]. Incidentally, they did not endure well the test of time either.

The first postulate, that the infectious agent should be present in diseased organisms, not healthy ones, was recognized as flawed by the author himself, when he subsequently identified asymptomatic carriers. The second, that the organism should be grown in pure culture, suffers from the problem of unculturable bacteria as well as viruses, which do not form pure cultures. And the third postulate, stating that the agent retrieved from an ill person should precipitate the disease in a healthy one, conflicts with the possibility of variable immune response and resistant individuals.

Yet Koch's proposals were received as honest and valuable by the academic world at his time and are cited till our days. Indeed, there is no evidence in his career that he was a plagiarist or reckless investigator. As regards earlier authors, Harvard astronomer Owen Gingerich points out that it was relatively frequent for past scientists to cherry-pick data as they fitted their theories. In those days, such was quite acceptable or at least was not stamped as malicious or fraudulent [21] (Fig. 1.1).

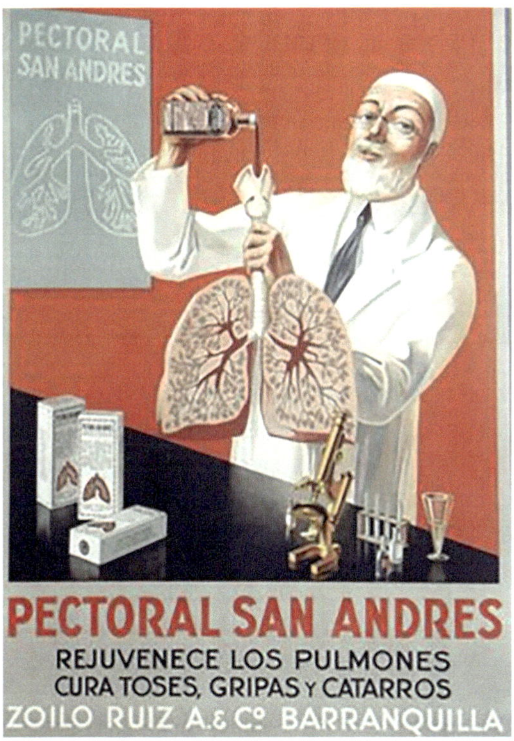

Fig. 1.1 Until the middle of the twentieth century, the majority of drugs in clinical use lacked proper clinical trials, and efficacy with occasional exceptions was questionable (Picture from Wellcome collection.org). Reproduction allowed by Creative Commons Attribution 4.0 International License (https://creativecommons.org/licenses/by/4.0/)

1.6 How Should Fraud Be Managed?

In the book by Ben-Yehuda and Oliver-Lumerman [22], which spans from the late nineteenth to early twenty-first centuries, about 750 incidents of falsification, fabrication, plagiarism, and other academic misconducts were documented. Repercussions, if any, mostly occurred in recent times and were institution-centered. Only in our times were scientific integrity rules and administrations created, research procedures updated, misconduct prevention emphasized, and postgraduate courses or training programs implemented, in some circumstances specifically aiming at previous offenders.

Less stringent actions targeted the perpetrators, with rare exceptions being at most reprimanded or temporarily excluded from the institution [23].

Nevertheless, in recent decades, even authorities and geniuses have not been exempted from integrity codes, and depending on the context, the country, and the consequences, severe penalties could affect those found guilty, encompassing permanent loss of job or university appointment, professional licensure, and even fines and imprisonment [22, 23].

Of course not all punished assume guilt or take it lightly. Vigorous reactions are not common yet they tend to grow. A recent example occurred with a New Zealand author, who had one article retracted and two more flagged with "expressions of concern." He announced that he is initiating legal process against the journal and its publisher [24].

References

1. Friberg J (1981) Methods and traditions of Babylonian mathematics: Plimpton 322, Pythagorean Triples, and the Babylonian triangle parameter equations. Historia Mathematica 8(3):277–318
2. Neugebauer O (1969) The exact sciences in antiquity. Dover Publications, New York
3. Woolf P (1981) Fraud in science: how much, how serious? Hastings Cent Rep 11(5):9–14

4. John LK, Loewenstein G, Preiec D (2012) Measuring the prevalence of questionable research practices with incentives for truth telling. Psychol Sci 23(5):524–532
5. Gongola J, Scurich N, Quas JA (2017) Detecting deception in children: a meta-analysis. Law Hum Behav. 41(1):44–54
6. Catano VM, Turk J (2007) Fraud and misconduct in scientific research: a definition and procedures for investigation. Med Law 26(3):465–476
7. Janket SJ, Meurman J, Diamandis EP (2019) Advocate cultivation of academic ethics: why is it necessary? F1000Res 30(8):1830
8. Artino ARJ, Driessen EW, Maggio LA (2019) Ethical shades of gray: international frequency of scientific misconduct and questionable research practices in health professions education. Acad Med 94(1):76–84
9. The European Code of Conduct for Research Integrity. ec.europa.eu/info/fundingtenders/opportunities/docs/2021-2027/horizon/guidance/european-code-of-conduct-for-researchintegrity_horizon_en.pdf. Accessed 10 Nov 2020
10. https://www.universityworldnews.com/post.php?story=20180905095203579. Accessed 5 May 2021
11. https://www.theguardian.com/science/2017/jun/27/profitable-business-scientific-publishing-bad-for-science. Accessed 5 May 2021
12. ICMJE. Recommendations for the conduct, reporting, editing, and publication of scholarly work in medical journals. Updated December 2019. icmje.org/icmje-recommendations.pdf. Accessed 10 Nov 2020
13. Retractionwatch.com. Accessed 10 Nov 2020
14. Retractionwatch.com/2012/02/27/the-first-ever-english-language-retraction-1756
15. Retractionwatch.com/2016/03/14/what-did-retractions-look-like-in-the-17th-century/
16. Claudius Ptolemy. newscientist.com/people/claudius-ptolemy. Accessed 10 Nov 2020
17. Kollerstrom N. The dark side of Isaac Newton. historicalnovelsociety.org/reviews/the-dark-side-of-isaac-newton-sciences-greatest-fraud/. Accessed 10 Nov 2020
18. Weeden NF (2016) Are Mendel's data reliable? The perspective of a Pea geneticist. J Hered 107(7):635–646
19. Koch R (1882) Die Atiologic der Tuberkulose (The etiology of tuberculosis). Berlin Klin Wochenschr 15:221–230
20. Henle FGJ. Von den Miasmen und Kontagien (On miasmata and contagion). Bull Hist Med 1840 1938: 907 910
21. Gingerich O (1993) The eye of heaven: Ptolemy, Copernicus, Kepler. Springer, USA
22. Ben-Yehuda N, Oliver-Lumerman A (2017) Fraud and misconduct in research: detection, investigation, and organizational response. University of Michigan Press, Ann Arbor, MI
23. Dal-re R, Bouter LM, Cuijpers P, Gluud C, Holm S (2020) Should research misconduct be criminalized ? Res Ethics 16(1–2):1–12
24. retractionwatch.com/2020/11/18/author-initiates-a-legal-process-against-a-journal-and-its-publisher-after-a-retraction-expressions-of-concern/

Research Integrity: The Roles of Academicians, Their Institutions, and Other Agencies

2

Janet D. Robishaw, David L. DeMets, Scott M. Alter, Joanne Krasnoff, Charles H. Hennekens, and Sarah K. Wood

Abstract

While education and clinical pursuits are subject to challenges in achieving and maintaining academic integrity, in recent decades, the majority of issues that have arisen concern scholarship and research.

In this chapter, we review the principles to guide academicians and their institutions with respect to the prevention and treatment of fraud as well as their collaborations with medical journals, the press, and industry. Finally, we discuss missed opportunities to achieve and maintain research integrity at two distinguished academic institutions.

J. D. Robishaw
Department of Biomedical Science, Charles E. Schmidt College of Medicine, Florida Atlantic University, FL, USA

D. L. DeMets
Department of Biostatistics and Informatics, University of Wisconsin School of Medicine & Public Health, Madison, WI, USA

S. M. Alter · J. Krasnoff · C. H. Hennekens (✉) · S. K. Wood
Charles E. Schmidt College of Medicine, Florida Atlantic University, FL, USA
e-mail: chenneke@health.fau.edu

2.1 Introduction

Most academic institutions aspire to achieve scholarship and research. In addition, most academic medical institutions have a tripartite mission that includes scholarship and research, education, as well as clinical. All missions in all academic institutions, whether medical or not, must be pursued with competence, compassion, character, integrity, dignity, and grace. While education and clinical pursuits are subject to challenges in achieving and maintaining integrity, in recent decades, the majority of issues that have arisen concern scholarship and research.

For the purposes of this chapter, the following definitions are adopted:

- Scholarship is work that leads to peer-reviewed commentaries, editorials, reviews, books, and chapters.
- Contract research is clinical science which is largely initiated by the funding agency in which the awardee enrolls subjects into multicenter investigations and the academic institution is remunerated based on each subject enrolled.

  Discovery research is basic or clinical science initiated by the investigator who writes proposals to obtain peer-reviewed funding from the government, private foundations, or industry that provide direct costs to the academician and indirect costs to the academic institution that are usually funneled to enhance the

J. Faintuch, S. Faintuch (eds.), *Integrity of Scientific Research*,
https://doi.org/10.1007/978-3-030-99680-2_2

future productivity of the discovery researcher. The primary goal is to test hypotheses and produce original articles or brief reports that are published in the peer-reviewed literature. These endeavors may include collaborators who are students, residents, and faculty. A subsidiary goal is to submit abstracts to national or local meetings to share the findings with other academics and secure their input about the findings.

There are increasing requirements for pre-medical and medical students as well as residents to be involved in scholarship and research activities, to learn the discipline, to enhance their curricula vitae, as well as to develop strong applications whether for medical school, residency, fellowship, or academic promotions. Most medical schools support scholarship and research opportunities for their students either within the curriculum or extracurricularly, as well as integrate teaching sessions that address research ethics and principles underlying research conduct. Academic institutions have the responsibility to provide appropriate education and development in the foundational tenets underlying the achievement and maintenance of research integrity.

Training on the importance of mentorship, authorship responsibility, and accountability as well as research misconduct and the roles and responsibilities of Institutional Review Boards (IRBs) and other research regulatory offices such as the Institutional Biosafety Committee (IBC), Institutional Animal Care and Use Committee (IACUC), and Office of Environmental Health and Safety (EHS) are also of increasing importance. Case-based learning used to highlight situations that have occurred intentionally and unintentionally at all levels of the academic enterprise should also be an adjunct to educate students, residents, fellows, and faculty wishing to pursue scholarship or research.

2.2 Prevention and Treatment of Academic Fraud

The achievement and maintenance of research integrity at academic institutions is both increasingly complex and an urgent necessity. This is due, at least in part, to the fact that there are perceptions and realities that fraud, misconduct, and fake news in the academic, medical, and social environment are increasing [1–4].

Few would disagree with Ben Franklin's statement that "an ounce of prevention is worth a pound of cure." In practice, however, there is wide debate about the application of this wisdom to healthcare systems [5]. This wide debate includes achievement and maintenance of research integrity at academic institutions, which sometimes is driven by short-term necessities rather than long-term prevention.

Whether treatment or prevention, we believe the hierarchy of responsibilities begins with the principal investigator (PI) and includes the academic institution, in particular, their Institutional Review Boards (IRBs), as well as the funding agency whether governmental, foundational, or industrial and, finally, the medical journals. On the one hand, academicians typically have the autonomy to choose to work within their areas of expertise as well as to devise their own research methods to achieve this goal.

Academicians hold a "sacred trust" that begins with their students, residents, and faculty and extends to the general public. Academicians are, and deserve to be, held in high esteem by the public, but this brings greater responsibilities. Of concern to academicians, this trust may be eroding because research misconduct appears to be on the rise for several decades [6]. It is important to note that misconduct does not mean honest differences of opinion or even honest errors. It implies fabrication, falsification, or plagiarism. Fabrication is defined as making up data, whereas falsification is defined as manipulating data, and plagiarism is the appropriation of the work of others.

The inability of academic institutions to achieve and maintain research integrity leads to "outside policing" by funding agencies and an increasingly distrustful general public. Failure to achieve these levels of oversight will cause financial, reputational, and legal consequences. We believe that academicians should have the primary responsibility for the identification, prevention, and treatment of misconduct in medical research. We also believe that federal agencies should not usurp the authority or responsibilities of academicians and their institutions. This implies, however, that academicians and their institutions are ready, willing, and able to do so. This, in turn, implies imperatives that are based both on ethics and self-interest [7]. We should not let the perfect be the enemy of the good ("le mieux est l'ennemi du bien," Voltaire, 1770). We believe that such leadership should begin with academicians not governmental or other agencies.

2.3 Collaborations of Academician and Academic Institutions with Medical Journals

The "sine qua non" of medical journals is the conduct of independent and rigorous peer review that include scientific accuracy and avoidance of misstatements or overstatements. Peer-reviewed journals should require every coauthor to have contributed in meaningful ways. In additions, authors should sign statements of attestation following the recommendations of the International Committee of Medical Journal Editors (ICMJE). Journals should ensure that research is free from commercial bias. In addition, disclosure policies for reviews and commentaries should be particularly rigorous because the authors are not publishing new findings, but are offering what are expected to be their unbiased opinions about the totality of evidence. Journals should ensure that all of their authors and reviewers as well as their own editorial boards should have no financial interests related to any manuscript. The implementation of these policies will facilitate independent peer review. Journals should allow readers full access to the disclosure statements of their authors, preferably online. We also concur with the ICMJE requirement for all investigators to register all randomized trials on clinicaltrials.gov. The registration of the trial before beginning ensures that the stated hypotheses are, indeed, a priori and not a posteriori and data derived. The registration also contributes to achieving and maintaining research integrity with respect to randomization, adherence, and follow-up of all participants in the trial. There is a crucial need for independent and multidisciplinary data monitoring committees (DMCs) [8, 9]. The primary role of independent DMCs in Phase 3 trials is to protect the safety of randomized subjects. To do so requires frequent monitoring of safety data and periodic monitoring of efficacy data. The utilization of these safeguards protects the patient, investigator, and sponsor with respect to the perception and reality of research integrity.

2.4 Collaborations of Academicians and Academic Institutions with the Press

Academicians must always avoid misstatements and overstatements to the press about their scholarship and research. While such misstatements and overstatements of benefit to risk ratios may increase publicity, academic promotion, and grant support, the clear and present dangers include misinforming and confusing colleagues and frightening patients and making it even more difficult to conduct the high-quality research necessary to test the hypothesis.

With respect to interactions with the press, all faculty members whether salaried, adjunct, affiliate, clinical, or visiting should be subject to the same stringent guidelines approved by the academic institutions. This is necessary for several cogent reasons. First, the press will not generally be aware of these distinctions when media coverage occurs for scholarship or research. Second, the academic institution will have reputational and legal liabilities. Each academic institution should have clear guidelines for speaking with the press or in the development of press releases.

With respect to the former, the general guideline is to seek and obtain prior approval from the media relations department of the institution before addressing any member of the press. As regards the latter, we believe that press releases should follow publications in peer-reviewed journals. Press releases accompanying preliminary analyses or even published abstracts should generally be discouraged. Academics must remain cognizant that the mere mention of their academic institution implies they are speaking on their behalf. It could be argued that even disclaimers may not be sufficient to avoid potential reputational and legal liability.

2.5 Collaborations of Academicians and Academic Institutions with Industry

Academicians and their institutions are increasingly dependent upon outside funding. By the end of the twentieth century, there had been remarkable gains that emanated from large-scale randomized trials of treatment and prevention of cardiovascular disease [10], many of which had been derived from industry funding. These landmark findings have resulted from equipoise and détente, namely, balanced and trusted interrelationships between the investigators, the sponsor, and the regulatory authorities [11].

We believe that to achieve such détente requires awareness of the different specific goals of these various factions. In our view, the industry has a primary goal to prove the efficacy of their drug to achieve approval by regulatory authorities. Regulatory authorities, in turn, have a primary goal to approve only drugs or devices with favorable benefit to risk ratios in order to minimize harm to the general public, so their major emphasis is safety in the context of a favorable benefit to risk ratio. The priority of academicians is to determine whether there is a favorable benefit to risk ratio of the intervention under investigation. This is achieved by maintaining their independence in the design, conduct, analysis, and interpretation of randomized trials of sufficient size and duration and with high adherence and follow-up. These circumstances will facilitate the most reliable quantitation of the benefit to risk ratio. These issues are relevant to device trials. In drug trials, dose is also a major consideration. Obstacles in a particular country may include changes in the medical care delivery system that decrease the influence of healthcare providers and increase the influence of legal and business interests [1].

2.6 Missed Opportunities at Two Outstanding US Academic Institutions

Duke University is one of the most outstanding academic institutions in the USA and world. Nonetheless, at Duke, an instance of academic fraud occurred in relation to developing genomic predictors for cancer risk assessment and treatment. After external challenges were made about the scientific validity of genomic marker research conducted in the laboratory of Joseph Nevins, MD, and his mentee Anil Potti, MD [12, 13], Duke commissioned an investigation in the fall of 2009 to assess whether the markers were sufficiently reliable to drive the selection of therapies in three randomized trials. In July 2010, the National Cancer Institute (NCI) called into question both the reliability of the biomarkers and Duke's 2009 investigation. Due, in part, to a failed audit, the National Cancer Institute (NCI) and Duke invited the Institute of Medicine (IOM) to investigate the research process in the Nevins-Potti lab, to identify flaws that may have led to the external challenges, and to make recommendations about how to avoid such mistakes in the future [14–16]. The Office of Research Integrity (ORI), within the US Department of Health and Human Services (DHHS), focused their report entirely on one individual faculty member and made no comment on the institution's responsibility and its failure to provide adequate oversight and investigation. These actions missed an important opportunity to emphasize the institution's critical responsibilities in oversight of research integrity and the importance of institutional transparency and accountability [4].

At Harvard, which is considered by some the beacon of academic institutions of the USA and world, several decades ago, academic fraud was detected initially in relation to fabricated data by John Darsee, MD, on experiments in cardiovascular disease in dogs. Later, it was uncovered that these activities were of long-standing duration as they had occurred at Emory his previous institution as well as Notre Dame, his undergraduate institution. The initial investigation was not of sufficient depth or breadth to uncover either the nature or the extent of the fraud, so this represented another missed opportunity [3].

The striking resemblances of these two instances decades apart included the lack of integrity of the investigator and a tendency of the institution to limit the scope of the initial investigations. These cases also highlight the need for courage and vigilance of the departmental and medical school authorities in the early initiation and completion of comprehensive and transparent investigations, as well as holding liable all individuals involved regardless of their positions in the hierarchy of the academic institution.

It may be of interest to note that DLD, the second author of this chapter, was a member of the committee that conducted the investigation at Duke. In addition, Professor Howard Morgan, the mentor of JDR, the first author of this chapter, was Chair and DLD a member of the committee that investigated Harvard.

2.7 Conclusions

In addressing the goal of achieving and maintaining academic integrity, the responsibilities must be shared. The primary responsibility should rest upon the principal investigator, including each and every collaborator in all scholarly or research activities. In addition, the institutions, especially those in a position of leadership, must play a significant role. Academic leaders in academic institutions must create and maintain an environment to maintain and enhance integrity and, as a corollary, maintain proper oversight. When, despite all efforts at prevention, the possibility arises of the occurrence of violations, especially academic fraud, the academicians and their institutions should respond rapidly and sufficiently to adequately diagnose and treat the issues whether the occurrence results from inadvertent or intentional misconduct [2].

Academicians, their institutions, the funding agency, and the medical journals should all ensure that there is a prompt, thorough, and transparent review process that results in appropriate accountability at all levels. All academicians as well as their faculties and staffs engaged in research must speak truth to power [17].

References

1. Robishaw JD, DeMets DL, Wood SK, Boiselle PM, Hennekens CH (2020) Establishing and maintaining research integrity at academic institutions: challenges and opportunities. Am J Med 133(3):e87–e90
2. DeMets DL (1997) Distinctions between fraud, bias, errors, misunderstanding, and incompetence. Control Clin Trials 18(6):637–650
3. DeMets DL (1999) Statistics and ethics in medical research. Sci Eng Ethics 5(1):97–117
4. DeMets DL, Fleming TR, Geller G, Ransohoff DF (2017) Institutional responsibility and the flawed genomic biomarkers at duke university: a missed opportunity for transparency and accountability. Sci Eng Ethics 23(4):1199–1205
5. Faust HS, Menzel PT (eds) (2011) Prevention vs. treatment: what's the right balance? Oxford University Press
6. Institute of Medicine (2012) Evolution of translational genomics: lessons learned and a path forward. National Academies Press
7. Dingell JD (1993) Shattuck lecture-misconduct in medical research. N Engl J Med 328(22):1610–1615
8. DeMets DL, Furberg CD, Friedman LM (eds) (2006) Data monitoring committees in clinical trials: a case studies approach. Springer
9. Ellenberg SS, Fleming TR, DeMets DL (2019) Data monitoring committees in clinical trials, 2nd edn. Wiley
10. Hennekens CH (1998) Increasing burden of cardiovascular disease; current knowledge and future directions for research on risk factors. Circulation 97(11): 1095–1102
11. Hennekens CH, DeMets DL (2011) Data and safety monitoring boards of randomized trials: emerging

principles and practical suggestions. Clin Invest 1(1): 53–57
12. Goldberg P (2009) A biostatistics paper alleges patient harm in two Duke clinical studies. Cancer Lett 35(36): 1–2
13. Baggerly K, Coombes K (2009) Deriving chemosensitivity from cell lines: forensic bioinformatics and reproducibility research in high-throughput biology. Ann Appl Stat 3:1309–1334
14. Institute of Medicine (IOM) (2012) Evolution of translation genomics: lessons learned and a path forward. National Academies Press
15. Goldberg P (2010) By defending Potti, Duke officials become targets of charges of institutional failure. Cancer Lett 26(28):1–2
16. Goldberg P (2011) IOM Committee will probe Duke scandal together with other "omics" case studies. Cancer Let 37(1):1–2
17. American Friends Service Committee. Speak truth to power: a quaker search for an alternative to violence. 1955.

Indictment or Information Can Lie: Post-Truth in Science

3

Mariella Scerri and Victor Grech

Abstract

Misinformation and false news are the provenance of post-truth, a recent phenomenon whereby debate is framed by appeals to emotion with repeated assertion of half-truths and outright lies. While dubbed a modern concept, misuse of information and fake news have their roots grounded in history. Indeed, scholars argue that even though certain features of post-truth were foreshadowed in earlier times, a combination of different factors has currently created a new set of circumstances which justify its designation as a post-truth era. This chapter will therefore take a closer look at the origins of post-truth and how it infiltrated science and medicine and negatively influenced debates about climate change, vaccine uptake and the COVID-19 pandemic. This chapter argues the importance of utilising Fairclough's critical discourse analysis (CDA) approach in analysing discourse which accompanies conspiracy theories and fake news. Only after a thorough examination of discourse can effective strategic planning to combat misinformation take place.

M. Scerri (✉)
University of Leicester, Leicester, UK

V. Grech
Faculty of Medicine and Surgery, Mater Dei Hospital, Msida, Malta
e-mail: victor.e.grech@gov.mt

Post-Truth in Science

> Beware of false knowledge; It is more dangerous than ignorance.—George Bernard Shaw.

3.1 Introduction

"False news is harmful to our community; it makes the world less informed, and it erodes trust. It's not a new phenomenon, and all of us — tech companies, media companies, newsrooms, teachers — have a responsibility to do our part in addressing it" [1]. This was part of a media campaign created recently by Facebook, to admonish and warn its users and encourage their familiarisation with a guide on the fake news phenomenon. This included scepticism towards headlines and checking story sources. Facebook alleges that it is actively engaged in fighting the proliferation of misinformation in main areas: to disrupt economic motives as fake news is financially driven; to build new products to stem the spread of misinformation; and to encourage people to make knowledge-based decisions regarding false news [1]. Misinformation and false news are the provenance of post-truth, a recent phenomenon whereby "debate is framed by appeals to emotion with repeated assertion of half-truths and outright lies" [2].

In 2016, the *Oxford Dictionary* named post-truth the word of the year after a 2000% spike in

J. Faintuch, S. Faintuch (eds.), *Integrity of Scientific Research*,
https://doi.org/10.1007/978-3-030-99680-2_3

usage in the preceding year. As a broad phrase, post-truth captures the *zeitgeist* [3].

3.2 Historical Roots of Fake News and Misinformation

The mid-1700s witnessed a spate of fake news, when at the peak of the Jacobite rebellion in Great Britain, in an attempt to subvert the establishment, printers incited false news reports claiming that King George II was ill. This fake information was picked up by other reputable printers and republished, making this piece of information indistinguishable from fact. Indeed, Attorney General Dudley Ryder responded to complaints in a letter:

> As the publication of such false news of his Majesty, has a tendency to disquiet the minds of his subjects, hurt public credit, and diminish the regard and duty which they owe him, I think the doing it with such views is an offence punishable at Common Law, and for which an indictment or information can lie. And the frequency of such publications is evidence of such wicked designs. But as every false report of this kind which may arise from mistake only cannot be charged as a crime, so it is very difficult to say how often it must be repeated in the paper to make it criminal [. . .] I don't know any method to prevent this practice but by prosecuting the offenders when they are guilty. [4]

Fast forward in time and two centuries later, the fake news of an alien attack on America on Sunday, 30 October 1938, was indeed prototypical. Broadcasting an adaptation of the 1898 novel *War of the Worlds* by HG Wells, the Columbia Broadcasting Network aired a major part of the plot over the radio as a series of breaking news alerts. The realistic attack alarmed many listeners and tricked them into believing that there was an alien invasion taking place [5]. Towards the end of the broadcast, doctors, nurses and soldiers were so panic-stricken that they reported for duty ready to fight the Martians. Police stations across the country responded to a multitude of calls, while newspaper reporters rushed to prepare special editions [5].

Indeed, scholars argue that even though certain "features of post-truth were foreshadowed in earlier times, a combination of different factors has currently created a new set of circumstances which justify its designation as a post-truth era" [6]. Ironically, fake news and misinformation flourished and propagated in the twenty-first century at a time when the Internet provides easy access to information. The large fact-free campaign during Brexit; the growing use of disinformation campaigns by politicians in Hungary, Russia and Turkey; and the deceit and lies surrounding the US presidential election by Trump gave rise to a "growing international trend where some feel emboldened to try to bend reality to fit their own opinions" [3]. The false messages that are relayed have common features; they insist that they have inside information; however, they do not provide a reference to support the alleged source. While the "motivation behind creating these messages is unclear, it might relate to attention seeking behaviour and conspiracist ideation" [7]. Political as well as cultural factors have been linked to the rise of post-truth where actual truth is relegated a secondary position.

3.3 The Origins of Post-Truth

Postmodernism and the emergence of the philosophical concept of relativism in the late twentieth century mark the inception of post-truth [8]. Postmodernism has been an enabler of power in "sustaining different regimes of truth", often obscure and discreet [6]. This has led "the general public to question objective facts and created a setting in which alternative facts are legitimated". The problem is aggravated when relativist views are normalised and make way to lies which can be excused as "alternative points of view" or "legitimate opinions" and turned into a maxim in which everyone is entitled to their own truth [9]. While postmodernist thinking has vocalised opinions of minority people in society, dismissal of facts and theories by powerful elites gives rise to widespread dissemination of inaccurate discourses and half-truths. This is compounded by the "development of digital communication tools and media which has led to a dramatic augmentation conducive to an information overload" [10]. The cyber revolution confronts the authority of conventional sources of information including scientific research [11].

The narrative of postmodernism and post-truth is well illustrated in George Orwell's *Nineteen Eighty-Four* wherein the State deliberately modifies and alters historic and archival records to match the propaganda of the day [12]. Of note is Orwell's entry in his diaries about war, "All propaganda is lies even when one is telling the truth" [13]. The BBC induction course at the start of his career helped him to understand propaganda machinations. Indeed, the stint at the BBC was influential in his inception of *Nineteen Eighty-Four*, and he came to realise "how politically ignorant the majority of people are, how uninterested in anything outside their immediate affairs" [14]. This provides a blank slate to governments giving them access to language manipulation in order "to serve political agendas and to re-shape language in the cause of a dominant ideology" [14].

Writing about the future is fraught with difficulties and uncertainty. The task for any writer is to sound the warning to his readers about what could happen. The clout of totalitarianism in *Nineteen Eighty-Four* is an entangled and complex motif, but the "novel's narration — with its texts within texts — also enacts its own phantasmagoria, a world where both everything is true and nothing is true" [15]. As Lynskey [16] avers, Orwell anticipates what Hannah Arendt [17] explains in *The Origins of Totalitarianism*, published a year after Orwell's demise: "The ideal subject of totalitarian rule is not the convinced Nazi or the convinced Communist, but people for whom the distinction between fact and fiction (i.e., the reality of experience) and the distinction between true and false (i.e., the standards of thought) no longer exist". Nowadays, this is in part due to circulating online misinformation.

For example, the events surrounding Donald Trump's impeachment which led to violent unrest at the Capitol in January 2021 are "part of a completely alternate reality" and have reached a point of "radicalisation" [18]. Wardle and Brookie, directors of the Atlantic Council's Digital Forensic Research Lab, highlight the negative repercussions that these conspiracy theories would have and how they will "outlast the Trump administration" [18]. The violent rhetoric on online platforms and the convergence of different types of conspiracy theories surrounding Trump's administration will have lasting effects and, unfortunately, will also infringe on other political and scientific topics.

In *A Democratic Staff Report Prepared for the Use of the Committee on Foreign Relations United States Senate,* Robert Menendez [19] argues that an even greater cause of concern is the latest model of digital authoritarianism which is likely to alter the digital domain in the near future. The Chinese Communist Party, in Orwellian fashion, is cultivating digital authoritarianism within China's borders through the development of an obtrusive, ubiquitous surveillance state that uses "emerging technologies to track individuals with greater efficiency and to bolster its censorship mechanisms to ascertain that information considered detrimental to the regime does not reach its citizens" [19]. Indeed, a fundamental respect for human rights is threatened; and in turn, such means of surveillance and population control can negatively influence campaigns worldwide [19]. Furthermore, the diversification of digital authoritarianism in China has far-reaching consequences for the free world at large because China is influencing and shaping its own version of a censored Internet in its own political image [19].

Despite China's authoritarian governing narrative, the country is a key player in major economic and political international spheres and is capable of "providing the communist regime with increased status among other nations" [19]. As journalist Richard McGregor [20] avers, China's manifesto—"the idea that authoritarian political systems are not only legitimate but can outperform Western democracies"—is a valid concern. The country's growing influence on the digital sphere "to promote an alternative model for the digital domain based on state control, seems to support European illiberal democracies" as is evidenced in Hungary and Poland [20].

Emerging innovative tools enable China to categorise and label individuals and pre-emptively take necessary action against those considered problematic to the regime [21, 22]. The banning of US social media platforms such as Facebook, Instagram, Twitter, WhatsApp, Pinterest and YouTube from China's servers [23] has two main advantages for the dictatorial state: censoring these platforms is successful in withholding information that would be considered derogatory to the Party and it also allows Chinese platforms with similar services to thrive and expand [19]. China's digital efforts need to be examined against a global backdrop and the ramifications this will have on other countries explored. "At best, China is selling digital technology that has remarkable capacity for surveillance and control to authoritarian or authoritarian-leaning countries with no second thought for the consequences. At worst, China is pairing its economic investment with aggressive outreach and training on Internet governance and domestic regulations to further inculcate authoritarian values and methods of social control" [19].

3.4 The Effect of Post-Truth on Science and Medicine

Unfortunately, the incursion of fake information, conspiracy theories and emerging digital authoritarianism have infiltrated science and medicine. While science constitutes the "last metanarrative with any significant cachet in the post-postmodern condition" [24], it continually falls prey to post-truth after certain significant events which occurred in the last several decades. Once respected for the authority of its method, scientific results are now openly questioned and doubted by squads of non-experts. Perceived as potentially offering a haven of truth for the general public, the distinguished probity of this pillar is being undermined by post-truth utterances [25].

One of the rudimentary challenges in fake news and conspiracy theories' identification in science is that it is prevalent in many different specialties of scientific knowledge. Norris [26] argues that scientific knowledge is complex and specialised, and this curtails the comprehension and acumen that can realistically be expected from any lay person. Moreover, scientific knowledge is in a constant state of flux. For this reason, scientific experts tend to rely on the knowledge of their peers to conduct their research, particularly in interdisciplinary research. Studies also show that misinformation causes people to abandon facts altogether. A study by van der Linden et al. [27] concludes that participants who were presented with both a persuasive fact and a similar piece of inaccurate information held tight to their original belief—the misinformation cancelled out the fact. The campaigns and conspiracy theories about climate change and against vaccination programmes are inarguably the greatest challenges that science and medical communities had to face in the last few decades. Indeed, the COVID-19 pandemic compounded an already fraught situation.

3.4.1 The Environmental Cost of Misinformation

Climate change is the biggest environmental story and arguably one of the toughest challenges to tackle misinformation. Not only is it scientifically complex but also politically perfidious, making it an impossible endeavour to correct and counteract the misinformation from media sources. Objectivity—a key component of climate change conventional equilibrium—led both journalists and the general public astray. The problem arose when firms dealing with public relations and scientists funded by the fossil fuel industry were the seat of doubt and misinformation about the reality of human-caused climate change. This led journalists to repeat the information in an attempt to be "balanced" [28], thereby leading the media to create an appearance of significant scientific debate over anthropogenic climate change, when, in fact, there was little disagreement. Such a disparate coverage, particularly evident up until 2004 [29], falsely framed climate change as a "debate" in the public eye [30]. Not long after Boykoff and Boykoff's [31] influential study, coverage changed to reflect more closely scientific consensus. By 2010, Block [32] noted that most journalists had stopped covering climate change as a scientific controversy.

Such a debate is particularly problematic because its repercussions can be great. When lay persons lack the necessary knowledge and skills to interpret data themselves, they seek guidance and knowledge from trusted experts. Climate sceptics create mistrust, suspicion and doubt, and while they may not be responsible for the conspiracy theories, theories abound that assert that climate scientists are faking their data to finance their research through funding [33]. Climate change conspiracy theories are uniquely harmful because they can steer public opinion and hamper efforts to adapt to and reduce the impact of global warming. To aggravate the situation, climate change can be considered as an example of "proportionality bias"; there is a large discrepancy between what is a large-scale event with enormous significance and what is portrayed by scientists and governments as a consequence of small every day events such as transportation.

This proportionality bias therefore becomes the seat of many conspiracy theories about climate change. Such theories typically postulate that there is no occurrence of global warming and instead claim that scientific findings are fabricated lies by researchers who have something to gain. Due to this common thread, they are often referred to collectively as the great global warming conspiracy. Douglas and Sutton [33] identify four of these conspiracy theories: scientists are involved politically; to get research funding; global warming is a green scam; and global warming is an attempt to promote nuclear power. As with other conspiracy theories, in these examples, the alleged conspirators are driven by either political power or money. However, comparing climate change conspiracy theories with other popular conspiracy theories reveals some crucial differences.

One notable difference is that people "prefer to believe that they are part of an enduring and moral social group that can be confident and strong in the face of conflict and threat" [34]. People want to be reminded that they are headed towards a bright future and that they should be optimistic about their prospects [35]. This "feel good" factor is part of the psychological appeal of climate change conspiracy theories. "Denial of climate change is likely to do a lot more for people's general sense of equanimity", and therefore, these conspiracy theories discredit the "overwhelming evidence that humans are contributing to the destruction of their own environment" [33].

Two mainstream explanations also contribute to an idiosyncratic feature of climate change conspiracy theories. The official version—that humans are inducing dangerous levels of climate change—is disputed and countered by others. Indeed, organisations such as Greenpeace [36] try to counter-argue and suggest that a thorough explanation is lacking. This allows industrialists to orchestrate and fund campaigns in which misinformation is presented to the public [37]. The other important elucidation is that other conspiracy theories are led by backroom political machinations. A classical case is the withholding of important information about greenhouse gas emissions in China which was removed from the summary for policy makers in the fifth Intergovernmental Panel on Climate Change assessment [38]. This raises concerns about the global politics involved in climate change—specifically, that the gravity of the situation and the culpability of humans are undermined.

Both conspiracy and counter-conspiracy theories are therefore represented as "insincere, and scientific data are represented as political tools rather than value-free observations of the world" [33]. This creates uncertainty and mutual distrust, which, when politicised, compromises an evidence-based approach of scientific findings by the general public [39]. Perhaps most corrosive is that the politicisation of climate change leads to a choice to be made between competing conspiracy theories.

3.4.2 Conspiratorial Beliefs of Anti-vaccination

Similarly, misinformation and conspiracy theories surrounding vaccination programmes create uncertainty in public discourse; "even worse conspiratorial ideation can lead to outright opposition to and rejection of the scientific method" [40]. Mistrust and conspiratorial beliefs are endemic in anti-vaccination groups, with those advocating the scientific and medical consensus often regarded as agents of some "ominous interest group bent on concealing the truth" [41]. Inarguably, the development of vaccines has been an important discovery in the history of medicine. However, many global regions have witnessed a decline in vaccination programmes, particularly the combined measles, mumps and rubella (MMR) vaccination [42].

This particular decline is attributed to the publication of Andrew Wakefield's article in *The Lancet* in 1998 in which he linked autism to MMR vaccination [43]. Although the article has since been retracted, the research discredited and the author is no longer permitted to practise medicine, the repercussions are still felt nowadays. In 2008, measles was declared to be endemic in the United Kingdom, 14 years after its spread was halted in the population [42], while MMR vaccination rates lie well below the recommended 95% uptake [42]).

Fundamental to the anti-vaccine conspiracy movement is the argument that large pharmaceutical companies and governments conceal information about vaccines to meet their own sinister objectives [44]. Anti-vaccine conspiracy theories therefore reflect suspicion and mistrust of scientific research and try to undermine vaccine efficacy and safety [44]. In particular, parents faced with the decision to have their children vaccinated may be more likely to seek vaccine-related information through the Internet than through their doctor or paediatrician [44]. Unfortunately, websites which propagate anti-vaccine conspiracy theories are usually some of the top "hits" [44]. Although many people are sceptical of anti-vaccine conspiracy allegations, recent research suggests group discussions on vaccination tend to promote such conspiracy theories [44].

3.4.2.1 The Role of the Social Media

The potential for disseminating harmful health-related information through social media has increased exponentially as "thought influencers" in the anti-vaccine movement attract a large following. A study conducted in Italy found an inverse correlation between MMR vaccine coverage and Internet search activity, Facebook posts and tweets [45]. In another study of 153 YouTube videos about immunisation, the results show that negative videos are more likely to receive a rating, have higher mean star ratings and have more views [46]. Vaccine objectors reject evidence-based information in favour of vaccination because misinformation is simply their version of information. Unfortunately, social media acts as a "postmodern Pandora's box", releasing arguments that are not easily dismissible. Once in circulation, misinformation is not easily retracted [47].

3.4.3 The Infodemic on COVID-19

The global COVID-19 pandemic is the perfect storm for aggravating a situation encumbered by misinformation and conspiracy theories. During a pandemic, the risk of contagion, alongside alarmist publicity stoked by social media, induces fear more than the disease itself [48], resulting in potential patients to panic and to succumb to unlikely offerings of prophylactic agents, treatments and cures [49]. The infodemic surrounding the COVID-19 pandemic has led the general public to question mask-wearing, to opt for fake treatment and to question vaccines as a safe preventive method. This was perpetuated by a plethora of news articles which have documented the insouciant attitude of several male leaders who diminished the importance of the "little flu" or claim it to be of "moderate risk". This incited citizens to imitate the diminishment discourse and encouraged them to flout local government guidelines for self-isolation, social distancing and mask-wearing.

The careless attitude of leaders such as Johnson, Putin, Bolsonaro and Trump is a spectacular feature of coronavirus emo-truth political performances and became a widely recognised sign of toxic masculinity in reporting around the world [50]. Emo-truth is a particular form of "aggressive masculine performance of trustworthiness, corresponding to a code for recognizing it, resulting in a legitimated status of the popular truth-teller, and at odds with more official scientific, institutional truth-tellers" [50]. Arguably, the most disturbingly spectacular performance of unmasked toxicity was waged by armed men, who stormed the Michigan (USA) state capitol to intimidate lawmakers before a vote to extend the lockdown. They alleged that their individual freedon was "threatened", and they opposed "government tyranny" manifest in lockdowns, social distancing and mask "imposition" [51]. These men demonstrated a disregard for truth claims about the seriousness and/or dangers of the virus or, perhaps more prominently, a disregard for the collective danger the virus poses.

3.4.3.1 Unproven Therapeutic Protocols

Recently, in the United States, media attention focused on the possible therapeutic benefit of chloroquine derivatives. Although they are not approved treatments for COVID-19, notable increases in inhalation exposures as well as exposures to bleach products and alcohol-based sanitisers are documented. In Madagascar, claims of a herbal coronavirus "cure" produced from the *Artemisia* plant were also circulating, compelling the WHO to state that the tonic is not evidence-based and is potentially toxic [52]. The WHO is also trying to combat the political polarisation and online misinformation that threaten to undermine COVID-19 vaccination programmes worldwide. Indeed, multiple sources of health information may fuel vaccine hesitancy [53]. As access to technology has improved, social media has attained global penetrance. Two lingering doubts in social media conspiracy theories about the safety of the COVID vaccines are the speed with which they have been developed and approved and whether the mRNA vaccines may alter DNA in some of the body's organs. Despite clinical trials revealing the safety and efficacy of coronavirus vaccines, a good proportion of society remain hesitant about being administered with the newly developed vaccines, thereby making the fight against the COVID-19 disease even more difficult.

3.5 A Critical Discourse Analysis Approach to Post-Truth

Identifying the root of conspiracy theories and the effect of post-truth in medicine is not enough, if the deleterious effects on science and medicine are to be counteracted. A proper understanding of how social media discourse is constructed becomes vital, as it will provide a deeper investigation of the invested meaning in post-truth. Norman Fairclough's critical discourse analysis (CDA) approach offers a framework which promises a deeper inquiry into a "communicative event" [54]. Fairclough's CDA has three basic properties: "it is relational, it is dialectical and it is transdisciplinary". CDA provides a point of contention between "discourse and other objects, elements or moments, as well as analysis of the internal relations of discourse" [54]. Narratives in social practice constructed in this transdisciplinary way allow for various "points of entry" for discourse analysis as well as sociological, economic, political and cultural analysis. The three-tiered framework, namely, the close textual and linguistic analysis in the first level, the analysis of discursive practices in the second level and the deep study of the social practice and sociocultural phenomena in the third level [54], is eminent in the framing and examination of conspiracy theories.

As meanings move from text to text, they are altered and therefore become open to transformation. This gesticulation of meaning called "mediation" involves "the constant transformation of meanings, both large scale and small, significant and insignificant, as media texts and texts about media circulate in writing, in speech and audiovisual forms, and as we, individually and collectively, directly and indirectly, contribute to their production" [55]. Movement of meanings involves both continuity and change which are both "contingent upon the nature of the events and texts that mediated meanings move into" [56]. Moreover, the possibility of transformation suggests that mediated meanings enter processes of meaning-making as part of the resources for meaning-making. Ultimately, these resources for meaning-making are both specific and general and concrete and abstract. Climate change and the coronavirus pandemic as specific events are surrounded by concrete representations regardless of whether these representations are conspiracy theories or evidence-based information. Indeed, media texts are a "class of texts which are specialised for moving resources for meaning-making between texts, and more abstractly between different social practices, fields, domains and scales of social life" [56].

Fairclough avers that social transformations in life are extensively "discourse led" in the sense that it is discourses which change first. As new discourses "enter and achieve salience or dominance in particular social fields or domains and at different social scales, [...], or are recontextualised within them, dialectical processes may ensue in which discourses are enacted in ways of acting" [54].

Social events (and texts) are shaped on the one hand by social practices and social structures and on the other hand by social agents. Any social media text has "causal effects on non-semiotic as well as semiotic elements of social life" [57]. It figures in three main ways as part of events: in acting, representing and identifying. The writing (the text) is part of the action, while it simultaneously represents aspects of the scientific world (such as climate change, the pandemic and vaccination) and identifies other social actors (the general public, patients and parents). As events do not exist independently but are "interconnected chains or, more loosely, networks which are in part chains or networks of texts" [54], participants in chains or networks of events orient to ways of chaining or networking which are parts of networks of social practices, including, semiotically, what Fairclough calls "genre chains" [54].

"Genres are regularly and predictably chained together such that meanings are moved and transformed along the chain, and recontextualised and transformed in regular ways in accordance with recontextualising principles" [54]. In this case, accurate information about climate change, vaccination programmes and COVID-19 measures is lost or thwarted along the chain. Therefore, understanding the semiotic principles of texts and how they are recontextualised to target specific groups is important for the scientific and medical communities as it will help them to come up with effective strategies to counter post-truth.

Post-truth in science and medicine reflects the discursive struggle operating between misinformation and the determination of truth. Fairclough's conception of social practice based on a Marxist tradition is particularly relevant here. Fairclough sees the development of power relations as a hegemonic struggle. He utilises the terms "ideology" and "hegemony" as a way to theorise changes in power relations in society. For Fairclough, hegemony is the power invested over a society as a whole; however, it is always partially attained and is more about winning consent through ideological means rather than domination. Discursive changes have their roots in tyrannical struggles where uneven power relations are located [58]. Fairclough's "understanding of ideology focuses on the implicit and unconscious materialization of ideologies that are manifest in individual and collective life". He uses Althusser's concept of interpellation—the idea that the individual internalises values and ideologies from the surrounding society and acts upon them—to explain how ideology constitutes the subject [58].

The two ideologies of a sustainable environment and good public health find resistance especially when interpellation comes from different ideological positions (such as misinformation and conspiracy theories), and this leads to uncertainty and confusion causing discursive change to begin. The controversies surrounding climate change, public trust in vaccines and the COVID-19 pandemic are not only a power struggle between the general public's comprehension of facts and science's factual knowledge. Disagreements in public between science and medical experts over evidence-based facts have also created confusion. This has left the public with difficult choices to make. The monopoly on truth which traditionally has been on the side of authorities backed up by a knowledgeable base has shifted and is now under pressure.

3.5.1 New Channels and Messages

The scientific and medical communities are now required to introduce new ways of communicating information to their populations and new ways of adjusting to the mechanisms of public decision-making. Authorities find their positions questioned in defining truth. Lay people are not concerned or perhaps unaware of the intricate webs of social structures, relations of power and the nature of social practices. What is exceedingly worrying is that their practices leave an impact on the hegemonic struggles around them, which in turn create the perfect medium for conspiracy theories to proliferate. Accurate and concise knowledge sharing is therefore paramount in shaping the public attitudes whether it is about measures needed to reduce the carbon footprint, attitudes towards vaccination or mitigation measures during a pandemic. Combating the plenitude of misinformation generated on social media and the Internet is not an easy feat; however, medical and scientific communities who are responsible for disseminating knowledge need to be cognizant of the power invested in discourse, if they are to counter-argue and reverse the adverse effects of conspiracy theories.

References

1. Working to stop misinformation and false news (2020) Facebook for media. https://www.facebook.com/formedia/blog/working-to-stop-misinformation-and-false-news. Accessed 5 Jan 2021
2. Tesich S (1992) A government of lies. The Nation:12–13
3. McIntyre L (2016) Post-truth. MIT Press, London
4. Black J (1987) The English Press in the eighteenth century. Routledge Revivals, London
5. Examples of fake news from history. The Social Historian. https://www.thesocialhistorian.com/fake-news/. Accessed 8 Jan 2021
6. Foroughi H, Gabriel Y, Fotaki M (2019) Leadership in a post-truth era: a new narrative disorder? Leadership 15(2):135–151. https://doi.org/10.1177/1742715019835369
7. Wang Y, MA MK, Torbica D (2019) Stuckler systematic literature review on the spread of health-related misinformation on social media. Soc Sci Med 240:112552
8. Gardner HE (2011) Leading minds: an anatomy of leadership. Hachette, London
9. Pomerantsev P (2016) Why we're post-fact. Granta Magazine. https://granta.com/why-were-post-fact/. Accessed 22 Jan 2021
10. Knight E, Tsoukas H (2018) When fiction trumps truth: what 'post-truth' and 'alternative facts' mean for management studies. Organ Stud 40:183–197
11. Gottfried J & Shearer E (2016) News use across social medial platforms 2016. Pew Research Center http://www.journalism.org/2016/05/26/news-use-across-social-media-platforms-2016/. Accessed 17 Jan 2021
12. Orwell G (1949/1984) Secker & Warburg, London
13. Orwell S & Angus I (1968) The collected essays, journalism and letters of George Orwell; Volume 2: my country right or left, 1940–1943 Secker and Warburg, London
14. Dale J (2020) What would nineteen eighty-four be like if it were written today? New Writing 17(1):19–30
15. Szalai J (2021) How Orwellian became an all-purpose insult. The New York Times. https://www.nytimes.com/2021/01/13/books/orwellian-1984.html. Accessed 12 Jan 2021
16. Lynskey D (2019) The ministry of truth. Doubleday, New York
17. Arendt H (1951) The origins of totalitarianism. Meridian Books, New York
18. Culliford E (2021) Online misinformation that led to Capitol siege is 'radicalization,' say researchers. Reuters https://www.reuters.com/article/us-misinformation-socialmedia-idUSKBN29H2HM. Accessed 18 Jan 2021
19. Menendez R (2020) The New Big Brother: China and the new digital authoritarianism. A democratic staff report prepared for the use of the committee on Foreign Relations United States Senate. https://www.foreign.senate.gov/imo/media/doc/2020%20SFRC%20Minority%20Staff%20Report%20-%20The%20New%20Big%20Brother%20-%20China%20and%20Digital%20Authoritarianism.pdf. Accessed 12 Jan 2021
20. McGregor R (2018) Xi Jinping's ideological ambitions. WSJ. https://www.wsj.com/articles/xi-jinpings-ideological-ambitions-1519950245. Accessed 16 Jan 2021

21. Mozur P (2019) One month, 500,000 face scans: How China is using A.I. to profile a minority. New York Times. https://www.nytimes.com/2019/04/14/technology/china-surveillance-artificial-intelligence-racial-profiling.html. Accessed 19 Jan 2021
22. Wang M (2019) China's algorithms of repression: reverse engineering a Xinjiang Police Mass Surveillance App. Human Rights Watch. https://www.hrw.org/video-photos/interactive/2019/05/02/china-how-mass-surveillance-works-xinjiang#:~:text=The%20Human%20Rights%20Watch%20report,of%20its%20%E2%80%9CStrike%20Hard%20Campaign. Accessed 18 Jan 2021
23. Pham S (2017) China adds Pinterest to list of banned sites. CNN. https://money.cnn.com/2017/03/17/technology/pinterest-banned-china/. Accessed 16 Jan 2021
24. Grech V (2011) Infertility in science fiction [dissertation]. University of Malta
25. Grech V (2017) Fake news and post-truth pronouncements in general and in early human development. EHD 115:118–120
26. Norris SP (1995) Learning to live with scientific expertise - toward a theory of intellectual communalism for guiding science teaching. Sci Ed 79(2):201–217. https://doi.org/10.1002/sce.3730790206
27. Van Der Linden S, Leiserowitz A, Maibach EW (2017) Scientific agreement can neutralize politicization of facts. Nat Hum Behav 2:2–3
28. Gelbspan R (2005) Disinformation, financial pressures, and misplaced balance. Nieman Reports, pp 77–79
29. Antilla L (2005) Climate of skepticism: US newspaper coverage of the science of climate change. Global Env Change 15:338–352
30. Boykoff M (2010) U.S. climate coverage in the '00s. extra!. http://fair.org/extra-online-articles/u-s-climate-coverage-in-the-00s/. Accessed 25 Jan 2021
31. Boykoff M, Boykoff J (2004) Balance as bias: global warming and the U.S. prestige press. Global Env Change 14:125–136
32. Block B (2010) Covering climate change. World Watch 23.2. http://www.worldwatch.org/node/6373. Accessed 25 Jan 2021
33. Douglas KM, Sutton RM (2015) Climate change: why the conspiracy theories are dangerous. Bull At Sc 71(2):98–106. https://doi.org/10.1177/0096340215571908
34. Wohl MJA, Branscombe NR, Reysen S (2010) Perceiving your group's future to be in jeopardy: extinction threat induces collective angst and the desire to strengthen the in group. Per Soc Psych Bull 36(7): 898–910
35. Feyginal JT, Goldsmith RE (2010) System justification, the denial of global warming, and the possibility of 'system-sanctioned change'. Pers Soc Psych Bull 36(3):326–338
36. Gibson C (2012) Koch brothers exposed: fueling climate denial and privatizing democracy. Greenpeace https://www.greenpeace.org/usa/koch-brothers-exposed-fueling-climate-denial-and-privatizing-democracy/. Accessed on 16 Jan 2021
37. Fischer D (2013, December 23) Dark money' funds climate change denial effort. Scientific American. https://www.scientificamerican.com/article/dark-money-funds-climate-change-denial-effort/. Accessed 16 Jan 2021
38. Howard BC (2014, July 3) Data deleted from UN climate report highlight controversies. Nat Geo. https://www.nationalgeographic.com/news/2014/7/140703-ipcc-climate-report-deleted-data-global-warming-science/. Accessed 18 Jan 2021
39. Cohen GL (2003) Party over policy: the dominating impact of group influence on political beliefs. J Pers So Psych 85(5):808–822
40. Lewandowsky S, Gignac GE, Oberauer K (2013) The role of conspiracist ideation and worldviews in predicting rejection of science. PLoS One 10(8): e75637
41. Jacobson RM, Targonski PV, Poland GA (2007) A taxonomy of reasoning flaws in the anti-vaccine movement. Vaccine 25(16):3146–3152
42. Health Protection Agency (2008) Measles figures soar. The Guardian https://www.theguardian.com/society/2008/feb/22/health. Accessed 20 Jan 2021
43. Burgess DC, Burgess MA, Leask J (2006) The MMR vaccination and autism controversy in United Kingdom 1998–2005: inevitable community outrage or a failure of risk communication? Vaccine 24:3921–3928
44. Kata A (2012) Anti-vaccine activists, Web 2.0, and the postmodern paradigm - an overview of tactics and tropes used online by the anti-vaccination movement. Vaccine 30:3778–3789
45. Aquino F, Donzelli G, De Franco E, Privitera G, Lopalco PL, Carducci A (2017) The web and public confidence in MMR vaccination in Italy. Vaccine 35: 4494–4498
46. Keelan J, Pavri-Garcia V, Tomlinson G, Wilson K (2007) YouTube as a source of information on immunization: a content analysis. JAMA 298:2482–2484
47. Stein RA (2017) The golden age of anti-vaccine conspiracies. Germs 7(4):168–170. https://doi.org/10.18683/germs.2017.1122
48. Debiec J (2020) Fear can spread from person to person faster than the coronavirus – but there are ways to slow it down. The Conversation https://theconversation.com/fear-can-spread-from-person-to-person-faster-than-the-coronavirus-but-there-are-ways-to-slow-it-down-133129. Accessed on 18 Jan 2021
49. Taylor S (2019) The psychology of pandemics: preparing for the next global outbreak of infectious disease. Cambridge Scholars, Cambridge
50. Harsin J (2020) Toxic white masculinity, post-truth politics and the COVID-19 infodemic. Eur J Cul Studies 10:1177
51. Berlatsky N (2020) Michigan gunmen stormed the statehouse and proved these protests have never really been about coronavirus. The Independent. https://www.independent.co.uk/voices/michigan-gunmen-

statehouse-coronavirus-protests-trump-working-people-a9494336.html. Accessed 20 Jan 2021
52. Coronavirus: caution urged over Madagascar's 'herbal cure' (2020) BBC News. https://www.bbc.com/news/world-africa-52374250. Accessed 16 Jan 2021
53. Vaccine confidence volatile, vulnerable to misinformation (2020) The Straits Times. https://www.straitstimes.com/world/vaccine-confidence-volatile-vulnerable-to-misinformation. Accessed 16 Jan 2021
54. Fairclough N (2013) Critical discourse analysis: the critical study of language. Taylor and Francis, New York
55. Silverstone R (1999) Why study the media? Sage, London
56. Lassen I, Strunch J, Vestergaard T (2006) Mediating ideology in text and image: 10 critical studies. John Benjamins Publishing Co., Amsterdam
57. Fairclough N & Jessop B (2010) Critical realism and semiosis. Research Gate https://www.researchgate.net/publication/313037147_Critical_realism_and_semiosis. Accessed 25 Jan 2021
58. Fairclough N (1992) Discourse and social change. Polity Press, Cambridge

Legislation on Research Misconduct: Rationales and Reflections—A Swedish Perspective

4

Kjell Asplund

Abstract

Prompted by the Macchiarini research fraud case, an Act on Research Misconduct was introduced in Sweden in January 2020. In accordance with international convention, it defines unlawful research misconduct as "...severe deviation from good research practice in the form of fabrication, falsification or plagiarism." To qualify as severe, the misconduct must be either intentional or caused by serious neglect. Universities are obliged to report suspected cases for investigation by a newly established governmental agency. The Act does not include regulation of penalties or sanctions. The expected consequences of the Act and its limitations are discussed. The crucial roles of universities and scientific journals in the prevention of misconduct remain, and improved European legislation to protect whistleblowers is described. Future directions in the struggle against research misconduct include use of technical advances to detect data and image manipulation.

Information in this chapter was modified from the author's Swedish popular science book—Kjell Asplund (2021) Fuskarna: Om Macchiarinis och andras svek mot vetenskapen (The frauds: On the betrayal of science by Macchiarini and others). Fri Tanke, Stockholm, Sweden

K. Asplund (✉)
Department of Public Health and Clinical Medicine, Umeå University, Umeå, Sweden

4.1 Introduction: The Macchiarini Case

Few scientists have had such a direct impact on Swedish legislation as the thoracic surgeon Paolo Macchiarini. His scientific fraud prompted an Act on Research Misconduct covering not only medical but all types of academic research. As a background to the Swedish regulation on research misconduct, I will, throughout the chapter, return to his story to illustrate why an Act on Research Misconduct was regarded as necessary.

In comparison with other major research frauds, Macchiarini did not stand out in terms of number of articles retracted because of research misconduct. But few other affairs have comprised such a stress test for large organizations.

In 2010, Paolo Macchiarini was appointed at the Karolinska Institute as guest professor and to the Karolinska University Hospital as consultant [6]. He had an international background with research in Paris and London, and he had been working as a thoracic surgeon in Florence, Barcelona, and Hannover. During his work in Barcelona, he was the first in the world to perform transplantation of a cadaver trachea. This achievement reported in *The Lancet* in 2008 received worldwide attention. The intervention was described as bold and innovative, actually visionary—a new, intriguing possibility to treat patients with advanced tumors and other severe tracheal diseases. During the twentieth century, several legendary surgeons had worked with

J. Faintuch, S. Faintuch (eds.), *Integrity of Scientific Research*,
https://doi.org/10.1007/978-3-030-99680-2_4

frontline technologies at the Karolinska. Macchiarini was recruited to follow in their footsteps. Stockholm was envisaged to be the central hub of an international network, the European Airway Institute.

In 2011, Paolo Macchiarini proceeded with the world's first transplantation of a synthetic trachea seeded with stem cells [6, 16]. It stirred great attention not only among his surgeon colleagues but also in the media; several TV companies were present to cover the story. During 2011–2013, Macchiarini transplanted another two patients with synthetic tracheas, one of them on two occasions. All three patients died after serious complications and prolonged severe suffering in intensive care. As the disastrous outcomes of these and other operations became apparent, his appointment as consultant at the hospital was terminated in 2013. He remained in his academic position at the Karolinska Institute until 2016, when the full story of his scientific fraud was uncovered.

Five months after the pioneering artificial trachea transplantation, Macchiarini and his team reported on the case in *The Lancet* [16]. Later scrutiny revealed that some of the content was fabricated and some falsified. Most importantly it was stated that the patient had recovered and had no remaining symptoms. New airway epithelium had been formed from the stem cells that the transplant had been seeded with. There was a statement that the local research ethics committee had approved the project. Nothing of this was true. At the end, not only *The Lancet* article but also another five of Macchiarini's articles were retracted by the respective journals.

4.1.1 Personal Experience

I was assigned by the hospital and its owner, the Stockholm County Council, to lead a team investigating the clinical aspects of Macchiarini's transplantation. After reviewing the medical charts and the relevant literature and interviewing more than 60 people involved, we contacted Macchiarini. He was low-key and appeared offended. He had flown in to Stockholm to tell us his version of what actually had happened, to put facts right. The message he conveyed: Sweden was a country that had treated him highly unfairly and he, being a world leading surgeon, had now left for new missions.

In our report, I and my investigation team concluded that he had been breaking several Swedish laws: the Ethics Review Act, the Drug Act, the Health Care Act, and the Patient Safety Act. Our many recommendations included improved employment procedures for combined academic and clinical positions, actions against group think, an education program on clinical trial regulations, a tighter system for introducing new treatments, and more determined reactions to whistleblowing [6].

This serious fraud case reached vast media attention, in Sweden and internationally. The dramaturgy required culprits. The senior researchers among Macchiarini's 25 co-authors lost research grants and prestigious positions in Swedish academia, e.g., memberships of the Nobel Prize awarding committee and the Royal Academy of Sciences. The President of the Karolinska Institute and the Chairman of the Board of the Institute were forced to resign, and the government discharged the Director General of the Swedish Higher Education Authority. Two heads of clinical departments at the Karolinska University Hospital were dismissed.

Macchiarini came to be a tough stress test for the Karolinska Institute, for the Karolinska University Hospital, but also for the Swedish research community at large. The process revealed critical deficiencies in the systems and compelled universities, tertiary healthcare institutions, as well as the government to oversee the scientific environment in order to detect and manage suspected fraud. Macchiarini also left a kind of cultural legacy. Research misconduct issues started to be discussed much more intensively among scientists, and the number of reports on suspected misconduct surged. At the governmental level, work on an Act on Research Misconduct was initiated.

4.2 Macchiarini's Legacy: New Legislation on Research Misconduct

In the second half of the 1960s, internal review boards, inspired by the Helsinki Declaration and focused on medical research only, had been established at medical faculties in Sweden. Similar review boards were later established in social sciences and humanities. In 2004, a Research Ethics Review Act and a system of six regional ethics review committees, with an appeal mechanism, were introduced [3].

In 2010, an advisory national expert committee for review of suspected research misconduct was established. Universities could turn to this committee for review of suspected cases. Taking the committee's opinions into consideration, the President of the respective university then took the final decision on whether there was research misconduct or not and decided on sanctions.

This system had obvious shortcomings, the most important being that universities only rarely brought matters of suspected fraud to the national expert committee; it seemed that the universities wanted to keep the investigations and decisions in-house—after all, public attention might tarnish the university's reputation. A second criticism was that the role of the national expert committee was only advisory with all decision-making power remaining at the universities.

One of the few matters brought to the committee was the Macchiarini case, possibly as a result of the intense criticism of how the Karolinska Institute handled it. The committee investigated six of the articles published by the Macchiarini group, involving 43 authors altogether. Misconduct was confirmed in all of the articles [17]. Based on the report by the expert committee, the President of the Karolinska Institute passed the decision that seven of the researchers were guilty of research misconduct. The majority of the other co-authors were found to be "blameworthy" (a connotation that was widely criticized). In general, senior co-authors were judged harder—with their experience, they were particularly responsible for the content of the articles; inexperienced junior collaborators could not have the same responsibility.

Under considerable media and political pressure, the government appointed one committee to investigate the legislation as to research misconduct and one to oversee the Research Ethics Review Act. The proposals of both inquiries have resulted in new legislation.

The Swedish Act on Research Misconduct has been in operation since 1 January 2020 [29] (Box 4.1).

Box 4.1 Key Items in the Swedish Act on Research Misconduct [29]

- Definition: A severe deviation in the form of fabrication, falsification, or plagiarism that is intentional or caused by serious neglect.
- Scope of action: Academic organizations, governmental agencies, regions and local communities, and companies, associations, and foundations with a considerable public ownership or influence. Not private organizations.
- Researcher responsibility: Good research practice, with the researcher's organization having the overriding responsibility.
- Suspected misconduct: Forwarded to the National Board for Assessment of Research Misconduct.
- Delators: Includes ordinary citizens; the report may be anonymous; the Board may their take own initiatives.
- Time limit: Cases older than 10 years are not investigated.
- Research misconduct with no intent or major neglect: Reported back to the accountable research organization. The same applies to lack of good research practice.

(continued)

Box 4.1 (continued)

- Research misconduct present: Agency decision forwarded to research funding organizations, authorities, and scientific journals.
- No regulation of penalties or other sanctions.

The definition of research misconduct follows the conventional FFP model (fabrication, falsification, plagiarism). In its document *The European Code of Conduct for Research Integrity*, ALLEA (All European Academies) has listed not only principles for good research practice but also what should be regarded as violations of research integrity [2]. This includes not only the three FFP forms of misconduct but also a considerable number of what ALLEA labels "unacceptable practices." Similar guidance is available in other countries, including the United States (e.g., [24]).

4.2.1 Limited Scope

In the discussions preceding the Act, the proposal was criticized for not embracing research misconduct in private companies and other non-public organizations. In Sweden, less than a third of all research is performed by public institutions and thus subject to the Act [30].

4.3 The National Board for Assessment of Research Misconduct

A new independent governmental agency, the National Board for Assessment of Research Misconduct, is established. Previously only university leaderships could initiate a notification to the national expert committee for investigation of suspected misconduct. This was voluntary and only occasionally used. Now, reporting to the new Board is mandatory for universities and other accountable public agencies. This means that in-house investigations of suspected research misconduct will no longer be performed, with reduced risk of vested interests influencing the investigation process and its conclusions. In addition, not only universities but anyone can report his or her suspicions.

The National Board is chaired by an experienced judge. In the Board, there is also broad academic representation (humanities, biomedicine, public health, natural sciences, and social sciences). In comparison with similar organizations in other countries, such as the Office of Research Integrity at the US Department of Health and Human Services [31], the mission of the Swedish organization is restricted to investigation of individual cases of suspected research misconduct. In policy-making or education, the Board only provides indirect guidance by the decisions it makes.

One of the first decisions of the Board was about a case that reached considerable attention in the academic world. Once again, the Karolinska Institute was in focus. When the President had to resign because of the Macchiarini affair, the Vice President stepped in as Acting President. The Board investigated the new Acting President for suspected research misconduct and found her guilty in four articles where she was the corresponding author. The Board reasoned that she had profound experience and had had several opportunities to scrutinize the manuscripts; as corresponding author, she had the ultimate responsibility for the content of the articles, concluding that there had been gross neglect on her part.

She appealed the Board's decision to an administrative court. Repeated research fraud was confirmed but the evidence for her personal responsibility for this was not convincing enough. She was therefore acquitted. This verdict was of particular interest, since it was the first formal juridical test of a decision by the new National Board.

4.3.1 Ongoing Impact

In comparison with the national system before the new law was introduced, the number of cases reported by the universities is much higher, indicating good adherence to the regulation on mandatory reporting. The great majority of reports have concerned life sciences research. In reviewing the decisions the Board has published so far, a few general observations on how it has interpreted the new Act can be made:

- *What definitions are used?* The Act on Research Ethics Review does not provide exact definitions of fabrication, falsification, or plagiarism. The Board therefore refers to definitions provided in guidelines and codes of research ethics, for instance, the European Code of Conduct for Research Integrity [2].
- *Which researcher(s) are investigated?* Even if coworkers may have been involved (and possibly have been manipulating data), only the responsible researcher(s) reported by the university or others has been subject to investigation. This is in accordance with Swedish administrative law practice; only matters that are reported to the Board are investigated (as opposed to criminal cases in which any potentially guilty person is investigated).
- *What delineations of research misconduct are used?* In addition to the FFPs (fabrication, falsification, and plagiarism), the misconduct must be "serious," defined as intentional or caused by gross negligence. An example of "intentional" is when the deviations have recurred in several publications. Fabrication or falsification of data or images is always considered a "serious deviation."
- *What is plagiarism?* It is not uncommon that text on methods, for instance, on instruments and data analyses, is directly copied from previous articles. When this is done without citing the source, it constitutes plagiarism. Plagiarism of the researcher's own previous publications without citing the source (self-plagiarism) is equated with other forms of plagiarism.
- *What defense is accepted?* A common response when someone is accused is that there may have been misunderstandings and mistakes, but this has not affected the main results or the conclusions and should not be regarded as misconduct. The Board has not accepted this reasoning.
- *What is reporting of results?* Reporting of results is considered to include both publication and "other types of public presentation." Thus, posters and other presentations at conferences are covered by the Act. Presentation of working material for critical in-house review before publication is not a violation of the law.
- *Acquittal.* If a researcher has not adhered to good research practice but this is not considered serious nor intentional, the accused researcher is acquitted.

4.4 Penalties Are Not Included

The Act does not include regulation on penalties or sanctions [29]. When the Board has reached a decision that research misconduct is confirmed, it is up to the university or other accountable organization to decide on sanctions. This may range from a warning or salary deduction to dismissal. Thus, there is still room for between-university variations in what consequences confirmed misconduct may have.

In very serious cases of misconduct, other legislation must be considered. Aspects like physical or mental injury to research subjects or economic criminality may be regarded. Again, the Macchiarini case could be illustrative. The ensuing legal process has been protracted. A previous investigation of Macchiarini for manslaughter was discontinued, but the case was reopened, the prosecution now being for severe physical abuse. When this was written (April 2022), a 3 weeks long trial was ongoing, 9-11 years after Macchiarini performed his synthetic tracheal transplantations. The accusation has been serious abuse and the prosecutor has been advocating a 5 year imprisonment sentence.

4.4.1 Applicability of Ordinary Legislation

The Swedish Criminal Code contains one chapter on fraud and other misconduct and one on falsification offenses [4]. They could hypothetically be applied in cases of research misconduct, but to the best of my knowledge, there is no example of a research fraudster being convicted in court on these grounds, neither has there been any conviction according to anti-corruption legislation.

In the public debate on criminality of any sort, a spinal reflex reaction could be to demand tough penalties. When it comes to breaking the law on research misconduct, the question emerges: Should the Swedish Act on Research Misconduct include regulations on penalties?

4.4.2 Incarceration for Ethical Failures

In other countries, there are examples of scientists who have been sentenced to prison for research misconduct, e.g., the stem cell scientist Hwang Woo-Suk in South Korea [9] and the anesthesiologist Scott Reuben [11] and the obesity and aging researcher Eric Pohlman [18], both in the USA. If Macchiarini is convicted for serious abuse, he will be the first in Sweden to be sentenced to prison, but that will not be primarily for research misconduct but for his transplantations as clinical procedures. The link to research misconduct is indirect only. Had the outcome of the first transplanted patient been correctly described in *The Lancet* article, it seems unlikely that the other two patients would have been transplanted.

4.4.3 Professional Versus Criminal Sanctions

In the Swedish debate, it has been argued that research misconduct should be regarded as a variant of honor crime and penalized by law [20], but I propose that scientists who have committed research misconduct are usually penalized sufficiently hard by the academic world. In serious cases, the fraudster has to leave his or her university and will have great difficulties in finding a new position. Research grants are retracted, and the researcher becomes an outcast in the academic world and loses all professional positions of trust. Rehabilitation is out of question, and there is usually no way back to academia. I see this as sufficiently deterrent punishment. Reconciliation is rarely a characteristic of the academic culture.

4.4.4 Psychological Profile

The personality traits of researchers found guilty of research misconduct have not been systematically studied; however, other problematic individuals in the academic world, cheating students, have been studied in detail [32]. They have the same target as other students: passing examinations, achieving good grades, being awarded by stipends, or simply being appreciated. But as opposed to other students, they think that their goals are superior to moral principles. How they reach the goal, if so by cheating, is of secondary importance—the moral barriers are not there.

Psychological profiling also shows that the risk of being caught for cheating is not very deterrent. Cheaters are, in general, more prepared to take high risks than non-cheaters. When reviewing anecdotal information from cases of research misconduct, it seems that many of the research fraudsters have a psychological profile similar to that of cheating students. If they are prepared to take very big risks, possible penalties may not be a major constraint.

4.4.5 High Rollers

Most researchers behind major scientific breakthroughs have been risk-prone. When entering a completely new research territory, a researcher cannot be absolutely sure about what is waiting: success or fiasco? "Bold" is an adjective that many use to describe how research should be; lack of boldness characterizes stagnant research. Most ordinary researchers trudge on in the non-risky main furrow. To build a research profile that stands out, an easy shortcut would be to publish fake results.

Making questions on research misconduct a legal matter may be two-edged. With a lawyer as Chair, the National Board will be strict in its judgments of evidence. This seems reassuring from the accused researcher's point of view. However, it may also mean that some of those who have actually faked information are acquitted because the weight of the evidence is queried. It may become a didactic problem to explain what an acquittal really means: There is no convincing evidence that violation of the law has been committed, i.e., there has been no serious or intentional fabrication, falsification, or plagiarism. This does not exclude that less serious violations and other forms of unacceptable behavior, for instance, those identified by ALLEA in its Code of Conduct for Research Integrity [2], could have occurred.

4.5 Prevention: How Can Universities and Journals Contribute?

The new Act addresses prevention only indirectly. For prevention, an improved mechanism for early detection would be needed. The Japanese anesthesiologist Yoshitaka Fujii fabricated data for two decades before his fraud was finally disclosed. The late disclosure contributed to the extraordinarily high number of articles (183) that had to be retracted [26]. Although questioned in the scientific literature since long, results published by the famous British psychology professor Hans Eysenck together with his coworker German sociologist Ronald Grossarth-Maticek were eventually disclosed as severely manipulated in 2019 only [21, 25]. Eysenck was spared the embarrassment; he died in 1997. In many cases of delayed exposure of research misconduct, whistleblowers backed by sturdy facts had been active, but they had been neglected by universities and authorities.

4.5.1 Zero Tolerance Culture

As I propose, the work of investigational and inspection bodies will have to be complemented by preventive activities by universities and scientific journals. The most serious forms of misconduct (fabrication, falsification, and plagiarism) often go hand in hand with other deviations from good research practice, behaviors that ALLEA calls unacceptable [2], although they are not illegal. A research environment in which results that do not support the hypothesis are hidden and evade publication, where the importance of own results are exaggerated and the results of others neglected, where inconvenient data points are moved or deleted, honorary authorship is accepted, and signals of possible misconduct are disregarded—that is a milieu that is prepared for more serious violations of research integrity. This is, to a large extent, a matter of leadership. Leaders of research groups, institutions, and universities are instrumental in creating environments where any step toward fraud, however small, is not tolerated. For the individual researcher, it is also a matter of self-respect: "Is this the kind of scientist I want to be?"

Structural preventive measures used in Sweden are more or less generic: education in good research practice and guidelines on research integrity. There is also some focus on storage of research data. When someone is suspected of research misconduct, is it then surprising that the original data tend to have disappeared? Data have typically been lost when offices have been moved or computers have been stolen. Techniques for storing laboratory protocols so that they cannot be changed in retrospect and good routines to file recordings and results are now widely used. This also provides protection for the researcher who is unfairly accused of misconduct.

4.5.2 The Role of Journals

4.5.2.1 Will Scientific Journals Report Suspected Cases of Research Misconduct to the New Board?

Some forms of research fraud such as plagiarism of a previously published image may be extremely difficult for a common reviewer to detect. In addition, there is an obvious risk of confirmation bias. A reviewer primarily examines internal validity of the research (and external validity, when appropriate). In all article reviews, these quality issues are relevant. Research misconduct on the other hand is relatively uncommon, and the reviewer's mind is usually not set to detect fraud (a variant of confirmation bias). The situation is very different if a reviewer or a committee has been assigned to investigate a case of suspected fraud; they actively look for facts confirming misconduct.

What adds to the problem is that many prestigious scientific journals have a fast-track system with little time for reviewers to probe manuscripts in depth. Many of the articles with fake data have been accepted after a fast-track procedure. This was how Paolo Macchiarini's article on the first-ever transplantation of a synthetic trachea in *The Lancet* [16] and Mandeep Mehra's and Sapan Desai's COVID-19 fake articles in *The Lancet* [22] and *New England Journal of Medicine* [23] were processed. Nevertheless, editors do apply technical tools to detect at least some forms of misconduct. Major journals now use computer programs to detect plagiarism.

4.6 Whistleblowing, a High-Risk Undertaking, Now with Improved Legal Protection

Whistleblowers are often neglected and questioned, sometimes humiliated and penalized. To fight research misconduct, the position of whistleblowers would need to be strengthened. This is now in progress in Europe. In 2019, the European Union decided on a Whistleblower Protection Directive [13]. It is targeted at both private and public sectors, meaning that it applies to whistleblowers also in academic settings.

Whistleblowers may be external or internal. External ones may have different angles of approach: journal editors, critical readers, colleagues or competitors in the same research field, and scientists who fail to reproduce published results. Trachea transplantation is a narrow research area. When the Belgian surgeon Pierre Delaere, one of Macchiarini's colleagues/competitors, was reached by the news from Stockholm that Macchiarini had performed the world's first transplantation of a synthetic trachea, he wrote to the Karolinska Institute [6]. This could not possibly work; he asked if research misconduct could have been involved. Not only did he write five insisting letters, but he also offered to come to Stockholm to present his criticisms and concerns. His warnings were dismissed by the ethics committee of the Karolinska Institute, perhaps because the committee regarded Delaere's observations as part of an academic conflict between competitors, rather than whistleblowing.

Three years later, four of Macchiarini's colleagues at the Karolinska Institute and Karolinska University Hospital collected detailed documentation on what they considered to be serious research misconduct in six articles with Macchiarini as senior author; three of the whistleblowers had been his co-authors. They were met with mistrust and threatened with reprisals. Had it not been for the massive attention their accusation had aroused at the *New York Times* [14], *Nature* [10], a Swedish TV documentary series entitled *The Experiments*, and other media, I am not certain that a detailed investigation would have been initiated. Slow reaction to whistleblowing is evident from many other conspicuous cases of research misconduct. Early signals are not taken seriously, and the final disclosure is delayed, often for many years, as in the Macchiarini case.

In reviewing a large number of instances of whistleblowing in medical research, it seems to me that a general pattern emerges: Whether or not whistleblowing will lead to a thorough investigation is a matter of power—the power based on the whistleblower's personal position and authority, the corresponding power of the accused, and the power entailed by solid facts.

4.6.1 The Sudbø Affair

The Norwegian cancer researcher Jon Sudbø leads the Scandinavian scoreboard of most retracted articles. He was eager to have esteemed US cancer researchers as co-authors, and they published in high-prestige journals. Camilla Stoltenberg, one of the top scientists at the Norwegian Public Health Institute and sister of the Prime Minister (later Secretary General of NATO), read a *Lancet* article by Sudbø and his collaborators [12]. She had good insights into the population registers that Sudbø was supposed to have used. She reacted: Could the remarkable findings be explained by research fraud? An ensuing investigation found that Sudbø had manipulated data ever since his time as PhD student. Stoltenberg's whistleblowing caused immediate actions because she was someone with scientific and formal personal power and her alarm had robust substance.

4.6.2 Hans Eysenck

Despite the fact that the famous British psychologist Eysenck's research frauds had long been well known and exposed in the scientific literature, it took more than two decades until his articles started to be retracted (a process that is still ongoing) [21, 25]. It seems that his stature as one of the giants in his field protected him. His defense against accusations of misconduct had a biblical tone, and upon rereading it, it gives me an impression of a sense of personal invulnerability.

Internal whistleblowers, often junior coworkers, usually lack formal power. This is, however, outweighed by the power of insight, deep knowledge of the conducted research, and its substance. Yet the outcome could still be dependent on the personal power of the accused.

4.6.3 The Hostile Atmosphere Against Whistleblowing

A US study in the 1990s of more than 200 whistleblowers in multiple organizations, not specifically in the academic or medical environment, showed dismal consequences: nine out of ten lost their jobs or were degraded, one in four ended up in court, and one in ten had attempted suicide [19]. An Australian study performed during the same period and also covering a spectrum of professionals ended up with similar although not quite as negative results [28]. The Australian authors described a typical pattern of the employers' reactions. They try to isolate the whistleblower by depicting him or her as incompetent, disloyal, and mentally unstable. A second strategy is to warn other employees to support the whistleblower and a third to avoid, at all means, to investigate the whistleblower's alarm.

A Swedish study on whistleblowing in social services, healthcare, schools, and churches, published in 2008 [15], confirmed this pattern. The whistleblowers themselves were also studied. They typically had long employments and were considered as reliable coworkers. The authors discerned organizational from personal consequences. It often resulted in actual changes in the organization, but this did not mean that the whistleblowers went free from reprisals.

As alluded to above, the European Union Whistleblower Protection Directive [13] was adopted in 2019 and is in the process of being implemented by national legislations of member states. In Sweden, national legislation to protect whistleblower already exists, but this is considered toothless. The EU Directive is substantially more targeted (Box 4.2). Moreover, adherence to the law will be supervised by a designated authority (The Swedish Work Environment Authority).

Box 4.2 Main Points in the EU Whistleblower Protection Directive Applied to National Legislation as Proposed by a Swedish Governmental Committee [5]

- Replaces existing Swedish Whistleblowing Act.
- Internal reporting (i.e., whistleblower) channels and procedures for employers, public and private, with 50+ workers; supervised by the Swedish Work Environment Authority.
- With few exceptions, no whistleblower liability for breaches of restrictions on disclosure of information.
- Information on reporting person or other individuals is protected.
- Only information of public interest protected. Information on reporting person's own working or employment conditions protected in exceptional cases.
- Knowingly false information is not protected.

4.7 Future Directions

4.7.1 Statistical Analyses to Detect Research Misconduct

In his hunt of medical research frauds, British anesthesiologist John Carlisle has been successful. Of the top ten scientists with most retracted articles on Retraction Watch's rather infamous ranking list [26], Carlisle has contributed to reveal six of them [1].

It all started with a conversation Carlisle had with colleagues in Torbay community hospital in South England on a series of articles published by the Japanese anesthesiologist Yoshitaka Fujii. Fujii had conducted randomized controlled trials on various drugs to reduce postoperative nausea and vomiting. The Torbay anesthesiologists found the results too good to be true. When Carlisle approached *Anesthesia & Analgesia*, one of the journals that had published the remarkable results, the editor challenged him: Why don't you scrutinize Fujii's research yourself? Carlisle took on the challenge.

He explored the statistics reported in Fujii's articles with the postulate that all biological data spontaneously vary, a "natural" variation [1]. However, published data appeared much more precise than if they had come from "natural" measurements. This was the start of a process that resulted in retraction of 183 of Fujii's articles, a world record [26].

Carlisle's method is statistical but he is not a professional statistician. His method is homemade and far from precise. It can identify articles that are suspected cases of research misconduct, yet, when reviewed more closely, many of these may prove to be accurate. When Carlisle himself evaluated his approach, he found that it resulted in a considerable number of precise hits but also that many of the shots ended up outside the target [8].

Statistical software that—in John Carlisle's spirit—is able to separate fabricated or beautified data from data based on real observations is now being developed. This may become a useful tool for detection and investigation of suspected cases of research misconduct to be used by inspection bodies, universities, and scientific journals.

4.7.2 Disclosure of Image Duplication and Manipulation

If statistics is John Carlisle's tool, the own eye is Elisabeth Bik's; she looks for image duplication and manipulation. After 15 years of gut microbiota research at Stanford University, this Dutch-American microbiologist took on a hobby: hunting, with research misconduct as prey [27]. When she and her colleagues reviewed 5000 articles with images in the biomedical field, they found that 4% were suspected to be manipulated [7]. Most frequent subjects to manipulation are microscopic pictures and electrophoreses. Photoshop has greatly expanded the manipulation options and techniques. Bik reports her observations to publishers, journal editors, and research institutions. Her scrutiny has resulted in more than 170 retractions of articles and more than 300 corrections published by the authors.

When Bik approaches authors of an article with suspected image manipulation, the most common response is something like: "Thanks for pointing out the error. We will send a correction to the journal and, by the way, the errors do not affect our conclusions" [27].

There is now software for detection of image manipulation ("digital image forensics"). According to Bik, the performance is less than optimal. It seems likely that artificial intelligence-based algorithms will improve the software precision. In analogy with John Carlisle and his statistical tools, sophisticated computer techniques may soon be used by inspection bodies and universities to detect and investigate cases of suspected image manipulation [27].

Conflicts of interest None.

References

1. Adam D (2019) The data detective. Nature 571:462–464
2. ALLEA (European Federation of Academies of Sciences and Humanities) The European Code of Conduct for Research Integrity (2017). Revised edition. allea.org/wp-content/uploads/2017/05/ALLEA-European-Code-of-Conduct-for-Research-Integrity-2017.pdf. Accessed 12 May 2021
3. Anonymous (2003) The ethical review act. onep.se/media/2348/the_ethical_review_act.pdf. Accessed 12 May 2021
4. Anonymous (2019) The Swedish criminal code. legislationline.org/download/id/8662/file/Sweden_criminal_code_am2020_en.pdf. Accessed 12 May 2021
5. Anonymous (2020) Ökad trygghet för visselblåsare [Improved protection of whistleblowers, summary in English]. SOU 2020:38. regeringen.se/rattsliga-dokument/statens-offentliga-utredningar/2020/06/sou-202038/. Accessed 12 May 2021
6. Asplund K, Blom N, Johansson K, et al (2016) The Macchiarini case. Investigation of the synthetic trachea transplantations at Karolinska University Hospital. Summary in English. sll.se/globalassets/1.-halsa-och-vard/bilagor---nyhet/bilagor-nyheter-2016/the-macchiarini-case-summary-eng.pdf?_t_id=jn7cuiNj-Rit3scHO1EqPDA%3d%3d&_t_uuid=4fsLCHhJQcCQKsxlbS0bwQ&_t_q=-macchiarini&_t_tags=language%3asv%2csiteid%3a8824540f-e7eb-44a0-94f3-c48a0b63798c%2candquerymatch&_t_hit.id=PublicWebv2_Cms_EPiSrvr_EpiDependencies_-Document/94&_t_hit.pos=3. Accessed 12 May 2021
7. Bik EM, Casadevall A, Fang FC (2016) The prevalence of inappropriate image duplication in biomedical research publications. MBio 7(3):e00809–e00816
8. Carlisle JB (2017) Data fabrication and other reasons for non-random sampling in 5087 randomised, controlled trials in anaesthetic and general medical journals. Anaesthesia 72:944–952
9. Choe S-H (2009, October 26) Disgraced cloning expert convicted in South Korea. *New York Times*. nytimes.com/2006/05/12/health/12iht-korea.1725751.html. Accessed 12 May 2021
10. Cyranoski D (2016, February 4) Artificial-windpipe pioneer under scrutiny again. *Nature* (News Article)
11. Dr Geoff (2017) Scott S Reuben – anesthesiologist who went to prison for faking pain control trials. drgeoffnutrition.wordpress.com/2017/11/19/scott-s-reuben-anesthesiologist-who-went-to-prison-for-faking-pain-control-trials/. Accessed 12 May 2021
12. Ekbom A, Helgesen G, Lunde T, et al (2006) Report by the investigation commission. Summary in English. web.archive.org/web/20060920200752/rikshospitalet.no/content/res_bibl/6633.pdf. Accessed 12 May 2021
13. European Union Directive (EU) 2019/1937. //eur-lex.europa.eu/legal-content/EN/TXT/PDF/?uri=CELEX:32019L1937&from=EN. Accessed 12 May 2021
14. Fountain H (2014, November 24) Leading surgeon is accused of misconduct in experimental transplant operations. *New York Times*. nytimes.com/2014/11/25/world/leading-surgeon-is-accused-of-misconduct-in-experimental-transplant-operations.html. Accessed 12 May 2021
15. Hedin U-C, Månsson S-A (2008) Repressalier mot kritiker i offentliga organisationer [Reprisals against critics in public organizations]. Summary in English. Socialvetenskaplig Tidskrift 3–4:276–294
16. Jungebluth P, Alici E, Baiguera S et al (2011) Tracheobronchial transplantation with a stem-cell-seeded bioartificial nanocomposite: a proof-of-concept study (Retracted). Lancet 378:1997–2004
17. Karolinska Institute (2016) External inquiry finds Paolo Macchiarini guilty of misconduct. news.ki.se/external-inquiry-finds-paolo-macchiarini-guilty-of-misconduct. Accessed 12 May 2021
18. Kintisch E (2006, June 28) Poehlman sentenced to 1 year in prison. *Science Magazine*
19. Lennane J (2012) What happens to whistleblowers, and why. Social Med 6:249–258
20. Lynøe N, Lambertz G (2013) Gör forskningsfusk straffbart! [Make research fraud punishable] Läkartidningen 110:CDAI

21. Marks DF (2019) The Hans Eysenck affair: time to correct the scientific record. J Health Psychol 24:409–420
22. Mehra MR, Desai SS, Ruschitzka F et al (2020a) RETRACTED: Hydroxychloroquine or chloroquine with or without a macrolide for treatment of COVID-19: a multinational registry analysis. Lancet S0140-6736:31180–31186
23. Mehra MR, Desai SS, Kuy SR et al (2020b) Cardiovascular disease, drug therapy, and mortality in Covid-19 (retracted). N Engl J Med 382(25):e102
24. National Academies of Sciences, Engineering, and Medicine, Committee on Responsible Science (2017) Fostering integrity in research. National Academies Press, Washington DC
25. Pelosi AJ (2019) Personality and fatal diseases: revisiting a scientific scandal. J Health Psychol 24: 421–439
26. Retraction Watch (2021) The Retraction Watch leaderboard. retractionwatch.com/the-retraction-watch-leaderboard/. Accessed 12 May 2021
27. Shen H (2020) Meet this super-spotter of duplicated images in science papers. Nature 581:132–136
28. Smith R, Brown AJ (2008) The good, the bad and the ugly: whistleblowing outcomes. In: Brown AJ (ed) Whistleblowing in the Australian Public Sector. Australian National University, Canberra, pp 109–136
29. Swedish Council for Higher Education (2019) Act on responsibility for good research practice and the examination of research misconduct (2019:504). uhr.se/en/start/laws-and-regulations/Laws-and-regulations/act-on-responsibility-for-good-research-practice/. Accessed 12 May 2021
30. Swedish Research Council. The Swedish Research Barometer (2019) The Swedish research system in international comparison [Summary in English]. vr.se/english/analysis/swedish-research-in-figures.html. Accessed 12 May 2021
31. US Department of Health & Human Services (2021) The Office of Research Integrity. ori.hhs.gov/. Accessed 12 May 2021
32. Williams KM, Nathanson C, Paulhus DL (2010) Identifying and profiling scholastic cheaters: their personality, cognitive ability, and motivation. J Exp Psychol Appl 16:293–307

Part II

Accountability of Scientific Teams

Publishing, Perishing, and the Infodemic of Fake Science

5

Alex W. Kirkpatrick and Thomas E. Randall

Abstract

The spread of misinformation and disinformation related to science and technology has impeded public and policy efforts to mitigate threats such as COVID-19 and anthropogenic climate change. In the digital age, such so-called fake science can propagate faster and capture the public imagination to a greater extent than accurate science. Therefore, ensuring the most reliable science reaches and is accepted by audiences now entails understanding the origins of fake science so that effective measures can be operationalized to recognize misinformation and inhibit its spread. In this chapter, we review the potential weaknesses of science publishing and assessment as an origin of misinformation; the interplay between science, the media, and society; and the limitations of literacy as an inoculation against misinformation; and we offer guidance on the most effective ways to frame science to engage non-expert audiences. We conclude by offering avenues for future science communication research.

5.1 Introduction

In our digital age, myriad media compete for the strained attention of polarized audiences [1]. Individuals are inundated with information of varying accuracy and quality and must ultimately decide for themselves which material to believe and which to cast off in an information-rich environment [2]. Meanwhile with the supremacy of the traditional media gatekeeper over the agenda of the day, even the truth-value of competing issues is waning, while the power of the public to steer societal discourse has dramatically increased through the ease with which unverified information can be shared at the touch of push button [3]. Within this heady mix of light-speed information, science stories are afforded less attention by traditional gatekeepers like newspaper editors [4]. The story itself, if covered, is likely to be addressed not by a specialist science reporter but a generalist who may inaccurately translate the language of science into an eye-catching and entertaining content that can compete against clickbait and the ubiquity of digital information and entertainment [4]. It is within this complex, diverse, and forever fluxing Petri dish that the bacteria of fake science breed.

A. W. Kirkpatrick (✉)
The Edward R. Murrow College of Communication, Washington State University, Pullman, WA, USA
e-mail: alex.kirkpatrick@wsu.edu

T. E. Randall
Info-Tech, London, ON, Canada

J. Faintuch, S. Faintuch (eds.), *Integrity of Scientific Research*,
https://doi.org/10.1007/978-3-030-99680-2_5

5.2 Shaky Science for Everyday Life

Science misinformation has generated widespread doubt concerning the efficacy of vaccines [5], the existence of climate change [6], and even the safety of relatively innocuous electromagnetic technologies such as smart meters [7] and 5G cellphone towers [8]. Such doubt has consequences in terms of public and policy behavior. During the COVID-19 pandemic, misinformation persisted in relation to the origins of the coronavirus, the importance of vaccines, and the credibility of medical professionals that attempted to communicate efficacy information publicly, with many still doubting the existence of the novel coronavirus even as the global death toll exceeded two million [9]. Doubts concerning the legitimacy of anthropogenic climate change has impeded public and policy efforts to mitigate serious environmental threats [6].

Misconceptions about electromagnetism and its association with cancer can even impede policy support for energy installation policies [7]. It has been argued that the power of misinformation has been enhanced and enabled during the term of US President Donald Trump whose administration helped to foster a post-truth era of alternative facts, fake news, and the routine distrust of experts, which has challenged the supremacy of the scientific method as the gold standard for knowledge production [10]. It is therefore vital that scientists and science communicators explore and attempt to understand how and why fake science becomes so attractive to audiences, how and why it spreads, and what responsibilities the scientific and academic communities have toward inhibiting its reach.

5.2.1 Science Publishing and Assessment

Much of the responsibility for ensuring the accuracy of science adopted by publics lies with the scientific community itself as generators of science, innovation, and change. Despite the ubiquitous presence of fake science in digital age, publics still afford the scientific community a significant level of support [11] and in general trust the results of scientific research [12] and defer to science and scientists themselves [13]. In other words, to differing extents, people expect and assume there are rigorous processes within the scientific community that ensure the quality and veracity of the science shared in the public sphere. Such practices include double-blind paper reviews by other experts [14], ensuring experimental validity [15], and publicly correcting erroneous information or outdated science [16]. It is partially a result of these vaunted processes that scientific research is considered gold standard for knowledge production [10]. But can the assumption that information from seemingly scientific sources is necessarily accurate block the spread of fake science?

5.3 Publication Weaknesses

There are worrying signals that indicate the review process and publication industry for the sciences, and thus the scientific process itself, is losing integrity. The pressures of a *publish-or-perish* mentality in academia can lead to suboptimal science and the proliferation of predatory journals. Under the guise of a robust scientific method, research produced through this process could be shared in good faith by publics who are unaware of declining credibility within the scientific publication industry. The *publish-or-perish* mentality results from a chain of pressure to succeed in academia, whereby the volume of academic publications acts as proxy for scientific productivity and ability [17]. Just as standardized test scores are presumed by some to reflect an individual's intelligence, journal publications are considered by many a reflection of an individual's ultimate academic attainment and intellect.

Building a strong résumé through multiple publications in high-ranking journals improves the authors' reputation. A reputation as a productive and capable publisher of empirical work then increases the authors' chances of attaining a full-time position at a higher education institution in an extremely competitive academic job market. In time, and with multiple well-cited articles attributed to them, the authors' chances of being afforded valuable funding and grants increase, further enhancing their potential to publish at quality journals and enhance their reputation and employability. Indeed, many universities and colleges demand a certain number of publications from their tenure-track professors, and, where this minimum output is not achieved, an academic risks losing their job and long-term livelihood [18]. This intense pressure to publish can eventually lead to fraud whereby the author intentionally distorts or exaggerates the results in order to publish seemingly groundbreaking research in prestigious journals [19]. Indeed, the rate of retraction in academic journals due to fraud has increased tenfold since 1975 [17].

Even when there are no intentions to deceive, pressure to publish tends to produce poorer methods and more false positives [20]. This has created a so-called replication crisis in several scientific disciplines, whereby the validity of published studies cannot be supported or replicated through following the original authors' methodology [21]. Yet often those published articles remain in the public sphere for public consumption and sharing. A questionable review process is further diminished by journals cherry-picking work, occasionally even hyping results and distorting refereeing in the competition for high status and profits from publication charges. Prestigious journals that claim accurate, valid, and robust review processes are still incentivized to weight positive and original results higher to increase readership ahead of null or replication studies [17]. Consequently, both authors and publishers could have reasons to game the system to their mutual advantage and to the pejoration of scientific veracity [22].

5.4 The Epidemic of Predatory Journals

Scientists of all career stages may be tricked into submitting [23] to questionable publications in an effort to improve their résumé [24]. Academics from low- and middle-income countries may be particularly vulnerable to predatory journals, as they attempt to overcome barriers to publication that exist in their nations due to a lack of resources and opportunities [25]. Such journals are counterfeit publications, abusing the open-access model whereby the author pays for publication and receives no compensation for the output, risking a diminishment rather than an enhancement of academic reputation [26]. These journals publish articles online with little or no peer review, low academic standards, and little credibility aiming to maximize profit from as many paying authors as possible.

Between January and May 2020, $33,807.41 was paid by authors for 350 articles published across 109 predatory journals on the topic of COVID-19 alone [27]. Given the already polarized nature of COVID-19 across political society, this amount of corrupted science further exacerbates the spread of misinformation, especially given how few non-academics are sufficiently knowledgeable about the relative quality of different journals.

5.5 Trustworthy Journal Classification

An independent international body to rate the quality of academic journals and note the predatory nature of some publishers could help to ameliorate the situation, rather than relying on metrics such as the impact factor, which is a measure of the frequency with which the average article in a journal has been cited in a particular year and speaks only indirectly to the quality of the journal. A standardized measure of quality, reach and review methodology could empower media and publics to determine the objective quality of the scientific information they receive rather than simply trusting or distrusting in the metonymy of science and its arcane processes.

5.6 Self-Correcting Science?

Even if the academic publishing culture is to evolve, some erroneous information will still emerge. A hallmark of the scientific method is to correct misinformation as quickly and as publicly as possible when it is discovered. Verifying and replicating each published study is not financially feasible and cannot keep pace with the rate of submissions at journals. Indeed, even if a study is retracted and discredited, the damage may already have been done. This is the legacy of Andrew Wakefield's study published in *The Lancet* [28] which reinforced a culture of anti-vaccination by falsely associating MMR jabs with occurrences of childhood autism. Though ten of the paper's 13 authors have since publicly renounced the study and *The Lancet* has retracted it, the paper remains a source for fueling conspiracies and, ultimately, the spread of fake science [29]. Once in the public sphere, the problem of fake science becomes even more complicated.

5.6.1 Science Through the Media

Most people do not peruse science, technology, engineering, and mathematics (STEM) into higher education and, as adults become dependent upon media resources to stay informed about the latest innovations, learn about science and orientate themselves within an increasingly technoscientific cosmos [30]. According to the National Science Board [31], while traditional mass media channels like radio, television, and magazines are still significant providers of STEM information to publics, the Internet is increasingly the primary source for such information. As concerns health innovations [32], it was found that participants in a study uniformly started with an Internet search. Indeed the Internet offers publics greater access to scientific information than at any point in history, as open-source databases are made more readily available and traditional science media transitions to digital formats. Due to the phenomenon of intermediality and the blurring of the lines between interpersonal, mass, and digital communication [33], our conception of media as an information source includes not only traditional mass media but the Internet and social media also.

5.7 The Pervasive Invasion of Electronic Media

Members of the public may have little recourse but to accept science as it is presented through media, even if that media is inaccurate or wilfully fake. Agenda-setting theory suggests that if media makes an issue salient through reporting, members of the public are more likely to be aware of that issue and to judge it as important to society [34]. In other words, media exposure has the potential to build awareness of accurate and inaccurate information alike, facilitating the spread of misinformation and increasing people's reliance on erroneous information when forming attitudes [35, 36]. Consumption of non-mainstream media (e.g., blogs and tabloids) has been correlated with acceptance of conspiracy theories, while consumption of more traditional sources (newspapers and network TV) has been found to be negatively associated with belief in some of the same conspiracies [37]. With members of the public increasingly turning to non-mainstream media [38], it is manifest that people risk increasingly being exposed to unverified and inaccurate information and conspiracy.

5.8 Illusory Truths

Traditional media outlets, at least, do abide by a code of ethics whereby they often devote space and effort to correcting misinformation [39]. *The New York Times*, for example, has a site dedicated to welcoming complaints about the accuracy of their reporting and republishing recently corrected articles [40]. Yet mere exposure to misinformation, even when that is labeled as false or inaccurate, still increases awareness of misinformation and risks instinctive acceptance of additional erroneous messages. Research on the illusory truth effect [41] has shown that people perceive an opinion to be widely held, after repeated exposure to it, even if expressed by a single source. Pennycook et al. [42] found that a single exposure to fake news headlines on Facebook increased subsequent perceptions of accuracy both immediately after exposure and after 1 week. This effect occurred even for less believable headlines, those labeled as contested by fact-checkers, or the information was counter-attitudinal inconsistent with the reader's political ideology [42].

Retracting misinformation may not do much to diminish its continued influence [43], while repeatedly identifying a claim as false might conduct people to accept it as true in the long term [44]. Sufficed to say, prevention is better than any cure at our disposal, and ensuring that the media receive and report accurate science in the first instance must be prioritized.

5.9 Reputable and Traditional Guidance

Use of mainstream broadcast and print media as an information source has been correlated with accurate beliefs about protection from infection, while use of conversational media outlets correlated with conspiracy theories about the seriousness of COVID-19 [45]. Synchronously, the platform through which individuals consume information also impacts how misinformation spreads. Social media sites are a common source of news information for the majority of Americans—53% of American adults get news from social media "often" or "sometimes" [38]. This is problematic when the algorithms that govern social media platforms, and what content users see, spread information not by way of accuracy but through profit maximization. When users click on additional content, the platform learns over time what material users engage most with. In turn, this model encourages "clickbait" material, keeping users online for longer stretches of time and increasing opportunities for further advertising.

5.10 User Engagement Tricks

The progressive polarization of clickbait material limits content access over time. Emphasis on emotionally charged article headlines coupled with the need to receive social feedback from the kinds of material users share could discourage people from considering the accuracy of information [42]. Therefore, it is unsurprising that social media companies have struggled with limiting the spread of misinformation during the 2016 and 2020 American elections and COVID-19 pandemic [38].

5.11 Trolls and Bots

These are actor or automated accounts that "use social media to start arguments, upset people, and sow confusion among users by circulating inflammatory and often false information online" [46]. Indeed, 20% of all tweets about the 2016 presidential election came from bots [46]. Such tweets were not random; these accounts are often funded and supported by unknown sources (the so-called dark money) with malevolent intent to spread misinformation. For instance, spending by undisclosed donors totaled $27 million in the 2016 election [46]. In a context of education inequity, media illiteracy, biased media sources, and the design of social media platforms, it is perhaps unsurprising to see the extent of science misinformation spread. Being beholden to media resources in order to learn about science post-formal education then becomes problematic when media themselves promote and propagate misinformation, prioritizing likes and digital traffic over accuracy. While social media sites have belatedly taken action against some of the most vociferous and influential sources of misinformation [47], it is probable that more pressure from science, society, and policymakers will be required if corporations like Facebook are to take any sustained and consistent action against all types of science misinformation. When mitigative actions such as fact-checking and open correction of misinformation are shown to be relatively ineffective at attenuating the spread of fake news, what can society do to empower publics themselves to recognize and reject fake science?

5.11.1 Knowledge Deficit Solutions

Media literacy can be understood as the ability to access, analyze, evaluate, and create messages in a variety of forms [48]. When individuals do not have the skills to evaluate media for its veracity, they are prone to accepting false information or resharing misinformation with others. Indeed, 23% of Americans have admitted to sharing fake science online [49], and one might safely assume that among the remaining 77% Americans, a portion will have shared fake science *unknowingly*. Specifically a science literacy deficit has often been blamed for poor public perceptions about science in general, further paving the way for the adoption of fake science [50]. The UK Bodmer Report [51] articulated a persistent and widely held perception on the behalf of the scientific community in the UK as to why publics sometimes hold negative beliefs about science and misperceive scientific risks and what constitutes a public, even. The espoused cause of such miscommunication was public ignorance, and the remedy was simply an amelioration of that ignorance through various educational methods now commonly grouped under the umbrella of the deficit model of science communication.

Bucchi [52] points out three broad and questionable assumptions on the part of the scientific community that underlie the deficit model. Firstly, it is assumed that publics and policymakers alike are misinformed about science and the issues raised through its development. Secondly, this misinformed state is exacerbated by a sensationalist media that does not cover technoscientific issues accurately or adequately, and, as a result, policymakers and publics often adopt irrational fears which fuel hostility toward accurate science and the scientific community. Lastly, and perhaps most importantly, is the assumption that these irrational fears (e.g., GPS microchips can be administered secretly through the COVID-19 vaccine) and the negative consequences thereof (e.g., hindrance of health research and diffusion of vaccines) can be attenuated by simply accurately informing the public about science, often via the media itself, through better management of the flow of mediated messages from the laboratory to the mind of the public [52].

5.12 Science Literacy Pitfalls

This paternalistic, downstream attitude and approach to science communication came under heavy criticism from the late 1980s onward, not least due to seeming ineffectiveness of efforts to raise science literacy, as manifested through public surveys over time in the USA and UK particularly [53]. Scholars soon began to provide evidence of the unfair characterization of the public as deficient and scientific knowledge as sufficient, even questioning the existence of a monolithic *general public* at all. Wynne [54] provided an early case study in 1992 into the reception of scientific knowledge by Cumbrian sheep farmers in Northern England in response to restriction introduced following the Chernobyl nuclear incident in 1986. Via in-depth interviews with the affected community, Wynne highlights the multiple ways in which scientific authorities disregarded the local knowledge of hill farming practices, displaying an acute ignorance for their social identities and cultural expertise that ultimately resulted in numerous bungled large-scale social interventions aimed at attenuating the spread of hazardous materials. These interventions, such as a ban suddenly placed on the movement and slaughtered sheep from affected areas (unnecessarily damaging the local economy), contributed to the mistrust the farmers developed for the scientific community, even generating conspiracy theories among the community that the botched attempt at preventing the absorption of radioactive cesium into peat soil, and the sheep themselves, was in some fashion a deliberate attempt to undermine the cultural identities of the Cumbrian hill farmers.

Public uptake of science does not depend wholly on scientific literacy but rather depends predominantly upon beliefs constructed as a function of the social networks with which they identify and value predispositions like trust [55]. Science literacy can even *enhance* rather than diminish the spread of fake science. A sample of US citizens was more likely to share a fake article vaunting the Large Hadron Collider's ability to produce black holes when they perceived themselves to have greater understanding of the results of scientific research. Those who were *less* confident in their ability were *less* likely to share the fake science online [56]. Indeed, intellectual humility has been associated with fact-checking and reduced likelihood that people will share COVID-19 misinformation [57].

5.13 Engagement Models

Contextual, dialogical approaches became a widespread remedy to the miscommunication between science and society from the late 1990s onward in an effort to upstream science policy decision-making by incorporating the contextual values and knowledge of varied publics and stakeholders (see [58]). This modern shift toward dialogue is perhaps best encapsulated by the publication of UK House of Lords Third Report on Science and Technology [59], which recognized the pejorative effect that the mishandling of and misinformation related to bovine spongiform encephalopathy in UK farming had had on public confidence in science. In doing so, the House of Lords Select Committee recommended two-way consultations between scientific authorities and publics both at the national and local levels, suggesting means of dialogue such as focus groups, citizen juries, Internet dialogues, and consensus conferencing, in an effort to incorporate contextual knowledge and public concerns into the trajectory of future research and policy.

5.14 Ongoing Pitfalls

Wynne [60] criticized new dialogue and engagement strategies for continually reinventing and repackaging public deficit explanations for public mistrust by persisting to imagine the public in reductive and simplistic ways and engaging these groups as if they truly reflected these *imagined* groups. Wynne points out that power imbalances are still inherent in the dialogue models, and he decries science's attempts to legitimize decisions through superficially democratic means. He also decries science's failure to recognize the increasing strains it puts on public trust in an ever more innovation-craving, technologically advanced global economy whereby publics have necessarily been forced to become more dependent on the scientific community to have their technoscientific needs met. Others point out that abandoning deficit thinking entirely ignores the requirement for sufficient level of scientific literacy to make informed decisions via deliberative processes [61].

5.15 Shared Initiatives

Perhaps in an effort to synchronously satisfy the need for both public literacy and dialogue, many recent science communication efforts (particularly in the UK) have emphasized participatory engagement [62]. Methods such as citizen science, open science, and crowdsourcing attempt to incorporate public expertise into the knowledge production process, often integrating new communication technologies like social media platforms, which allow for spontaneous public engagement in a number of different ways [63]. Some now see creative interaction *not* as simply a method of legitimizing science and policy but a major goal of science communication and a process capable of engendering a positive relationship between science and society regardless of quantitative outcome measures such as knowledge and affect [64].

What is so far unclear is the extent to which domain-specific knowledge or positive engagement with a specific science or innovation can truly diminish the spread of fake science. Trust and perceived literacy can easily amplify the spread of misinformation through social systems, particularly when misinformation and disinformation are presented in the familiar format of science communication. Perhaps then, scientists and their institution should get to know their enemy in this infodemic and establish why fake science is often more attractive to audiences than the real thing.

5.15.1 Framing the Story of Science

Science communication to publics beyond the scientific community refers to the art and technique of informing, influencing, and motivating individual, institutional, and public audiences with the dissemination of science information [65]. Poor science communication risks hindering scientific credibility in relation to other contextual forms of knowledge and enhancing the spread and acceptance of fake science. Therefore, scientists have an important role to play for limiting the spread of fake science. As experts in their field, scientists must be staunch gatekeepers, better managing the flow of empirical data from the lab through the media and to the public [66]. Moreover, scientists must uphold the scientific method for deciding what information is valid enough to be published within their community and communicated beyond their disciplines. However, if proper scientific governance, publishing and retraction guidelines, and peer review were sufficient to combat the spread of fake science, misinformation and misconceptions about science would not hold sway over the public imagination the way it has and does still. An appreciation of message effects and media psychology can potentially help scientists and their institution better frame the science emerging from their labs as well as ensuring its empirical validity.

5.16 To Win, Understand Your Enemy

A proven strategy for preventing the spread of fake science is to utilize narratives and tell stories using language that is engaging and easily understandable to general audiences rather than attempting to communicate abstract statistics without contextualizing data within the broader social environment [67, 68]. Some point to the relative persuasive appeal of emotionally driven storytelling over rational systems of thought and providing concrete episodic examples versus overarching themes, citing omnipresence of religious and superstitious thought across human history and cultures [69]. In their review of narrative persuasion [70], they provide evidence to support the propositions that absorption in a narrative and identification with it, or responses to characters in a story, can even suppress counterarguing where the information offered is counterattitudinal.

5.17 Persuasive Language: Ethos, Pathos, and Logos

Narrative may be particularly useful when attempting to counteract fake science or challenge widespread misconceptions such as those surrounding the efficacy of MMR vaccines [71]. Health communication studies in particular have effectively operationalized narrative forms of persuasion to achieve positive health outcomes across a number of health domains such as blood pressure [72] and childhood tooth decay [73]. Scientists may benefit from understanding not only the power of narrative but that narrative and context will likely be foisted on their results by the media that filter their results for public consumption regardless. It may behoove science and our broader society should scientists themselves better define the narrative and context associated with their results rather than deferring to journalists to package their findings.

Related to narrative and presenting persuasive, episodic, concrete examples to audiences is the concept of psychological distance. Concretizing arcane concepts through narrative can synchronously be viewed as reducing psychological distance to the relevant objects, people, or events. Psychological distance refers to the mutually associated dimensions of perceived temporal (e.g., how soon could COVID-19 affect me?), geographic (e.g., does COVID-19 affect my local region or just faraway someplace?), and social distances (e.g., does COVID-19 affect my age group or people similar to me?) and the dimension of hypotheticality (e.g., is contracting COVID-19 purely hypothetical or conversely very probable?) [74]. By reducing perceived distance to threats like climate change, one can prime a heightened perception of risk and consequently influence proactive behaviors like support for pro-environmental policies.

5.18 Immediacy and Personal Relevance

Kirkpatrick [56] details two experiments during which the lexical concreteness or abstractness of a fake science news headline was manipulated. Lexically concrete, specific messages primed psychological proximity to the threat over more general, abstract forms of communication, and, in turn, this perceived proximity heightened the perception that the threat was greater and the likelihood that the message recipient would share the science misinformation online [56].

Contextual models arise in large part in recognition of the diverse, interactive, context-dependent ways in which varied publics form attitudes and perceptions of science and technology, and not simply as results of their field-specific knowledge. In fact, perceptions of risk-benefit, individual differences, attitude, and value predispositions are the most often reported determinants of the public acceptance of new technologies [75]. Recent studies even hint at the important role that fear, anger, surprise, and disgust can play in informing not only judgments made as to the quality of scientific research but also subjective and objective understanding of science [76]. Risk-benefit perceptions in particular constitute an important contextual system that people rely on when making decisions about their environment and have therefore been studied widely in relation to the diffusion of science and technology. In a telephone survey [77], as perceptions of risks associated with nanotechnology increased, belief in its usefulness decreased, as did support for the industry surrounding the science's implementation.

5.19 Institutional Commitment

Interdisciplinary communication projects that include experts from academic disciplines such as communication, rhetoric, and computational sciences are necessary, so that scientific output might be appropriately translated into common languages and presented attractively via digital methods. These are stories that raw statistics cannot tell without interpretation, and stories that scientists have often neglected to tell appropriately much less *sell* to audiences.

References

1. Feezell JT, Wagner JK, Conroy M (2020) Exploring the effects of algorithm-driven news sources on political behavior and polarization. Comput Hum Behav 106626
2. Walter N, Ball-Rokeach SJ, Xu Y, Broad GM (2018) Communication ecologies: analyzing adoption of false beliefs in an information-rich environment. Sci Commun 40(5):650–668
3. Goode L (2009) Social news, citizen journalism and democracy. New Media Soc 11(8):1287–1305
4. Dunwoody S (2021) Science journalism: prospects in the digital age. In: Bucchi M, Trench B (eds) Routledge handbook of public communication of science and technology. Routledge
5. Lahouati M, De Coucy A, Sarlangue J, Cazanave C (2020, July 15) Spread of vaccine hesitancy in France: What about YouTube™? *Vaccine* [Internet]. [cited 2020 Jul 20]. http://www.sciencedirect.com/science/article/pii/S0264410X20309026
6. Maertens R, Anseel F, van der Linden S (2020) Combatting climate change misinformation: evidence for longevity of inoculation and consensus messaging effects. J Environ Psychol 70:101455

7. Hmielowski JD, Kirkpatrick AW, Boyd AD (2020) Understanding public support for smart meters: media attention, misperceptions, and knowledge. J Risk Res:1–17
8. Greenspan RL, Loftus EF (2021) Pandemics and infodemics: research on the effects of misinformation on memory. Hum Behav Emerg Technol 3(1):8–12
9. Chu H, Yuan S, Liu S (2021) Call them COVIDiots: exploring the effects of aggressive communication style and psychological distance in the communication of COVID-19. Public Underst Sci 30:240–257. https://doi.org/10.1177/0963662521989191
10. McIntyre L (2018) Post-truth [Internet]. MIT Press; [cited 2021 Apr 20]. 240 p. The MIT Press essential knowledge series. https://mitpress.mit.edu/books/post-truth
11. Funk C, Hefferon M, Kennedy B, Johnson C (2019) Trust and mistrust in Americans' views of scientific experts [Internet]. Pew Research Center Science & Society. [cited 2019 Dec 29]. https://www.pewresearch.org/science/2019/08/02/trust-and-mistrust-in-americans-views-of-scientific-experts/
12. Gauchat G (2012) Politicization of science in the public sphere: a study of public trust in the United States, 1974 to 2010. Am Sociol Rev 77(2):167–187
13. Howell EL, Wirz CD, Scheufele DA, Brossard D, Xenos MA (2020) Deference and decision-making in science and society: how deference to scientific authority goes beyond confidence in science and scientists to become authoritarianism. Public Underst Sci 29(8): 800–818. https://doi.org/10.1177/0963662520962741
14. Tennant JP, Ross-Hellauer T (2020) The limitations to our understanding of peer review. Res Integr Peer Rev 5(1):6
15. Carver J, VanVoorhis J, Basili V. Understanding the impact of assumptions on experimental validity. In: Proceedings 2004 international symposium on empirical software engineering, 2004 ISESE '04. 2004. pp 251–60
16. Bode L, Vraga EK, Tully M (2020) Correcting misperceptions about genetically modified food on social media: examining the impact of experts, social media heuristics, and the gateway belief model. Sci Commun 43. https://doi.org/10.1177/1075547020981375
17. Grimes DR, Bauch CT, Ioannidis JPA (2018) Modelling science trustworthiness under publish or perish pressure. R Soc Open Sci 5(1):171511
18. Coggburn JD, Neely SR (2015) Publish or perish? Examining academic tenure standards in public affairs and administration programs. J Public Aff Educ 21(2): 199–214
19. Yan J, MacDonald A, Baisi L-P, Evaniew N, Bhandari M, Ghert M (2016) Retractions in orthopaedic research: a systematic review. Bone Jt Res 5(6): 263–268
20. Smaldino PE, McElreath R (2016) The natural selection of bad science. R Soc Open Sci 3(9):160384
21. Shrout PE, Rodgers JL (2018) Psychology, science, and knowledge construction: broadening perspectives from the replication crisis. Annu Rev Psychol 69(1): 487–510
22. Hopf H, Krief A, Mehta G, Matlin SA (2019) Fake science and the knowledge crisis: ignorance can be fatal. R Soc Open Sci 6(5):190161
23. Clemons M, de Costa e Silva M, Joy AA, Cobey KD, Mazzarello S, Stober C et al (2017) Predatory invitations from journals: more than just a nuisance? Oncologist 22(2):236–240
24. Cobey KD, Lalu MM, Skidmore B, Ahmadzai N, Grudniewicz A, Moher D (2018) What is a predatory journal? A scoping review. F1000Research [Internet] 7. https://www.ncbi.nlm.nih.gov/pmc/articles/PMC6092896/
25. Forero DA, Oermann MH, Manca A, Deriu F, Mendieta-Zerón H, Dadkhah M et al (2018) Negative effects of "predatory" journals on global health research. Ann Glob Health 84(4):584–589
26. Beall J (2012) Predatory publishers are corrupting open access. Nat News 489(7415):179
27. Vervoort D, Ma X, Shrime MG (2020) Money down the drain: predatory publishing in the COVID-19 era. Can J Public Health 111(5):665–666
28. Wakefield A, Murch S, Anthony A, Linnell J, Casson D, Malik M et al (1998) RETRACTED: Ileal-lymphoid-nodular hyperplasia, non-specific colitis, and pervasive developmental disorder in children. Lancet 351(9103):637–641
29. Bowes J (2016) Measles, misinformation, and risk: personal belief exemptions and the MMR vaccine. J Law Biosci 3(3):718–725
30. Hallman WK (2017) What the public thinks and knows about science—and why it matters. In: Jamieson KH, Kahan DM, Scheufele D (eds) The Oxford handbook of the science of science communication. Oxford University Press, Oxford, pp 61–72
31. National Science Board (2018) Science and Technology: Public Attitudes and Public Understanding [Internet]. National Science Board (Science & Engineering Indicators 2018). https://nsf.gov/statistics/2018/nsb20181/report/sections/science-and-technology-public-attitudes-and-understanding/highlights
32. Sundstrom B (2016) Mothers "google it up:" extending communication channel behavior in diffusion of innovations theory. Health Commun 31(1): 91–101
33. Rice RE (2017) Intermediality and the diffusion of innovations. Hum Commun Res 43(4):531–544
34. McCombs M (2018) Agenda-setting. In: The Blackwell encyclopedia of sociology [Internet]. American Cancer Society, pp 1–2. https://onlinelibrary.wiley.com/doi/abs/10.1002/9781405165518.wbeosa025.pub2
35. Gentzkow MA, Shapiro JM (2004) Media, education and anti-Americanism in the Muslim world. J Econ Perspect 18(3):117–133

36. Garrett RK, Weeks BE, Neo RL (2016) Driving a wedge between evidence and beliefs: how online ideological news exposure promotes political misperceptions. J Comput-Mediat Commun 21(5): 331–348
37. Stempel C, Hargrove T, Stempel GH (2007) Media use, social structure, and belief in 9/11 conspiracy theories. Journal Mass Commun Q 84(2):353–372
38. Shearer E, Mitchell A (2021) News use across social media platforms in 2020 [Internet]. Pew Research Center's Journalism Project. [cited 2021 Apr 20]. https://www.journalism.org/2021/01/12/news-use-across-social-media-platforms-in-2020/
39. Coddington M, Molyneux L, Lawrence RG (2014) Fact checking the campaign: how political reporters use twitter to set the record straight (or not). Int J Press 19(4):391–409
40. The New York Times (2021) Corrections [Internet]. *The New York Times*. [cited 2021 Apr 27]. https://www.nytimes.com/section/corrections
41. Hasher L, Goldstein D, Toppino T (1977) Frequency and the conference of referential validity. J Verbal Learn Verbal Behav 16(1):107–112
42. Pennycook G, Cannon TD, Rand DG (2018) Prior exposure increases perceived accuracy of fake news. J Exp Psychol Gen 147(12):1865–1880
43. Thorson E (2016) Belief echoes: the persistent effects of corrected misinformation. Polit Commun 33(3): 460–480
44. Skurnik I, Yoon C, Park DC, Schwarz N (2005) How warnings about false claims become recommendations. J Consum Res 31(4):713–724
45. Jamieson KH, Albarracín D (2020 Apr) The relation between media consumption and misinformation at the outset of the SARS-CoV-2 pandemic in the US. Harv Kennedy Sch Misinformation Rev [Internet] 1(3) https://misinforeview.hks.harvard.edu/article/the-relation-between-media-consumption-and-misinformation-at-the-outset-of-the-sars-cov-2-pandemic-in-the-us/
46. Iyengar S, Massey DS (2019) Scientific communication in a post-truth society. Proc Natl Acad Sci 116(16):7656–7661
47. Bond S. Facebook (2021) Widens ban on COVID-19 vaccine misinformation in push to boost confidence [Internet]. *NPR.org*. [cited 2021 Apr 28]. https://www.npr.org/2021/02/08/965390755/facebook-widens-ban-on-covid-19-vaccine-misinformation-in-push-to-boost-confiden
48. Livingstone S (2004) Media literacy and the challenge of new information and communication technologies. Commun Rev 7(1):3–14
49. Barthel M, Mitchell A, Holcomb J (2016) Many Americans believe fake news is sowing confusion [Internet]. Pew Research Center's Journalism Project. [cited 2021 Apr 20]. https://www.journalism.org/2016/12/15/many-americans-believe-fake-news-is-sowing-confusion/
50. Feinstein N (2011) Salvaging science literacy. Sci Educ 95(1):168–185
51. The Royal Society (1985) The public understanding of science [Internet]. London, The Royal Society, p 46. https://royalsociety.org/~/media/Royal_Society_Content/policy/publications/1985/10700.pdf
52. Bucchi M (2008) Of deficits, deviations and dialogues: Theories of public communication of science. In: Bucchi M, Trench B (eds) Handbook of public communication of science and technology [Internet], 1st edn. Routledge, London, UK, pp 71–90. https://www.taylorfrancis.com/books/e/9780203928240
53. Miller S (2001) Public understanding of science at the crossroads. Public Underst Sci 10(1):115–120
54. Wynne B (1992) Misunderstood misunderstanding: social identities and public uptake of science. Public Underst Sci 1(3):281–304
55. Takahashi B, Tandoc EC (2016) Media sources, credibility, and perceptions of science: Learning about how people learn about science. Public Underst Sci 25(6): 674–690
56. Kirkpatrick AW (2021) The spread of fake science: lexical concreteness, proximity, misinformation sharing, and the moderating role of subjective knowledge. Public Underst Sci 30(1):55–74
57. Koetke J, Schumann K, Porter T (2021) Intellectual humility predicts scrutiny of COVID-19 misinformation. Soc Psychol Personal Sci 13:277–284
58. Rowe G, Frewer LJ (2005) A typology of public engagement mechanisms. Sci Technol Hum Values 30(2):251–290
59. House of Lords Select Committee on Science and Technology. Science and Technology [Internet]. (2000) London, UK: UK Parliament. [cited 2020 Feb 24]. Report No.: 3. https://publications.parliament.uk/pa/ld199900/ldselect/ldsctech/38/3801.htm
60. Wynne B (2006) Public engagement as a means of restoring public trust in science – hitting the notes, but missing the music? Public Health Genomics 9(3): 211–220
61. Miller JD (2004) Public understanding of, and attitudes toward, scientific research: what we know and what we need to know. Public Underst Sci 13(3): 273–294
62. Einsiedel EF (2014) Publics and their participation in science and technology: Changing roles, blurring boundaries. In: Bucchi M, Trench B (eds) Routledge handbook of public communication of science and technology [Internet], 2nd edn. Routledge, pp 125–139. https://www.taylorfrancis.com/books/e/9780203483794
63. Hetland P (2016) Models in science communication policy. Nord J Sci Technol Stud 2(2):5
64. Wilkinson C, Weitkamp E (2016) Creative research communication: theory and practice. Manchester University Press, Manchester, 347 p
65. US National Center for Health Statistics, editor (2012) Healthy People 2010: final review. Hyattsville, MD: U.S. Dept. of Health and Human Services, Centers for

Disease Control and Prevention, National Center for Health Statistics; 1 p. (DHHS publication)
66. Tattersall A (2018) New research must be better reported, the future of society depends on it [Internet]. The Conversation. [cited 2021 Apr 21]. http://theconversation.com/new-research-must-be-better-reported-the-future-of-society-depends-on-it-87407
67. Chen M, Bell RA (2022) A meta-analysis of the impact of point of view on narrative processing and persuasion in health messaging. Psychol Health 37(5): 545–562
68. Goldstein CM, Murray EJ, Beard J, Schnoes AM, Wang ML (2020) Science communication in the age of misinformation. Ann Behav Med 54(12):985–990
69. Epstein S (1994) Integration of the cognitive and the psychodynamic unconscious. Am Psychol 49(8): 709–724
70. Slater MD, Rouner D (2002) Entertainment—education and elaboration likelihood: understanding the processing of narrative persuasion. Commun Theory 12(2):173–191
71. Duchsherer A, Jason M, Platt CA, Majdik ZP (2020) Immunized against science: narrative community building among vaccine refusing/hesitant parents. Public Underst Sci 29(4):419–435
72. Houston TK, Allison JJ, Sussman M, Horn W, Holt CL, Trobaugh J et al (2011) Culturally appropriate storytelling to improve blood pressure. Ann Intern Med 154(2):77–84
73. Balint KE, Das E, Stel G, Hoppener M (2021) Can a funny story about tooth brushing decrease plaque scores in children? A longitudinal field experiment. Health Commun 37(7):802–812
74. Trope Y, Liberman N (2011) Construal level theory. In: Lange PAMV, Kruglanski AW, Higgins ET (eds) Handbook of theories of social psychology: collection: Volumes 1 & 2. SAGE, London, pp 118–134
75. Gupta N, Fischer ARH, Frewer LJ (2012) Socio-psychological determinants of public acceptance of technologies: a review. Public Underst Sci 21(7): 782–795
76. Drummond C, Fischhoff B (2020) Emotion and judgments of scientific research. Public Underst Sci 00(0):1–16
77. Cacciatore MA, Scheufele DA, Corley EA (2011) From enabling technology to applications: the evolution of risk perceptions about nanotechnology. Public Underst Sci 20(3):385–404

6 Policies and Ethical Challenges in Social Science Research

Henry Poduthase and Lisa Garza

Abstract

Scientific misconduct is adversely affecting the entire scientific community and threatens the trust of the scientifically published research outcomes. These behaviors in the forms of fabrication, falsification, or plagiarism must be seen seriously by the research community. This chapter discusses about various scientific misconducts in social science research in general and provides details about the core values of research, various forms of scientific misconduct, and ethical challenges of social science research. Furthermore, the authors discuss about various steps in place to address these fraudulent activities in the current system. Finally, this chapter talks about the importance of understanding fraudulent scientific behaviors and the various measures that should be in place to ensure quality in research.

Only when certain events recur in accordance with rules or regularities, as in the case of repeatable experiments, can our observations be tested—in principle—by anyone.... Only by such repetition can we convince ourselves that we are not dealing with a mere isolated 'coincidence,' but with events which, on account of their regularity and reproducibility, are in principle inter-subjectively testable.—Karl Popper [1, p. 45]

6.1 Introduction

Social science research is the scientific inquiry encompassing an extensive range of systematic activities that are poised toward creating new knowledge or improving the existing knowledge and understandings. The effectiveness and authenticity of this scientific knowledge, for the major part, rest on how honestly, fairly, and accurately the researcher has undertaken the research. When the researchers have shown scientific dishonesty and lack of integrity, the impact falls upon the entire scientific community as well as that on the consumers of this information.

According to the US Department of Agriculture, "Scientific integrity is the condition resulting from adherence to professional values and practices when conducting, reporting, and applying the results of scientific activities that ensures objectivity, clarity, and reproducibility, and that provides insulation from bias, fabrication, falsification, plagiarism, inappropriate influence,

H. Poduthase (✉)
West Texas A&M University, Canyon, TX, USA
e-mail: hpoduthase@wtamu.edu

L. Garza
Department of Psychology, Sociology, and Social Work, West Texas A&M University, Canyon, TX, USA
e-mail: lgarza@wtamu.edu

J. Faintuch, S. Faintuch (eds.), *Integrity of Scientific Research*,
https://doi.org/10.1007/978-3-030-99680-2_6

political interference, censorship, and inadequate procedural and information security" [2].

6.2 Core Values of Research

Fundamentally, research should be rooted in various commonly accepted norms, the prominent ones being those of communalism, universalism, disinterestedness, and organized skepticism [3]. While communalism promotes the common ownership of scientific knowledge, universalism promotes the freedom for any scientist to contribute to the existing knowledge. Disinterestedness points toward the cautious behavior expected from the scientist to avoid personal gain and to stand for the good of the scientific knowledge. Finally, organized skepticism promotes the role other scientists could critically examine the authenticity of the newly produced scientific information.

The conceptualization of the core values of research identified these six values: (1) objectivity, (2) honesty, (3) openness, (4) accountability, (5) fairness, and (6) stewardship. When researchers are aware about these core values and perform scientific inquiry within the parameters of ethical principles and values, there will be exceptional creation of information [3].

Objectivity Objectivity refers to scientific claims, methods, and results which should be independent of all influences and personal biases of the scientists themselves. Ideally, the core of scientific inquiry should be to develop or improve scientific knowledge in the specific area of research [4].

Honesty As a universal human value, honesty has a deep impact in scientific inquiry. It is the responsibility of the researchers to be honest in the freedom they enjoy advancing knowledge, and this is actually the foundation of all other values.

Openness Researchers should be transparent in presenting all the relevant information that led them to come into a conclusion or outcome. This is critical in understanding what the researcher concluded and if the results could be replicated or verified [5].

Accountability It directly refers to the trustworthiness of the research conducted as well as the willingness to demonstrate the validity of the work implemented. The researchers should be responsible for the work, statements, actions, data collected, and end results.

Fairness "Fairness in adhering to explicit criteria and processes reinforces a system in which the core values can operate and trust among the parties can be maintained" [3].

Stewardship This is the commitment of the researchers to their fellow researchers and aspiring researchers. A good stewardship implies being helpful and providing service to the fellow researchers and the institution.

6.2.1 Scientific Misconduct

Scientific misconduct could be either intentional or incidental. In both cases, the result is directly or indirectly affecting the scientific community as well as that of the consumers of the research. Such behavior sometimes overlaps with *SPIN*. SPIN refers to "specific intentional or unintentional reporting that fails to faithfully reflect the nature and range of findings and that could affect the impression the results produce in readers" [6]. Even if the researcher began with a good intention, data manipulation or faulty data analysis eventually creates an outcome that is highly influenced by biases. Of course unintentionally created scientific errors might also influence the final report. When the spin occurs, if it is based on the vested interests of the author due to financial, intellectual, academic, or other personal biases, that becomes more complicated as this is a deliberate act by the researcher [7].

6.2.1.1 A Protean Question

According to the US Office of Science and Technology (OSTP), research misconduct is the "fabrication, falsification, or plagiarism in proposing, performing or reviewing research or in reporting research results." Lalchchandama [8] states about Yoshihiro Sato, a Japanese bone health researcher who plagiarized, fabricated data, and forged authorships. The extent of his scientific fraud activities spans from 1996 to 2013, leading to the retraction of more than 60 studies from the scholarly literature. When Grey and his colleagues at the University of Auckland in New Zealand analyzed Sato's work, they found that university investigations into research misconduct on many occasions are inadequate, opaque, and poorly conducted [9]. Hence, there should be procedures in place to understand, evaluate, and restrict or eliminate scientific misconduct in research.

6.3 Understanding Scientific Misconduct in Social Sciences

The existing policies and procedures in the scientific community emphasize accepted training practices such as the Collaborative Institutional Training Initiative (CITI), Institutional review board (IRB) approval, and peer review.

6.3.1 Falsification

Falsification is regarded as one of the cardinal sins of research where research materials, equipment, or processes are manipulated, and data are omitted, or results are not accurately represented in the final report [10]. There are several elements to falsification, from the beginning of the research to the final report. We must understand the prevalence of falsification from the tenure pressure experienced by the professors to the financial biases of various institutions. The fear of negative results to a statistical testing would lead the authors to manipulate the data so as to prove their hypothesis. One of the major challenges of social science theories is that they are often loosely constructed and could be stretched to fit various sets of experimental results [11]. Thus, falsification is a major challenge for social science research.

6.3.2 Fabrication

Scientists are seekers of truth who pursue scientific methodologies to come into a conclusion. However, there are several scandals including that of the South Korean geneticist Hwang Woo-Suk's fake stem cell lines and that of the physicist at Bell Labs in Murray Hill (NJ, USA) Jan Hendrik Schon's duplicated graphs, which depict how well data could be fabricated and later published in reputed journals that theoretically have several steps in place to identify fraudulent behaviors [12]. In the research conducted by Daniele Fanelli (2009), it was found that 2% of the scientists admit serious forms of misconduct. When researchers deliberately present and report fake or non-existent procedures, data, and findings, they are committing fabrication [13].

6.3.3 Plagiarism

Citing the source is an essential component in research when the author provides that details in a particular paper are taken from another publication. This is a proper way to acknowledge the work of others and respect their contribution. Citing gives the reader an idea about the emergence and the background of the researcher's idea development. Plagiarism is defined as using someone's ideas, statements, or writing style without recognizing the original contributor [14]. There are several forms of plagiarism [15]:

1. Direct form—Complete or partial copy of text without mentioning the source
2. Mosaic form—Borrowing assorted ideas and opinions from other source without properly acknowledging it

3. Self-plagiarism—Reusing own work without stating about the primary source

6.4 Some Ethical Challenges to Social Research

A sociology doctoral student was eager to participate in a disaster field research study—a project of great importance and one where she had the privilege of working with faculty who were well known in their respective field of study. She thought she was well prepared to be in the field and was excited and ready to collect data, conduct interviews, and make observations. She understood the challenges of research and the code of ethics researchers abide by; however, within a couple of days, she found all she had learned in her research methods classes would be tested. "Susie," a subject who was to be observed came to her and was clearly in need of immediate assistance. Susie's face, hair, and clothes were dirty and messy—she was upset and exhausted as she and her family needed food, water, clean clothing, and a place to sleep.

Susie was desperate as she had been turned away from other agencies and had been looking for supplies—anything to help the family who needed resources during a difficult and devastating disaster. The graduate student's personal values of compassion and social justice immediately kicked in as she attempted to assist Susie. She was later informed that her role was that of a "researcher" who was collecting data, and she was instructed that she could not assist. She reported being devastated and discouraged. Her commitment to humanity and social justice came through, and despite all she had learned in the classroom, she struggled to understand her role as a researcher.

This scenario demonstrates just one of an array of challenges and ethical pitfalls that social scientists encounter when collecting data—bias, concern for participants, and issues of validity to name a few.

6.4.1 Influence of Bias

When collecting data, it is imperative that the researcher maintain objectivity. This might create challenges when we are pulled in or encounter unexpected circumstances concerning what we are studying. The graduate student is now feeling for Susie. That interest in our chosen subject could potentially lead to issues of bias and a temptation to distort what we are studying, in other words, the possibility of "going native." How we phrase questions on a survey, how we ask questions on an interview schedule, our facial expressions, and our tone, all can give participants clues on our personal beliefs about that topic, thus possibly impacting the validity of our research. While it is almost impossible for researchers to maintain perfect objectivity, if we are attentive of how and why we are influenced by our social and political views, we can avoid bias in our research.

Of course humanitarian questions and other unforeseen matters can be pressing and for moral and ethical reasons should not be overlooked, however, preferentially by external persons or agencies not connected with the research project.

Another specific example of bias in qualitative research often occurs when observing social interactions or social phenomena. Researchers may record not only what they are observing but may inadvertently record their personal interpretation and connect that to the data they are collecting. We should remain mindful when recording observations that we record just that—what we have observed, not what we think we have observed.

6.4.2 Scientific Validity

According to Bordens and Abbott [16], validity can be defined as the extent to which one is measuring what one intends to measure. Validity can be examined in several ways, for example, construct, internal, and external validity. When conducting a study, the researcher carefully follows the steps in the research process: developing questions; reviewing of the relevant literature; formulating hypotheses, methodology, or research design; sampling; collecting data; analyzing the data; and reporting findings. One could say that these steps are a bit easier when the study is quantitative in nature rather than qualitative. When qualitative research has been chosen, the design can be more forgiving. As Chenail [17] reminds us, changing the research question once we are in the field is acceptable; however, he points out that the researchers must be aware they have adjusted their study. This endeavor may become the moment that bias will creep into a well-intended project.

Referring to the opening scenario, if the graduate student takes into consideration Susie's situation, then the construction of the research questions and the goal of the research have become impacted or potentially changed. It is imperative to self-reflect throughout the research process and to ensure research questions align with the methodology that has been chosen and be certain to adhere to the goals and objectives of the study. Incidental or unplanned events such as that have to be addressed in ways as transparent, ethical, and effective as possible, without interfering with or derailing the protocol.

6.4.3 Maintaining Methodological Rigor

Lincoln and Guba [18] maintain the reliability or "trustworthiness" of qualitative research is achieved using the following criteria: credibility, transferability, dependability, and confirmability. Credibility refers to the confidence in the findings of the research. Transferability simply refers to the consistency in findings of the current study to similar studies. Dependability can be described as the findings being consistent if repeated. Finally, confirmability ensures the findings are shaped by the data collected from the participants, not the research bias. Referring to the opening scenario, in order for the eager graduate student to maintain methodological rigor, she will need to consider persistent observations, triangulation, and peer briefing. Being aware of one's reactions to research participants or subjects and having knowledge of one's own beliefs will not hamper the management of emergencies and will ensure validity and reliability of the study.

6.4.4 Reactive Nature of Field Research

Another possible pitfall the researchers may find themselves in is the reaction the participants may have to the research. The participants may find themselves attempting to give information they think we want to hear. Looping back to the opening situation, was Susie truly in dire need? Researchers cannot make assumptions that participants are passive and that their behavior or responses are a normal representative [16].

6.4.5 Ethical Protection of Participants

It must be ensured that participants fully understand the research and the implications on them. The informed consent document should be clear, detailed, and easy to read, and participants need to be fully aware that they have autonomy and can leave the study at any time. Confidentiality and anonymity of course are to be strictly maintained.

6.4.5.1 Ethical Repercussions on the Investigator

What happens when the research topic itself or the participants illicit an emotional reaction in the one collecting the data? The relationship between the researcher and participants may very well be impacted by the race, ethnicity, gender, and age of the one collecting the data [19]. Too often when conducting research, the experiences of the researcher are scarcely addressed. This relationship poses an important question: can the researcher, if negatively impacted by the participants or the topic itself, leave the study? According to Denzin and Giardina, the answer to this question is yes [20]. If the struggle between continuing the study and the individual ethics of the researcher is significant, it is acceptable for that individual to withdraw.

6.5 Measures in Place to Address Scientific Misconduct

The first formal governmental definition of research misconduct was published in the Public Health Service Interim Policy of 1985 in the United States. However, the Health Research Extension Act of 1985 required government and universities to ensure more aggressive approach in identifying research misconduct and to eliminate them. Over the years, major checks were put in place to identify and eliminate fraudulent behavior within the scientific research field [21].

6.5.1 Institutional Review Board (IRB)

According to the US Food and Drug Administration (FDA), "IRB is an appropriately constituted group that has been formally designated to review and monitor biomedical research involving human subjects. In accordance with FDA regulations, an IRB has the authority to approve, require modifications in (to secure approval), or disapprove research." Hence, IRB protects the rights and welfare of the research participants and ensures all necessary steps are taken during the process of data collection to create a safe or low risk for the participants.

6.5.2 Training

Providing training is essential in creating and reinforcing the fundamentals of research ethics and values. A training program organized by the Collaborative Institutional Training Initiative (USA) in the year of 2000 is an excellent example of such an effort. The Course Catalog organized by CITI covers multiple topics about ethics and compliance and is subscribed by over 2500 organizations [22].

6.5.3 Peer Review

Peer review conducted by scientific journals is critical as papers are evaluated by an independent expert who can analyze and critique about the research output of the author. The mostly blind peer reviewers check the manuscript for accuracy, validity of the methodology, and also ethics and would recommend appropriate revisions or rejection of the paper.

6.6 Concluding Remarks

There are several historical evidences of unethical research in the United States. The most discussed is the 1932 Tuskegee syphilis study. This protocol was funded by the US Public Health Service, and 400 black men were the victims of untreated syphilis in an infamous episode of scientific misconduct. It eventually triggered the creation of the National Center for Bioethics in Research and Health Care at Tuskegee University (Tuskegee, AL, USA). Disadvantaged populations are mostly exploited in these questionable research studies. When we come into the realm of falsification, fabrication, and plagiarism, instances are quite widespread around the world. In social sciences, the likelihood of conducting fraudulent behavior is high, as social scientists in most part deal with qualitative variables rather than quantifiable ones. By their nature, these are less easy to track and exactly replicate, as circumstances could somewhat change from one setting to another. Hence, fundamental values like objectivity, honesty, openness, fairness, accountability, and stewardship should be the guiding principles for every researcher. Furthermore, institutional steps in place should not be neglected, in order to ensure that suspected misconduct is promptly identified and investigated.

References

1. Popper K (1959) The logic of scientific discovery. Hutchison, London

2. U.S. Department of Agriculture (n.d.) Scientific integrity and research misconduct. usda.gov/our-agency/staff-offices/office-chief-scientist-ocs/scientific-integrity-and-research-misconduct#:~:text=Research%20misconduct%20%2D%20a%20subset%20of,or%20in%20reporting%20research%20results. Accessed 22 May 2021
3. National Academies of Sciences, Engineering, and Medicine (2017) Fostering integrity in research. The National Academies Press, Washington, DC
4. Reiss J, Sprenger J (2020) Scientific objectivity. In The Stanford encyclopedia of philosophy (winter edition). https://plato.stanford.edu/cgi-bin/encyclopedia/archinfo.cgi?entry=scientific-objectivity. Accessed 22 May 2021
5. Levin N, Leonelli S, Weckowska D, Castle D, Dupre J (2016) How do scientists define openness? Exploring the relationship between open science policies and research practice. Bull Sci Technol Soc 36(2):128–141
6. Boutron I, Ravaud P (2018) Misrepresentation and distortion of research in biomedical literature. Proc Natl Acad Sci 115(11):2613–2619. https://doi.org/10.1073/pnas.1710755115
7. Fletcher RH, Black B (2007) "Spin" in scientific writing: scientific mischief and legal jeopardy. Med Law 26(3):511–525
8. Lalchchandama K (2020) The making of scientific ethics-lessons from unethical conducts. Sci Vis 20 (3):118–131
9. Grey A, Bolland M, Gamble G, Avenell A (2019) Quality of reports of investigations of research integrity by academic institutions. Res Integrity Peer Rev 4: 3. https://doi.org/10.1186/s41073-019-0062-x
10. U.S. Office of Science and Technology Policy (1999) Fact sheet. clintonwhitehouse3.archives.gov/WH/EOP/OSTP/html/9910_20_2.html. Accessed 22 May 2021
11. Folger R (1989) Significance tests and the duplicity of binary decisions. Psychol Bull 106:155–160. https://doi.org/10.1037/0033-2909.106.1.155
12. Saunders R, Savulescu J (2008) Research ethics and lessons from Hwanggate: what can we learn from the Korean cloning fraud? J Med Ethics 34(3):214–221
13. Bos J (2020) Fabrication and cheating. In: Research ethics for students in the social sciences. Springer, Cham. https://doi.org/10.1007/978-3-030-48415-6_5
14. Masic I (2014) Plagiarism in scientific research and publications and how to prevent it. Materia sociomedica 26(2):141–146. https://doi.org/10.5455/msm.2014.26.141-146
15. Begovic E (2014) My view on plagiarism. Acta Informatica Medica 22(2):145–146
16. Bordens KS, Abbott BB (2018) Research design and methods: a process approach, 10th edn. McGraw-Hill, New York, NY
17. Chenail RJ (2011) Ten steps for conceptualizing and conducting qualitative research studies in a pragmatically curious manner. Qual Rep 16(6):1713–1730
18. Lincoln YS, Guba EG (1985) Naturalistic inquiry. Sage Publications, Newbury Park, CA
19. Aluwihare-Samaranayake D (2012) Ethics in qualitative research: a view of the participants' and researchers' world from a critical standpoint. Int J Qual Methods 11(2):64–81
20. Denzin NK, Giardina MD (2007) Decolonizing and politics of knowledge: ethical futures in qualitative research. Left Coast Press, Walnut Creek, CA
21. The Office of Research Integrity (ORI 2.0) – Historical background. ori.hhs.gov/historical-background. Accessed 22 May 2021
22. CITI Program (n.d.). about.citiprogram.org/. Accessed 22 May 2021

Pseudoscience During the COVID-19 Pandemic

7

Yutori Takai and Kenji Matsui

Abstract

The COVID-19 pandemic remarkably changed the landscape of scientific research. Preprint and open-access publication culture has been accelerated, low-quality research has been disseminated, and scientists have been forced to counteract the pseudoscience amid the infodemic. This paper discusses scientific discourses that were originally legitimate scientific research but have been "pseudoscientified" amid the pandemic by the media and politicians. This phenomenon, called pseudoscientification of science, had structural backgrounds which existed before the pandemic. Through an examination of examples, we identified their structural problems: constant pressure on researchers to publish, eye-catching press releases, and report approval systems that give the appearance of completely proven effectiveness. Through such measures, politicians cherry-pick the results of scientific studies and try to retain public support by obtaining emergency approvals for drugs; such misconducts have led to the pseudoscientification of science during the pandemic. To counteract these issues, we need to first solve these structural problems.

7.1 Introduction

The COVID-19 pandemic accelerated preprint and open-access publication, and low-quality research has been disseminated worldwide. In particular, the retractions that occurred in *The Lancet* and *The New England Journal of Medicine* (*NEJM*), two prestigious medical journals, were widely received as a warning about the state of publishing that has neglected the various procedures to ensure the credibility of science. In addition, COVID-related pseudoscientific discourse was widely spread by the high-speed information distribution network. As the "pseudoscience" has spread through social sites, the situation has been described as the "first post-truth pandemic," and the resulting social response has been recognized as a public health issue that was just as important as the spread of coronaviruses [1]. These events lead us to insist on what has always been obvious: science, not pseudoscience, and high-quality science, not low-quality science.

The alarm may cloud the existence of a structural background for understanding the current state of science. The naivety of such a warning

Y. Takai
Gunma University, Ishikawa, Japan
e-mail: yutori-t@gunma-u.ac.jp

K. Matsui (✉)
Division of Bioethics & Healthcare Law, The National Cancer Center, Tokyo, Japan
e-mail: kematsui@ncc.go.jp

J. Faintuch, S. Faintuch (eds.), *Integrity of Scientific Research*,
https://doi.org/10.1007/978-3-030-99680-2_7

stems from the assumption that science and pseudoscience can be easily dichotomized, whereas the existence of a "demarcation problem" has become common sense in the current philosophy of science. Rather we would like to discuss the scientific discourses that are originally a legitimate kind of scientific research but have been "pseudoscientified" under the pandemic conditions by the media and politicians. COVID-related research was often distorted, exaggerated, and circulated as a kind of pseudoscience through the manipulation of politicians advocating (or criticizing) the fight against the pandemic and through the mass media interested in easy-to-understand innovative findings. This fact teaches us that the boundary between science and pseudoscience can be blurred by the way they are interpreted and communicated.

Furthermore, the perception of an "emergency" situation that the warning guideline assumes masks the structural problems involved in science that existed prior to the pandemic. These problems are the pressure on scientists to publish, which are plagued by a constant lack of funding; the resulting exhaustion of the peer review system; the overflow of predatory journals; and the society's expectation of instant practical benefits. The COVID-19 pandemic conditions have expanded and exposed these structural problems. Thus, if we understand the retraction cases mentioned earlier as exceptionally created by the "emergency situation," we will not be able to see the real problem.

7.2 The COVID-19 Pandemic and Non-science

The COVID-19 pandemic is considered "the first pandemic of the post-truth era" and has spawned a plethora of pseudoscientific discourse falsifying science [2, 3]. As the World Health Organization (WHO) warned, dealing with pseudoscience is becoming a public health issue [4]. The discourse about the origin of the virus stands out. Despite mounting evidence to the contrary, the pseudoscientific belief that viruses are man-made is still circulating and believed by a significant population in some countries [5, 6]. In addition, conspiracy theories claimed that 5G radio waves caused the spread of the infection [7]. Many unscientific discourses based on the so-called traditional medicine and non-Western medicine were also produced during the pandemic. Since COVID-19 patients often spontaneously heal, it is easy to claim that such medicine works [8]. Similar cases came about from the so-called "Western" medicine, too. For example, a paper from India pointing out the similarity between the human immunodeficiency viruses (HIV) and the new coronavirus was published on the preprint server *bioRxiv*, however immediately criticized and withdrawn [9, 10].

The rapid expansion of non-peer-reviewed preprints has led to the circulation of many low-quality researches worldwide; to some extent, science itself has broken down the line between what is and is not science [11, 12].

There has long been a shared consensus in the field of philosophy of science that it is not possible to conceptually organize a strict division between what is and is not science [13]. Yet, various pragmatic classifications continue to be attempted [14]. The mass preprint publication framework broke down the standards of scientific practice in a different way, as science itself gave up some of its standards and opened the door to pseudoscience. COVID-19-related research retractions became prominent [15], and the conditions that distinguish science from non-science are partly threatened from within [16].

7.3 The Pseudoscientification of Science

In this paper, the term "pseudoscientification of science" refers to the phenomenon where researches that aren't unscientific in themselves have been circulated as pseudoscientific discourse, because they have been transformed and conveyed through the relationship of political forces and mass media and their respective interests.

7.3.1 Hydroxychloroquine

On March 28, 2020, the US Food and Drug Administration (FDA) granted emergency approval to the anti-malarial drug hydroxychloroquine for the treatment of COVID-19, even though there was no strong evidence for the efficacy of the drug at the time. This authorization is rumored to be a response to political pressure from the White House [17]. The only evidence was a paper "published" on YouTube on March 16, and March 20, 2020. It is true that the decision was not necessarily "unscientific," since the United States does not require a rigorous scientific basis within certain contexts. However, it was criticized by scientists [18, 19]. In celebration of this emergency approval, President Trump called the drug a "game changer" [20].

It was not considered a problem that the efficacy and clinical risk/benefit assessment of hydroxychloroquine had not been fully evaluated; therefore, in principle, that was not pseudoscience. The problem was that the President's assessment of hydroxychloroquine as a "game changer" resulted in a pseudoscientific discourse. The FDA revoked the approval in June [21]. The so-called game changer could not change the game.

7.3.2 Plasma Transfusions

For plasma that was also granted an emergency approval, President Trump said at a press conference with the FDA Commissioner Stephen Hahn, "It reduces mortality by 35%," and he called it a "breakthrough" in the treatment of COVID-19 [22]. Even the team who conducted the study on which the prescription was based became puzzled, not knowing where the "35%" figure came from. As a matter of fact, the "35%" was forcibly created by comparing a subgroup of the study subjects with other groups within the trial [23]. It was obvious cherry-picking [24]. Again, the original investigation was not unscientific. The problem is that politicians attempted to impress the public with "great strides." This is a classic case of pseudoscientification trying to meet the public's expectations for the development of a cure [18].

7.3.3 Remdesivir

In October 2020, Gilead's remdesivir was approved for use as a therapeutic agent in the European Union and the United States. The FDA never consulted a group of outside experts and approved its use by relying only on trial data, which was convenient for Gilead. The FDA did not take seriously the data from the WHO study that cast doubt on the efficacy of remdesivir, which was released just before approval [25]. The WHO strongly expressed concern; however, the FDA did not take up the matter [26]. President Trump praised it in the press, and the CEO of Gilead was also present. In this case, somewhat different from the "cherry-picking" within a single study, as in the case of hydroxychloroquine, a specific study was selected among a variety of others. Science is probabilistic in nature and is open to correction. Therefore, it is possible for studies with conflicting conclusions to be scientific. However, one should not consider only certain favorable protocols.

7.3.4 Povidone-Iodine Disinfectant

In August 2020, Dr. Akifumi Matsuyama announced the results of a study conducted by the Osaka Habikino Medical Center, showing that povidone-iodine's virus-killing effect is also applicable against the new coronavirus. The Governor of Osaka Prefecture was also present at the press release of the results of this wonderful "research," and he confidently recommended that citizens gargle with povidone-iodine [27]. However, the study was not randomized, as gargling with a liquid containing povidone-iodine was compared with historical data from people who did not undergo the intervention, and criticism promptly followed [28]. The Osaka Governor exaggerated the results of the study, stating that "It sounds hard to believe, but it's true. A study found that povidone-iodine solution could be effective in treating corona" [29]. Although the original study did not immediately support clinical application and never supported gargling as a preventive measure, the prefectural governor repeatedly made statements as if he had received scientific evidence for preventive measures. This press resulted in negative social effects such as increased sales of povidone-iodine all over Japan. Since the disinfectant also kills normal bacteria in the oral cavity, there were concerns about negative clinical effects as well [29]. At the end, the encouragement of gargling with povidone-iodine was thoroughly criticized and forgotten, and the Osaka Governor, Yoshimura, apologized on a television program for the way he made the announcement in April 2021.

7.3.5 Further Events

In addition to these paradigmatic cases, there are many reports of similar political intervention in science worldwide, such as the sudden posting of a document on the Center for Disease Control and Prevention (CDC) official website, despite strong resistance from scientists, stating that "people without COVID-19 symptoms do not need to get tested for the coronavirus" [30], and the rumored political intervention at the Scientific Advisory Group for Emergencies (SAGE), the major UK scientific agency [31]. In response to these cases, *Nature* issued an unusual document expressing strong concerns about political intervention in science [32]. Even if a research is of good quality and not pseudoscientific, it can be turned into pseudoscience when it is reframed by political intervention or exaggerated by mass media and social networking services (SNSs).

7.4 The Structural Problem of Science

The "pseudoscientification of science" was a problematic event that emerged during the pandemic; however, the soil that enabled its growth was structural and existed even before. It is a consequence of interconnected contexts, i.e., the public's immediate expectations from science, the political intervention in science to gain public political support, and the system of the scientific community that tries to maintain its own activities by responding to the interests of society. These contexts were enhanced by the following three elements: (1) preprint and mass publication culture, (2) reframing of scientific results, and (3) fast-track approval system for drugs.

7.4.1 Preprint and Mass Publication Culture

Researchers are under constant pressure to publish, which has created a climate for preprint publication [33]. The emerging culture prioritizes publishing results as soon as possible, even if there is uncertainty about them, and bringing them to the attention of others, even though they have not been peer-reviewed or revised by experts. Such a mass publishing culture has invited predatory journals that publish low-quality "peer-reviewed" papers and enabled industrialized journals that are built on the high cost of submission fees and open access [34]. Since peer review is often a form of volunteer labor, the academic community, exhausted by lack of funding, is increasingly deprived of the energy to devote to peer review [35], even though this problematic situation existed before the pandemic [36].

Clearly, the culture of preprint publishing was underway and expanded before the pandemic, and it is not necessarily an inappropriate culture in and of itself. However, there is no doubt that it has undermined the basic assumption of science that peer-reviewed publications in journals are "scientific results," and this has allowed politicians to gain public support at any given moment, even if these findings have not been sufficiently reviewed by the scientific community.

7.4.2 Press Releases of Research Results

Faced with the constant need to acquire funds, some researchers and institutions opted for a way to attract public attention through sensational media broadcasting. In part, such releases reflect the difficult financial situation that certain institutes are in, as evidenced by the Korean stem cell fraud in 2006 and the Japanese STAP (stimulus-triggered acquisition of pluripotency) cell fraud, both of which were subjected to lavish broadcasts on mass media [37]. Such media strategy was in concert with the mass media's desire to communicate new and easily understandable research results to the public. The researchers used and were used by the media [38].

The demand for transparency in research stems from the idea of clinical trial registries, originally aimed at making negative data visible and critically examining the data; however, it also includes the idea of providing the public with free access to research information [39]. The generalization of the strategy of social networking sites (SNSs) has created an unhealthy relationship between the public's growing expectations from science and the scientific community's attempts to meet them. With this pandemic, the public expected science to develop a cure for COVID-19, and science was quick to respond by revealing its findings to the public as much as possible. However, such a natural stance on the part of science has resulted in scientists cutting corners or exaggerating their research results, as exemplified in the case of the povidone-iodine study in Osaka. The press release culture with its motto of simplicity became an easy target for exploitation by politicians [40].

7.4.3 Drug Approval System in Crisis

Emergency drug approval applies to products for which there is no sufficient scientific basis; however, in the absence of any existing treatments, the decision has provided hope for the public. This can be understood as an unintended amplification, by the FDA's approval system, of the public's expectation of immediate and innovative research from science, an expectation that existed even before the pandemic. Politicians understood the public's expectations and accordingly tried to utilize it to secure the support that had been lost due to the failure to control the spread of the infection. The approval body, which should authorize drugs purely on scientific evidence, has been caught between the public's high expectations, the politicians who want to meet those expectations and gain support, and the scientific community that strongly opposes approval without evidence.

In addition, pharmaceutical companies' profits are also deeply involved in approval systems. With regard to the remdesivir case, there are suspicions that Gilead withheld data from the WHO study that was detrimental to the company, to gain approval in the European Union. Gilead denied this and explained that they had not received sufficient data from the WHO; however, the WHO claimed that they had completely handed over the trial data [25]. In any case, such difficulties and controversies exist in normal times as well, not just during pandemics.

7.5 Conclusion and Implications

Many scientists have voiced concerns about all these rapid and deeply disturbing changes and argued that the quality of science should always be maintained [12], as the pandemic has turned scientific research from a marathon into a sprint racing event [15], generating a "pandemic exceptionalism" [41].

This exceptionalism also applies to the relationship between science and politics as, in some cases, politicians violently intervened in science.

Many researchers have kept an eye on the FDA and similar approval bodies, hoping for strictly evidence-based, independent decisions.

The fact that approval agencies are caught up among diverse interests is not unique to the pandemic era. Politics and science have always been inseparable [32], and politicians have always made use of the science that suits their agenda [42]. Approval agencies must be more independent than ever to prevent the exploitation of citizens' health and property for the benefit of private companies and politicians. A higher level of scientific review body beyond national approval bodies would be ideal in cases of global pandemics.

Of note is the phrase "follow the science" that was uttered by politicians worldwide during the pandemic. While it may seem to respect science at first glance, there is a political decision at work as to which "science" is selected and which is dismissed [43].

What is needed is to change the structural background of science: first, to reduce the pressure on the scientific community to publish and to create an environment where scientists can spend their time on research with stable financial resources rather than short-term competitive funding. This will allow scientists to halt the expansion of mass and preprint publishing and selectively release scientific results that have undergone a more accurate review process in the community. This will also reduce the risk of scientific results being cherry-picked or exaggerated by those with political agendas.

Next, the relationship between society (the public) and the scientific community must be reconsidered. Directly related to the funding issue mentioned above, it is clearly unfavorable for research institutes to conduct eye-catching press releases and sometimes embellish their research results. With the addition of SNSs and many web-based media platforms to traditional mass media, people now have direct access to the latest scientific research results. However, understanding and interpreting the results of research requires special scientific literacy, and not all studies provide knowledge that can be immediately applied to clinical practice. In an age when the "latest research findings" go viral on SNSs with short, catchy phrases, it is very important that people understand their content, namely, to promote scientific literacy[44].

Acknowledgments This manuscript was funded by the Grants-in-Aid for Scientific Research [(B) #19H03868] from Japan Society for the Promotion of Science and by a grant from Japan Agency for Medical Research and Development [19190204].

References

1. Shelton T (2020) A post-truth pandemic? Big Data Soc 7(2):1–6
2. Cinelli M, Quattrociocchi W, Galeazzi A, Valensise CM, Brugnoli E, Schmidt AL, Zola P, Zollo F, Scala A (2020) The COVID-19 social media infodemic. Sci Rep 10(1):16598
3. Kasapçopur Ö (2020) Science and pseudoscience during the COVID-19 pandemic. Turk Pediatri Ars 55(4): 335–336
4. WHO (2020, February 8) Director-General's remarks at the media briefing on 2019 novel coronavirus. who.int/director-general/speeches/detail/director-general-s-remarks-at-the-media-briefing-on-2019-novel-coronavirus%2D%2D-8-february-2020. Accessed 30 May 2021
5. Nie JB (2020) In the shadow of biological warfare: conspiracy theories on the origins of COVID 19 and enhancing global governance of biosafety as a matter of urgency. J Bioeth Inq 17(4):567–574
6. Thomson C (2021, May 1) More Canadians think COVID-19 is a 'biological weapon' than believe the Ogopogo exists: poll, *TRYCITY news*. tricitynews.com/local-news/poll-finds-more-canadians-think-covid-19-is-a-biological-weapon-than-believe-the-ogopogo-exists-3683594. Accessed 30 May 2021.
7. Schraer R, Lawrie E (2020, April 15) Coronavirus: Scientists brand 5G claims 'complete rubbish.' *BBC News*. bbc.com/news/52168096. Accessed 30 May 2021
8. Aramesh K (2020, June 2) The COVID-19 pandemic and the problem of pseudoscience. *bioethics. net*. bioethics.net/2020/06/the-Covid19-pandemic-and-the-problem-of-pseudoscience/. Accessed 30 May 2021
9. Jia D (2020, February 14) Scientists slam Indian study that fueled coronavirus rumors. *NIKKEI ASIA*. asia.nikkei.com/Spotlight/Caixin/Scientists-slam-Indian-study-that-fueled-coronavirus-rumors. Accessed 30 May 2021

10. Sharma M, Scarr S, Kelland K (2020, February 19) Speed science; the risks of swiftly spreading coronavirus research. *Reuters*. graphics.reuters.com/CHINA-HEALTH-RESEARCH/0100B5ES3MG/index.html. Accessed 30 May 2021
11. Calcaterra G, Bassareo PP, Nodari S, Barilla' F, Di Franco A, Romeo F (2020) An expression of concern on research during the Covid-19 pandemic. J Cardiovasc Med 21(10):838–839
12. Scheirer W (2020) A pandemic of bad science. Bull At Sci 76(4):175–184
13. Laudan L (1982) Commentary: science at the bar—causes for concern. Sci Technol Hum Values 7(4): 16–19
14. Resnik DB (2000) A pragmatic approach to the demarcation problem. Stud Hist Phil Sci 31:249–267
15. Bramstedt KA (2020) The carnage of substandard research during the COVID-19 pandemic: a call for quality. J Med Ethics 46(12):803–807
16. Fortaleza CMCB (2021) Emergency science: epistemological insights on the response to COVID-19 pandemics. Infect Control Hosp Epidemiol 42(1): 120–121
17. Saag MS (2020) Misguided use of hydroxychloroquine for COVID-19: the infusion of politics into science. JAMA 324(21):2161–2162
18. Singh JA (2020) COVID-19: science and global health governance under attack. *S Afr Med J* 110(6):445–446
19. Thomson K, Nachlis H (2020) Emergency use authorizations during the COVID-19 pandemic: lessons from hydroxychloroquine for vaccine authorization and approval. JAMA 324(13):1282–1283
20. Thomas K (2020, April 17) Trump calls this drug a 'game changer.' Doctors aren't so sure. *New York Times*. nytimes.com/2020/04/17/health/trump-hydroxychloroquine-coronavirus.html. Accessed 30 May 2021
21. Food and Drug Administration (2020, June 15) Coronavirus (COVID-19) update: FDA revokes emergency use authorization for chloroquine and hydroxychloroquine. fda.gov/news-events/press-announcements/coronavirus-Covid19-update-fda-revokes-emergency-use-authorization-chloroquine-and. Accessed 30 May 2021
22. Lovelace B Jr (2020, August 24) Scientists doubt convalescent plasma touted by Trump is a 'breakthrough' coronavirus treatment. *CNBC*. cnbc.com/2020/08/24/scientists-doubt-convalescent-plasma-touted-by-trump-is-a-breakthrough-coronavirus-treatment.html. Accessed 30 May 2021
23. Kupferschmidt K, Cohen J (2020) In plasma OK, critics see politics, not science. Science 369(6507): 1038–1039
24. Chatlani S (2020, September 8) San Diego scientists call out FDA commissioner over 'cherry picked' COVID-19 plasma data. *KPBS Public Media*. kpbs.org/news/2020/sep/08/scientists-butt-heads-fda-over-misleading-breakthr/. Accessed 30 May 2021
25. Cohen J, Kupferschmidt K (2020, October 28) The 'very, very bad look' of remdesivir, the first FDA-approved COVID-19 drug. *Science*. sciencemag.org/news/2020/10/very-very-bad-look-remdesivir-first-fda-approved-Covid19-drug. Accessed 30 May 2021
26. Oakes K (2020, November 19) WHO backs off on remdesivir as FDA issues another EUA. *Regulatory Affairs Professionals Society*. raps.org/news-and-articles/news-articles/2020/11/who-backs-off-on-remdesivir-as-fda-issues-another. Accessed 30 May 2021
27. Reidy G (2020, August 4) Gargle medicine can fight coronavirus, Osaka governor says, citing limited data. *The Japan Times*. japantimes.co.jp/news/2020/08/04/national/gargle-medicine-coronavirus-osaka-governor-hirofumi-yoshimura/. Accessed 30 May 2021
28. Reidy G (2020, August 5) Experts shoot down Osaka governor's claim that gargling helps beat virus. *The Japan Times*. japantimes.co.jp/news/2020/08/05/national/osaka-hirofumi-yoshimura-gargling-coronavirus/. Accessed 30 May 2021
29. Tabata H, Matsumoto K (2020, August 5) Osaka Gov. urges use of iodine gargle solution for COVID treatment, but experts wary. *The Mainichi*. mainichi.jp/english/articles/20200805/p2a/00m/0na/016000c. Accessed 30 May 2021
30. Mandavilli A (2020, September 17) CDC virus testing guidance was posted against scientists' objections. *New York Times*. nytimes.com/2020/09/17/health/coronavirus-testing-cdc.html. Accessed 30 May 2021
31. Abbasi K (2020) Covid-19: politicisation, "corruption," and suppression of science. BMJ 371:m4425
32. Nature Editorial, Science and politics are inseparable (2020, October 8) *Nature*. media.nature.com/original/magazine-assets/d41586-020-02797-1/d41586-020-02797-1.pdf. Accessed 30 May 2021.
33. Kaiser J (2017) The preprint dilemma. Science 357(6358):1344–1349
34. Green J (2018) Critical analysis, credibility, and the politics of publishing in an era of 'fake news'. Crit Public Health 28(2):129–131
35. Bell K, Green J (2020) Premature evaluation? Some cautionary thoughts on global pandemics and scholarly publishing. Crit Public Health 30(4):379–383
36. Teixeira da Silva JA (2020) An alert to COVID-19 literature in predatory publishing venues. J Acad Librariansh 46(5):102187
37. Moore A (2006) Bad science in the headlines. Who takes responsibility when science is distorted in the mass media? EMBO Rep 7(12):1193–1196
38. Cassidi J (ed) (2006) Science, risk and the media—do the front pages reflect reality? Social Market Foundation, London
39. Dal-Ré R, Bracken MB, Ioannidis JP (2015) Call to improve transparency of trials of non-regulated interventions. BMJ 350:h1323

40. Camargo KR Jr (2020) Trying to make sense out of chaos: science, politics and the COVID-19 pandemic. Cad Saude Publica 36(4):e00088120
41. London AJ, Kimmelman J (2020) Against pandemic research exceptionalism. Science 368(6490):476–477
42. Hahn MB (2020) The battle between politics and science is costing us a timely victory over the COVID 19 pandemic. J Am Osteopath Assoc 120(11):806–807
43. Stevens A (2020) Governments cannot just 'follow the science' on COVID-19. Nat Hum Behav 4(6):560
44. Laugksch RC (2000) Scientific literacy: a conceptual overview. Sci Educ 84(1):71–94

Part III

Research Ethics

8 Lysenkoism: A Fine Line Between Formation of Scientific News and Disinformation

Agata Strządała

Abstract

Lysenkoism, a set of convictions about biology created by Lysenko, had its origins in botany and horticulture technics, such as vernalization and grafting, incorporating several general biological ideas (about nongenetic heredity and reinterpretation of Darwinism) which stood in contrast with contemporary science. Because of its totalistic ambitions and its promise of modifying nature to the benefit of socialist countries, it expanded its reach to zoology and science of farm animals. Statements on the promise of "grafting" or "vegetative reproduction" of farm animals were not unusual. The text examines the process whereby the involved scientific institutions started to disinform public opinion and reviews the conditions essential in turning production of scientific news to fake news in general.

8.1 Introduction

Lysenkoism was based on the idea of inheritance of acquired characteristics and rejection of standard genetics and cell theory. It was introduced firstly in the Soviet Union (USSR) during Stalinism (1927–1953), and after 1948, it was enforced in some other communist states (e.g., Poland, Czechoslovakia, China)—leading to elimination or suspension of research and teaching of genetics [49, 42, 44, 50]. In places like Italy [35] or Japan [47], Lysenkoist ideas were willingly adopted by groups of leftist or antifascist political convictions. Albeit the term is derived from Trofim Denisovich Lysenko (1898–1976), some of its concepts were developed by others, such as Olga Lepeshinskaya's (1871–1963) spontaneous emergence of life from inanimate matter or Ivan Vladimirovich Michurin's (1855–1935) horticulturalist ideas, retrospectively incorporated in it. At the time of its prominence, the nomenclature New Biology, Lysenko-Michurinism, or Creative Darwinism was used. The term Lysenkoism is generally reserved to the political and historical discourse about the phenomenon.

8.1.1 Fundamentals of the New Biology

It strongly opposed the Mendelian laws of inheritance and also genetics based on the discoveries by Thomas H. Morgan (1866–1945) and Hermann J. Müller (1890–1967). It openly denied August Weismann's experiments on mice tails (1834–1914) that disproved the theory of inheritance of acquired characteristics (tails grew back in the next generations, despite the fact that parents had them cut). According to the New

A. Strządała (✉)
Department of Medical Humanities and Social Science, Wroclaw Medical University, Wroclaw, Poland
e-mail: agata.strzadala@umw.edu.pl

J. Faintuch, S. Faintuch (eds.), *Integrity of Scientific Research*,
https://doi.org/10.1007/978-3-030-99680-2_8

Biology, traits acquired during parents' life are inherited by offspring, and environmental conditions have a primary impact on shaping the biological characteristics of living creatures. What made Lysenkoism unique regarding its predecessors (Lamarckism and epigenetics) was its official status as the only correct trend in science and the fact that it took over the scientific discourse with increasing support from political authorities.

8.1.2 Politicization of Science

Use of science in the "moralization" of society is common enough. Nonetheless, Lysenkoism took it one (or more) step(s) farther, thanks to political authorities (including Stalin himself), till it reached its zenith in the 1940s. Gradually classic genetic concepts were banned from scientific and educational institutions, such as schools and universities.

The Second World War served as an additional impetus to the development of "national science" corresponding with the pressing demand to make a difference not only on the battlefield but also in the economy [43]. Ideas from Lysenkoism were swiftly implemented in agriculture without convincing evidence of its effectiveness, and opponents were persecuted. One of them, the famous Russian geneticist Nikolai Vavilov (1887–1943), was arrested and died in prison in 1943. Other scientists toed the party line out of fear or straightforward conformism while looking for career opportunities. Many were actually convinced that Lysenkoism could solve contemporary problems such as famine, assuming it would improve production of food and induce prosperity of socialist nations. Willingly or not, scholars in Soviet countries worked in a reality in which the New Biology played a significant role.

8.2 Lysenkoism and Michurinism

The New Biology is frequently associated and mixed up with Ivan Michurin's works. Lysenko considered himself to be a Michurin disciple. Nonetheless, Lysenko actually reinterpreted some concepts founded by the Russian gardener in line with the New Biology and generalized them to support its claims. Michurinism was not so much the foundation of the New Biology but had rather retrospectively incorporated therein. Most notably, Michurin's disapproval of Mendel's law and his conviction of the influence of the environment on heredity traits were included [39]. It is worth noting that reluctant reception of Mendel's law at the time was quite common among horticulturalists and not specific only to Michurin.

Notwithstanding Michurin's limited contribution to scientific theory, his achievements as a grower included production of new and interesting hybrids. Michurin's lifelong passion toward horticulture led him to work on new varieties of fruits (apples, cherries, plums, peaches). As a main method, he used hybridization of distant species (e.g., apple and pear, almond and peach), originating from different climates, which were far more resistant to cold. Eventually, this procedure led to significant geographic changes in cultivation of some trees, moving them much farther north. This procedure was later called the "Michurin acclimatization principle" [39].

His consistent success in horticulture led him to believe that nature can be shaped almost freely by humans by controlling the environmental factors. Moreover, he believed that such environmentally induced phenotypic changes can be transferred to the next generation. Lysenko later fully accepted this idea claiming that by controlling the conditions of upbringing, one can "manually" manage heredity. Michurin's works were the product of passionate and creative effort, interpreted post-factum in the spirit of Lamarckism. Lysenko took over his achievements and treated them as proof for Michurin's theoretical conjectures.

8.3 The Main Components of Lysenkoism, Their Implementation, and Social Consequences

8.3.1 Vernalization

As son of peasants and due to turbulent turns of history, Lysenko never received full and systematic scientific education, yet he finally graduated from the Institute of Agrotechnology in Kiev. Thanks to the new political and social reality in Soviet Russia, he had an opportunity to develop and publicize his own works (and results of cooperation with his father) [41]. In 1927, he adapted some existing technics and concepts of cultivation of plants such as vernalization and added to them new interpretations without empirical or experimental confirmation.

Vernalization is a procedure whereby exposure to cold induces flowering. This phenomenon had been already described in the middle of the nineteenth century and established in the second decade of the twentieth century [32]. As assumed by Lysenko, conditioning of seeds to cold would induce development, as well as productivity and resistance to cold. Lysenko later added the conviction that a vernalized state can be hereditary, thus claiming that he could permanently change winter seeds into spring seeds by soaking them in cold water [32]. Per his conjecture, this procedure could enhance food production in the harsh climate of Siberia and Central Asia, where frosty winters were causing shortages of food and famine for people and farm animals.

8.3.2 Food for the Masses and Execration for Scientists

Famine in parts of the USSR, particularly in Ukraine, was predominantly a result of brutal collectivization of farming, war against wealthy peasants/kulaks (deportations, imprisonment, executions), and systematic confiscations of crops, which led to destruction of the existing socioeconomic system of farming [37]. Poor infrastructure and transportation and inefficient distribution of food further exacerbated the issues. Famine was actually endemic in certain territories of Russia, similar to some parts of China and India. To cope with these issues, an ambitious plan had been proclaimed by Stalin to introduce the New Biology in agriculture, and Lysenko was a key player [46].

Despite doubts and criticism from leading USSR geneticists, such as Vavilov, the plan was put into practice. Lysenko not only rejected the idea of genes and genetics, believing that somatic cells have direct impact on heredity, but, when unable to deliver promised outcomes, blamed geneticists of obstruction and sabotaging. This resulted in further and more brutal persecution of opponents and marginalization of genetics and traditional biology.

8.3.3 Creative Darwinism and Close Planting

Later in his career, Lysenko reinterpreted Darwinism by questioning the concept of struggle for survival as influenced by Malthusianism (Malthus theories were criticized by Lenin) [10]. Lysenko concluded that intraspecies competition does not exist and that the high density of population of the same species will induce altruism and cooperation, as if belonging to the same social class. Practical implication was a technic of close planting. In 1948, Stalin introduced this procedure to transform the steppes, steppe-forests, and deserts of South Russia and Central Asia into productive forests and agricultural areas. Nonetheless, due to lack of natural resources for the large amount of seeds planted too close to each other and harsh environmental conditions, seedlings were dying out, and the action of forestation failed [46].

Another "innovation" by Lysenko was a new concept of genesis of species, based on the idea of sudden transformation due to environmental factors. The theory of "stadial development" claimed that in the plant's life cycle, transformation from one phase to the next occurs rapidly and dramatically due to "breaking the heredity." Hence, rye could transform into wheat if favorable conditions are met, and under adverse influences, wheat would change into weed during one generation.

8.3.4 Spontaneous Generation

Lysenko also incorporated the Lepeshinskaya theories of spontaneous emergence of life from inanimate matter, by virtue of a "vital substance," in opposition to Rudolf Virchow's (1821–1902) discovery that cells are a result of division of other pre-existing cells. The implementation of Lysenko's ideas predictably turned out to be not only farfetched but also futile and catastrophic. All these facts notwithstanding, some of his technics such as close planting or introduction of new species were implemented also in other countries notably in China, as a part of Mao's Great Leap Forward [51], with the same failures. Close planting led to waste of resources, further deteriorating the Chinese food supply, a country on the brink of a great famine [33]. Hunger in reality resulted from communist collectivization of farmland, inefficient transportation, poor land and water management, as well as intentional extermination of sparrows (blamed for stealing grains) leading to fast breeding of insect devoid of natural enemies. Ironically, a greater catastrophe was averted due to local authorities' realization of the shortcomings of close planting and hence abandoning this technic. It was preserved only on some fields reserved for official visitations and propaganda purposes [33].

8.4 Lysenkoism in Poland

8.4.1 Proclamation and Implementation

The Polish Communist Party and state institutions were openly and actively involved in promoting the New Biology after the Second World War. Shortly after becoming a one-party state in 1948, the New Biology started to be introduced in science on an institutional level [3, 42, 44]. Both the adoption of Lysenkoism in 1948 (beginning of Stalinism in Poland) and its later rejection in the People's Republic of Poland were strictly dependent on political shifts [45, 48, 54]. Lysenkoism was not a grassroots movement. Scholars who did not comply were ostracized, censored, and banned from teaching genetics, and their research was discontinued.

The starting point of Lysenkoism was a conference *On The New Genetics* in Warsaw (March 30, 1949) including representatives of the Central Committee of the Polish United Workers' Party, together with Jerzy Albrecht (1914–1992), head of the Department of Propaganda, Culture and Education [9]. The conference was organized and led by Polish zoologist Jan Dembowski (1889–1963), a deputy in Polish Parliament and later its Marshal (1952–1957). Additionally, Dembowski held the position of deputy chairman of the Council of State and a founder and first president of the Polish Academy of Sciences (PAN) [52].

When Wincenty Lesław Wiśniewski (1904–1958), a zoologist and eminent researcher of Polish waters, commented on Dembowski's presentation on Lysenko's theory noting its weaknesses, Dembowski accused him of being in favor of a compromise between classical genetics and the New Biology. He stressed that Morgan's and Weisman's theories were finished and terminated any public exchange of viewpoints on genetics for years [9].

Dembowski played a very important role in reestablishing Polish science in postwar times, as hundreds of Polish scholars were executed directly or sent to concentration camps during Nazi Germany occupation of Poland (1939–1945) or were subsequently killed by the NKVD during the Soviet invation of Poland (1939-1940) [31, 52]. Dembowski was not only a main propagator of Lysenkoism, but also an influential politician, therfore his personal involvement in the New Biology had tremendous consequences for Polish scientific life

The Polish United Workers' Party, the Marxist Naturalists' Circle (later Marxist Naturalists' Association), and the Copernicus Society with its magazines *Kosmos* and *Wszechświat* actively joined the campaign of the New Biology [52].

Staring from 1948, Polish science has developed dependence on Soviet science. A series of conferences devoted to the New Biology were of key importance for the popularization of this trend in science [1, 2, 13]. The teaching of genetics was officially banned at schools and universities at that time [52].

One tragic figure was the Polish biologist Stanisław Skowron (1900–1976). In pre-war times, he had been an advocate of eugenics, and before the advent of the New Biology, he had published students' books about genetics. However, in light of the new political situation, perhaps out of fear, he radically converted to the New Biology becoming one of its leaders [40].

8.4.2 Genesis: Internal Factors

The New Biology before 1948 was disregarded by Polish science. Unlike eugenics, a known and discussed idea (albeit neither popular nor widely accepted in interbellum Poland [38]), the New Biology had been restricted only to a narrow circle around Jan Dembowski. The genesis of Dembowski's views originated from his contact with Russian scientists in pre-war times, his political view, and his scientific specialization: protozoology. From the very beginning of his career, Dembowski kept in close touch with the scientific community in Russia. Dembowski was born in St. Petersburg and spent his childhood and youth in Tambov in the central part of Russia, where he graduated from high school. In the years 1907–1912, he studied at the University of St. Petersburg. Between 1912 and 1914, he was an assistant of prof. Valentin A. Dogiel (1882–1955) in the Department of Zoology at the local university. Later, he continued his studies in Vienna and in 1918 came to Poland and began working at the Nencki Institute in Warsaw. Dembowski obtained doctorate in 1920 and habilitated in 1922; he conducted research also outside of Poland in France, Italy, and the USA.

Dembowski political affiliations gravitated toward Soviet version of socialism. Dembowski, who spent the war in Vilnius, started there his collaboration with Soviet communists. He became a member of the Union of Polish Patriots, an organization which became a Soviet agency in Poland, introducing the communist system in the country. In 1944, he moved to Moscow for 3 years, where he held the position of a scientific attaché at the Polish Embassy [34].

Additionally, owing to his interest in protozoa, Dembowski developed the conviction that environmental conditions shape the nature of living things. Interestingly, protozoa change their characteristics dramatically during different phases of their life and inhabit diverse environments corresponding to their life cycle. Protozoa are also able to reproduce asexually (without genetic exchange between organisms by binary fission) and exchange genetic material without reproduction. These specific features of protozoa were compatible with Lysenko's claims on the crucial role of environment in heredity, the ability to transform from one form of life to another, the stadial development and grafting, as well as the vegetative reproduction of plants. Already in interbellum times, Dembowski had been a convinced epigeneticist.

Dembowski's first work promoting epigenesis was published in 1927 [7]. Although he did refer to the achievements of Morgan and Weismann, he concluded that the role of epigenesis would increase and that it would set a new direction in biology. His book was strongly politicized and stylized in a manner of Marxism-Leninism, and Dembowski categorically claimed that epigenesis is a historical necessity. His work included statements that will later become a permanent elements of the style of Creative Darwinism, i.e., a new theory of ontogenesis (the dependence of ontogenesis on appropriate external conditions) and the belief that biological phenomena can be controlled and freely manipulated by humans [52].

His next publication, *Principles of General Biology,* was even more radical. Dembowski in it strongly supported the idea of heredity of acquired characteristics and criticized the idea of division between the germplasm and somatic cell lines [8]. Furthermore, he advocated the idea of animals' ability to regenerate due to totipotent abilities of extracellular "creative substance" which can be preserved at all stages of ontogenesis. Dembowski failed to notice the reality of stem cells, but rather interpreted it in line with Lepeshinskaya's concept of "creative substance." He also rejected the idea that genes can be located in chromosomes.

Dembowski's books had been met with harsh criticism. Polish botanist Bolesław Hryniewiecki (1875–1963) accused him of subjectivism and selectivity in the choice of presented facts and theories. Hryniewiecki also pointed out

Dembowski's basic errors in botany, his complete ignorance in genetics, concluding that Dembowski statements contradicted the well-established and generally accepted current state of knowledge [12]. Thereafter, Dembowski faced the consequences of publicizing his convictions and lost his position at the Nencki Institute. His career was put on hold, and as a result, he moved to Vilnius and was hired by the University of Stefan Batory [34].

Against this background, his cooperation with Russian Lysenkoists and Soviet political organizations as well as his future role in promoting the New Biology is hardly surprising. Finally, in 1948, equipped with the political and institutional power as member and Marshal of Parliament and organizer of scientific institutions in postwar Poland, he became a major designer of Lysenkoism. Up to this point, Dembowski's views were rather unique in Poland and by no means representative of the majority of Polish biologists. Personal, scientific, and political origins of Lysenkoism in Poland were united in Dembowski's persona. Nevertheless, Dembowski's research, teaching, and organizational efforts in science cannot be reduced to his role in Lysenkoism. Moreover, the campaign of the New Biology would not be possible without the positive responses from other scholars.

8.4.3 The New Biology in Zoology

Notwithstanding its botanic foundation (such as grafting or vernalization), Lysenkoism expanded its reach into zoology and especially into the improvement of farm production, a high social and political issue at the time. It was believed that application of Lysenkoist ideas to farm animals had the potential to help economic growth [3].

Some Polish zoologists and farm animal scientists actively responded to the political order to conduct research on large animals according to the New Biology. This was done in clear contrast with their own knowledge about genetics and physiology of pigs, chickens, or cows [15–21, 30]. Any direct and open opposition, as well as continuation of genetic research, was retorted by intimidation, censorship, and ostracism [2, 13]).

Cold rearing of cattle was treated as an equivalent of vernalization; it was expected that raising cows in cold weather would induce higher milkiness and better resistance against pathogens. The hereditary changes according to Lysenkoists would be caused also by intense milking (the more you milk, the more milk cows will produce in the future), whereas reduced milking was thought to lead to atrophy of glands in next generations [2, 22–24]).

Polish zoologists were referring to M.F. Ivanov works and especially to the Ukrainian white steppe pigs, which he introduced. Lysenkoists also claimed that chromosomal determination of sex is a false conviction; instead, they stated that sex is determined only by temperature or metabolism (nota bene including mammals). Additionally, vegetative reproduction was believed to be possible and beneficial in the context of enhanced production of farm animals. Skowron argued that vegetative reproduction is plausible in case of pigs or cows and that parents' cytoplasm influence positive, adaptive hereditary changes in their offspring. What is more, Skowron claimed that multiple males' spermatozoids can be directly involved in conception, hence causing hereditary changes of progeny (additional spermatozoids of a second male used for fertilization played a role of a "mentor" according to grafting technic transferred to zoology) [5, 25–30, 53].

In Poland, soon after a political pressure subsided in the mid-1950s, Lysenkoism was abandoned. It was not only criticized as a set of ideas but also its implementation was exposed and ridiculed [4, 6, 11, 14]. However, its long-term consequences, such as dependence on Soviet science, prioritizing practice before basic research, politicization of science, censorship, and diminished cooperation with non-socialist countries, played a significant role for decades to come. Professional and personal development of many scholars was disturbed as well.

Due to its practical ineffectiveness, Lysenkoism in farming was ceased on the experimental level. However, blind reliance on scientific discoveries originating from the Eastern Bloc and putting practical use as priority before diligent basic research was concluded were still destructive and harmful. For example, the introduction of Sosnowsky's hogweed (*Heracleum sosnowskyi*) in Poland, while not a part of Lysenkoism but still followed a similar trend. In the late 1950s, it was believed that Sosnowsky's hogweed, originally inhabiting the area of Caucasus, could be used in medicine, due to its coumarin content. Then, because it was growing and developing faster in a mild Polish climate, there were attempts to use it as an animal feed, and farmers were encouraged to plant it. Additionally, it was planted in state-owned experimental stations. As a consequence, the plant becomes common, especially in Southern Poland. Unfortunately, Sosnowsky's hogweed was not suitable as animal feed and caused serious chemical burns in animals and humans. As of now, Sosnowsky's hogweed is considered in Poland an invasive, dangerous plant. Growing it is strictly forbidden under law, and a lot of effort is being made to eradicate it. Nonetheless, due to its prevalence in Poland, it is responsible for many cases of chemical burns every year [36]. To this day, many still name the plant ironically a "Stalin's revenge."

8.5 Conclusions

Utopian ambitions, politicization of science, and direct support and "instructions" from the state are not phenomena specific for Lysenkoism; rather, they are common characteristics of science as a social institution. Nonetheless, the study of the New Biology stresses a few factors which may lead to scientific ideas becoming disinformation, fake news, and in breach of research ethics. Noteworthy factors are:

1. Science is financially and organizationally dependent on an exclusive source of support (state, corporation, industry).
2. It responds to urgent and serious social and political demands (due to famine, war, economic crisis, natural catastrophe, and epidemy).
3. Society enjoys only limited freedom of speech and civic liberties.
4. Brutalization of social life and common use of violence.

While the two former factors typically ignite scientific progress and direct it in line with social need, they may also lead to the narrowing of the focus of scientific activities and limiting diversity in science. Such diversity plays an important role in balancing scientific discourse and institutions with social and ethical values. For example, Dembowski's personal involvement into the New Biology would have no implication for the scientific life in Poland if free exchange of thoughts had been preserved as it was in pre-war times. Most notably, absence of civil liberties and brutalization of life may remove the ethical boundaries of scientific research, discourse, and its applications.

References

Primary Sources

1. Anonymous (1952) Konferencja naukowa poświęcona nauce Pawłowa (Sprawozdanie stenograficzne) (Scientific Conference Dedicated to Pavlov's Science. (Shorthand Report). Państwowy Zakład Wydawnictw Lekarskich, Warszawa
2. Anonymous (ed) (1951) Materiały konferencji agrobiologów, biologów, medyków w Kuźnicach 27.12. 1950-06.01.1951. (Proceedings of the conference of agrobiologist, biologist and physician in Kuźnice), vol 1. Warszawa, Książka i Wiedza
3. Anonymous (1953) Nauki przyrodnicze wiele zawdzięczają Stalinowi... (Natural Sciences Owe a Lot to Stalin). Kosmos 3:3–5
4. Ber A (1956) Pawłowizm pod lupą (Pavlovism Under The Magnifying Glass). Po Prostu 23:1–5
5. Bilewicz S, Kamiński Z (1952) Rozród u ssaków (Mammalian Reproduction). Postępy Wiedzy Rolniczej 5(17):35–47
6. Bóbr (Jan. 1956) Zmowa milczenia (Conspiracy of silence). Po Prostu 4:1,6
7. Dembowski, Jan, 1927a. Szkice biologiczne (Biological overview). Lwów.
8. Dembowski J (1927b) Zasady biologii ogólnej (Principles of general biology). Wydawnictwo M. Acta, Warszawa
9. Dembowski J (1949) "O nowej genetyce" (About the new genetics). In: Dembowski J (ed) O nowej genetyce i dyskusja uczonych polskich o teorii

Miczurina-Łysenki (About the new genetics and discussion of polish scholars about the Michurin-Lysenko's Theory). Książka i Wiedza, Warszawa, p 9–66
10. Drozdowicz AE (1953) Recenzja "Zagadnienia twórczego darwinizmu. Materiały kursu biologii w Dziwnowie 7 V-7 VIII 1952" (Review: the issue of creative darwinism. proceedings of the course in biology in Dziwnów, 7 July–7 August 1952). Kosmos 3:69–73
11. Grodziński W (1956) Oczyścić pole ofensywy (Clean up the offensive field). Po Prostu 4:6
12. Hryniewiecki B (1928) Recenzja książki Jana Dembowskiego "Zasady biologii *ogólnej*" Review of the book by Jan Dembowski, principles of general biology. Kosmos Seria B 53:246–248
13. Jaczewski T (1951) Konferencja biologiczna w Kuźnicach (Biological conference in Kuźnice). Życie Nauki 3:234–241
14. Kuźnicki L (1955) Przerwijmy zmowę milczenia. Darwinizm a łysenkizm (Let's break the conspiracy of silence. Darwinism and Lysenkoism). Po Prostu 42–43:4
15. Malinowski E (1951) Problem gatunku (The topic of the species). In: Unknown (ed) Materiały konferencji agrobiologów, biologów i medyków w Kuźnicach (Proceedings of the conference of agrobiologist, biologist and physician in Kuźnice), vol vol. 1. Książka i Wiedza, Warszawa, pp 287–325
16. Marchlewski T (1951a) Agrobiologiczne momenty w produkcji zwierzęcej (Agrobiological moments in the animal production). Postępy Wiedzy Rolniczej 2(10):67–82
17. Marchlewski T (1951b) Krytyczny przegląd badań nad drozofilą (Critical review of the research on Drosophila). In: Unknown (ed) Materiały konferencji agrobiologów, biologów i medyków w Kuźnicach (Proceedings of the conference of agrobiologist, biologist and physician in Kuźnice), vol vol. 2. Książka i Wiedza, Warszawa, pp 397–425
18. Marchlewski T (1954) Objawy wegetatywnego rozszczepienia u zwierząt (Signs of vegetative cleavage in animals). Kosmos 2(7):224–225
19. Michajłow W (1949) Dyskusja uczonych polskich o teorii Miczurina-Łysenki (Discussion of Polish scholars about the Michurin-Lysenko's theory). In: Dembowski J (ed) "O nowej genetyce" i dyskusja uczonych polskich o teorii Miczurina-Łysenki (About the new genetics and discussion of polish scholars about the Michurin-Lysenko's theory). Książka i Wiedza, Warszawa, pp 87–95
20. Michajłow W (1953) O metodzie dialektycznej w parazytologii (About dialectical method in parasitology). Kosmos 3(4):10–24
21. Michajłow W (1952) Stadialność rozwoju – prawem ogólnobiologicznym (Stadial development as a law in general biology). Postępy Wiedzy Rolniczej 5(17): 25–34
22. Olbrycht T (1949) Dyskusja uczonych polskich o teorii Miczurina-Łysenki (Discussion of Polish scholars about the Michurin-Lysenko's theory). In: Dembowski J (ed) "O nowej genetyce" i dyskusja uczonych polskich o teorii Miczurina-Łysenki (About the new genetics and discussion of Polish scholars about The Michurin-Lysenko's theory). Książka i Wiedza, Warszawa, pp 99–109
23. Pająk JZ (1949) Agrobiologia w zootechnice (Agrobiology in zootechny). Biologia w Szkole. 4(8): 73–91
24. Petrusewicz K (1951) Wstęp (Introduction). In: Unknown (ed) Materiały konferencji agrobiologów, biologów, medyków w Kuźnicach 27.12. 1950-06.01.1951. (Proceedings of the conference of agrobiologist, biologist and physician in Kuźnice), vol vol. 1. Książka i Wiedza, Warszawa, pp 1–2
25. Raabe Z (1953) Na marginesie artykułu St. Skowrona "O tak zwanych prawach Mendla" (In the margin of St. Skowron's article "about so called Mendel's laws"). Kosmos 1(2):42–46
26. Skowron S (1951a) Co wiemy o dziedziczności? (What do we know about heredity?). Katowice.
27. Skowron S (1951b) Z zagadnień regeneracji (The issue of regeneration). In: Unknown (ed) Materiały konferencji agrobiologów, biologów i medyków w Kuźnicach (Proceedings of the conference of agrobiologist, biologist and physician in Kuźnice), vol vol. 2. Książka i Wiedza, Warszawa, pp 307–346
28. Skowron S (1952) Z zagadnień teorii stadialnego rozwoju (The Theory of Stadial Development). Postępy Wiedzy Rolniczej 1(13):3–11
29. Skowron S (1953) W odpowiedzi na artykuł Z. Raabego (Answering to the Z. Raabe's article). Kosmos 3(4):63–65
30. Skowron S, Fidelus (Jan. 1954) Z zagadnień zapłodnienia u zwierząt (The Topic of Insemination in Animals). In: Michajłow W, Petrusewicz K (eds) Materiały konferencji młodej kadry biologów w Kortowie (Proceedings of the conference of youth biologists in Kortów). Państwowe Wyd. Rolnicze i Leśne, Warszawa, pp 299–306

Secondary Sources

31. Albert Z (1975) Kaźń profesorów lwowskich - lipiec 1941/studia oraz relacje i dokumenty zebrane i oprac. przez Zygmunta Alberta (The Murder of Lwów Professors in July 1941). Wrocław: Wydawnictwo Uniwersytetu Wrocławskiego
32. Amasino R (2004) Vernalization, competence, and the epigenetic memory of winter. Plant Cell 16(10): 2553–2559. https://doi.org/10.1105/tpc.104.161070
33. Becker J (1998) Hungry ghosts. Mao's secret famine. Holt Paperback, New York

34. Brzęk G (2007) Historia zoologii w Polsce do drugiej wojny światowe (History of zoology in Poland until the Second World War). Polska Akademia Umiejętności, Kraków
35. Cassata F (2012) The Italian communist party and the "lysenko affair" (1948–1955). J Hist Biol 45(3): 469–498. https://doi.org/10.1007/s10739-011-9286-4
36. Chmielewski J, Czarny-Działak M, Dziechciaż M, Uściński P, Bąk-Badowska J, Florek-Łuszczki M, Szpringer M (2017) Barszcz Sosnowskiego – Roślina Zagrażająca Zdrowiu Człowieka. Medycyna Środowiskowa - Environ Med 20(3):55–59
37. Davis RW (1980) The socialist offensive: the collectivization of soviet agriculture. Harvard University Press, Cambridge
38. Gawin M (2003) Rasa i nowoczesność. Historia polskiego ruchu eugenicznego (1880–1952) (Race and modernity. History of the Polish eugenic movement (1880–1952)). Wydawnictwo Neriton, Instytut Historii PAN, Warszawa
39. Goncharov NP, Savel'ev NI (2016) Ivan V. Michurin: on the 160th anniversary of the birth of the Russian Burbank. Russ J Genet Appl Res 6:105–127. https://doi.org/10.1134/S2079059716010068
40. deJong-Lambert W (2009) From Eugenics to Lysenkoism: the evolution of Stanisław Skowron. Hist Stud Nat Sci 39(3):269–299. https://doi.org/10.1525/hsns.2009.39.3.269
41. deJong-Lambert W (2012) The Cold War politics of genetics research: an introduction to the Lysenko Affair. Springer, Dordrecht-New York
42. deJong-Lambert W (2008) The new biology: Lysenkoism in Poland. VDM Verlag, Saarbrücken
43. Krementsov N (1997) Stalinist science. Princeton University Press, Princeton, NJ
44. Köhler P (2008) Łysenkizm w botanice polskiej (Lysenkoism in Polish Botany). Kwartalnik Historii Nauki i Techniki 53(2):83–161
45. Köhler P (2013a) Łysenkizm w Polsce na tle ówczesnej sytuacji politycznej (Lysenkoism in Poland in Terms of The Political Situation of The State). In: Köhler P (ed) Studia and Łysenkizmem w polskiej biologii (Studies in Lysenkoism in Polish biology). Kraków, Instytut Pamięci Narodowej, pp 25–70
46. Köhler P (2013b) Zarys historii łysenkizmu w ZSRR (An outline of the history of Lysenkoism in the USSR). In: Köhler P (ed) Studia and Łysenkizmem w polskiej biologii (Studies in Lysenkoism in Polish biology). Kraków, Instytut Pamięci Narodowej, pp 25–70
47. Lida K (2015) A controversial idea as a cultural resource: the Lysenko controversy and discussions of genetics as a 'democratic' science in postwar Japan. Soc Stud Sci 45(4):546–569. https://doi.org/10.1177/0306312715596460
48. Paczkowski, Andrzej (1999) Terror und Überwachung: die Funktion des kommunistischen System in Poland von 1944 bis 1956. transl. Hanna Labrenz-Weiss. Berlin: Der Bundesbeauftragte für die Unterlagen des Staatssicherheitsdienstes der Ehemaligen Dt. Demokratischen Republik
49. Orel V (1992) Jaroslav Krizenecky (1896–1964), tragic victim of Lysenkoism in Czechoslovakia. Q Rev Biol 67(4):487–494
50. Roll-Hanson NILS (2005) The Lysenko effect: the politics of science. Humanity Books, New York
51. Schneider L (2003) Biology and revolution in twentieth-century China. Rowman and Littlefield, Lanham, MD
52. Strządała A (2011) Łysenkizm w polskiej zoologii w latach 1948–1959. Wdrażanie Łysenkizmu (Lysenkoism in Polish Zoology in 1948–1959. Implementation of Lysenkoism). In: Płonka-Syroka B (ed) Gra możliwości. Studia z historii medycyny i farmacji XIX i XX wieku (The game of possibilities. studies from the history of medicine and pharmacy in the 19th and 20th centuries). Akademia Medyczna im. Piastów Śląskich we Wrocławiu, Wrocław, pp 185–212
53. Strządała A (2012) Zasadnicze komponenty łysenkizmu (Lysenkoism in Polish zoology: the main components of the new biology). In: Płonka-Syroka B, Płonka-Syroka A (eds) Obraz świata jako konstrukt kultury (The image of the world as a cultural construct). Arboretum, Wrocław, pp 59–81
54. Werblan A (2009) Stalinizm w Polsce (Stalinism in Poland). Towarzystwo Wydawnicze i Literackie, Warszawa

9 Misconduct and Consent: The Importance of Informed Consent in Medical Research

Marton Gergely, Fida K. Dankar, and Saed Alrabaee

Abstract

Human subject research in the medical arena offers an indisputable contribution to society. However, all medical research needs to be conducted through maintaining strict levels of research standards and by complying with applicable regulations and guidelines. Despite this, medical research misconduct has been around for centuries, with countless examples of gross ethical oversights and morally flawed decisions made by both medical practitioners and researchers. One area of such misconduct is within that of informed consent. Despite the continually evolving rules and regulations surrounding informed consent in medical research, countless breaches are present surrounding the topic. In this paper, we discuss the origins of informed consent, the regulations and guidelines surrounding it, the common types of informed consent necessary to be gathered, and the potential pitfalls therein. We conclude with an overview of a selection of breaches in informed consent in the area of medical research and their likely reasons.

M. Gergely · F. K. Dankar (✉) · S. Alrabaee
United Arab Emirates University, College of IT, Al Ain, United Arab Emirates
e-mail: mgergely@uaeu.ac.ae; fida.dankar@uaeu.ac.ae; salrabaee@uaeu.ac.ae

9.1 Introduction

Medical research involving human subjects is laboriously regulated worldwide and must adhere to national and international norms, industry standards, and operational regulations [1, 2]. Deviation or incompliance with these standards, ethical principles, and regulations will not only hinder scientific progress and jeopardize the public perception of scientific outcomes but also can be considered misconduct or fraud [3].

The US Public Health Service defines medical research misconduct as the "fabrication, falsification, or plagiarism in proposing, performing, or reviewing research, or in reporting research results" [3]. These categories are further expanded by George and Buyse [3] as:

1. Fabrication: Making up data or results and recording or reporting them.
2. Falsification: Manipulating research settings, materials, equipment, calibration, or process or changing or omitting data or results which may affect the results' accuracy as presented in the research report.
3. Plagiarism: Confiscation of others' ideas, processes, results, or words without giving appropriate credit.

The issue of misconduct and fraud with regard to this paper's focus, informed consent in medical research, falls succinctly under the category of "falsification," or the manipulation of research settings or processes. As fabrication revolves

J. Faintuch, S. Faintuch (eds.), *Integrity of Scientific Research*,
https://doi.org/10.1007/978-3-030-99680-2_9

around the research results and data therein, this does not concern itself with the research process. Similarly, plagiarism is devoted to the copying of others and thus does not apply to informed consent. However, the category of falsification deals directly with the malicious modification or alteration of research settings and their respective documents (e.g., forging informed consent forms in order to conduct research). As such, the falsification (or intentional misuse) of informed consent in medical research is, ipso facto, considered by the US Public Health Service definition as a form of medical misconduct.

Such misconduct in informed consent has been around for over a century in general medical care and for many decades in research protocols, with countless examples of gross ethical oversights and morally flawed decisions made by both medical practitioners and researchers [4–6]. Further complicating matters is today's modern notion of informed consent and the advances in data gathering methods and long-term storage options (with the advent of nearly infinite database capacities, there is no need to ever delete patient data; instead, it can be continually reused for newer and newer studies). Despite these advances in research methodology, international regulations concerning informed consent have remained relatively static and unchanged for some time.

Informed consent still relies heavily on antiquated legal statutes such as the 1946 Nuremberg Code, the 1964 Declaration of Helsinki, the 1979 Belmont Report, and the 1991 Common Rule (naturally, it should be noted that the Declaration of Helsinki and the Common Rule have been continually revised, up to 2013 and 2018 respectively; however, they still maintain the basic underpinnings of their original form) [4, 6]. Since the idea of informed consent has eclipsed its classical form, it begins to beg the question:

- "Are today's informed consent gathering rules and regulations enough to contain and curb misconduct in the ever-evolving and data-hungry landscape of modern medical research?"

9.2 Background to Informed Consent

The idea of informed consent was brought about by countless violations of ethical and moral standards in both medical care and medical research, spanning over the last 100 years [1]. To complicate matters, the need of informed consent has changed due to advances in medicine, as well as the type of data being gathered, further complicating how physicians and researchers should interact with their patients and subjects. As such, informed consent was born from unfortunate yet fully preventable events in two distinct areas: medical care and human subject research [7].

9.2.1 The History of Informed Consent

Informed consent in medical care is generally attributed to a result of physicians' authoritarian (and perhaps overprotective) approach to patient treatment [4]. Despite rules dating back to as early as 1665 (wherein physicians and surgeons were required to obtain a patient's consent prior to treatment under the Duke of York's Laws of 1665, the founding laws for New York), there have still been myriad court cases due to adverse effects of medical procedures that patients opposed, did not approve, or were not provided with adequate information about [4–6].

9.2.2 Historic Consent Litigations in Medical Care

The first major court case regarding informed consent can be traced back to the 1905 case of Pratt v. Davis, where a surgeon removed an unknowing woman's uterus and ovaries, in order to treat her epilepsy [4, 8]. Here, the court ruled that a "physician or surgeon, however skillful [. . .] [cannot] violate without permission the bodily integrity of his patient" [8]. This step, albeit small, was the first true advance toward modern informed consent. Another turning point was in the 1914 case of Schloendorff v. Society of New York Hospital [4–6, 9], where, despite the patient's explicit refusal to a procedure, the surgeons proceeded. Furthermore, complications arose from the surgery, and the doctors were forced to amputate some of the patient's fingers as a result. In this case, the court ruled that "Every human being of adult years and sound mind has a right to determine what shall be done with his own body [. . .]" [6, 9]. A third notable court case is the 1957 case of Salgo v. Leland Stanford Jr. University Board of Trustees [4, 5], wherein the term "informed consent" was used for the first time. In this case, a patient awoke permanently paralyzed after a routine procedure. The patient had never even been informed that such a risk could exist, and as such, the ruling of this case caused modern medical care to take note that failure to disclose medical risks or alternative treatments could be cause for legal action [5].

When humans are left to their own volition in the void of a formalized definition of informed consent, it is not merely medical care-related misconduct that arises. Over the past century, there have been numerous research studies in the medical field with disastrous outcomes that have even been deemed as human rights violations. These, in turn, have spurred the creation of numerous legal statutes broadly circumscribing the notion of informed consent as well as the creation of the Belmont Report and Institutional Review Boards.

9.2.3 Breaches of the Consent Principle in Scientific Research

The first event was the Nazi's medical experimentation throughout World War II [4, 5, 10]. Nazi physicians tortured and killed non-consenting concentration camp prisoners in the name of scientific research. In 1946, these atrocious acts were heard in a trial known as "The Medical Case" which ultimately led to the introduction of the Nuremberg Code and, much later, indirectly led to the creation of the Declaration of Helsinki.

Another famous example that arose due to the lack of proper research scrutiny was the Stanford Prison Experiment of the 1971 [10]. Herein, a group of students were asked to act as prisoners and guards in an improvised prison environment, ultimately taking the roles to heart and resulting in prolonged psychological trauma for the subjects [11].

The third notable unethical human subject research study was the Tuskegee Syphilis Experiment, which was brought to light immediately after the Stanford Prison Experiment [4, 10]. In this experiment, 600 impoverished African Americans were studied regarding the effects of untreated syphilis over a 40-year span, despite being told the experiment would last only a few months. The subjects were not informed if they actually had the disease and were not offered any medicine (despite there being a known cure) [12]. This massive human rights violation, along with the aforementioned Stanford Prison Experiment, led to the creation of the Federal Policy for the Protection of Human Subjects (also known as the Common Rule), as well as the 1979 Belmont Report, both of which, in essence, required oversight processes for research involving human subjects to be conducted by governing bodies (which later became Institutional Review Boards or Ethical Review Boards) and detailed basic ethical principles in human subject research [4, 10].

9.3 The Definition of Informed Consent

Informed consent must include three vital "pillars" [7]:

1. Study information: Disclosure of all information, including any and all risks about a study to the participants regardless of the effect they may have on a subject's willingness to participate in the study.
2. Subject's comprehension and understanding: The mental capacity of the subjects should be approached in such a way as to fully cognize the information given to them by the researchers, as well as understand the risks and benefits of their participation.

3. Voluntary participation: Necessity of a subjects' consent to be free and voluntary (both for entering and withdrawing from a study), with no caveats, coercion, or influence from the investigator.

These three pillars can be combined to form an overarching modern-day definition of the notion of informed consent: "Informed consent is the full disclosure of the nature of the research and the participant's involvement, adequate comprehension on the part of the potential participant, and the participant's voluntary choice to participate" [7, 13].

These three pillars are easily addressable in the classical model of informed consent gathering (i.e. one consent per study, per subject). However, the manner of collecting informed consent in today's medical research landscape has evolved significantly, and it has become increasingly difficult to ensure that all three of the abovementioned pillars are addressed prior to gaining a research subject's consent.

9.4 Types of Consent and Their Collection

In general, modern informed consent for medical research is gathered for one of the three purposes:

1. For an individual's participation in a single study (such as a clinical trial).
2. For an individual's participation in a biomedical database to be used for multiple undefined research studies (participation in multiple studies).
3. For the return (and permitted sharing) of medical information related to the subject.

Traditionally, consent was gathered for specific studies, wherein a subject would agree (or decline) to participate in a study after learning the risks and benefits of the research (e.g., clinical trials, surveys, etc.). As these studies were most often done in person, gaining consent was relatively simple and comprehensible. The boundaries of these studies were generally clear, and as such, consent for participation was rather straightforward.

Modern medical research relies heavily on data analytics and on large databases of multiple information types from various observational and instrumental sources such as health data, research data, and data from health devices/services. This data is often used in multiple studies, for varied research purposes, and is often shared widely with other investigators and over an indefinite time period. This stands in contrast to the traditional, in-person, single study research mentioned in the first type of consent. Nonetheless, informed consent still needs to be taken every time such data is used for research. This problem of retroactive (or proactive) informed consent gathering creates a sundry of new issues that have barely been defined in modern medical guidelines and legal statutes.

Finally, the third type of consent ties neatly in with the first and second and is often the most forgotten category of consent: the consent for the return (and permitted sharing) of a patient's medical information. Ever since the Tuskegee Syphilis Experiment of 1971, it has become highly unethical to not disclose needed medical information to a subject or to refuse to share medical findings with permitted third parties (such as a subject's chosen medical provider). For example, if during a study the researchers discover that a subject has an actionable (i.e., treatable) disease, it is their ethical obligation to relay this information to the research subject should they choose to receive it [13].

Even when studies allow for the return of information back to participants, the recommendation is to return "actionable results" only, that is, results associated with some kind of available preventive strategy. However, these recommendations exclude the individuals from the decision, reducing their autonomy and subsequently the "voluntariness" pillar. Moreover, in such cases, the subject should also be able to share these findings with their chosen healthcare provider, without any major hurdles. However, despite these ethical obligations, many research studies still often restrict what information can and should be given back to subjects or shared with permitted third parties [13] (Fig. 9.1).

9.5 Recent Advances and Their Effect on Consent Collection

As the three pillars of informed consent are addressed, a multitude of new issues and problem surface. These are information sharing, overwhelming of the subjects, subject vulnerability, and emergency research. Each of these can hinder proper informed consent gathering and needs to be addressed in every research scenario.

First, the sharing of information needs to be considered depending on the subjects' needs. As illustrated in Fig. 9.1, the pillar of Information can be subdivided into not just the information of the study, as previously defined, but also information related to a patient, as well as who this information can be shared with, what information should be shared, and under what circumstances. In some cases, it is essential for subjects participating in medical research to receive vital information about their health [13]. It may also be necessary to share them with the subject's medical providers, so that they may be able to seek advice and treatment.

Second, the overwhelming of subjects are very real phenomena, as detailed by Bester et al. [14]. The authors argue that "there are medical situations in which the information involved in making a decision is of such a nature that the decision-making capacity of a patient is overwhelmed by the sheer complexity or volume of information at hand. In such cases a patient cannot attain the understanding necessary for informed decision making, and informed consent is therefore not possible" ([14], p. 871). In these cases, it becomes the researchers' duty and obligation to convey the study information in such a way that the subject is able to understand it without becoming overloaded. This can be done in many ways, for example, with the aid of a surrogate decision-maker (having someone else makes the decision for the subject) or through the use of shared decision-making (where the physician or researcher deliberates with the patient, assisting them to understand topics, until a decision can be reached) [14]. If this is not employed correctly, both the Information and Comprehension pillars of informed consent are jeopardized, as portrayed in Fig. 9.1.

Third, it is vital to protect vulnerable subjects, as this can severely affect their comprehension of research information and their voluntary participation (as shown in Fig. 9.1). Vulnerable subjects can be defined as people who lack the capacity to consent or may be exposed to coercion to participate in medical research such as children, prisoners, mentally impaired, sick individuals desperately seeking medical treatment, or poor individuals needing money from medical trials [15]. In these cases, researchers must take the utmost care in ensuring they are not inadvertently coercing individuals or pressuring them to participate in research. An example of this would be the aforementioned Nuremberg medical experiments, where prisoners were forced to participate in clinical trials and medical experiments.

Fourth, when it comes to emergencies (such as the current global COVID-19 pandemic), there are cases when it may be acceptable to forgo following the three pillars of informed consent when conducting time-sensitive necessary research. However, how and when this can be done is still a matter of debate. For example, Title 21 of the Code of Federal Regulations that governs the Food and Drug Administration of America (FDA) details the requirements for waiving informed consent under certain emergency circumstances (which can be permitted with prior Institutional Review Board approval) [15].

As with the aforementioned Tuskegee Syphilis Experiment, even under necessary circumstances, research can be done ethically or unethically. To this day, the Tuskegee case has had negative effects on the black community and their initiative in taking vaccines (such as the new COVID-19 vaccines) for the fear of a repeat of such a travesty [16, 17]. In fact, since 2005, it is estimated that only two-thirds of published cases in emergency research were even registered with ethics committees [15]. As such, these emergency situations often invalidate a subjects' voluntariness (as seen in Fig. 9.1) in participating in a

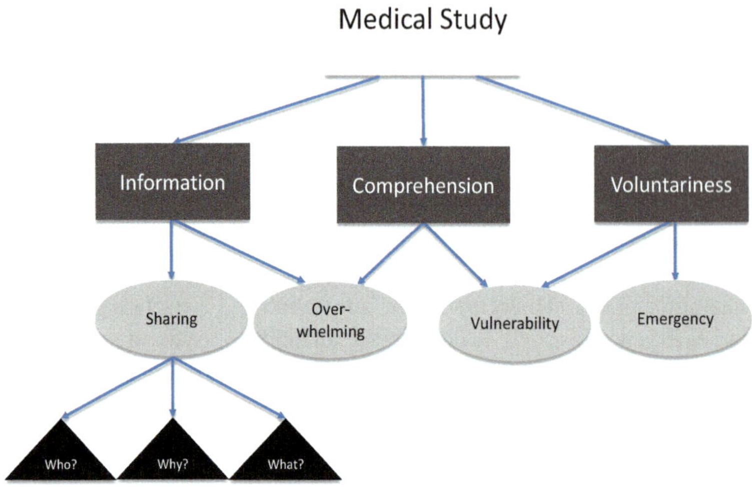

Fig. 9.1 Potential pitfall areas in informed consent collection

research study, as they may not even know that they are participating in medical research or feel an undue need to participate.

9.5.1 Broad and "Opt-Out" Consent

While the above issues could affect the ability to properly collect informed consent for medical research, another large issue arising is the ability to do all of this in a continued fashion for multiple, continuous, ongoing, or changing studies. To deal with this modern consent dilemma, the majority of modern data warehouses practice "broad" consent [18]. Broad consent authorizes all future and unforeseeable data uses without any specific information about any future research, nor any possibility for further oversight [19], and it is usually collected once at data collection time. As such, it grants the data holder full governance over participants' data.

Another common consent is the "opt-out." It assumes broad consent for research by default, unless individuals explicitly opt out. These consent models were introduced to avoid having to contact data subjects every time the data is used in a research study, which is seen as a big hurdle. However, these consents tend to diminish subjects' autonomy and information at the advantage of research and science. To compensate for the downgrading of consent, institutional review boards and ethics review committees have been tasked to provide research oversight, by reviewing each data usage/sharing instance to ensure that it is not problematic. However, the extent of oversight and control from these boards is subjective and varies from one implementation to another.

Although these consents are instituted in certain countries (e.g., broad consent in the new revision of the US Common Rule), they are not seen as ethically acceptable. As shown by Hayden, they can increase the possibility of research misconduct by taking advantage of people's vulnerabilities [20] (such as their lack of comprehension of the risks and their need for treatment).

9.6 Examples of Misconduct in Informed Consent

While born out of varying necessities, each of the major research ethics documents (the 1946 Nuremberg Code, the 1964 Declaration of Helsinki, the 1979 Belmont Report, and the 1991 Common Rule) serves as a slightly different statement of ethical principles for medical research involving human subjects or identifiable human material and data [4, 6]. Nevertheless, all of these international efforts are focused on one common goal: to achieve reliable research practices and outcomes that do not violate human rights and privacy and contribute significantly toward advancing the science of our well-being. These principles, standards, regulations, and laws were all placed to govern research practices and therein combat inherent misconduct and fraud.

Naturally, breaches in ethics and morals regarding informed consent fall under these many guidelines discussing misconduct and fraud in medical research. Such fraud and misconduct have been an archaic thorn in the side of medical research, yet they still persist, even in today's modern medical world. Unfortunately, most informed consent cases are settled quickly and quietly by funding agencies and the corporations responsible for the research. As such, the majority of these cases rarely make it into the public eye [21]. Table 9.1 reviews a selection of these cases (in chronological order) and elaborates on them in terms of their location, the violations of informed consent, their respective consequences, and the pillars (and/or sub-domains) of informed consent that were broken.

When reviewing Table 9.1, three interesting patterns arise:

First, it is interesting to note that the majority of the violations listed above occur in third world and impoverished locations. As noted by Nair [32], it is significantly cheaper to conduct clinical trials in these regions, and as regulations are more relaxed, companies often fail to seek adequate consent in these cases. Even though only 1.4% of clinical trials done worldwide take place in India, as evidenced in the table above, there are numerous examples of publicly reported violations of informed consent and research misconduct. This could be simply because there are an extraordinary number of such cases in third world countries like India, but more likely it is because most informed consent cases in the Western world are settled swiftly and silently [21].

Second, the violations were most often of Information and Comprehension. These two constructs are inextricably intertwined, as subjects not only need to be provided with the requisite information regarding a study but also have the adequate mental faculties to process and understand this information. In the majority of these cases, researchers did not provide their subjects with the full story of the research at hand. Due to this, subjects agreed to participate in these trials without full Information (and Comprehension) of the risks involved. As such, it can be said that these participants agreed to participate in the research under false pretenses. As mentioned prior, hiding information from subjects is a severe offense in research and, as evidenced in Table 9.1, can lead to numerous consequences such as dismissal or suspension of the researcher, legal investigations along with heavy fines, and even the dismantling of the research organization.

Third, the only sub-domain of informed consent repeatedly violated was Vulnerability. Researchers utilized children, mentally ill patients, and impoverished individuals needing free access to medicine, among others, to gain easy access to research subjects for their research purposes. Of course, this is a grave violation of informed consent, and researchers should take care to ensure that they are not taking advantage of any vulnerable individuals or groups, as their true "voluntariness" may be subject to questioning.

9.7 New Perspectives

The advances of big biomedical data are posing new challenges to informed consent and are creating a friction between proponents of open data and proponents of informed consent. The former are calling for a relaxed approach on consent (such as broad consent), while the latter are calling for a refocus and reform on informed consent. New research in the area suggests alternatives to consent such as "community-based governance" which recognizes people's willingness to engage in risky activities for the benefit of others and thus shifts the decision from the individual to the community [33]. Other alternatives offer participants complete control over the usage of their data

Table 9.1 Examples of consent breaches in medical research

Case and source	Description	Consequences	Pillar of informed consent violated
Trovafloxacin trials [22] Kano, Nigeria, 1996	Parent consent was not sought, and preliminary ethical approval was not provided. A total of 11 deaths occurred in both arms (190 participants), besides cases of brain damage and paralysis	Two initial lawsuits in the USA were redirected to Nigeria. In 2007, Nigeria filed charges of $7 billion and subsequently ordered the arrest of eight former directors of Pfizer. Pfizer denies any wrongdoing	Information, comprehension, and voluntariness. As children were the subject of the trial, the sub-domain of vulnerability is also violated
Cariporide trial [22] Argentina, 1997	None of the 137 patients fully consented. The signatures were forged, or participants ignored the nature of the study. Thirteen patients died	A criminal lawsuit was started against Aventis, and the company decided not to apply for US license	Information, comprehension, and voluntariness
Bezwoda trials [23, 24] South Africa, 2000	Prof. Werner Bezwoda failed to submit his protocols for ethical review or to collect informed consent from participants	Dismissal of Prof. Werner Bezwoda from the university and publishing an article negating his reported outcomes	Information, comprehension, and voluntariness
NDGA trials [22] India, 2000	A drug was tested without adequate preclinical investigation, informed consent, and alternative treatment, on 26 cancer patients; there were 2 deaths	The research was interrupted for 6 months. The Indian Council of Medical Research conducted an inquiry; however, the results were not made public. Johns Hopkins University barred the involved scientist from serving as principal investigator on any future human research	Information and comprehension
New York Foster Care Children trials [22] USA, 2002	Clinical trials with HIV-infected children at the New York City Agency for Children's Services (ACS). Only ACS signed the consent forms. Participants became severely ill with life-threatening side effects	The trials were halted in 2002. An investigation confirmed noncompliance with legal regulations	Information, comprehension, and voluntariness. Vulnerability (pediatric patients) was also violated
Hepatitis E vaccine trial [22] Nepal, 2003	A vaccine was investigated by US Walter Reed Institute and GlaxoSmithKline on 2000 Royal Nepalese Army soldiers. They were poor and risked coercion by superiors	Post-trial access to the vaccine was requested in case of approval, before further studies. Apparently GSK did not respond	Information, comprehension, voluntariness, and vulnerability
Nevirapine PMTCT trials [22] Uganda, 2003	Consent about protocol changes was not collected, wrong doses were given, and adverse effects were not disclosed to participants	In 2004, the FDA recommended to stop using the drug with certain patients and issued warnings	Information and comprehension
Letrozole trials [22] India, 2003	Letrozole (Sun Pharmaceuticals) was investigated for an off-label indication (breast cancer in postmenopausal women), without informed consent	After complaints, a clarification letter was sent to relevant professionals	Information and comprehension
Risperidone trials [22] India, 2003	Risperidone (Johnson & Johnson) was tested for mania against placebo, depriving participants from baseline medications. Consent was also inadequate	Johnson & Johnson assured that consent had been obtained	Information, comprehension, and voluntariness

(continued)

Table 9.1 (continued)

Case and source	Description	Consequences	Pillar of informed consent violated
VGV-1 trials [22] China, 2003	HIV-positive patients were given an investigational drug with inadequate consent and information about risks, personal expenses were not reimbursed, and results were not made available to them	The Institutional Review Board of the National Center for AIDS/STD Prevention and Control recommended that trials must be better explained and expenses should be paid	Information and comprehension
TGN 1412 trials [22] UK, 2003	An apparently safe investigational drug caused potentially fatal multi-organ failure in six volunteers. Informed consent failed to inform potential risks and had other inadequacies	The clinical trial was suspended	Information and comprehension
Tenofovir trials on HIV transmission [22] Cameroon, 2005	400 tested sex workers received inadequate information about risks, in a language many did not speak	The trials were cancelled in March 2005	Information and comprehension
SFBC Miami Test Centre [22] USA, 2005	Poor immigrants were enrolled in studies with inadequate information and were not always allowed to leave the protocol. Payment was retained till the end, as a form of coercion	SFBC denied the allegations; however, the test center eventually closed	Information, comprehension, and voluntariness
ART treatment interruption trials [22] Uganda and Zimbabwe, 2006	Patients were desperate to get free treatment, some risks were not disclosed, and during treatment interruption, several deaths occurred	Participants at increased risk of adverse events including death were not appropriately informed	Information, comprehension, voluntariness, and vulnerability
Bhopal Memorial Hospital and Research Centre trials [25, 26] *India, 2010*	The ethics committee did not appropriately consider the risks. *Some subjects were told by doctors that they always did trials for new drugs, and there was "nothing to worry about"*	The hospital made more than 10 m rupees (£140,000) from British, US, and French drug companies for carrying out the trials	Information, comprehension, voluntariness, and vulnerability
HPV trials [25, 27] India, 2011	Participating students (children) were not informed that the vaccine was a trial, not a compulsory government program, and entailed risks	After many had been vaccinated, the government eventually stopped the trials	Information, comprehension, voluntariness, and vulnerability
Indore trials [25, 28, 29] India, 2011	Mentally ill participants were engaged in clinical studies without informed consent	Suspension of the responsible doctors for 3 months	Information, comprehension, voluntariness, and vulnerability
Covaxin trials [30, 31] India, 2020	A coronavirus vaccine was applied without adequate information especially regarding the use of placebo in half of the cohort	A potentially questionable trial that could undermine the quality of the data and the trust in the vaccine	Information, comprehension, voluntariness, vulnerability, and emergency

without any reliance on third parties [34]. Autonomy is achieved through the usage of blockchains, and access to individual-level data proceeds according to individual's permissions that can be updated and revoked in real time.

References

1. Gupta A (2013) Fraud and misconduct in clinical research: a concern. Perspect Clin Res 4(2):144–147
2. Polanin-Huk J, Huk J, Filip R (2010) Fraud and misconduct in clinical research. J Pre-Clin Clin Res 4(2): 158–160
3. George SL, Buyse M (2015) Data fraud in clinical trials. Clinical Investigation 5(2):161–173
4. Capron AM (2018) Where did informed consent for research come from? J Law Med Ethics 46:12–29
5. Green DST, MacKenzie CR (2007) Nuances of informed consent: the paradigm of regional anesthesia. HSS J 3:115–118
6. Wolf SM, Clayton EW, Lawrenz F (2018) The past, present, and future of informed consent in research and translational medicine. J Law Med Ethics 46:7–11
7. Dankar FK, Gergely M, Dankar SK (2019) Informed consent in biomedical research. Comput Struct Biotechnol J 17:463–474
8. Pratt v Davis (1905). https://www.ravellaw.com/opinions/967d63e3e4aceb0c7ccc62e78af2d543. Accessed 23 March 2021
9. Schloendorff v. Soc'y of N.Y. Hospital (1914). https://www.ravellaw.com/opinions/13aa93a26f08eeced5264b5eaa468369?query=Schloendorff%20v.%20Soc%27y%20of%20N.Y.%20Hospital%2C%201914. Accessed 23 March 2021
10. Plant R, Pons AP (2006) MIS research involving human subjects: processes and IRB requirements. Commun Assoc Inf Syst 17:1–46
11. Haney C, Banks WC, Zimbardo PG (1973) A study of prisoners and guards in a simulated prison. Naval Res Rev 30:4–17
12. Jones JH, Tuskegee Institute (1981) Bad blood: the Tuskegee syphilis experiment. Free Press, New York
13. Dankar FK, Gergely M, Malin B, Badji R, Dankar SK, Shuaib K (2020) Dynamic-informed consent: a potential solution for ethical dilemmas in population sequencing initiatives. Comput Struct Biotechnol J 18:913–921
14. Bester J, Cole CM, Kodish E (2016) The limits of informed consent for an overwhelmed patient: clinicians' role in protecting patients and preventing overwhelm. AMA J Ethics 18(9):869–886
15. Nandra R, Brockie AF, Hussain F (2020) A review of informed consent and how it has evolved to protect vulnerable participants in emergency care research. Effort Open Reviews 5:73–79
16. Reeves J (2021) Vaccine skepticism lurks in town famous for syphilis study. AP News. https://apnews.com/article/coronavirus-vaccine-skepticism-us-eae3315d2dfecd6443b56a2dc7bcb887. Accessed 23 March 2021
17. Vaidya A (2020) How Tuskegee experiment puts extra burden on black physicians during pandemic. Becker's Hospital Review https://www.beckershospitalreview.com/hospital-physician-relationships/how-tuskegee-experiment-puts-extra-burden-on-black-physicians-during-pandemic.html. Accessed 23 March 2021
18. Dankar FK, Ptitsyn A, Dankar SK (2018) The development of large-scale de-identified biomedical databases in the age of genomics—principles and challenges. Hum Genomics 12(19):1–15
19. Mittelstadt BD, Floridi L (2016) The ethics of big data: current and foreseeable issues in biomedical contexts. Sci Eng Ethics 22(2):303–341
20. Check Hayden E (2013) Privacy protections: the genome hacker. Nature 497(7448):172–174
21. Fuchs RE, Robinson SG (2018) Strict liability for lack of informed consent in clinical trials. Troutman Pepper Hamilton Sanders LLP. https://www.troutman.com/insights/strict-liability-for-lack-of-informed-consent-in-clinical-trials.html. Accessed 23 March 2021
22. Weyzig F, Schipper I (2008) SOMO briefing paper on ethics in clinical trials #1: examples of unethical trials. Centre for Research on Multinational Corporations (SOMO) https://www.somo.nl/examples-of-unethical-trials/. Accessed 23 March 2021
23. Crown J (2000) breast-cancer treatment put into question. Lancet 355(9217):1828–1829
24. Weir C, Murray G (2011) Frauds in clinical trials: detecting it and preventing it. Significance 8(4): 164–168
25. Bhakuni H (2020) Informed consent to clinical research in India: a private law remedy. Medical Law International 20(3):256–283
26. Lakhani N (2011) From tragedy to travesty: drugs tested on survivors of Bhopal. Independent. https://www.independent.co.uk/news/world/asia/tragedy-travesty-drugs-tested-survivors-bhopal-6262412.html. Accessed 23 March 2021
27. Bagla P (2013) Indian Parliament comes down hard on cervical cancer trial. Science. https://www.sciencemag.org/news/2013/09/indian-parliament-comes-down-hard-cervical-cancer-trial. Accessed 23 March 2021
28. Chatterjee S (2017) MCI suspends 8 Indore doctors' registration over unethical drug trial. News18 https://www.news18.com/news/india/mci-suspends-8-indore-doctors-registration-over-unethical-drug-trial-1482739.html. Accessed March 23, 2021
29. Krishnan V (2012) MP govt to act against 11 Indore doctors. Mint. https://www.livemint.com/Politics/NxmwzRufEq2Efjcl888OFJ/MP-govt-to-act-against-11-Indore-doctors.html. Accessed 23 March 2021
30. Bhuyan A (2021) How covaxin trial participants in Bhopal were misled. The Wire. https://science.

thewire.in/health/peoples-hospital-bhopal-covaxin-clinical-trials-exploitation-ethics-ground-report/. Accessed 23 March 2021

31. Mitra E, Hollingsworth J (2021) More than a dozen slum residents in an Indian city say they thought they were being vaccinated. They were part of clinical trials. CNN https://edition.cnn.com/2021/02/25/asia/india-vaccine-trials-covid-ethics-intl-dst-hnk/index.html. Accessed 23 March 2021
32. Nair A (2015) Clinical research: regulatory uncertainty hits drug trials in India. The Pharmaceutical J. https://pharmaceutical-journal.com/article/feature/clinical-research-regulatory-uncertainty-hits-drug-trials-in-india. Accessed 23 March 2021
33. Prainsack B, Buyx A (2013) A solidarity-based approach to the governance of research biobanks. Med Law Rev 21(1):71–91
34. Grishin D, Raisaro JL, Troncoso-Pastoriza JR et al (2021) Citizen-centered, auditable and privacy-preserving population genomics. Nat Comput Sci 1: 192–198

Research Biobanks and External Researchers Under the European General Data Protection Regulation: Between Controller-Processor Relationship and Joint Controllership

10

Andrea Parziale and Ciara Staunton

Abstract

The growing involvement of biobanks in international collaborative research creates invaluable opportunities but raises critical ethical and legal issues, e.g. on the use and sharing of health and genetic data. The appropriate governance of biobanks is key to ensuring research integrity and public trust in the methods and findings of the research. A crucial part of this requires identifying who is the controller, joint controller, or processor of personal data under the European General Data Protection Regulation (GDPR). These qualifications determine who is liable for compliance with different data protection provisions and how data subjects can implement their rights.

This chapter maps the issues related to the GDPR qualifications of biobanks and external researchers. In particular, it outlines whether and under what conditions biobanks and external researchers are in a controller-processor relationship or joint controllers and the respective operational implications for biobanks. Finally, it provides some general practical advice to research biobanks.

10.1 Introduction

The global landscape of biobanks has grown exponentially over the years [1], and biobanking has emerged as a critical reference for biomedical research in the fight against the novel coronavirus, for instance [2]. Indeed, the number of biobanks across the globe continues to multiply, with their global market being expected to grow at a compound annual rate of 6.2% between 2021 and 2028, exceeding 100 billion US dollars by 2028 [3].

Simultaneously, the role and modus operandi of research biobanks have changed significantly. While biosample and data collections were originally confined to individual researchers' laboratories, over time biobanks have become complex infrastructures for multi-institutional research projects, thanks to the advances in digitalisation and other technologies [4]. The emergence of national and international biobank networks and increasingly numerous collaborations with other private and public institutions have boosted the development and valorisation of research biobanks [5].

A. Parziale
Institute for Biomedicine, Eurac Research, Bolzano, Italy

Luiss Guido Carli University, Rome, Italy
e-mail: andrea.parziale@maastrichtuniversity.nl

C. Staunton (✉)
Institute for Biomedicine, Eurac Research, Bolzano, Italy

School of Law, Middlesex University, London, UK
e-mail: ciara.staunton@eurac.edu

J. Faintuch, S. Faintuch (eds.), *Integrity of Scientific Research*,
https://doi.org/10.1007/978-3-030-99680-2_10

10.1.1 Commercial, Scientific, and Public Health Implications

These developments entail an ever-stronger drive to sample and data sharing across different institutions, companies, and countries [6]. To accelerate both health research and its translation into the practice of medicine, funding agencies increasingly encourage and sometimes require biobanks to share data and collaborate with national and international institutions, including economic operators (e.g. pharmaceutical companies) [7].

The growth of biobanks and their involvement in international collaborative research create invaluable opportunities for scientific and medical progress but raises significant ethical and legal issues, partly on the use and sharing of health and genetic data [8]. Issues of consent, return of unsolicited results, risk of stigmatisation and discrimination, and community engagement are just some of the ethical issues that must be considered [9]. The appropriate governance of biobanks must also be addressed to ensure the integrity of the research and the public's trust in the methods and findings of the research. A crucial part of this requires identifying the respective duties and responsibilities of all the actors involved in personal data processing for multi-institutional (and multinational) scientific and medical research projects. Important in this is Regulation (EU) 2016/679 on the protection of natural persons with regard to the processing of personal data and on the free movement of such data (General Data Protection Regulation, GDPR).

10.1.2 Managing and Operative Roles

There are certain key individuals listed in the GDPR. As defined in Article 4, they include a controller who "determines the purposes and means of the processing of personal data" and a processor who "processes personal data on behalf of the controller". There is also the possibility of joint controllers under Article 26, whereby "two or more controllers jointly determine the purposes and means of processing". However, it is not always clear whether biobanks and external researchers are joint controllers or in a controller-processor relationship.

10.2 Data Protection in Research Biobanks: Exemptions and Safeguards

Within research biobanks, appropriate data governance is key to ensuring the integrity of the research and the public's trust in the use of the samples and data. In this respect, compliance with the GDPR is of primary importance.

The GDPR adopts a principle-based approach to the processing of personal data (Article 5). At the outset, personal data must be processed in a lawful, fair, and transparent manner towards the data subjects. In addition, data must be "collected for specified, explicit and legitimate purposes and not further processed in a way incompatible with those purposes" (purpose limitation). Furthermore, data must be limited to what is necessary for such purposes (data minimisation) and must not be kept longer than necessary (storage limitation). Finally, data must be accurate and processed in a secure way. Adherence to these principles is likely to change data management practices for biobanks, and importantly, it is the controller that must ensure and demonstrate compliance with all these principles (accountability principle, Article 5(2) GDPR).

10.2.1 Rules for Sensitive Data

In addition to this, the GDPR generally prohibits the processing of certain "special categories of data" (Article 9(1) GDPR), that includes health-related and genetic data (Article 9(1)), unless the processing falls within one of the exceptions as set out in Article 9(2). For research biobanks, they may process "sensitive" data if the data subjects give explicit consent; it is performed in the legitimate activities of not-for-profit bodies (with appropriate safeguards); the data subject made the data public; it is necessary for reasons of substantial public interest and this public interest, that can include public health, is based on EU or Member State law that also provides for appropriate safeguards; and it is necessary for research, with a basis in EU or Member State law and with appropriate safeguards. It is for the controller to identify the basis on which to process sensitive data and ensure that there are appropriate safeguards in place to respect the fundamental rights and interests of the data subject.

10.2.2 Rightful Disclosures and Protections to Biobank Participants

The GDPR also regulates data subjects' rights. These include the right to information, which includes information on the identity of the controller, the contacts of the Data Protection Officer, the purpose and legal basis of the processing, the recipients of the personal data, and, if applicable, the data transfers to a third country or an international organisation (Article 13 GDPR); the right to access personal data concerning them that are processed by the controller (Article 15 GDPR); the right to rectification of inaccurate personal data concerning them (Article 16 GDPR); the right to the erasure of personal data concerning them ("right to be forgotten", Article 17 GDPR), e.g. if the data are no longer necessary for the purposes of the processing or the data subject withdraws their consent and there is no other lawful basis for the processing; and the right to object to the processing of personal data when based on a public or legitimate interest (Article 21 GDPR).

10.2.3 Scientific Research Exemptions

If personal data is processed for scientific research purposes, the GDPR allows for derogations from several processing requirements and data subjects' rights. In particular, personal data may be retained for longer than necessary and used for a purpose not specified at data collection if it is to be used for research (Article 5(1) (c) and (e) GDPR). The rights to access (Article 15), rectification (Article 16), erasure (Article 18), and object (Article 21) may also be derogated if they are likely to render impossible or seriously impair the research (Article 89(2) GDPR) [10].

However, these derogations and an exemption to the general ban on the processing of personal information can only apply if safeguards "for the rights and freedoms of the data subject" are in place (Article 89 GDPR). Article 89 requires such safeguards to be "appropriate" and "in accordance with the GDPR" and "ensure that technical and organisational measures are in place in particular in order to ensure respect for the principle of data minimisation".

Further guidance on potential safeguards may be found in soft law tools, international treaties, and other legal instruments regulating the use of health research data [11]. These instruments point out that clear and transparent governance procedures that oversee the use of data are crucial to protecting data subjects' rights and promoting trustworthiness [11]. These governance procedures should include policies on data transfer, feedback of findings, storage of data, withdrawal of consent, re-contact of data subjects, access requests of data subjects, intellectual property, and commercial exploitation of research results [11].

10.2.4 Conflicting Interests of Biobank Stakeholders and External Parties

As an appropriate safeguard, research biobanks' policies should also appropriately regulate access requests from third parties [12]. Access policies may be either open-access or controlled-access. In open-access models, data is made available for users without any constraint. Initially supported by the Human Genome Project and numerous funders, these models have raised concerns regarding the re-identifiability risk inherent in health and genetic data [13] and the possibility of exploitation and inequitable data use [14]. Indeed, unrestricted open access risks perpetuating inequalities in knowledge production and distribution [14]. Notably, such policies may not be legally compliant with data protection regulations worldwide, which increasingly afford special protection to health and genetic data.

10.2.5 Supervising Biobank Access

When "sensitive" data are shared, controlled-access models are, therefore, required [15]. Such models require clear and transparent criteria and procedures for data users' qualifications and the review procedure [16]. They generally operate through Data Access Committees (DACs) that review data access requests from external researchers for approval or disapproval [16]. Since there are no widely accepted guidelines and criteria, the organisation and functioning of DACs tend to vary significantly across biobanks. In principle, DACs are called to promote data sharing and protect the rights of data subjects and communities [17]. To this end, DACs assess the consistency of the proposed research with the original consents and evaluate potential harms for data subjects and communities, such as misuse of data, breaching participant confidentiality, and group discrimination and stigmatisation [17]. Given the complexity of these assessments, DACs should include a reasonable number of components covering relevant scientific and ethical-legal expertise [18]. DACs, especially those of institutions and consortia, usually publicly share information on the assessment criteria on their web pages or in their respective publications. Transparent access procedures are crucial to promoting trustworthy data sharing [19].

10.2.6 Transfer Agreements

Once a DAC approves an access request, it is a standard practice that the biobank and the research partner conclude the so-called access agreements [16]. This is to legally bind the parties to ensure that research participants' data is adequately protected during the execution of the research. Such contracts usually include Data Transfer Agreements (DTAs) and, if biological samples are also shared, Material Transfer Agreements (MTAs). In specifying the conditions of data and biosample transfers, DTAs and MTAs must clarify the different roles of the parties regarding specific data processing operations. The GDPR differentiates the roles of controller and processor. Since these roles have distinct duties and responsibilities regarding data processing, effective and transparent data governance requires that it is clear from the outset who is the controller and who is the processor.

10.3 Research Biobanks as Controllers or Processors

The legal qualification of the parties as controllers or processors depends on their decision-making role regarding the purpose and means (i.e. the *why* and *how*) of a specific data processing. This emerges from the respective definitions under Article 4(7) and (8) GDPR. Under Article 4(7), a controller is the natural or legal person or other body that determines the processing purposes and means of the processing. They will also decide on the lawful basis for the processing of personal data and overall responsibility for compliance with GDPR. Under Article 4(8), the processor is, instead, defined as a natural or legal person or other body which processes personal data "on behalf" of the controller.

Further guidance on these definitions may be found in Opinion 1/2010 of the Article 29 Data Protection Working Party [20] ("Opinion"). Although this document was published under the previous Data Protection Directive, it continues to be relevant under the GDPR as the legal definitions of controller and processor have not changed. The Opinion clarifies that controller and processor are independent functional concepts to be determined based on factual analysis, regardless of the terms used in contractual arrangements.

A controller is defined by the concrete ability to decide on the purposes and the essential elements of the means of the processing. These latter include the elements "traditionally and inherently reserved to the determination of the controller, such as 'which data shall be processed?', 'for how long shall they be processed?', 'who shall have access to them?'". Conversely, more practical methodological issues regarding the means may be delegated. This role can stem either from a legal entitlement or just "factual influence". The CJEU's case law has confirmed this broad concept of controllership in *Google Spain* (C-131/12), *Wirtschaftsakademie* (C-210/16) [21], and *Fashion ID* (C-40/17) [22].

10.3.1 External Data Processing

The Opinion also clarifies that the processor is a legal entity separate from the controller and processing personal data on behalf of the controller. "Separate from the controller" means that the processor must be an external organisation. The EDPB's Guidelines 07/2020 on the concepts of controller and processor in the GDPR [23] ("Guidelines") confirm that the controller's employees are not processors because they process data within the controller's entity. Instead, acting "on behalf" of the controller means that the processor acts in the interest of the latter. Thus, the processor must follow the decisions and instructions of the controller concerning the purpose and the essential elements of the means of the processing. Also, when processing personal data, the processor must not pursue any interest of their own. Although enjoying some flexibility, the processor infringes the GDPR if it goes beyond the controller's instructions to the point that it decides on the purpose and means of the processing.

In sum, the definitions of controller and processor are context-dependent and require a case-by-case assessment. In principle, a research biobank qualifies as a controller when it determines the purpose and means of the processing. Therefore, a biobank that collects biosamples and associated data and reuses them in research will qualify as a controller. As such, the biobank must ensure, and be able to demonstrate, compliance with the data protection principles (Article 5 GDPR). Secondly, it must facilitate the exercise of the data subjects' rights (Article 14(2) GDPR) and fulfil the relevant obligations (Articles 15–21 GDPR). It must also implement appropriate technical and organisational measures to ensure, and be able to demonstrate, compliance with data protection law (Article 24 GDPR).

10.3.2 Privacy Protection at All Levels

Furthermore, both when determining the purpose of the processing and during the processing itself, the biobank must adopt appropriate technical and organisational measures (e.g. pseudonymisation) to implement the data protection principles effectively (e.g. data minimisation), integrating the necessary safeguards in the processing to meet the GDPR requirements and safeguard the data subjects' rights ("privacy by design" Article 25(1) GDPR). Finally, the biobank must adopt appropriate technical and organisational measures to ensure that, by default, only the personal data necessary for the purpose of the processing are processed ("privacy by default", Article 25(2) GDPR).

10.3.3 Hefty Financial Punishments

Violation of the GDPR obligations may expose the biobank to compensation claims from data subjects (Article 82(2) GDPR), unless it demonstrates that it was in no way responsible for the event causing the damage (Article 82(3) GDPR). The biobank is also subject to administrative fines up to ten million euros (or 2% of the global turnover if it is an undertaking) by the competent supervisory authority (Article 83 GDPR).

Conversely, if an external entity acts as a controller and a biobank processes personal data on its "behalf", the biobank qualifies as a processor. As such, the biobank must fulfil the obligations outlined in Article 28 GDPR. In particular, the biobank can process data only following the controller's documented instructions. It must also bind the persons authorised to process data to confidentiality and take the technical and organisational measures necessary for the safety of the processing. In addition, the biobank must assist the controller in fulfilling the controller's obligations to respond to data subjects' requests for exercising their rights and in the event of a data breach. Furthermore, it must delete or return, at the choice of the controller, all data to the controller after providing its services. Finally, the biobank cannot engage a "sub-processor" unless authorised by the controller (Article 28(2) GDPR).

As a processor, the biobank is liable for the damages caused by the processing only if it has violated one of the processor-specific obligations outlined by the GDPR or the controller's lawful instructions (Article 82(2) GDPR). However, like the controller, the processor is exempt from liability if it is not "in any way responsible for the event giving rise to the damage" (Article 82(3) GDPR). Finally, the biobank is subject to administrative fines in case of noncompliance (Article 83 GDPR).

10.4 Controller-Processor Relationship

When acting as a controller, the biobank may resort to a processor (Article 28 GDPR). For instance, a biobank that conducts its own research may contract a consultant or some other kind of third party to analyse biosamples and data on its behalf. This is a clear controller-processor relationship, in which the third party acts as the biobank's processor.

In such circumstances, the biobank must use "only processors providing sufficient guarantees to implement appropriate technical and organisational measures" (Article 28(1) GDPR). To this end, the biobank should consider the processor's technical expertise, reliability, and resources (Recital 81 GDPR). In addition, the processing by a processor must be regulated by a contract or other legal act between the controller and processor, which must be concluded in writing (e.g. in electronic form) (Article 28(3) GDPR). The contract must provide, among other things, that the processor processes data only "on documented instructions from the controller, including with regard to transfers of personal data to a third country" (Article 28(3) GDPR). The controller and processor may choose to negotiate their contract or rely on standard contractual clauses, adopted either by the European Commission or a national supervisory authority (Article 28(7) and (8) GDPR). The controller-processor relationship may be regulated in the DTA or MTA.

Thus, in a controller-processor relationship, the biobank exerts strict control over the processor. In principle, this ensures that the processor will not misuse the data and the trust relationship between the biobank and the research participants is preserved. On the other hand, the biobank remains the entity primarily responsible for GDPR compliance, including safeguards and technical and organisational measures (*ex plurimis*, Articles 5, 24, and 25 GDPR).

Conversely, the processor is liable only for violating either the (few) GDPR direct processor obligations or the biobank's instructions (Article 28 GDPR). If the processor processes data beyond the controller's instructions, the processor will be considered a controller regarding that processing (Article 28(10) GDPR) and will incur relevant GDPR sanctions.

10.5 Joint Controllership

When a biobank and another entity engage in the same data processing, joint controllership may also arise. Under joint controllership, two or more parties jointly determine the purpose and means of specific processing operations (Article 26 GDPR). The Opinion clarifies that a substantive and functional approach should be taken to assess joint controllership. In particular, joint controllers participate in determining either the purposes or the essential elements of the means of the processing. Contractual agreements may be indicative of joint controllership, but the factual circumstances of the relationship between the parties should always be considered.

The Guidelines further clarify that this joint participation may consist of a "common" or "converging decision". To assess whether a converging decision is in place, "an important criterion is that the processing would not be possible without both parties' participation in the sense that the processing by each party is inseparable, i.e. inextricably linked. The joint participation needs to include the determination of purposes on the one hand and the determination of means on the other hand".

Accordingly, joint controllership likely arises when a biobank shares data with another institution as part of a collaborative research project. In this case, the biobank and the other institution draft the research project together, thus jointly determining the purposes and means of the processing. This is confirmed by the Guidelines, which provide an example specifically regarding joint research projects. The example concerns several institutes that decide to participate in a joint research project, submit personal data to a platform to this end, and use such data to conduct the research. The Guidelines state that all the institutes qualify as joint controllers because they have co-decided the purpose and means of the processing, i.e. the joint research and the use of the platform, respectively.

10.5.1 External Collaborations

More nuanced is the situation in which a biobank shares data with an external researcher conducting its own independent research. In controlled-access models, the biobank strictly instructs the processing by the external researcher in a way a controller would do with a processor. However, the external researcher does not seem to process data merely "on behalf" of the biobank. Instead, it primarily defines the purpose of the processing by drafting the research protocol in the first place. In contrast, the biobank's DAC typically only accepts or rejects the respective access request.

The Guidelines' "converging decision" concept may help clarify this "grey area" in terms of joint controllership as both the external researcher's access request and the DAC's decision seem indispensable for the processing to take place. However, it would be helpful if the EDPB provided more granular guidance on research activities in which the stakeholders' roles are more differentiated than in collaborative research.

When joint controllership does arise, the biobank and the external research must transparently allocate their respective responsibilities for compliance with the GDPR (Article 26(1) GDPR). In particular, they must regulate the exercise of data subject's rights and the information obligations under Articles 13 and 14. Although the GDPR does not specify the legal form of this arrangement, the Guidelines recommend resorting to legally binding acts, e.g. a contract. Regardless of the terms of the joint controllership arrangement, the data subjects can exercise their rights under the GDPR in respect of each of the controllers (Article 26(3) GDPR). Therefore, data subjects' requests for withdrawing their consent or cancelling their data can always be addressed either to the biobank or the external researcher.

Article 26 provides that data subjects can exercise their rights "against each of the controllers" but does not mention supervisory authorities. Also, the joint and several liability regime outlined in Article 82 is explicitly limited to damages compensation and, therefore, does not apply to administrative penalties. The CJEU's case law seems to confirm this. The Court concedes that joint controllers are not necessarily involved in all stages of the processing. Therefore, the "level of responsibility of each of them must be assessed with regard to all the relevant circumstances of the particular case" (C-210/16 *Wirtschaftsakademie*, para 43). This means that an entity "cannot be considered to be a controller in the context of operations that precede or are subsequent in the overall chain of processing" in which it has no influence (C-40/17 *Fashion ID*, para 74).

Acknowledgements The authors are grateful to Deborah Mascalzoni (Institute for Biomedicine, Eurac Research; Centre for Research Ethics and Bioethics, Department of

Public Health and Caring Sciences, Uppsala University) for her comments on an earlier draft of this chapter.
The authors declare financial support by CHRIS-2D (FESR1102).

References

1. Hewitt R, Watson P (2013) Defining biobank. Biopreserv Biobanking 11(5):309–315
2. Simeon-Dubach D, Henderson MK (2020) Opportunities and risks for research biobanks in the COVID-19 era and beyond. Biopreserv Biobanking 18(6):503–510
3. Biobanks market size, share & trends analysis report by product, by service (cold chain, lab processing), by biospecimen type, by biobank type (virtual, physical), by application, by region, and segment forecasts, 2021–2028. https://www.grandviewresearch.com/industry-analysis/biobanks-industry
4. Coppola L, Cianflone A, Grimaldi AM, Incoronato M, Bevilacqua P, Messina F, Baselice S, Soricelli A, Mirabelli P, Salvatore M (2019) Biobanking in health care: evolution and future directions. J Transl Med 17(1):172. https://doi.org/10.1186/s12967-019-1922-3
5. Simeon-Dubach D, Roehrl MH, Hofman P, Puchois P (2020) Enhancing cooperation between academic biobanks and biomedical industry: better mutual understanding and new collaborative models are needed. Biopreserv Biobanking 18(2):144–149
6. Bull S, Bhagwandin N (2020) The ethics of data sharing and biobanking in health research. Wellcome Open Res 5:270. https://doi.org/10.12688/wellcomeopenres.16351.1
7. Caulfield T, Burningham S, Joly Y, Master Z, Shabani M, Borry P, Becker A, Burgess M, Calder K, Critchley C, Edwards K, Fullerton SM, Gottweis H, Hyde-Lay R, Illes J, Isasi R, Kato K, Kaye J, Knoppers B, Lynch J et al (2014) A review of the key issues associated with the commercialization of biobanks. J Law Biosci 1(1):94–110. https://doi.org/10.1093/jlb/lst004
8. Mascalzoni D (ed) (2015) Ethics, law and governance of biobanking. Springer. https://www.springer.com/gp/book/9789401795722
9. Ashcroft JW, Macpherson CC (2019) The complex ethical landscape of biobanking. Lancet Public Health 4(6):e274–e275. https://doi.org/10.1016/S2468-2667(19)30081-7
10. Staunton C (2021) Individual rights in biobank research under the GDPR. In: Slokenberga S, Tzortzatou O, Reichel J (eds) GDPR and biobanking. law, governance and technology series, vol vol 43. Springer, Cham. https://doi.org/10.1007/978-3-030-49388-2_6
11. Staunton C, Slokenberga S, Mascalzoni D (2019) The GDPR and the research exemption: considerations on the necessary safeguards for research biobanks. Eur J Hum Genet 27:1159–1167. https://doi.org/10.1038/s41431-019-0386-5
12. Staunton C, Slokenberga S, Parziale A, Mascalzoni D (2022) Appropriate Safeguards and Article 89 of the GDPR: Considerations for Biobank, Databank and Genetic Research. Front Genet 13:719317.https://doi.org/10.3389/fgene.2022.719317
13. Homer N, Szelinger S, Redman M et al (2008) Resolving individuals contributing trace amounts of DNA to highly complex mixtures using high-density SNP genotyping microarrays. PLoS Genet 4:e1000167
14. Maxmen A (2021) Why some researchers oppose unrestricted sharing of coronavirus genome data. Nature 593:176–177. https://doi.org/10.1038/d41586-021-01194-6
15. Mascalzoni D, Bentzen HB, Budin-Ljøsne I et al (2019) Are requirements to deposit data in research repositories compatible with the European Union's general data protection regulation? Ann Intern Med 170:332–334. https://doi.org/10.7326/M18-2854
16. Shabani M, Chassang G, Marelli L (2021) The impact of the GDPR on the governance of biobank research. In: Slokenberga S et al (eds) GDPR and Biobanking. Springer, pp 45–60. https://doi.org/10.1007/978-3-030-49388-2_4
17. Cheah PY, Piasecki J (2020) Data access committees. BMC Med Ethics 21:12. https://doi.org/10.1186/s12910-020-0453-z
18. Murtagh MJ, Blell MT, Butters OW et al (2018) Better governance, better access: practising responsible data sharing in the METADAC governance infrastructure. Hum Genomics 12:24. https://doi.org/10.1186/s40246-018-0154-6
19. Shabani M, Knoppers BM, Borry P (2015) From the principles of genomic data sharing to the practices of data access committees. EMBO Mol Med 7:507–509. https://doi.org/10.15252/emmm.201405002
20. Article 29 Data Protection Working Party (2010) Opinion 1/2010 on the concepts of "controller" and "processor", 00264/10/EN WP 169. https://ec.europa.eu/justice/article-29/documentation/opinion-recommendation/files/2010/wp169_en.pdf
21. Berger M, Eisendle D (2019) Web 2.0 and the concept of 'data controller': recent developments in EU data protection law. Indian J Law Technol 15(1):20–39
22. Primož G (2019) Fashion ID GmbH & Co. KG v Verbraucherzentrale NRW e.V.: more control, more data protection for website visitors? (C-40/17 Fashion ID): case note. Eur Data Protect Law Rev 4:579–582
23. EDPB (2021) Guidelines 07/2020 on the concepts of controller and processor in the GDPR. https://edpb.europa.eu/system/files/2021-07/eppb_guidelines_202007_controllerprocessor_final_en.pdf

11 Research Involving Human Participants and Their Biological Material: A South African Perspective

Melodie Labuschaigne and Magda Slabbert

Abstract

Southern Africa, including South Africa, is one of the most genetically divergent regions in the world, with highly sought-after human biological material. The South African legal framework governing research involving human participants and their biological material is complex and multilayered. Despite protection afforded by the South African Constitution and the National Health Act and regulations to research participants, including the scope of activities permitted with regard to their biological materials, ambiguities, legal gaps and discrepancies remain, pointing to a fragmented and often contradictory legal situation. Further clarity is needed on the place of broad consent vis-à-vis the transfer, exchange and sharing of human biological material. Whilst permitted in the case of the transfer of and further use of human biological material under research ethics committee oversight, the role of broad consent in the processing of personal information under the Protection of Personal Information Act is still unsettled. It is hoped that this would be clarified by a Code of Conduct for Health Research (in progress) under the same act.

M. Labuschaigne (✉) · M. Slabbert
Department of Jurisprudence, School of Law, University of South Africa, Pretoria, South Africa
e-mail: slabbmn@unisa.ac.za; slabbm@unisa.ac.za

Abbreviations

ASSAf	Academy of Science of South Africa
DNA	Deoxyribonucleic acid
GN	Government notice
HBM	Human biological material
HREC	Health Research Ethics Committee
NHA	National Health Act 61 of 2003
NMTA	National material transfer agreement
POPIA	Protection of Personal Information Act 4 of 2013
RNA	Ribonucleic acid
RSA	Republic of South Africa
SA	South Africa
UK	United Kingdom
USA	United States of America

11.1 Introduction

Southern Africa is widely regarded as one of the most genetically divergent regions in the world and hence a very valuable source of genetic information for many researchers across the globe [1]. As a result, the region, including South Africa, has become a popular target for its human biological samples, leading to rightful

J. Faintuch, S. Faintuch (eds.), *Integrity of Scientific Research*,
https://doi.org/10.1007/978-3-030-99680-2_11

concerns regarding the long- and short-term benefits of these predominantly foreign studies for South Africa. A specific concern relates to the use and export of human biological materials that have led to calls for the protection of South Africa's genomic sovereignty [2].

In 2018, the Wellcome Sanger Institute, a major genome research centre in Hinxton, United Kingdom, was accused of commercialising a gene chip developed by using DNA donated by African donors. With concerns raised by, among others, the University of Stellenbosch and University of KwaZulu-Natal in South Africa, it was alleged that some of the material transfer agreements in this study seemingly did not provide for commercialisation, nor did participants' consent to restrict commercialisation. Although an external legal investigation found no "unlawful exploitation of scientific work" or a breach of contract and intellectual property rights, the 75,000 gene arrays stored at Sanger have not been used and expired at the end of 2019 [3]. The possible fallout of this incident is that the trust that has been established with African institutions and people may have been harmed in the process.

Potential exploitation of South Africa's human biological material is not limited to external entities or researchers only. Earlier in 2015, the South African health insurer, Discovery, entered into a partnership with the Human Longevity Incorporation, which would provide genetic testing to Discovery members for $250 (US) per person. However, as legal scholars at the time observed, this initiative not only posed significant problems regarding the transfer or exploitation of biological samples and data out of South Africa but also brought the lack of protection for potential donors in South Africa into sharper focus [8]. In return for a reduced-cost genetic test, Discovery members' samples and deidentified information would be transferred to and stored with the Human Longevity Incorporation (belonging to Craig Venter) in the United States of America. Under the ruse of clinical care and the lofty aim of advancing medical research, but capitalising on South Africa's inadequate regulatory framework relating to the use, storage and export of biological samples at the time, the apparent aim was to generate one of the world's biggest databases for whole genome, phenotype and clinical data. Consequently, thousands of Discovery members' (RSA citizens) saliva samples were sent to the United States to further this aim.

This example is one of the many where the one-way flow of samples out of South Africa, with questionable benefit to the local population, impacts negatively on the development of local infrastructure, capacity and expertise [8], not to mention South Africans' concerns relating to control over their own samples [5] or collective genetic heritage [7]. Further concerns that arise in the context of the export of human biological material from South Africa relate to issues of informed consent, intellectual property rights, benefit-sharing and community engagement [6].

The purpose of this chapter is to discuss the regulation of the removal, use and transfer of human biological materials in South Africa, with specific focus on the statutory framework, including the recently introduced standard material transfer agreement for human biological materials (SNMTA), published in Government Notice 719, *Government Gazette* 41781 of 20 July 2018 [15], as well as the export and further use of human biological material generally. The chapter will also touch on challenges regarding research participants' informed consent to the further use and sharing of their biological materials.

11.2 Legal Framework Governing the Use and Transfer of Human Biological Material

11.2.1 Constitution of the Republic of South Africa, 1996 [12]

Any law or conduct which is inconsistent with the Constitution is invalid (Section 2 of the Constitution). The Constitution entrenches informed consent as a requirement for the participation in medical research through Section 12(2)(c) which states that "everyone has the right to bodily and psychological integrity, which includes the right not to be subjected to medical or scientific experiments without their informed consent". The word "experiment" resembles the wording "no one shall be subjected without his free consent to medical or scientific *experimentation*" (emphasis added) used in the International Covenant on Civil and Political Rights of the United Nations of 16 December 1966, generally regarded as referring to research.

Regulation on the use of human tissue (referred to as "human biological material" in this chapter, denoting a more comprehensive category of human tissue in South African law) may be done through national legislation, such as the United Kingdom's Human Tissue Act 2004 (c 30) [24]. In South Africa, the use and control of human biological material are governed by Chapters 8 and 9 of the National Health Act 61 of 2003 [16], which repealed and replaced the former Human Tissue Act 65 of 1983 (the latter modelled after the United Kingdom's Human Tissue Act) [13].

11.2.2 National Health Act 61 of 2003 [16]

Chapter 8 of the National Health Act (hereafter NHA), titled "Control of use of blood, blood products, tissue and gametes in humans", came into effect in 2012 after the promulgation of the NHA regulations in Proclamation No. 11, published in *Government Gazette* 3081 of 27 February 2012 [17].

"Human tissue" is defined in law in South Africa in Section 1 of the NHA as "human tissue, and includes flesh, bone, a gland, an organ, skin, bone marrow or body fluid, but excludes blood or a gamete". In regulations published in terms of the NHA (discussed below), the term "human biological material", referring to a broader category of human tissue that includes gametes and ostensibly human blood and blood products as well, is used. Both categories of "human tissue" and "human biological material" are relevant in context of health research. Biological materials (and related data) are collected through different purposes, e.g. specifically for research purposes, incidentally to diagnostic or therapeutic purposes or for a combination of purposes, which may include the intention of possible future research use.

"Health research" is defined in the NHA as "any research which contributes to knowledge of, amongst others, the biological, clinical, psychological or social processes in human beings; human pathology; the cause of disease and the development or new application of pharmaceuticals and medicines, and the development of new applications of health technology" (Section 1).

Only an "authorised institution" (meaning an institution designated as such by the Minister of Health in terms of the NHA) may acquire or use tissue lawfully imported or removed from the body of a deceased or living person (Section 54 of the NHA). No person may remove tissue from the body of another living person unless it is done with the written consent of the person from whom the tissue is removed, granted in the "prescribed manner" and in accordance with "prescribed conditions" (Section 55). "Prescribed conditions" or "prescribed manner" would refer to those outlined in regulations, according to the definition of "prescribed" in the NHA.

The use of such removed tissue, extracted from a living person, is limited to "such medical or dental purposes as may be prescribed" in terms of Section 56 of the NHA. The Act again does not elaborate on what the prescribed purposes are, but these are stated in regulations (discussed below). The same section further prohibits the removal of tissue from a mentally ill person as well from a minor (below the age of 18 years) if it is not replaceable by natural processes, unless Ministerial consent has been granted. Only a registered medical practitioner or dentist may remove or use tissue (Section 59), and if it is removed from a living person, it must be done in a hospital or an authorised institution (Sections 59 and 58 accordingly). Such tissue may only be donated to a hospital, a university, an authorised institution, a medical practitioner and a dentist or a tissue bank, in accordance with Regulations Regarding the General Control of Human Bodies, Tissue, Blood, Blood Products and Gametes.

Section 60 of the NHA makes it an offence to sell or trade in human tissue. If human tissue is imported or exported, the importer or exporter may receive payment, but this is limited to cover only the costs involved with the importation or exportation. A person who donates tissue may only be reimbursed for the reasonable costs incurred by him or her to provide such a donation. Thus, compensation for providing human biological material for research is restricted to reasonable costs incurred with the donation. To receive any form of compensation, financial or otherwise, is unlawful and a criminal offence in terms of the NHA.

11.2.2.1 Cadaver Tissue

The regulation of the donation of tissue by deceased persons is addressed in Section 62, which states that such tissue may only be donated for purpose of health research (Section 64). The NHA authorises the Minister of Health in Section 68 to make regulations to further explain the provisions in Chapter 8. This has been done by way of several sets of regulations published on 2 March 2012 in *Government Gazette* 35099. Relevant to health research are the regulations relating to the use of human biological material (Government Notice (GN) R177) [18]; the import and export of human tissue, blood, blood products, cultured cells, stem cells, embryos, foetal tissue, zygotes and gametes (GN R181) [19]; tissue banks (GN R182) [20]; and the general control of human bodies, tissue, blood, blood products and gametes (GN R180) [21].

Regulations relating to research with human participants were published in terms of Chapter 9 of the NHA, in *Government Gazette* 3800 of 19 September 2014 (GN R 719) [22]. To date, no regulations relating to biobanking or biobanks specifically have been promulgated. Chapter 9 of the NHA provides for the establishment of both a National Health Research Committee (Section 69), tasked inter alia with advising the Minister on health research priorities, and a National Health Research Ethics Council (Section 72), which, among others, is responsible for the development of guidelines for the functioning of health research ethics committees, currently contained in the second edition of the Department of Health's 2015 Guidelines, titled, *Ethics in Health Research: Principles, Processes and Structures* [23], which describe the primary role of health ethics research committees to protect the interests of research participants.

11.2.2.2 The Health Research Ethics Committee

An institution that conducts health research must either have its own health research ethics committee (hereafter HREC) or must have access to such a committee (Section 73 of the NHA). There are currently 46 research ethics committees registered with the Council, comprising mostly university-based research ethics committees, those representing various other bodies involved in health research and one committee belonging to a private company. The 2015 Department of Health Ethics Guidelines are legally binding on HRECs and apply to all research involving human tissue or human biological material, as well as data derived from these. The legal force of the Ethics Guidelines originates from regulations 2(a) and 6(b) of the 2014 regulations relating to research with human participants, referred to above, that provide that all research with human participants, their samples and derived data must at a minimum comply with the 2015 Ethics Guidelines issued by the Department of Health.

11.2.2.3 Informed Consent and Assent

"Research or experimentation" on a living person in terms of Section 71 of the NHA may only be conducted "in the prescribed manner" and with the written consent of the person after he or she "has been informed of the objects of the research or experimentation and any possible positive or negative consequences on his or her health". Concerning a minor (e.g. a child under the age of 18 years), the same provision stipulates that therapeutic research may only be conducted if it is in the minor's best interest and with the consent of the minor's parents or guardian, and if the said minor is capable of understanding what is to be conducted, with the said minor's consent as well. In the case of nontherapeutic research involving a minor, Ministerial consent is required, in addition to the consent of the parent or guardian, or that of the minor. The Minister will not give consent if the same research can be conducted on adults, or if there will be no specific benefit for the minor.

11.2.3 Regulations Promulgated Under the National Health Act

Health research requires that both biological material and data be made accessible to researchers but also requires the willingness of research participants to contribute their biological material and data. This requires significant trust from participants but also confidence in a regulatory model that protects both the interests and rights of participants and researchers. Unfortunately, as the discussion below will show, an analysis of a range of regulations relevant in this context reveals an ambiguous and often contradictory picture, which has not been satisfactorily resolved by the recent addition of the mandatory national standard material transfer agreement.

As alluded to earlier, contrary to the NHA's reference to "human tissue", the NHA's regulations relating to the use of human biological material govern a broader category of "biological material", defined as "material from a human being including DNA, RNA, blastomeres, polar bodies, cultured cells, embryos, gametes, progenitor stem cells, small tissue biopsies and growth factors from the same" (Section 1). Echoing more or less similar requirements laid down in the NHA regarding the removal of human biological material from living or deceased donors, these regulations further determine that in the case of the removal of biological material from a deceased person, where no consent has been provided by the deceased whilst alive, the spouse, partner, major child, parent, legal guardian, major brother or major sister of the deceased may consent to the removal, in this order of listing (Section 4). The regulations clarify in Section 5 the "medical and dental purposes" for the use and removal of human biological material referred to in the NHA, which are stated as (a) DNA, RNA and chromosome-based genetic testing, (b) health research (c) training or (d) archaeological, medical or heritage value studies relating to DNA obtained from human genetic material.

Moreover, a human biological material register should be kept by those using the material (Section 12); the confidentiality of all records must be guaranteed; and written consent is required where the information is disseminated to other institutions and for long-term storage of genetic material, stem cells or research findings (Section 13). If used for research purposes (Section 13), identifiable or related health information generated from the use of the material should be anonymous or anonymised. Despite providing more granularity to the NHA provisions, these regulations do not regulate biobanking research using human biological material adequately, as the regulations, though referring to the transfer of information, are silent on the transfer and further use of human biological material (hereafter HBM). Ad hoc collection of biospecimens also occurs in clinical (for diagnostic purposes) and academic settings (for teaching and research purposes), sometimes in the absence of established governance or quality control processes. For this reason, references to biobanking should consider that a limited understanding of a biobank will not suffice, as there are a range of other stakeholders involved in the collection, storage and sharing of biospecimens, regardless of the context.

11.2.3.1 Tissue Banks vs. Biobanks

The regulations elating to tissue banks provide little direction on biobanking activities. These regulations define a "tissue bank" in Section 1 as "an organisation, institute or person that provides or engages in *one or more services* involving cells and/or tissue from living or deceased individuals *for transplantation* purposes" (emphasis added). Although tissue banks are generally considered banks for the therapeutic purposes relating to tissue transplantation, contrary to a limited understanding of biobanks whose purpose appears to be research directed, it is peculiar that Section 17 of the same tissue bank regulations refers (with regard to the requirements pertaining to tissue bank activities) to "the research and development of tissue samples". This ambiguity and conflation of the role of tissue and biobanks, in the absence of clear biobanking regulations, creates a legal vacuum in the landscape of tissue regulation [4].

The definitions of human tissue in the NHA and its tissue bank regulations are also different. The regulations relating to tissue banks define tissue as "a functional group of cells" that "is used collectively in regulations to indicate both cells and tissue". Whilst the NHA's tissue definition makes no reference to human cells and excludes gametes and blood, the tissue bank regulations' definition appears confusing and misaligned with that of the main Act. Similar discrepancies exist in relation to the definition of "cells" in the NHA and other regulations [4]. Whilst researchers may not fault these inconsistencies, differing descriptions exacerbate already existing textual ambiguities, creating legal interpretational hurdles in complex scenarios involving research using human samples.

11.2.3.2 Participant Co-ownership of the Results and Intellectual Rights

An interesting feature of the Regulations Regarding the General Control of Human Bodies, Tissue, Blood, Blood Products and Gametes is the reference in regulation 26 to "exclusive rights" in tissue by any person who lawfully acquires tissue by virtue of any provision of the NHA and its regulations. The inverse result of this provision is that a person who provides his or her biological material for use in a research project is deemed to have transferred "exclusive rights" in the HBM to the research organisation that receives the material for use in research. This result is in contrast with the ownership of the results of the research funded by public funds in terms of the Intellectual Property Rights from Publicly Financed Research and Development Act 51 of 2008 [14]. Section 4 (1) of this Act determines that a recipient of such funding has the right to register and own the intellectual property emanating from the research and may even co-own intellectual property rights with a private entity under certain conditions in terms of Section 15(2). However, since research participants contributing their biological material for research are not the recipient of the funding and do not conduct the actual research, they are effectively excluded from becoming co-owners of any intellectual property emanating from publicly funded research using their biological material [9].

11.2.3.3 Compensation for Damages

It is instructive that the regulations relating to research with human participants, which give legal force to the 2015 Department of Health Ethics Guidelines, in addition to unpacking the requirement of informed consent, also require that provision be made for compensation for research-related injury, for more than minimal risk research, as well as that representatives from participating communities, where research is carried out, be consulted (Section 3). The same section requires that research results be disseminated in a timely and competent manner to the participants and the community. One limitation of these regulations is that despite the comprehensive description of informed consent of participants, the regulations do not provide for participant's consent prior to the export or transfer of their HBM outside South Africa. Moreover, the regulations do not provide for benefit-sharing by communities whose HBMs are used in research.

Another set of NHA regulations pertinent to health research is the regulations relating to the import and export of human tissue, blood, blood products, cultured cells, stem cells, embryos, foetal tissue, zygotes and gametes. No person may import or export any tissue without a permit requested in writing from the Department of Health (Section 2), and only samples obtained under the terms of the NHA may be exported. If the Director-General from the Department of Health is satisfied that the information submitted in support of an application meets the set requirements, a permit will be issued. Applicants need to demonstrate in writing that the tissue was donated according to the NHA and that the procured tissue will be used in terms of the Act. Section 7 of these regulations requires that registers of imports and export be kept and be provided to the Department at intervals not exceeding 6 months. The latter requirement, prior to the introduction of the material transfer agreement discussed next, was the only form of oversight over the movement of HBM from South Africa to other destinations, yet does not protect the rights of research participants following the export of their samples, not to mention protecting the confidentiality and privacy of their health information.

11.2.3.4 Participant Rights in Case of Exported Samples

Prior to the introduction of the national material transfer agreement in 2018, donors' consent for the export of their HBM was lacking. As far as the export and transfer of data related to their HBMs are concerned, the Protection of Personal Information Act 4 of 2013 (hereafter POPIA) provides some protection in Section 72 relating to the transfer of personal information regarding a research participant (as data subject). This provision requires the consent of the data subject and a binding agreement with the third party to whom information is transferred, the latter providing "an adequate level" of data protection for the transfer of personal information. If the recipient country does not provide an adequate level of protection, prior authorisation for the transfer from the SA Information Regulator is necessary. It is still unclear in South Africa what would qualify as an "adequate level" of protection. Although enacted in 2013, POPIA provisions affecting consent of research participants came into effect on 1 July 2020. It is interesting to note that at the time of writing, the Academy of Science of South Africa (hereafter ASSAf) is in the process of drafting a Code of Conduct for Research in South Africa in terms of Chapter 7 of POPIA, which would provide practical guidance on how the obligations and conditions for lawful processing of personal information can be applied and complied with in the research sector.

11.3 Standard Material Transfer Agreement for Human Biological Materials (SNMTA)

The South African research community participates in many international research collaborations, such as H3Africa ([26], https://h3africa.org), B3Africa ([25], www.b3Africa.org) and MalariaGEN ([27], www.malariagen.net), which cannot function without the sharing of HBM and related data. The multiple uses of samples and data require alternative consent models in addition to specific consent, such as broad and tiered consent, referred to in the 2015 Department of Health Guidelines. The Guidelines endorse the use of broad consent to "allow for future and secondary uses of data, in line with the opportunities to use such data in advancing knowledge to improve health" (2015, p. 31). The Guidelines advise that ethics approval be obtained for all future research, whilst, as mentioned, the regulations relating to research involving human participants are silent on the issue of participants' consent prior to, and for, the transfer or export of biological material. Bar the provisions of the material transfer agreement and on the face of the NHA and its regulations alone, legal requirements relating to tissue or HBMs are limited to obtaining consent prior to the removal of the material, as well as using it in the manner outlined in the NHA and regulations. This shortcoming has led to an unsatisfactory and fragmented regulatory impasse, alleviated to some extent by the subsequent promulgation of the SNMTA in 2018. The need for a SNMTA template has also been borne out by the historical exploitation of South African HBM, the lack of benefit-sharing returning to those donating the biological material, as well as the reality that non-South African material transfer agreements are inadequate in addressing unique legal concerns specific to South Africa [9].

According to the notice published with the standard NMTA of HBMs, the SNMTA applies to "all the providers and recipients of the biological material for use in research or clinical trials under the auspices of the Health Research

Ethics Committees", which means in every instance where HBM is transferred for the purpose of research or clinical trials from the provider to the recipient, where at least either of the institutions is located in South Africa.

The NMTA refers to an HREC (required by law to be registered with the SA National Health Research Ethics Council in terms of Section 73 of the NHA), the providing institution (provider) and the recipient institution (recipient), as parties to the agreement. The feasibility of the role of the HREC as a party to the SNMTA has led to opposing scholarly views, notably by Labuschaigne and Mahomed [4] and Thaldar et al. [9], which mainly relate to questions regarding the contractual capacity and juristic personhood of HRECs. The SNMTA is subject to a suspensive condition that it will only take legal effect (a) once the HREC has approved the project of which the SNMTA forms part and (b) the relevant HREC has approved this agreement (Clause 3.5). The SNMTA's definition of "Materials" includes both HBMs *and* data, whereas the notice to the schedule that contains the SNMTA makes reference to "biological materials", which ostensibly refer to HBMs exclusively. This curious discrepancy creates an anomaly with regard to the inclusion of data relating to HBMs in the transfer of the HBMs. However, with the enactment of the POPIA in toto from 1 July 2021, the data associated with HBM in the realm of health research would be governed by POPIA and not the SNMTA.

11.3.1 Institutional Responsibilities and Donor Rights

The SNMTA provides in Clause 3.3 that the providing institution (providing the HBMs) is regarded as the custodian of the materials, a notion that has been supported with regard to the storage of HBMs elsewhere [10], particularly because it allows for the oversight of materials from the initial collection of the materials up to the dissemination of the research findings and because it promotes principles of transparency, fairness to research participants and shared responsibility among all research stakeholders [11]. The donor of HBM remains the owner of his or her material in terms of Clause 3.3 under the NMTA, until such time the material is destroyed. As stated already, the Regulations Regarding the General Control of Human Bodies, Tissue, Blood, Blood Products and Gametes provide in Section 26 for the acquisition of "exclusive rights" in HBMs by any person or institution who lawfully acquires tissue by virtue of any provision of the NHA and its regulations. In order to reconcile this provision with the NMTA's "ownership" of donors in their HBMs in Clause 3.3, Thaldar et al. [9] suggest that the ownership provided for in the NMTA be regarded as *nominal* or symbolic ownership without any rights in the HBM or the associated data. He argues that research participants' rights vis-à-vis the research institution collecting their HBMs, which include, among others, the right to refuse that the HBM be used for further research beyond the initial consent and the right to request the destruction of his or her HBMs, are not dependent on the donor being "owner" of the donated HBM ([9], p.9). In fact, references to ownership in this context may compound legal expectations with regard to fair benefit-sharing and the rights and duties of all parties.

The SNMTA also outlines the obligations of the contract parties, e.g. the provider (Clause 4), recipient of the materials (Clause 5) and the HREC (Clause 6). In the case of the provider, obligations stated are those relating to obtaining export permits and authorisations, informing the donor where materials have become identifiable or obtaining informed consent for further uses of the materials. The recipient must protect and maintain the confidentiality of the material (e.g. HBMs and data) and may only transfer material listed in an annexure to the NMTA and, if intending to do so, with HREC approval. If any material becomes identifiable, the recipient must inform the provider without delay.

11.3.2 Benefit-Sharing vs. Financial Benefits

The SNMTA also provides for the sharing of benefits that must be discussed and negotiated between the provider and recipient before the transfer of the material (Clause 7). A benefit-sharing agreement should be appended to the SNMTA in Annexure B. Clear national guidelines on benefit-sharing are required to guide researchers to the various models that are available, which would not fall foul of Section 60 of the NHA prohibiting payment for HBMs, except to reimburse reasonable costs relating to the donation of the HBM.

11.3.3 Consent for Primary vs. Secondary Use of Biomaterials

Clause 10 addresses informed consent issues and requires the provider to obtain the donor's informed consent for the transfer of the materials to the recipient, as well as furnishing the HREC with the donor consent and protocol for approval, including those relating to secondary uses involving the material. The NMTA's definition of informed consent appears to describe dynamic consent through its reference to it being an "on-going information sharing process which allows a Donor to consent to participate and determine whether and how their Materials will be utilised". We submit that this flexible informed consent requirement is not inconsistent with that of the regulations relating to research with human participants and should be welcomed in the context of the transfer of HBMs.

11.3.4 Legal Conflicts

The SNMTA's dispute settlement clause (e.g. 11) directs that in the event that a dispute cannot be resolved amicably, any party can institute action in accordance with South African law in a South African court, unless the parties agree to resolve such dispute by arbitration in terms of a separate arbitration agreement. This clause is criticised by Thaldar et al. [9] who argue that it would be more fair to determine that the jurisdiction of the provider be the applicable legal system and place of adjudication for disputes arising from the SNMTA, regardless of whether through litigation or arbitration, as this would promote better access to justice for research participants, as they are more likely to be domiciled in the same country as the provider.

11.3.5 Intellectual Property Rights

As far as intellectual property rights are concerned, the SNMTA stipulates in Clause 12 that these will be handled in accordance with relevant and applicable laws and third-party agreements, which, as per the law of contract, would need to be appended to the NMTA with specific reference in the SNMTA to their existence and binding force with regard to the respective parties. Current challenges by research ethics committees registered with the National Health Research Ethics Council, with the SNMTA's implementation in practice, have led to a revision of the SNMTA by the NHREC of the SNMTA. The National Department of Health is currently considering the revised SNMTA.

References

Journal Articles

1. Choudhury A, Aron S, Sengupta D et al (2018) African genetic diversity provides novel insights into evolutionary history and local adaptations. Hum Mol Gen 27(R2):R209–R218. https://doi.org/10.1093/hmg/ddy161
2. Ledford H (2010) Africa yields two full human genomes. Nat 18 463(7283):943–947. https://doi.org/10.1038/463857a
3. Kokstad E (2019) Major U.K. genetics lab accused of misusing African DNA. Afr Sci Stud 30:1. https://doi.org/10.1126/science.aba0343
4. Labuschaigne M, Mahomed S (2019) Regulatory challenges relating to tissue banks in South Africa: impediments to accessing healthcare. S Afr J Bioethics

Law 12(1):27–31. https://doi.org/10.7196/SAJBL.2019.v12i1.674
5. Moodley K, Sibanda N, February K et al (2014) 'It's my blood': ethical complexities in the use, storage and export of biological samples: perspectives from South African research participants. BMC Med Ethics 15(4):1–7. https://doi.org/10.1186/1472-6939-15-4
6. Nienaber A (2011) Consent to and authorisation of the export and use of human biological specimens for future research – perspectives from three African countries. CILSA 44(2):254–225. https://www.jstor.org/stable/23253098
7. Slabbert MN, Pepper MS (2010) "A room of our own?" Legal lacunae regarding genomic sovereignty in South Africa. Tydskrif vir Hedendaagse Romeins-Hollandse Reg (J Contemp S Afr Roman-Dutch Law) 73:432e50
8. Staunton C, Moodley K (2016) Data mining and biological sample exportation from South Africa: a new wave of bioexploitation under the guise of clinical care? S Afr Med J 106(2):136–138. https://doi.org/10.7196/SAMJ.2016.v106i2.10248
9. Thaldar D, Botes M, Nienaber A (2020) South Africa's new standard material transfer agreement: proposals for improvement and pointers for implementation. BMC Med Ethics 21(85):1–13. https://doi.org/10.1186/s12910-020-00526-x
10. Verlinden M, Nys H, Ectors N et al (2016) Qualitative study on custodianship of human biological material and data stored in biobanks. BMC Med Ethics 17:15. https://doi.org/10.1186/s12910-016-0098-0
11. Yassin R, Lockhart N, Del-Riego MG et al (2010) Custodianship as an ethical framework for biospecimen-based research. Can Epidemiol Biomarkers Preven 19(4):1012–1015. https://doi.org/10.1158/1055-9965.EPI-10-0029

Statutes and Regulations

South Africa

12. Constitution of the Republic of South Africa, 1996
13. Human Tissue Act 65 of 1983
14. Intellectual Property Rights from Publicly Financed Research and Development Act 51 of 2008
15. Material transfer agreement for human biological materials (GN 719), Government Gazette 41781 of 20 July 2018
16. National Health Act 61 of 2003
17. Proclamation No. 11, Government Gazette 3081 of 27 February 2012
18. Regulations relating to the use of human biological material (GN R177), Government Gazette 35099 of 2 March 2012
19. Regulations relating to the import and export of human tissue, blood, blood products, cultured cells, stem cells, embryos, foetal tissue, zygotes and gametes (GN R181), Government Gazette 35099 of 2 March 2012
20. Regulations relating to tissue banks (GN R182), Government Gazette 35099 of 2 March 2012
21. Regulations relating to the general control of human bodies, tissue, blood, blood products and gametes (GN R180), Government Gazette 35099 of 2 March 2012
22. Regulations relating to Research with Human Participants (GN R 719), Government Gazette 3800 of 19 September 2014
23. Republic of South Africa (2015) Guidelines: ethics in health research: principles, processes and structures, 2nd edn. National Department of Health, Pretoria

United Kingdom

24. Human Tissue Act 2004 (c 30)

Websites (Organisations)

25. B3Africa (2015). www.b3africa.org. Accessed 12 Apr 2021
26. H3Africa (2010). https://h3africa.org. Accessed 12 Apr 2021
27. MalariaGEN (2005). www.malariagen.net. Accessed 12 Apr 2021

Part IV

Research Misconduct

Ethical and Legal Risks of Artificial Intelligence in Radiology

12

Tugba Akinci D'Antonoli

Abstract

In radiology, artificial intelligence is endowed with many potential practical applications, both at the individual level and considering the profession in general. However ethical and legal conflicts could exist, and attention to the patient rights are paramount in this context. Given the novelty of the field and the not yet fully mapped challenges ahead, it is essential to count with the continuing input of radiologists along with ethicists, as more and more domains are impacted by these valuable and even disruptive advances.

Abbreviations

AI	Artificial intelligence
CE	Conformité Européenne
CLAIM	Checklist for Artificial Intelligence in Medical Imaging
CONSORT	Consolidated Standards of Reporting Trials
CT	Computed tomography
DL	Deep learning
EQUATOR	Enhancing the Quality and Transparency of Health Research
EU	European Union
FDA	US Food and Drug Administration
FL	Federated learning
GANs	Generative adversarial networks
GDPR	General Data Protection Regulation
GPT-3	Generative Pre-trained Transformer 3
HIPAA	Health Insurance Portability and Accountability Act
ML	Machine learning
MRI	Magnetic resonance imaging
NIH	National Institutes of Health
PACS	Picture archiving and communication systems
PHI	Protected health information
RANZCR	Royal Australian and New Zealand College of Radiologists
SPIRIT	Standard Protocol Items: Recommendations for Interventional Trials
STARD	Standards for Reporting of Diagnostic Accuracy Studies
USA	United States of America

T. Akinci D'Antonoli (✉)
Department of Pediatric Radiology, University Children's Hospital Basel, Basel, Switzerland
e-mail: tugba.akincidantonoli@usb.ch

J. Faintuch, S. Faintuch (eds.), *Integrity of Scientific Research*,
https://doi.org/10.1007/978-3-030-99680-2_12

12.1 Introduction

> There's nothing artificial about AI. It's inspired by people, it's created by people, and—most importantly—it impacts people. It is a powerful tool we are only just beginning to understand, and that is a profound responsibility.—Fei-Fei Li [1].

Artificial intelligence (AI) aims to mimic human intelligence [2]. Machine learning (ML), or the ability to learn from data without specific programming, and deep learning (DL), which copies neural network arrangements to classify data and arrive at conclusions, are important within this field [2].

12.2 Imaging Diagnosis

A significant number of algorithms in medicine have been cleared by the Food and Drug Administration (FDA) in the USA and got Conformité Européenne (CE) mark in European Union (EU) [3]. Such applications include detection of atrial fibrillation and diabetic retinopathy [4]. In radiology the market is very substantial as demonstrated by the majority of proposals so far [3].

One common utilization is workflow prioritization, for life-threatening emergencies such as intracranial hemorrhage in computed tomography (CT) scans [4, 5]. Waiting time and problem reporting can be substantially shortened [5]. Risk scoring is another tedious yet potentially lifesaving technique. A high-accuracy deep learning model for mammography has been developed [6]. Image quality optimization, lesion diagnosis and characterization, as well as segmentation are additional major fields in progress [7].

12.3 Legal and Ethical Risks of Using AI

Minority populations and as diverse targets and results as possible are essential in the development stages, so that the algorithm bypasses distortions and prejudices. A legal AI program in the USA systematically attributed worse scores to African Americans with minor or no criminal history, when compared to white Americans with a major criminal history [8]. Deadly accidents with self-driving cars managed fully by AI are rare; however they do occur, even with highly advanced resources [9].

Another type of risk arises due to the vulnerability of AI algorithms to adversarial attacks. Adversarial attacks are deceiving inputs that force the AI systems to make an error, i.e., upcoding diagnosis and procedures that allow the highest billing. The aim of such an attack could be to increase revenue. Recent studies already showed the feasibility of adversarial attacks in medicine [10, 11]. In another study, researchers have investigated modifying images with generative adversarial networks (GANs) to inject or remove features in mammography images [12]. As a result, researchers were successful in deceiving the radiologists: The modified images were indistinguishable from the unmodified mammograms [12].

Moreover, using AI-powered worklists prioritizing algorithms can come with unexpected drawbacks. The false-negative examinations can be pushed too far down the worklist, below the true- and false-positive examinations, and can cause delays in some patients to access lifesaving treatments [13].

Liability that arises from AI's shortcomings or misguidance is a risk, similarly to automation bias, i.e., the tendency to favor AI-generated decisions, overlooking eventual errors [14]. Due to large amounts of data fed into such AI-based models, we are unable to fully understand all of the hidden paths of decisions. Who should be held responsible for the undesirable results, the user or the producer of the algorithm [15]? Physicians who adhere to the standard of care are protected by law [16], and penalties might apply when physicians choose AI-made nonstandard care medical decisions [16, 17].

In a survey study, researchers examined the liability judgments of potential jurors in a case in which a patient was harmed by a physician's decision to accept or reject AI recommendations for standard or nonstandard care [18]. The jury pool view surprisingly favored the use of AI in precision medicine. Thus, when physicians receive an AI recommendation for standard care, they have a legal incentive to accept it. If nonstandard care is advised and thus rejected, they do not necessarily free themselves from liability.

12.4 Possible Preventive and Corrective Interventions

The ethical framework for AI applications in radiology should always incorporate the core principles of biomedical ethics, emphasizing specific issues that can be easily compromised by AI applications, namely, transparency and accountability [20, 21].

12.4.1 Autonomy and Data Privacy

Medical imaging data usually contains patient demographic features and eventually clinical details [22]. In the USA, the Health Insurance Portability and Accountability Act (HIPAA) and, in Europe, the General Data Protection Regulation (GDPR) mandate anonymization and, if possible, pseudonymization (a de-identification procedure) as optional in the GDPR [23]. Automatic software is available to anonymize or de-identify images, yet their accuracy is not fully demonstrated [24]. To this aim, each image should be checked by a human reviewer [24]. Such a tedious image-per-image control was performed before the release of the US National Institutes of Health (NIH) ChestX-ray8 dataset, which consists of 100,000 images from 30,000 patients [25], and DeepLesion dataset, comprising 32,000 CT images from 4400 unique patients [26], by the NIH Clinical Center. However AI algorithms might still be able to reveal the identity of the participants [27].

New resources that can de-identify the data using the ChestX-Ray14 and COVID-19 datasets have been announced [27]. The problem remains when body contours and facial recognition can be reconstructed [28, 29]. Such rescued images of the face from cranial MRI could be compared with photos from social media or other sources [30]. Distorting facial features or removing surface soft tissue data from cranial radiographs could be helpful [31].

Institutional review boards or ethics committees should be contacted to ensure compliance with the patient consent, and access to data on a need-to-know basis, with data tags identifying each user, is similarly recommended [32].

12.4.1.1 Clinical Research and Modalities of Informed Consent

In many countries consent is not required for scientific use of information stored in data banks, even though journal articles could need it [22]. Consent forms do not always address secondary use of data. A "tiered" consent form is an alternative, indicating how and what data to share [22]. As images may be used repeatedly for training and validation of algorithms, continuous use should be mentioned. Permanent communication between researchers and participants via dynamic consent is also convenient [33], or perhaps control of future use of data by meta-consent [34].

12.4.1.2 Stakeholders

Institutions, radiology clinics, and research teams are currently the owners of the images; however participants often want to keep the rights, deciding for instance conditions of access by third parties, or even demanding financial compensation. The downside could be barriers and delays in AI adoption. One sharing framework considers the entire process as a form of public good as long as no entity directly profits from the data, and useful knowledge results [35].

12.4.1.3 Confidentiality Risks

Data breaches and cyberattacks threaten patient privacy [29, 36, 37]. Long-term projects like the UK Biobank regularly test Internet technology systems, conduct security checks by experts, remove individual identifiers, do not allow any form of de-anonymization, and strictly restrict as well circulation or exchange of personal data [38].

12.4.1.4 Federated Learning (FL)

Within this pathway models are shared instead of patient data to keep training data private and to enable secure collaborations among multiple institutions [24]. Although FL can provide data privacy, training examples could be reconstructed through model inversion [39]. Therefore, privacy-preserving measures should still be necessary [39].

12.4.2 Beneficence, Nonmaleficence, and Justice

AI methods could reduce the time to recognize clinical emergencies and save lives. Nevertheless failures and biases could have the opposite effect, discriminating subjects and delaying diagnosis [40]. Personal or professional views of fairness should not be imposed on people or on AI algorithms. Competing moral concerns should always be acknowledged and recognized. Conflicts may arise, for example, in the allocation of resources: Each physician might prioritize his or her patients rather than the next one who really needs it first. This could occur either because of unawareness of other patients' information or because of cognitive biases. Whatever the reason, such moral conflicts could be inherited by an AI algorithm, which in turn perpetuates the same biased allocation [20]. Discrimination against African Americans in a widely used healthcare allocation algorithm in the USA has been denounced. It was driven by the wrong endpoint, namely, profit: The algorithm was not trying to predict future healthcare needs, but rather healthcare costs [41]. Since white Americans tend to have higher healthcare premiums than African Americans, white Americans were assigned a higher score. Within the context of breast cancer, by including diverse data, it was possible to create a model that was equally accurate for white and African American women compared to the traditional Tyrer-Cuzick model, in which African American women were not fairly represented [6].

12.4.2.1 Continuous Self-Learning Algorithms

Such models should require feedback loops and continuous surveillance for possible bias, without overlooking new insights for ongoing optimization [37, 42]. By addressing the differences between "locked" and "adaptive" algorithms and requiring "product life cycle reporting" for "adaptive" algorithms, the new FDA framework can help address these challenges [43–45]. Policies also need to be put in place on how new true-positive findings should be handled in a retrospective analysis or new false-positive findings should be addressed in a prospective setting. In this context, guidelines similar to those for adverse drug reaction reporting could be adapted and developed after careful examination of end-user feedback loops [38].

12.4.2.2 Commercial and Other Nonscientific Applications

Insurance companies could stratify patients according to risk of disease occurrence or predicted treatment costs for maximum profit, endangering the ethics of AI employment [19, 35, 37, 42]. A non-healthcare misconduct example was the Volkswagen scandal, whereby an algorithm was programmed to fake low nitrogen oxide emission levels to meet US standards, when the actual emission was 40 times higher [37, 42].

12.4.3 Transparency and Accountability

Artificial intelligence systems should be clearly structured, and the intent of the system should be available to all users [46]. Malfunctions or harms need to be traceable and the decision-making process should be verifiable. If legal liability applies, it should be attributed to a responsible party [20, 32]. Communication about the criteria for patient selection and sharing of the algorithm code is required [2, 4]. Even before routine use, all methods demand validation, whereas data and codes need also to be shared [4].

12.4.3.1 Major Disasters

The tragic Boeing 737 MAX airplane accidents, even though without direct healthcare implications, demonstrated what can happen when an automated system triggers an action without human involvement. Systems should be endowed with warnings when they are engaged and have an override function [47].

12.4.3.2 Challenges and Competitions

These are relatively common in the realm of AI and could foster breakthroughs, yet transparent ranking guidelines or standardization of challenge designs is advisable [48], in order to contribute to translational applications [48, 49]. Censorship of algorithm codes is only defensible within specific circumstances, like the Generative Pre-trained Transformer 3 (GPT-3), an automatic language generator able to mimic human-written text [50]. As GPT-3 can generate misinformation in the wrong hands, the code should not be shared.

12.5 AI Ethical Codes

Traditional methods of regulating a product, research, or development may not be suitable for adoption in AI technologies since different phases of implementation require different measures. The recommended approach to governance consists of two steps: before and after implementation [51]. Before implementation (ex ante phase), the risks must be foreseen, and precautions must be taken, whereas after implementation (ex post phase), the damage that has occurred must be assessed retrospectively, and a solution must be sought for it [51]. Ex ante regulation could be hampered by AI systems that require little to no physical infrastructure, are designed without deliberate coordination, are executed through multi-institutional and multinational collaboration, or have potential threats that would be difficult for outside observers to detect. Due to the autonomous nature of AI, ex post regulation could be inefficient as well, due to foreseeability and oversight challenges [51].

Nevertheless regulation of AI applications is constantly introduced and updated, such as the *Ethics Guidelines for Trustworthy AI* by the European Commission [46], the *Asilomar AI Principles* [52], and the *Montreal Declaration for Responsible AI* [40]. As many as 84 different guidelines are clustered around the classic pillars of autonomy/freedom, privacy, beneficence, nonmaleficence, justice/fairness, transparency, responsibility, trust, dignity, sustainability, and solidarity [53]. The most represented principle was transparency, whereas sustainability, dignity, and solidarity were underrepresented compared to others [53]. This could serve as a cautionary tale for us in radiology not to overlook any crucial ethical principles.

12.5.1 Regulatory Procedures

AI-driven devices are submitted to the FDA in the USA, whereas the European approval process is carried out by accredited agencies that can issue a CE mark [3], on the basis of safety, performance, reliability, and additional requirements depending on the level of risk (e.g., clinical applications are high-risk). The FDA handles AI containing products as standard medical devices and applies regulations according to their features [3].

To support the development of safe and effective medical devices that use AI algorithms, the FDA first released a framework proposal [45] and later an action plan in direct response to stakeholder feedback on that framework [43]. The paper is titled *Artificial Intelligence/Machine Learning-Based Software as a Medical Device and,* as the title suggests, proposes a regulatory framework for emerging AI technologies that treats them as medical devices. To date, FDA-approved algorithms are primarily so-called locked algorithms that do not continuously learn with each use. The manufacturer can modify these "locked" algorithms from time to time by training them with new data.

In contrast, there is massive potential for improvement in so-called adaptive or continuous learning algorithms, which can learn independently without manual modification, through use in the real world. To regulate AI algorithms, FDA plans to take the following actions: a tailored regulatory framework that accounts for future modifications; good machine learning pathways to establish best practices; patient-centered approach that incorporates transparency for users; regulatory science efforts to reduce bias and increase generalizability; and a total product lifecycle approach to evaluate real-word performance [43]. This audit will require a premarket review that includes the algorithm performance, anticipated modifications, and the manufacturer's ability to control the risks of modifications [44, 54].

12.5.2 AI Applications in Radiology

The Canadian Radiology Society published some time ago a white paper on ethical and legal issues related to AI in radiology, focusing primarily on privacy and data sharing issues [29]. Afterward the Royal Australian and New Zealand College of Radiologists (RANZCR) prepared *Ethical Principles for AI in Medicine* [55] relying on safety, privacy, and protection of data, avoidance of bias, transparency and explainability, application of human values, decision-making on diagnosis and treatment, teamwork, accountability for decisions made, and governance. Around the same time, a multi-society statement entitled *Ethics of AI in* Radiology was published, which is a joint effort of the American College of Radiology, the European Society of Radiology, the Radiological Society of North America, the Society for Imaging Informatics in Medicine, the European Society of Medical Imaging Informatics, the Canadian Association of Radiologists, and the American Association of Physicists in Medicine [14]. This statement approaches the topic on three levels: the ethics of data, the ethics of algorithms, and the ethics of practice.

Both the multi-society and RANZCR papers highlight the value of patient privacy and support every effort to safeguard and maintain the security of patient data. In addition, both papers discuss avoiding bias and prejudice by using diverse data derived from both the general population and minority groups. Both have similar definitions for the role of AI in the decision-making process. The RANZCR paper limits the role of AI to an advisory position, leaving the ultimate decision to mutual communication between the patient and physician [55]. Similarly, when AI is part of the process, the multi-society paper advises that the decision is made after having transparent communication between healthcare providers, physicians, and patients [14]. The RANZCR paper assures "shared liability" among the manufacturer, practitioner, and management that allowed the use of AI [55]. While the multi-society paper supports the concept of "shared liability," it first calls for the precise definition of the "AI-assisted standard of care" because in routine practice, the person who complies or fails to comply with the "standard of care" is liable [14]. Finally, both agree that radiologists should be involved in the development of a code of ethics for AI applications in radiology.

12.5.3 Guidelines Concerning AI Research

The Radiological Society of North America/RSNA recently organized a checklist designed for authors and reviewers of *Radiology: Artificial Intelligence* journal. The Checklist for Artificial Intelligence in Medical Imaging (CLAIM) includes classification, image reconstruction, texture analysis, and workflow optimization, with the goal of improving the reproducibility and generalizability of AI research [56]. The Enhancing the Quality and Transparency of Health Research (EQUATOR) network and the Standards for Reporting of Diagnostic Accuracy Studies-Artificial Intelligence (STARD-AI) Steering Group deserve emphasis [57, 58]. The STARD-AI is complementary to the other new AI extensions of the EQUATOR network program, such as Consolidated Standards of Reporting Trials (CONSORT-AI) and Standard Protocol Items: Recommendations for Interventional

Trials (SPIRIT-AI) [59]. It is being developed in close collaboration with stakeholders, clinicians, computer scientists, journal editors, researchers, trialists, industry leaders, regulators, funders, policymakers, and patient groups as well as groups from typically underrepresented regions such as Asia and Africa, to ensure that the AI extension to STARD is considered to be globally applicable [58].

12.6 Future Perspectives

When picture archiving and communication systems (PACS) became available in the 1990s [60], radiologists feared professional irrelevance as images were widely available to general clinicians. When AI algorithms exceeded human diagnostic performance, the same threats were echoed [61]. Nobody is sure about the disruptive potential of AI in radiology in the future, yet chances are that experts will always be required [62]. This should be a safeguard for professional ethics and integrity as well, in face of accidental "misconduct" or "bias" by the machine algorithm.

Training of residents is one of the preliminary steps, contributing to knowledgeable and responsible utilization of AI by means of academic and clinical activities [63]. As knowledge levels increase, medical attitudes toward AI shift from anxiety to more positive views [64], and the public should be informed as well [65].

Acknowledgments I would like to thank Xenia Deligianni, Francesco Santini, and Joe Cocker for their helpful comments.

References

1. Hempel J (2018) Fei-Fei Li's quest to make AI humanity. Wired 12:1–34. wired.com/story/fei-fei-li-artificial-intelligence-humanity/. Accessed 24 Apr 2021
2. Tang A, Tam R, Cadrin-Chênevert A, Guest W, Chong J, Barfett J, Chepelev L, Cairns R, Mitchell JR, Cicero MD, Poudrette MG, Jaremko JL, Reinhold C, Gallix B, Gray B, Geis R, O'Connell T, Babyn P, Koff D, Ferguson D, Derkatch S, Bilbily A, Shabana W (2018) Canadian association of radiologists white paper on artificial intelligence in radiology. Can Assoc Radiol J 69:1–16
3. Muehlematter UJ, Daniore P, Vokinger KN (2021) Approval of artificial intelligence and machine learning-based medical devices in the USA and Europe (2015–20): a comparative analysis. Lancet Digit Heal 3:e195–e203
4. Topol EJ (2019) High-performance medicine: the convergence of human and artificial intelligence. Nat Med 25:44–56
5. O'Neill TJ, Xi Y, Stehel E, Browning T, Ng YS, Baker C, Peshock RM (2021) Active reprioritization of the reading worklist using artificial intelligence has a beneficial effect on the turnaround time for interpretation of head CT with intracranial hemorrhage. Radiol Artif Intell 3:e200024
6. Yala A, Lehman C, Schuster T, Portnoi T, Barzilay R (2019) A deep learning mammography-based model for improved breast cancer risk prediction. Radiology 292:60–66
7. Hosny A, Parmar C, Quackenbush J, Schwartz LH, Aerts HJWL (2018) Artificial intelligence in radiology. Nat Rev Cancer 18:500–510
8. Angwin J, Larson J, Mattu S, Kirchner L (2016) Machine bias. ProPublica. propublica.org/article/machine-bias-risk-assessments-in-criminal-sentencing. Accessed 24 April 2021
9. Shaw J (2019) Artificial intelligence & ethics. *Harv Mag* 44–49,74. harvardmagazine.com/2019/01/artificial-intelligence-limitations. Accessed 24 Apr 2021
10. Finlayson SG, Chung HW, Kohane IS, Beam AL (2019) Adversarial attacks against medical deep learning systems. arXiv: 1804.05296v3
11. Finlayson SG, Bowers JD, Ito J, Zittrain JL, Beam AL, Kohane IS (2019) Adversarial attacks on medical machine learning. Science (80-) 363:1287–1289
12. Becker AS, Jendele L, Skopek O, Berger N, Ghafoor S, Marcon M, Konukoglu E (2019) Injecting and removing suspicious features in breast imaging with CycleGAN: a pilot study of automated adversarial attacks using neural networks on small images. Eur J Radiol 120:108649
13. O'Connor SD, Bhalla M (2021) Should artificial intelligence tell radiologists which study to read next? Radiol Artif Intell 3:e210009
14. Geis JR, Brady A, Wu CC, Spencer J, Ranschaert E, Jaremko JL, Langer SG, Kitts AB, Birch J, Shields WF, van den Hoven van Genderen R, Kotter E, Gichoya JW, Cook TS, Morgan MB, Tang A, Safdar NM, Kohli M (2019) Ethics of artificial intelligence in radiology: summary of the joint European and North American multisociety statement. Insights Imaging 10:101
15. Schmelzer R (2019) What Happens when self-driving cars kill people? *Forbes*. forbes.com/sites/cognitiveworld/2019/09/26/what-happens-with-self-driving-cars-kill-people/?sh=154a1e5f405c. Accessed 24 Apr 2021

16. Sullivan HR, Schweikart SJ (2019) Are current tort liability doctrines adequate for addressing injury caused by AI? AMA J Ethics 21:160–166
17. Price WN, Gerke S, Cohen IG (2019) Potential liability for physicians using artificial intelligence. JAMA 321:2281–2282
18. Tobia K, Nielsen A, Stremitzer A (2021) When does physician use of AI increase liability? J Nucl Med 62: 17–21
19. The National Commission for the Protection of Human Subjects of Biomedical and Behavioral Research (1978) The Belmont report ethical principles and guidelines for the protection of human subjects of research. hhs.gov/ohrp/regulations-and-policy/belmont-report/read-the-belmont-report/index.html. Accessed 24 Apr 2021
20. Floridi L, Cowls J, Beltrametti M, Chatila R, Chazerand P, Dignum V, Luetge C, Madelin R, Pagallo U, Rossi F, Schafer B, Valcke P, Vayena E, Corporation N, Europe D, Affairs G, Ethics H (2018) AI4People – an ethical framework for a good AI society: opportunities, risks, principles, and recommendations. Minds Mach 32:1–24
21. Kohli M, Geis R (2018) Ethics, artificial intelligence, and radiology. J Am Coll Radiol 15:1317–1319
22. Kotsenas AL, Balthazar P, Andrews D, Geis JR, Cook TS (2021) Rethinking patient consent in the era of artificial intelligence and big data. J Am Coll Radiol 18:180–184
23. The European Commission (2016) Directive (EU) 2016/1148 of the European Parliament and of the Council of 6 July 2016 concerning measures for a high common level of security of network and information systems across the Union. Off J Eur Union 6: 30. https://eur-lex.europa.eu/legal-content/EN/TXT/PDF/?uri=CELEX:32016L1148&from=EN. Accessed 24 Apr 2021
24. Langlotz CP, Allen B, Erickson BJ, Kalpathy-Cramer-J, Bigelow K, Cook TS, Flanders AE, Lungren MP, Mendelson DS, Rudie JD, Wang G, Kandarpa K (2019) A roadmap for foundational research on artificial intelligence in medical imaging: from the 2018 NIH/RSNA/ACR/The Academy Workshop. Radiology 291:781–791
25. Wang X, Peng Y, Lu L, Lu Z, Bagheri M, Summers RM (2017) Chest X-ray8: hospital-scale chest X-ray database and benchmarks on weakly-supervised classification and localization of common thorax diseases. arXiv:1705.02315 [cs.CV]
26. Yan K, Wang X, Lu L, Summers RM (2018) DeepLesion: automated mining of large-scale lesion annotations and universal lesion detection with deep learning. J Med Imaging 5:036501
27. Packhäuser K, Gündel S, Münster N, Syben C, Christlein V, Maier A (2021) Is medical chest x-ray data anonymous? arXiv: 2103.08562v1
28. Prior FW, Brunsden B, Hildebolt C, Nolan TS, Pringle M, Vaishnavi SS, Larson-Prior LJ (2009) Facial recognition from volume-rendered magnetic resonance imaging data. IEEE Trans Inf Technol Biomed 13:5–9
29. Jaremko JL, Azar M, Bromwich R, Lum A, Alicia Cheong LH, Gibert M, Laviolette F, Gray B, Reinhold C, Cicero M, Chong J, Shaw J, Rybicki FJ, Hurrell C, Lee E, Tang A (2019) Canadian association of radiologists white paper on ethical and legal issues related to artificial intelligence in radiology. Can Assoc Radiol J 70:107–118
30. Schwarz CG, Kremers WK, Therneau TM, Sharp RR, Gunter JL, Vemuri P, Arani A, Spychalla AJ, Kantarci K, Knopman DS, Petersen RC, Jack CR (2019) Identification of anonymous MRI research participants with face-recognition software. N Engl J Med 381:1684–1686
31. Mazura JC, Juluru K, Chen JJ, Morgan TA, John M, Siegel EL (2012) Facial recognition software success rates for the identification of 3D surface reconstructed facial images: implications for patient privacy and security. J Digit Imaging 25:347–351
32. Liew C (2018) The future of radiology augmented with Artificial Intelligence: a strategy for success. Eur J Radiol 102:152–156
33. Finlay T, Collett C, Kaye J, Bentzen HB, Beck S, D'Abramo F, Javaid MK, Teare HJA, Caenazzo L, Felzmann H, Mascalzoni D, Budin-Ljøsne I, Jones E, Simpson A, Katić V (2017) Dynamic consent: a potential solution to some of the challenges of modern biomedical research. BMC Med Ethics 18:1–10
34. Ploug T, Holm S (2016) Meta consent – a flexible solution to the problem of secondary use of health data. Bioethics 30:721–732
35. Larson DB, Magnus DC, Lungren MP, Shah NH, Langlotz CP (2020) Ethics of using and sharing clinical imaging data for artificial intelligence: a proposed framework. Radiology 295:675–682
36. Pesapane F, Volonté C, Codari M, Sardanelli F (2018) Artificial intelligence as a medical device in radiology: ethical and regulatory issues in Europe and the United States. Insights Imaging 9(745–753):38
37. Char DS, Shah NH, Magnus D (2018) Implementing machine learning in health care—addressing ethical challenges. N Engl J Med 378:981–983
38. Akinci D'Antonoli T (2020) Ethical considerations for artificial intelligence: An overview of the current radiology landscape. Diagnostic Interv Radiol 26:504–511
39. Li W, Milletarì F, Xu D, Rieke N, Hancox J, Zhu W, Baust M, Cheng Y, Ourselin S, Cardoso MJ, Feng A (2019) Privacy-preserving federated brain tumour segmentation. Lect Notes Comput Sci (including Subser Lect Notes Artif Intell Lect Notes Bioinformatics) 11861 LNCS:133–141. arXiv: 1910.00962v1
40. de Montréal U (2018) Montreal declaration for a responsible development of artifical intelligence.

nouvelles.umontreal.ca/en/article/2017/11/03/montreal-declaration-for-a-responsible-development-of-artificial-intelligence/. Accessed 24 Apr 2021

41. Obermeyer Z, Powers B, Vogeli C, Mullainathan S (2019) Dissecting racial bias in an algorithm used to manage the health of populations. Science 366:447–453
42. Vayena E, Blasimme A, Cohen IG (2018) Machine learning in medicine: addressing ethical challenges. PLoS Med 15:e1002689
43. IMDRF SaMD Working Group (2021) Artificial intelligence/machine learning (AI/ML)-based software as a medical device (SaMD) action plan. https://www.fda.gov/media/145022/download. Accessed 24 Apr 2021
44. Harvey HB, Gowda V (2020) How the FDA regulates AI. Acad Radiol 27:58–61
45. FDA (2019) Proposed regulatory framework for modifications to artificial intelligence/machine learning (AI/ML)-based software as a medical device (SaMD) https://www.fda.gov/media/122535/download. Accessed 24 Apr 2021
46. High-Level Expert Group on Artificial Intelligence (2019) Ethics guidelines for trustworthy AI. https://digital-strategy.ec.europa.eu/en/library/ethics-guidelines-trustworthy-ai. Accessed 24 Apr 2021
47. Mongan J, Kohli M (2020) Artificial intelligence and human life: five lessons for radiology from the 737 MAX disasters. Radiol Artif Intell 2:e190111
48. Maier-Hein L, Eisenmann M, Reinke A, Onogur S, Stankovic M, Scholz P, Arbel T, Bogunovic H, Bradley AP, Carass A, Feldmann C, Frangi AF, Full PM, van Ginneken B, Hanbury A, Honauer K, Kozubek M, Landman BA, März K, Maier O, Maier-Hein K, Menze BH, Müller H, Neher PF, Niessen W, Rajpoot N, Sharp GC, Sirinukunwattana K, Speidel S, Stock C, Stoyanov D, Taha AA, van der Sommen F, Wang CW, Weber MA, Zheng G, Jannin P, Kopp-Schneider A (2018) Why rankings of biomedical image analysis competitions should be interpreted with care. Nat Commun 9:1–13
49. Prevedello LM, Halabi SS, Shih G, Wu CC, Kohli MD, Chokshi FH, Erickson BJ, Kalpathy-Cramer J, Andriole KP, Flanders AE (2019) Challenges related to artificial intelligence research in medical imaging and the importance of image analysis competitions. Radiol Artif Intell 1:e180031
50. Brown TB, Mann B, Ryder N, Subbiah M, Kaplan J, Dhariwal P, Neelakantan A, Shyam P, Sastry G, Askell A, Agarwal S, Herbert-Voss A, Krueger G, Henighan T, Child R, Ramesh A, Ziegler DM, Wu J, Winter C, Hesse C, Chen M, Sigler E, Litwin M, Gray S, Chess B, Clark J, Berner C, McCandlish S, Radford A, Sutskever I, Amodei D (2020) Language models are few-shot learners. arXiv: 2005.14165v4
51. Scherer MU (2016) Regulating artificial intelligence systems: risks, challenges, competencies, and strategies. Harv J Law Technol 29:354–400
52. Asilomar Conference (2017) Asilomar AI principles. In: Princ. dev. conjunction with 2017 Asilomar Conf. [Benevolent AI 2017]. https://futureoflife.org/ai-principles/. Accessed 24 Apr 2021
53. Jobin A, Ienca M, Vayena E (2019) The global landscape of AI ethics guidelines. Nat Mach Intell 1:389–399
54. Wiens J, Saria S, Sendak M, Ghassemi M, Liu VX, Doshi-Velez F, Jung K, Heller K, Kale D, Saeed M, Ossorio PN, Thadaney-Israni S, Goldenberg A (2019) Do no harm: a roadmap for responsible machine learning for health care. Nat Med 25:1337–1340
55. The Royal Australian and New Zealand College of Radiologists (2019) Ethical principles for artificial intelligence in medicine. ranzcr.com/college/document-library/ethical-principles-for-ai-in-medicine. Accessed 24 Apr 2021
56. Mongan J, Moy L, Kahn CE (2020) Checklist for artificial intelligence in medical imaging (CLAIM): a guide for authors and reviewers. Radiol Artif Intell 2: e200029
57. Bossuyt PM, Reitsma JB, Bruns DE, Gatsonis CA, Paul P, Irwig L, Lijmer JG, Moher D, Rennie D, De HCW, Kressel HY, Rifai N, Golub RM, Altman DG, Hooft L (2015) STARD 2015: an updated list of essential items for reporting diagnostic accuracy studies. Radiology 277:826–832
58. Sounderajah V, Ashrafian H, Aggarwal R, De Fauw J, Denniston AK, Greaves F, Karthikesalingam A, King D, Liu X, Markar SR, McInnes MDF, Panch T, Pearson-Stuttard J, Ting DSW, Golub RM, Moher D, Bossuyt PM, Darzi A (2020) Developing specific reporting guidelines for diagnostic accuracy studies assessing AI interventions: the STARD-AI steering group. Nat Med 26:807–808
59. Ibrahim H, Liu X, Rivera SC, Moher D, Chan A-W, Sydes MR, Calvert MJ, Denniston AK (2021) Reporting guidelines for clinical trials of artificial intelligence interventions: the SPIRIT-AI and CONSORT-AI guidelines. Trials 22(1):11
60. Hall FM (1998) Perils of PACS. Radiology 213:307–308
61. Langlotz CP (2019) Will artificial intelligence replace radiologists? Radiol Artif Intell 1:e190058
62. Mazurowski MA (2019) Artificial intelligence may cause a significant disruption to the radiology workforce. J Am Coll Radiol 16:1077–1082
63. Wiggins WF, Caton MT, Magudia K, Glomski SA, George E, Rosenthal MH, Gaviola GC, Andriole KP (2020) Preparing radiologists to lead in the era of artificial intelligence: designing and implementing a focused data science pathway for senior radiology residents. Radiol Artif Intell 2:e200057

64. Huisman M, Ranschaert E, Parker W, Mastrodicasa D, Koci M, Pinto de Santos D, Coppola F, Morozov S, Zins M, Bohyn C, Koç U, Wu J, Veean S, Fleischmann D, Leiner T, Willemink MJ (2021) An international survey on AI in radiology in 1,041 radiologists and radiology residents part 1: fear of replacement, knowledge, and attitude. Eur Radiol 31(9):7058–7066
65. Kitamura FC, Marques O (2021) Trustworthiness of artificial intelligence models in radiology and the role of explainability. J Am Coll Radiol 18(8):1160–1162

13 Typhoid Fever as a Biological Nazi Weapon in the Terezin Jewish Ghetto

Kateřina Horáčková

Abstract

The paper is a historical overview study presenting the specific position of the Theresienstadt Jewish ghetto among other internment facilities, the living conditions of ghetto prisoners as the main predictor of the emergence of diseases and the targeted use of typhoid fever infection as a biological weapon against a vulnerable population (extremely underfed and undernourished civilian prisoners) at the end of World War II. Technically this was not scientific misconduct, as those events occurred far away from research laboratories or academic institutions in the context of a war. Nevertheless they fit a pattern of widespread medical omission and in many instances of active collaboration with the atrocities of the Nazi regime, which led to a criminal trial against members of the medical profession in Germany after World War II.

13.1 Introduction

In world history the Nazi Holocaust against Jews is the most horrific evidence of atrocities ever inflicted on man by man. It has been only 76 years since World War II and the Holocaust ended. Some of those who passed through the walls of Jewish ghettos or concentration camps are still with us, those who experienced the brutality of the Nazi guards, those who hid in various shelters under new identities or those who were torn from their mother's or father's arms and taken by train to a foreign country.

In the territory of former Czechoslovakia (now Czech Republic), the fortress town of Theresienstadt (Terezin) is a place most associated with World War II. On 20 January 1942, senior representatives of Nazi Germany, led by Reinhard Heydrich, met at a conference in Wannsee to discuss the "Final Solution of the Jewish Question". As the Nazis were not ready to carry out the physical liquidation of Jews at this stage of their plan, they chose a "ghettoization" of the Jewish population into artificially created towns (ghettos) as an interim solution, in order to accelerate the physical liquidation of Jews [1, 2]. Joseph Goebbels, Hitler's minister of propaganda, referred to ghettos as a "territory of death". The ghettos were to serve as a transfer station for the final extermination in concentration or labour camps [3].

Theresienstadt was built at the end of the eighteenth century by order of Holy Roman Emperor

K. Horáčková (✉)
Faculty of Health Sciences, Nursing Department, University of Pardubice, Pardubice, Czech Republic
e-mail: katerina.horackova@upce.cz; https://fzs.upce.cz/fzs/katerina-horackova

J. Faintuch, S. Faintuch (eds.), *Integrity of Scientific Research*,
https://doi.org/10.1007/978-3-030-99680-2_13

Joseph II to prevent the advance of Prussian troops towards Prague. The fortress town (main fortress) served as a Jewish ghetto, and the small fortress became a prison for the Prague section of the secret Nazi police (Gestapo), and also for some ghetto inhabitants [2, 4].

The ghetto was governed by Jewish "self-administration", which actually followed strict orders by the Nazis. Protectorate gendarmes (Czech nazi militias) helped paramilitary German Nazi Schutzstaffel (SS) troops guard the city. Different from the SS, the gendarmes mostly felt for the prisoners and tried to help them or keep them in touch with the outside world [4, 5].

13.2 A Model Concentration Camp

In addition to accommodating Jews and redistributing them further to extermination camps, Theresienstadt also served a role in the official Nazi propaganda. It was designed to be a "model Jewish settlement", a demonstration that Hitler, and thus all of Nazi Germany, treated Jews kindly and humanely. German propaganda spoke of Theresienstadt as a spa town, as "proved" by manipulated films, which were important instruments influencing public opinion. This illusion was also deliberately created for the members of the International Committee of the Red Cross. Another feature was a high number of world-renowned doctors, scholars and scientists from all over Europe, a group of prominent individuals (some carrying Nobel Prizes or other important awards) whose disappearance would not go unnoticed.

Theresienstadt was not surrounded by barbed wire fences, SS troops were not too conspicuous on the streets, and the prisoners were not punished as brutally as in other concentration camps, which may have created a certain sense of freedom in its inhabitants, which most of them only realized when transported to other concentration camps. A final difference was the absence of gas chambers and inhumane medical experiments, which were common practices in other ghettos and concentration camps [6, 7].

Theresienstadt was unique in its rich cultural life with lectures, readings, recitations, theatre performances, concerts and operas. Although illegal, it was possible to perform educational and sporting activities for children. These activities were often a manifestation of resistance against fascism and Nazi power and, from a psychological point of view, had a preventive effect on the development of psychosomatic diseases. Also health care in the ghetto was organized at a level that cannot be compared with other ghettos or concentration camps [7, 8].

13.3 Living Conditions and Health Troubles

The ghetto was typical for a high morbidity and mortality rate. Diseases of the digestive tract, dying of "old age", respiratory abnormalities, diseases of the blood and heart, infectious and nervous problems, malignancies, urinary and genital tract disorders and violent death were the most common causes of mortality. Frequent transports of new prisoners were full of sick, old, blind patients, as well as pregnant women. Overcrowding (the peak was recorded in September 1942, with about 1.6 square metres of accommodation space per person), tragic accommodation and sanitary conditions, the scarcity of drinking water and poor-quality food, inadequate heating during the winter and the psychological strain caused infectious diseases (which often reached epidemic proportions), as well as new, previously unknown diseases, whose causes were widely debated by doctors in Theresienstadt [8, 9].

13.4 Healthcare Facilities

These conditions gave rise to a new branch of medicine called "Theresiana". Due to its specific factors, the symptomatology of known diseases was quite different. The health of the prisoners was looked after by doctors, nurses and auxiliary staff recruited from prisoners. Theresienstadt had a main hospital, a secondary hospital, a children's hospital, a hospice for dying patients and a nursing home. Health care included many outpatient clinics of various specialists, such as those of a urologist, dermatologist, dentist, etc. Medical and surgical procedures, especially in the early days of the ghetto, were carried out in the most primitive manner. Gradually, a great development could be observed, especially thanks to the dedication of the medical staff themselves [9, 10].

During the period of the ghetto's operation (1941–1945), many infectious diseases were recorded. Some of them reached epidemic proportions. Although the official statistics list diseases of the digestive tract and respiratory tract as the most common cause of death among the prisoners, it should be pointed out that infectious diseases of these systems were also included. Therefore, they can be considered the main cause of death in the ghetto. The most relevant infections were typhoid fever, diphtheria, scarlet fever, lung tuberculosis, encephalitis, measles, rubella, mumps, whooping cough and epidemic typhus. At the end of the war, epidemiology changed for worse, at the same time that atrocities committed by the Nazis became more common [9, 11].

13.5 The Small Fortress Prison

Very harsh conditions prevailed. The showers and washrooms often did not work, and if they did only cold water was available. Not being able to wash was part of the punishment. The living conditions led to an enormously high morbidity rate. The hospital in the small fortress was designed for 60 people but accommodated 190 patients. There were often as many as three patients on one straw bed. Every morning before the roll call, corpses were piled up in front of the hospital entrance. Before despatching them to the crematorium, the guards ripped out their gold teeth and crowns [12, 13]. Solitary confinement cells were set up for patients with infectious diseases. From the autumn of 1944 on, prisoners with infectious diseases were left in mass cells among the other prisoners. The air in the cells was unbreathable and there was no ventilation.

13.6 Female Prisoners

In the women's ward, pregnant women also gave birth, in absolutely insane conditions. The baby's umbilical cord was tied with non-sterilized, ordinary trousers' drawstring. Often after the birth, the mother was taken to the hospital in the city of Litoměřice, where the child was taken away from her and never returned. The mother was usually transferred to a concentration camp in Eastern Europe, from which practically no one came back either. In November 1944, one of the women's cells was occupied by still surviving women from the hastily liquidated concentration camps (Auschwitz, Christianstad, etc.), in face of the advance of Allied armies. It was called the "cell of death". There were no bunk beds, no toilets and no running water, the women were denied medicine, and the female doctor had no access. Thanks to the solidarity of other prisoners, the women on death row received help, and many survived [12, 13].

Theresienstadt was officially liberated on 8 May 1945. When the first Red Army troops arrived a day later, the situation was complicated by a typhus epidemic [2].

13.7 Epidemic Typhus in Terezin

This is an acute high-fatality infectious disease typical of the prison environment as well as of wars, caused by the intracellular microorganism *Rickettsia prowazekii*, and transmitted by lice. During World War I, it had already triggered tens of thousands of fatalities, with the name of trench fever. Factors influencing the emergence and spread include poverty (destitution, hunger, lack of clothing), overcrowding, overpopulation and mixing of populations, inadequate personal hygiene (lack of soap, hot water, inability to wash), lack of disinfectant solutions, lack of vaccines and medicine, stress, torture, fear and abundance of infected lice [21, 22]. The mortality rate is 10–40% if untreated. The older the person, the greater the likelihood of death [3, 16]. Soldiers as well as prisoners of war need to be bathed, deloused and disinfected. Clothes are first disinfected with steam and then with gas and finally treated with delousing powder [16].

In the Nazi concentration camps, epidemic typhus posed a great danger, despite theoretically available delousing measures. The science of the time was limited, and epidemic typhus was essentially treated with two non-curative drugs: lobeline, a nicotinic alkaloid that promoted breathing, and Sympatol (synephrine) a sympathomimetic alkaloid that stimulated the heart. None of them helped much [1, 3, 16].

13.8 Jewish Guilt

Historically the association of epidemic typhus with Jews had been one of the roots of antisemitism in Germany but also in France and other European countries. Jews were often denounced as carriers of lice and thus of typhus and other diseases, as depicted in propaganda posters. The Third Reich held the Jews responsible for the poor health of the German nation [1, 3]. At the same time, typhus was associated with biological warfare as part of the Nazi genocide.

The Theresienstadt ghetto was spared from typhus until April 20, 1945, although the lice infestation of the population was a serious problem. Especially in 1942 and 1943 (at the height of the ghetto's overcrowding), lice infestation spread throughout the ghetto and predominantly affected the elderly, malnourished and sick prisoners. Until that date a single case of typhus had been recorded, in January 1944 [17, 18].

13.9 Endemic Louse-Borne Diseases

A very important part of the healthcare system in this ghetto, different from other concentration camps with practically no health assistance at all, was the disinfection group, which was successful throughout the internment. The typhus epidemic at the end of the war was not the fault of this group. With the increasing arrival of new prisoners, the capacity of the delousing station was no longer sufficient, and therefore more space had to be made available. Moreover deloused prisoners almost immediately came into contact with their fellow lice-infested colleagues, as isolation in the cramped conditions was impossible [19].

Significant improvements were made with the use of tiny amounts of hydrogen cyanide, called Zyklon B, the same human killing gas used in higher doses at the death chambers of Auschwitz, where two million Jews are annihilated. It was employed to bathe the prisoners, follow on with manual delousing and at the same time gas the quarters. Moreover flies, bedbugs and fleas had also to be exterminated, so it was necessary to gas some of the blocks from time to time. Due to the location of the ghetto near the Eger River, the prisoners faced another problem, which was rats. The conditions for their breeding were very favourable due to the overcrowding of the city [5, 17, 20].

13.10 Other Jewish Ghettos

The environmental conditions elsewhere were incomparably worse. Not less than 450,000 Jews were forcibly moved into the Warsaw ghetto (2.6 km^2), with a mean of over seven persons in rooms or other house dependencies, lacking space on the floor to lie down. Many had to sleep on the outside often suffering fatal freezing. The Lodz ghetto covered a similar area (2.5 km^2), and 250,000 Jews and Roma (Gypsies) were sent to it, with still severe overcrowding. Unlike Theresienstadt, tuberculosis and epidemic typhus were highly prevalent all the time in those places. At least three epidemics of typhus occurred in the Warsaw ghetto. More than 100,000 people suffered from this disease, and 25,000 of them died. The epidemics were eradicated but always at the cost of a large number of victims. Delousing practices were humiliating and repressive, as they meant standing naked often in the freezing cold, being washed and having the clothes treated with harsh chemicals that often seriously damaged people's health [3, 21–23].

13.11 The End of the War Calamity

The situation in the Theresienstadt ghetto tragically deteriorated at the very end of the war when a typhus epidemic broke out in full force. The main source of the infection was the miserable starving and lice-infested prisoners from other concentration camps, which were being evacuated by the Nazis as Allied troops advanced in Europe, and forcibly transferred to places farther from the front like Theresienstadt. Between 20 April and 1 May 1945, the Germans moved in approximately 13,000, seriously ill, emaciated and lice-infested prisoners with virtually no quarantine [16, 17]. It was to be expected that among this mass of prisoners, there would be those infected with typhus. It can therefore be assumed that this infection was deliberately introduced into Theresienstadt in this way. At that time, the total population of Theresienstadt was about 17,000 people.

13.12 The 11th Hour "Death Transports"

The name transport was often inappropriate, as trains were scarce in Europe and reserved for other military priorities. Prisoners were typically obligated by SS troops to walk during many days from one country to another, without food or shelter, during which thousands succumbed out of hunger, cold and exhaustion (death marches). The first motivation for these transports in the very final phase of the war was to cover up the evidence of Nazi atrocities in the concentration camps, by gradually disposing of the prisoners during the journey, or by liquidating them without a trace in an area not immediately threatened by the arrival of the liberators. The second reason was the desperate and fanatical desire to preserve this slave force for further use inside the Reich (German territory), in the development of "miracle weapons" that were to change the course of the war at the last moment [3, 24, 25].

13.13 Purposeful Introduction of the Epidemic

The epidemic was preliminarily spread to the ghetto by prisoners from the small fortress, at the end of 1944 [26]. As the war front approached, for reasons that are not yet clear, the order was given to open the small fortress and thus to mix prisoners from both parts of Theresienstadt, which promptly became a breeding ground for infectious and parasitic diseases. In addition to epidemic typhus, dysentery, scabies and tuberculosis had to be dealt with. The scale and mortality of these events were enormous for a comparatively small camp. On the peak week, there were 745 infected and 130 dead. Theresienstadt was eventually closed and a 14-day quarantine was declared [17, 26].

13.14 Scorched Earth Policy

It is more than likely that the occupation authorities, especially the SS troops operating in the ghetto, may have attempted either of their own volition or on the orders of a higher decision-making structure, to complete the process of the so-called Final Solution of the Jewish Question in the Protectorate of Bohemia and Moravia (parts of the Czech Republic under direct Nazi rule). This step together with the consideration of the potential spread of the disease even beyond the ghetto borders would correspond to the "scorched earth" method applied by the SS troops, especially during the retreat.

A final circumstance that reinforces the hypothesis of the use of epidemic typhus as a biological weapon for revenge was the already dehumanized behaviour and attitudes of prisoners, especially those newly arrived from the death marches and extermination concentration camps. These people, after many years of suffering and absolute loss of dignity, defied any social order or discipline. Violence or murder posed no threat to them [17]. Thus the Nazis knew that these people would resist any process of resocialization and would probably even attempt to escape, which would lead to spreading the epidemic beyond the walls of Theresienstadt to the civilian population.

13.15 Rescue of the Survivors

On 2 May 1945, Paul Dunant, a member of the International Red Cross, arrived in Theresienstadt and took it under the protection of the organization. In Prague the epidemiologist MUDr. Karel Raška launched the Czech Help Action to help the Theresienstadt prisoners. Although the initiative was organized by the Czech Medical Association, it was a result of a cooperation of a number of organizations. A large-scale humanitarian action started and a fund-raiser for Theresienstadt was announced [6, 27]. With the cooperation of former medical prisoners and Czech and also Soviet doctors from the Red Army, it was possible to separate the healthy inmates from the sick and to carry out basic hygiene measures to prevent further spread of epidemic typhus [6, 28].

References

1. Trunk I (1972) Judenrat: the Jewish Councils in Eastern Europe under Nazi occupation. University of Nebraska Press, Lincoln, USA
2. Votoček O (1980) Terezín (Terezin). Prague, Odeon
3. Baumslag N (2005) Murderous medicine: Nazi doctors, human experimentation, and typhus. Greenwood Publishing Group, Santa Barbara, CA
4. Kárný M (1991) Konečné řešení: Genocida českých židů v německé protektorátní politice (Final Solution: Genocide of Czech Jews in German Protectorate politics). Academia, Prague
5. Adler HG (2007) Terezín 1941–1945 Tvář nuceného společenství I. Dějiny (Theresienstadt 1941–1945: The face of a Forced community I. History), Brno, Barrister & Principal
6. Chládková L (2005) Terezínské ghetto (Terezin Ghetto). V Ráji, Prague
7. Hydráková A, Krejčová H, Svobodová J (1996) Prominenti v ghettu Terezín 1942–1945 (Celebrities in the Terezín ghetto 1942–1945). ÚSD AV ČR, Prague
8. Makarova E, Makarov S (2002) Kuperman V. Univerzita přežití. Osvětová a kulturní činnost v terezínském ghettu 1941–1945 (University of Survival. Educational and cultural activities in the Terezin ghetto 1941–1945). G plus G, Prague, Czech Republic
9. Lagus K, Polák J (1964) Město za mřížemi (The city behind bars). Naše vojsko, Prague
10. Adler HG (2006) Terezín 1941–1945. Tvář nuceného společenství. II.díl – Sociologie (Theresienstadt 1941-1945: The face of a Forced community I. Sociology). Barrister & Principal, Brno, Czech Republic
11. Prochnik R (1945) Statistika – Židé v Terezíně". (Statistics – "Jews in Terezín") Collection of the Jewish Museum in Prague. Prague, Czech Republic
12. Kulišová T (1964) Malá pevnost Terezín (the small fortress Terezin). Naše vojsko, Prague
13. Kypr P (1950) Malá pevnost Terezín (The small fortress Terezin). Mír, Prague
14. Votýpka J, Kolářová I, Horák P (2018) O parazitech a lidech (About parasites and humans). Triton, Prague
15. Raška K (1954) Epidemiologie (Epidemiology). Státní zdravotnické nakladatelství, Prague
16. Dobson M (2009) Nemoci: příběhy nejnebezpečnějších zabijáků historie. (Disease: the extraordinary stories behind history's deadliest killers). Slovart, Prague
17. Pick J, Polák R, Pacovský J (1947) Terezín očima hygienika. Zpráva z terezínského koncentračního tábora o boji proti hmyzu a skvrnitému tyfu (Theresienstadt in the eyes of a hygienist: the report from the concentration camp in Terezín on the fight against insects and typhus). Lékařské knihkupectví a nakladatelství, Prague, Czech Republic
18. Grodin MA (2014) Jewish medical resistance in the holocaust. Berghahn, New York, NY
19. Weisz GM (2010) Dr. Fleck fighting fleck typhus. Soc Stud Sci 40:145–153
20. Acta Theresiania sv.1 2003. Denní rozkazy rady starších a Sdělení židovské samosprávy v období 1941–1945 (Orders of the day by the Council of the Elders and Reports of the Jewish self-government of Theresienstadt 1941-1945). Prague, Sefer
21. Altman LJ (2011) The Warsaw Ghetto uprising. Striking a blow against the Nazis. Enslow Publisher, Berkeley Heights, NJ
22. Weisz GM, Albury WR (2013) Ghetto medicine: the special case of Ghetto Lodz, 1940-44. Isr Med Assoc J 15:203–208
23. Lang HJ (2014) Ženy z bloku 10. Lékařské pokusy v Osvětimi (Women in barrack 10: medical experiments in Auschwitz), Prague, Euromedia Group
24. Blodig V (1995) Poslední dny války, první měsíce míru. Terezín v roce osvobození (The last days of the War, the first month of peace. Terezín in the year of liberation). Terezínské listy: Proceedings of Terezin Memorial, Prague, Czech Republic, Ústí nad Labem: Severočeské nakladatelství
25. Hes M (2015) A brány se otevřely: osvobození 1945: Dachau – Osvětim – Terezín (And the gates opened ...: liberation 1945: Dachau – Auschwitz – Terezin). Epocha, Prague
26. Langhamerová M, Vajskebr J (2009) Malá pevnost Terezín 1940–1945 (The Small Fortress Terezin). V ráji, Prague

27. Jarošová J (1974) Terezín v roce 1945 (Terezin in 1945). In: Collective of authors. Terezínské listy. Proceedings of the Terezin Memorial. Prague, Czech Republic

28. Lukeš A (2008) Svatý týden v Terezíně (Graduation in Terezin ghetto). Terezínské listy: Proceedings of Terezin Memorial, Prague, Czech Republic, Naše vojsko

14 Fraud, Misconduct, and Unethical Practices in Biomedical Research in China

Ruipeng Lei and Renzong Qiu

Abstract

The first part of the chapter is to present the concept of research misconduct and the general situation about research integrity and research misconduct in China. The second part is to briefly account for several notorious cases about fraud, misconduct, and unethical practices in China, such Golden Rice trials, unproved and unregulated stem cell therapy, head-to-body transplantation, He Jiankui incident, and paper mill. The third part is trying to analyze the causes or reasons of these cases in China.

R. Lei
School of Philosophy, Huazhong University of Science and Technology, Wuhan, China

Center for Bioethics, Huazhong University of Science and Technology, Wuhan, China

The Hastings Center, Garrison, NY, USA

WHO Ethics & COVID-19 Expert Group, Geneva, Switzerland

R. Qiu (✉)
Center for Bioethics, Huazhong University of Science and Technology, Wuhan, China

The Hastings Center, Garrison, NY, USA

Institute of Philosophy, Chinese Academy of Social Sciences, Beijing, China

International Institute of Philosophy, Paris, France

14.1 The Concept of Research Misconduct in China

The concept of research misconduct in China is not consistent between different sectors. For example, in the interim regulations on the dealing with research misconduct promulgated by the Minister of Science and Technology, research misconduct is defined as [1]:

1. Providing false information in respect of the title, resume, and research basis of the relevant personnel
2. Plagiarism or plagiarizing the scientific research results of others
3. Fabricating or falsifying scientific research data
4. Violating the regulations on informed consent and protection of privacy in research involving human subjects
5. Violating the norms for protecting experimental animals
6. Other misconduct in scientific research

However, in the Norms of Academic Publishing promulgated by National General Administration of Press and Publications, research misconduct is defined as including plagiarism, fabrication, falsification, inappropriate authorship, duplicate or multiple submission, and overlapping publication [2].

We believe that these two definitions of research misconduct are both too wider than due [3, 4]. In a report [5], we first recommend that

J. Faintuch, S. Faintuch (eds.), *Integrity of Scientific Research*,
https://doi.org/10.1007/978-3-030-99680-2_14

research misconduct should be defined as fabrication, falsification, or plagiarism in the course of a project proposal, research review, or reporting of research results, and we can use the word "fraud" to refer to these triples. Fabrication refers to making fake data or results; falsification is the manipulation of research materials, equipment, or processes to alter or omit data or results so that the research does not match the research records; plagiarism is the appropriation of someone else's ideas, research processes, research results, or written expressions without in any way acknowledging the contributions of the author. This definition would align with current international definitions, such as those of the United States [6] and the European Union [7].

Misconduct in scientific research is a kind of action that violates scientific research ethics with bad nature and serious consequences. The action in it is intentional and deliberate, which corrupts the core value of science—honesty. Such a definition can enable us to concentrate our energy and time, and use limited resources to focus on the filing, investigation, and treatment of research misconduct, and avoid the expansion of the blame of research misconduct to innocents. Secondly, we recommend that we have to strictly distinguish research misconduct from other ethical violations and "honest errors."

14.2 Unethical Practices vs. Research Misconduct

Attention should be paid to strictly distinguishing the boundary between research misconduct and other acts that violate scientific ethics, for instance:

- Protection of human subjects and experimental animals, which is handled by other regulations and institutions.
- Fraud in her/his CV by a researcher does not belong to the research misconduct directly related to scientific research. It can be dealt with by the human resources department according to the rules that are formulated by the department.
- "Inappropriate authorship," "duplicate or multiple submission," and "repeated publication" may deemed as violating ethics. The nature of these conducts is still controversial, and should not be defined as research misconduct due to their indirect relation with scientific research itself.
- "The marks of origin not complying with norms" should not be generally defined as research misconduct (plagiarism). The core of plagiarism in research misconduct is to take the ideas and achievements of others as one's own without acknowledging the contributions of the authors in any way, that is, it is intentional and deliberate. Acknowledgement of an author's contribution can be done in a number of ways, e.g., in the form of notes or quotations, in the title of the paper, notes in the text, citation marks, and text in context.
- Distinguishing between "original articles" or "research articles" that report the author's own research results and "review articles" that report the results of others, including review and survey or overview. For review articles, if the literature in an article has been marked, it is inappropriate to use the term "mark not enough" or "not complying with the norms." It would expand the blame of research misconduct to innocents.
- Honest errors are not research misconduct. The main difference is whether there is intentional subjectivity. Human beings are not saints; how can they be free from mistake? To err is human. Scientific research is to explore the unknown; therefore mistakes are inevitable (including negligence, carelessness, etc.), and should not be judged too harshly. Therefore, a clear distinction must be made between "intentional" and "unintentional" [5].

We suggest to use the term "unethical practices" to refer to all other wrongdoings in scientific research rather than research misconduct.

Since the reform and openness policy, scientific research has been heavily invested by the government, and lots of scientific programs of international collaboration have been flourishing. Almost at the same time, the research misconduct and other unethical practices have been rampant in many fields of scientific research, in spite of other fields of scientific research such as space science where scientists have been upholding research integrity. In the first issue of 2007 Bulletin of Oversight Committee, National Natural Science Foundation, 13 misconduct cases were reported. One institute submitted six fraudulent applications.

14.3 The Hanxin Digital Computer Chip Production

The most notorious case of research misconduct is the Hanxin Laboratory and Organization. On January 17, 2006, a whistleblower on Tsinghua University website accused Jin Chen, then the Dean of the School of Microelectronics, Shanghai Jiaotong University, of fraud for the Hanxin No. 1 chip he invented. On Tsinghua University website, 80% of the users supported the whistleblower. In August 2002 Chen, who had been an engineer working at the US Motorola company, bought ten MOTO-freescale 56,800 chips from the United States. He had the logo and letters MOTO on the chips removed and stamped Hanxin No. 1 and the Hanxin logo on them.

Via various connections he obtained faked documents of "domestic designed" (from Jiaotong University), "domestic produced" (from Shanghai Zhongxin International), "domestic packaged" (from Shanghai Yuwei Science & Technology), and "domestic tested" (from Shanghai Center for Integrated Circuits Design). He used every effort to convince the officials of MOST (Ministry of Science and Technology), MOII (Ministry of Information Industry), and SCDR (State Commission on Development and Reform) they were true. Then he invited well-known experts at integrated circuits to attend an evaluation meeting, the conclusion of which was: "No. 1 Hanxin is a high memory large-scale integrated circuit at internationally advanced level."

The first high memory digital signal processor/DSP chip of which China possessed intellectual property rights was officially born. In the same way, he updated Hanxin No. 1 to No. 2, No. 3, No. 4, and up to No. 5. During 3 years he used fabricated technical data to apply for funds to Shanghai Commission on Science & Technology, MOST, MOII, Shanghai Commission on Development & Research, SCDR, etc. 40 times and obtained nearly CNY 100 million (=USD 64 million). Chen did everything personally in editing technical data and relevant materials, submitting application forms, and organizing question and answer meetings and public relation activities with officials and experts. Part of the funds entered into his pocket or his accounts in US banks.

His company was awarded the prizes of "High-Tech Enterprise," "Soft Enterprise," and "Innovation," and he himself was awarded prizes such as "President's Prize of Shanghai Jiatong University," "Leader of Shanghai Scientific & Technological Innovation," "Shanghai Top Ten IT New Upstarts," "Shanghai Top Ten Distinguished Youth," and others. The title of "Yangtze River Scholar" was awarded by the Ministry of Education. He also used the faked chips to apply for 12 patents which were approved. On May 12, 2006, Shanghai Jia Tong University released a bulletin on adjudication of the fraud of Hanxi series of chips. The university decided to remove Chen Jin from the post of Dean of the School of Microelectronics and from professorship. MOST decided to end his research projects, require him to return the funds back, and cancel his qualification to take future national research projects. The Ministry of Education decided to withdraw his title "Yangtze River Scholar," cancel his qualification to enjoy special allowances, and require him to return the funds back. The SCDR decided to end the projects of high-tech industrialization and require him to return the funds back [8].

14.4 Clinical Trials of Golden Rice

On August 1, 2012, a paper was published online about the genetic engineering of golden rice (*Oryza sativa*) to increase its level of β-carotene [9]. In December 2002, the US Institute of Diabetes and Digestive and Kidney Diseases had approved a project by the lead author, Guangwen Tang (at the time working at Tufts University, Boston, USA), to carry out a comparative study into the effectiveness of β-carotene in golden rice vs. spinach (employing as control a pure β-carotene/oil capsule), at providing vitamin A to children. In September 2003 a co-author of the article, Shian Yin, who was the principal investigator/PI of the Chinese team, signed a collaborative agreement with the Zhejiang Academy of Medical Sciences (ZAMS). In November 2003, the Institutional Ethics Review Committee (IERC) of ZAMS approved the project, which was to be conducted in China. Tufts University's Institutional Review Board (IRB) similarly approved the study in 2003. Then, in August, 2004, Tufts University signed a collaborative research memorandum with ZAMS in which Tang was PI and Yin Wang (in addition to Shian Yin) was PI of the Chinese team. In 2008, the site of the project was changed to the Central Primary School of Jiangkou Township, Hengyang County, in Hunan Province, and the project was merged with a previous project of Yin's, which had studied the effectiveness of β-carotene in vegetables converting to vitamin A.

The study started on May 20, 2008, and ended on June 23, 2008. On May 29, 2008, Tang took golden rice cooked in the United States with her to China without making the legally required report to the authorities. On June 2 Tang heated golden rice and mixed it with ordinary cooked rice, and distributed this to the children without informing either them or their parents. On June 2, 2008, the Institutional Review Board (IRB) of Tufts University approved the Chinese version of the consent form, which mentioned only golden rice and did not explain that it was genetically modified rice (GMR), whereas the consent forms issued in 2003 and 2008 had made it clear that the golden rice was GMR. As the original project was merged with another one, and the trial site was also changed, it should have been submitted to the IERC for review and renewed approval. In collusion with Yin, Wang stole ZAMS' stamp and put it on a fabricated Tufts document to extend the validity of the 2003 review.

In July 2008 someone in a Chinese regulatory department who had heard that Tufts University was conducting golden rice trials in China asked the authors about them; they lied and replied that the trials had not started, whereas in fact they had already finished. There was no report before, during, or after to the director of their institution (ZAMS). On May 22, 2008, the project team convened a meeting with the schoolchildren's parents or guardians to provide information, but they did not mention that they would use GMR. In the consent form, the words "Rice rich in β-carotene" were deliberately used. Nor did the form highlight any potential risks of ingesting such rice. This was intentional concealment of the truth. The parents or guardians were not given the complete consent form, but only the last page. This consent was therefore not valid. Eventually all three Chinese PIs were disciplined by China's Ministry of Health.

14.5 Critical Analysis of the Trial

The unethical practices in the case include the following: cheating the regulatory body, illegally bringing in GMR from abroad, and hiding the true information (GMR) from research participants and their guardians. However, we believe that the case also indicates the incompetence of the IERC of ZAMS. The IERC should have considered the following points: (1) Does the protocol meet the ethical requirement of specially protecting children as vulnerable group? (2) How should informed consent to GMR trials be validly obtained, given that such trials are the subject of fierce controversy in China? (3) Is international collaboration equitable and just? Does it aim to solve health problems in China? Or does it aim just to collect data to resolve a debate in a sponsor country? It seems clear that the IERC never raised these questions nor asked the PIs to answer these questions, but instead simply approved the project by rubber-stamping it [10, 12].

14.6 Unproven and Unregulated Stem Cell Therapy

The golden rice trial was local, but the case of unproven and unregulated stem cell therapy is something that affected the whole country. Since 2005 there has been a notorious unruly situation for unproven and unregulated stem cell therapy, which attracted many foreign patients who had lost hope in other medical therapies to engage in a form of medical tourism to mainland China and even reaching into neighboring countries. It is estimated that more than 500 medical institutions and biotech companies were involved. This lasted for about 7 years until 2012 and may still exist "underground." No positive outcome was reported. In turn physical, mental, and economic harms have been inflicted on about at least 100,000 cheated patients, and billions of CNY have been earned as profits by the physicians, scientists, and companies involved. This example of bad science and bad ethics carried out by Chinese physicians, scientists, and regulators demonstrates a number of deficiencies that must in future be avoided, including the following:

- A lack of scientific underpinnings: Not as the routine hematopoietic cell therapy, stem cell treatment has not been supported by scientific evidence yet, neither by robust basic research.
- A lack of research has been carried out to identify specific stem cell products and to have their quality tested by authoritative third parties.
- A lack of validly assessed preclinical (lab, animal) studies have been carried out.
- A lack of randomized controlled trials observing fundamental ethical requirements have been reported in this context.
- The procedures were carried out under the pretense that they were routine practice, deliberately confusing experimental therapy with treatment.
- Practitioners obtained invalid consent from patients by cheating, using exaggerated, fabricated, and misleading advertisements.
- Practitioners exploited desperate patients who were eagerly seeking magic medical treatment.

The interventions with stem cell therapy (SCT) which seem miraculous tools however were eventually disproved involving a cluster of troubles stemming from pervasive medical commercialism, degeneration of medical professionalism, diseased health system and policy, as well as prevailing corruption in medical care and the society at large [11–13].

14.7 Head-to-Body Transplantation

Head-to-body transplantation (HBT) was devised by Sergio Canavero (Turin, Italy), subsequently partnering with Xiaoping Ren at Harbin Medical University (China). The protocol HEAVEN (head anastomosis venture, 2013) was designed to offer a new life to terminally ill patients with irreversible somatic (non-neurologic) diseases. It consisted in preserving just the head of the patient, connecting it to the healthy body from a cadaver donor. As the head is the site of wisdom, memory, and personality, the candidate would literally be born again in the new organism.

The number one obstacle was reconstruction and refunctionalization of the sectioned spinal cord, something achieved only in simple animal models, never in higher mammals or in man. Ischemia of the recipient's brain before vascular reanastomosis could have consequences. Moreover, immune rejection could precipitate serious damages, even with modern immune suppressive therapy.

Along the years, Canavero and Ren announced scores of animal experiments and a few in cadavers, reportedly with success. Given the limited provided details, no other center was able to replicate the results. The plan was eventually aborted (2018) by Harbin Medical University [12, 14].

14.8 He Jiankui Incident

In 2019, Dr. He Jiankui from the Southern University of Science and Technology (Shenzhen, China) announced to have created genome-edited (CRISPR-Cas9 technique) twin girl babies. Thirty fertilized eggs were harvested and 70% were gene-edited, from seven couples interested in having children, where the husband was HIV positive and the wife HIV negative. The experiment was considered scientifically and medically inappropriate, thus unethical and even illegal. The CCR5 gene which was knocked out has many positive functions, such as the prevention of infectious diseases, including flu. There are many safe and effective methods to prevent HIV in the offspring, as the route of HIV transmission is quite limited. Moreover the protein which helps HIV to enter into the nucleus is produced both by the CCR5 and the CXCR4 gene. The intervention is actually prohibited in China, where genetic manipulation of human gametes, zygotes, and embryos for reproductive purposes is not allowed, whereas HIV carriers are not candidates to assisted reproductive technologies. He was sentenced to 3 years in prison [15, 16].

14.9 Paper Mill

A large black market for fraudulent or plagiarized papers eventually published in indexed journals has been pointed out in China [17]. Fabricated images including Western blot bands have also been the subject of harsh debates [18]. According to famous Dutch scientific integrity consultant Elisabeth Bik, authorships on ready-to-submit or already-accepted papers are offered for substantial money. As many as 412 papers were sold to physicians in China [19].

14.10 General Underlying Mechanisms

14.10.1 Conflict of Interest

Scientific research and bioethical research in particular are closely linked with capital and market operations in contemporary times. Trials of new medicines or equipment (used for diagnosis or treatment) are mostly financed by pharmaceutical or medical equipment industries. At the same time, the companies build close relationships with scientists and physicians who may buy shares of these companies, be invited to be member or even chair of a variety of their expert or executive committees with good pay, or receive a range of benefits including luxury meals, hotels, and bribes (money). When scientists or physicians put their own interests or those of the companies from which they get gains over those of patients/research participants, they became *ji gong jin li* (eager for quick success and instant benefits). The result is driving the scientists or physicians to commit misconducts such as fabrication, falsification, and plagiarism or cause unacceptable harms to research participants or patients.

14.11 Perverse Incentives by the Government or Local Agencies: The Han Case

Chunyu Han is a molecular biologist at Hebei University of Science and Technology in Shijiazhuang, China, who co-authored a paper in *Nature Biotechnology* in 2016 describing how an enzyme called NgAgo could edit genomes nearly as well as the widely used CRISPR-Cas9 gene editing tool. The paper was retracted in 2017. However soon after the initial publication, Han was made vice president of the Hebei Association of Science and Technology, and his university planned to invest 224 million yuan (US$33 million) into a research center for gene editing with Han's team at its core [12]. There is no evidence proving that Han committed research misconduct, yet the attitude of *ji gong jin li* would lead to misconduct or other unethical practices sooner or later. The He Jiankui incident is another convincing proof.

By promoting scientific discovery and invention, the Ministry of Education permits young teachers to temporarily leave their teaching posts and run business in order to use the profits to promote scientific research, and then apply their scientific results into business, thus creating a virtuous circle. He was eager to get profits and believed that he would get a future Nobel Prize for his genome-edited babies, notoriously violating ethical norms and legal stipulations [12, 14, 15].

14.12 Failure of Self-Discipline Alone

An unavoidable result of the collaboration between scientists and industry is that their self-discipline is no longer effective albeit it is still needed. "The ass is dominant over the head" (*pigu jueding naodai*) is an expression popular in a Chinese website which means that a person's decision is made by her/his position. Faced with conflicts of interests, scientists tend to consider more their own or the company's interests, when they have close interactions with this company. That is why research must be regulated and supervised by regulatory bodies as well as the community (the public). Another reason for ineffectiveness of self-discipline is that the ethical issues in scientific and biomedical research have become so complicated, intriguing, and even subtle that scientists have difficulty to grasp them, without professional ethical knowledge and skills.

14.13 Regulatory Gap

The key reason of research misconduct and other unethical practices being so rampant is the existence of a regulatory gap. Regulations on research misconduct and other unethical practices are often lacking or inconsistent, as well as improperly implemented. Many cases such as in the so-called stem cell therapy are not investigated and handled, or they are dropped without any penalty. Even in some regulations on scientific research, there is no article on penalty [20].

On the other hand, the regulations on the academic promotion of physicians require published research articles. The majority of physicians lack the necessary skills or are too busy in clinical practices, so they have to buy articles from a paper mill. The central government is now considering to overturn the requirement of published research articles when physicians apply for academic promotion, and several local governments have already done it, such as the Qinghai Municipal Government [21].

14.14 Incompetence of Supervising Bodies

Now almost all universities established an office of research integrity. However, their personnel are not well trained. Some do not know the difference between research and review articles, and confuse acknowledged citations with plagiarized ones, so innocent authors can be blamed for committing plagiarism [3, 4, 14, 15].

14.15 Ethics Dumping

Trials which are deemed questionable or even prohibited in developed countries are sometimes transferred to developing countries where there is a regulatory gap. These cases are called ethics dumping. When protocols of international collaboration are reviewed, attention must be given to the risks of ethics dumping [22].

References

1. Minister of Science and Technology (MOST), Interim Regulations on Handling Research misconduct in Implementing National Plans of Science and

Technology (2007). gov.cn/ziliao/flfg/2006-11/10/content_438430.htm. Accessed 6 Sept 2021
2. National General Administration of Press and Publications (NGAPP): Norms of Academic Publishing: The Definition of Research misconduct in Journals (2019). sohu.com/a/329168225_120051578. Accessed 6 Sept 2021
3. Qiu R (2020, July 4) First the definition should be robust in dealing with research misconduct. Health Newspaper
4. Qiu R (2020, July 14) Which conduct can be defined as misconduct? Health Newspaper.
5. Lei R, Qiu R, Jia P (2020) Policy recommendations on the definition of research misconduct and its determination. An internal report to Ministry of Education, Ministry of Science and Technology, and National Health Commission, submitted to these Ministries on July 3, 2020
6. NIH (2019) Research misconduct – definitions. grants.nih.gov/research_integrity/definitions.htm. Accessed 6 Sept 2021
7. ALLEA (2017) The European Code of Conduct for Research Integrity. allea.org/wp-content/uploads/2017/05/ALLEA-European-Code-of-Conduct-for-Research-Integrity-2017.pdf. Accessed 6 Sept 2021
8. Qiu R (2007) Research integrity issues in China-Mainland, presented at the World Conference on Research Integrity: fostering responsible research. Lisbon, Portugal, September 17–19, 2007
9. Tang G, Hu Y, Yin S, Wang Y, Dallal GE, Grusak MA, Russell RM (2012) β-carotene in golden rice is as good as β-carotene in oil at providing vitamin A to children. Am J Clin Nutr 96(3):658–664
10. Zhai X, Qiu R (2020) The Declaration of Helsinki in China: an example of the tension between international guidelines and native cultural values. In: Schmidt U, Frewer A, Sprumont D (eds) Ethical research: declaration of Helsinki, and the past, present and future of human experimentation. Oxford University Press, Oxford, pp 443–467
11. Lei R, Qiu R (2019) Fight for regulation and its enforcement on stem cell therapy in China. Int J Cell Sci Mol Biol 6(2):555682 h
12. Lei R, Zhai X, Zhu W, Qiu R (2019) Reboot ethics governance in China. Nature 569:184–186
13. Qiu R (2013) The unruly situation of 'stem cell therapy': ethical and regulatory issues in clinical translation of stem cells (in Chinese). Sci Soc 3(1):8–15
14. Lei R, Qiu R (2020a) Impassable scientific, ethical and legal barriers to body-to-head transplantation. Bioethics 34(2):172–182
15. Lei R, Qiu R (2020b) Ethical and regulatory issues in human gene editing: Chinese perspective. Biotechnol Appl Biochem 67(6):880–891
16. Zhai X, Lei R, Zhu W, Qiu R (2019) Chinese bioethicists respond to the case of He Jiankui. thehastingscenter.org/chinese-bioethicists-respond-case-jiankui/. Accessed 6 Sept 2021
17. Hvistendahl M (2013) China's publication bazaar. Science 2013 342(6162):1035–1039
18. Christopher J (2018) Systematic fabrication of scientific images revealed. https://doi.org/10.1002/1873-3468.13201. https://febs.onlinelibrary.wiley.com/doi/full/10.1002/1873-3468.13201. Accessed 6 Sept 2021
19. Bik E (2020) The Tadpole Paper Mill. scienceintegritydigest.com/2020/02/21/the-tadpole-paper-mill/. Accessed 6 Sept 2021
20. Ministry of Science and Technology (MOST)/ National Health Commission (NHC) (2003) The guiding ethical principles on human embryonic stem cell research. http://www.lawinfochina.com/display.aspx?lib=law&id=3346&CGid=. Please also see: Cheng L. Qiu R. Duan H. et al: Ethics, China already has clear stem cell guidelnes. Nature 440, 992(2006). https://www.nature.com/articles/440992b
21. Sohu (2017, April 4) Good news for physicians! No longer need research articles when applying promotion. sohu.com/a/134582059_443462. Accessed 6 Sept 2021
22. Qiu J (2019, February 2) No dumping. Please. Recent events highlight an unpleasant scientific practice: ethics dumping. Rich-world scientists conduct questionable experiments in poor countries. The Economist. economist.com/science-and-technology/2019/02/02/recent-events-highlight-an-unpleasant-scientific-practice-ethics-dumping. Accessed 6 Sept 2021

15 Misrepresentation of Scientific Figures

Davis M. Leaphart, Apryl C. Cronley, and Thomas E. Brothers

Abstract

Issues with research quality and reproducibility have grown in the past few decades. Increasing statistical complexity, availability of "user-friendly software," and lack of proper statistical training have been identified as possible causes in this multifactorial problem. Issues also stem from scientific misrepresentation, which has evolved from blatant fraud to new tactics such as selectively reporting outcomes and HARKing in order to produce "positive findings." These veiled attempts, whether purposeful or not, may be driven by researchers' desires for publication or other secondary gains. Any published error erodes the faith in scientific findings and the research community. Different solutions and shortcomings have been reviewed to help educate readers and researchers alike combat issues in their own work.

D. M. Leaphart · A. C. Cronley
College of Medicine, Medical University of South Carolina, Charleston, SC, USA
e-mail: leaphard@musc.edu; cronley@musc.edu

T. E. Brothers (✉)
Department of Surgery, Division of Vascular Surgery, Medical University of South Carolina, Charleston, SC, USA
e-mail: brothete@musc.edu

15.1 Introduction

The scientific method has grown over time to test the theories of man. As study design, measurements, and calculations have become more complex, the rules for reporting research findings, like the science they are employed to serve, have evolved over time. Carl Gauss (1777–1855) wrote in the book *Theoria motus corporum coelestium in sectionibus conicis solem ambientium* (Theory of the Motion of the Heavenly Bodies Moving about the Sun in Conic Sections): "(numbers) are all to a certain extent approximate only, the results of all calculations performed by the aid of these numbers can only be approximately true" [1].

With the invention and adoption of the slide rule, numbers were rounded inter-operationally to the most precise increment available on the instrument while performing calculations [2]. This led to limited reporting of significant digits and rules for uniform data representation. With increased computing power, we have witnessed more rigorous data analysis and greater research throughput than ever before [3].

Utilizing user-friendly software, researchers feel empowered to do statistical analyses beyond their statistical knowledge while not fully understanding the nuances and application of different statistical tests; as a result, poor research quality and reproducibility have risen [4]. To combat the fallibility in present-day research quality, guidelines on topics ranging from research design

J. Faintuch, S. Faintuch (eds.), *Integrity of Scientific Research*,
https://doi.org/10.1007/978-3-030-99680-2_15

to data reporting have been published [5]. Even so, there is a growing body of evidence suggesting that scientific misrepresentation remains widespread, likely secondary to external pressures for more significant findings and publications [6].

15.1.1 Scientific Publications

The widespread use of research publications in health care are instrumental in modifications of best practices and standards of care. However, the ethical principles fundamental to scientific research are arguably mistakenly assumed to be abided by and verified. Publications are presumed to be reported entirely and completely in a transparent, accurate fashion that has been reviewed and confirmed by a third-party peer reviewer. While many assume biomedical research to be objective in nature, there is often failure to fully appreciate the subjectivity present.

When writing a manuscript, researchers have a broad range of options when it comes to reporting, analyzing, presenting, and interpreting the data. Every step of the process has the potential for introducing bias, both conscious, such as in fraud, and unconscious, such as when investigators lack proper statistical training and unknowingly misinterpret their conclusions. These biases influence investigators to misrepresent data and results and ultimately draw unfounded conclusions in an attempt to report their work in a more favorable light, making publication and citation more likely. This is often referred to as spin [7].

15.2 Fraud

There have been many notable incidents of scientific fraud such as the now infamous and retracted case series linking MMR vaccines and autism [8]. While the fallout from this underpowered and extremely biased study continues to rage among skeptics and conspiracists, the majority of scientific research has pivoted to more subtle forms of misrepresentation such as emphasizing secondary outcomes, selectively reporting findings, or emphasizing intermediate results rather than defined endpoints [9].

These hidden attempts to change a study's message may be different than blatant data fabrication, but the amount of fraud is increasing and is done with the same intent of secondary gain [10]. Many of the tactics and methods of deception are similar between accidental and purposeful misrepresentation, but *intention* distinguishes between fraud and an error [11]. For example, selective reporting becomes fraud when authors purposefully elect not to report data because it weakens their arguments.

15.2.1 Manipulated Images

In Matthew Walker's book *Why We Sleep* [12], a previously published graph was intentionally adjusted by removing some data points (example figures recreated in Fig. 15.1), demonstrating association between fewer sleep hours and likelihood of injury without disclosing the change [13]. While the doctored figure is more convincing for the author's argument, his intentional lack of disclosure in cherry-picking data highlights this attempt to misrepresent the true findings.

Instances of intentional misconduct were estimated to be roughly half as common as unintentional mistakes based on a few hundred retracted journal articles indexed with MEDLINE [14]. However when fraud is exposed, all of an author's publications should be questioned for validity, as 50% of fraudulent first authors are repeat offenders [15]. It takes close to 4 years from publication to retraction [10], and intentional fraud takes more time and investigation to be uncovered and retracted compared to honest mistakes [14]. Journals recognize a responsibility in policing their published authors, and larger high impact factor journals retract fraudulent articles quicker than their less cited counterparts. This may be driven in part by the higher incidence of fraud in high impact journals [10]. It seems that researchers are compelled to lie about data and results in order to publish in a more prestigious journal.

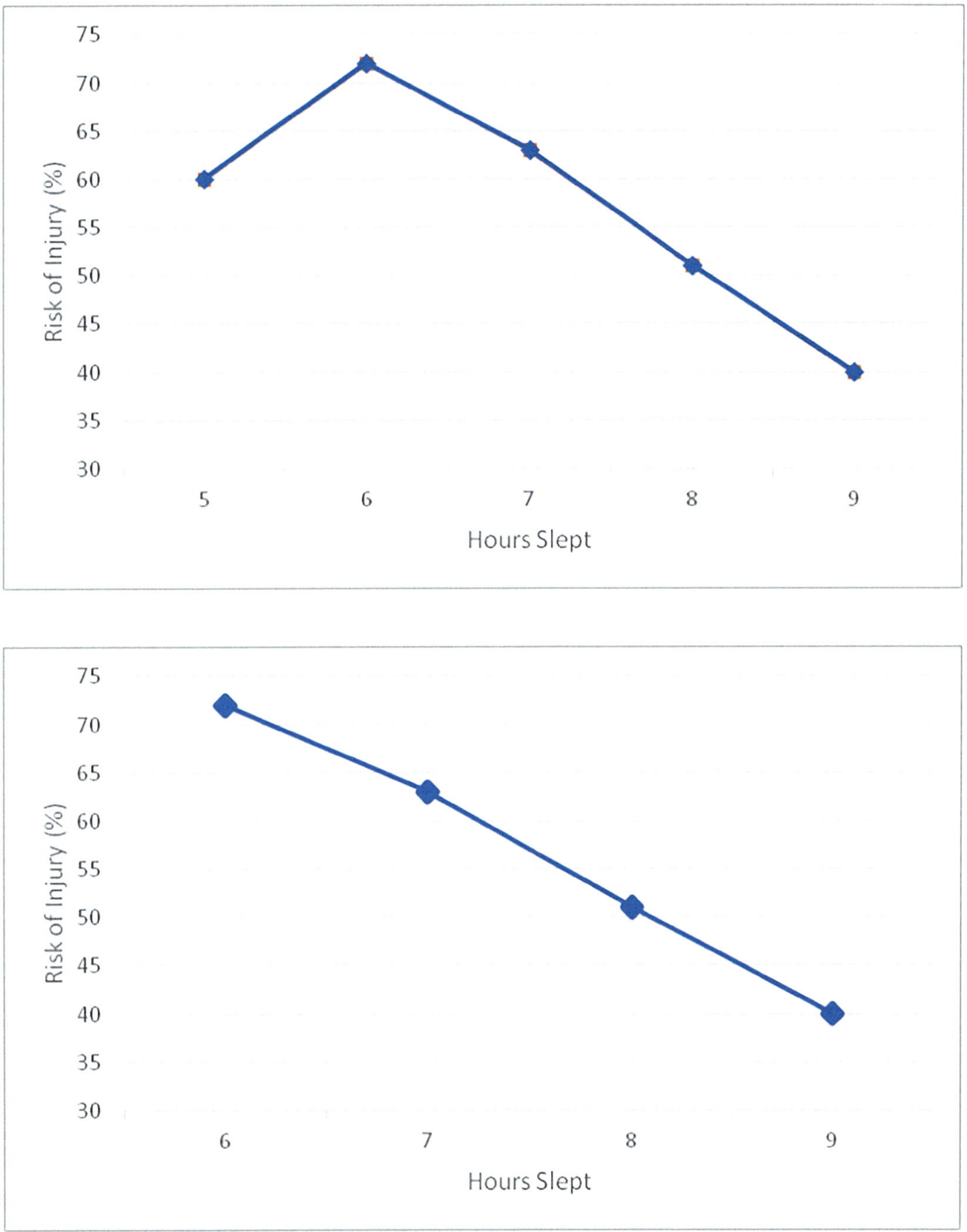

Fig. 15.1 Example of intentionally groomed figures to emphasize desired outcomes. Upper image: original findings. Lower image: selective (manipulated) presentation of the curve

15.2.2 The Role of Whistleblowers

There are different enforcement entities combating scientific fraud, like the US Office of Research Integrity/ORI; however research colleagues are more likely to uncover questionable studies. High-level fraud is discrete and requires a focused eye and collateral knowledge to understand the research, methods, and hypothesis being studied. Monitoring data for fraud during studies is increasingly important when multiple centers and providers are participating. Even if fraud is not suspected, data and statistical monitoring ensures anomalies are caught quickly so that corrective action can be implemented without affecting the endpoints or timeline [11].

15.2.3 Rule of Thumb

Benford's Law, which assesses the probability that a numeral 1–9 is the first digit, assuming a random distribution, can be used for quick randomization checks [16]. While this method shows promise for random tasks like randomizing patients to different protocol arms, nonrandom distributions such as age won't follow the same frequency pattern. New curves for digit frequency can be generated for population data, and individual subsets can be compared to the population for discrepancies.

15.2.4 Data Collision

Repeats of a patient's information that should be fairly and randomly distributed should be investigated for discrepancies [11]. Even with lack of consensus regarding data monitoring, it is still paramount for colleagues, investigators, co-investigators, and authors to report scientific misconduct and fraud, when they see it. Means for doing so is provided by institutionally based offices of research integrity, or the national Office of Research Integrity, where reporting can take place anonymously without concern for penalty or retribution. It is also the publishing journal's obligation to ensure transparency by providing clear criteria and processes for retracting published papers. This should include but not be limited to the ways in which retraction notices are sent as well as the investigatory results that predicated the retraction.

Still, discerning between misconduct and unintentional error can be impractical, and it is nearly impossible to determine if an author's error was due to lack of understanding, unconscious bias, or actual willingness to mislead or misrepresent. Well-conducted research can involve conflicting data, differences in experimental design, or differing interpretation of information [17].

15.2.5 Conflicting Scientific Approaches

David O'Neill accused New York University (NYU) and NYU professor Nina Bhardwaj, his former supervisor, of research misconduct related to statistical analysis [18]. O'Neill claimed in his wrongful dismissal lawsuit that the analysis was deceptive because it was chosen after the data had been collected, a form of *hypothesizing after the results are known*, with the acronym HARKing [17]. While NYU feels that this episode highlights differences in opinion on how to conduct studies, they stated that no wrongdoing was found on the part of Bhardwaj [19].

O'Neill felt that his dismissal for unprofessionalism was retaliatory and provided insight into the often-hierarchical structure institutions employ. Many academic and research institutions leave researchers hesitant to report scientific misconduct, especially considering those reporting the misconduct are often subordinates. In addition, this situation highlights the uncertainties surrounding misconduct, error, and differences in scientific opinion [17]. This case outlines debated practices such as choosing a statistical method after the results are known to conspicuously achieve a desired result [20].

15.3 Honest Errors

> The truth is that the problem of poor statistical reporting is long-standing, widespread, potentially serious, concerns mostly basic statistics, and yet is largely unsuspected by most readers of the biomedical literature [21].

Fraud and misconduct are based on intention to mislead [11]. Honest errors include the use of unconventional research methods, mistakes in statistical analysis, misunderstood conclusions, or proposition of controversial hypotheses [17]. Honest errors and accidental misrepresentation of scientific figures may not carry the same connotation as intentional fraud, but negligence to follow convention is sloppy and may have unintended consequences [22]. Authors and publishers alike may commit and permit the publishing of basic statistical errors [23], varying from incorrectly reporting the correct number of significant figures for a mean to not indicating which statistical software was used in the analysis. This lack of statistical quality has been a focus of medical literature for some time [24] and has grown to include other disciplines. While these careless errors may seem inconsequential, journals publishing more errors are correlated to lower impact factors, possibly indicating fewer peer citations and acceptance by the community [25].

15.3.1 Sloppy Statistics

Up to 50% of published psychology investigations have at least one *p*-value that does not match with the corresponding statistical parameters of the sample [26]. As analysis complexity increases, the percentage of *p*-values that are incongruent increases, indicating more errors in difficult statistical analysis [27]. The impact of these discrepancies is hard to predict and is multifactorial depending on topic interest, number of citations, relevance to future studies, and how long it takes to uncover the error. One simple approach is to count the number of citations a retracted article has received as a surrogate for the continued relevance and discussion [28]. For example, *Primary Prevention of Cardiovascular Disease with a Mediterranean Diet* has received 2757 citations and continues to grow with new citations despite a notice of retraction posted on the electronic study [29]. Similarly, *Effects of Omega-3 Polyunsaturated Fatty Acids on Inflammatory Markers in COPD* has continued to receive positive citations 13 years after retraction, even with "RETRACTED" added to the study's electronic title [30, 31]. Worse, 96% of new citations after the retraction fail to mention that the original study was retracted for data fabrication [30].

15.3.2 Article Checklists

Given that researchers, reviewers, and publishers not uncommonly fail to recognize problems prior to publication, it makes sense for initial efforts to standardize the reporting process. CONSORT, which provides reporting guidelines for randomized clinical trials (RCTs), has developed over the previous 30 years and has evolved their recommendations and gained approval from most major medical journals (CONSORT). The EQUATOR Network formed in the early 2000s after the first statement from CONSORT. While CONSORT limits their guidelines to RCTs, the EQUATOR Network recognized that other publication types, i.e., case reports, observational studies, guidelines, etc., were also plagued with reporting issues [32]. Since inception the Network has compounded guidelines in reporting for many different types of studies and has grown into a worldwide network with most journals supporting the easily accessible published guidelines.

15.3.3 Publication Loopholes

Unfortunately, author guidelines may not address all the necessary information being reported, and journals vary in their "Instructions to Authors" and review process, leading to opportunities for errors in statistical representation by authors. Among a small subset of six surgery journals, four did not have published guidelines for data presentation or reporting [25].

15.3.4 Overlooked or Neglected Guidelines

Researchers may not be aware that guidelines exist or, alternatively, refuse to adopt them [33, 34]. By adopting a set of external guidelines, such as from the EQUATOR Network or CONSORT, or by developing journal-specific reporting guidelines that must be followed prior to submission, journals can force statistical and data reporting to fit a defined standard. For journals that have adopted a specific set of guidelines, sample guidelines should be provided to authors along with attestation forms that researchers must acknowledge prior to journal submission.

Some argue that checklists are not adequate to combat the issues at hand, as no studies have validated published guidelines [35]. Journals that helped write the STROBE guidelines did not actually perform better on the STROBE statistical quality checklist [35]. Likewise, journals that mandated that a STROBE checklist on confounding variables be completed did not score better on statistical review [36]. However, few studies have implemented a checklist requirement at submission with quality commentary proof on complete and disputed line items alike [37]. As reported by *Nature*, 49% of surveyed individuals felt checklists subjectively improved the quality of their publications [38].

There are different methods for internal review ranging from spot checking calculations to an entire statistical review by trained statistician. The *statcheck* program (statcheck.io), which compares *p*-values to other statistical parameters such as the sample size, and the GRIM (granularity-related inconsistency of means) method (peerj.com/preprints/2064v1/), which calculates possible averages based on sample size, have been used to look for overlooked statistical errors in psychology journals and are freely available to all [26, 39].

15.4 Scientific Misrepresentation

Among the scientific community, *spin* has been described as the biggest threat to objective research [9] and has become increasingly prevalent in published work, occurring in almost half of all cardiovascular RCTs [40]. While this *spin* sometimes is directed to intentionally mislead the reader for secondary gain, at other times it is accidental or subconscious [7, 41].

15.5 Misinterpretation of Conclusions

In rare diseases or new scientific frontiers there may be a limited amount of quality data, potentially impairing evidence-based analysis of new contributions. A historical example is when Dr. Duncan MacDougall reported the weight of a human soul being 21 g by measuring body mass before and after death in six patients [42]. While there were other flaws and assumptions made in this early work, the lessons of small sample sizes and investigator subjectivity hold true. Randomized controlled trials enroll hundreds of participants, ensuring statistical power and accurate population representation while attempting to mitigate random error. While such trials are considered the golden standard in clinical research, there still exists opportunity to misinterpret findings, besides other study designs that introduce more opportunity for systemic error and investigator bias. For example, propensity matching and pooling data sets for large meta-analyses introduce assumptions and alterations to the original objective data and may lead to incorrect conclusions [43].

The addition of a statistical reviewer has been shown to increase the quality of statistical reporting [35, 44, 45]. Insufficiently trained physicians and reviewers often attempt analysis using "user-friendly" statistical software or reviewing analysis far beyond their statistical understanding [36].

Peer reviewers should be screened in a standardized fashion for necessary statistical skills to recognize errors. Unless reviewers are directed to do a detailed review of the statistics, they may assume someone else prior to submission or the publisher would conduct that review. This could represent a potential, although likely unintentional, gap in the review process.

15.5.1 The Relevance of Hindsight

Recognizing mistakes can often be accomplished only in hindsight. The $CHADS_2$ method is a classic example. It was initially developed for risk stratification in atrial fibrillation and was validated with years of data [46, 47]. A decade later a Danish group reviewed those with $CHADS_2$ scores of 0 or 1 and found an approximately four times higher risk of stroke than previously reported [48]. This led to the development of the CHA_2DS_2-VASc score accounting for three variables that were previously not used. More than 17 years after the initial $CHADS_2$ system was published, another iteration that included female sex, a modifier in higher-risk cases, was recommended [49]. While the initial study was a quality protocol, it simply overlooked variables that increase stroke risk at low-risk $CHADS_2$ levels.

15.6 Tactics for Misrepresentation

Journals have been biased toward only publishing positive studies, further reinforcing the notion that a study must provide positive results to be deemed a success [6]. There are 2.2–4.7 increased odds of publication when the outcomes are statistically significant [50]. Identical manuscripts have been submitted to reviewers that only differed in the conclusion as statistically significant or not. Those without significant findings were recommended for publication less often and were found to have more edits [51]. This emphasis has led to increasing statistically significant studies, termed positive bias [51], and various adaptations to increase the likelihood of finding significance.

Statistically significant results are 2.5 times more likely to have their study protocols completely reported [52], while drug toxicities are not as extensively reported when the primary endpoint is positive in chemotherapy trials [53]. This highlights the cost to transparency in an attempt to gain publications and peer approval. Whether leaving out negative results or not defining protocols, these omissions may be influenced by funding. Those with commercial funding were three times more likely to attempt to spin the results [53, 54].

15.6.1 Endpoints

CONSORT has rules for defining and reporting the primary endpoints in RCTs [55], along with guidelines on reporting study changes, including protocols, analysis, and endpoints. Prior to these, 62% of published RCTs in Denmark had undisclosed changes in primary endpoints [52]. Endpoint manipulation in review articles and meta-analyses has grown, with 30% of studies including adjusted endpoints [56]. Publishing these adjusted endpoints misrepresents the original hypothesis and fails to change the true-life outcomes.

Secondary endpoints are treated in a similar fashion, with more emphasis placed on reporting positive secondary endpoints in the absence of primary endpoint significance. In a sample of oncology RCTs, 59% of trials were found to use secondary endpoints and subgroups as positive findings when the primary endpoint was not met [53].

15.6.2 HARKing

Hypothesizing is a part of the scientific method to address unexpected findings and further refine conclusions. Hypothesizing after the results are known, or HARKing, occurs when presenting a post hoc explanation for known data as a priori hypothesis or not presenting a true priori hypothesis after it is disproven [20, 57]. Instead of testing a hypothesis, HARKing confirms conclusions without the proper study enrollment and thorough control for confounding variables, possibly introducing deviations from the study's goal or Type I error [20]. Close to half of researchers self-report this practice [57].

15.6.3 P-Hacking

A commonly employed tactic is to test multiple hypotheses and different types of analysis on the same data to find the statistical significance, dubbed p-hacking [58]. While some support p-hacking as a means to find new associations, almost as if finding a needle in a haystack, true confirmation with predefined endpoints and a priori hypotheses should be employed to confirm a novel association.

15.6.4 Preregistration of Scientific Protocols

All endpoints and hypotheses for every study should be registered in specialized sites. This may prove difficult for observational studies and institutional retrospective analyses, but a mandatory holiday between registration and submission may discourage HARKing. Sharing deidentified data allows others to verify the findings and allows for secondary analysis [59]. In addition to deidentified databases which can house a repository of patient information, adding a second statistician or internal statistical reviewer who is blinded to the conclusions for quick analysis can help minimize bias. While this might require more resources, money, and time, it ensures quality work prior to submission and may help reduce the number and types of required revisions prior to publication. Additionally, as seen in the evolution of the $CHADS_2$ scoring system, sharing data allows for more retrospective analysis and addition of new hypotheses encompassing different variables for future study.

References

1. Gauss CF (1809) Theoria motus corporum coelestium in sectionibus conicis solem ambientium. Hamburg, Perthes F and Besser IH
2. Moore S. The Lost Art of Signficant Digits: More digits don't always mean more accuracy. qualitydigest.com/inside/statistics-column/lost-art-significant-digits-042417.html. Accessed 5 Oct 2021
3. Landhuis E (2016) Scientific literature: information overload. Nature 535(7612):457–458
4. Begley CG, Ioannidis JPA (2015) Reproducibility in science. Circ Res 116(1):116–126
5. CONSORT (2010). consort-statement.org/about-consort/history. Accessed 25 Jan 2021
6. Fihn SD (2019) Combating misrepresentation of research findings. JAMA Netw Open 2(5):e192553-e
7. Boutron I, Ravaud P (2018) Misrepresentation and distortion of research in biomedical literature. Proc Natl Acad Sci 115(11):2613
8. Wakefield AJ, Murch SH, Anthony A, Linnell J, Casson DM, Malik M et al (1998) Ileal-lymphoid-nodular hyperplasia, non-specific colitis, and pervasive developmental disorder in children. Lancet 351(9103):637–641
9. Al-Marzouki S, Roberts I, Marshall T, Evans S (2005) The effect of scientific misconduct on the results of clinical trials: a Delphi survey. Contemp Clin Trials 26(3):331–337
10. Fang FC, Steen RG, Casadevall A (2012) Misconduct accounts for the majority of retracted scientific publications. Proc Natl Acad Sci U S A 109(42): 17028–17033
11. George SL, Buyse M (2015) Data fraud in clinical trials. Clin Investig (Lond) 5(2):161–173
12. Walker MP. Why we sleep: unlocking the power of sleep and dreams; 2017.
13. Gelman A, Guzey A (2020) Statistics as squid ink: how prominent researchers can get away with misrepresenting data. Chance 33(2):25–27
14. Nath SB, Marcus SC, Druss BG (2006) Retractions in the research literature: misconduct or mistakes? Med J Aust 185(3):152–154
15. Steen RG (2011) Retractions in the scientific literature: do authors deliberately commit research fraud? J Med Ethics 37(2):113
16. Carlton Collins J. Using Excell and Benford's law to detect fraud. journalofaccountancy.com/issues/2017/apr/excel-and-benfords-law-to-detect-fraud.html. Accessed 5 Oct 2021
17. Resnik DB, Stewart CN Jr (2012) Misconduct versus honest error and scientific disagreement. Account Res 19(1):56–63
18. Marris E (2010) Statistics spark dismissal suit. Nature 467:260
19. Abramson SB (2011) Differing opinion, not misconduct. Nature 470(7335):465

20. Kerr NL (1998) HARKing: hypothesizing after the results are known. Personal Soc Psychol Rev 2(3): 196–217
21. Lang TSM (2006) How to report statistics in medicine: annotated guidelines for authors, editors, and reviewers, 2nd edn. American College of Physicians, Philadelphia, PA
22. Nuijten MB (2016) Preventing statistical errors in scientific journals. Europ Sci Ed 42(1):8–10
23. Lang T (2004) Twenty statistical errors you can find in biomedical research activities. Croatian Med J 45(4): 361–370
24. Gardner MJ, Bond J (1990) An exploratory study of statistical assessment of papers published in the British Medical Journal. JAMA 263(10):1355–1357
25. Leaphart DM, Cronley AC, Brothers TE (2019) Errors in reporting significant figures in surgical and medical journals in 2017. J Surg Res 243:567–573
26. Nuijten MB, Hartgerink CH, van Assen MA, Epskamp S, Wicherts JM (2016) The prevalence of statistical reporting errors in psychology (1985–2013). Behav Res Methods 48(4):1205–1226
27. Georgescu C, Wren JD (2017) Algorithmic identification of discrepancies between published ratios and their reported confidence intervals and P-values. Bioinformatics 34(10):1758–1766
28. Estruch R, Ros E, Salas-Salvadó J, Covas M-I, Corella D, Arós F et al (2013) Primary prevention of cardiovascular disease with a mediterranean diet. N Engl J Med 368(14):1279–1290
29. Nieminen P, Carpenter J, Rucker G, Schumacher M (2006) The relationship between quality of research and citation frequency. BMC Med Res Methodol 6(1): 42
30. Schneider J, Ye D, Hill AM, Whitehorn AS (2020) Continued post-retraction citation of a fraudulent clinical trial report, 11 years after it was retracted for falsifying data. Scientometrics 125(3):2877–2913
31. Matsuyama W, Mitsuyama H, Watanabe M, Oonakahara K-i, Higashimoto I, Osame M et al (2005) Retracted: effects of omega-3 polyunsaturated fatty acids on inflammatory markers in COPD. Chest 128(6):3817–3827
32. EQUATOR (2016). equator-network.org/about-us/history/#:~:text=The%20EQUATOR%20Network%20held%20its,reviewers%2C%20medical%20writers%20and%20funders. Accessed 11 Apr 2021
33. Howell VSA, O'Leary JD, Donnell CM (2015) The effect of the SQUIRE (standards of quality improvement reporting excellence) guidelines on reporting standards in the quality improvement literature. BMJ Qual Saf 24(6):400–406
34. Bastuji-Garin S, Sbidian E, Gaudy-Marqueste C, Ferrat E, Roujeau JC, Richard MA et al (2013) Impact of STROBE statement publication on quality of observational study reporting: interrupted time series versus before-after analysis. PLoS One 8(8):e64733
35. Dexter F, Shafer SL (2017) Narrative review of statistical reporting checklists, mandatory statistical editing, and rectifying common problems in the reporting of scientific articles. Anesth Analg 124(3):943–947
36. Pouwels KB, Widyakusuma NN, Groenwold RH, Hak E (2016) Quality of reporting of confounding remained suboptimal after the STROBE guideline. J Clin Epidemiol 69:217–224
37. Mascha EJ, Vetter TR (2017) The statistical checklist and statistical review: two essential yet challenging deliverables. Anesth Analg 124(3):–719, 721
38. Nature - Anonymous (2018) Checklists work to improve science. Nature 556:273–274
39. Brown NJL, Heathers JAJ (2017) The GRIM test: a simple technique detects numerous anomalies in the reporting of results in psychology. Soc Psychol Personal Sci 8(4):363–369
40. Khan MS, Lateef N, Siddiqi TJ, Rehman KA, Alnaimat S, Khan SU et al (2019) Level and prevalence of spin in published cardiovascular randomized clinical trial reports with statistically nonsignificant primary outcomes: a systematic review. JAMA Netw Open 2(5):e192622-e
41. Yavchitz A, Boutron I, Bafeta A, Marroun I, Charles P, Mantz J et al (2012) Misrepresentation of randomized controlled trials in press releases and news coverage: a cohort study. PLoS Med 9(9):e1001308
42. MacDougall D (1907) The soul: hypothesis concerning soul substance together with experimental evidence of the existence of such substance. Am Med 2:240–243
43. Cote MP, Lubowitz JH, Rossi MJ, Brand JC (2018) Reviews pooling heterogeneous, low-evidence, high-bias data result in incorrect conclusions: but heterogeneity is an opportunity to explore. Arthroscopy 34(12): 3126–3128
44. Cobo E, Selva-O'Callagham A, Ribera J-M, Cardellach F, Dominguez R, Vilardell M (2007) Statistical reviewers improve reporting in biomedical articles: a randomized trial. PLoS One 2(3):e332
45. McClean M, Silverberg JI (2015) Statistical reporting in randomized controlled trials from the dermatology literature: a review of 44 dermatology journals. Br J Dermatol 173(1):172–183
46. Gage BF, van Walraven C, Pearce L, Hart RG, Koudstaal PJ, Boode BS et al (2004) Selecting patients with atrial fibrillation for anticoagulation: stroke risk stratification in patients taking aspirin. Circulation 110(16):2287–2292
47. Gage BF, Waterman AD, Shannon W, Boechler M, Rich MW, Radford MJ (2001) Validation of clinical classification schemes for predicting stroke: results from the National Registry of Atrial Fibrillation. JAMA 285(22):2864–2870
48. Olesen JB, Torp-Pedersen C, Hansen ML, Lip GY (2012) The value of the CHA2DS2-VASc score for refining stroke risk stratification in patients with atrial fibrillation with a CHADS2 score 0-1: a nationwide cohort study. Thromb Haemost 107(6):1172–1179
49. Nielsen PB, Skjøth F, Overvad TF, Larsen TB, Lip GYH (2018) Female sex is a risk modifier rather than a

risk factor for stroke in atrial fibrillation: should we use a CHA(2)DS(2)-VA score rather than CHA(2)DS(2)-VASc? Circulation 137(8):832–840
50. Dwan K, Altman DG, Arnaiz JA, Bloom J, Chan A-W, Cronin E et al (2008) Systematic review of the empirical evidence of study publication bias and outcome reporting bias. PLoS One 3(8):e3081
51. Emerson GB, Warme WJ, Wolf FM, Heckman JD, Brand RA, Leopold SS (2010) Testing for the presence of positive-outcome bias in peer review: a randomized controlled trial. Arch Intern Med 170(21):1934–1939
52. Chan AW, Hróbjartsson A, Haahr MT, Gøtzsche PC, Altman DG (2004) Empirical evidence for selective reporting of outcomes in randomized trials: comparison of protocols to published articles. JAMA 291(20): 2457–2465
53. Vera-Badillo FE, Shapiro R, Ocana A, Amir E, Tannock IF (2013) Bias in reporting of end points of efficacy and toxicity in randomized, clinical trials for women with breast cancer. Ann Oncol 24(5): 1238–1244
54. Reynolds-Vaughn V, Riddle J, Brown J, Schiesel M, Wayant C, Vassar M (2020) Evaluation of spin in the abstracts of emergency medicine randomized controlled trials. Ann Emerg Med 75(3):423–431
55. Schulz KF, Altman DG, Moher D (2010) CONSORT 2010 statement: updated guidelines for reporting parallel group randomized trials. Ann Intern Med 152(11):726–732
56. Kirkham JJ, Altman DG, Williamson PR (2010) Bias due to changes in specified outcomes during the systematic review process. PLoS One 5(3):e9810
57. Rubin M (2017) When does HARKing hurt? identifying when different types of undisclosed post hoc hypothesizing harm scientific progress. Rev Gen Psychol 21(4):308–320
58. Nelson LD, Simmons J, Simonsohn U (2018) Psychology's renaissance. Annu Rev Psychol 69(1): 511–534
59. Ohmann C, Banzi R, Canham S, Battaglia S, Matei M, Ariyo C et al (2017) Sharing and reuse of individual participant data from clinical trials: principles and recommendations. BMJ Open 7(12):e018647

Part V

Academic Institutions

Myths About Race and Racism on University Campuses

16

Dana Strauss, Monnica T. Williams, Muna Osman, and Jade Gallo

Abstract

Despite recent increased interest in equity, diversity, and inclusion in post-secondary education, racism persists on university campuses as it does throughout Western society, fueled by malignant myths and falsehoods. These myths operate such that universities, intended as places to inspire intellectual growth and development, promote misinformation and create a toxic environment for students of colour. This chapter addresses some of these harmful myths including (1) the right to "academic freedom", even if it means enacting racism, and (2) the notion that racial microaggressions and systemic racism do not exist on campuses, (3) that tokenism is as good as inclusion, and (4) that psychedelic communities have transcended racism. Due to the increasing necessity of post-secondary education, dispelling racist myths is critical for the eradication of racism and the evolution of universities into spaces that are truly inclusive, comfortable, and capable of inspiring growth for all students.

16.1 Introduction

In 2021, we watched a 4-week trial of former Minneapolis police officer Derek Chauvin, for a murder that the whole world witnessed on social media. As such, it comes as no surprise that racism is still rampant on university campuses. Although diversity and inclusion have become hot topics in post-secondary institutions, idealizing these values is not the same as practising them. Predominantly White institutions (PWIs) boast of their cultural centres and clubs, learning communities for minoritized groups, and "distinct" and "inclusive" course selection to prospective and incoming students, specifically those of colour, as a means of appealing to a wider range of applicants. Over recent years, there has been an increase in students of colour attending PWIs, drawn in through exaggerated marketing of inclusion and the ease of finding a community within a community. However, claims of diversity and inclusion are often only surface level, revealing many myths and misconceptions surrounding inclusion and other issues related to race and racism. This chapter aims to illuminate some of these harmful myths in hopes of facilitating meaningful advancements toward

D. Strauss · M. T. Williams (✉) · M. Osman
University of Ottawa, School of Psychology, Ottawa, ON, Canada
e-mail: dstra012@uottawa.ca; Monnica.Williams@uOttawa.ca; mosma039@uottawa.ca

J. Gallo
Department of Psychological Sciences, University of Connecticut, Storrs, CT, USA
e-mail: jade.gallo@uconn.edu

J. Faintuch, S. Faintuch (eds.), *Integrity of Scientific Research*,
https://doi.org/10.1007/978-3-030-99680-2_16

real equity, diversity, and inclusion at post-secondary institutions.

Racism, an ideology that has taken form through harmful prejudices, attitudes, and behaviours toward another ethnicity or phenotype rooted in White supremacy, is a global issue that is systemic in nature [1]. It has bled into all aspects of life, making it almost impossible for those affected by it to escape. This rings true throughout academia, especially in post-secondary institutions, affecting students, faculty, staff, and executives alike [2]. Most PWIs have long legacies of racial prejudice and racism embedded within the institution through policies and social traditions which continue to create uncomfortable environments for students of colour to this day [3]. For example, until 2003, one of the United States' biggest universities, the University of Mississippi "Ole Miss", had a mascot named Colonel Reb that represented a White confederate from the Old South. Undoubtedly, this created an unwelcoming environment for Black students and other students of colour (e.g. [4, 5]). Likewise, until recently, Ryerson University in Toronto, Canada, still boasted a large statue of its founder, Egerton Ryerson, who, in 1847, recommended the establishment of the infamous Canadian residential school system which served as a tool for cultural genocide against Indigenous peoples [6, 7], although at the time he may have believed these actions were in their best interests [8].

PWIs usually also have few faculty of colour and an underrepresentation of people of colour (POC) on school boards, making it difficult for students to connect with a sense of community on campus that exists beyond their peers. Milem, Chang, and Antonio [9] call this phenomenon "compositional diversity". Poor compositional diversity makes it harder for students of colour to find guidance, mentorship, and security with someone who may understand their experiences and adversities as a person of colour [10].

McClain and Perry [10] also studied retention rates in students of colour affected by their campus racial climate using compositional diversity along with four other factors: institutional historical legacy of inclusion or exclusion, psychological climate (how each member of the community perceives and experiences the institution as a whole), behavioural climate (interaction among varying groups and individuals and the quality of intercultural relationships), and structural diversity (administration and operations within the institution). These five key components of campus racial climate highlight the marginalization and exclusion of students of colour and might explain the lower retention rates for students of colour compared to non-Hispanic White students. Since school culture strongly impacts students' aspirations and beliefs, a White-centric culture on college campuses hinders students of colour [11]. In contrast, a culturally inclusive environment creates a sense of belonging within a group or community; thus, it is necessary for colleges to have programmes, spaces, and resources that contribute to creating a culturally inclusive environment. A campus environment that is intentionally inclusive and welcoming of all people and cultures may dramatically increase retention rates for students of colour [3, 10].

Since education is a principal component of anti-racist interventions, the core curriculums at all levels of education should include courses on culture, decolonization, and anti-racism. Also, these teachings should be integrated into all classes. For example, in introduction to psychology, cultural differences in symptomatology should be discussed for each disorder taught. Part of this education should include dispelling harmful and misleading myths about inclusion and issues related to race and racism. This chapter will discuss some of these pervasive myths, namely, the myth that (1) the "right to academic freedom" has nothing to do with racism, (2) racial microaggressions and systemic racism do not exist here, (3) inclusion and tokenism are the same thing, and (4) psychedelic communities have transcended racism. These myths become exposed when problems arise that challenge the status quo. Through dispelling these myths, universities may begin to dismantle systems and policies grounded in White supremacy and foster a truly inclusive campus environment, which, in turn, may decrease mental health issues and increase retention rates among students of colour.

16.2 Myth 1: The Right to Academic Freedom

On September 23, 2020, a White professor at the University of Ottawa used the n-word in class, sparking a student-led Black Lives Matter campus sit-in. The professor's subsequent suspension led to a heated debate over the right to "academic freedom" and a fission within the University. *Le Journal de Montréal* responded by criticizing the University's decision to suspend the professor and likening the University of Ottawa's BIPOC Caucus to the deadly coronavirus all while denying that systemic racism exists in Canada [12, 13].

In a letter sent to his undergraduate students on February 1, 2021 (the first day of Black History Month), Dr. François Chapleau, a White, francophone professor at the University of Ottawa, defended his approval of a letter (signed by 34 other White francophone professors) in support of the professor's use of the n-word in class. He argued that "there are about as many 'sensitivities' as there are members. Who will decide if a term or an idea should be part of the 'code' or require a punitive action?" ([14], p. 6).

Academic freedom is a sacred value cherished by academic institutions. It is often compared to basic human rights and freedoms and as a universal ideal [15, 16]. However, there are many definitions of this concept, and its advocates vary substantially in whether the "academic" boundary of this freedom is emphasized. On one hand, some academics equate freedom to the absence of any responsibility and accountability to their institutions or the public, advocating that academics be granted substantial latitudes that many other professions are not granted—including the right to perpetrate racism. On the other hand, some academics recognize the professional obligation they have to their profession and society to practice responsible and ethical science.

Within psychology, the American Psychological Association (APA) has explicitly outlined in policy their commitment to the "free and responsible practice of science" [17]. Internationally, regulatory bodies such as the International Union of Psychological Science (IUPsyS) and the International Council for Science (ICSU), to which APA is affiliated, adhere to principles of the universality of science. These principles include (a) the free and responsible practice of science, (b) equitable access to science, and (c) the responsibility of scientists to society to promote benefits and prevent the harms associated with their work [17]. It is clear from these principles that APA and its international counterparts recognize and are committed to the ethical considerations in the implementation of academic and scientific freedom and uphold appropriate accountability both within and outside academia.

According to the APA dictionary of psychology, academic freedom, "within an institution of learning, [refers to] the liberty of a teacher to educate, of a student to study, and of both to express opinions, particularly regarding controversial topics (e.g. morals, religion, politics), without interference from or punishment by government, school officials, teachers, or community groups" [18]. This definition heavily focuses on concepts of liberty of teachers, the expression of opinions, and the lack of interference from broader society. Furthermore, there is no explicit mention of the responsibility of professors, institutions, or society to place appropriate and justified limits on academic freedom. Although APA's policy is to uphold the "free and responsible" research and practice of psychology, this definition does not mention the ethical responsibility of academic professionals to promote benefits and prevent harms, as well as the responsibility of science to broader society. This lack of conceptual clarity can imply the boundary between academic freedom and racism is somehow blurred.

Although many were swayed by the eloquent words of those who conflated racism and academic freedom, the occurrence and aftermath of this event surrounding the n-word at the University of Ottawa was particularly harmful to Black, Indigenous, and people of colour (BIPOC) on campus. Defending the use of the racial slur created a hostile campus environment because it invalidated the lived experiences of racism and discrimination and the toll that racial trauma takes on mental health. By calling racial slurs and microaggressions "sensitivities", professor Chapleau is effectively minimizing the trauma endured by BIPOC at a majority White campus. Further, his question of who will decide if a term or idea should be part of a "code" of prohibited language is exclusionary, implying that the racialized groups that are most affected are not qualified to make this decision.

This past year was marked by universities, organizations, and academic associations around the world publishing statements in solidarity with Black and other racialized groups to express their commitment to equity and racial justice. For example, the Canadian Association of University Teachers [19] expressed their solidarity with Black communities and their commitment to dismantle racism on campuses and in academic spaces. Only months after their commitment, when given the opportunity to illustrate their solidarity to Black communities, CAUT made a formal statement that was in complete contradiction to their expressions of solidarity, as they joined and in support of an academic using the n-word in class [20]. Even after the professor apologized for this offensive act, the CAUT reaffirmed their support of the use of a racist slur in an academic setting, citing academic freedom as the justification.

Both national and international regulatory bodies of psychological science are clearly committed to the values of academic responsibility and ethical scholarship, yet when examined more closely, there is a lack of equitable and responsible practices to manifest these commitments. In these examples, the unquestioned right of academics to speak freely, even when they might be perpetuating racial prejudice by using slurs, is prioritized above harm reduction for Black students, solidarity with Black communities, and anti-racist commitments. In line with systemic racism, the liberties, comforts, and "sensitivities" of White professors are designated as more important than the harms and mental distress experienced by BIPOC members of the university community [3]. The concept of academic freedom must be understood as including both responsibility and accountability to ensure university campuses are places where BIPOC students, faculty, and staff are valued and respected.

16.3 Myth 2: Racial Microaggressions and Systemic Racism Do Not Exist Here

Although racial slurs are generally understood to be racist, racial microaggressions are less well understood, and their validity has even been disputed by prominent academics [21]. Racial microaggressions have been described as "brief, everyday exchanges that send denigrating messages to people of colour because they belong to a racial minority group" ([22], p. 273). In a recent article entitled "Microaggressions: Clarification, Evidence, and Impact", Williams [23] clarifies common misconceptions associated with microaggressions, namely, what they are and the harms they can cause. Williams [24] explains that while microaggressions may be entirely unintentional on the part of the perpetrator, the cumulative effects on the victims can potentiate long-lasting and devastating mental health outcomes including stress, anxiety, depression, posttraumatic stress disorder (PTSD), obsessive-compulsive disorder (OCD), substance use, alcohol abuse, and suicide, as well as physical ailments such as hypertension and impaired immune response.

Although much of the microaggression research to date has been conducted in the United States, racial microaggressions are an international issue. Willis [25] writes about the racial and gender microaggressions encountered by Black American women on study abroad programmes in the British Isles, the Mediterranean, and West Africa. Many other studies to date have also documented the experiences of BIPOC Americans studying abroad [25–27]. Data from six Western European countries (Austria, Belgium, France, Germany, the Netherlands, and Sweden) showed that non-Muslims hold negative attitudes toward the headscarf worn by Muslims [28]. A study by Burdsey [29] documented racial microaggressions experienced by British Asian players in English men's first-class cricket. Miller et al. [30] surveyed adoptive parents and adopted adolescents of colour in France about their experiences with prejudice and feeling different, and they found that 63% of the adopted adolescents interviewed were upset by rude and/or racist comments made by complete strangers.

Microaggressions are also common on college campuses. Researchers have documented racial microaggressions experienced by Aboriginal and East and South Asian international students at multiple Canadian universities [31, 32]. Baker [33] studied the prevalence of racial microaggressions reported by a sample of 170 White, first-year university students in St. John's, Newfoundland and Labrador, Canada. Baker's [33] research revealed that despite Sue's [22] assertion that members of disempowered groups are more likely to accurately perceive whether a racist event has transpired compared to those with racial/White privilege and power, "Whites are acutely aware of racism, discrimination, and prejudice" as well ([33], p. 376). Baker [33] also found a trend of passivity among the White students such that they did not call out the racist behaviour they witnessed even when they recognized that it was wrong and unwarranted, and even when the victim was their friend. In his book, *White Guys on Campus: Racism, White Immunity, and the Myth of "Post-Racial" Higher Education*, [34] draws attention to the racial microaggressions perpetrated by White male undergraduate students in American colleges and highlights the underlying ideologies, policies, and practices that perpetuate racism, otherwise known as systemic racism.

Moreover, contrary to the myth widely believed by many Canadians, systemic racism is not confined to the United States. For example, in June of 2019, a Black male University of Ottawa student was carded and left to sit for hours in handcuffs on the street for skateboarding on campus [35]. The student described this incident as "humiliating" and an act of "blatant racism", adding "uOttawa security used their authority to harass and demean me" [35]. In January of 2021, the University of British Columbia (UBC) silenced an Indigenous professor for voicing her concerns about 12 teaching candidates in a mandatory course on Indigenous Education. After receiving complaints from the 12 students as well as an anonymous letter from one of the students' parents, the administration deleted the professor's interim reports which included her assessment that certain students in the class displayed tendencies of White supremacy rendering them unfit to educate students on Indigenous topics and perspectives. The administration further instructed the professor to also delete her copies and to "never speak of the content of the interim reports to anyone" [36]. The professor has since been placed on administrative leave and has not been offered any further contract work as a sessional instructor [36].

In February of 2021, one of the authors became aware that students at Bard College in Berlin, Germany, were subjected to explicit racism when a professor suggested his class have a debate about which races are more intelligent. To add insult to injury, the associate dean of the college offered three options to the students who voiced concerns over the professor's behaviour, none of which included accountability for the professor or the college to its students: (1) the students could take up their concerns directly with the offending professor, (2) the students could have a mediated conversation with the professor, and (3) the associate dean could speak privately to the professor ensuring the anonymity of the concerned students. While these options may seem fair at first glance, they do not account for the power imbalance in the situation. And for option 3, it is not appropriate to seek the perspective of only one of the parties, especially the one with power and motivation not to be seen as racist. Further, it is always wrong to centre the voice of the oppressor (the powerful, in this case the professor) over the voice of the oppressed (in this case, the students). The institutional racism enacted by the associate dean and coupled with the initial overt racism perpetrated by the professor compounds the harm done to the students. Since the institution knew about the event, they should have taken the initiative to investigate it without putting students in a vulnerable position [3, 37].

16.4 Myth 3: Tokenism and Inclusion Are the Same Thing

Inclusion is defined as including a person of colour because one truly values that individual's talents and skills and/or principles of equity, diversity, and inclusion (EDI); in contrast, tokenism is the act of including a person of colour merely to feign EDI values [37]. Tokenism can also involve homogenizing diverse groups based on race and/or the assumption that the perspective of one member of a marginalized group is representative of the perspective of their entire group [38, 39]. Like systemic racism and racial microaggressions, tokenism is ubiquitous. It can manifest in many different forms and spaces. Here we provide a few examples to better illustrate some of the ways in which tokenism may present itself.

Despite her extensive scholarship on racism and cultural diversity and inclusion within the field of psychology (e.g. [37, 40, 41]), the second author of this chapter (MTW) shares her own personal examples in which she has been tokenized. In one such instance, she was asked to serve on a hiring committee that she was not qualified to serve on purely because she is a person of colour. When she explained to her department chair that she was not qualified to evaluate neuroscience candidates (her expertise being in psychopathology instead), the response was simply that the dean wanted a person of colour on the committee. Not only was this decision illogical and unfair to prospective applicants, but it was also hurtful to her because it meant that her skills and expertise were not being valued or recognized, and instead she was reduced to the colour of her skin. Further, the other committee members knew she had no useful scholarly contributions to make in the deliberation process by virtue of her lack of expertise in this hiring area, which made her feel less competent, humiliated, as well as tokenized.

In another instance, she was an invited panellist for a psychology conference in Washington, DC. The panel was titled *Being a Mentor in the Era of #Metoo*. As the only person of colour on the panel, she felt a responsibility to draw attention to issues related to the #Metoo movement that solely impacted people of colour. One important issue which felt particularly relevant to her at the time was the historical pattern of White women victimizing men of colour, often resulting in life-threatening consequences for the accused man of colour in the name of White women's virtue. As she was speaking on this important issue, the moderator abruptly cut her off, and as swiftly as the mic was taken from her, so too were her concerns ignored and her uncomfortable words forgotten. In her place, a White woman was given a platform to describe her own experience of victimization. While this event illuminates the ways in which White feminism functions as White supremacy, at the core, unmistakably, is tokenism. If the conference organizers were not tokenizing her, but instead actually valuing inclusion, perhaps she would not have been publicly humiliated and shamed into silence [42].

The final example of tokenism was learned the day before writing this piece. One of the only Black students in the graduate psychology programme at the University of Ottawa shared that she is related to the Black student pictured on the University of Ottawa website. Despite their significant underrepresentation in the uOttawa graduate psychology programme (based on observation as the University does not collect any data on the race or ethnicity of its students), Black, Indigenous, and people of colour (BIPOC) are still used to give the illusion of diversity at universities. This is not an isolated incident either; unfortunately, tokenism is commonplace at many Western universities [43]. An embarrassing attempt at promoting racial diversity at the University of Wisconsin-Madison came to light in the fall of 2000, when it was revealed that a photograph featured on the front of their undergraduate application booklet had been altered to add the image of a Black student among a sea of White faces at a sporting event.

There have been many calls to action with suggestions for how to implement real diversity and inclusion. For example, in addition to implementing equitable hiring practices that include hiring more faculty of colour, Williams and Kanter [37] suggest creating a department subdivision dedicated to cultural diversity science. First and foremost, prioritizing cultural diversity science is ethical, equitable, and inclusive. Further, it will likely also attract diverse faculty on the basis of expertise rather than race, solving the issue of "diversity hires" [37]. As with racial microaggressions and systemic racism, tokenism is harmful to people of colour and to the field of psychology as a whole. People of colour cannot truly feel or be valued, and the field will not truly be inclusive or equitable until we stop tokenizing people of colour to feign EDI values and instead prioritize real equity, diversity, and inclusion.

16.5 Myth 4: Psychedelic Communities Have Transcended Racism

Psychedelic communities have long been perceived as eclectic groups of young people, flouting the expectations of society and embracing love, peace, and liberal values. Indeed, college campuses were the origins and epicentre of psychedelic awareness in the 1960s, and college campuses today remain a relatively safe haven for exploration and self-discovery with all manner of chemical agents. Nonetheless, a discussion of psychedelics and the psychedelic community may seem out of place in this chapter centred around prejudice at universities; therefore, we preface this section with the following: it is important to draw attention to the myth that psychedelics cure racism; thus, the psychedelic community has transcended racism, particularly in the wake of the resurgence of psychedelic research at universities across the globe and considering the historical exclusion of Black, Indigenous, and people of colour from psychedelic research [44, 45]. We believe it is necessary to deconstruct this myth so that academic psychedelic programmes and research initiatives will make a concerted effort to decolonize and prioritize inclusivity.

2020 was a tough year to say the least. The coronavirus wreaked havoc around the world, hitting Black and Brown communities the hardest [46, 47]. The year 2020 also bore witness to some of the worst police violence to emerge in the public sphere. The cold-blooded, callous killings of George Floyd and Breonna Taylor (along with many more Black men, women, and children) at the hands of police shook the world and brought international attention to the Black Lives Matter (BLM) movement. Protests against police brutality reverberated around the world, and organizations in the West responded with statements of solidarity with BLM. Despite resounding international support for BLM, individuals and organizations throughout the West subscribing to conservative and fringe right-wing politics voiced their opposition, and the psychedelic community was no exception.

Labate and Buchanan [48] analysed social media responses within the psychedelic community to statements of support for BLM issued by psychedelic organizations including the Multidisciplinary Association for Psychedelic Studies (MAPS), Chacruna, Psychedelics Today, San Francisco Psychedelic Society, Psychedelic Support, Lucid News, DoubleBlind, and Psychedelic Seminars. After reviewing the profiles of the commenters, they found that the comments that were hostile toward or highly critical of BLM/anti-racist initiatives or the psychedelic organization's support of these initiatives were authored almost exclusively by White men while those that refuted these comments included men and women of varying cultural backgrounds. Labate and Buchanan [48] categorized the negative comments into eight categories: anti-BLM, divisive, ideological, infiltration, spiritual bypassing, "race does not exist", systemic racism denial, and virtue signaling. They also found that the more mainstream psychedelic organizations (e.g. MAPS) harboured more of these hostile/negative comments than smaller organizations.

It is not surprising that there was criticism of BLM/anti-racist initiatives and psychedelic organizations' support for these initiatives as denials of the existence of systemic racism and criticism of BLM permeate all of society. Rather, what is surprising is the apparent myth that the psychedelic community is diverse, inclusive, and devoid of racism simply by virtue of the fact that its members use psychedelics. While psychedelics have been known to inspire mystical experiences of love and connection, the belief that psychedelics on their own can eradicate racism is foolish and indicative of one's own privilege and limited understanding of racism. In addition, a quick glance at the board of directors of most psychedelic organizations will reveal the lack of diversity and inclusion fostered by the psychedelic community, and the mere presence of comments opposing BLM or support for BLM evidences the racism that persists within the psychedelic community.

This myth is not just surprising; more importantly, it is also extremely harmful because it insulates and upholds racism and White supremacy within the psychedelic research community. By creating the facade that there is no racism/diversity problem, White people can continue to ignore their own implicit biases and the ways in which White Supremacy benefits them and harms BIPOC, thus perpetuating systemic racism. Psychedelics are currently being explored to heal racial trauma for people of colour, and some scholars have even suggested that psychedelics might also help to heal the racism entrenched in the perpetrators and in all people, a product of generations of pathological racial socialization [49]. Nonetheless, at present, White people are primarily the beneficiaries of psychedelic medicine, while people of colour are stigmatized as the drug users [44, 50].

This myth is especially important to acknowledge and deconstruct in the wake of the resurgence of Western psychedelic research [45]. As psychedelics reenter the university campus, it is vital that we understand the gravity of such a myth and consider the repercussions it will have on who will be represented and included in future research and resulting treatments. Western psychedelic science has historically excluded and even criminalized Black, Indigenous, and other racialized groups who have traditional ties to these plant medicines, and for whom without their maintenance of these plant medicines and their contributions to the field of psychedelic research, the field would not be where it is today [23]. Thus, it is the responsibility of the academic and psychedelic communities to come together with Indigenous healers and prioritize the inclusion of BIPOC in this psychedelic resurgence to change the course of psychedelic research and treatments to be what they say they are already: diverse, inclusive, and transcendent of racism [51].

16.6 Conclusion

This chapter is not intended to be a comprehensive examination of all racist myths. Many other harmful myths stemming from racist ideologies prevail within university campuses and society at large, such as that of the "model minority", "colorblindness", and the notion that "universal" access to health care cures health disparities. Although this chapter does not cover these myths and many more, it is necessary to dispel all racist myths in order to create a safe and nurturing environment for all students on campus and, ultimately, to build a safer and more inclusive society for all.

There are several reasons why these racist myths persist. First, the myths of meritocracy and beliefs in a just world, as two complimentary myths, contribute to sustaining racism. Many White people prefer to believe they have properly earned all their advantages and privileges. When we believe our success is a direct result of our effort, we deny White privilege as the unearned benefits given to White people. Similarly, belief in a just world suggests people deserve what happens to them whether good or bad [1, 52, 53]. Out of psychological self-protection, people want to believe the world is a fair place. However, holding these beliefs means we deny discrimination and feel mistreatment and inequities are justified and deserved [52]. Second, people lack historical knowledge of racism, colonialism, and genocide and so are unable to understand current issues as they do not have the historical context needed to put everything into perspective.

As the popularity and necessity of higher education increases, so does the importance of dispelling these myths, along with the many others that exist on university campuses. Through much organizational and policy changes, proper inclusivity on college campuses will strengthen retention rates, enhance experiences, and provide more comfortability among students of colour for a greater sense of community [3]. Racism, unfortunately, exists throughout all stages and walks of life and must be tackled one system at a time. Racism in the university atmosphere is particularly harmful as it discourages students and can cause traumatic experiences, and it should not be tolerated.

References

1. Haeny A, Holmes S, Williams MT (2021) The need for shared nomenclature on racism and related terminology. Perspect Psychol Sci 16(5):886–892. https://doi.org/10.1177/17456916211000760
2. Coleman MJ (2005) Racism in academia: the white superiority supposition in the "unbiased" search for knowledge. Eur J Polit Econ 21(3):762–774. https://doi.org/10.1016/j.ejpoleco.2004.08.004
3. Williams MT (2019) Adverse racial climates in academia: conceptualization, interventions, and call to action. New Ideas Psychol 55:58–67. https://doi.org/10.1016/j.newideapsych.2019.05.002
4. Desai SR, Abeita A (2017) Institutional microaggressions at a Hispanic serving institution: a Diné (Navajo) woman utilizing tribal critical race theory through student activism. Equity Excell Educ 50(3):275–289
5. Williams MT, Skinta MD, Kanter JK, Martin-Willett-R, Mier-Chairez J, Debreaux M, Rosen DC (2020) A qualitative study of microaggressions against African Americans on predominantly White campuses. BMC Psychol 8(111):1–13. https://doi.org/10.1186/s40359-020-00472-8
6. Griffith J (2018) Off to school: filmic false equivalence and Indian residential school scholarship. Hist Stud Educ/Revue d'histoire de l'éducation 30(1):69–83
7. Truth and Reconciliation Commission of Canada (2015) Canada's residential schools: the history, Part 1 Origins to 1939. The final report of the Truth and Reconciliation Commission of Canada. The Truth and Reconciliation Commission of Canada. http://www.trc.ca/assets/pdf/Volume_1_History_Part_1_English_Web.pdf
8. Smith D (2017, July 5) Egerton Ryerson doesn't deserve an anti-Indigenous label. The Globe and Mail. https://www.theglobeandmail.com/opinion/egerton-ryerson-doesnt-deserve-an-anti-indigenous-label/article35558895/
9. Milem J, Chang M, Antonio A (2005) Making diversity work on campus: a research-based perspective. Association of American Colleges and Universities. https://www.aacu.org/sites/default/files/files/mei/MakingDiversityWork.pdf Accessed 23 Feb 2021
10. McClain KS, Perry A (2017) Where did they go: retention rates for students of color at predominantly White institutions. Coll Stud Aff Leadership 4(1):3
11. McClafferty KA, McDonough PM, Nuñez AM (2002) What is a college culture? Facilitating college preparation through organizational change. Paper presented at the annual conference of the American Educational Research Association, New Orleans, Los Angeles, 1–5 April
12. Bombardier D (2020) Les croisés de l'antiracisme: the crusaders of anti-racism. J Montreal. https://www.journaldemontreal.com/2020/10/16/les-croises-de-lantiracisme. Accessed 2 Feb 2021
13. Facal J (2020) Vol au-dessus d'un nid de coucous: flight over a cuckoo's nest. J Quebec. https://www.journaldequebec.com/2020/12/10/vol-au-dessus-dun-nid-de-coucous. Accessed 2 Feb 2021
14. Chapleau F (2021) Letter to my students. https://www.dropbox.com/scl/fi/h7bbx6esu5uw7srqwoj9y/Letter-to-my-students.docx?dl=0&rlkey=np5nyt7smruiuadakvpsv9tbp. Accessed 21 Jan 2021
15. Gottfredson LS (2010) Lessons in academic freedom as lived experience. Personal Individ Differ 49(4):272–280. https://doi.org/10.1016/j.paid.2010.01.001
16. Horn M (1999) Academic freedom in Canada: a history. University of Toronto Press. https://doi.org/10.3138/9781442670570
17. American Psychological Association (2016) Resolution on the free and responsible practice of science, freedom of movement of scientists, and APA international engagement. http://www.apa.org/about/policy/free-responsible-practice.aspx. Accessed 20 Feb 2021
18. VandenBos GR (2007) APA dictionary of psychology. American Psychological Association
19. Canadian Association of University Teachers (2020) Statement on anti-racism protests. https://www.caut.ca/latest/2020/06/statement-anti-racism-protests. Accessed 20 Feb 2021
20. Friesen J (2020) Ottawa professor says she regrets using slur in discussion on reappropriated terms. The Globe and Mail, A8
21. Lilienfeld SO (2017). Microaggressions? Prejudice remains a huge social evil but evidence for harm caused by microaggression is incoherent, unscientific and weak. Aeon. https://aeon.co/essays/

why-a-moratorium-on-microaggressions-policies-is-needed. Accessed 5 Feb 2021
22. Sue DW, Capodilupo CM, Torino GC, Bucceri JM, Holder A, Nadal KL (2007) Racial microaggressions in everyday life: implications for clinical practice. Am Psychol 62(4):271–286. https://doi.org/10.1037/0003-066X.62.4.271
23. Williams MT, Reed S, George J (2020) Culture and psychedelic psychotherapy: ethnic and racial themes from three Black women therapists. J Psychedelic Stud 4(3):125–138. https://doi.org/10.1556/2054.2020.00137
24. Williams MT (2020b) Psychology cannot afford to ignore the many harms caused by microaggressions. Perspect Psychol Sci 15(1):38–43. https://doi.org/10.1177/1745691619893362
25. Willis TY (2015) "And still we rise…": Microaggressions and intersectionality in the study abroad experiences of Black women. Front Interdiscip J Study Abroad 26(1):209–230. https://doi.org/10.36366/frontiers.v26i1.367
26. Goldoni F (2017) Race, ethnicity, class and identity: implications for study abroad. J Lang Identity Educ 16(5):328–341
27. Talburt S, Stewart MA (1999) What's the subject of study abroad? Race, gender, and "living culture". Mod Lang J 83(2):163–175
28. Helbling M (2014) Opposing Muslims and the Muslim headscarf in Western Europe. Eur Sociol Rev 30(2):242–257. https://doi.org/10.1093/esr/jct038
29. Burdsey D (2011) That joke isn't funny anymore: Racial microaggressions, color-blind ideology and the mitigation of racism in English men's first-class cricket. Sociol Sport J 28(3):261–283
30. Miller LC, de Montclos M-OP, Matthews J, Peyre J, Vaugelade J, Baubin O, Chomilier J, de Monleon J-V, de Truchis A, Sorge F, Pinderhughes E (2020) Microaggressions experienced by adoptive families and internationally adopted adolescents in France. Adopt Q 23(2):135–161. https://doi.org/10.1080/10926755.2020.1719253
31. Clark DA, Kleiman S, Spanierman LB, Isaac P, Poolokasingham G (2014) 'Do you live in a teepee?' aboriginal students' experiences with racial microaggressions in Canada. J Divers High Educ 7(2):112–125. https://doi.org/10.1037/a0036573
32. Houshmand S, Spanierman LB, Tafarodi RW (2014) Excluded and avoided: racial microaggressions targeting Asian international students in Canada. Cult Divers Ethn Minor Psychol 20(3):377–388. https://doi.org/10.1037/a0035404
33. Baker J (2017) Through the looking glass: white first-year university students' observations of racism in St. John's, Newfoundland and Labrador. Canada Sociol Inq 87(2):362–384. https://doi.org/10.1111/soin.12165
34. Cabrera NL (2019) White guys on campus: Racism, White immunity, and the myth of "post-racial" higher education. Rutgers University Press. https://doi.org/10.36019/9780813599106
35. Gillis M, Crawford B, Laucius J (2019) 'Humiliating': Black uOttawa student handcuffed in campus carding incident. The Ottawa Citizen. https://ottawacitizen.com/news/local-news/humiliating-black-uottawa-student-cuffed-in-campus-carding-incident
36. Junos K (2021) UBC prof condemns school for allegedly burying teacher candidate reports. https://www.citynews1130.com/2021/01/31/ubc-professor-faculty-of-education-reports-destroyed/. Accessed 21 Jan 2021
37. Williams MT, Kanter JW (2019) Promoting diversity and inclusion on college campuses. In: Williams MT, Rosen DC, Kanter JW (eds) Eliminating race-based mental health disparities: Promoting equity and culturally responsive care across settings. Oakland, California, pp 243–276
38. Nadal KL (2011) The racial and ethnic microaggressions scale (REMS): construction, reliability, and validity. J Couns Psychol 58(4):470–480. https://doi.org/10.1037/a0025193
39. Weber A, Collins SA, Robinson-Wood T, Zeko-Underwood E, Poindexter B (2018) Subtle and severe: microaggressions among racially diverse sexual minorities. J Homosex 65(4):540–559. https://doi.org/10.1080/00918369.2017.1324679
40. Williams MT (2020a) Microaggressions: clarification, evidence, and impact. Perspect Psychol Sci 15(1):3–26. https://doi.org/10.1177/1745691619827499
41. Williams MT, Skinta MD, Martin-Willett R (2021) After Pierce and Sue: evidence for a revised racial microaggressions taxonomy. Perspect Psychol Sci 16(5):991–1007. https://doi.org/10.1177/1745691621994247
42. Williams MT (2021) How White feminists oppress Black women. In: Labate B, Cavnar C (eds) Psychedelic justice: on gender, diversity, sustainability, reciprocity, and cultural appropriation. Synergetic Press, Santa Fe, NM
43. Robinson SJ (2013) Spoke tokenism: Black women talking back about graduate school experiences. Race Ethn Educ 16(2):155–181. https://doi.org/10.1080/13613324.2011.645567
44. Michaels TI, Purdon J, Collins A, Williams MT (2018) Inclusion of people of color in psychedelic-assisted psychotherapy: a review of the literature. BMC Psychiatry 18(245):1–9. https://doi.org/10.1186/s12888-018-1824-6
45. Rochester J, Vallely A, Grof P, Williams M, Chang H, Caldwell K (2021) Entheogens and psychedelics in Canada: proposal for a new paradigm. Can Psychol, Advance Online Publication. https://doi.org/10.1037/cap0000285
46. Fortuna LR, Tolou-Shams M, Robles-Ramamurthy B, Porche MV (2020) Inequity and the disproportionate impact of COVID-19 on communities of color in the United States: the need for a trauma-informed social

justice response. Psychol Trauma 12(5):443–445. https://doi.org/10.1037/tra0000889
47. Sequist TD (2020) The disproportionate impact of COVID-19 on communities of color. NEJM Cata Innov Care Delivery 1(4). https://doi.org/10.1056/CAT.20.0370
48. Labate B, Buchanan N (2020) Hate & social media in psychedelic spaces. https://chacruna.net/psychedelic-community-social-media-racism/. Accessed 17 Feb 2021
49. Buchanan N, Smith D (2021) Debunking racist myths in the psychedelic community: a conversation with NiCole Buchanan and Darron Smith. Chacruna Institute for Psychedelic Plant Medicines. https://www.crowdcast.io/e/debunking-racist-myths/1. Accessed 17 Feb 2021
50. Jahn ZW, Lopez J, de la Salle S, Williams MT (2021) Racial/ethnic differences in prevalence for hallucinogen use by age cohort: findings from the 2018 National Survey on Drug Use and Health. J Psychedelic Stud 5 (2):69–82. https://doi.org/10.1556/2054.2021.00166
51. George JR, Michaels TI, Sevelius J, Williams MT (2020) The psychedelic renaissance and the limitations of a White-dominant medical framework: a call for indigenous and ethnic minority inclusion. J Psychedelic Stud 4(1):4–15. https://doi.org/10.1556/2054.2019.015
52. Rezrazi A, Gangloff B (2020) Relations among gender, status and various dimensions of the belief in a just world and justice sensitivity. Cogn Brain Behav XXIV (2 (June)):93–121. https://doi.org/10.24193/cbb.2020.24.06
53. National Inquiry into Missing and Murdered Indigenous Women and Girls (2019) Reclaiming power and place: the final report of the national inquiry into missing and murdered Indigenous women and girls. National Family Advisory Circle. https://www.mmiwg-ffada.ca/final-report/

Harassment as Scientific Misconduct 17

Erika Marin-Spiotta, Linda Gundersen, Rebecca Barnes, Meredith Hastings, Blair Schneider, and Janet Stemwedel

Abstract

Sexual and other types of identity-based harassment, discrimination, and exclusionary behaviors have been identified as barriers to diversifying science, technology, engineering, mathematics, and medicine (STEMM) fields. By creating hostile environments, disrupting career progression, and sowing distrust, these behaviors affect the integrity of scientific and medical practice. Improving workplace climate requires effective policies and codes of conduct as well as cultural change, and one way to do that is treat harassment and other hostile behaviors with at least the same level of concern as data fabrication, falsification, and plagiarism. Moving away from a purely legal framework to an ethical framework would also be effective toward addressing bullying and other abusive behaviors that have harmful impacts yet are not considered unlawful. We present arguments for expanding current definitions of responsible conduct of research and highlight several case studies of program interventions from professional associations, academic institutions, and funding agencies.

Author is retired from USGS at the time of publication.

E. Marin-Spiotta (✉)
University of Wisconsin-Madison, Madison, WI, USA
e-mail: marinspiotta@wisc.edu

L. Gundersen
U.S. Geological Survey, Ocean View, DE, USA

R. Barnes
Colorado College, Colorado Springs, CO, USA
e-mail: rbarnes@coloradocollege.edu

M. Hastings
Brown University, Providence, RI, USA
e-mail: Meredith_Hastings@brown.edu

B. Schneider
Kansas Geological Survey, Lawrence, KS, USA
e-mail: blair.schneider@ku.edu

J. Stemwedel
San Jose State University, San Jose, CA, USA
e-mail: janet.stemwedel@sjsu.edu

17.1 Introduction

Sexual and other types of harassment, discrimination, and exclusionary behaviors like bullying create hostile workplace environments. A 2018 US National Academies of Sciences, Engineering, and Medicine (NASEM) report concluded that sexual harassment, which affects 50–60% of women at all levels in academia, is a barrier to the retention of women in science, technology, engineering, mathematics, and medicine (STEMM) [1]. In 2019, only 16% of women in the USA who held bachelor's degrees in science and engineering were employed in science and engineering occupations, compared to 35% of men. More women than men with science and engineering degrees were employed in related occupations, including healthcare (37% vs 25%) and in

J. Faintuch, S. Faintuch (eds.), *Integrity of Scientific Research*,
https://doi.org/10.1007/978-3-030-99680-2_17

non-science and engineering jobs (47% vs 39%) [2]. Women overall made up 38.5% of the academic STEMM doctoral degree workforce in 2019; people of color of all genders made up 8.9% [2].

Low diversity in STEMM is maintained by low retention rates. More women (56%) than men (44%) left their neuroscience PhD programs before completion, for example, and students from underserved and minoritized racial and ethnic groups left at the same rate than they entered [3]. These patterns are not unique to one field. Persistent low representation and discrimination contribute to experiences of isolation and increased identity-based harassment [4–6]. Scientists who experienced negative workplace climate reported skipping professional activities, including conferences and field research [7], and considering leaving their disciplines [8].

Sexual harassment is prevalent even in the biological sciences which have the greatest proportion of women. In 2017 more women enrolled in US medical schools than men [9]. Despite this, 66% of women medical faculty reported being left out of professional advancement opportunities, and 30% reported being sexually harassed by a peer or superior [10]. One reason for this is that women become less visible in more senior positions. Women make up only 14.7% of full professors and 9.25% of department chairs in academic medicine [11].

17.1.1 Harassment vs Misconduct

Despite the pervasiveness and detrimental effects of sexual and other types of harassment, discrimination, and bullying on the STEMM workforce, these behaviors are excluded from definitions of research ethics and responsible conduct of research. Current definitions of research misconduct by US federal agencies are constrained to data fabrication, falsification, and plagiarism (hereafter FFP). Until recently, most research ethics trainings did not consider harassment, bullying, or discrimination.

High-profile cases and increasing public awareness of sexual harassment in higher education and other STEMM sectors have drawn attention to the need for more effective policies. Marin-Spiotta et al. [12] argued that when the academic community violates its social contract to provide learning and working environments that are free from harassment, it risks further eroding public trust in higher education and in the scientific enterprise, which is especially concerning in a time of intense political debate on public funding of research [12]. In recognition of the damaging effects of sexual harassment on scientific integrity and workforce retention, the 2018 NASEM report concluded that sexual harassment be considered equally important as scientific misconduct [1]. Tenbrunsel et al. [13] proposed reframing sexual harassment from a legal issue to an ethical issue as a strategy for its prevention [13]. Marin-Spiotta [14] called for STEMM ethics training that reveals how power dynamics and organizational structures allow harassment, bullying, and other exclusionary behaviors to persist. Page et al. [15] argued that framing sexual harassment as misconduct places responsibility on academic institutions [15]. Mello and Jagsi [16] advocated for gender bias and harassment in the medical profession to be handled as matters of professional ethics and for individuals to practice their ethical obligation to intervene against incidences of inequities and harassment [16].

17.1.2 The Scientific Integrity Environment

The definition of research or scientific misconduct has changed in the last three decades, and there is precedence for the consideration of the harmful effects of sexual harassment on scientific integrity. We provide recommendations for moving forward to reduce the harmful impact of harassment, bullying, and discrimination on the STEMM enterprise and highlight the role of professional associations, academic institutions, and funding agencies. Whereas this chapter relies on data from the USA, where the authors are employed, the problems addressed in this chapter are not restricted to the USA (see [17]), and the recommendations can be applied to other countries and regions. Similarly, the focus of the chapter is STEMM, yet most recommendations are applicable to other disciplines.

17.2 Sexual Harassment in the Workplace and Academia

17.2.1 Definitions and Factors Contributing to Sexual Harassment

Psychologists have identified three categories of behavior under sexual harassment: sexual coercion, unwanted sexual attention, and gender harassment [1, 18]. Sexual coercion occurs when sexual activity, bribes, or extortion are used in return for benefits related to education or employment. Unwanted sexual attention can include repeated, nonreciprocal requests for a date or sex, unwanted physical advances or touching, or sharing of unsolicited material of a sexual nature. Gender harassment is the most common type of sexual harassment in the workplace yet the least recognized and represents behaviors that denigrate, objectify, exclude, or display hostility toward someone because of their gender. Indeed, research in academia confirms findings from other workplaces that gender harassment is most common and rarely exists in isolation of other forms of harassment [1, 19].

In the USA, sexual harassment becomes illegal when enduring it becomes a condition for continued employment and/or when it is severe or pervasive enough to interfere with one's work performance. We note that behaviors that are not recognized under legal frameworks in the USA and elsewhere still have harmful impacts on individual health and wellness, on professional satisfaction and outcomes, and on workplace climate, with consequences at all levels of an organization.

17.2.1.1 Indifference to Abuse

Factors that contribute to the prevalence of sexual and gender harassment include organizational climate, perceived tolerance of hostile behaviors, gendered perceptions of the profession or discipline, and workplace representation [20, 21]. In US universities, and especially in STEMM, severe power imbalances created by current systems of advising, funding, and career promotion, along with current legacies of historical exclusion, and a culture of impunity facilitate the persistence of harassment [22]. Large power differentials exist within relationships among individuals of different rank among staff, faculty, students, trainees, medical residents, and postdoctoral researchers. In the apprenticeship model of training, early-career individuals depend on a senior person for access to resources for completing professional degrees and for securing employment. Trainees are hence particularly vulnerable. Of the 21% of graduate students who reported being bullied in an international study, almost 50% identified their supervisor and more than 30% identified another academic staff as the perpetrators [23]. Fear of retaliation is one the most common reasons why people do not pursue formal complaints [1].

17.2.1.2 Fierce Challenges to Success

In research environments, the "publish-or-perish" drive and intense competition for resources and ideas, which foster noncollaborative environments where people are taught to prioritize data over people's well-being, contribute to the tolerance of abusive behaviors. In the medical profession and academia, hazing-style practices set up expectations that sustaining abuse creates toughness of character. These behaviors have significant negative effects on mental health and wellness, with professional consequences [24, 25].

17.2.2 Prevalence and Impacts of Harassment

Academia is reported to have the second highest incidence rate of sexual harassment (58%), behind the military (69%), compared to near 45% for government and private-sector workplaces in the USA [26]. Fifty to 60% of women faculty have been sexually harassed

[1]. A 2020 study revealed that 82.5% of women faculty and 65.1% of men faculty in one medical school had been sexually harassed by other faculty, staff, or students in the last 12 months and 64.4% of women and 44.1% of men faculty were sexually harassed by patients or patient families [27].

17.2.2.1 Junior Personnel

Trainees and those in precarious positions are especially vulnerable due to power dynamics. An international survey of >6000 PhD students revealed that 39% experienced gender discrimination, 33% racial discrimination or harassment, and 15% sexual harassment [25]. In one US university, 20% of undergraduate and graduate women students in science, >25% of women engineering students, and >40% of medical students reported being sexually harassed by faculty or staff [28]. Hsiao et al. [29] found that 52% of medical students, 31% of residents/fellows, and 25% of faculty respondents had experienced sexual harassment in the last 12 months. In the UK and Australia, 40–50% of students in higher education have experienced sexual harassment [30, 31].

Sexual harassment has documented negative effects on the career success of women, especially, but not exclusively. Most research on sexual harassment has focused on women, resulting in lesser recognition of its effects on men. Women of color, transgender and nonbinary individuals, as well as people who identify as disabled or as having a disability experience the greatest rates of harassment and exclusionary behaviors.

17.2.2.2 A Wide Range of Consequences

Impacts of harassment range across a spectrum of psychological, epidemiological, physical, and emotional well-being. Sexual and other types of harassment experiences decrease job satisfaction and increase the intention to leave careers [32–35]. Individuals who experienced sexual harassment at work exhibited decreased work productivity and withdrawal from work or work-related events [36]. Researchers have reported a significant decrease in mental health and emotional well-being including depression, anxiety, and even post-traumatic stress disorder [33, 37–39].

17.2.2.3 Professional Repercussions

Harassment affects other aspects of career advancement, including promotion and job retention. Women who self-report experiences of sexual harassment in the workplace were negatively stereotyped and were less likely to receive a recommendation for promotion [40]. Individuals who reported sexual harassment faced retaliation from coworkers and supervisors, including being fired [41]. Targets of sexual harassment had greater rates of career disruption (i.e., changing jobs) and more financial stress [42].

17.2.2.4 Disproportionate Impacts

Research guided by an intersectional framework reveals disproportionate impacts on groups already underserved in STEMM. The experiences and consequences of sexual harassment on career advancement differ for women of color as compared to white women [7, 32, 38, 43]. Sexual and racial harassment have interactive effects on psychological well-being and occupational outcomes [39, 44]. Transgender, gender nonconforming, and LGBTQ scientists who identify as women experienced greater hostility than cisgender men, and LGBTQ physicists of color faced greater levels of discrimination than their white peers [45, 46]. Harassment and discrimination have clear negative impacts on individual careers and on the STEMM enterprise.

17.3 Ethical Perspectives on Harassment, Bullying, and Discrimination

17.3.1 History of Research Misconduct Definition

The 2018 NASEM report on sexual harassment briefly summarizes the history of the current definition of research misconduct in the USA [1]. In the 1980s, the US National Science Foundation (NSF) and the Public Health Service tried to develop standard definitions to address growing recognition of misconduct (see [47]). These definitions included FFP and phrases that related to "other serious deviations" from accepted practices [48]. During this time, NSF used this definition to successfully prosecute numerous cases, including those involving sexual harassment [49]. A concern that the "other serious deviations" language might be used to restrict novel approaches to science led to consideration of revising the definition of scientific misconduct [50].

In their 1992 report on *Responsible Science*, the National Academy of Sciences (NAS) reviewed the various definitions of research misconduct and removed the "other serious deviations" phrase, arguing that the definition should only include fabrication, falsification, and plagiarism. The report stated that "...sexual harassment and financial mismanagement are illegal behaviors regardless of whether scientists are involved, but these actions are different from misconduct in science because they do not compromise, in a direct manner, the integrity of the research process" [51].

FFP were incorporated into the federal definition of research misconduct released by the US Office of Science and Technology Policy [52] and are the current accepted definition for federally funded research. The national misconduct policies of at least 22 other countries also include FFP [53]. The most recent (2017) NAS report *Fostering Integrity in Research* upheld this definition of research misconduct [54]. It includes sexual harassment with "other misconduct" that affects the integrity of research, but states that it is not unique to conducting research, and therefore should not be included in the definition. A 2019 working group report to the US National Institutes of Health (NIH) included harassment, bullying, and discrimination under a separate category of professional misconduct (https://acd.od.nih.gov/documents/presentations/12122019ChangingCulture_Report.pdf).

17.3.2 Sexual Harassment and Discrimination Damage Scientific Integrity

Sexual and other types of harassment create a hostile and unfair environment that negatively affects the research conducted by individuals and groups, damaging health, scientific careers, collaboration, performance, productivity, and thus scientific quality and integrity [1]. The 2018 NASEM report explicitly addressed the effects of sexual harassment on accountability, stewardship, and fairness, three of the six core values of research integrity.

17.3.2.1 Discrimination

Gender discrimination has been documented at all stages of the academic and research enterprise, from the initial employment process to tenure and promotion, resources allotted for startup, and honors and recognition. Gender bias in peer review and publications [55–57] has detrimental effects on the research process. Gender and racial disparities in awarding of funding can limit the advancement of science [58–62].

17.3.2.2 Broader Definitions

Buzzelli [49] detailed a case that included sexual harassment, sexual coercion, and rape of women students by a professor, who further withheld research data, advising, and resources [49]. The NSF helped investigate and resolve this case. The author defended the broader definition of research misconduct so that NSF could make sure that behaviors that affected research practice, such as the appropriate training and mentoring of students, protection of intellectual property, and prevention of hostile research environments, were covered [49]. According to Buzzelli [49], the phrase "other serious deviation from accepted practices" in the definition of RCR refers to "professional ethics standards" and would not preempt novel research approaches or applications [49].

A limitation of approaches like those taken by the original OSTP (2000) definition, the NAS 1992 and 2017 reports, and arguments such as those by Resnik in 2019 [63] is that they are not rooted in ethical arguments but in what affects the integrity of a very narrow definition of the scientific process. These efforts are focused on an easily administered, easily understood definition, rather than the complex reality of the behaviors that affect scientific integrity and the ethical norms needed to protect that integrity.

17.3.2.3 Recent Initiatives

Directives on establishing scientific integrity policies in the federal government released by the Obama administration embraced a broad definition that included ensuring a "culture of scientific integrity" [64]. Many federal agencies responded by including codes of ethics that addressed professional behavior, honesty, respect, coercion, and intimidation [65, 66] (csldf.org/resources/scientific-integrity/). Some agencies, however, explicitly direct allegations of certain kinds of ethical and criminal breaches such as sexual harassment and issues of "fraud, waste, and abuse" in the conduct of science to other existing policies and procedures.

The 2018 NASEM report stated that: "A powerful incentive for change may be missed if sexual harassment is not considered equally important as research misconduct, in terms of its effect on the integrity of research" [1]. In a follow-up article, Benya [67], the study director and co-editor for the report, reiterated the need for science organizations to consider sexual harassment under research misconduct [67]. The NSF (2018), NIH (2020), and National Aeronautics and Space Administration (2020) created new rules related to the reporting of sexual harassment and grantees, but as Benya [67] noted, consistent policy, not a "patchwork of rules," is needed.

In 2021, US President Biden directed the OSTP and federal agencies to review current scientific integrity policies and consider whether they "support scientists and researchers of all genders, races, ethnicities, and backgrounds; and advance the equitable delivery of the Federal Government's programs" (https://www.whitehouse.gov/briefing-room/presidential-actions/2021/01/27/memorandum-on-restoring-trust-in-government-through-scientific-integrity-and-evidence-based-policymaking/).

17.3.3 Redefining Research Misconduct to Include Harassment and Discrimination

Many members of the scientific and ethics communities have expressed support for including sexual harassment and other discriminatory behaviors in the definition of research misconduct [14, 67–70]. Steneck [71] argued that the continued use of a narrow definition for FFP misses serious questionable practices which may constitute the majority of misconduct in science [71]. Sexual and other types of harassment can be considered as an ethical issue from the perspective of the perpetrator, who commits an act of harm, and from the perspective of a workplace or organization's response to the harmful action [72]. Tenbrunsel et al. [13] identified sexual harassment as a violation of care, fairness, loyalty, authority, and purity, based on Graham et al.'s [73] moral foundations theory [13, 73]. In the medical profession, Mello and Jagsi [16] asserted that sexual harassment and gender bias violate a bedrock ethical principle of respect for others, including patients and other healthcare workers [16].

17.3.3.1 An Ethical Framework

Researchers have argued that framing sexual harassment as an ethical violation is more likely to increase its recognition as a costly unethical activity with harmful impacts to individuals and organizations and to motivate corrective action [13, 72]. O'Leary-Kelly and Bowes-Sperry [72] applied Jones's [74] dimensions of moral intensity [74] to the context of sexual harassment. They identified awareness of the magnitude of consequences, both for the target of the behavior, and the perpetrator, and degree of social consensus as important factors in influencing the likelihood of ethical decision-making. Perceived tolerance through weakness of policies and lack of enforcement, as well as perceived complicity of leadership, are known to contribute to the frequency of sexual harassment in an organization and often magnify the harm of the initial violation.

An ethical framework would also mainstream feminist ethics of care approaches to current practices in STEMM (see [75]). An ethics of care approach to sexual harassment prevention training, rather than the predominant legal compliance approach, is expected to be more effective at changing perceptions of the harm of sexual harassment [76]. Framing sexual harassment as an ethical issue may also allow for different resolution approaches, which may encourage more reporting than the current legal framework under Title IX policies in the USA [13].

17.3.3.2 Long-Term Abusive Profile

Another important reason for considering sexual harassment under research misconduct is that rather than a one-time lapse of judgment, research on perpetrators of sexual harassment shows persistent engagement in unethical behavior. Of 300 reported cases of sexual misconduct by faculty, 50% involved repeat offenders [77]. In the biomedical and health sciences, an analysis of 125 faculty accused of sexual misconduct found that 59% of them had been involved in harassing behavior of more than 2 years' duration and 61% had targeted more than one individual, with 18% more than five individuals [78]. These numbers are likely an underestimate, given paucity in reporting and confidentiality rules that limit public access to information.

Sekreta [79] asserted that by "excluding sexual harassment from the definition of "science misconduct" the federal government has reinforced the notion that sexual harassment affects neither the integrity of scientific research nor accepted scientific norms" [79]. Greater recognition that harassment negatively affects the ability of persons of all genders, including men, to perform their professional obligations may garner wider support for a broader definition of research misconduct.

17.4 Preventive and Corrective Policies and Strategies

We recommend that government research and funding agencies consider a common code of ethical conduct and a consistent, broader definition of research misconduct including but not limited to the following: identity-based discrimination; harassment; bullying; coercion; intimidation; unfair peer review; withholding data and methodology; neglectful, abusive, or exploitive supervision in research; irresponsible publication practices of editors, reviewers, and authors; inadequate institutional policies for protecting researchers and research integrity; gross negligence by persons in their professional activities; and vandalism, including tampering with research experiments or instrumentation [54].

17.4.1 A Scientific Integrity Net

This implementation will require accessible policies with clear and detailed definitions and procedures; confidential due processes that protect both complainant and respondent; and fair and thorough evaluation of evidence by a panel of experts, an appeals process, and appropriate, fair, and consistent sanctions and penalties. These will need to be coupled with best practices for education, adoption of the NASEM recommendations for changing academic and research environments, strengthening leadership and mentoring standards, and collaboration across the various university and agency offices that handle sexual harassment, research misconduct, discrimination, ethics, and related functions, to create transparent and robust processes for handling violations.

Following their 2018 report, the NASEM created an ongoing Action Collaborative that brings together leaders and experts from academia and the broader STEMM community to identify evidence-based strategies and actions to prevent sexual and gender harassment (chapter coauthor Barnes is a member). In their first year of work

(2019–2020), the Action Collaborative produced a rubric for organizations to assess their current prevention, response, remediation, and evaluation programs and policies around sexual harassment, including the explicit recommendation to move beyond current legal frameworks to address workplace climate (https://www.nationalacademies.org/our-work/action-collaborative-on-preventing-sexual-harassment-in-higher-education#sl-three-columns-ae23481a-cc59–4050-946a-1f17dbbb54cc). The recommendations explicitly state that sexual harassment should be treated as a violation of research integrity.

17.4.2 Role of Professional Associations

Professional associations can take a leadership role in setting cultural expectations in a discipline [12, 14]. Associations have an important responsibility to address harassment and other exclusionary behaviors in the first place because these occur at conferences [80, 81]. Another reason is because societies administer awards and honors in the discipline and can either incentivize abusive behaviors by failing to include them in the evaluation process or actively discourage and penalize the full spectrum of misconduct by wielding effective codes of conduct. Mello and Jagsi [16] identified areas where medical associations have clear expectations for intervention to disrupt risky behaviors and recommend these are expanded to include sexual harassment and discrimination [16].

Whereas scientific societies have not always paid attention to research ethics [82], most professional associations in STEMM and other disciplines today have ethical codes of conduct to guide behavior at society events and publications. Van Valey and Hillsman [83] provided a historical perspective of professional ethics in the practice of social science research and recognized that association codes of conduct provide guidelines for responsible professional self-regulation within a discipline and can serve as "guideposts" for academic institutions [83]. Frankel [84] outlined a role for associations to handle issues of misconduct within the scientific community, rather than by government [84]. While not advocating for a broader federal definition at the time, Resnik [63] recognized that associations have a strong influence on professional ethics and career development and thus should lead the way in redefining research misconduct [63].

17.4.2.1 Case Study from the Earth and Space Sciences

The American Geophysical Union (AGU) is an international professional society with >60,000 members from >130 countries. AGU's 2020 strategic plan includes excellence, integrity, respect, diversity, collaboration, and scientific education and outreach as its principal values. In 2012, the AGU created a new scientific integrity policy and code of conduct [85] that broadened the definition of misconduct and focused on professional behavior, adapting the Singapore Statement (wcrif.org/guidance/singapore-statement) as the core and the US Department of Interior Scientific Procedures Handbook process for dealing with allegations of misconduct (https://doi.gov/sites/doi.gov/files/elips/documents/305%20DM%203_%20Handbook%20-%20Scientific%20Integrity%20Procedures.pdf).

In response to a high-profile case of a serial sexual harasser in the space sciences community, AGU convened a call for action at their 2015 annual meeting with leadership from the AGU, American Astronomical Society, Association for Women Geoscientists (AWG), and Earth Science Women's Network (ESWN) (author note: multiple authors of this chapter serve or recently served on the leadership of these organizations) [12]. AGU co-hosted a NSF-funded workshop in Washington, D.C., with leaders from the scientific community, government, and professional associations and recommended that the definition of research misconduct be expanded to include sexual harassment [86].

Subsequently, AGU established a task force to review its ethics policy (author note: Marin-Spiotta and Gundersen served on this task force) and in 2017 adopted a new policy that includes harassment, discrimination, and bullying as misconduct. The policy includes harassment related to ethnic origin, race, religion, citizenship, language, political persuasion, sex, gender identity, sexual orientation, disability, age, or economic class [87, 88]. The code of conduct applies to AGU members, leaders, and staff, regardless of where the incident occurs, and also applies to anyone attending an AGU program.

AGU now requires award nominators to indicate if they are aware of nominees having violated the expanded code of ethical conduct. Before an award or leadership position is announced, nominees are required to self-disclose whether they have violated the code or been under investigation for violations of related policies at their own institutions. The new code of ethics includes a range of sanctions for violations and has been used to rescind a scientific award [89]. Many organizations in the geosciences and across disciplines now have adopted similar policies [90].

AGU has launched a suite of programs to address ethics, harassment, discrimination, bullying, diversity, equity, and inclusion, including releasing annual ethics reports of its activities; examining its own practices such as awards, publication, and peer review to reduce bias; and establishing the AGU Ethics and Equity Resource Center.

17.4.3 Changes in Academia and the Workplace

A lot has been written about the limitations of current legal frameworks and administrative processes in place to respond to incidences of sexual and other types of harassment and discriminatory behaviors in academia [1, 15, 91]. Broadening the definition of academic misconduct policies to include not only how we treat data but how we treat people [14] would go a long way toward shifting perceptions about the cost of allowing these behaviors to persist. Strategies to reduce the impact of harassment, bullying, and discrimination need to be incorporated into required responsible conduct of research training [14, 92], and that prevention and response training follow evidence-based practices [93].

17.4.3.1 Ethical Climate

Academia needs to face perverse incentives in its reward structures that contribute to the hiring and promotion of serial harassers and create environments for abuses of power. A new hiring policy at the University of Wisconsin-Madison adopted in 2018 asks whether applicants have been the subject of harassment investigations or complaints. These conversations need to be included into all levels of career advancement and promotion, including tenure. Clancy et al. [93] call for "brave leadership" for creative solutions that address the systemic conditions that allow for harassment and gendered and racialized disrespect to manifest in higher education [93].

17.4.3.2 Opposition to Change

Institutional policies to address sexual harassment in higher education have often been met with resistance by faculty, who claim infringement on academic freedom and faculty rights [13, 94]. As Eyre [95] discussed, this framing contributes to legitimizing oppressive practices and marginalizes feminist discourses in academia [95]. It also ignores the damage that tolerating behaviors that create hostile workplace climates, like sexual harassment, have not only on individual lives and careers, but on the integrity of academic pursuits. We acknowledge that policies about professional conduct have a troubling history of discriminating against, in particular, Black people and their responses to racism in education and in the workplace. Broadening the definition of professional conduct therefore needs to be done with care.

17.4.3.3 Further Initiatives in the Earth and Space Sciences

The ADVANCEGeo Partnership (serc.carleton.edu/advancegeo) is an example of a collaborative approach to address sexual harassment and hostile workplace climates led by geoscientists and social scientists from multiple US universities and geoscience societies (AGU, AWG, ESWN). Funded by the NSF in 2017, the program's main organizational change intervention is a community-based model for bystander intervention education to equip faculty and academic leaders with the tools to recognize and interrupt behaviors as they happen, as well as support those targeted. The work is informed by a feminist ethics of care and an intersectional framework, necessary to address persistent inequities in the earth and space sciences [22]. The program promotes ethical codes that identify harassment, bullying, and discrimination as research misconduct.

17.4.4 Role of Funding Agencies and Foundations

Funding agencies in the USA and UK recently have revised their policies on sexual harassment, recognizing its effects on scientific integrity [70, 96, 97]. In 2018, the NSF announced a new policy that requires awardee institutions to notify NSF of findings and investigations regarding violations by investigators of policies and codes of conducts around sexual harassment, other forms of harassment, or sexual assault (83 FR 47940, 2018). In 2019, the NIH apologized that "it has taken so long to acknowledge and address the climate and culture that has caused such harm" (nih.gov/about-nih/who-we-are/nih-director/statements/update-nihs-efforts-address-sexual-harassment-science). In 2019, the NAS adopted a new policy to revoke membership, once a lifetime honor, for violations of its code of conduct and in 2021 used it for the first time to expel a known serial sexual harasser [98].

17.4.4.1 Support for Those Victimized

These types of policies would be more effective if they did not rely on individual institutional policies and procedures for reporting and conducting investigations, and also include safeguards for protection from retaliation. Furthermore, funders and institutions would do well to increase independent funding for early-career researchers, students, and trainees to diffuse power dynamics and also create pathways for supporting those whose career progress has been slowed, stalled, or derailed by being exposed to abuse, including mechanisms for re-entry into the STEMM workforce.

17.5 Conclusions

An ethical framework to research misconduct would recognize that tolerating abusive behaviors that create hostile workplace climates has consequences not only on the intended targets but on the entire scientific and research community. There is also little recognition of the emotional labor and lost productivity involved in surviving, responding to, reporting and investigating harassment and supporting its targets, which disproportionately falls on groups that already experience identity-based discrimination. This displacement of labor and responsibility is also unethical.

We note that recognition of the harm of harassment on scientific integrity through its effects on targets, victims or survivors should never be construed to justify their removal from STEMM, instead the focus should be on perpetrators and the organizations that tolerate harm.

We also acknowledge that policies about professional conduct have a troubling history of racializing professionalization to discriminate against, in particular, Black people and their responses to discrimination and racism in education and in the workplace. Including harmful behaviors under ethical conduct needs to be done with care and in recognition of how policies have discriminated in the past.

Acknowledgments Support for this work was provided by the US National Science Foundation ADVANCE program (Human Resources Division Awards: 1725879, 1726021, 1725650, 1726163, 1725424).

References

1. National Academies of Sciences, Engineering, and Medicine (2018) Sexual harassment of women: climate, culture, and consequences in academic sciences, engineering, and medicine, Washington, DC, The National Academies Press. https://doi.org/10.17226/24994. Accessed 5 May 2021
2. National Science Foundation, National Center for Science and Engineering Statistics (2019) Women, minorities, and persons with disabilities in science and engineering: Alexandria, VA. Report No.: 19–304. nsf.gov/statistics/wmpd. Accessed 5 May 2021

3. Sved A Report of survey of neuroscience graduate, postdoctoral, & undergraduate programs (academic year 2010–2011). Society for Neuroscience; 2011. sfn.org/Careers/Higher-Education-and- Training/Neuroscience-Training-Program-Survey. Accessed 5 May 2021
4. Ong M, Wright C, Espinosa L, Orfield G (2011) Inside the double bind: a synthesis of empirical research on undergraduate and graduate women of color in science, technology, engineering, and mathematics. Harvard Ed Rev 81(2):172–209
5. Willis TY, Mattheis A, Dotson B, Brannon L-J, Hunter M, Moore A et al (2019) "I Find Myself Isolated and Alone": Black women's experiences of microaggressions at an Hispanic-Serving Institution. J Women Gend Higher Ed 12(2):186–204
6. Ramos D, Yi V (2020) Doctoral women of color coping with racism and sexism in the academy. Int J Doct Sci 15:135–158
7. KBH C, KMN L, Rodgers EM, Richey C (2017) Double jeopardy in astronomy and planetary science: Women of color face greater risks of gendered and racial harassment: gender, race, and harassment in astronomy. J Geophys Res Planets 122(7):1610–1623
8. Aycock LM, Hazari Z, Brewe E, Clancy KBH, Hodapp T, Goertzen RM (2019) Sexual harassment reported by undergraduate female physicists. Phys Rev Phys Educ Res 15(1):010121
9. Association of American Medical Colleges. More women than men enrolled in U.S. medical schools in 2017 (AAMC News). news.aamc.org/ press-releases/ article/applicant-enrollment-2017. Accessed 5 May 2021
10. Jagsi R, Griffith KA, Jones R, Perumalswami CR, Ubel P, Stewart A (2016) Sexual harassment and discrimination experiences of academic medical faculty. JAMA 315(19):2120–2121
11. Epstein NE (2017) Discrimination against female surgeons is still alive: where are the full professorships and chairs of departments? Surg Neurol Int 8:93
12. Marin-Spiotta E, Schneider B, Holmes MA (2016) Steps to building a no-tolerance culture for sexual harassment. Earth and Space Science News /EOS. 97
13. Tenbrunsel AE, Rees MR, Diekmann KA (2019) Sexual harassment in academia: ethical climates and bounded ethicality. Annu Rev Psychol 70(1):245–270
14. Marín-Spiotta E (2018) Harassment should count as scientific misconduct. Nature 557(7704):141–141
15. Page T, Bull A, Chapman E (2019) Making power visible: "slow activism" to address staff sexual misconduct in higher education. Violence Against Women 25(11):1309–1330
16. Mello MM, Jagsi R (2020) Standing up against gender bias and harassment — a matter of professional ethics. N Engl J Med 382(15):1385–1387
17. Bondestam F, Lundqvist M (2020) Sexual harassment in higher education – a systematic review. Eur J Higher Ed 10(4):397–419
18. Gelfand MJ, Fitzgerald LF, Drasgow F (1995) The structure of sexual harassment: a confirmatory analysis across cultures and settings. J Vocat Behav 47:164–177
19. Rosenthal MN, Smidt AM, Freyd JJ (2016) Still second class: sexual harassment of graduate students. Psychol Women Q 40:364–377
20. Fitzgerald LF, Drasgow F, Hulin CL, Gelfand MJ, Magley VJ (1997) Antecedents and consequences of sexual harassment in organizations: a test of an integrated model. J Appl Psychol 82(4):578–589
21. Topa Cantisano G, Morales Domínguez JF, Depolo M (2008) Perceived sexual harassment at work: meta-analysis and structural model of antecedents and consequences. Span J Psychol 11(1):207–218
22. Marín-Spiotta E, Barnes RT, Berhe AA, Hastings MG, Mattheis A, Schneider B et al (2020) Hostile climates are barriers to diversifying the geosciences. Adv Geosci 53:117–127
23. Lauchlan E (2019) Nature survey 2019 report, London: Shift Learning. figshare.com/s/74a5ea79d76ad66a8af8?file=18543281. Accessed 5 May 2021
24. Evans TM, Bira L, Gastelum JB, Weiss LT, Vanderford NL (2018) Evidence for a mental health crisis in graduate education. Nat Biotechnol 36(3): 282–284
25. Woolston C (2019) PhDs: the tortuous truth. Nature 575:403–406
26. Ilies R, Hauserman N, Schwochau S, Stibal J (2003) Reported incidence rates of work-related sexual harassment in the united states: using meta-analysis to explain reported rate disparities. Pers Psychol 56(3):607–631
27. Vargas EA, Brassel ST, Cortina LM, Settles IH, Johnson TRB, Jagsi R (2020) #MedToo: a large-scale examination of the incidence and impact of sexual harassment of physicians and other faculty at an Academic Medical Center. J Women's Health 29(1):13–20
28. Swartout K (2018) University of Texas Climate Survey. report on the University of Texas System Campus Climate Survey. In: Sexual harassment of women: climate, culture, and consequences in academic sciences, engineering, and medicine [Internet]. Washington, DC, USA: The National Academies Press. https://doi.org/10.17226/24994. Accessed 5 May 2021
29. Hsiao CJ, Akhavan NN, Singh Ospina N, Yagnik KJ, Neilan P, Hahn P et al (2021) Sexual harassment experiences across the academic medicine hierarchy. Cureus 13(2):e13508
30. Australian Human Rights Commission. Change the course: national report on sexual assault and sexual harassment at Australian Universities 2017, p 254. humanrights.gov.au/our-work/sex-discrimination/publications/change-course-national-report-sexual-assault-and-sexual. Accessed 5 May 2021
31. National Union of Students and The 1752 Group (2018) Power in the academy: staff sexual misconduct

in UK higher education. London, p 55. 1752group.com/power-in-the-academy-report/. Accessed 5 May 2021
32. Cortina LM, Fitzgerald LF, Drasgow F (2002) Contextualizing Latina experiences of sexual harassment: preliminary tests of a structural model. Basic Appl Soc Psychol 24(4):295–311
33. Willness CR, Steel P, Lee K (2007) A meta-analysis of the antecedents and consequences of workplace sexual harassment. Pers Psychol 60(1):127–162
34. Pina A, Gannon TA (2012) An overview of the literature on antecedents, perceptions and behavioural consequences of sexual harassment. J Sex Aggress 18(2):209–232
35. Sojo VE, Wood RE, Genat AE (2016) Harmful workplace experiences and women's occupational well-being: a meta-analysis. Psychol Women Q 40(1): 10–40
36. Schneider KT, Swan S, Fitzgerald LF (1997) Job-related and psychological effects of sexual harassment in the workplace: empirical evidence from two organizations. J Appl Psychol 82:401–451
37. Avina C, O'Donohue W (2002) Sexual harassment and PTSD: is sexual harassment diagnosable trauma? J Trauma Stress 15(1):69–75
38. Woods KC, Buchanan NT, Settles IH (2009) Sexual harassment across the color line: experiences and outcomes of cross- versus intraracial sexual harassment among Black women. Cultur Divers Ethnic Minor Psychol 15(1):67–76
39. Buchanan NT, Settles IH, Wu IHC, Hayashino DS (2018) Sexual harassment, racial harassment, and well-being among asian american women: an intersectional approach. Women Ther 41(3–4):261–280
40. Hart CG (2019) The penalties for self-reporting sexual harassment. Gend Soc 33:534–559
41. Hesson-Mcinnis MS, Fitzgerald LF (1997) Sexual harassment: a preliminary test of an integrative modell. J Appl Soc Psychol 27(10):877–901
42. McLaughlin H, Uggen C, Blackstone A (2012) Sexual harassment, workplace authority, and the paradox of power. Am Sociol Rev 77(4):625–647
43. Wasti SA, Cortina LM (2002) Coping in context: sociocultural determinants of responses to sexual harassment. J Pers Soc Psychol 83(2):394–405
44. Buchanan NT, Bergman ME, Bruce TA, Woods KC, Lichty LL (2009) Unique and joint effects of sexual and racial harassment on college students' well-being. Basic Appl Soc Psychol 31(3):267–285
45. Atherton TJ, Barthelemey RS, Deconinck W, Falk ML, Garmon S, Long E, et al. (2016) LGBT climate in physics: building an inclusive community. College Park, MD: American Physical Society. aps.org/programs/lgbt/upload/LGBTClimateinPhysicsReport.pdf Accessed 21 May 2021
46. Richey CR, Lee KMN, Rodgers E, Clancy KBH (2020) Gender and sexual minorities in astronomy and planetary science face increased risks of harassment and assault. Bull AAS 51(4)
47. Gross C (2016) Scientific misconduct. Annu Rev Psychol 67(1):693–711
48. Price AR (1994) Definitions and boundaries of research misconduct: perspectives from a federal government viewpoint. J Higher Ed 65(3):286
49. Buzzelli D (1993) The definition of misconduct in science: a view from NSF. Science 259(5095):584
50. Schachman H (1993) What is misconduct in science? Science 261(5118):148–149
51. National Academy of Sciences (1992) Responsible science: ensuring the integrity of the research process: volume I. National Academies Press (US), Washington (DC). nap.edu/catalog/1864/responsible-science-ensuring-the-integrity-of-the-research-process-volume. Accessed 21 May 2021
52. Office of Science and Technology Policy. Federal Research Misconduct Policy. Executive Office of the President 2000 Federal Register 65(235): 76260–76264. ori.hhs.gov/federal-research-misconduct-policy. Accessed 5 May 2021
53. Resnik DB, Rasmussen LM, Kissling GE (2015) An international study of research misconduct policies. Account Res 22(5):249–266
54. National Academies of Sciences, Engineering, and Medicine (2017) Fostering integrity in research. National Academies Press, Washington, D.C.. [cited 2021 Apr 25]. nap.edu/catalog/21896
55. Kaatz A, Gutierrez B, Carnes M (2014) Threats to objectivity in peer review: the case of gender. Trends Pharmacol Sci 35(8):371–373
56. Helmer M, Schottdorf M, Neef A, Battaglia D (2017) Research: gender bias in scholarly peer review. elife 6: e21718
57. Hagan AK, Topçuoğlu BD, Gregory ME, Barton HA, Schloss PD (2020) Women are underrepresented and receive differential outcomes at ASM journals: a six-year retrospective analysis. MBio 11(6):e01680–e01620
58. Ginther DK, Schaffer WT, Schnell J, Masimore B, Liu F, Haak LL et al (2011) Race, ethnicity, and NIH research awards. Science 333(6045):1015–1019
59. Shen H (2013) Despite improvements, female scientists continue to face discrimination, unequal pay and funding disparities. Nature 495:22–24
60. Head MG, Fitchett JR, Cooke MK, Wurie FB, Atun R (2013) Differences in research funding for women scientists: a systematic comparison of UK investments in global infectious disease research during 1997–2010. BMJ Open 3(12):e003362
61. Witteman HO, Hendricks M, Straus S, Tannenbaum C (2019) Are gender gaps due to evaluations of the applicant or the science? A natural experiment at a national funding agency. Lancet 393(10171):531–540
62. Taffe MA, Gilpin NW (2021) Racial inequity in grant funding from the US National Institutes of Health. elife 10:e65697
63. Resnik DB (2019) Is it time to revise the definition of research misconduct? Accountab Res 26(2):123–137

64. Holdren JP. Scientific integrity: fueling innovation, building public trust The White House Blog. 2010. obamawhitehouse.archives.gov/blog/2010/12/17/scientific-integrity-fueling-innovation-building-public-trust-ostp. Accessed 5 May 2021
65. Government Accountability Office (2019, April) Scientific Integrity Policies: Additional Actions Could Strengthen Integrity of Federal Research. United States Government Accountability Office. Report No.: GAO-19-265. gao.gov/assets/gao-19-265.pdf. Accessed 5 May 2021
66. Gallo ME (2021, April) Federal scientific integrity policies: a primer, p 17. (Congressional Research Service). Report No.: R46614. crsreports.congress.gov/product/pdf/R/R46614/5. Accessed 5 May 2021
67. Benya F (2019) Treating sexual harassment as a violation of research integrity. Issues Sci Technol 35(2): 56–59
68. Stemwedel JD (2016) What can be done to address harassment in science? Forbes. [cited 2021 May 3]. forbes.com/sites/janetstemwedel/2016/01/29/what-can-be-done-to-address-harassment-in-science/?sh=4867f0675294. Accessed 5 May 2021
69. Anonymous (2018) Science suffers from harassment. Sci Am 318(3):8
70. Atkin E (2018) Why scientists accused of sexual misconduct can still get government grants. The New Republic. newrepublic.com/article/146733/scientists-accused-sexual-misconduct-can-still-get-government-grants. Accessed 5 May 2021
71. Steneck NH (2006) Fostering integrity in research: definitions, current knowledge, and future directions. Sci Eng Ethics 12(1):53–74
72. O'Leary-Kelly AM, Bowes-Sperry L (2001) Sexual harassment as unethical behavior: the role of moral intensity. Hum Resour Manag Rev 11(1–2):73–92
73. Graham J, Nosek BA, Haidt J, Iyer R, Koleva S, Ditto PH (2011) Mapping the moral domain. J Pers Soc Psychol 101(2):366–385
74. Jones TM (1991) Ethical decision making by individuals in organizations: an issue-contingent model. Acad Manag Rev 16:366–395
75. Edwards R, Mauthner M (2012) Ethics and feminist research: theory and practice. In: Ethics in qualitative research, 2nd edn. SAGE Publications Ltd, London, pp 14–28
76. White TI (1998) Sexual harassment: trust and the ethic of care. Bus Soc Rev 100–101(1):9–20
77. Cantalupo NC, Kidder WC (2018) A systematic look at a serial problem: sexual harassment of students by university faculty. Utah Law Rev 3:117
78. Espinoza M, Hsiehchen D (2020) Characteristics of faculty accused of academic sexual misconduct in the biomedical and health sciences. JAMA 323(15): 1503–1505
79. Sekreta E (2006) Sexual harassment, misconduct, and the atmosphere of the laboratory: the legal and professional challenges faced by women physical science researchers at educational institutions. Duke J Gender Law Policy Spring:115–138
80. Settles IH, O'Connor RC (2014) Incivility at academic conferences: gender differences and the mediating role of climate. Sex Roles 71(1–2):71–82
81. Biggs J, Hawley PH, Biernat M (2018) The academic conference as a chilly climate for women: effects of gender representation on experiences of sexism, coping responses, and career intentions. Sex Roles 78(5–6):394–408
82. Iverson M, Frankel MS, Siang S (2003) Scientific societies and research integrity: what are they doing and how well are they doing it? Sci Eng Ethics 9(2): 141–158
83. Van Valey TL, Hillsman ST. The code of ethics of the american sociological association. asanet.org/sites/default/files/savvy/images/asa/docs/pdf/CodeofEthics.pdf. Accessed 5 May 2021
84. Frankel MS (2000) Scientific societies as sentinels of responsible research conduct. Proc Soc Exp Biol Med 224(4):216–219
85. Gundersen LC, Townsend R. Formulating the American Geophysical Union's scientific integrity and professional ethics policy: challenges and lessons learned. In: Geoethics: Ethical challenges and case studies in Earth Sciences , Elsevier, USA, 2017. p. 83–93. pubs.er.usgs.gov/publication/70193730. Accessed 5 May 2021
86. Wendel J (2016) AGU-sponsored workshop targets sexual harassment in the sciences. Eos 20:97
87. McPhaden M (2017) American Geophysical Union adopts and implements a new scientific integrity and professional ethics policy. In: Gundersen LC (ed) Special Publications. John Wiley & Sons, Inc., Hoboken, NJ
88. Williams BM, McEntee C, Hanson B, Townsend R (2017) The role for a large scientific society in addressing harassment and work climate issues. Ann Geophys 60:11
89. Reardon S (2018) Geoscience society rescinds award to top seismologist after ethics investigation. Nature 16
90. Schneider B, Holmes MA, Marin-Spiott E (2018) Sexual harassment in the sciences: response by professional societies. American Geosciences Institute. Report No.: 125. americangeosciences.org/geoscience-currents/sexual-harassment-sciences-response-professional-societies. Accessed 5 May 2021
91. Cantalupo NC, Kidder WC (2019) Systematic prevention of a serial problem: sexual harassment and bridging core concepts of bakke in the #MeToo Era. UC Davis Law Rev 52(57):2349–2405
92. Gundersen LC (ed) (2017) Appendix B: resources and references for scientific integrity, ethics and geoethics. In: Special Publications. Hoboken, NJ: John Wiley & Sons, Inc
93. Clancy KBH, Cortina LM, Kirkland AR (2020) Opinion: use science to stop sexual harassment in higher education. Proc Natl Acad Sci U S A 117(37): 22614–22618

94. Candido AH (1997) A right to talk dirty?: academic freedom values and sexual harassment in the university classroom. Univ Chicago Law School Roundtable 4(1):45
95. Eyre L (2000) The discursive framing of sexual harassment in a university community. Gender Ed:293–307
96. Witze A (2016) How should science funders deal with sexual harassers? Nature 530:138–139
97. Else H (2018) Report harassment or risk losing funding, says top UK science funder. Nature 557:149
98. Ghorayshi A, Vergano D 2021, May 27 In an unprecedented move, astronomer Geoff Marcy has been kicked out of the prestigious National Academy of Sciences. Buzzfeed News buzzfeednews.com/article/azeenghorayshi/geoff-marcy-national-academy-sciences-expelled. Accessed 5 May 2021

Sexual Harassment in Orthopedic Training: Personal Perspectives and Outlooks for the Future

18

Rivka Ihejirika-Lomedico and Joseph Zuckerman

Abstract

Sexual harassment, discrimination, and misconduct have become a topic of great attention and importance in orthopedic training. In this chapter we will focus on the incidence and scope of harassment in orthopedics, its inception, cultural norms within the field, and perpetrators of the problem. Furthermore we will discuss solutions and potential barriers to implementing these solutions. Personal perspectives from orthopedic attendings and residents are woven into objective data to demonstrate the reality and complexity of the dilemma.

Abbreviations

AAOS American Academy of Orthopaedic Surgeons
AOA Alpha Omega Alpha
PC Politically correct

R. Ihejirika-Lomedico · J. Zuckerman (✉)
NYU Langone Orthopedic Hospital, New York, NY, USA
e-mail: Joseph.Zuckerman@nyulangone.org

18.1 Background

Sexual harassment is pervasive internationally. The World Health Organization reports that over 30% of women have experienced sexual violence and over 75% of women have experienced sexual harassment globally [1, 2]. While sexual violence predominantly occurs in a domestic setting, sexual harassment is frequently encountered in the workplace. Sexual harassment in the workplace has recently been brought to the forefront of media attention with numerous high-profile figures being exposed for sexual misconduct and abuse. Tarana Burke, a social activist, coined the phrase "me too" on her Myspace page in 2006 to bring activism to sexual abuse and harassment. In 2017, that phrase and the movement went viral as the #MeToo gained star power from celebrities such as Alyssa Milano calling on women to post on social media if they had experienced such abuses [3]. The impact of the #MeToo movement was to bring a wave of reckoning through all industries, particularly professional ones, as people were forced to grapple with the open secret of sexual harassment in their workplaces and instituting consequences for "bad" behavior [4, 5].

Medicine has long been a silo where information, good or bad, is not shared with others. Problems within a department can fester and create a hostile and malignant work environment. Intimidation, harassment, and discrimination, while often rampant, can frequently be ignored

J. Faintuch, S. Faintuch (eds.), *Integrity of Scientific Research*,
https://doi.org/10.1007/978-3-030-99680-2_18

even when reported. The #MeToo movement has created a trickle-down effect causing those in academic medicine to question how pervasive and impactful sexual harassment truly is [6, 7].

While sexual harassment has garnered the most notoriety on a wide scale, some argue that other forms of abuse such as bullying and discrimination are worse as they create a culture of intolerance and stifle adequate intellectual growth in medicine. Furthermore, gender harassment potentiates sexual harassment in academic medicine. An environment where women are treated as if they don't belong can make them feel as though the injustices they suffer are an acceptable part of joining a club that is not meant for them [8, 9].

The Royal Australasian College of Surgeons defines inappropriate conduct in the workplace succinctly [10]:

> Discrimination: treating a person with an identified attribute or personal characteristic less favorably than a person who does not have the attribute or personal characteristic (e.g. gender, age, religious belief, political belief, pregnancy, breastfeeding, disability, impairment, marital status, family responsibilities, sexual orientation, race and cultural background).

> Bullying: A behavior or pattern of behaviors that a reasonable person would expect might victimize, humiliate, undermine, or threaten a person to whom the behavior is directed.

> Sexual Harassment: unwelcome sexual advances, request for sexual favors and other unwelcome conduct of a sexual nature, by which a reasonable person would be offended, humiliated or intimidated. Sexual harassment may include, but is not limited to sexual innuendo; sexually explicit or offensive jokes; graphic verbal commentaries about an individual's body; sexually degrading words used to describe an individual; pressure for sexual activity; persistent requests for dates; intrusive remarks, questions or insinuations about a person's sexual or private life; unwelcome sexual flirtations, advances or propositions; and unwelcome touching of an individual, molestation or physical violence such as rape.

> Harassment: unwanted, unwelcome or uninvited behavior that makes a person feel humiliated, intimidated or offended. Harassment can include racial hatred and vilification, be related to disability or the victimization of a person who has made a complaint.

Gender discrimination or harassment occurs when individuals are treated less favorably than others or with unwanted behavior based on their gender. Within orthopedics, women often describe this as not being given equal opportunity to participate in surgical or academic opportunities as their male colleagues or being told they aren't strong enough to complete a task regardless of their demonstrated ability.

Siller et al. distill these forms of harassment and abuse into the term "psychological violence" or deliberate behavior that compromises a person's mental capacity and fortitude. In their interview-based study of women in academic medicine in Australia, they found that the hierarchy of medicine preys on the more vulnerable physicians, most often young women [11]. Harassment is exceedingly common in academic medicine affecting over 80% of female faculty [5]. Over time this repeated psychological violence causes a reduced capacity to work, ranging from increased time off from work to depression, anxiety, reduced academic productivity, and ultimately burnout. Hu and colleagues outline the complex issue of burnout clearly: "Burnout, a syndrome of emotional exhaustion, cynicism, and reduced effectiveness at work, has been linked to poor health, alcoholism, depression, and suicide in physicians. Burnout has adverse effects on patient care and the physician workforce, since burned-out physicians are more likely to report that they have made medical errors, more frequently reduce their work efforts, change jobs, or leave the field of medicine. The prevalence of burnout appears to be higher among surgeons, trainees, and women than in other groups." In their study of over 7000 residents, they found that burnout was associated with increased suicidal ideation [12]. The consequence of harassment of all forms in the workplace leads to demoralization and mental health concerns [13].

> I feel like I have to compromise myself and my morals to make it so that I could pass. I feel worn down. I have been holding it in but it is tiring. I haven't reached the breaking point yet but it takes a toll mentally.—Female medical student

> It is almost like a cycle for me where I go through periods of deep depression often set off by the littlest things. Maybe I am getting pimped in the OR and someone snickers when I get a question wrong. Or patients' family member will see me walking down the hallway and come up and ask me if I can change their mother's bedpan. It's like the straw that broke the camel's back. I have thought about dropping out numerous times. I have wished I wouldn't wake up from falling asleep on occasion. I haven't sought professional help because I usually dig myself out of it after a day or two. Nobody knows how hard it is.—Female orthopedics resident

18.2 Harassment in Orthopedic Surgery

Is there a problem of harassment or psychological violence in orthopedic surgery? In a 2019 survey study of members of the American Academy of Orthopaedic Surgeons (AAOS) specifically addressing discrimination, harassment, and bullying, researchers found that over 80% of women reported being on the receiving end of these types of misconduct [14].

> We expected as the girls in my residency to get sprayed down with a pulse lavage so that we could get mopped up at the end of the case. They would pat us dry on our chests. It wasn't private just between the two of us and secret. It was in the operating room with everybody watching so there was a pressure just to take it in stride and laugh it off but it's one of those things that makes you uncomfortable. How bad is it at the end of the day? Frankly if you go through your educational and work life feeling a little uncomfortable all the time then it must impact your ability to learn.—Female orthopedics attending

Orthopedic surgery is evolving at a glacial pace. As a specialty that started out entirely comprised of men (like all fields in medicine), it is slowly becoming more heterogenous. The AAOS conducts a census every 2 years which is referred to as the Orthopedic Practice in the United States Survey. In 2018, 92.3% of practicing orthopedic surgeons were male. As a comparison, in 2018 orthopedic surgery residency trainees were 15% female nationwide. While this does represent a substantial increase from the number of licensed and practicing attending orthopedic surgeons, this number is still the lowest of any other medical specialty [15–17]. Why is the rate of change so slow and why hasn't gender representation increased to the same extent as it has in other specialties?

Van Heest argues that gender diversity remains stagnant in orthopedic surgery because female medical students choose specialties based on exposure and mentorship during medical school at higher rates than their male student colleagues. Orthopedic education in medical school is minimal compared to other specialties and must be sought out intentionally with elective rotations. Furthermore, orthopedic faculty mentorship may not be as readily present, affable, or interested as other specialties [17–19]. These factors along with the "culture" of orthopedic surgery may have built a psychological barrier for recruiting women to the field. In that context then, what is the culture of orthopedics?

> Orthopedics has a culture of toxic masculinity to some extent... Although as a field it is willing to repudiate any instance of sexual harassment immediately, it definitely is a place where sexual harassment can occur because of the underlying masculine culture that has been prevalent for a long time.—Male orthopedics resident

> It's like a fraternity at an Ivy League school. It is very hard to get into. The people on the inside hold themselves in high regard even though they may be no smarter than those in different organizations. It's highly secretive. There are rituals that may involve punishment and physical discomfort as part of the initiation and that members take great pride in having gone through. Senior members like to inflict the same tortures on junior members in a vicious ongoing cycle.—Female medical student

> Orthopedics is a boy's club. It's what I picture being in a locker room is like except for on a much grander scale. The only difference is that I am allowed to be there now and allowed to see it all. Sometimes it's not harassment against me but hearing the conversations that people have amongst themselves which are very uncomfortable. For example, I've heard someone tell a male resident who abstains from alcohol, "It's too bad that you're sober when you go out. Imagine how many drunk women you could sleep with if you just had one drink." Basically insinuating that it is okay to sleep with an incapacitated woman if you could say in

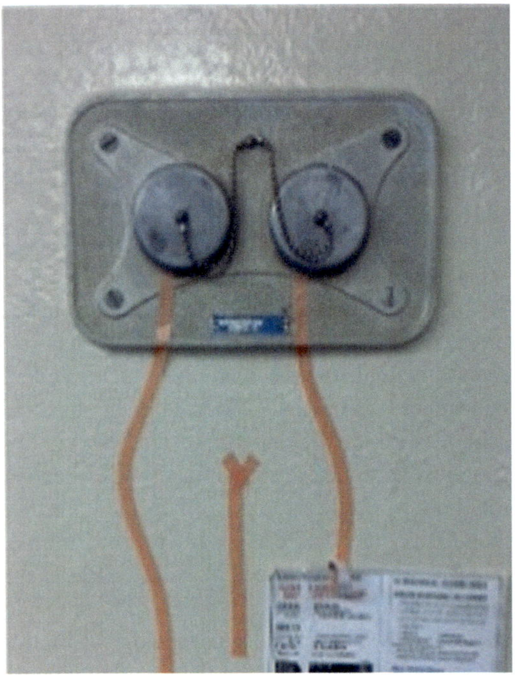

Fig. 18.1 Contributed by attending

any way that you were drinking too.—Female orthopedics resident

As an attending I walked into the operating room where those big valves stick out on the wall someone had taken pink tape and made it into a female figure with the valves as the breast with a little V shape for the vagina.—Female orthopedics attending (Fig. 18.1)

Despite stereotypes of orthopedics being a "boy's club" where harassment is known to occur, why do women continue to enter the field? Do women find harassment to be acceptable? Do they realize that harassment is common when they are medical students or are they naive?

There was a white male attending who was the head of one of the orthopedic departments at the hospital where I was doing my away rotation. The moment he saw me walk into the clinic work room he looked me up and down then grinned and remarked "boy I wish I could have dated you when I was a younger, it would have pissed my parents off so much". As a black female amongst a group of all white residents and this older man I didn't know what to say. I felt naked. Like an exotic animal on display for amusement. Nobody said anything to him, including me. Later after he left the room the residents apologized to me and said he was known to be a creepy pervert. The sad thing is I ranked that program number 1 and ended up matching there in the end.—Female orthopedics resident

I was first exposed to orthopedic surgery in college. My experience was never an issue of sexual harassment or gender bias or anything like that; I just didn't like the orthopedic surgeon who I worked with. I thought he was an *sshole and therefore I assumed that all orthopedists were *ssholes. He was a fraternity brother type of guy who worked out a lot, yelled at the nurses and was very degrading to the operating room staff. I thought to myself, wow, I do not want to be like this guy.—Female orthopedics resident

I became an orthopedic surgeon because I got really good board scores and I liked sports. It is not something that I was working towards and I did not really know what I was signing up for. When everyone said it's really hard, I told them I haven't done anything that's not hard and I want to do the hardest thing. I encourage everybody to be thoughtful of their decision-making and to be very inquisitive, ask lots of people lots of questions. Orthopedic surgery to me was like I found my crew. I found jocks. I really enjoyed the ribald and I never felt targeted by it. I love making stupid jokes that were off color and that was the fun of ortho.—Female orthopedics attending

Kisel et al. described the term "hidden curriculum" to be the insidious education of bourgeoning physicians on clinical ward rotations to accept dehumanizing and inappropriate verbal and physical treatment. They argue that this treatment begins in medical school as social proprietary behavior breaks down. Patients are lain bare for multiple eyes to see, poke, and prod [20]. The most intimate aspects of their personal histories and sexual lives are exposed and discussed with, at best an impassive clinical air and at worst judgment and disdain.

On arthroplasty rotations during our indications conferences when you see a patient with a large penis or breast implants on the x-ray everyone in the room will snicker. That goes from the medical students to the senior attendings. There is always some comedy in the human figure.—Male orthopedics resident

The ubiquitous, relentless nature of irreverence to the human form is part of what causes providers to become desensitized slowly over time. As patients are deconstructed, the boundaries of social norms among professionals also begin to wither [21–23].

Surveys of gender and sexual harassment in medical school have shown that this type of behavior occurs more often on surgical rotations than on medical rotations. An anonymous questionnaire survey of medical students at 14 different medical schools by Nora et al. revealed that women experience greater instances of harassment than men. Female medical students have long been told they should pursue more "female" or "family-friendly" specialties such as family medicine, pediatrics or obstetrics, and gynecology. Both men and women reported an increasing prevalence of these behaviors in academic hospitals than in community hospitals or outpatient offices [24].

Evidence suggests that women are less sensitive than men to gender and sexual harassment. Stratton et al. conducted a survey study of over 1027 medical students and found that although men experienced fewer instances of gender and sexual harassment than women, those few instances had a greater negative impact on their desire to pursue a particular specialty. In their study, men stated that gender bias and harassment were a main deterrent from pursuing obstetrics and gynecology, while women reported harassment in general surgery had less of an impact on their willingness to pursue the field. The authors also speculated that harassment is not always perceived negatively. "Within surgical training, certain intimidating and harassing behaviors may be perceived by faculty and house staff as functional educational tools... Seemingly unwelcome environments may also attract students who, in response to gender discrimination and sexual harassment behaviors encountered on some clerkships, actually strengthen their resolve to impart a distinctly male or female perspective within that discipline... We cannot rule out the possibility that a small number of students may, in fact, be attracted to specialties or programs that they associate with gender discrimination and sexual harassment" [25].

18.3 Perpetrators: Superiors, Support Staff, and Patients

Who are the perpetrators of harassment and discrimination in orthopedics? When asked this question, many immediately look to superiors or peers in their past who have been harassers as they feel the power dynamic made them powerless to act or speak out.

> I had an attending who used to put his arm around me and one time he got his hand caught in my coat because he was trying to touch my butt. Everyone has their own line of what they can laugh off and when that line gets crossed. For me that line wasn't crossed to the point where I felt like I needed to report somebody. Many women feel like if they are the squeaky wheel about something like that not only does it reflect poorly on them, but if the desired outcome is not met then there may be retaliation. The most typical outcome is that nothing is done. The biggest fear is a fear of reprisal and it is not unfounded.—Female orthopedics attending

> I was walking outside when I ran into a co-resident taking a jog. As I reached down to tie my shoelaces I noticed him take his phone out quickly and then put it away. I asked him if he took a photo of me and he said no with an incredulous face. About month later when I was sitting next to him in a conference, he was scrolling through his text communications with another resident trying to show me an old x-ray and scrolled past the photo of me from that day, bent over and exposed. I was humiliated.—Female orthopedics resident

> One of the really old anesthesia attendings came into the operating room with a patient in the room and said, "Damn Dr. [Talento], I don't mean to offend but you are a fox." I was humiliated and didn't know how to respond.—Female orthopedics resident

> The only time I felt weirded out was by my chief residents when I was an intern. One had voiced multiple times that he was very interested in sleeping with me despite the fact that he was engaged and the other one voiced... that he wanted to sleep with me despite the fact that he was married... The married person's wife actually contacted me on Facebook and said stop f*cking my husband... I told her that she should speak to her husband. I spoke to him about it and he brushed it off... It kind of rolled into the second-year residency end of year roast where the rumor that we had slept together was then publicized to the residency program... I confronted him about this... and I asked him if he told people we had slept together and he

said "oh yeah, totally." I said, "but we never slept together so why did you tell people that". He said, "oh, I just thought it was fun."—Female orthopedics resident

This one attending used to bro out with the team because it was all guys and except me. He would find inappropriate [sexual] memes and gifs and show all the other guys on the team and laugh. In the middle of case I was scrubbed into they somehow began talking about Harvey Weinstein and how he attacked girls. He said something along the lines of "when I was in college I drank a lot and did some things but you don't see me all over the news." Essentially implying that he had done things to women and harassed them. I felt so uncomfortable because I couldn't leave because I was in the middle of surgery but I just stopped talking and stepped back. I didn't know what to say.

I felt really uncomfortable on call alone with him. I waited until multiple other females had gone through that same rotation after me before I said something because I did not want to be labeled as the person who blows the whistle, or as the person who's too PC (politically correct) and nobody can joke around her.—Female orthopedics resident

Much is written about misconduct when coming from professionals of the same rank or a superior status. However, many also complain of sexual harassment from support staff in the operating room or hospital environment.

People noticing my body happens all the time. After I had a baby, at least 3 or 4 ancillary staff members looked me up and down on different occasions and remarked on how good my body looked. When an attending anesthesiologist did the same thing, leering and joking about my figure in front of my male co-resident, I didn't even notice but my colleague was shocked. He said he couldn't believe someone would say something like that to me. I had to do a double take to even remember what was said. I think I get comments so often that they go in one ear and out the other. I wouldn't even fathom reporting that type of behavior even though I know it is not right.—Female orthopedics resident

There was an assistant who was in the operating room who had inappropriate pet names for all the female residents, nurses and sometimes attendings as well. He called them sweetie or honey and would push up his body against them or invade their personal space. I saw it all the time and it just seemed like it was the way things flowed. None of the females that he was doing this to said anything so it never seemed that it was my place to intervene. There are a lot of gray zones. He was always pushing up against them but the nature of the work is very physical so it would be hard to say "Hey, you were moving up right against her as you were helping to move that patient." ... He eventually said something to the wrong person. He behaved in his normal manner with a new attending of authority who had a zero-tolerance policy. She escalated it quickly. A lot of people over the years had allowed him to cross the line habitually in small ways... After she escalated this, many other people came out of the woodwork saying that they had all been verbally and/or physically sexually assaulted by him and he was fired swiftly.—Male orthopedics resident

The discussion on workplace harassment falters when the focus rests largely on unprofessional conduct between employees. Much verbal and physical sexual harassment and gender harassment occur from patients [12]. These problems can go beyond gender-based microaggressions and sexual harassment to full physical assault. These situations are exceedingly common, estimated to affect up to 75% of the physician workforce in some instances, and challenging to regulate [26]. Physicians who experience repeated verbal and/or physical abuse from patients have higher rates of burnout and its sequelae [27, 28].

Patients constantly make advances on me. The most benign ones are from people commenting on how pretty I am or how young I look. It may seem nice but it is infuriating. I feel like it undermines my credentials as I get whittled down to a young pretty face rather than the surgeon it has taken me years to become.—Female orthopedics resident

At the VA there was a representative who was coming in to show medical compression stockings and there was an older veteran who she was showing the compression stockings to. He began to ask her what the compression stockings looked like when she was wearing them, does she ever wear just compression stockings and nothing else and then he asked her if she would sit on his lap and try on the compression stockings. It was really uncomfortable, but he was an 80-year-old guy and she was smiling. I didn't know how uncomfortable she was and I didn't really know what to say. Both him and the representative laughed it off but it was clear it was inappropriate.—Male orthopedics resident

One of the harder parts of navigating the space of harassment from patients is aftermath when you have to interact with the patient and you know they're not a good person but you still have to give them care. I was an intern at one of our hospitals with a female senior resident on the service. After rounds a patient commented to me how big her breasts were and how she was so good-looking. He asked me if I had ever slept with her. I didn't really know what to do because I wanted to build rapport with the patient but I also could not condone that sort of thing being said about my superior. I told him that we shouldn't talk about that and the patient let it go.—Male orthopedics resident

18.4 Gender Harassment: The Ultimate Burden?

If you're a woman and you made it to your thirties and you've not been sexually harassed then you have been living in a bubble. Unfortunately sexual harassment is part of the status quo and in a lot of ways I'm thankful nothing worse has ever happened to me but at the same time why is it so normalized? That's a more frightening thing to think about.—Female orthopedics resident

While sexual harassment is generally regarded as unacceptable, gender harassment or discrimination has become a unique challenge in orthopedics. This viewpoint was stated by multiple interviewees who in fact felt that sexual harassment was so ubiquitous it was not seen as a problem when it occurs in the workplace. To them, more sinister and undermining is the issue of gender harassment.

Sexual harassment is something that we all actually get in our daily lives and sometimes I wonder if gender discrimination is worse.... I really believed that orthopedic surgery was a meritocracy in that once you got in it didn't matter what your sex or gender was or your color but there is a true undercurrent of subtle, insidious discrimination that belies that. There is a tendency for like to gravitate towards like and because our field is so homogeneous you have to intentionally include people who do not look like the norm. That requires intentional effort and it often times doesn't happen. I don't think it's evil or malicious, but it definitely happens more than I thought it did. It often happens on the margins but that's where careers are made and opportunities happen. It happens on the golf course we don't get invited to. It's thoughtless inclusion rather than intentional exclusion.—Female orthopedics attending

When I was a third-year resident I was doing a total hip with a PA and an older male attending. He looked at me when the patient was in the room, not yet asleep yet, and he asked me, "How are we going to do this case?" I said, "What do you mean?" He said, "Well, we need a fourth person." I said "Why? We can just do this case you and I. Felicity (the PA) is here to help hold hook." Then he asked, "Who is going to help reduce the hip?" I looked at him and I remember being very confused and telling him that I could reduce the hip with a very incredulous look on my face. I had never been in a situation where anyone said I couldn't do something because I was a woman physically. I thought that was incredibly rude, inappropriate and unprofessional in so many ways. I was so upset that I left the room to gain composure. I didn't want to come back and I thought to myself, f*** you if you're going to treat me like that, your resident who is here to help you and you don't even know how to operate then you need to check yourself. The best part of the case was that I didn't even end up getting to do anything. He screwed up everything because he's a terrible surgeon. I just stood there watching him and in the end he couldn't reduce the hip. He looked around and asked if there was someone that they could call to come in the room to help reduce the hip. I responded "Why don't I give it a shot". I got it in right away on the first try because in the end it's not about strength it is about technique.—Female orthopedics resident

When I was an intern at County Hospital I was changing a wound VAC on a man who had an infected hip and his neighbor in the next bed over kept saying "nurse, nurse, nurse." While focusing on my work, I kept telling him, "I am not your nurse" and that I was in the middle of doing something. The patient then got out of bed and grabbed my ass to get my attention. Mind you I was fully involved in changing the other patient's complex dressing so my hands were occupied but I dropped everything and whirled around and shouted at him, "I am not your nurse, do not touch me, you are getting out of here today! If you are good enough to get out of bed you can go home." It's not disrespectful to call an orthopedic resident a nurse but it feels that way when it happens over and over again. It's that kind of microaggression where it's not the first nurse who says it, or the first patient who says it, or the first attending who says it but when it's the 20th time someone calls you by the wrong role that you finally snap. People react to my frustration and say "wow, why did you react like that, it's not an insult to be called a nurse." The reality is that it is not who I am, I am a doctor.—Female orthopedics resident

18.5 Special Considerations

How do we combat negative societal norms entering the professional workspace? Many women feel that within orthopedics there is a double standard. Women have to work harder to get recognized, are disciplined more severely, and have to compromise their work-life balance to be seen as a team player.

> What's really disappointing to me is the double standard. There was attending who was having sex with a nurse on hospital property and the nurse ended up "leaving" while the attending is still here working, seemingly without any repercussions... There is also a male pediatric surgeon who operated on both a mother and child then was caught having sex with the mom in his office while the child was sitting outside the clinic room unaccompanied. That guy is still hired. I think if it had been me, I wouldn't have a job anymore.—Female orthopedics resident

> I slept with a [medical device] representative and everyone judged me and was shocked but when my male co-residents do it all the time no one cares.—Female orthopedics resident

> There was a junior female orthopedic resident who slept with a senior male resident. Everyone including myself judged the female resident. The male resident got away with it and was unscathed even though he was the Senior resident in the position of power manipulating the situation.—Female orthopedics resident

Pregnancy during residency has long been a taboo. Historically in all male residency programs, having children was of minimal consequence to other members of the team. The male resident's wife would manage everything, and some would take a few days off at most; some report even just a few hours off for the delivery.

> During fellowship interviews [this 2021 cycle] one of the fellows boasted of the greatness of the program. My wife was induced at the hospital I was working at and it was great! I scrubbed cases until she was dilated enough then came in to watch the delivery and went right back to work afterwards.—Female orthopedics resident

The mentality which encourages that type of approach to parental leave is not unique.

> We are taught so young don't get pregnant. In your fellowship don't give us a bad name by doing something like that and getting pregnant. It's actually not harassment it's discrimination.—Female orthopedics attending

> You know sometimes women are perpetrators of a culture of abuse... At my medical school... there was a resident who got pregnant in her second year of residency... and ended up having to go on bedrest half way through. It put a toll on the program because all of her male colleagues had to cover for her. I remember when she was a fifth year resident, she got pregnant again and went out on maternity leave and everyone gave her a hard time. They said she should've chosen a different specialty because it wasn't advisable to get pregnant during orthopedic residency and she should've known that coming in. I took that with me and when I was a resident at my own residency program a few years later I said something similar about a female resident to one of my attendings and he scolded me. He told me I shouldn't think that way and if I did I should keep it to myself. Nobody should tell a woman when the right time for her to get pregnant is. It is a personal choice. That really shocked me. I was embarrassed to realize I was a part of the problem.—Female orthopedics attending

> I had friends who were told explicitly not to get pregnant but it was always clear implicitly that if you got pregnant you would be hurting your classmates and your program and it was discouraged. I think that is changing and I even had some male co-residents who took paternity leave. That demonstrates the change in mindset at some institutions. The reality is it's way better to have children in your residency because of your pocket book. The opportunity cost is huge when you have children as an attending and you have to take time off or have complications.—Female orthopedics attending

> People don't talk about fertility preservation but they're happy to tell you not to get pregnant until you would be technically a geriatric mother. The longer you wait to even preserve your fertility, the lower your chances are.—Female orthopedics attending

Many interviewees also spoke of the problems faced by double minorities. In a survey study of AAOS members, African-American surgeons reported experiencing higher rates of bullying, discrimination, and harassment than any other ethnic group [14]. Coupled with increased gender harassment, it stands to reason that African-American females may be subject to more harassment than Caucasian females, further discouraging them from joining the field of orthopedics.

> Being a minority puts you at risk for being harassed in different ways. I think that race plays a role in that a lot of what minorities face day to day is analogous to what women and members of the LGBTQ community feel more so than others with regard to sexual harassment in orthopedics specifically. I think race has to be part of the conversation when you talk about sexual harassment in the workplace. There's one way that it affects white women and there's a different way that it affects women of color and if you don't address that you're missing part of the conversation. What I have seen is that women who are minorities face a double threat of sexual harassment and racism compounded often. I only deal with one of those things I can't imagine if I had to deal with both.—Male orthopedics resident

> I always knew I wanted to do a surgical specialty that was very tactile but choosing ortho was hard for me. I had never seen a black attending orthopedist and certainly never a black female orthopod in any level of training. Support staff at my medical school sneered at me after I told them I was interested in ortho. I wasn't sure if it was because of my race or my gender. Even the ortho program director at my home program barely made time for me and would not write me a letter of recommendation despite the fact that I had great board scores and was AOA [Alpha Omega Alpha].—Female orthopedics resident

18.6 Solutions: Response and Responsibility

In response to reports of sexual harassment and abuse in medical school, Recupero et al. recommended that all medical schools have an established chain of reporting for medical students to notify the administration of harassment behavior and a systematic method of addressing any issue [29]. In many medical schools, there is someone dedicated to receive complaints who is not directly affiliated with the medical school such as an ombudsman. However, in residency programs this reporting process often becomes less uniform, less impartial, and more problematic. For example, the program director, chairman, or other supporting faculty often take the role of the overseer. This can present an issue as residents may feel these individuals have their own interests at stake and may want to protect the perpetrator or the institutional name and/or may be the perpetrator themselves [30].

Binder and colleagues discuss the problem of "covert retaliation" where victims who report abuse are often silently not promoted, not given the ability to do the same work as others in the workplace, and are excluded from participation on teams. The knowledge that their report will not be anonymous and that they may have to continue to work with the perpetrator after reporting leads to underreporting of harassment. They recommend two methods of negotiating with harassment reporting to reduce retaliation: group reporting and bystander reporting. Many people accused of harassment have had complaints from other women in the past. Gathering together those who have been abused to file a joint complaint against a perpetrator strengthens the veracity of their complaints and diminishes the likelihood of any negative fallout on an individual woman who registers the complaint. Additionally, sexual harassment often occurs in a public space with others watching. A more senior uninvolved witness should take it on themselves to report on the victim's behalf knowing that the likelihood of retribution to them will be low [6, 31].

Perhaps most importantly, individuals themselves need to have a plan ahead of time to combat episodes of harassment. Within the workplace among superiors or colleagues, one junior faculty advises to remove all emotionality from the encounter.

> Laughing it off is not the right approach when something doesn't feel right. The trick is to remove your emotion from it and just respond, "Dude, that's not cool" or if it's someone more senior "I appreciate that you were trying to be funny but I didn't find it very humorous" or "I see what you're trying to say and I'm not upset but moving forward let's not talk about that". If you can learn to impart your feedback in a non-emotional non-confrontational way then it will be easier to engage.—Female orthopedics attending

Many residents appreciated and enjoyed that, during times where they were being disrespected by a patient on rounds, a senior resident stepped in and spoke up for them. However, a senior female attending warns of the pitfalls of relying on others to stand in to support junior residents without forethought:

> I wouldn't want an attending to assume they have to speak for me. A large part of the message is that mentors need to say something but the reality is that they need to be supportive of how women want to handle these problems... Sometimes you have to speak up for yourself because you're the only one in the room. If you are on a team-based rotation where there is one team leader then you may need to figure out ahead of time how you're going to handle it. For example, if you are on a team that only has one female, the leader of that team should ask the minority person how they would like these situations to be handled... They are very likely to occur in patient encounters. "Would you like me to speak up for you? Would you like to speak up for yourself in that moment? If it's a really hostile patient I recommend that we leave the room and give a patient some time to calm down or would you like to be removed?"... Another way is to have a signal so if you're really feeling threatened, then everyone knows that they can have your back and that they won't be demeaning by speaking up for you. The whole goal is to convince the patient you are there to take care of them, qualified to take care of them and that you have the same goals.—Female orthopedics attending

Hock et al. developed a training protocol to help resident and attending physicians manage sexual and gender harassment coming from patients. They found that while numerous training modules existed to help attending physicians intervene when residents were being harassed by patients, there were no methods to teach physicians how to advocate for themselves while alone with patients. In their training model, they used female resident facilitators who had experienced sexual harassment themselves from patients to teach male and female residents and faculty members. Using role play as a physician being harassed by a patient, they emphasized the following eight key points in effective physician-patient dialogue [32]:

1. Address patients' inappropriate behavior with the word "I".
   (a) "I don't appreciate your comments about my gender."
2. Speak about the "behavior" that the patient did, not about the patient being good or bad themselves.
   (a) "What you said was unkind." Versus "You are unkind."
3. Challenge the "effect" of what the patient said on you, not what their "intent" was.
   (a) "I don't think you intended to hurt my feelings but I found what you said very hurtful."
4. Educate them on equality
   (a) "I have gone through the same training as all the male physicians and request that you treat me with equally."
5. Appeal to their well-being and medical care
   (a) "Your comments make me feel uncomfortable which makes it challenging to give you the best care possible because I feel unsafe."
6. Don't fall into the trap of laughter or minimizing.
7. Tell them what you prefer them to do or say.
   (a) "I am your physician, please call me Dr. X."
   (b) "Talking about my physical appearance is inappropriate and I will not tolerate it. Let's focus on what brought you to the hospital today."
8. Leave the room if you need to and tell the patient why.
   (a) "Your behavior is inappropriate. I am leaving the room because I do not feel comfortable"

This overall strategy gives victims of harassment verbal tools to combat inappropriate behavior immediately rather than waiting for someone else to intervene or going along with the negative patient encounter. When addressing inappropriate physical contact from patients, options are fewer, and the swiftest resolution is often to extricate oneself immediately from the encounter.

18.7 Endgame: Creating a Culture of Intolerance

Many believe that increasing representation of females is the key to reducing harassment in medicine [33, 34]. But is there a magic number to reach where women are represented enough?

According to the Mansfield Rule, once 30% of a group is represented in a field, they have reached a state where they no longer represent a niche and their needs begin to become part of the normative culture [19]. In line with those aims, there are pipeline programs such as the Perry Initiative, Ruth Jackson Orthopaedic Society, and Nth Dimensions which strive to increase the number of females and underrepresented minorities in orthopedics [35, 36]. These organizations each target different age sectors to spark an interest in medicine for high school and college students and in orthopedics specifically for those already in medical school.

> Even though it's been incredibly slow, our numbers are changing. We need to hit that magical 30% to really make a lot of change more quickly. I think more residency programs have had more female residents, forcing them to think about it and create policies about it... The me-too movement has highlighted that sexual harassment is present in every profession to some degree. That put some people on notice but it also sparked a lot of discussions... I think we have to have people who are invested in changing it so we need to change our numbers.—Female orthopedics attending

Gender discrimination, bias, and harassment are not isolated to orthopedics or the United States [37]. In a systematic meta-analysis, Sprow et al. reviewed global literature on gender microaggressions in surgery and found that female attendings and trainees internationally reported "being treated like second-class citizens" where their opinions were less valued than men regardless of their surgical competence. Additionally, while orthopedics is one of the least gender diverse surgical sub-specialties, more gender diverse specialties such as general surgery suffered the same problems [38, 39]. This begs the question: if increasing female representation hasn't changed the prevalence of gender harassment in general surgery, how can it change harassment in orthopedic surgery? Many are skeptical that psychological violence will reach an end in academic medicine.

> [Do you think there will be a time when harassment is not the status quo?] I don't know if that is ever possible because humans are animals and there's some parts of our nature that are instinctive natures that are within us. Yes, we have conscience and the ability to make decisions but there are underlying drives that have evolutionarily provided an advantage. I think realistically harassment will always be there on some underlying level. It's also not black and white people sexually harass each other across genders and sexes.—Female orthopedics resident

Although women are the most frequent victims of psychological violence, they can also be perpetrators. Men and women must both stand together against inappropriate behavior in the workplace for any change to be made.

Many have expressed a concern that zero-tolerance policies are unrealistic and create a hostile work environment where amity suffers [40]. Burnout in orthopedics occurs not only because of harassment but also because of the physical and emotional demands of the work [41]. Not only does zero tolerance put everyone on high alert, but it may cause people to disengage from participation in mentorship and education altogether.

> We spend a lot of time at work to have to tiptoe around our colleagues. It's not great [harassment] but if we have a zero-tolerance policy for everything then life can get restrictive. You have to like what you do in life and at work you have to have some level of camaraderie. You have to think, "Is the behavior that I'm performing or experiencing enhancing or undermining camaraderie" and go from there.—Female orthopedics attending

> Everyone has a different threshold for what's okay and what is not okay. I feel like there is an entire generation of orthopedic surgeons who are checked out of mentoring because it's not worth it... A female resident came up to me and asked, "[Dr. Refugio] would you be my mentor? I have already reported three other mentors for bad behavior but you're awesome." I thought to myself, I don't want to work with you.—Female orthopedics attending

Perhaps a realistic ideal isn't the cessation of workplace harassment altogether with a zero-tolerance policy but people feeling empowered to speak up in the moment without concern for retribution.

> I think what we want to strive for rather than a zero-tolerance policy, is a work environment where everybody knows what their [own] line is and people feel empowered to speak up knowing that there will be no reprisals. If someone says, "hey that makes me feel uncomfortable," everyone

should be able to respond and navigate that space with the person. A space where there is no judgement and the person who said the comment doesn't feel negatively towards a person who corrected them... I think our vision right now is all or nothing but we need to understand how to live comfortably grey.—Female orthopedics attending

Harassment and discrimination in the workplace is not just a female problem perpetrated by males. Almost everyone interviewed for this chapter, male and female, admitted that they had been perpetrators of abuse at some time in their training to people of both genders. Fighting long-held internal biases requires deep reflection and intentional action to succeed. Many are unaware of their role in the cycle of abuse until they are specifically told about the harm their actions have caused. Individually, the onus lies on victims to speak up and on perpetrators to listen. Institutionally, residency programs should strive to create safe, anonymous, and objective mechanisms for reporting. Nationally, leaders must set an example that harassment, slurs, bullying, discrimination, and abuse are unacceptable. Inertia and institutional behavioral changes will be the biggest hurdles to eradicating harassment and discrimination, but what each of us can do to change the culture is to challenge ourselves and our colleagues to defend each other and speak up. Change doesn't happen overnight; it starts with people taking ownership of fostering an environment of mutual respect.

Acknowledgments Thank you to all those who took the time to interview anonymously for this chapter. Your insights were invaluable.

Note from the Authors Numerous anonymous interviews were conducted to gain insight into harassment behavior that occurs in the medical profession. In each instance, these personal narratives and perspectives are italicized. To maintain the anonymity of contributing physicians and medical students, the names of people, hospitals, and institutions have been changed. Any similarity in name to people living or deceased is purely coincidental. Personal perspectives contained in this chapter are meant solely to show the diversity of opinions regarding harassment in orthopedics and do not necessarily reflect the opinions of the authors.

References

1. World Health Organization (2018) Violence against women prevalence estimates, p 2021
2. Xepoleas MD, Munabi NCO, Auslander A, Magee WP, Yao CA (2020) The experiences of female surgeons around the world: a scoping review, vol 18. Human Resources for Health, BioMed Central Ltd
3. Kaufman MR, Dey D, Crainiceanu C, Dredze M (2019) #MeToo and Google inquiries into sexual violence: a hashtag campaign can sustain information seeking. J Interpers Violence. 886260519868197
4. Suk J, Abhishek A, Zhang Y, Ahn SY, Correa T, Garlough C et al (2019) #MeToo, networked acknowledgment, and connective action: how "empowerment through empathy" launched a social movement. Soc Sci Comput Rev
5. Vargas EA, Brassel ST, Cortina LM, Settles IH, Johnson TRB, Jagsi R (2020 Jan 1) #MedToo: a large-scale examination of the incidence and impact of sexual harassment of physicians and other faculty at an academic medical center. J Women's Health 29(1):13–20
6. Bates CK, Jagsi R, Gordon LK, Travis E, Chatterjee A, Gillis M et al (2018) It is time for zero tolerance for sexual harassment in academic medicine. Acad Med 93:163–165. Lippincott Williams and Wilkins
7. Choo EK, van Dis J, Kass D (2018) Time's up for medicine? only time will tell. N Engl J Med 379(17): 1592–1593
8. Amoli MA, Flynn JM, Edmonds EW, Glotzbecker MP, Kelly DM, Sawyer JR (2016 Sep 1) Gender differences in pediatric orthopaedics: what are the implications for the future workforce? Clin Orthop Relat Res 474(9):1973–1978
9. Dzau VJ, Johnson PA (2018) Ending sexual harassment in academic medicine. N Engl J Med 379(17): 1589–1591
10. Crebbin W, Campbell G, Hillis DA, Watters DA (2015) Prevalence of bullying, discrimination and sexual harassment in surgery in Australasia. ANZ J Surg 85(12):905–909
11. Siller DH, Beck-Rabanser C, Hochleitner PDM, Exenberger DS (2021) "Not a woman-question, but a power-question": a qualitative study of third parties on psychological violence in academic medicine. Work Heal Saf 69(1):41–49
12. Hu Y-Y, Ellis RJ, Hewitt DB, Yang AD, Cheung EO, Moskowitz JT et al (2019 Oct 31) Discrimination, abuse, harassment, and burnout in surgical residency training. N Engl J Med 381(18):1741–1752
13. David EA, Stephens EH, Kane L (2020) Mistreatment and burnout in surgical residency training. N Engl J Med 382(6):581–583
14. Balch Samora J, Van Heest A, Weber K, Ross W, Huff T, Carter C (2020 Dec 15) Harassment, discrimination, and bullying in orthopaedics: a work environment and culture survey. J Am Acad Orthop Surg 28(24):e1097–e1104

15. Cannada LK (2016 Sep 1) Women in orthopaedic fellowships: what is their match rate, and what specialties do they choose? Clin Orthop Relat Res 474(9):1957–1961
16. Chambers CC, Ihnow SB, Monroe EJ, Suleiman LI (2018 Sep 5) Women in orthopaedic surgery: population trends in trainees and practicing surgeons. J Bone Joint Surg Am 100(17):e116
17. O'Connor MI (2016) Medical school experiences shape women students' interest in orthopaedic surgery. Clin Orthop Relat Res 474(9):1967–1972
18. Rohde RS, Wolf JM, Adams JE (2016) Where are the women in orthopaedic surgery? Clin Orthop Relat Res 474(9):1950–1956
19. Van Heest A (2020) Gender diversity in orthopedic surgery: we all know it's lacking, but why? Iowa Orthop J 40(1):1–4
20. Kisiel MA, Kühner S, Stolare K, Lampa E, Wohlin M, Johnston N et al (2020 Dec 10) Medical students' self-reported gender discrimination and sexual harassment over time. BMC Med Educ 20(1):503
21. Schoenefeld E, Marschall B, Paul B, Ahrens H, Sensmeier J, Coles J et al (2021 Dec) Medical education too: sexual harassment within the educational context of medicine – insights of undergraduates. BMC Med Educ 1:21(1)
22. Whicker E, Williams C, Kirchner G, Khalsa A, Mulcahey MK (2020 Nov 1) What proportion of women orthopaedic surgeons report having been sexually harassed during residency training? A survey study. Clin Orthop Relat Res 478(11):2598–2606
23. Witte FM, Stratton TD, Nora LM (2006) Stories from the field: Students' descriptions of gender discrimination and sexual harassment during medical school. Acad Med 81(7):648–654
24. Nora LM, McLaughlin MA, Fosson SE, Stratton TD, Murphy-Spencer A, Fincher RME et al (2002) Gender discrimination and sexual harassment in medical education: perspective gained by a 14-school study. Acad Med 77(12 I):1226–1234
25. Stratton TD, McLaughlin MA, Witte FM, Fosson SE, Nora LM (2005) Does students' exposure to gender discrimination and sexual harassment in medical school affect specialty choice and residency program selection? Acad Med 80(4):400–408
26. Wu Y, Jiang F, Ma J, Tang Y-L, Wang M, Liu Y (2020 Jan 29) Experience of medical disputes, medical disturbances, verbal and physical violence, and burnout among physicians in China. Front Psychol 11: 556517
27. Ames SE, Cowan JB, Kenter K, Emery S, Halsey D (2017 Jul 19) Burnout in orthopaedic surgeons: a challenge for leaders, learners, and colleagues: AOA critical issues. J Bone Joint Surg Am 99(14):e78
28. Wu Y, Jiang F, Ma J, Tang YL, Wang M, Liu Y (2021) Experience of medical disputes, medical disturbances, verbal and physical violence, and burnout among physicians in China. Front Psychol 11:556517. https://doi.org/10.3389/fpsyg.2020.556517
29. Recupero PR, Heru AM, Price M, Alves J (2004) Sexual harassment in medical education: Liability and protection. Acad Med 79:817–824. Lippincott Williams and Wilkins
30. Best CL, Smith DW, Raymond JR, Greenberg RS, Crouch RK (2010) Preventing and responding to complaints of sexual harassment in an academic health center: a 10-year review from the medical university of South Carolina. Acad Med 85(4):721–727
31. Binder R, Garcia P, Johnson B, Fuentes-Afflick E (2018) Sexual harassment in medical schools: the challenge of covert retaliation as a barrier to reporting. Acad Med. Lippincott Williams and Wilkins 93: 1770–1773
32. Hock LE, Barlow PB, Scruggs BA, Oetting TA, Martinez DA, Abràmoff MD et al (2021) Tools for responding to patient-initiated verbal sexual harassment: a workshop for trainees and faculty. MedEdPORTAL J Teach Learn Resour 17(1):11096
33. Zuckerman JD (2020) CORR Insights®: what proportion of women orthopaedic surgeons report having been sexually harassed during residency training? A survey study. Clin Orthop Relat Res 478(11): 2607–2609
34. Zuckerman JD (2018 Aug 1) What's important: diversity in orthopaedic surgery. J Bone Joint Surg Am 100(15):1351–1352
35. Lattanza LL, Meszaros-Dearolf L, O'Connor MI, Ladd A, Bucha A, Trauth-Nare A et al (2016 Sep 1) The Perry initiative's medical student outreach program recruits women into orthopaedic residency. Clin Orthop Relat Res 474(9):1962–1966
36. Mason BS, Ross W, Ortega G, Chambers MC, Parks ML (2016 Sep 1) Can a strategic pipeline initiative increase the number of women and underrepresented minorities in orthopaedic surgery? Clin Orthop Relat Res 474(9):1979–1985
37. Phillips SP, Webber J, Imbeau S, Quaife T, Hagan D, Maar M et al (2019 Jan 1) Sexual harassment of Canadian medical students: a national survey. EClinicalMedicine 7:15–20
38. Sprow HN, Hansen NF, Loeb HE, Wight CL, Patterson RH, Vervoort D et al (2021) Gender-based microaggressions in surgery: a scoping review of the global literature. World J Surg 5(5):1409–1422
39. Van Heest AE, Fishman F, Agel J (2016) A 5-year update on the uneven distribution of women in orthopaedic surgery residency training programs in the United States. J Bone Jt Surg Am Vol 98(15):e64
40. Rihal CS, Baker NA, Bunkers BE, Buskirk SJ, Caviness JN, Collins EA et al (2020) Addressing sexual harassment in the #MeToo era: an institutional approach. Mayo Clin Proc 95:749–757. NLM (Medline)
41. Fitzgerald CA, Smith RN, Luo-Owen X, Turay D, Ferrada P, Han J et al (2019) Screening for harassment, abuse, and discrimination among surgery residents: an EAST multicenter trial. Am Surg 85(5):456–461

Institutional-Level Tracking to Combat Mistreatment of Medical Students, Residents, and Fellows

19

Jessica Hane and Taj Mustapha

Abstract

Mistreatment is prevalent in academic medicine resulting in personal and professional harm to individuals. Medical trainees that experience mistreatment have higher rates of burnout, suicidal ideation, and attrition. Additionally, mistreatment has deleterious effects on clinical team function and patient care. It is essential that academic institutions track mistreatment, as well as take appropriate prevention and mitigation steps. The following chapter outlines the prevalence of mistreatment among medical trainees, types of mistreatment, and current tracking tools. A framework for determining the optimal intervention and entity best positioned to intervene based on the type of mistreatment is presented. The process of mapping measures of mistreatment to both type and level of intervention is necessary to create a more equitable training environment.

J. Hane (✉) · T. Mustapha
Departments of Medicine and Pediatrics, University of Minnesota Medical School, Minneapolis, MN, USA
e-mail: jhane@umn.edu; must0035@umn.edu

19.1 Introduction

Mistreatment is commonplace in academic medicine. The term is broad, and studies show that perceptions of mistreatment vary based on academic rank, specialty, gender, and race [1–4]. While typically referring to discrimination and harassment, mistreatment can also include microaggressions, bullying, and a hostile workplace. Mistreatment is prevalent not only in medical school but at all stages of physicians' careers [5, 6].

19.1.1 Students

In a landmark study at a major medical school, 46.4% of all respondents and 80.6% of seniors reported abuse while enrolled in medical school [5]. Unfortunately, medical students today are still experiencing abuse despite tracking, prevention, and mitigation efforts [5]. Since 1991, the Association of American Medical Colleges (AAMC) has surveyed graduating medical students about their experience of mistreatment. Data from 27,504 medical students (representing 72.1% of graduating medical students in the United States) was consistent with 35.4% experiencing at least one type of mistreatment [6]. Medical students that identified as female, underrepresented minority, multiracial, Asian, lesbian, gay, or bisexual reported a higher prevalence [6].

J. Faintuch, S. Faintuch (eds.), *Integrity of Scientific Research*,
https://doi.org/10.1007/978-3-030-99680-2_19

19.1.2 Residency and Fellowship

Trainees interact with within-specialty attendings and peers, as well as outside-of-specialty attendings and trainees, interdisciplinary staff, and patients. The intersection between multiple professional cultures and unique specialty-specific microcultures makes tracking and preventing mistreatment more challenging. The prevalence of harassment and discrimination during medical training in a large meta-analysis was 59.4% [7], though one study showed 91% of trainees reporting that they have experienced one form of harassment or discrimination during their medical training [8]. The most common form of mistreatment was gender discrimination, followed by verbal harassment [7].

19.1.3 Gender-Specific Problems

There is robust evidence that women experience sexual harassment and gender-based discrimination at higher rates than male colleagues [9–11]. Nearly half of women in medical school report experiencing some form of sexual harassment [12]. Within the academic medical faculty, 30% of women have personally experienced sexual harassment, and of those women, 47% reported these experiences negatively affected their career path [13].

19.1.4 Racial and Ethnic Minorities

Underrepresented in medicine (UIM) is defined by the AAMC Executive Committee as those racial and ethnic populations that are underrepresented in the medical profession relative to their numbers in the general population, and UIM trainees report higher rates of mistreatment and discrimination based on race [6]. In systems that reinforce imposter syndrome, provoked stereotype threat, implicit bias, microaggressions, and harassment, higher burnout rates are noticed [12]. This cannot be divorced from the fact that nationwide, 18.9% of all dismissed residents are Black, despite representing only 4.7% of that population [14]. In every specialty analyzed by the ACGME, Black residents were overrepresented in dismissals. There are also increased burnout rates in UIM faculty, who face unique challenges pertaining to their environment that contribute to increased rates of emotional exhaustion and depersonalization [13].

19.1.5 Long-Lasting Consequences

Mistreatment can be linked to negative mental health outcomes, including depression, anxiety, and burnout [2, 15]. Female physicians who experienced sexual or gender-based harassment in medical training were more likely to report a history of depression or suicide attempts [16]. Mistreatment in training is linked to lower career satisfaction, increased cynicism, and decreased commitment to a profession [12, 17]. Medical students who experience mistreatment are less likely to plan careers in academic medicine. Students who reported mistreatment experienced adverse effects for a month or more, and 16.2% reported that it would "always affect them" [5].

19.1.6 The Benefits of Diversity

Teams composed of diverse personnel operate with increased creativity, promote cross-cultural competence, and make accurate diagnoses twice as often as teams of homogenous identities. Overall racial, ethnic, and gender diversity is an institutional driver of excellence [18]. In clinical medicine in particular, representative diversity (racial, ethnic, and gender diversity reflective of the patient population) is linked to greater patient satisfaction [19], improved patient engagement in healthcare decisions [20], and improved health outcomes and health equity [21, 22]. While recruiting for diversity is important, developing a culture of inclusion is critical to reaping the benefits of diversity in teams, as well as retention of a diverse workforce. It is well known from other fields that a lack of organizational justice or institutional fairness leads to a significant amount of employee attrition, as well as personal and professional harm.

19.2 Types of Mistreatment

The following are functional categories that are supported by the literature in terms of definitions, tracking mechanisms, and interventions.

19.2.1 Microaggression

It was coined by psychologist Chester Pierce in 1977 to describe the "subtle, stunning, often automatic, and nonverbal exchanges which are 'put-downs' of Blacks by offenders" [23]. The definition was expanded by Sue et al. [24] to include "brief and commonplace daily verbal, behavioral, or environmental indignities, whether intentional or unintentional, that communicate hostile, derogatory, or negative racial slights and insults toward people of color." We've gradually come to understand that microaggressions also target individuals based on gender, race, sexuality, disability, and/or other marginalized status. Microaggressions are typically broken down into four categories: microassaults, microinsults, microinvalidations, and environmental microaggressions (Table 19.1).

19.2.2 Sexual Harassment

Sexual harassment is characterized by requests for sexual favors, unwelcome sexual advances, sex-based threats, and verbal or physical behavior that is sexual in nature. Examples of behaviors include sexually suggestive jokes or comments, requests for sexual favors or unwelcome sexual advances, negative comments about pregnancy or motherhood, microaggressions, sexual assault, or quid pro quo harassment, which translates to "something for something," like giving the employee something, such as a promotion, in exchange for a sexual favor.

19.2.3 Humiliation and Shame

This type of mistreatment is often the product of a culture that promotes blame and shame, though it can also be the result of individual bullying behavior [25]. Bullying may reflect unmanageable work and stress on the part of the perpetrator, and/or lack of community and empathy between perpetrator and victim [26].

19.2.4 Threatened with Physical Harm or Otherwise Intimidated

It often persists due to the power differential inherent in the academic medical environment.

Table 19.1 Definitions of types of microaggressions

Type	Definition [24, 55]	Example
Microassault	A verbal remark or behavior to harm an individual, typically a conscious effort	"I don't want a f** taking care of my child"
Microinsult	A rude or insensitive comment that demeans an individual's identity or heritage, often an unconscious effort	"You are surprisingly good at procedures for a woman"
Microinvalidation	A verbal remark or behavior that excludes or minimizes the thoughts, feelings, and lived experience of another person, often an unconscious effort	"Your people must be so proud of you for getting into medical school"
Environmental microaggression	A systemic or environmental manifestation of microassaults, microinsults, and microinvalidations	The presence of imagery that depicts only white people in the role of physician

This behavior can represent individual pathology and possibly a culture tolerant of such pathology.

19.3 Current Tools to Track Mistreatment

19.3.1 Culture Survey

Employees should evaluate if an organization's ideals align with its behavior. This type of survey attempts to gauge the perceptions, norms, beliefs, and values within an institution. The results can show if there is a predisposition for mistreatment within a specific environment. It may also shed light on the effectiveness of institutional policy and codes of conduct.

19.3.2 Climate Survey

Typically a questionnaire is administered to employees to gauge perceptions about the work environment. These types of surveys are highly variable in scope and focus and ideally should be tailored to the needs of an organization. In the medical environment, a complicating factor is the cultural heterogeneity between specialties. Despite being at the same university, a woman in a majority-male training program, such as surgery, is likely to experience more discrimination than in pediatrics [27]. Most climate surveys do not parse out differences between these microcultures and lack the specificity to recognize if issues are due to the intersection of cultures instead of a problematic single microculture [28, 29].

19.3.3 AAMC and ACGME Survey

19.3.3.1 Mistreatment Survey

Mistreatment surveys provide information about prevalence, types, and sources of mistreatment in the academic training environment. They are typically anonymous and may be administered at the program or institutional level. One such example is the annual mistreatment survey administered to all residents and fellows at the University of Minnesota [30] (Fig. 19.1). The survey uses a modified version of the mistreatment questions in the Association of American Medical Colleges Graduation Questionnaire and a question from the Racial Microaggressions Scale. Li et al. [8] administered a mistreatment survey to residents in ten emergency medicine programs to understand their experience with verbal abuse and threats, physical threats and attacks, sexual harassment, and racial harassment.

19.3.4 Focus Group

Focus groups allow a small group of individuals to participate in a facilitated discussion about their experiences in the work environment. These groups are usually optional and only include a sample of a larger population of trainees. Some trainees may not feel comfortable sharing their experiences in this environment. The responses are typically kept anonymous and aggregated such that individual concerns are not necessarily addressed unless separately reported by the individual.

19.3.5 Individual Reporting

Individual reporting is time intensive and burdensome, placing full responsibility on the individual experiencing mistreatment. Additionally, this requires trust in the system that there won't be retaliation and that positive change will be made in response to the report. A study of gender-based discrimination among medical trainees found that the most common reason for not filing a formal report was the belief that nothing would happen (65%), being too busy (37%), not knowing how to file a report (29%), and fear of negative consequences (25%) [31]. The burden of reporting also requires that an incident meet threshold of severity and rarity to justify the effort of the individual. Commonplace but impactful mistreatment, namely, microaggressions, are rarely reported through incident reporting mechanisms [32, 33]. This may explain the finding that the majority men (86%) and women (96%) have experienced or observed gender-based discrimination, but less than 5% formally reported those experiences [31].

A

- Have you been subjected to name calling or offensive remarks related to your gender, gender identity, sexual orientation, race, ethnicity or religion?
- Have you felt uncomfortable, undermined, or insulted by someone responding to your gender, gender identity, sexual orientation, race, ethnicity, or religion? (These actions and comments are collectively termed "microaggressions.")
- Have you been present when someone made offensive remarks, jokes or comments, not directed at you or anyone present, related to gender, gender identity, sexual orientation, race, ethnicity, or religion? (For example, someone telling a racist, misogynistic or homophobic joke where they assumed no one present would be offended?)
- Have you been subjected to unwanted sexual advances, requests for sexual favors or otherwise subjected to sexual harassment?
- Have you been threatened with physical harm or otherwise intimidated?
- Have you been humiliated or shamed?

B	Never	Once	More than once	Prefer not to answer
From patients/clients or their companions?	○	○	○	○
From faculty within your program?	○	○	○	○
From other faculty?	○	○	○	○
From interprofessional team members not in your program (e.g. nurses, pharmacists, social workers)?	○	○	○	○
From trainees (residents/fellows) within your program?	○	○	○	○
From other trainees (residents/fellows) in other training programs?	○	○	○	○

Fig. 19.1 Survey components. (**a**) Details the six question terms. (**b**) Shows the matrix response possibilities that were presented with each question stem. Figure originally published in Mustapha et al. [30]

19.4 Preventing, Addressing, and Mitigating Mistreatment

19.4.1 Actionable and Consequential Code of Conduct

The goal of a code of conduct is to transparently express expectations of behavior. Codes of conduct are variable in scope and may pertain to a specific job class, or to all people operating within a given environment. A common problem though is that not all codes of conduct are actionable—meaning they may not have accompanying procedures regarding actions to take if violated. Further, even actionable codes of conduct may not be consequential, meaning the action taken for violating the code may not be significant enough to correct, deter, or prevent recurrence of behavior. Codes of conduct that are both actionable and consequential can help to address mistreatment in the forms of microassaults (explicit derogatory remarks) [24], sexual harassment, bullying through humiliation, and threats of physical harm [34].

19.4.2 Bystander Training

Bystander training raises awareness about the different types of mistreatment, how mistreatment affects the whole care team, and how to intervene should someone witness mistreatment. An introductory training is typically 1-h long and provides specific scripts and role-playing scenarios to practice intervening. Advanced trainings and refresher courses involve group members generating specific examples and working through potential interventions. While the initial advanced training is typically expert facilitated, subsequent refreshers may be facilitated by a group lead during regular meetings and can be as short as 15 min each. Bystander training can help to address mistreatment in the forms of offensive remarks, microaggressions, sexual harassment, bullying through humiliation, and threats of physical harm [35, 36].

19.4.3 Meaningful and Accessible Individual Reporting System

The goal of individual reporting is to identify specific individuals whose behavior either violates a known code of conduct or is otherwise contributing to an unsafe work environment. The outcome of individual reports is largely based on the procedures that accompany codes of conduct or other Human Resources regulations, such as Title IX in US Federal legislation. Reported behavior that cannot be linked to codes of conduct or regulations often is not addressed at all, especially if the victim does not self-identify. Individual reporting is often limited by the individual's belief that nothing will happen, a burdensome reporting process, or fear of negative consequences [31]. To combat all forms of mistreatment, individual reporting should be easily accessible and user-friendly and have an option for confidentiality.

19.4.4 Victim Support Services

The type of service necessary will vary based on the experience of mistreatment; potential services include counseling, support groups, peer support, and legal services. Ideally, victim support services are available for all types of mistreatment.

19.4.5 Implicit Bias Training

Implicit bias refers to learned or automatic thoughts, attitudes, and stereotypes. There are numerous approaches to implicit bias training, but one recently proposed framework is through transformative learning theory (TLT) [35]. The essential components of TLT include the experience, critical reflection, dialogue, and action (behavioral change). This method encourages learners to not only recognize biases but also learn how to manage them. Implicit bias training can be most helpful to address mistreatment in the forms of offensive remarks, microaggressions, sexual harassment, humiliation, and shame.

19.4.6 External Review for High-Stakes Evaluations

The goal of external review for high-stakes evaluations is to allow individuals who are experiencing mistreatment to be fairly and rigorously evaluated in their work. External reviews may consist of direct observations, or review of material (in the case of medical learners, this could be case notes), by a diverse panel [37, 38]. All assessment tools should strive to evaluate objective criteria and performance to reduce bias—but as this is already a goal in developing medical performance assessments, it was not specifically mentioned as an intervention [39]. External review can help to mitigate the effects of all types of mistreatment.

19.4.7 Restorative Justice Mechanisms

This approach is distinguished from more general conflict resolution in that it acknowledges that there is an offender and a victim. The restorative justice approach promotes mutual understanding of harm by all involved, creates an environment that encourages offenders to take responsibility rather than minimize harm, and explores ways to rebuild trust [40]. Currently, a restorative justice model, one that favors rehabilitation over punitive measures, is not commonplace in academic medical centers. Restorative justice mechanisms best address mistreatment in the forms of offensive remarks, sexual harassment, humiliation and shame, and threats of harm, though it can also be utilized for microaggressions [41] (Box 19.1).

Box 19.1 Employing restorative justice [42]
In 2014, four female dental students at Dalhousie University filed a sexual harassment complaint due to offensive and discriminatory material posted about them by male students on a private Facebook group, the "Gentleman's Club." Multiple Facebook posts in the group reflected "misogynistic, sexist, and homophobic attitudes." The female students who filed the complaint decided to pursue a restorative justice process. The members of the Facebook group agree to full participation in an investigation and restorative justice process.

> Restorative justice provided us with a different sort of justice than the punitive type most of the loudest public voices seemed to want. We were clear from the beginning, to the people who most needed to hear it, that we were not looking to have our classmates expelled as 13 angry men who understood no more than they did the day the posts were uncovered. Nor did we want simply to forgive and forget. Rather, we were looking for a resolution that would allow us to graduate alongside men who understood the harms they caused, owned these harms, and would carry with them a responsibility and obligation to do better.—Female participants from the class of DDS2015 in *Report from the Restorative Justice Process at the Dalhousie University Faculty of Dentistry* [42]

During the investigation, the Faculty of Dentistry suspended the men from clinical duties and did not allow them to attend class with other classmates. The investigation and subsequent restorative process took nearly 5 months and included information sessions, individual and group interviews, group meetings, seminars, and workshops. The members of the Facebook group spent nearly 150 h in sessions and training. Additionally, the men planned the event "Day of Learning" for dental students and other stakeholders to process the harm caused, learn more about the restorative justice, and consider a path forward. The process focused on community building, inclusion and

(continued)

Box 19.1 (continued)

equality, professionalism and ethics, curriculum and program structure, and reporting processes and conflict resolution. Ultimately, the Academic Standards Class Committee (ASCC) and female complainants concluded that the men involved in the Facebook group had remediated their behavior and met the standard of professionalism required for graduation. This extensive process allowed a path forward for involved parties and robust dialogue about the types of culture and climate that tolerate discrimination.

19.5 Proposed Framework

Addressing the mistreatment of medical trainees requires an aggressive, multimodal approach at multiple levels in the complex academic training environment. Work teams in academic medicine are fluid by design, and each profession, such as nursing, pharmacy, and social work, has its own professional culture [43]. Medical teams represent an intersection between these different professional cultures and unique specialty-specific microcultures. Trainees and faculty interact with within-specialty attendings and trainees, as well as outside-of-specialty attendings and trainees, interdisciplinary staff, and patients. There have been many attempts to effectively track, prevent, and mitigate mistreatment. Despite well-meaning efforts, many interventions, particularly those that target a single institutional level, are not effective [44].

Transparency in result reporting is critical, as perceived tolerance for mistreatment is an important predictor of mistreatment occurring in a given institution [12]. For issues uncovered across all programs, such as patient mistreatment of trainees and faculty, the larger institution needs to support and direct interventions to address the problem. Similarly, for issues uncovered only within a single program, and only involving that program's faculty or trainees, that single program is clearly best positioned to act. For every result in between, however, determining the appropriate entity to address is more complex (Fig. 19.2).

Consider the following matrix structure for understanding the complex academic hospital system, where the reporting structures for domains, clinical service lines, and departments interact with each other perpendicularly. Clinical service lines (CSLs) are groupings of services across populations and disease groups (e.g., Emergency Room, Operating Suite, Med-Surg Units). Domains are environments of care where multiple anatomical specialties and disease states are addressed (e.g., Maternal-Child Health, Hospital Medicine, Surgical Specialties). Departments are traditional academic discipline groupings (e.g., Department of Surgery, Department of Medicine, Department of OB-GYN, Department of Emergency Medicine). To understand the entity best suited to intervene on mistreatment, it is necessary to consider the relationship between these three main organizational structures. It is essential to consider the relationship between organizational structures and groups of individuals reporting mistreatment, as well as the relationship between organizational structures and potential sources of mistreatment. Figure 19.3 shows the organizational structures mapped to the sources of mistreatment over which they have significant control and/or influence (Box 19.2).

Box 19.2 Who should take responsibility to address mistreatment?

Case 1: Trauma Surgery, Orthopedics, Urology, and General Surgery all report a significant amount of bullying from faculty not in their program. All programs spend time in the Operating Suite, and all programs are part of the surgical service line. Given that faculty report up through the service line, it would make most sense

(continued)

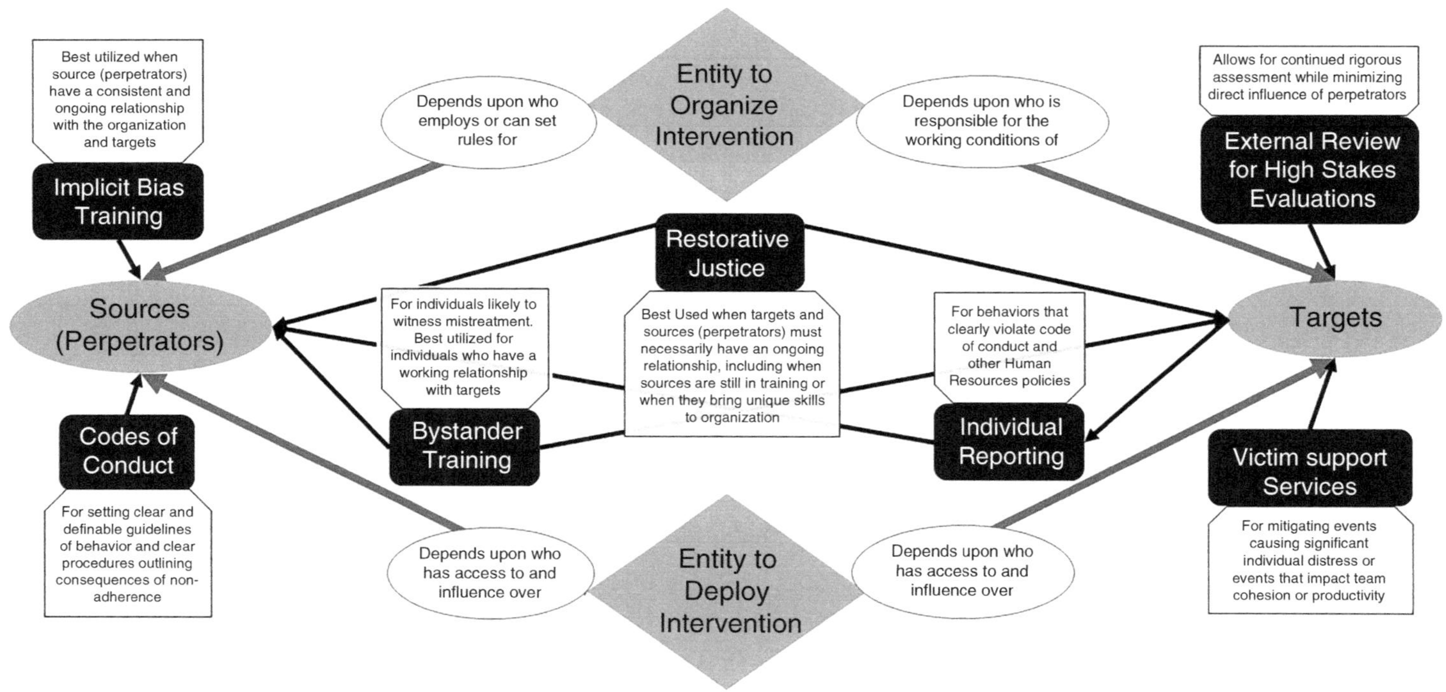

Fig. 19.2 Overall concept map showing who is responsible for addressing mistreatment and what interventions can be used. The gray arrows show the relationship between entities (or levels) within the organization and the sources and targets of mistreatment. The types of interventions that might be used to address mistreatment are shown, as well as their relationship to targets and sources. Of note, the entity to organize the intervention could also be the entity to deploy the intervention if the relevant relationships overlap

Box 19.2 (continued)

for an intervention to be organized by the clinical service line.

Case 2: Pediatric Surgery, General Pediatrics, Pediatrics Infectious Disease, and Pediatric Hospital Medicine Fellows report significant microaggressions from interprofessional team members. While all these programs operate within the same service line, they also operate within the same domain of the pediatric inpatient units. As the domain has most influence over interprofessional team members, in this case, it would be most effective for the domain to organize the intervention.

Case 3: A mistreatment survey reveals that multiple training programs within a single department (e.g., the Internal Medicine Residency Program, the Gastroenterology Fellowship Program, the Hematology/Oncology Fellowship Program, and the Infectious Disease Fellowship Program) show a significant prevalence of bullying from attending physicians. In this scenario, the Department of Medicine should organize a department-wide intervention to address a culture of bullying.

19.5.1 Coordinated and Harmonious Organizational Oversight

Historically, clinical service lines and domains have rarely been utilized to address mistreatment experienced by trainees and faculty. Mapping demonstrates the many circumstances under which coordinated institutional efforts would be the most effective. Additionally, mapping demonstrates the entity best positioned to intervene in each situation and shifts responsibility away from the proximal units (individual programs) where mistreatment is reported. It is essential that clinical service lines, domains, and departments share in the responsibility of ensuring a fair work environment.

19.6 Determining How to Intervene

Microaggressions, which are particularly difficult to describe and report, require clear policies and codes of conduct. Further interventions such as implicit bias training [35], external reviews of bias in performance evaluations [45], and the presence and accessibility of restorative justice mechanisms are necessary to effect microaggressions [36, 46, 47]. If pervasive, employing climate surveys that specifically ask about inclusivity can shed light on additional areas for intervention [48].

Direct offensive remarks, or microassaults, are more clearly identifiable by witnesses and are more amenable to bystander intervention [36, 49]. The presence of indirect offensive remarks suggests that the perpetrator interprets the climate and culture to be safe and tolerant of such remarks [36]. Similar to direct offensive remarks, indirect offensive remarks and witnessing microaggressions can be addressed by accessible reporting mechanisms and enforceable codes of conduct, as well as bystander training.

19.6.1 Sexual Harassment

Responding to sexual harassment in the workplace requires a multimodal approach. Interventions including clear and enforceable policies, accessible reporting mechanisms, restorative justice mechanisms, bystander training, and implicit bias training could all be beneficial [36, 50]. If pervasive, employing a climate survey regarding sexual harassment may be useful [51].

19.6.2 The Cycle of Abuse

The behaviors of humiliation, shame, threatening, and intimidation often represent a "cycle of abuse" in which medical students who are mistreated go on to be physicians who mistreat others [52, 53]. Climate surveys can demonstrate organizational contributors to a culture of shame

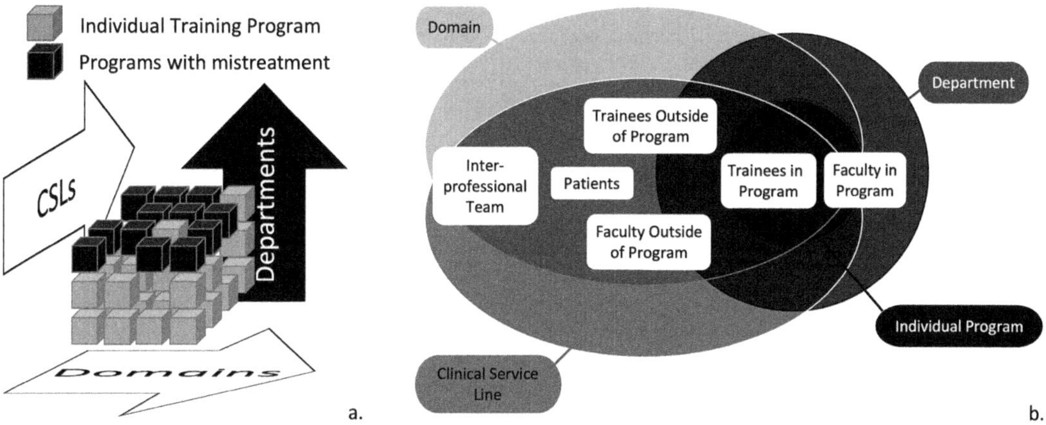

Fig. 19.3 Conceptual models to determine group(s) best positioned to organize and deploy intervention. CSLs (clinical service lines) are groupings of services across populations and disease groups (e.g. cancer care, neurosciences, primary care). Domains are environments of care where multiple anatomical specialties and disease states are addressed (e.g. perioperative, emergency department, hospitalists). Departments are traditional academic discipline groupings. There are two separate relationships to consider: (**a**) The relationship between organizational structures and groups of individuals reporting mistreatment. Shown is a pattern of findings that might prompt intervention beyond individual programs (e.g. department). (**b**) The relationship between organizational structures and potential sources of mistreatment. Understanding these relationships helps to determine which organizational entity or structure is best positioned to organize and deploy interventions

and reveal areas for intervention [54]. These types of behaviors can be mitigated by clear and enforceable policies and codes of conduct, as well as individual reporting mechanisms. Restorative justice mechanisms are important to address the "cycle of abuse" and provide an opportunity for the victim and perpetrator to move forward after the incident [47, 48] (Box 19.3).

Box 19.3 Putting it all together

An annual mistreatment survey demonstrates that residents are experiencing microassaults from patients on the psychiatry unit. Several individual reports reveal derogatory comments made by both patients and staff. Additionally, a climate survey demonstrates a higher level of bullying and harassment among interprofessional colleagues within the psychiatry department compared to other departments. A multimodal approach is necessary to address mistreatment within the psychiatry unit and department. Addressing mistreatment only at the level of the psychiatry residency is unlikely to lead to meaningful change. The Behavioral Health clinical service line should employ actionable mistreatment policies, as well as a patient code of conduct. The Inpatient Psychiatry domain will also need to establish clear procedures in response to the mistreatment policies and ensure staff are trained on those procedures. The domain should also consider holding implicit bias training for all staff. Finally, the Department of Psychiatry should provide bystander training to residents and staff as well as support services.

References

1. Cook DJ, Liutkus JF, Risdon CL, Griffith LE, Guyatt GH, Walter SD (1996) Residents' experiences of abuse, discrimination and sexual harassment during residency training. CMAJ 154(11):1657–1665

2. Yau BN, Chen AS, Montgomery KB, Dubuque N, McDowelle DM (2021) An internal perspective: the psychological impact of mistreatment. Acad Psychiatry 45(3):308–314
3. Babaria P, Abedin S, Berg D, Nunez-Smith M (2012) "I'm too used to it": a longitudinal qualitative study of third year female medical students' experiences of gendered encounters in medical education. Soc Sci Med 74(7):1013–1020
4. Richardson DA, Becker M, Frank RR, Sokol RJ (1997) Assessing medical students' perceptions of mistreatment in their second and third years. Acad Med 72(8):728–730
5. Silver HK, Glicken AD (1990) Medical student abuse: incidence, severity, and significance. JAMA 263(4): 527–532
6. Hill KA, Samuels EA, Gross CP, Desai MM, Sitkin Zelin N, Latimore D et al (2020) Assessment of the prevalence of medical student mistreatment by sex, race/ethnicity, and sexual orientation. JAMA Intern Med 180(5):653–665
7. Fnais N, Soobiah C, Chen MH, Lillie E, Perrier L, Tashkhandi M et al (2014) Harassment and discrimination in medical training. Acad Med 89(5):817–827
8. Li SF, Grant K, Bhoj T, Lent G, Garrick JF, Greenwald P et al (2010) Resident experience of abuse and harassment in emergency medicine: ten years later. J Emerg Med 38(2):248–252
9. Hu Y-Y, Ellis RJ, Hewitt DB, Yang AD, Cheung EO, Moskowitz JT et al (2019) Discrimination, abuse, harassment, and burnout in surgical residency training. N Engl J Med 381(18):1741–1752
10. Lu DW, Lall MD, Mitzman J, Heron S, Pierce A, Hartman ND et al (2020) #MeToo in EM: a multicenter survey of academic emergency medicine faculty on their experiences with gender discrimination and sexual harassment. West J Emerg Med 21(2):252–260
11. de Bourmont SS, Burra A, Nouri SS, El-Farra N, Mohottige D, Sloan C et al (2020) Resident physician experiences with and responses to biased patients. JAMA Netw Open 3(11):e2021769
12. National Academies of Sciences, Engineering, and Medicine (2018) Sexual harassment of women: climate, culture, and consequences in academic sciences, engineering, and medicine. The National Academies Press, Washington, DC
13. Jagsi R, Griffith KA, Jones R, Perumalswami CR, Ubel P, Stewart A (2016) Sexual harassment and discrimination experiences of academic medical faculty. JAMA 315:2120–2121
14. McDade WA (2019) Accreditation C for GME. Integrating a sustainable diverse and inclusive GME workforce. In: AAMC learn serve lead, Phoenix, AZ
15. Cook AF, Arora VM, Rasinski KA, Curlin FA, Yoon JD (2014) The prevalence of medical student mistreatment and its association with burnout. Acad Med 89(5):749–754
16. Frank E, Brogan D, Schiffman M (1998) Prevalence and correlates of harassment among US women physicians. Arch Intern Med 158(4):352–358
17. Thomas NK (2004) Resident burnout. JAMA 292: 2880–2889
18. Association of American Medical Colleges (AAMC) (2009) Striving toward excellence: faculty diversity in medical education. Washington, DC. yumpu.com/en/document/read/22797053/. Accessed 4 May 2021
19. Cooper-Patrick L, Gallo JJ, Gonzales JJ, Vu HT, Powe NR, Nelson C et al (1999) Race, gender, and partnership in the patient-physician relationship. JAMA 282(6):583–589
20. Saha S, Komaromy M, Koepsell TD, Bindman AB (1999) Patient-physician racial concordance and the perceived quality and use of health care. Arch Intern Med 159(9):997–1004
21. Alsan M, Garrick O, Graziani GC (2019) Does diversity matter for health? experimental evidence from Oakland. Am Econ Rev 109:4071–4111
22. Smedley BD, Stith AY, Nelson AR (2003) Unequal treatment: confronting racial and ethnic disparities in health care. The National Academies Press, Washington DC. pubmed.ncbi.nlm.nih.gov/25032386/. Accessed 4 May 2021
23. Pierce CM, Carew JV, Pierce-Gonzalez D, Wills D (1977) An experiment in racism. Educ Urban Soc 10(1):61–87
24. Sue DW, Capodilupo CM, Torino GC, Bucceri JM, Holder AMB, Nadal KL et al (2007) Racial microaggressions in everyday life: implications for clinical practice. Am Psychol 62(4):271–286
25. Sklar DP (2019) Recognizing and eliminating shame culture in health professions education. Acad Med 94: 1061–1063
26. Aryee S, Sun LY, Chen ZX, Debrah YA (2007) Antecedents and outcomes of abusive supervision: test of a trickle-down model. J Appl Psychol 92(1): 191–201
27. Pololi LH, Brennan RT, Civian JT, Shea S, Brennan-Wydra E, Evans AT (2020) Us, too. Sexual harassment within academic medicine in the United States. Am J Med 133(2):245–248
28. Wear D, Keck-McNulty C (2004) Attitudes of female nurses and female residents toward each other: a qualitative study in one U.S. Teaching Hospital. Acad Med 79(4):291–301
29. Gjerberg E, Kjølsrød L (2001) The doctor-nurse relationship: how easy is it to be a female doctor co-operating with a female nurse? Soc Sci Med 52(2):189–202
30. Mustapha T, Ho Y, Andrews JS, Cullen MJ (2019) See no evil, hear no evil, stop no evil: institutional-level tracking to combat mistreatment of residents and fellows. J Grad Med Educ 11(5):601–605
31. McKinley SK, Wang LJ, Gartland RM, Westfal ML, Costantino CL, Schwartz D et al (2019) "Yes, I'm the Doctor": one department's approach to assessing and

addressing gender-based discrimination in the modern medical training era. Acad Med 94:1691–1698
32. Sue DW (2010) Microaggressions in everyday life: race, gender, and sexual orientation. John Wiley & Sons, Hoboken, NJ
33. Silver JK, Rowe M, Sinha MS, Molinares DM, Spector ND, Mukherjee D (2018) Micro-ineq Med 10:1106–1114
34. Ashkanasy NM, Jackson CRA (2000) Organizational culture and climate. In: Handbook of organizational culture and climate. Sage Publications, Inc, Thousand Oaks, CA, pp 398–415
35. Sukhera J, Watling CJ, Gonzalez CM (2020) Implicit bias in health professions. Acad Med 95(5):717–723
36. Banyard VL, Moynihan MM, Cares AC, Warner R (2014) How do we know if it works? Measuring outcomes in bystander-focused abuse prevention on campuses. Psychol Violence 4(1):101–115
37. Newkirk P, Diversity, Inc (2019) The failed promise of a billion-dollar business. Bold Type Books, New York, NY
38. Giuliano L, Levine DI, Leonard J (2011) Racial bias in the manager-employee relationship: an analysis of quits, dismissals, and promotions at a large retail firm. J Hum Resour 46(1):26–52
39. Steinke P, Fitch P Minimizing bias when assessing student work. files.eric.ed.gov/fulltext/EJ1168692.pdf. Accessed 4 May 2021
40. Acosta D, Karp DR (2018) Restorative justice as the rx for mistreatment in academic medicine: applications to consider for learners, faculty, and staff. Acad Med 93(3):354–356
41. Karp DRSMGT (2018) Restorative justice in universities: case studies of what works with restorative responses to student misconduct. In: The Routledge international handbook of restorative justice. Routledge, New York, NY. Chapter 17
42. Llewellyn J, MacIsaac J, MacKay M. Report from the restorative justice process at the Dalhousie University Faculty of Dentistry. whistler.bibliocommons.com/v2/record/S49C126606603. Accessed 4 May 2021
43. Zelek B, Phillips SP (2003) Gender and power: nurses and doctors in Canada. equityhealthj.com/content/2/1/1. Accessed 4 May 2021
44. Fried JM, Vermillion M, Parker NH, Uijtdehaage S (2012) Eradicating medical student mistreatment. Acad Med 87(9):1191–1198
45. Carbado DW, Gulati M (2000) Working identity. Cornell Law Rev 85(5):1270–1278
46. Offermann LR, Basford TE, Graebner R, Degraaf SB, Jaffer S (2013) Slights, snubs, and slurs: leader equity and microaggressions. Equal Divers Incl 32(4): 374–393
47. AAMC webinars and online courses: restorative justice for academic medicine (RJAM): Using RJ to respond to mistreatment and bias (tier II). aamc.elevate.commpartners.com/products/restorative-justice-for-academic-medicine-rjam-using-rj-to-respond-to-mistreatment-and-bias-tier-ii-november-21-2019. Accessed 4 May 2021
48. Person SD, Jordan CG, Allison JJ, Fink Ogawa LM, Castillo-Page L, Conrad S et al (2015) Measuring diversity and inclusion in academic medicine: the diversity engagement survey. Acad Med 90(12): 1675–1683
49. Coker AL, Bush HM, Cook-Craig PG, DeGue SA, Clear ER, Brancato CJ et al (2017) RCT testing bystander effectiveness to reduce violence. Am J Prev Med 52(5):566–578
50. Justice beyond bars: exploring the restorative justice alternative for victims of rape and sexual assault. thefreelibrary.com/Justice+beyond+bars%3A+exploring+the+restorative+justice+alternative...-a0414824835. Accessed 4 May 2021
51. Estrada AX, Olson KJ, Harbke CR, Berggren AW (2011) Evaluating a brief scale measuring psychological climate for sexual harassment. Mil Psychol 23(4): 410–432
52. Major A (2014) To bully and be bullied: harassment and mistreatment in medical education. Virtual Mentor 16:155–160
53. Hunt D, Barzansky B, Migdal M (2009) When bad things happen in the learning environment. Virtual Mentor 11:106–110
54. Priesemuth M, Schminke M, Ambrose ML, Folger R (2014) Abusive supervision climate: A multiple-mediation model of its impact on group outcomes. Acad Manag J 57(5):1513–1534
55. Minikel-Lacocque J (2013) Racism, college, and the power of words. Am Educ Res J 50(3):432–465

Part VI

Sensationalistic Science

Misleading Research of So-Called Alternative Medicine

20

Edzard Ernst

Abstract

Research into so-called alternative medicine (SCAM) faces many problems. Practitioners of SCAM try to persuade consumers that, for several reasons, research in their field is not all that important. The amount of research into SCAM is relatively small and its quality is often very poor. Crucially, SCAM researchers are often deeply biased towards their therapy and seem incapable of critical assessments. Adverse effects of SCAM are thus both under-investigated and under-reported. These and many further problems contribute to much confusion and mislead consumers about the value of SCAM. Such confusion can lead to ill-advised therapeutic decisions potentially endangering public health.

20.1 Introduction

The area of healthcare which we will discuss in this chapter has many different names; they may have subtly different meanings but are often employed synonymously.

E. Ernst (✉)
University of Exeter, Complementary Medicine Group, Peninsula Medical School, Exeter, UK
e-mail: e.ernst@exeter.ac.uk

- Fringe medicine is rarely used today. It denotes the fact that the treatments under this umbrella are not in the mainstream of healthcare.
- Unorthodox medicine is based on the fact that medical orthodoxy tends to shun most of the treatments under this umbrella.
- Unconventional is wide open to misunderstandings: any innovation in medicine might initially be called unconventional.
- Traditional medicine describes the fact that most of the modalities in question have been used for centuries and thus have a long tradition of usage.
- Alternative medicine is the term everyone seems to know and which is therefore most commonly employed in non-scientific contexts.
- Complementary medicine accounts for the fact that the treatments tend to be used by patients in parallel with conventional medicine.
- Complementary and alternative medicine (CAM) describes the phenomenon that many of the treatments can be employed either as a replacement of or as an adjunct to conventional medicine.
- Holistic medicine denotes the fact that practitioners often pride themselves to look after the whole patient—body, mind and spirit.
- Natural medicine describes the notion that many of the methods in question are natural.

J. Faintuch, S. Faintuch (eds.), *Integrity of Scientific Research*,
https://doi.org/10.1007/978-3-030-99680-2_20

- Integrated or integrative medicine is the term used for adding alternative modalities to routine care.

Of all these terms, alternative medicine seems by far the most popular one. Yet, I prefer using the term 'so-called alternative medicine' (SCAM) [1]. It accounts for two important facts: [1] if a treatment does not work, it cannot possibly serve as an adequate alternative; and [2] if a therapy does work, it should be part of conventional medicine. Thus, there cannot be an alternative medicine, as much as there cannot be an alternative chemistry or physics. Some advocates find the term and its abbreviation derogatory, but my decision to use it in most of my current writings was inspired by one of the best-known champions of alternative medicine, Prince Charles: in his book *Harmony*, he repeatedly speaks of 'so-called alternative treatments' [2].

20.2 SCAM Enthusiasts Are Less than Enthusiastic About Rigorous Research

After seeing the often-negative results of the rigorous scientific tests of their favourite therapy, SCAM enthusiasts tend to be disappointed. Deep down, they 'know' (are convinced) that their SCAM works. But why does the research suggests otherwise? This question almost inevitably leads to a state of cognitive dissonance—inconsistent thoughts caused by discrepancies between belief and fact—which causes psychological discomfort and therefore needs to be resolved. And the resolution which many SCAM supporters find is disarmingly simple: SCAM defies science! Their SCAM, they conclude, is too subtle, too holistic, too individualised, too special to be squeezed into the straitjacket of what they call 'conventional science'.

In a clinical trial, they claim, everything must be standardised and reduced to measurable endpoints. But, in 'real life', they do not standardise! On the contrary, they tailor their treatments according to the special needs of each patient. Clinical trials ignore or disallow this and therefore, such tests are no adequate tools for evaluating their SCAM. Consequently, these SCAM enthusiasts reject 'conventional' science and decide that other types of evidence are preferable.

Enthusiasts of SCAM often feel strongly about these issues, and they make sure that the public knows their concerns. Countless articles have therefore convinced consumers that they have a point: science is too insensitive a tool to be applied to SCAM. A recent example is this statement from the UK 'All-Party Parliamentary Group for Integrated Healthcare [3]' about guidelines by NICE, the UK National Institute for Health and Care Excellence:

> NICE guidelines are too narrow and do not fit well with models of care such as complementary, traditional and natural therapies, and should incorporate qualitative evidence and patient outcomes measures as well as RCT evidence. Complementary, traditional and natural healthcare associations should take steps to educate and advise their members on the use of Measure Yourself Medical Outcome Profiles (MYMOP), and patient outcome measures should be collated by an independent central resource to identify for what conditions patients are seeking treatment, and with what outcomes.

The experience of hundreds of years, SCAM enthusiasts feel, tells us that SCAM works, and it weighs more than any scientific assessment.

20.3 Experience Is Not Evidence

What they tend to forget is that experience is often complex and rarely easy to interpret. When a patient experiences a symptomatic improvement after SCAM, both the practitioner and the patient assume this to be the result of the treatment. This assumption ignores several important facts. Two events—the treatment and the improvement—that follow each other in time are not necessarily causally related. Even the most superficial scan of the possibilities discloses several reasons for clinical improvement, other than the treatment itself:

- The natural history of the condition
- Regression towards the mean
- The placebo effect
- Concomitant treatments
- Social desirability

Such phenomena can contribute to or determine the clinical outcome in such a way that inefficacious or even mildly harmful treatments appear to be efficacious. In other words, the prescribed treatment is only one of the several influences on the clinical outcome, and even the most impressive clinical experience can be totally misleading.

The multifactorial nature of any clinical response requires that we control for all the factors that might determine the outcome other than the SCAM. Ideally, we would need to create a situation or an experiment where two groups of patients are exposed to the full range of factors, and the only difference is that one group does receive the SCAM, while the other one does not. And this is precisely the model of a controlled clinical trial.

20.4 Clinical Trials

There are hundreds of clinical trials of SCAMs where the holistic, subtle, individualised nature of the treatment was fully accounted for. Here are just two examples from my own research:

- **Spiritual healing** is a popular complementary and alternative therapy; in the UK almost 13,000 members are registered in 9 separate healing organisations. The present randomised clinical trial [4] was designed to investigate the efficacy of healing in the treatment of chronic pain. One hundred and twenty patients suffering from chronic pain, predominantly of neuropathic and nociceptive origin resistant to conventional treatments, were recruited from a pain management clinic. The trial had two parts: face-to-face healing or simulated face-to-face healing for 30 min per week for 8 weeks (part I) and distant healing or no healing for 30 min per week for 8 weeks (part II). The McGill Pain Questionnaire was predefined as the primary outcome measure, and sample size was calculated to detect a difference of eight units on the total pain rating index of this instrument after 8 weeks of healing. VASs for pain, SF36, HAD scale, MYMOP and patient subjective experiences at week 8 were employed as secondary outcome measures. Data from all patients who reached the predefined mid-point of 4 weeks (50 subjects in part I and 55 subjects in part II) were included in the analysis. Two baseline measurements of outcome measures were made, 3 weeks apart, and no significant differences were observed between them. After eight sessions there were significant decreases from baseline in McGill Pain Questionnaire total pain rating index score for both groups in part I and for the control group in part II. However, there were no statistically significant differences between healing and control groups in either part.
- **Homeopathy** is frequently used to treat asthma in children. In the common classical form of homeopathy, prescriptions are individualised for each patient. There has been no rigorous investigation into this form of treatment for asthma. Methods: In a randomised, double-blind, placebo-controlled trial [5], the effects of individualised homeopathic remedies were compared with placebo medication in 96 children with mild to moderate asthma as an adjunct to conventional treatment. The main outcome measure was the active quality of living subscale of the Childhood Asthma Questionnaire administered at baseline and follow-up at 12 months. Other outcome measures included other subscales of the same questionnaire, peak flow rates, use of medication, symptom scores, days off school, asthma events, global assessment of change and adverse reactions. Results: There were no clinically relevant or statistically significant changes in the active quality of life score. Other subscales, notably those measuring severity, indicated relative improvements, but the sizes of the effects were small. There

were no differences between the groups for other measures.

Such studies were designed to minimise all possible sources of bias and confounding. They had a control group, i.e. a group of patients who received a treatment that is different from the SCAM which is being tested. At the end of the treatment period, we could therefore compare the effects of our SCAM with those of another intervention, a placebo or no treatment at all.

The over-riding principle of any clinical trial is always the same: we want to make sure that we can reliably determine whether the SCAM was the cause of the clinical outcome. Causality is evidently the key here; and the crucial difference between clinical experience and scientific evidence is that the former does not allow conclusions about causality, while the latter does. What SCAM practitioners and patients witness can have a myriad of causes; what scientists observe in a well-designed trial is caused by the treatment per se.

20.5 The Quantity of Research into SCAM

At the time of writing these lines (May 2021), a MEDLINE search for 'complementary and alternative' produced a total of more than 27,000 articles on the subject. For 1993, the year when I started being a full-time SCAM researcher at the University of Exeter, Medline listed just 259 such papers. The numbers subsequently increased steadily and reached its peak in 2020, when ~1500 articles were listed in this category.

Many consumers will think that these figures are impressive, perhaps even higher than expected. However, when we put them into perspective, they might feel like thinking again. In the light of the total number of MEDLINE-listed articles in several other (randomly chosen) fields, the number of SCAM papers seems deplorably low:

- Rheumatology: >136,000
- Pharmacology: >6,587,000
- Psychiatry: >672,000
- Paediatrics: >989,000
- Immunology: >1,836,000

Not all the 27,000 SCAM papers mentioned above are original research articles. I estimate that only about 10–20% of them report original research; the rest is a mixture of opinion pieces, editorials, viewpoints, letters to the editor, etc.

20.6 The Quality of Research into SCAM

While the amount of research into SCAM is less than impressive, the main problem is the quality. Much of conventional clinical research is of dubious quality, but most of research in SCAM is outright dismal. Reports of research should fully describe which questions were addressed and why, what was done, what was shown and what the findings mean [6]. In SCAM, these conditions are hardly ever met. Since 2012, I publish a blog [7] where I regularly analyse recent SCAM research, and I am never short of truly appalling pseudoscience to criticise. Perhaps the ubiquity of serious flaws in SCAM studies is best demonstrated by the fact that almost every systematic review of SCAM ever published includes a passage where the authors discuss that their conclusion was severely limited, due to the poor quality of the primary research.

20.7 Problems Encountered When Conducting SCAM Research

Evidently, this is due to several factors. In most countries, SCAM research funding is on a very small scale. For instance, only 0.08% of the UK National Health Service research budget go towards SCAM research [8]. It seems obvious that lack of plausibility of many SCAMs deters scientific review committees from defining SCAM as a priority. A vicious circle is then set in motion: little plausibility means no funds, therefore no research, therefore little plausibility.

Clinical trials of SCAM are often more expensive than the already costly studies of conventional medicine. This is because many SCAM treatments are therapist-led (and therapists need to be paid, of course), effect sizes are often small (requiring large sample sizes), and therapeutic effects may appear only after long treatment periods (requiring long treatments and follow-up periods in clinical trials). Collectively, these factors mean greater expense.

A further aggravating factor is that, for most SCAM modalities, intellectual property cannot be protected; thus, commercial investments are rarely forthcoming. This shortage of SCAM research funds has three important consequences:

- It prevents relevant projects from happening.
- It hinders the development of a research infrastructure comparable to that of conventional medicine.
- Crucially, it keeps well-trained career scientists from entering the field.

Many SCAM therapies (e.g. acupuncture, massage therapy, spinal manipulation) are physical by nature, which creates methodological challenges. What, for instance, is an acceptable 'placebo' control for a trial of massage treatments? Like several other areas of conventional medicine (e.g. physiotherapy, surgery, psychotherapy), blinding patients in clinical trials can be difficult or even impossible. Thus, the highest level of scientific rigour cannot always be realised in trials of SCAM.

20.8 SCAM Researchers Might Be the Biggest Problem of Them All

The most serious obstacle to high-quality research is, however, the attitude of the SCAM researcher. The typical SCAM researcher has little or no scientific background; she has been drawn to SCAM research foremost by her personal devotion to SCAM. In other words, she is a SCAM enthusiast and her hope is that her research will prove her favourite SCAM to be safe and effective. This emotional baggage weighs heavily on the research; it results in a conflict of interest that is almost never declared and often much more powerful than any financial conflicts could ever be.

20.8.1 Unbiased Protocols

In my own research team at Exeter, I have witnessed this lamentable phenomenon over and over again, and I have seen how it impacts on the quality of SCAM research. If a researcher's starting point is to prove SCAM, this represents a fundamentally different stance from that of a true scientist whose ambition is to test SCAM as rigorously as possible. The latter approach is, of course, the only correct one, while the former attitude can even amount to an abuse of science.

We have repeatedly shown that the percentage of negative findings (results that fail to demonstrate that SCAM is effective) published in such journals is zero or very close to zero [9]. On my blog [1], I have published a series of articles satirically called 'The Alternative Medicine Hall of Fame' where I highlight the work of prominent SCAM researchers. These assessments show that many SCAM researchers have exclusively published positive findings and never drawn a single negative conclusion related to their often implausible SCAM. This is, I think, the strongest evidence currently available about the worrying influence of the SCAM researcher's bias on the research outcome.

20.9 The Safety of SCAM

Virtually all forms of SCAM have the potential to cause adverse effects, some of which can be serious, such as the following:

- Herbal remedies can be toxic, contaminated or adulterated [10, 11].
- .Dietary supplements and herbal remedies can interact with prescribed drugs [12].
- An acupuncture needle can penetrate a vital organ [13].

- Chiropractic spinal manipulation can cause a stroke and death [14].

These direct risks can all cause fatalities and, due to the absence of post-marketing surveillance, their frequencies are essentially unknown [15]. However, by far the biggest danger of SCAM is its indirect risks. They relate to the advice and lack of medical competence of certain SCAM providers [16]. Even a seemingly harmless SCAM like homeopathy can turn out to be life-threatening, if it is used as an alternative therapy for a serious disease [17, 1].

20.10 Proving the Ineffectiveness of SCAM

In discussions with SCAM enthusiasts, one argument comes up with unfailing regularity. Science, they claim, has not proven the ineffectiveness of their form of SCAM. We therefore ought to be fair and give it the benefit of the doubt.

Science is an inadequate tool for proving a negative; therefore, it has not, and never will, conclusively prove that SCAM is ineffective. Nobody can conclusively disprove that extraterrestrial life exists or that a multitude of teapots made of chocolate is not currently circulating distant planets. Similarly, you cannot conclusively disprove that homeopathy (or any other SCAM) does not work. This is neither theoretically nor practically possible.

We can, for instance, conduct a clinical trial, and that might fail to demonstrate the effectiveness of homeopathy (or any other SCAM), but this is not a conclusive proof of its ineffectiveness. And even if we did consider it to be a conclusive proof, a homeopath can always claim that the result was false-negative because:

- The remedy was not well-chosen
- Its potency was suboptimal
- The patients were wrong
- The illness was not responding
- The treatment period was too short or too long

Armed with a negative trial, all we can ever state with confidence is that the study failed to show the effectiveness of SCAM. A SCAM practitioner would be entirely correct in arguing that this might be so, but it does not prove that his SCAM does not work. The absence of evidence is not the same as evidence of absence. And the many SCAM proponents who regularly remind us of this truism are, of course, correct.

20.10.1 Evidence-Based Therapy

In healthcare we do not normally bother about proving the ineffectiveness of a therapy. We are under the ethical, moral and legal obligation to use, whenever possible, those treatments that are supported by a positive proof of effectiveness. Our default position must therefore be that any treatment which lacks positive proof must be considered ineffective until there is sufficient evidence to show otherwise. This position is not one that is aimed against SCAM. It is necessary to safeguard the interest of the patient; responsible healthcare professionals must protect patients from ineffective treatments.

References

1. amazon.co.uk/SCAM-So-Called-Alternative-Medicine-Societas/dp/1845409701. Accessed 4 May 2021
2. amazon.co.uk/Harmony-New-Way-Looking-World/dp/0007348037. Accessed 4 May 2021
3. cnhc.org.uk/sites/default/files/Downloads/PGIH-Report.pdf. Accessed 4 May 2021
4. Abbot NC, Harkness EF, Stevinson C, Marshall FP, Conn DA, Ernst E (2001) Spiritual healing as a therapy for chronic pain: a randomized clinical trial. Pain 91(1-2):79–89
5. White A, Slade P, Hunt C, Hart A, Ernst E (2003) Individualised homeopathy as an adjunct in the treatment of childhood asthma: a randomised placebo controlled trial. Thorax 58(4):317–321
6. Glaszlou P, Altman DG, Bossuyt P et al (2014) Reducing waste from incomplete or unusable reports of biomedical research. Lancet 393(9913):267–276
7. edzardernst.com. Accessed 4 May 2021
8. Ernst E (1996) Regulating complementary medicine. Only 0.08% of funding for research in NHS goes to complementary medicine. BMJ 313(7061):882

9. Ernst E, Pittler MH (1997) Alternative therapy bias. Nature 385(6616):480
10. Posadzki P, Watson LK, Ernst E (2013) Adverse effects of herbal medicines: an overview of systematic reviews. Clin Med (Lond) 13(1):7–12
11. Posadzki P, Watson LK, Ernst E (2013) Contamination and adulteration of herbal medicinal products (HMPs): an overview of systematic reviews. Eur J Clin Pharmacol 59(3):295–307
12. Ernst E (2012) Interactions between drugs and supplements: the tip of an iceberg? Int J Clin Pract 66:1056–1078
13. Melchart D, Weidenhammer W, Streng A, Reitmayr S, Hoppe A, Ernst E, Linde K (2004) Prospective investigation of adverse effects of acupuncture in 97 733 patients. Arch Intern Med 164(1):104–105
14. Stevinson C, Ernst E (2002) Risks associated with spinal manipulation. Am J Med 112(7):566–571
15. Ernst E (2011) Fatalities after CAM: an overview. Br J Gen Pract 61(587):404–405
16. Ernst E (2011) Anthroposophy: a risk factor for non-compliance with measles immunization. Pediatr Infect Dis J 30(3):187–189
17. Ernst E (2007) Popular health advice: entertainment or risk factor? Br J Gen Pract 57(538):415

21 False Information and Mandatory Pediatric Vaccination

Giulia Spina, Sarah Barni, and Elena Bozzola

Abstract

Many different factors influence vaccination adherence, such as availability, distribution, and access, but also confidence, compliance, convenience, and other emotional aspects that can lead to vaccine hesitancy. Globalization, social networks, and mass media play a great role in divulgation concerning vaccines. On the Internet, there can also be found false, partial, sometimes frightful, and misleading informations, which usually have a great power of influence on people, because they strike vivid emotions. Bots, trolls, and conspiracy rumors frequently undermine belief in vaccine reliability. The risk-benefit ratio of vaccines is too positive to let false information and vaccine hesitancy prevail. Vaccines are a great tool that can prevent illnesses and protect the most fragile people. Mandatory vaccination has been a valuable strategy to enhance vaccination coverage; however full population commitment has to be mobilized in order to achieve success in vaccination.

G. Spina · S. Barni · E. Bozzola (✉)
Unit of Pediatrics, IRCCS Bambino Gesù Children's Hospital, Rome, Italy
e-mail: giulia.spina@opbg.net; sarah.barni@opbg.net; elena.bozzola@opbg.net

List of Abbreviations

ADEM	Acute disseminated encephalomyelitis
CVST	Cerebral venous sinus thrombosis
DIC	Disseminated intravascular coagulation
EMA	European Medical Agency
MMR	Measles-mumps-rubella
no-vax	No vaccination
SAGE	Strategic Advisory Group of Experts
SSPE	Subacute sclerosing panencephalitis
WHO	World Health Organization

21.1 Introduction

Many different factors can influence vaccination adherence such as availability, distribution, and access. However, even among high-income countries where vaccination is guaranteed and widely available, there is not total compliance. Nowadays with the Internet and modern globalization, people can virtually know everything.

Then, why would anyone avoid vaccination? Together with reliable, referenced informations, there can also be found false, partial, sometimes frightful, and misleading informations, which usually have a greater influence on people and are therefore frequently debated by the media.

J. Faintuch, S. Faintuch (eds.), *Integrity of Scientific Research*,
https://doi.org/10.1007/978-3-030-99680-2_21

21.2 Origins of Myths About Vaccination Consequences

In the late 1990s, a research paper, not well planned and with deceiving conclusions, suggested a link between the measles vaccine and the development of autism. It was first published in *The Lancet* but then retracted. It is probably the origin of the biggest controversy ever involving immunization, which drove the publication of an incredible number of works regarding measles vaccine [1].

Many following studies demonstrated the absence of any link between measles vaccines and autism. A recent American meta-analysis involving a total of 1,266,327 children confirmed no association between measles-mumps-rubella (MMR) vaccination and autism [2]. Anyhow it reinforced previous movements of "no-vax" (no vaccination), conducting to a sudden increase in measles cases in several parts of the world [1]. The false myth of measles vaccine and autism has been used by the mainstream media for its social appealing and frightful meaning; moreover, it had huge consequences, undermining immunization strategies and changing public's perception about vaccine safety.

Vaccination uptake in the UK dropped to levels as low as 79.9% in 2003–2004 [3]. In 2017, a great outbreak struck Minnesota (USA), mainly involving the Somali American population, where misinformation about MMR vaccine and autism had spread [4]. Another measles outbreak has been observed in Disneyland in Anaheim, California (USA), where half of the measles cases occurred in people eligible for vaccination who intentionally remained unvaccinated [5].

Romania had 7570 measles cases in October 2016, followed by Italy in early 2017, with 5006 cases, and then Germany with 981 ones [6]. In Italy, due to the political debate among no-vax groups, the deferral in imposing immunization, and the lack of clinical skills in early diagnosis, the 2017–2018 measles epidemic caused more than 8000 cases and 15 deaths [7].

21.3 The Role of Internet and Social Media

Two of the biggest differences between social media are Internet-based applications from one side and traditional media from the other one, where there is no editorial oversight nor fact check. Contents are often personal opinions and sold as if they were evidence [8]. Anyone can shout any kind of thought and people can decide to believe it or not. Facebook, Instagram, Twitter, LinkedIn, Pinterest, Snapchat, YouTube, Reddit, and TikTok are some of the different content-type platforms that enable interaction among users, who can even decide to remain anonymous or to have fake accounts.

Social media allow users to create virtual communities of people with shared beliefs and interests. This mechanism leads to self-select streams of content relevant to them, at the same time rejecting content which they do not agree with [9].

21.3.1 Dangerous Algorithms and Artificial Intelligence

In 2020 on Netflix, a documentary film called *The Social Dilemma* was released. Beyond the fiction, it explained how these social media use specific algorithms, in order to try to keep the user on their platforms. They constantly suggest contents and advertising which are more likely to be enjoyed by the users. This process leads to self-reinforcement and contributes to ideological isolation. The more one looks for anti-vaccination information, the more the social media finds out about it, strengthening polarization. As a matter of fact, there are significant public health apprehensions raised by anti-vaccination messaging on these platforms and the resulting vaccine hesitancy.

Since 2020 with the epidemic emergency of Sars-Cov-2, social media users started to write and share posts debating on the real existence of the virus, or on its intentional creation in Chinese laboratories, and then arguing with the use of facial mask, social distancing, and finally the reliability of vaccines.

Moreover, social media can spread misinformation by using vivid narratives and powerful imagery. The success of such methods can be described through the "fuzzy-trace theory" whereby individuals assimilate information by two methods: through precise memories which embrace all specific details and through gist memories which hold the crucial meaning only, in a broad way. Choices are typically made using the gist memories. Moreover, social media posts stating gist are more prone to be shared. On the opposite side, quantitative information presented by evidence-based medical literature on pro-vaccine platforms may be less evoking [10].

21.4 Vaccine Hesitancy

In 2019 the WHO listed "vaccine hesitancy" as one of the top ten threats to world health. Vaccine hesitancy can be a global risk because it may lead to under-coverage and reappearance of preventable diseases, which can have a high mortality index, especially in those with immunodeficiencies who cannot get vaccinated but who can rely on herd immunity.

The WHO SAGE working group [11] has adopted the following definition for it: "Vaccine hesitancy refers to delay in acceptance or refusal of vaccination despite availability of vaccination services. Vaccine hesitancy is complex and context specific, varying across time, place and vaccines. It is influenced by factors such as complacency, convenience and confidence." Vaccine hesitancy is a well-defined and analyzed phenomenon which doesn't always follow a logical path, because it is much more related to personal beliefs and misperceptions. The SAGE working group tried to decodify it with the "*3 Cs*" model: *c*onfidence, *c*omplacency, and *c*onvenience.

21.4.1 The "*3 Cs*" Theory

Confidence refers to trust in the efficacy and safety of vaccines; belief in the organization that supplies them, including the health services and health professionals; and approval of the policy's reasoning that decides on the needed vaccines.

Complacency occurs when the risks of vaccine-preventable diseases are perceived as low, and therefore vaccination is not considered a necessary preventive action. Immunization programs' accomplishments may, paradoxically, result in complacency and eventually in hesitancy. In fact, when people no longer face symptoms and complications of vaccine-preventable diseases, that's because of the vaccination coverage. On the contrary, it's more common to highlight vaccinations' adverse reactions. Therefore, some people decide not to get their children vaccinated, because they perceive vaccinations more harmful than infectious diseases.

Vaccination convenience is also a significant factor, and it includes physical accessibility, cost, geographical availability, and capacity to understand (language and health literacy). The quality of the service (real or perceived), delivered in a specific cultural context that should be appropriate and comfortable, also impacts the decision to be vaccinated.

A recent review [12] classified into five main concepts parents' justification for their vaccination hesitancy: risk conceptualization, mistrust, alternative health beliefs, view on parental responsibility, and information on vaccination. Although these concepts have emerged separately, they can also be inextricably linked.

21.4.2 The Five Steps of Vaccination Hesitancy

Risk conceptualization: parents may think of vaccinations as a risk for their children's health. They are worried to be poisoning the children, or they have concerns about the potential side effects such as disabilities, deafness, mental damage, coma, or death. Some parents fear autism, polio, or cancer, despite doctors' reassurances. Others believe that vaccines are able to compromise the immune system. Some parents think that a healthy diet and a fit lifestyle can avoid getting diseases that vaccines target. Some hesitant parents believe that in a "civilized country," some vaccinations are not needed, for example, measles vaccine. They often declare that these diseases are not so life-threatening to merit vaccination [12].

Mistrust: parents may be suspicious toward vaccine-related persons and institutions such as pharmaceutical companies, researchers, health practitioners, and even the media. They suspect that pharmaceutical economic interests affect vaccine research. Developing doubts about the integrity of these companies then leads to a lack of trust in the evidence presented to support vaccine efficacy and safety. Some parents think that medical practitioners are influenced by the interests of vaccine manufacturers and therefore that their information is not impartial.

Alternative health beliefs: some parents don't want to interfere with the natural course of events, preferring that children develop immunity through the disease; other parents reject the injection of substances they don't know in their children, worried that they can contain toxic elements.

Religious faiths: vaccination is seen as an act that would interfere with the divine actions. Many parents feel that it is up to God to decide if their children should be ill or not and that vaccination is contrary to that faith-based belief.

Parents' information on vaccination: parental decision-making is often affected by not having enough information about vaccination. Many of them blame their general practitioners. Hesitant parents also think about the media as misleading: they feel it is often contradictory and biased.

21.5 Conspiracy Beliefs

Vaccine hesitant beliefs and conspiracy theories have several elements in common. They both tend to simplify the complexity of reality, including doubt of scientific research, refusal of science, or acceptance of alternative explanations of scientific data, thus creating a new mainstream of beliefs and practices. Vaccine conspiracy beliefs are associated with lower education. Nevertheless the exact relation of education and vaccine hesitancy is not clear: studies have shown mixed results associating both lower and higher education [13].

The core vaccine conspiracy theory is that big pharmaceutical companies and governments alter data about vaccines to achieve their own maleficent intents [14]. People who believe it think that pharmaceutical factories earn a great profit from vaccines and that they would do everything in their power in order to protect this high economical income: they fabricate or falsify data on the efficacy of vaccines and erase evidence on serious side effects by persuading corrupt scientists [15].

These vaccine conspiracy views are not a rare, remote, or psychopathological occurrence. A national American survey reported that up to 56% of the interviewed people partially or fully endorsed such beliefs [16].

Hesitant parents are mostly used to surf on anti-vaccine-oriented web pages [17], which unsurprisingly present information in a very emotional, vivid, and moving way, often including parental testimonies regarding bad experiences after taking vaccines or pictures of children described as injured by vaccines [18]. In fact, false information finds support, not because it is considered reliable but because, on the unlikely chance, it is correct, the consequences would be dreadful. And this fear for a yet unrealistic scenario can block people from doing basic logical things such as vaccination.

21.5.1 Fake News, Bots, and Trolls

Sometimes misinformation and conspiracy theories are deliberately spread on the media [19]. A recent research [20] has shown that Russian bots and troll farms have pushed a great amount of anti-vaccination messages on Western social media. Bots are autonomous programs that promote content over and over, and trolls are individuals who falsify their identities with the intention of promoting disagreement. Russian trolls and Twitter bots post contents about vaccination at higher rates than an average user. They give equal attention to pro- and anti-vaccination arguments: this is a strategy called "amplification." It promotes quarrel across a range of controversial topics: it's not so important to shout against vaccination, as to keep feeding the debate and raise the doubt about efficacy and safeness of vaccination. Eventually, it may undermine public health and raise political discontent, leading social media users to question long-standing scientific consensus regarding vaccines and government politics.

Therefore, bots and trolls are deeply tangled in the online public health argument, and this is very important to know, especially because bots, trolls, or cyborgs are not easy to identify. Russian trolls and sophisticated bots promote both pro- and anti-vaccination chronicles, in order to breed political discord [21]. On the other hand, accounts that are known to dispense malware and commercial content are more likely to spread anti-vaccination messages, as clickbait to drive up advertising income and expose users to malware. In this case bot networks and spambots are primarily designed to deliver anti-vaccination messages for business [20].

Most recently, in the SARS-CoV-2 pandemic era, a lot of theories have been published on social media: the novel coronavirus was engineered in a lab in China; wealthy elites intentionally spread the virus to gain power and profit; COVID-19 is no worse than the flu; the "Black Lives Matter" protests led to increased transmission; any vaccine will be unsafe and a bigger risk than getting COVID-19 infection; COVID-19 was caused by 5G cellphone networks; and many others.

21.6 Is the Risk-Benefit Ratio of Vaccines Nowadays Positive?

The natural course of an infection, for example, measles, has a quite high percentage of mortality and complications, in this case affecting especially the central nervous system.

Acute disseminated encephalomyelitis (ADEM) is a measles's complication that usually occurs within 1–4 weeks after the skin rash. It causes an autoimmune demyelination, fever, seizures, and neurological deficits [22].

Its occurrence is 1 in 1000 cases [23]. Febrile seizures and encephalomyelitis that may occur after measles vaccination have an incidence of 1 in 2000–3000 and 1 in 1,000,000, respectively [24].

Moreover, measles can cause subacute sclerosing panencephalitis (SSPE) that occurs in 1 in 10,000–100,000 cases, 5–10 years after the acute illness. SSPE is more frequent in children infected with measles virus before the age of 2 years. SSPE causes seizures, progressive deterioration of cognitive and motor function, and eventually death within a few years [23]. The measles vaccine cannot cause SSPE, as studies have shown; on the contrary, successful measles vaccination programs can potentially eradicate SSPE, through the elimination of measles [25].

Vaccines can yet cause anaphylaxis, but its occurrence is generally very rare: with the measles vaccine, anaphylaxis occurs in 1–3.5 per 1,000,000 [24].

Measles's mortality was more than 2,000,000 deaths per year, before the broad use of vaccines. It then decreased to slightly more than 100,000 in 2015 [23]. During 2000–2015, increased coverage worldwide led to a 75% decrease in measles incidence and a 79% reduction in estimated measles mortality. During this period, an estimated 20.3 million deaths were prevented thanks to vaccination [26].

21.6.1 Herd Immunity

When enough people get vaccinated for a specific infectious disease, the vaccination coverage within that population reduces the incidence of the disease and can indirectly prevent the spreading of it even among not vaccinated people [27]. The minimum threshold for herd immunity varies depending on many factors. Firstly, the infectiousness of the disease but also the social and environmental context become important: maternal immunity, variation in age at vaccination, age differences in infection risk, seasonal changes, geographic heterogeneity, and social structure [28]. In ideal circumstances of vaccines delivered at random and individuals mixing at random, herd immunity thresholds are given by this simple formula: $(1-1/R0)$ [29], where R0 is the average number of people who will contract the disease from one person who has it. It usually applies to a population previously free of infection and not vaccinated. Ideally 100% of the population should get vaccinated, which is unlikely, but the closer we get to that number, the more fragile people (e.g., immunodeficient, children with genetic conditions, people who have contraindication to receive vaccination) can take advantage of general population immunization. For measles, which is highly contagious, minimum coverage is set at 95%.

21.6.2 Infection Elimination

The ultimate target is the disappearance of the infection as it occurred with smallpox: its global eradication was certified by the World Health Organization in 1978, and since then there has been no case of smallpox anywhere in the world, thanks to extensive vaccination programs [30]. Likewise for poliomyelitis, only two countries still present wild poliovirus transmission: Pakistan and Afghanistan [31]. Next on the list of feasible elimination is measles: in 2010, the World Health Assembly started a program for the eradication of measles to be achieved by 2015. Unfortunately, it hasn't been achieved yet, and the only way to persevere is to keep vaccinating.

21.7 Should Vaccination Be Mandatory?

In Europe 11 countries out of 31 (35.4%) have mandatory vaccinations for at least one condition out of diphtheria, tetanus, pertussis, hepatitis B, poliovirus, *Haemophilus influenzae* type b, measles, mumps, rubella, and varicella, regarding children under 18 months. Italy and Latvia have ten mandatory vaccines. Some countries (Bulgaria, Croatia, Czech Republic, France, Hungary, Poland, and Slovakia) have as many as nine vaccines which are compulsory among children [32].

Taking as an example Italy until June 2017, only four vaccines were mandatory (polio, diphtheria, tetanus, and hepatitis B). But over 5 years (2011–2015), there has been a drastic decrease in the immunization coverage: of just 5.3% for the measles vaccine, for example. This led Italy to be ranked as the sixth highest risk worldwide for measles cases in 2017. After 2017 the number of compulsory vaccines rose to 10. This new law is giving good results: preliminary data showed that almost one-third of the previously unvaccinated children born in 2011–2015 have been immunized. In addition, polio and measles vaccine uptake has improved by 1% and 2.9%, respectively, and by even more in selected regions [33].

In the USA a child cannot attend state schools if not regularly vaccinated, unless a valid medical exemption is provided. In the past some states accepted even non-medical exemptions, such as personal moral or religious reasons. Nowadays, new laws have been issued in order to limit this scenario. In California, for example, non-medical exemptions have been completely removed; in Michigan objecting parents are requested to attend immunization educational courses.

Australia has in place the laws of "No Jab, No Pay" and "No Jab, No Play": when families do not fully vaccinate their children ("No Jab"), they do not obtain the financial assistance payments ("No Pay"), nor are their children allowed to access kindergarten ("No Play") [34].

More than 100 countries around the world have national mandatory vaccination policies. Extreme penalties, such as loss of child custody, are relatively rare. Much more common are financial and educational penalties. Evidence has demonstrated that mandatory vaccination and fines are associated with higher vaccination coverage [35, 36]. In countries where immunization coverage is low, or where outbreaks occur, probably only the implementation of mandatory vaccination will save children's lives [37].

21.7.1 Ethical Conflicts

Mandatory vaccination is usually justified when three requirements are met: the infectious disease is a serious threat to public health (high mortality and/or morbidity); the utility of mandatory vaccination is high (if herd immunity is achieved on a voluntary basis, then mandatory vaccination is not required); coercion is proportionate (restriction of liberty should be as minimal as necessary to achieve the goal) [34].

21.7.2 Multi-pronged Approach

Mandatory vaccination cannot be the only answer. Large-scale communication intervention programs are required. The general population needs to be given the right tools to understand the topic in every aspect, in order to decide freely but consciously. The war against vaccine hesitancy needs to be won and not set aside by mandatory policies.

21.8 How to Promote Effective Knowledge

Communication plays a fundamental role in every successful immunization strategy.

In our contemporary era, we need to take into account mass and digital communication to be the most effective: mass media should be integrated with public health policies in order to empower people and advocate on such relevant topics.

The improved accessibility of the Internet and digital tools needs to be used by scientific communities to promote information to the general population. Scientific divulgation about immunization usually focuses on providing evidence-based facts or statistics to drive decisions. However, judgments concerning immunization are also driven by emotions and by specific local environments. Too often health-care providers tend to ignore the emotional part of public perception, missing the opportunity to establish an effective dialogue. A positive dialogue-based approach, which takes into account the interlocutor's fears and emotion, has been demonstrated to be very successful [1].

Vaccine-hesitant parents might work on rational reasoning, invest time and effort to understand vaccination, but eventually they make their choice of avoidance based on emotional reasoning. That's the reason why attempts to persuade hesitant parents using numbers or statistical data are not successful in increasing vaccine uptake. The perception of vaccine-related risks is more of a feeling than a rational thought [13]. There is evidence that attitudes can be reversed. For example, vaccine-hesitant college students rethought their position after interviewing individuals who had experienced the sequelae of vaccine-preventable diseases. For some, offering explanations may be enough, while others may require a motivational interviewing approach that of course can cost a significant amount of time for the practitioner [38].

21.8.1 Advertising Resources

Certainly, there is a growing need for enhancing digital communication strategies: online information needs should be monitored, multidisciplinary groups should be involved in the communication process, and new contents that balance facts with emotionally positive messaging should be created and then spread on multiple digital channels. Finally, more research in the field of effective communication for immunization should be promoted.

A crucial moment is listening to the public. Monitoring public opinion on vaccines through social media is essential to plan the right strategies and to prepare the most appropriate and pertinent content to elucidate doubts and fears. Moreover, not only health professionals should be engaged in this battle: associations and communities can be essential in spreading information to ethnic, religious, and other subgroups. Indeed, it is extremely important to be able to tailor and adapt divulgation to the social context and background beliefs of the target groups. Sometimes non-digital tools must be used too, in order to reach a wider audience that doesn't use digital devices [1].

The president of the American Academy of Pediatrics asked for partnerships with Google, Facebook, and Pinterest in order to make sure that parents using their platforms would interface with only science-based information. In reply, Facebook stated that groups and pages sharing anti-vaccine misinformation would be removed [39].

21.9 Ongoing Debates and Future Perspectives

Sometimes political and business interests jeopardize public health for their profit, and this cannot be allowed. The most recent episode just happened with the advent of COVID-19 vaccines. Mass media, social media, and also politicians kept talking, not fully informed, about this delicate topic, and eventually the administration of one of the first authorized vaccines, by AstraZeneca, has been suspended in more than a dozen countries, especially European ones. Even though the European Medical Agency (EMA) confirmed the safety of the vaccine, the media fuss and political forces succeeded in interrupting vaccination causing a tremendous delay, which means more infected cases and eventually deaths. Moreover, this led people to fear and mistrust vaccinations and to increase vaccine hesitancy tendency.

EMA then came out with an extraordinary review stating that over 20 million people received the vaccine and only 7 cases of blood clots in multiple blood vessels (disseminated intravascular coagulation, DIC) and 18 cases of CVST (cerebral venous sinus thrombosis) were reported. A causal link with the vaccine could not be proven [40].

Moreover, during the COVID-19 pandemic, non-COVID-19 immunization coverage declined. Some immunization services were discontinued, scheduled vaccinations were postponed due to the containment measures in place, and especially the fear of contracting the virus going into medical structures reduced the access [41]. Now, full commitment has to be put in the cause of vaccines: immunization coverage levels have to rise again, and trust in vaccination has to be reinforced if not rebuilt.

References

1. Bozzola E, Spina G, Tozzi A, Villani A (2020) Global measles epidemic risk: current perspectives on the growing need for implementing digital communication strategies. Risk Manag Healthc Pol 13:2819–2826
2. Jain A, Marshall J, Buikema A, Bancroft T, Kelly JP, Newschaffer CJ (2015) Autism occurrence by MMR vaccine status among US children with older siblings with and without autism. JAMA 313(15):1534–1540
3. NHS Digital (2017) Childhood vaccination coverage statistics, England, 2016–17. digital.nhs.uk/data-and-information/publications/statistical/nhs-immunisation-statistics/childhood-vaccination-coverage-statistics-england-2016-17. Accessed 10 Feb 2021
4. Olive JK, Hotez PJ, Damania A, Nolan MS (2018) The state of the anti-vaccine movement in the United States: a focused examination of nonmedical exemptions in states and counties. PLoS Med 15(6): e1002578
5. Clemmons NS, Gastanaduy PA, Fiebelkorn AP, Redd SB, Wallace GS, Centers for disease control and prevention (CDC) (2015) Measles—United States, January 4-April 2, 2015. MMWR Morb Mortal Wkly Rep 64(14):373–376
6. Coombes R (2017) Europe steps up action against vaccine hesitancy as measles outbreaks continue. BMJ 359:j4803
7. Lo Vecchio A, Montagnani C, Krzysztofiak A, Valentini P, Rossi N, Bozzola E, Castelli Gattinara G, Magurano F, Guarino A, Galli L, Italian Society for Pediatric Infectious Diseases Measles Study Group (2020) Measles outbreak in a high-income country: are pediatricians ready? J Ped Infec Dis Soc 9(4):416–420
8. Meleo-Erwin Z, Basch C, MacLean SA, Scheibner C, Cadorett V (2017) "To each his own": discussions of vaccine decision-making in top parenting blogs. Hum Vaccin Immunother 13(8):1895–1901
9. Schmidt AL, Zollo F, Scala A, Betsch C, Quattrociocchi W (2018) Polarization of the vaccination debate on facebook. Vaccine 36(25):3606–3612
10. Puri N, Coomes EA, Haghbayan H, Gunaratne K (2020) Social media and vaccine hesitancy: new updates for the era of COVID-19 and globalized infectious diseases. Hum Vaccin Immunother 16(11): 2586–2593
11. MacDonald NE, SAGE Working Group on Vaccine Hesitancy (2015) Vaccine hesitancy: definition, scope and determinants. Vaccine 33(34):4161–4164

12. Díaz Crescitelli ME, Ghirotto L, Sisson H, Sarli L, Artioli G, Bassi MC, Appicciutoli G, Hayter M (2019) A meta-synthesis study of the key elements involved in childhood vaccine hesitancy. Public Health 180:38–45
13. Tomljenovic H, Bubic A, Erceg N (2020) *It just doesn't feel right*–the relevance of emotions and intuition for parental vaccine conspiracy beliefs and vaccination uptake. Psychol Health 35(5):538–554
14. Jolley D, Douglas KM (2014) The effects of anti-vaccine conspiracy theories on vaccination intentions. PLoS One 9(2):e89177
15. Hviid A, Hansen JV, Frisch M, Melbye M (2019) Measles, mumps, rubella vaccination and autism: a nationwide cohort study. Ann Intern Med 170(8): 513–520
16. Oliver JE, Wood T (2014) Medical conspiracy theories and health behaviors in the United States. JAMA Intern Med 174(5):817–818
17. Dannetun E, Tegnell A, Hermansson G, Giesecke J (2005) Parents' reported reasons for avoiding MMR vaccination. A telephone survey. Scand J Prim Health Care 23(3):149–153
18. Betsch C, Renkenwitz F, Betsch T, Ulshofer C (2010) The influence of vaccine-critical websites on perceiving vaccination risks. J Health Psychol 15(3):446–455
19. Wilson SL, Wiysonge C (2020) Social media and vaccine hesitancy. BMJ Glob Health 5(10):e004206
20. Broniatowski DA, Jamison AM, Qi S, AlKulaib L, Chen T, Benton A, Quinn SC, Dredze M (2018) Weaponized health communication: twitter bots and Russian trolls amplify the vaccine debate. Am J Public Health 108(10):1378–1384
21. MacKinnon A (2020) Russian disinformation takes on coronavirus, pointing a finger at the United States. mediawell.ssrc.org/2020/02/14/russian-disinformation-takes-on-coronavirus-pointing-a-finger-at-the-united-states/. Accessed 13 Feb 2021
22. Bozzola E, Spina G, Valeriani M et al (2021) Management of pediatric post-infectious neurological syndromes. Ital J Pediatr 47(1):17
23. Moss WJ (2017) Measles. Lancet 390(10111): 2490–2502
24. World Health Organization (2015) Immunization safety surveillance. iris.wpro.who.int/handle/10665.1/12620. Accessed 20 Feb 2021
25. Campbell H, Andrews N, Brown KE, Miller E (2007) Review of the effect of measles vaccination on the epidemiology of SSPE. Int J Epidemiol 36(6): 1334–1348
26. Patel MK, Gacic-Dobo M, Strebel PM, Dabbagh A, Mulders MN, Okwo-Bele JM, Dumolard L, Rota PA, Kretsinger K, Goodson JL (2016) Progress toward regional measles elimination–worldwide, 2000–2015. MMWR Morb Mortal Wkly Rep 65(44):1228–1233
27. Fine P, Eames K, Heymann DL (2011) "Herd immunity": a rough guide. Clin Infect Dis 52(7):911–916
28. Fine PE (1993) Herd immunity: history, theory, practice. Epidemiol Rev 15(2):265–302
29. Gomes MGM, Corder RM, King JG et al (2020) Individual variation in susceptibility or exposure to SARS-CoV-2 lowers the herd immunity threshold. Preprint. medRxiv 2020.04.27.20081893
30. Meyer H, Ehmann R, Smith GL (2020) Smallpox in the post-eradication era. Viruses 12(2):138
31. World Health Organization (2020) who.int/news/item/25-08-2020-global-polio-eradication-initiative-applauds-who-african-region-for-wild-polio-free-certification. Accessed 22 Feb 2021
32. Bozzola E, Spina G, Russo R, Bozzola M, Corsello G, Villani A (2018) Mandatory vaccinations in European countries, undocumented information, false news and the impact on vaccination uptake: the position of the Italian pediatric society. Ital J Pediatr 44(1):67
33. Burioni R, Odone A, Signorelli C (2018) Lessons from Italy's policy shift on immunization. Nature 555(7694):30
34. Savulescu J, Giubilini A, Danchin M (2021) Global ethical considerations regarding mandatory vaccination in children. J Pediatr 231:10–16
35. Gravagna K, Becker A, Valeris-Chacin R, Mohammed I, Tambe S, Awan FA, Toomey TL, Basta NE (2020) Global assessment of national mandatory vaccination policies and consequences of non-compliance. Vaccine 38(49):7865–7873
36. Vaz OM, Ellingson MK, Weiss P, Jenness SM, Bardají A, Bednarczyk RA, Omer SB (2020) Mandatory vaccination in Europe. Pediatrics 145(2): e20190620
37. Hadjipanayis A, Dornbusch HJ, Grossman Z, Theophilou L, Brierley J (2020) Mandatory vaccination: a joint statement of the Ethics and Vaccination working groups of the European Academy of Paediatrics. Eur J Pediatr 179(4):683–687
38. Hotez PJ, Nuzhath T, Colwell B (2020) Combating vaccine hesitancy and other 21st century social determinants in the global fight against measles. Curr Opin Virol 41:1–7
39. The Lancet Child Adolescent Health (2019) Vaccine hesitancy: a generation at risk. Lancet Child Adolesc Health 3(5):281
40. European Medicines Agency (2021) COVID-19 vaccine AstraZeneca: benefits still outweigh the risks despite possible link to rare blood clots with low blood platelets. ema.europa.eu/en/news/covid-19-vaccine-astrazeneca-benefits-still-outweigh-risks-despite-possible-link-rare-blood-clots. Accessed 19 Mar 2021
41. Odone A, Bucci D, Croci R, Riccò M, Affanni P, Signorelli C (2020) Vaccine hesitancy in COVID-19 times. An update from Italy before flu season starts. Acta Biomed 91(3):e2020031

Part VII

Clinical Trials

Regulatory, Ethical and Political Challenges of Experimentation with LSD on Human Subjects

22

Kristof Janos Bodnar and Szabina Peter

Abstract

The inventor of LSD, Albert Hoffmann, himself referred to his most famous discovery as his 'problem child'. In the first part of this chapter, we outline the past regulatory, ethical and political issues that surrounded this much researched and debated agent until it ended up as a Schedule I substance, falling under the strictest regulations worldwide. We believe that the ethical, legal and political aspects are equally important to understand what had led to decade-long loss of interest in experimentation on human subjects with LSD. Having explored these issues, we turn our attention to the present research landscape and the future challenges and prospects of trials with this substance. We focus mainly on the research and bioethics aspects; besides we make some explanatory remarks on the past and present political and cultural/socio-critical perspectives that shaped and still affect public and professional discourse on the subject matter.

K. J. Bodnar (✉) · S. Peter
Department of Behavioral Sciences, Faculty of Medicine, University of Debrecen, Debrecen, Hungary
e-mail: bodnarjanos@med.unideb.hu

22.1 LSD: A Dangerous Hallucinogenic Agent or a Tool for Brain Investigation?

Lysergic acid diethylamide (commonly known as LSD, more precisely, LSD-25) is a synthetic ergot alkaloid, first synthesized by Albert Hoffmann in 1938, but 5 years passed until he unintentionally discovered its consciousness-altering properties on the 19th of 1943 (which date is remembered by the admirers of his discovery annually as the 'Bicycle Day') [1]. Scholarly discussion in scientific journals began soon after; Swiss psychiatrist Werner Stoll published the first paper on the effects of the substance on a small cohort of healthy volunteers and patients in 1947, and Joel Elkes was also an early pioneer in conducting a psychiatric experiment with LSD in 1951 [2–4]. Even in these two early instances, an interesting ethical perspective appeared: both Stoll and Elkes took a dose of LSD before they decided to give it to the study participants. The following years brought even more burgeoning scientific interest in this substance.

By 1963, researchers described the effects of the new substance in more than a thousand publications, most of them written on the experimental therapeutic applications of it. The majority of these were assessing it as a substance applied in the frame of 'psychedelic assisted psychotherapy' and involved as much as 40,000 participants [5]. These publications contained data of 120 LSD trials including more than 1700

J. Faintuch, S. Faintuch (eds.), *Integrity of Scientific Research*,
https://doi.org/10.1007/978-3-030-99680-2_22

participants, with remarkable federal funds provided for them—over USD4 million [6, 7].

The studies from this period—i.e. roughly before 1963—addressed the basic questions of novel substances research: pharmacological properties, toxicity and safety in pre-clinical and clinical settings and the therapeutic effectivity of the drug as an adjuvant to psychiatric and psychological treatments. Some of these questions were abandoned relatively quickly by the scientific community (like Abramson's 'model psychosis' hypothesis or investigating LSD's potential in treating schizophrenic patients [8]), while others—most notably regarding addictions and anxiety of terminally ill patients—the mental health professionals deemed as fruitful perspectives. Methodologically speaking, all in the latter experimental clinical applications or controlled trials applied LSD as an 'adjunct to psychotherapy'.

This decision, on the one hand, stemmed from the drug's supplier's (Sandoz) official indication of *Delysid* (the brand name for Sandoz LSD-25) for 'analytical psychotherapy' [1] and, on the other, because the early attempts to achieve psychological benefits from the usage of LSD without any form of counselling did not succeed at all [9]. This, as we will see later on, posed particular challenges to create proper study designs in the light of the strict evaluation standards of the new 1962 FDA amendments. Two dominant versions of these therapies were widespread—the so-called *psycholytic* and *psychedelic* ones: In the former, a relatively smaller amount of LSD (50–200 μg) was administered to the participants for several months, once or twice per week, prior to a therapy session. The latter consisted of using a single, greater dosage (400–800 μg), for which event the therapists prepared their patients through psychotherapy ongoing for weeks, on a daily basis [10].

Based on reviews of the literature on these studies [6, 10–12], two fields seemed to promise significant success: reducing pain, depression and anxiety of terminal patients and treating alcoholics. These studies involved more than a hundred patients, and—at least compared to the standards of research regulation *before* 1962—investigators assessed the application of LSD to be safe in the study groups, at the same time being effective in this adjuvant manner.

22.2 Ethical and Regulatory Constraints

The Food and Drug Administration (FDA) accepted a new set of regulations on proof of efficacy and safety of clinical trials to test new drugs—the so-called Drug Amendments of 1962. This affects both how researchers could evaluate the above-mentioned ocean of data from trials before the 1970s and how contemporary researchers needed to change the design and conduct of their studies onwards. Given the lack of regulations strictly prescribing particular designs, experimentation prior to 1962 with LSD mostly consisted of empirical studies on small populations, usually lacking control of any sort, and dominantly included therapy resistant, severely diseased participants [13, 14]. Retrospective (re)evaluation of studies from this period raises the following issues.

According to recent researches and reviews [15, 16], out of these hundreds of studies, only ten publications passed the muster of current research ethics and methodology standards: 'The quality of the great majority of the clinical trials did not conform to modern standards, with a non-randomized control group or without control group itself'—concludes a 2020 systematic literature review assessing data from 1949 on the therapeutic benefits of LSD. Besides, the authors conclude that 'it was observed a significant effect of LSD in four studies performed. However, this effect was related to quality of life and general health in some of the studies, with no clear improvements in alcohol abstinence' [15].

Naturally, we might lament whether the scarcity of qualitatively proper studies is indeed an (more) objective indicator of the real efficacy of adjuvant LSD therapy, a true answer to the high hopes of many researchers from this early period of experimentation. Or should we rather agree with Matthew Oram, who dedicated much scholarly effort to examine whether LSD and most prominently LSD-assisted psychotherapy trials were hardly tailored or tailorable to meet the requirements of the novel regulations [17]? This question is particularly important not only to fairly judge the research history of the substance but to guide the design and conduct of future human experimentation, even to shed light on why general FDA (or FDA-type) regulations might not fit all type of drug trials. In order to fully grasp Oram's concerns, we must examine two factors—the special attributes of LSD used as an adjuvant to psychotherapy and the concerns about (most notably, placebo) controls in these trials.

The first problem is best understood as an intrinsic property of not only LSD assisted psychotherapy but any assisted therapy in which neither part (the substance or the therapy) is superior than the other, nor would be able to achieve the effect of its components' combination. With regard to LSD, Albert Kurland, prominent investigator of several such studies, described this phenomenon as follows: 'We definitely do not see LSD being utilized here as a medication in the usual sense. Rather, it is conceived of as an activating mechanism that brings about a unique experience (. . .) upon which the therapeutic intent and structuring is focused' [18]. This is further complicated by the fact that these therapies were of a psychoanalytic or humanistic kind, in which the qualitative components, subjective feelings and interpretations are substantial.

In this sense, evaluation of such therapies—as according to the FDA type of regulations—and the adjuvant substances applied in them is much more challenging than the behaviourist therapies, whose effectivity could be assessed comparatively easier in quantitative terms [8]. Placebo, regarded as the 'golden standard' means of study control, was also deemed problematic in the case of LSD—many investigators arguing that blinding on either side was hardly achievable, caused by the easily observable and experienced signs of LSD influence. When actively controlled 'with ephedrine sulfate, in 19 of 20 cases the therapist correctly guessed which type of drug was administered to the patient, and 20 mcg of LSD was considered too low a dose' [15]. These factors combined, in the reading of Oram, may have contributed to several false-negative outcomes in the trials after 1962 [14].

However, at least regarding the placebo problem, it is worthy to note that some recent researches specially aim at further exploring such questions. The research of Olson and colleagues concludes that contrary to the above arguments, psychedelic effects might indeed be triggered in subjects by passive placebo means as well, thus suggesting that previous claims on the impossibility of placebo control might come from improper design [19]. This in our understanding might be a consequence of the particularly pivotal role 'set' and 'setting' play in the psychedelic experiences—therapeutic, experimental or recreational sort of 'trips' alike [16]. Consequently, to achieve more adequate design in future trials, we find it cardinal to further elaborate on these issues of set, setting and its role in the subjective experience and in the (objectively) measurable efficacy of LSD-assisted psychotherapies [20].

At the same time, it is equally important to see that comparatively fewer research investigated the direct—i.e. not therapy-adjuvant—biological and psychological effects of LSD [21]. However, it is worthy to note that basic biological properties concerning safety and toxicity were thoroughly investigated both on healthy and patient populations in that period, and in light of both early experiments, recent ones and meta-analyses [16], the following conclusion can be drawn: 'Classic hallucinogens have very low physiological toxicities, with no evidence of resulting organ damage or neuropsychological deficits even at very high doses' [11].

A significant drop in the number of trials between the end of the 1960s and the millennium occurred. The more it is important to surmise on this question, because in the midst of a scientific trend often labelled as 'psychedelic renaissance', there seems to be an idea. 'There is a common narrative among those who follow the history of psychedelic research that all clinical studies in the USA with this class of drugs terminated in 1976 and did not recur again until Rick Strassman began his dimethyltryptamine (DMT) clinical investigations in 1990' [6, 8]. However, even if the number of studies after 1965 indeed has dropped from 70 to 9, it should not be seen a deliberate—governmental or other politically induced—backlash or prohibition of human experimentation per se.

We agree with Oram on that this is a multi-causally explainable issue, stemming from the following reasons: First of all, after the 1962 restrictions, researchers aim to adhere to those voicing criticism about the methodological and ethical shortcomings of many contemporary trials more and more often [8]. Secondly, not only had Sandoz stopped the distribution of LSD in the August of 1965, but the following year they decided to withdraw their IND from the FDA—worthy to add that they never in the firsts place conducted systematic phase 1 or properly broad 2 and 3 studies: 'There were no data to submit in an NDA—or a sponsor to submit it' [8]. Thirdly, as Hoffmann himself put it, 'LSD abuse have occurred (. . .) completely beyond the control of Sandoz (reaching) the scale of a serious threat to public health' [1]. The forms and sources of these abuses were widespread: from therapeutic/research ones to the ones resulting from 'street use', linked to the rebelling youth or even to the highest governmental circles.

22.3 Political Aspects, Episodes of Fraud and Misconduct

We risk to claim that only a handful of other substances' experimental, therapeutic and 'recreational' usage was (and is) surrounded so heavily by ethical and political concerns as LSD's many applications, all along from its invention. This political aura sometimes stemmed from and was interwoven with its adjuvant psychotherapeutic use, often resulting from the applied humanistic therapeutic approaches' inherent socio-critical character [19]. A talking example for this is R. D. Laing's cultural, political and professional impact in the 1960s on the anti-psychiatric movement and his radical reinterpretation of mental normality and pathology—in which, according to some theoreticians [20], his admiration of LSD played a crucial role. He reflected on and criticized psychiatry's role in the political and social domain—its exploitive and abusive potential—and the social and political domains' role in the mental disorder individual and communal perception and treating. These ideas, taken together with his explicit views on LSD's overall therapeutic potential—as a "spiritual laxative" for patients, for individuals and for society *en general*—might have contributed to his labelling as an 'acid communist' [20].

Worthy to note here is that this label was taken up again more recently by the 'psychedelic socialists' [22]—a movement which could be best described as follows: 'It sought to change and raise the consciousness of singular people and the whole society, be that through the creative use of psychedelic chemicals, aesthetic experiments in music and other arts, social and political revolution, or all of the above' [23]. Even Michel Foucault described his 'acid trip' as a particularly meaningful experience which, according to some scholars, had an impact on his philosophy and contributed to the elaboration of his concepts of biopolitics and the 'care of the self' [24, 25].

Based on these insights, and on the often described 'spiritual experiences' of unity and the erosion of the self/ego—which can contribute to a change of the personality and *political views* [26, 27]— we might risk the conclusion that, even in the frame of properly designed adjuvant therapeutic session, LSD seems to carry an intrinsic 'political' character [22]. Is this character truly an intrinsic one? Whether and to what extent do the drug's biological properties have a role in this? Or does it have more (or exclusively) to do with extra-drug factors—such as the set-setting, the therapies' social and political 'characteristics'?

Unfortunately, the ethical and political landscape of early LSD research was not free of scandals and controversial practices: Of course, in the early 1950s, when scientific knowledge was sparse on the effects of LSD either on healthy or patient populations—and even *Delysid*'s manual *suggested* that researchers should self-administer the substance so to *familiarize* themselves with the effects—requirements of full disclosure and proper informing of the subjects could not have been fully met. Nevertheless, as recreational usage quickly became widespread due to the unrestricted legal and growing counter-cultural status of the drug, some researchers—most notably Richard Alpert (aka. Ram Das) and Timothy Leary, lecturers and therapists at Harvard University then—were accused of giving LSD (and other psychedelic substances) to their students [2].

Even if these experiments and the 'recruitment' did not violate any contemporary ethical norms, nor broke the law, still, voluntariness, safety protection and proper scientific methodology could easily have been compromised in them (similar though certainly less steep concerns were and are voiced towards Laing's own experimental therapies [28]). These researches were problematic even by those days' measures, for they often took place in informal frames—parties, concerts, non-controlled gatherings of various sorts. According to numerous commentators, even if these were fueled by the best intentions of investigators like Leary and Alpert, they surely contributed to the quasi-stigmatized status of LSD experiments to a certain extent [2, 19]. Even more contributed to public and professional mistrust the infamous MKUltra project (whose ill reputation still pertains—enough only to think about its depiction in the popular TV series *Stranger Things*).

This CIA-sponsored, covert project purposed to investigate drugs—LSD included—as means of mind-control, 'truth-serum' and other, potentially for military and intelligence aims (usually without any success). However, the tragic death of Dr. Frank Olson is a reminder of the serious ethical flaws of the project: he killed himself in 1953, after he was secretly dosed with LSD at a party [29]. However, these dangers certainly stemmed from the—political and professional—abuse or neglective conduct of these 'studies', and we, based on our review of the literature, agree with Das and his colleagues: 'Classic hallucinogens have very low physiological toxicities. (...) For the large majority of participants, the most relevant safety concern is the potential for dangerous and erratic behavior resulting from the intense subjective experiences with these drugs' [11].

Though general assessments of its safety and its potential therapeutic benefits—even if drawn from methodologically flawed studies—were known by the regulatory authorities [3], the US government had put LSD into Schedule 1 by means of the CSS, followed by its UN Convention regulation in 1971. This certainly contributed to a dearth of LSD studies in the upcoming decades [30]—until a significant interest returned into the researchers' community to further elaborate on the therapeutic potentials, harm reduction perspectives and the regulatory challenges they concern.

22.4 Future Perspectives

According to Das and his colleagues, these novel researches, by relying on more rigorous ethical and methodological means, attempt to provide 'a clearer characterization and definition of their [the psychedelic substances'] principal effects on the mind and brain' [11]. These studies are diverse and broad in their scope of interest: they range from fundamental pharmacokinetical ones through general neurological and psychiatric and psychological ones—including those not conducted in the 'classical era', viz. placebo-controlled trials on healthy participants to explore general psychological effects—to quite 'exotic ones' [6], only to mention two of this latter kind: The first is the much media attention generating microdosing study of Szigeti and his colleagues in which they applied 'a self-blinding setup procedure that enabled self-experimenters, who microdose on their own initiative using their own psychedelic, to implement placebo control and randomization without clinical supervision' [31]. The second is the study of Nour and colleagues on the possible association of political orientation—more precisely, a liberal, anti-authoritarian one—and psychedelic usage [32].

The first we deem important from a future regulatory and ethics perspective: it reflects on a 'grassroots' phenomenon with scientific rigor and methodology—namely, on the growing trend in the USA and Europe: people—more and more often in a publicly admitted way—take microdosages of psychedelics (most often LSD) for recreational, self-exploring or even 'productivity-optimizing' purposes [33, 34]. This study foreshadows the possibility of using means of 'citizen science' to explore LSD-related medical or psychological questions which would be much more difficult to explore on such a broad scale. By this methodology researchers could reach out for users whom could not be reached otherwise—largely due to the present legal and regulatory concerns regarding these substances in the vast majority of the world's countries. This can be a relevant aspect if a harm reduction-minded regulatory and political viewpoint becomes more accepted and widespread—naturally, researchers must do everything to accurately lay down and maintain the particular methodological (e.g. of self-blinding) and ethical concerns (e.g. of protecting anonymity) of any such studies.

The study of Nour and his colleagues raises interesting implications regarding informed consent [35]: whether and to what extent does such a possibility of a significant change in the personality—possibly resulting in the change of political preference—pose particular challenges to the informing process? Besides, should this 'liberalizing effect' be proven (which is certainly not the case yet and is disputed by professionals [36]), would it pose a novel type of ethical concern—say, that no governmental funds of a democratic country should be provided for studies possibly biasing participants into a given political direction? (It is hard not to think, to a certain extent, of *A Clockwork Orange*'s Ludovico therapy in this respect). Putting sci-fi-type concerns aside, the still strong link between politics and—recreational, experimental, therapeutic—LSD usage demands careful considerations on many (more tangible) levels as well.

As Fuentes and colleagues put it: 'LSD remains one of the most stigmatized and legally restricted agents among psychoactive substances. It is still included in Schedule I of the United Nations classification of drugs, restricting its use in research and making it difficult to potentially use it as a therapeutic tool in medicine' [15]. We believe that the 2019 ethical review's insights still apply today: this scheduling 'creates a challenge to the scientific validity of clinical LSD trials (...) The policy defines controlled substances in Schedule 1 as having no "accepted medical use"—in that way, by the same token, the scheduling itself obstructs efforts to investigate whether or not it actually has medical uses' [6, 37].

22.5 Ethical Guidelines for Future Research and Clinical Use

Granted that the experience (the 'psychedelic trip') is certainly one of a kind, the assessment of its therapeutic potentials and risks should involve the same means as in the case of 'magic bullet' type substances which act through their direct biological mechanisms or other, non-hallucinogen psychoactive drugs (like SSRIs) [35]. That is, its particularity should be explored by relying on the generally accepted research methodology and ethical norms—at the same time addressing the peculiarities of such experiments in the design, conduct and follow-up of these trials. What are these special ethical considerations?

Both Johnson and his colleagues [38], with regard to psychedelics *en general*, and Bodnar and Kakuk, focusing particularly on LSD trials, provided a systematic summarization of such issues, as Pilecki and his colleagues sum up: 'There is a consensus on the need for psychotherapeutic support for people undergoing treatment using psychedelics. In clinical trials, this typically takes the form of preparatory sessions, therapeutic support during dosing sessions and a few integration sessions after each dosing session' [39]. The importance of follow-up in any properly designed study should be evident, however, given the possibly overwhelming nature of the experience—the more for those who did not have previous psychedelic experiences—it demands researchers to place further emphasis on this issue regarding LSD trials. We believe that follow-up psychological counselling sessions should be made available—free of charge—for trial participants, even if they are healthy volunteers or if the experiment itself did not contained psychotherapy per se. Naturally, it seems evident that this psychological counselling should be done by properly trained therapists—which leads us to a related problem.

As mentioned earlier, even Sandoz promoted the idea that researchers should take LSD so to familiarize themselves with the peculiar consciousness-altering properties of the drug. This should be considered together with the literature-wide consensus on the utmost importance of 'set' and 'setting' for the individual experience—and we believe that one of the most crucial factors of the setting is the therapist/investigator herself. Thus, an ethical challenge arises: If recent regulations do allow for participants to take LSD in the context of a trial, and from a patient safety perspective, the best would be to have investigators with personal experience leading their 'trip', should researchers also be allowed to use LSD for these scientific purposes? The same questions apply, from a harm reduction perspective, to those counselors/therapists, who aim to provide professional help for those who seek integration therapy for their LSD experiences (either they went through those by participating in a trial or by illegal use). Should there be official trainings provided for such integration therapists in the frame of which their LSD usage would be accepted as legal? Again, all this has to do with the often described 'ineffable' nature of the mystical type/psychedelic experiences [15]—which issue itself demands a special, extended consent [35]—therefore that best could be understood and conveyed (during the informing process, the 'trip' and the follow-up counselling session) by someone having personal experience.

22.6 Conclusion

These trials would be necessary not only because of the potential therapeutic breakthroughs many believe they indeed promise [40]—most often in the field of mental health—but from a public (mental) health perspective, too, via a harm reduction viewpoint. We hypothesize that the 'psychedelic renaissance' does not only refer to an increase in the number of trials and scientific investigations, but more and more news emerge about a spread in the—grey-zoneish—microdosing, often openly admitted by those using it for recreation and exploration, even to boost their performance and productivity. All in all, we believe that it is no longer in the best interest of the public, from a harm reduction perspective to label this growing number of half/open advocates of LSD users as delinquents or 'junkies'; therefore we find it imperative to provide trustworthy, scientifically valid information about the effects of LSD usage—recreational and therapeutic ones as well.

As a conclusion, we would like to quote a pioneer of early LSD scientific research, Humphrey Osmond: 'A good doctor knows his limitations and does not set himself up as an all-knowing guru, social scientist, or a universal problem solver. (...) Limitations must be understood' [41].

References

1. Hofmann A (1983) LSD, my problem child: reflections on sacred drugs, mysticism, and science. J.P. Tarcher, Los Angeles
2. Dyck E (2015) Flashback: psychiatric experimentation with LSD in historical perspective. Can J Psychiatry 50(7):381–388
3. Novak SJ (2004) LSD before Leary: Sidney Cohen's critique of 1950s psychedelic drug research. In: Tracy SW, Acker CJ (eds) Altering American consciousness: the history of alcohol and drug use in the United States, 1800–2000. University of Massachusetts Press, Amherst, pp 87–110
4. Stoll W (1947) Lysergsäure-diäthyl-amid, ein Phantastikum aus der Mutterkorngruppe. Schweiz Arch Neur 60:1–2
5. Grinspoon L, Bakalar BJ (1981) The psychedelic drug therapies. Curr Psychiatr Ther 20:275–283
6. Bodnar KJ, Kakuk P (2019) Research ethics aspects of experimentation with LSD on human subjects: a historical and ethical review. Med Health Care Philos 22: 327–337. https://doi.org/10.1007/s11019-018-9871-9
7. Pollan M (2015) The trip treatment. New Yorker
8. Bonson KR (2018) Regulation of human research with LSD in the United States (1949–1987). Psychopharmacology 235:591–604. https://doi.org/10.1007/s00213-017-4777-4
9. Hagenbach D, Werthmuller L (2011) Mystic chemist: the life of Albert Hofmann and his discovery of LSD. Synergetic Press, Santa Fe

10. Dyck E (2015) LSD A new treatment emerging from the past, psychedelic medicine: a re-emerging therapeutic paradigm. CMAJ 187(14):1079–1080
11. Das S, Barnwal P, Ramasamy A, Sen S, Mondal S (2016) Lysergic acid diethylamide: a drug of 'use'? Ther Adv in Psychopharmacol 6(3):214–228 https://doi.org/10.1177/2045125316640440
12. Passie T (1997) Psycholytic and psychedelic therapy research 1931–1995: a complete international bibliography. Laurentius Publishers, Hannover
13. Oram M (2014) Efficacy and enlightenment: LSD psychotherapy and the Drug Amendments of 1962. J Hist Med Allied Sci 69(2):221–250
14. Oram M (2016) Prohibited or regulated? LSD psychotherapy and the United States Food and Drug Administration. Hist Psychiatry 27(3):290–306
15. Fuentes JJ, Fonseca F, Elices M, Farré M, Torrens M (2020) Therapeutic use of LSD in psychiatry: a systematic review of randomized-controlled clinical trials. Front Psych 10:943. https://doi.org/10.3389/fpsyt.2019.00943
16. Luoma JB, Chwyl C, Bathje GJ, Davis AK, Lancelotta R (2020) A meta-analysis of placebo-controlled trials of psychedelic-assisted therapy. J Psychoactive Drugs 52(4):289–299. https://doi.org/10.1080/02791072.2020.1769878
17. Oram M (2018) The trials of psychedelic therapy LSD psychotherapy in America. John Hopkins University Press, Baltimore
18. Kurland AA, Unger S, Shaffer JW, Savage C (1967) Psychedelic therapy utilizing LSD in the treatment of the alcoholic patient: a preliminary report. Am J Psychiatry 123(10):1202–1209. https://doi.org/10.1176/ajp.123.10.1202
19. Olson DE (2020) The subjective effects of psychedelics may not be necessary for their enduring therapeutic effects. ACS Pharmacol Transl Sci 4(2):563–567. https://doi.org/10.1021/acsptsci.0c00192
20. Richert L, Dyck E (2020) Psychedelic crossings: American mental health and LSD in the 1970s. Med Humanit 46:184–191
21. Sessa B (2018) The 21st century psychedelic renaissance: heroic steps forward on the back of an elephant. Psychopharmacology 235(2):551–560. https://doi.org/10.1007/s00213-017-4713-7
22. Zutshi V (2018) The political significance of LSD. https://www.opendemocracy.net/en/transformation/political-significance-of-lsd/. Accessed 10 May 2021
23. Gilbert J (2017) Psychedelic socialism, OpenDemocracy
24. Dean M, Zamora D (2021) The last man takes LSD foucault and the end of revolution. Verso, London
25. Wade S (2019) Foucault in California. Heyday, New York
26. Lyons T, Carhart-Harris RL (2018) Increased nature relatedness and decreased authoritarian political views after psilocybin for treatment-resistant depression. J Psychopharmacol 32(7):811–819. https://doi.org/10.1177/0269881117748902
27. Nour MN, Evans L, Robin LC-H (2017) Psychedelics, personality and political perspectives. J Psychoactive Drugs 49(3):182–191. https://doi.org/10.1080/02791072.2017.1312643
28. Paton M (2015) RD Laing: was the counterculture's favourite psychiatrist a dangerous renegade or a true visionary. https://www.independent.co.uk/life-style/health-and-families/rd-laing-was-the-counterculture-s-favourite-psychiatrist-a-dangerous-renegade-or-a-true-visionary-a6755021.html. Accessed 10 May 2021
29. Mashour GA (2007) From LSD to the IRB: Henry Beecher's psychedelic research and the foundation of clinical ethics. Int Anesthesiol Clin 45(4):105–111
30. Haave P, Pedersen W (2020) The promise and demise of LSD psychotherapy in Norway. Soc Hist Med. https://doi.org/10.1093/shm/hkz128
31. Szigeti B, Kartner L, Blemings A, Rosas F, Feilding A, Nutt DJ, Carhart-Harris RL, Erritzoe D (2021) Self-blinding citizen science to explore psychedelic microdosing. elife 10:e62878. https://doi.org/10.7554/eLife.62878
32. Nour MM, Evans L, Nutt D, Carhart-Harris RL (2016) Ego-dissolution and psychedelics: validation of the ego-dissolution inventory (EDI). Front Hum Neurosci 10:269. https://doi.org/10.3389/fnhum.2016.00269
33. Leonard A (2015) How LSD microdosing became the hot new business trip rolling stone. https://www.rollingstone.com/culture/culture-news/how-lsd-microdosing-became-the-hot-new-business-trip-64961/. Accessed 10 May 2021
34. Nye C (2017) Microdosing: the people taking LSD with their breakfast. https://www.bbc.com/news/health-39516345. Accessed 10 May 2021
35. Smith WR, Sisti D (2020) Ethics and ego dissolution: the case of psilocybin. J Med Ethics. Published Online First: 27 May. https://doi.org/10.1136/medethics-2020-106070
36. Johnson MW, Yaden DB (2020) There's no good evidence that psychedelics can change your politics or religion. Scientific American
37. Nutt D (2010) Drug harms in the UK: a multicriteria decision analysis. Lancet 376(9752):1558–1565
38. Johnson M, Richards W, Griffiths R (2008) Human hallucinogen research: guidelines for safety. J Psychopharmacol (6):603–620. https://doi.org/10.1177/0269881108093587
39. Pilecki B, Luoma JB, Bathje GJ (2021) Ethical and legal issues in psychedelic harm reduction and integration therapy. Harm Reduct J 18:40. https://doi.org/10.1186/s12954-021-00489-1
40. Nutt D, Carhart-Harris R (2021) The current status of psychedelics in psychiatry. JAMA Psychiat 78(2):121–122. https://doi.org/10.1001/jamapsychiatry.2020.2171
41. Osmond H, Siegler M (1973) Notes on orthomolecular psychiatry and psychotherapy. Orthomolecul Psychiat 2, 3:120, 118–126

Controversies with Clinical Trial Regulations in Low- and Middle-Income Countries

23

Bastian Rake

Abstract

Low- and middle-income countries are increasingly selected as locations for international clinical trials. The consequences have been debated within and beyond the academic literature. Low- and middle-income countries are also referred to as nontraditional countries. This chapter reviews the extant literature related to this debate and identifies the three main—partially interrelated—concerns: a lack of contribution to building up local innovation capacity, a lack of local regulatory capacity, and the prevalence of questionable clinical research practices and fraud. A call for long-term capacity-building efforts of local researchers and regulators in nontraditional countries is a priority.

23.1 Introduction

Pharmaceutical and biomedical research and development (R&D) activities have traditionally been conducted predominantly in North America, Western Europe, and a small number of other countries outside these regions, such as Japan and Australia. More recently pharmaceutical and biomedical research has become increasingly internationalized. Clinical trials are no exception to this development. Clinical trials, essential activities to assess the safety and efficacy of new drugs, vaccines, or medical devices, are increasingly conducted in low- and middle-income countries [1–4]. They are also labeled as nontraditional countries [3].

The increasing global footprint of clinical research has given rise to controversial debates about the scientific, ethical, and economic consequences associated with this development. In particular, the potential consequences for individuals from nontraditional countries who are participating in clinical trials have generated controversy. They may not be able to obtain access to the treatment under investigation at an affordable price after the completion of the trial. Noncompliance with internationally accepted principles of research ethics and good clinical practices may undermine the protection of the populations, which are generally seen as more vulnerable [5, 6]. Different or insufficient trial regulation and approval procedures could interfere with compliance with ethical standards and generally accepted research practices in conducting clinical trials and in reporting the corresponding results [7].

B. Rake (✉)
Maynooth University, Maynooth, Ireland
e-mail: bastian.rake@mu.ie

J. Faintuch, S. Faintuch (eds.), *Integrity of Scientific Research*,
https://doi.org/10.1007/978-3-030-99680-2_23

23.2 Advantages of Conducting Trials in Nontraditional Countries

23.2.1 For Trial Sponsors

Innovation infrastructure, which itself is determined by the availability of scientific and technological knowledge and other factors and conditions supporting R&D activities, is deemed essential for candidate countries [8, 9]. Labor cost advantages and the availability of well-trained and often English-speaking workforces should also be taken into account [10, 11]. In addition, nontraditional countries have developed a sufficient number of high-standard healthcare facilities comparable to Western countries [12]. The considerable and growing number of publications in biomedical and related areas originating in countries such as China and India demonstrates that they have built a knowledge base that is relevant and supportive for conducting clinical trials [13].

Clinical trial sponsors seek to speed up the trial duration and to lower the recruitment costs by ensuring that a sufficiently large number of participants are available at the host location [14]. This condition is generally met in many nontraditional countries where large parts of the population have limited access to (high-quality) healthcare. Furthermore, many (potential) clinical trial participants from low- and middle-income countries are “treatment naïve,” i.e., they have not previously received any treatment against a specific medical indication. This is favorable from the sponsors’ perspective, as the probability of statistically significant effects in favor of the sponsor’s expectations increases. At the same time, being treatment naïve reduces the risk of unintended interactions with ongoing medical treatments or influencing factors [14].

The opportunity to obtain local market approval is relevant, and a sample of participants who are representative of the population of the target market is often a precondition. The growing size of local market business opportunities is hence another factor that makes nontraditional countries more attractive [15, 16].

23.2.2 For Host Countries

Nontraditional countries may have a vital interest in testing new medications with local populations, as this is a chance to ensure that the treatments under investigation are safe and effective in different populations and in disease environments that correspond to local conditions [17, 18]. Nevertheless large proportions of the local population have limited access to (high-quality) healthcare [19]. This has been seen as an opportunity to provide trial participants access to novel treatments and healthcare more generally that would otherwise not be available to them [11]. The additional inflow of financial and other resources, such as medical equipment, that can be used for diagnosis and care beyond a particular trial should not be overlooked [19].

Governments may see participation in international clinical trials as a chance to improve national innovative capacity, i.e., a country’s long-term ability to produce commercially relevant innovations [20], even if the trials predominantly address disease areas with a global prevalence or mostly in high-income countries [2]. Finally, governments and policymakers support the creation of employment opportunities for often highly qualified graduates with medical, health, or science degrees [19].

23.2.3 Ethical Controversies

The protection of trial participants is an important aspect of medical research ethics and is closely linked to the notion of informed consent. When making this decision, (potential) participants should decide freely without coercion and with a full understanding of all information—particularly information regarding the potential benefits and risks—that is relevant in the context of a specific trial [21]. Researchers have questioned whether the principle of informed consent is followed or whether these trials exploit vulnerable populations. Large segments of the local population may be illiterate or not educated enough to fully understand trial-related information, are not very familiar with medical interventions in general, or live in poverty and are often in need of access to healthcare [22, 23]. Against this backdrop, it has been questioned whether they participate as real volunteers or due to economic or health needs.

Full Disclosure of Risks

Some clinical trials have raised concerns among academics, practitioners, and the general public, typically when participants either died or were injured during the trial and where a causal link to the treatment under investigation could not be ruled out. A well-known example is the death of a 10-year-old girl in Kano, Nigeria, who participated in a trial sponsored by a large Western pharmaceutical company during a meningitis epidemic. The case became well known, as several lawsuits in Nigeria and the United States addressed allegations that the trial sponsor failed to secure informed consent [24].

Long-Term Access

Lacking access to the new treatment after completion of the trial is another key concern [25]. Even if treatments are introduced to the local markets, they are often not available at prices that are affordable for large parts of the population [18]. At the same time, global public and private research have remained focused on diseases that are more prevalent in high-income countries [26]. Interest in diseases that are highly prevalent in nontraditional countries but not in high-income counties has been rather low [27, 28]. Thus local populations would have rather limited benefits from participating in clinical trials and however bear the associated risks [6]. Even if the trial addresses diseases with a local prevalence, there are concerns that the selection of subjects focuses more on their comparability to patients in the big and, hence, more profitable markets in (Western) high-income countries than on the specificities and diversity of local populations [29]. This focus on Western needs may imply that, even if a new treatment is tested in countries such as China or India, the results may provide only limited insights for the safety and efficacy of this treatment for local populations and may pay limited attention to specific local conditions.

Enticing Compensations or Payments

High payments, compared to the salaries that trial participants could normally achieve, may undermine their critical judgment and therefore invalidate their consent [30]. However, empirical evidence suggests that individuals use payment information as an indicator of the associated risk [31–33]. Another line of argumentation argues that financial incentives may not make participants' informed consent invalid but motivates vulnerable individuals to participate in activities that they would normally perceive as unacceptable [34].

Gelinas et al. [35] emphasize that generally "compensation rates should be set in line with local norms and practices." High financial incentives for trial participants are not per se an issue even if economically vulnerable individuals are the target audience. However, to be acceptable, the research should follow strict ethical research standards, such as securing approval by research ethics boards and safeguarding informed consent [35].

Contagious Diseases and Pandemics

Human infection challenge studies are those in which individuals are intentionally infected with a microorganism to study the clinical course or to test treatment options or vaccines [36]. In nontraditional countries they can support the production of knowledge regarding a specific disease that affects their populations. However compliance with strict ethical standards is particularly important, notably with regard to subsequent follow-up and treatment, as an illness will be triggered by the challenge, and participants cannot be simply abandoned and left alone [36].

23.2.4 Regulatory Controversies

To ensure that all the different actors involved in conducting clinical trials follow strict ethical standards, appropriate legal regulations must be in place, and compliance with these regulations must be monitored and enforced. The reality is, however, rather complex. The leading nontraditional clinical trial countries show a variety of regulation approaches ranging from codified comprehensive regulations to their virtual absence. Differences may involve mandatory insurance, compensable injuries, claim compensation processes, and the role of research ethics committees [37].

The mere existence of regulations that aim to protect trial participants is insufficient to guarantee their protection, as regulatory authorities may lack appropriate resources, including a sufficient number of trained staff to oversee compliance and to sanction violations of local regulations or international standards. Historically, regulatory bodies in many nontraditional trial countries were underdeveloped; for instance, drug decisions were made based on approval in Western countries [38].

Non-acceptable Trials

Nontraditional countries with lax regulations could be a magnet for protocols refused by more developed countries [11, 39]. Some nontraditional countries, including China, India, and Mexico, introduced reforms to strengthen their regulatory bodies, yet deficiencies may remain [40–43]. Against this background, the literature has argued that policymakers face ethical tradeoffs. Stricter regulations and better-equipped regulators may improve the protection of trial participants but may, at the same time, decrease a country's attractiveness as a location for conducting clinical trials so that the local population may not benefit from such involvement [44].

Inadequate External Oversight

The US Food and Drug Administration (FDA) is conducting inspections of foreign (non-US) trial sites and attempting to improve its ability to monitor trials conducted in nontraditional countries; nevertheless conducting such oversight depends on its budget, which is largely determined by the US Congress. With the available budget being constrained, the FDA has inspected only a limited number of foreign trial sites throughout the last decade [45]. In addition, trial sponsors may use loopholes in the regulations across countries to conduct trials without a registration in the United States or another Western country and to avoid inspections by the FDA or other Western regulators [46].

23.2.5 Knowledge Transfer Pitfalls

The positive impact of clinical trials for improving national innovative capacity may be exaggerated. The involvement of researchers from nontraditional countries in trial-related publications has remained modest [47, 48]. Regulatory changes that are aimed, as in the Indian case, to improve knowledge transfer to local actors through the facilitation of late-stage clinical trials seem to have limited benefits for domestic science systems. More specifically, Rake and Haeussler [48] find that the potential for learning and knowledge transfer is limited, as the knowledge produced in late-stage clinical trials is only used to a limited extent by domestic researchers. In line with these findings, Reid and Ramani [49] emphasize that the potential of learning from conducting clinical trials is limited and that investments in more basic biomedical research are likely to be more advantageous in terms of increasing the potential for knowledge transfer and building up national innovative capacity.

23.3 Nontransparent Research, Fraud, and Misconduct

Pre-registration

Efforts to increase the availability of clinical trial information in public registries have been undergoing for a considerable time. In September 2004, the International Committee of Medical

Journal Editors (ICMJE) announced that (prospective) trial registration will be made a precondition for publication in its member journals, which encompass many of the world's most prestigious biomedical journals. Thereafter, the regulatory environment and the corresponding legislation were amended in favor of (prospective) trial registration in the United States and internationally [50]. The compliance with the registration requirements of clinical trials is in general quite high and has increased the transparency of clinical research and the accessibility of information regarding relevant ongoing research for biomedical researchers, medical doctors, patients, and other interested parties [51]. However, recent empirical evidence indicates that the increase in transparency may not apply to all countries equally, as there are at least some cases in which trial sponsors do not follow domestic laws and do not register their trials in nontraditional countries' domestic databases, as would be required by the corresponding domestic regulations [52].

The lack of compliance with regulatory requirements asking for the full and timely disclosure of trial results remains a controversial issue across trial registries and the trials registered in these registries. Hence, the issue of nondisclosure of trial results is not restricted to trials conducted in nontraditional countries [53, 54].

Transparent Result Sharing and Questionable Outcomes
The quality and trustworthiness of the reported trial results have also been under scrutiny. A large percentage of reported outcomes and clinical trial data originating in nontraditional countries are of poor quality or subject to fraud or other forms of academic misconduct [55]. Carlisle [56] found that false and fatally flawed clinical trial data frequently occur in trial-related manuscripts from nontraditional countries, including China, South Korea, India, and Egypt. An investigation of China's State Food and Drug Administration (CFDA) raised concerns of data manipulation and data quality in approximately 80% of more than 1600 new drug applications that had been reviewed [57]. In 2015 the European Medicines Agency recommended the suspension of the market authorization of more than 700 generic drugs in the European Union due to serious concerns regarding the validity of the corresponding clinical trials conducted by an Indian company [58]. Similarly, the US Food and Drug Administration has found evidence of fraud, questionable research practices, or violations of good clinical research practices in trials conducted in nontraditional countries [59].

Selective Reporting and Inadequate Designs
Clinical trials conducted in nontraditional countries are often partially or fully contradictory to those obtained in traditional countries [60–62]. Traditionally, many medical colleges in countries such as India have focused on patient care, instead of laboratory investigations and clinical trials. Consequently there has been a lack of adequate institutional support, specialized facilities, and role models, as most academics in these institutions would at best occasionally publish in (international) peer-reviewed journals [63]. This institutional context may have led to a lack of understanding of the international norms of authorship, of designing and conducting (clinical) research, and of research ethics [64, 65]. While academics in these environments have not been well prepared for attempting to publish in international peer-reviewed journals, institutional changes such as making publications a prerequisite for academic promotion or financial incentives for publications raised the expectation to do so [65, 66]. At the same time, the internationalization of clinical research has increased the division of labor, which itself is associated with higher levels of academic misconduct, questionable research practices, or errors [67].

23.4 Future Perspectives

Improvements in national innovation capacity are a prerequisite for enabling nontraditional countries to conduct quality research that addresses local health needs. This increase in national innovative capacity is unlikely to be achieved through conducting additional (late-stage) clinical research but instead requires investments in domestic scientific and technological capabilities, including research facilities such as laboratories but also training for researchers and PhD students. These investments will also enhance a country's ability to use and build upon existing research originating from other countries and make the country a more attractive location for multinational companies, encouraging these companies to conduct more (basic) research and more clinical development locally [48].

White [39] suggests that regulatory bodies in traditional clinical trial countries, such as the US Food and Drug Administration, should increase inspections of foreign trial sites. Inspections by Western regulatory bodies may force trial sites in nontraditional clinical research countries to develop capacities to meet international standards and requirements. However, inspections by the FDA or other (Western) regulatory authorities may not lead to capacity building on the side of local regulators in nontraditional clinical research countries and may lead to increasing dependence on (Western) regulatory bodies. To build up local capacity for overseeing clinical trials, agreements between the FDA or other Western regulatory bodies and local authorities in nontraditional countries regarding joint inspections and information exchange may be a promising route, particularly if it includes training and competence development opportunities. The successful agreement between the FDA and the European Medicines Agency may serve as a role model for these types of agreements [39].

Blockchain-based systems offer the opportunity to empower regulators in traditional and nontraditional countries to validate statistical results and to run analyses that are independent of trial sponsors' influences. A precondition for this opportunity is that the raw data are captured correctly in the blockchain as early as possible [68].

In the case of the publication of questionable clinical research, White [39] argues in favor of improved information exchange between regulatory bodies and academic journals so that erroneous or fraudulent research can be either retracted or is not published in the first place. While regulatory authorities such as the FDA disregard questionable or fraudulent clinical trial data in their decision-making processes, there is no comprehensive information exchange system among journals and regulatory authorities [59]. This information exchange seems to be important, as peer review is unlikely to detect all erroneous or fraudulent research.

Cutter [69] suggest that journals' peer-review process is complemented by thorough (statistical) checks of the data and results by professional reviewers or other appropriate checks conducted by the journals, as this task requires extensive work that cannot be expected from volunteer reviewers. Strengthening academia's quality control mechanisms may help to avoid the publication of erroneous, fraudulent research but does not necessarily lead to increased capacity building in nontraditional countries that would help (clinical) researchers from these countries to avoid errors and follow generally accepted research practices. Hence, there is again a need for local capacity building. However, this capacity building should be accompanied by a review of the local incentive mechanisms that should put increasing emphasis on the integrity of (clinical) research.

When deciding about these regulatory changes and investments, policymakers face the challenge that tighter regulations and empowering regulatory authorities may make a country less attractive for conducting clinical trials [44, 70]. However, policymakers should not overemphasize these concerns for two reasons. First, scandals around questionable clinical research and increased scrutiny by (Western) regulatory authorities are also likely to reduce the attractiveness of a country for international clinical trials. Given the high drug development cost, (Western) sponsors may not want to conduct clinical research in locations that have a bad reputation due to allowing questionable clinical research practices, as this may increase scrutiny and endanger market approval. Second, it is questionable whether clinical trials that will not be conducted because of stricter regulations are those that would have supported competence building and positive health outcomes in host countries.

References

1. Cantner U, Rake B (2014) International research networks in pharmaceuticals: structure and dynamics. Res Policy 43:333–348
2. Haeussler C, Rake B (2017) The changing geography of clinical research: a critical analysis of its drivers. Ind Corp Chang 26:285–310
3. Thiers FA, Sinskey AJ, Berndt ER (2008) Trends in the globalization of clinical trials. Nat Rev Drug Disc 7:13–14
4. Drain PK, Parker RA, Robine M, Holmes KK (2018) Global migration of clinical research during the era of trial registration. PLoS One 13:e0192413
5. Lang T, Siribaddana S (2012) Clinical trials have gone global: is this a good thing? PLoS Med 9:e1001228
6. Nundy S, Gulhati CM (2005) A new colonialism? — Conducting clinical trials in India. New Eng J Med 352:1633–1636
7. Khin NA, Yang P, Hung HMJ, Maung-U K, Chen Y-F, Meeker-O'Connell A et al (2013) Regulatory and scientific issues regarding use of foreign data in support of new drug applications in the United States: an FDA perspective. Clin Pharmacol Ther 94:230–242
8. Varsakelis NC (2006) Education, political institutions and innovative activity: a cross-country empirical investigation. Res Policy 35:1083–1090
9. Fagerberg J, Srholec M (2008) National innovation systems, capabilities and economic development. Res Policy 37:1417–1435
10. Haakonsson SJ, Jensen PDØ, Mudambi SM (2013) A co-evolutionary perspective on the drivers of international sourcing of pharmaceutical R&D to India. J Econ Geogr 13:677–700
11. Maiti R, Raghavendra M (2007) Clinical trials in India. Pharmacol Res 56:1–10
12. Crone RK (2008) Flat medicine? Exploring trends in the globalization of health care. Acad Med 83
13. White K (2019) Science and engineering publication output trends: 2017 shows U.S. output level slightly below that of China but the United States maintains lead with highly cited publications. nsf.gov/statistics/2019/nsf19317/. Accessed 5 Sept 2021
14. Petryna A (2009) When experiments travel: clinical trials and the global search for human subjects. Princeton University Press, Princeton, NJ
15. Gassmann O, Han Z (2004) Motivations and barriers of foreign R&D activities in China. R&D Manag 34: 423–437
16. Epstein RJ (2007) Growth of the Asian health-care market: global implications for the pharmaceutical industry. Nat Rev Drug Disc 6:785–792
17. Wilson JF, Weale ME, Smith AC, Gratrix F, Fletcher B, Thomas MG et al (2001) Population genetic structure of variable drug response. Nat Genet 29:265–269
18. Glickman SW, McHutchison JG, Peterson ED, Cairns CB, Harrington RA, Califf RM, Schulman KA (2009) Ethical and scientific implications of the globalization of clinical research. New Eng J Med 360:816–823. https://doi.org/10.1056/NEJMsb0803929
19. Varawalla N, Jain R (2011) Chapter 7–Clinical trials in India. In: Chin R, Bairu M (eds) Global clinical trials. Academic Press, San Diego, CA, pp 119–157
20. Furman JL, Porter ME, Stern S (2002) The determinants of national innovative capacity. Res Policy 31:899–933
21. O'Neill O (2002) Autonomy and trust in bioethics. Cambridge University Press, Cambridge
22. Glantz LH, Annas GJ, Grodin MA, Mariner WK (1998) Research in developing countries: taking "benefit" seriously. Hast Cent Rep 28:38
23. Fisher JA (2009) Medical research for hire: the political economy of pharmaceutical clinical trials. Rutgers University Press, New Brunswick, NJ
24. Annas GJ (2009) Globalized clinical trials and informed consent. New Eng J Med 360:2050–2053
25. Chang PL (2017) The abandoned stakeholders: pharmaceutical companies and research participants. J Bus Ethics 143:721–731
26. Yegros-Yegros A, van de Klippe W, Abad-Garcia MF, Rafols I (2020) Exploring why global health needs are unmet by research efforts: the potential influences of geography, industry and publication incentives. Health Res Pol Sys 18:47
27. Barrenho E, Miraldo M, Smith PC (2019) Does global drug innovation correspond to burden of disease? The neglected diseases in developed and developing countries. Health Econ 28:123–143
28. Pécoul B, Chirac P, Trouiller P, Pinel J (1999) Access to essential drugs in poor countries: a lost battle? JAMA 281:361–367
29. Merz S (2020) Global trials, local bodies: negotiating difference and sameness in Indian for-profit clinical trials. Sci Technol Hum Values. 0162243920963813
30. Largent EA, Fernandez LH (2017) Paying research participants: regulatory uncertainty, conceptual confusion, and a path forward. Yale J Health Policy Law Ethics 17:61–141
31. Leuker C, Samartzidis L, Hertwig R, Pleskac TJ (2020) When money talks: judging risk and coercion in high-paying clinical trials. PLoS One 15:e0227898
32. Mantzari E, Vogt F, Marteau TM (2014) Does incentivising pill-taking 'crowd out' risk-information processing? Evidence from a web-based experiment. Soc Sci Med 106:75–82
33. Cryder CE, John London A, Volpp KG, Loewenstein G (2010) Informative inducement: study payment as a signal of risk. Soc Sci Med 70:455–464
34. Millum J, Garnett M (2019) How payment for research participation can be coercive. Am J Bioeth 19:21–31

35. Gelinas L, White SA, Bierer BE (2020) Economic vulnerability and payment for research participation. Clin Trials 17:264–272
36. Jamrozik E, Selgelid MJ (2020) Human infection challenge studies in endemic settings and/or low-income and middle-income countries: key points of ethical consensus and controversy. J Med Ethics 46:601
37. Chingarande GR, Moodley K (2018) Disparate compensation policies for research related injury in an era of multinational trials: a case study of Brazil, Russia, India, China and South Africa. BMC Med Ethics 19:8
38. Kremer M (2002) Pharmaceuticals and the developing world. J Econ Perspect 16:67–90
39. White CM (2020) Current system of overseeing drug trials in developing countries by the FDA is dangerous. Ann Pharmacother 54:928–932
40. Rodríguez FS (2010) Governing ethical clinical research in developing countries: exploring the case of Mexico. Sci Public Policy 37:583–596
41. Zhou Q, Chen X-Y, Yang Z-M, Wu Y-L (2017) The changing landscape of clinical trial and approval processes in China. Nat Rev Clin Oncol 14:577–583
42. Jayaraman K (2012) India mulling stricter laws to curb unethical trials. Nat Med 18:182
43. Gupta YK, Padhy BM (2011) India's growing participation in global clinical trials. Trends Pharmacol Sci 32:327–329
44. Aguilera B, DeGrazia D, Rid A (2020) Regulating international clinical research: an ethical framework for policy-makers. BMJ Glob Health 5:e002287
45. Hull D (2015) Reining in the commercialized foreign clinical trial. J Legal Med 36:367–401
46. Department of Health and Human Services (2010) Challenges to FDA's ability to monitor and inspect foreign clinical trials. ntrl.ntis.gov/NTRL/dashboard/searchResults/titleDetail/PB2010114711.xhtml. Accessed 5 Sept 2021
47. Hoekman J, Frenken K, Zeeuw D d, Heerspink HL (2012) The geographical distribution of leadership in globalized clinical trials. PLoS One 7:e45984
48. Rake B, Haeussler C (2019) Did relaxing clinical trial regulation enhance the stock of scientific knowledge in India? Not necessarily. PLoS One 14:e0210163
49. Reid SE, Ramani SV (2012) The harnessing of biotechnology in India: which roads to travel? Technol Fore Soc Change 79:648–664
50. Dickersin K, Rennie D (2012) The evolution of trial registries and their use to assess the clinical trial enterprise. JAMA 307:1861–1864
51. Viergever RF, Li K (2015) Trends in global clinical trial registration: an analysis of numbers of registered clinical trials in different parts of the world from 2004 to 2013. BMJ Open 5:e008932
52. Kumari S, Mohan A, Saberwal G (2020) Hidden duplicates: 10s or 100s of Indian trials, registered with ClinicalTrials.gov, have not been registered in India, as required by law. PLoS One 15:e0234925
53. Anderson ML, Chiswell K, Peterson ED, Tasneem A, Topping J, Califf RM (2015) Compliance with results reporting at ClinicalTrials.gov. New Eng J Med 372: 1031–1039
54. DeVito NJ, Bacon S, Goldacre B (2020) Compliance with legal requirement to report clinical trial results on ClinicalTrials.gov: a cohort study. Lancet 395:361–369
55. Resnik DB (2003) From Baltimore to Bell Labs: reflections on two decades of debate about scientific misconduct. Account Res 10:123–135
56. Carlisle JB (2021) False individual patient data and zombie randomised controlled trials submitted to anaesthesia. Anaesthesia 76:472–479
57. Woodhead M (2016) 80% of China's clinical trial data are fraudulent, investigation finds. BMJ 355:i5396
58. Watson R (2015) EU to suspend hundreds of drug authorisations after doubts over clinical studies in Hyderabad. Brit Med J 350:h574
59. Seife C (2015) Research misconduct identified by the US Food and Drug Administration: out of sight, out of mind, out of the peer-reviewed literature. JAMA Intern Med 175:567–577
60. Zhang D, Freemantle N, Cheng KK (2011) Are randomized trials conducted in China or India biased? A comparative empirical analysis. J Clin Epidemiol 64:90–95
61. Desai K, Carroll I, Asch S, Hernandez-Boussard T, Ioannidis JP (2019) Extremely large outlier treatment effects may be a footprint of bias in trials from less developed countries: randomized trials of gabapentinoids. J Clin Epidemiol 106:80–87
62. Panagiotou OA, Contopoulos-Ioannidis DG, Ioannidis JPA (2013) Comparative effect sizes in randomised trials from less developed and more developed countries: meta-epidemiological assessment. Brit Med J 346:f707
63. Ray S, Shah I, Nundy S (2016) The research output from Indian medical institutions between 2005 and 2014. Curr Med Res Pract 6:49–58
64. Chen W, Xing Q-R, Wang H, Wang T (2018) Retracted publications in the biomedical literature with authors from mainland China. Scientometrics 114:217–227
65. Elango B, Kozak M, Rajendran P (2019) Analysis of retractions in Indian science. Scientometrics 119: 1081–1094
66. Franzoni C, Scellato G, Stephan P (2011) Changing incentives to publish. Science 333:702
67. Walsh JP, Lee Y-N, Tang L (2019) Pathogenic organization in science: division of labor and retractions. Res Policy 48:444–461
68. Wong DR, Bhattacharya S, Butte AJ (2019) Prototype of running clinical trials in an untrustworthy environment using blockchain. Nat Commun 10:917
69. Cutter GR (2017) To catch a crook, you might try statistics. Nat Rev Neurol 13:9–10
70. Porter G (2018) Regulating clinical trials in India: the economics of ethics. Dev World Bioeth 18:365–374

Dangers of FDA Oversight of Clinical Trials in Developing Countries

24

C. Michael White

Abstract

Conducting more clinical trials outside the United States (US) intended to support Food and Drug Administration (FDA) approval of drugs poses a major challenge. There are ethical dilemmas where far away sites are seldom inspected and not to the same extent as US sites. There is also the risk of coercion, limited participant protections, and a reduced ability of participants to be compensated if harmed. There are prominent examples of ethical lapses or where lax oversight led to questionable research data being used in submissions for regulatory approval around the globe. This is combined with an FDA increasingly tied to user fees from the drug companies in exchange for more timely New Drug Application reviews that makes it harder to find malfeasance. Taken together, this convergence of risks should give us pause.

Abbreviations

ECGs	Electrocardiograms
EMA	European Medicines Agency
FDA	Food and Drug Administration
HIV	Human immunodeficiency virus
IND	Investigational New Drug
NDA	New Drug Application
OIG	Office of the Inspector General
PDUFA	Prescription Drug User Fee Act
US	United States
WHO	World Health Organization

C. M. White (✉)
University of Connecticut, Storrs, CT, USA
e-mail: Charles.White@uconn.edu

24.1 Introduction and Background

Until the passage of the Pure Food and Drug Act of 1906, there was little oversight of medications in the Unites States (US) [1]. To comply with this act, medications had to meet minimal manufacturing standards for strength and consistency of the active ingredient, purity of the product, and the quality of the manufacturing. However, medications did not have to prove that they were effective or safe for use in humans. The Food, Drug, and Cosmetic Act of 1938 established the Food and Drug Administration (FDA) under the US Department of Health and Human Services. One of the FDA's many charges under this legislation was to evaluate prescription and over-the-counter medications to determine if they had an appropriate balance of efficacy and safety for use in the US [1].

The preclinical and clinical trial process delineated by the FDA for manufacturers is rigorous, logically sequenced, and vital to protecting the health of the research participants and yielding the information needed to determine the

J. Faintuch, S. Faintuch (eds.), *Integrity of Scientific Research*,
https://doi.org/10.1007/978-3-030-99680-2_24

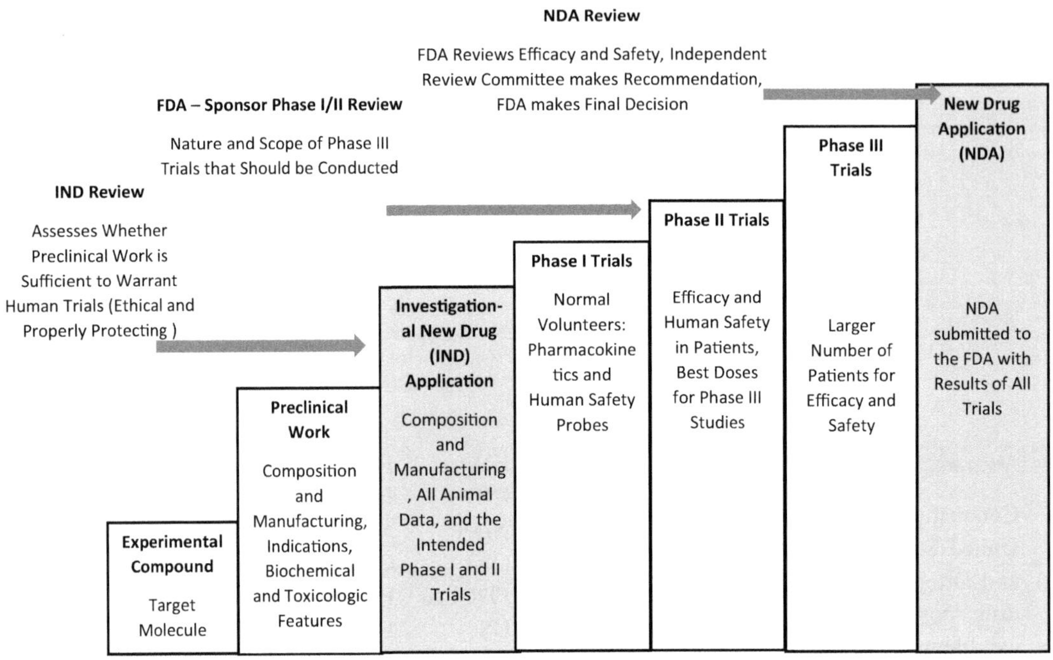

Fig. 24.1 FDA drug approval process. Legend: *FDA* Food and Drug Administration, *IND* Investigation New Drug, *NDA* New Drug Application. This is the normal step-wise process leading from a drug candidate to an FDA-approved drug. Modified from fda.gov/media/82381/download

potential benefits and harm that the drug could cause to US consumers [2]. Figure 24.1 describes the process from target compound identification to full drug approval by the FDA [2].

There are myriad reasons to relocate clinical trials away from the US and into developing countries, but they carry with them some limitations, risks, and ethical dilemmas.

24.2 Current Trends in the Geographical Migration of Clinical Trial Sites

The rate of research growth in the US has significantly eroded over the past 40 years [3, 4]. While FDA-regulated trials abroad have consistently increased every year in the past decade, clinical trials in the US have declined by 5.5% [5]. Before 1985, it was almost impossible to find foreign clinical investigators conducting trials relevant for FDA approval [3, 4]. However, by 2000, 22% of investigational new drug/IND clinical investigators resided outside the US, and by 2013, 43% resided abroad. The rate of research site growth in different geographic regions is displayed in Fig. 24.2 [3].

From virtually no research data coming from non-US sites in the 1970s and early 1980s, by 2008 78% of research subjects were not from the US, 80% of applications for drugs and biologics included data from non-US trials, and 8% of new drug applications/NDA's included data were derived from sites entirely outside of the US [3, 4]. We are rapidly approaching the time where most drugs that are approved for use by US citizens by the FDA will have their efficacy and safety assessed based on a dataset with a paucity of US participants.

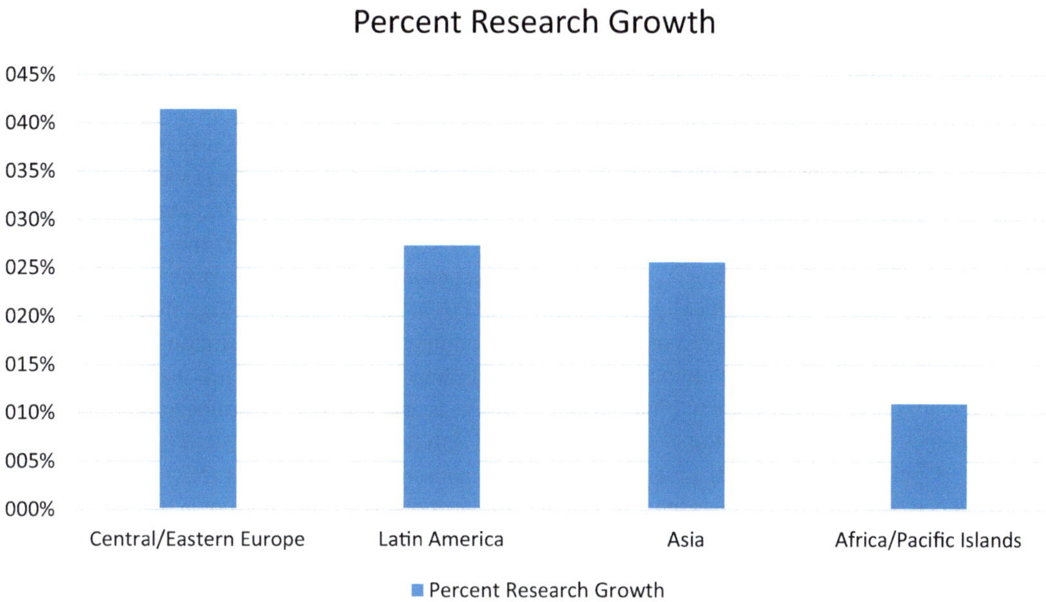

Fig. 24.2 Geographic assessment of percentage research site growth in developing countries [3, 4]. Legend: This table shows the areas of the world where drug research is growing at the fastest rate

24.3 Market Forces Driving International Migration of Clinical Trials

24.3.1 Cheaper Labor Costs

Be it cellphones, T-shirts, pharmaceuticals, or clinical trials, cheaper labor costs are incurred when they are manufactured or conducted in developing countries. For commodities, some of these savings are attenuated by the expense associated with packaging and shipping them across the globe to US consumers. With clinical trials however, the study data can be sent rapidly and inexpensively through online portals or even via email [6, 7].

24.3.2 Faster Enrollment Rates

Faster enrollment rates at overseas sites are a major incentive to move clinical trials abroad. If you can complete studies faster, you preserve more time on your patent life to profit from the sales of your drug [5]. An average of 505 subjects are recruited at a single clinical trial site in a developing country as compared to 75 subjects in a US clinical trial site [5].

However, to what extent are recruitment and enrollment faster in developing countries because the populations are more enthusiastic about supporting research or more densely populated, as compared to the coercion experienced by study subjects, investigators, and clinical practitioners?

24.3.2.1 Financial Incentives

Given economies of scale, the value of a dollar in the US is radically different than in developing countries; however this is not figured into the FDA guidance or decision-making. Over 2.6 billion citizens in the world live on less than $2 a day, with ~85% of citizens in some countries living below the international poverty line [5]. Regardless of the country, the FDA considers a payment from a sponsor to a clinical investigator to be "significant" if it is in excess of $25,000 [8]. When the FDA believes that the integrity of the data may be compromised by significant financial interests, they can audit the data or reject the trial in its entirety [8]. Is it not a much more significant enticement if an investigator in a developing country is given $20,000, as compared to one in the US receiving $25,001?

24.3.2.2 Participant Protection

Vulnerable populations in the US enjoy extra human subject protections, but there are no regulations whereby impoverished people in developing countries are specifically identified and given additional protections as a result [9]. For example, to get a 6-month extension for drug exclusivity on the US market, pharmaceutical companies can acquire additional data in the pediatric population, but special protections are not mandated by the FDA for research conducted in orphanages within developing countries [10]. Can the 4-year-old orphan really fail to give assent if the orphanage director needs the money to better their facility, take in other deserving children, or to keep the facility running? A similar ethical dilemma is faced by both parents and children in large families who are struggling with poverty, where incentives for participation in a clinical trial could be an important source of family income.

24.3.2.3 Intellectual Restrictions

Subjects in developing countries may have limitations in literacy or numeracy, making them unable to truly understand the potential risks of receiving experimental therapy or choosing an experimental regimen over a therapy that has already been established [5, 10]. It is easier to be coerced into participating in research if subjects cannot determine if the benefits to the individual or society are worth the risks they are taking on as a result.

24.3.2.4 Access to Legal Assistance

The dual potential for coercion and limited literacy/numeracy to distort a human subject's ability to gauge the relative benefits and risks of trial participation is magnified, when those same people have little to no legal recourse in their countries if they were subsequently harmed [5, 10].

24.3.3 FDA Rules Create Loopholes for Research Conducted in Developing Countries

As highlighted in Fig. 24.1, human subject research in the US cannot begin without the filing of an IND application [2, 9]. The FDA reviews this IND to determine if the animal studies conducted suggest that the drug is safe enough for initial human testing. The IND review also determines if the proposed Phase I and II clinical trials have enough safeguards to protect human subjects and whether the drug can be consistently and safely manufactured [9]. The FDA derives the authority to review this IND data because in order for a US clinical trial to begin, the drug travels from the manufacturing plant across state lines to the clinical sites [7, 8]. This interstate commerce is under the auspices of the federal government. However, what happens if the drug is manufactured outside of the US and is shipped to clinical trial sites in developing countries? In this paradigm, there is now no US interstate transportation of products, and the FDA does not have the same authority [8].

In these cases, the FDA allows clinical trial data to be included in an NDA without the company submitting an IND, describing what clinical trials will be conducted in Phase I and II of the drug approval process [11]. There are two main implications of this huge loophole. First, clinical trials can be conducted in humans based on pre-clinical studies that the FDA would have felt were insufficient to ensure research subject well-being. The FDA is wholly unaware whether a non-IND disclosed trial has been conducted until it is concluded and included in the company's NDA. Once it is included in the NDA, the FDA can retrospectively assure that the study adhered to good research practices or choose to inspect study records if they have concerns. Any problems identified would only lead to the trial data being discounted or rejected, not protecting the subjects that were already subjected to the questionable research methods. In this way, research subjects in the US are, by law, prospectively protected by the FDA, while those in developing countries receive inferior retrospective assessments [8]. As we will see in the next section, the FDA has limited ability and timeframe in which to perform these inspections as well.

24.3.3.1 Discarded or Mishandled Trials

Secondly, it is possible for a company to conduct a trial without FDA's knowledge, and they do not have to divulge the study results if they do not suit the company's desires [7]. The company can simply repeat the study after tweaking the dose, more selectively choosing the study population, or changing the times when outcomes are assessed. This ability to selectively include or discard trials is anathema to the logical continuum between the IND and NDA review processes which are critical to assuring a drug is efficacious and safe for use in US consumers. For trials conducted in the US, the FDA knows ahead of time what trials are going to be done, what methods will be used, and what outcomes will be assessed because they are divulged in the IND. The FDA then closes the loop and wants to see that data in the NDA before rendering a decision about its approval for use in the US.

If a Phase I or II clinical trial has a high number of adverse events or severe events, the FDA will want the company to explore the reasons why it happened, and this can shape subsequent contraindications or precautions in the official product labeling. If these studies are not included in the NDA, the document seems stronger than the totality of the clinical trial data would otherwise suggest, and the chances for approval are greater, while the restrictions to use are less onerous [7]. All of these things are good for profitability but not good for the patient or the payors for healthcare products.

24.3.4 The FDA Cannot Inspect Foreign Sites Like They Should

Before the human immunodeficiency virus (HIV) epidemic, the FDA was funded through federal appropriations (US taxpayer dollars), but the time from NDA submission to drug rejection or approval was exceedingly long [12]. After intense pressure from HIV activists, the funding mechanism was changed to a hybrid model, with funding from appropriations and user fees from the companies seeking to sell drugs in the US through the passage of the Prescription Drug User Fee Act (PDUFA) of 1992. By 2019, only 55% of the entire FDA's budget ($5.7 billion) was provided by federal appropriation [13]. More so, 64% of the FDA's funding for the oversight of drugs was paid by user fees, and the FDA's ability to collect these user's fees is tied explicitly to how rapidly they conduct NDA reviews [13]. As a result, the FDA is under intense pressure to rapidly decide whether to approve a new drug after submission of an NDA [14]. The review period from NDA submission to decision has been reduced from 19.1 months in 1993 to 8.5 months in 2015 [14].

Going through these trials carefully during this brief period of review is very difficult for the FDA. If the company submits their proposed clinical trials in the IND, the FDA can choose to inspect clinical trial sites while those trials are in process, or they can inspect their documentation after trial completion [14]. However, if the trial from a developing country is not disclosed until the NDA is filed, the FDA loses the ability to prospectively inspect the sites and must trust that the study data they receive in their retrospective review is truly representative of what occurred. Even when trials in developing countries were disclosed in INDs before they commenced, the frequency and intensity of FDA inspections were markedly different for domestic and foreign sites [3, 4]. According to the OIG, a domestic site was disproportionately more likely [odds ratio 15.87 (95% CI 9.69–25.99)] to be inspected by the FDA than a foreign one [3]. To date, there are dozens of countries where clinical trials have been conducted under INDs that have never been inspected by the FDA [3].

24.3.4.1 Expenses and Logistics of Inspection

Inspections in remote areas of the globe can cost up to $40,000 to complete, and the total inspection length is limited to 7 days, regardless of the amount of time it takes to travel there [3, 4]. Limiting the total inspection time in a developing country's clinical research site to the same duration as for US-based clinical site inspections is grossly unequal because once the inspectors arrive, they may or may not speak the native language and, if not, often rely on the site's personnel to chaperone them around the site and allow communication between the scientists or participants and the inspectors. The slower the pace, the less actual inspecting that can be done. It is apparent that the time available to ensure ethical compliance and cross-check data in medical records or original data sheets to the final submitted data is woefully inadequate [3, 4].

The FDA proposed increasing the standard user fees charged to manufacturers to allow greater oversight of foreign clinical trial sites [8]. The manufacturers inexplicably argued that clinical trial oversight is not part of the human drug review process, and no new revenue was included in the final PDUFA agreement in 2017 for this purpose. Since this agreement is in force until 2022, between now and then, only Congressional appropriation can create parity for US and non-US clinical sites [8].

24.3.4.2 Reduced Timeframe

To make matters worse, 81% of new drugs benefited from "Fast-Track," "Accelerated Approval," and "Priority Review" in 2018. This means that the FDA's ability to tease out clinical trial issues after the fact is becoming exceedingly difficult, unless the level of malfeasance is absurdly high [15]. Additionally, with the FDA's willingness to allow one rather than two or more pivotal trials for approval, that remaining pivotal trial must be of impeccable quality [15].

24.3.4.3 Good Manufacturing Practices

This situation is analogous to the FDAs oversight of manufacturing plants for prescription drugs [16]. Manufacturing has shifted largely overseas, and the FDA lacked the funding to adequately inspect these sites. When dangerous breeches of good manufacturing practices came to light due to whistleblowers in a few overseas plants, the FDA stepped up international inspections identifying malfeasance in many plants. These deviations from good manufacturing practices placed the US public at substantial risk from receiving substandard medications [16, 17].

We also know that the FDA does not have the resources to properly assess dietary supplements coming into the US, and there are many examples of products that were adulterated or contaminated as a result [18]. Without accountability there will always be companies involved in the research, manufacture, or sale of drugs and dietary supplements willing to cut corners, falsify or manipulate results, and as such place the public at risk for financial gain. In the case of clinical trials, it may not be the pharmaceutical companies themselves who are promoting malfeasance but the clinical research organizations they are contracting with.

24.4 Specific Examples of Clinical Trial Issues Overseas

The ethical and data quality issues associated with regulating clinical trials in developing countries is common among the FDA, World Health Organization (WHO), European Medicines Agency (EMA), and those countries themselves [7].

After the cyclo-oxygenase inhibitor rofecoxib was banned in the US for increasing the risk of cardiovascular events, two Indian clinical research organizations still agreed in principle to conducting research on rofecoxib in study subjects in a "Dateline NBC" sting operation in 2012 [19]. This shows how the company's responsibility to ethical conduct was secondary to securing research funding.

Trovafloxacin was an antibiotic with concerning animal data suggesting hepatotoxicity and bone/joint adverse effects [5]. In 1996, a clinical research site in Kano, Nigeria, was established after an outbreak of meningitis. The company claimed it had secured approval from an ethics committee, but evidence suggests there was no ethics committee in place at the time of stated approval and the approval letter was backdated. Doctors Without Borders set up a camp nearby to treat patients with an established treatment free of charge, but participants were not informed there was another treatment alternative for them [5].

Participants were not asked for their consent in the subjects' native language of Hausa, and no consent forms were filed [5]. Once the trial began, they did not properly assess for adverse events, even those known from pre-clinical work. Many significant adverse events occurred that might have alarmed the FDA, but they were not divulged. Up to 20% of subjects even experienced significant protocol deviations. Trovofloxacin was approved in 1997 but subsequently removed from the US market in 2001, when Phase IV post-marketing surveillance found excessive risks of hepatotoxicity [5]. This type of conduct is not just bad for the company that sponsors it; it also reflects badly on the US in the minds of those residing in that developing country.

China's version of the Food and Drug Administration (cFDA) led a 1-year intense investigation to verify the details of clinical trial dossiers supplied to support drug approvals in their country [20]. In 2016, they reported that 81% of the 1622 NDAs contained data from clinical trials with important integrity issues. This included many instances of discrepancies between trial data from the original record books and the submitted files, selective reporting of data, and not reporting the adverse events the patients experienced. In several cases, when they checked the submitted data versus data in the charts, there was no raw data available, strongly suggesting that the data was fabricated [20].

The cFDA was able to methodically piece together the real nature of the clinical trial research data submitted to them because they did not have an accelerated timeline, travel time to the clinical sites and expense were not barriers, there were no language barriers, and they had the authority to act [20]. Because the FDA oversight of foreign-conducted clinical trials lacks in these areas, it is unlikely that the FDA could be nearly as effective detecting this malfeasance abroad in developing countries [20].

While such widespread issues have not been reported internally in other developing countries, the most likely explanation is not that they lack similar ethical breeches but that their country's regulatory agencies have not undertaken a deep dive to uncover it [7]. This would be analogous to the whistleblower complaint against Ranbaxy where manufacturing deviations from FDA requirements were cited that then resulted in increased FDA inspections throughout India and China resulting in numerous violations through many companies [16, 17]. However in the case of clinical trials, revealing this level of malfeasance did not result in the FDA or other regulatory bodies conducting rigorous evaluations of non-US-conducted clinical trials.

24.4.1 World Health Organization/WHO and European Medicines Agency/EMA Inspections

In 2014, the WHO issued their second highest level of violation to an Indian-based clinical research organization [28]. The WHO team described problems with record integrity, subject safety, and poor-quality assurance in a study assessing the combination of lamivudine, zidovudine, and nevirapine for the treatment of HIV. The investigative team found ~66% of pre-study electrocardiograms (ECGs) were duplicates, suggesting actual ECGs may not have been performed or were substituted when the subject's own ECGs were not to the liking of investigators. When they assessed the files, the WHO team found forms that were blank, suggesting that they would be filled in later. The WHO ended up rejecting the study given the pervasive and serious nature of these issues.

The EMA has found serious data integrity issues as well. An Indian-based clinical research organization was found to have manipulated ECGs and ECG data between 2008 and 2014. In response to two such events in India, EMA recommended suspending the sale of approximately 700 drugs that were approved based on studies conducted by these research organizations [28].

In 2017, European inspectors determined that there were widespread data issues with studies conducted by an Indian clinical research organization, and the EMA recommended that studies for manufacturers such as Aurobindo, Sandoz, Sanofi, and Mylan be suspended until new studies were completed [21]. While these examples are in Indian-based clinical research organizations, it doesn't mean that these things are not occurring elsewhere in the world, just that given the issues that were known, increased scrutiny from the EMA occurred in India.

24.4.2 Other International Examples

Data quality issues have also been noted in major clinical trials. The TOPCAT trial assessed the impact of spironolactone versus placebo in patients with heart failure with preserved ejection fraction [22]. Overall, no significant benefits were seen with spironolactone, but heterogeneous findings were found between patients recruited in the Americas and those in Russia. In 2017, it was revealed that concentrations of spironalactone's active metabolite, canrenone, were undetectable in 30% of Russian subjects randomized to active treatment versus only 3% of subjects in the US and Canada. So, the reason for the discrepancy was because many subjects in Russia were not actually receiving the drug according to the study protocol [22].

The TRUE-AHF trial was published in 2017 and assessed the efficacy of the experimental drug ularitide in acute heart failure [23, 24]. Upon scrutiny, 17% of the population enrolled in the trial should not have been eligible for inclusion. This protocol breech, having $\geq$3 ineligible patients enrolled at a clinical site, only occurred in 2% of US sites but 63% of sites in the Estonia, Poland, the Czech Republic, and Serbia [23, 24].

The ARISTOTLE trial compared apixaban to warfarin in patients with atrial fibrillation. The FDA looked carefully at Chinese sites after one of them found much greater survival in the apixaban arm than with warfarin, which was out of line with the average site [25]. Upon inspection of that site and numerous others in China, the FDA excluded data from a total 24 Chinese sites with questionable data or data proven to be fraudulent. Publications resulting from the ARISTOTLE trial did not mention the controversy, even ones that came out over a year afterward. The data used in these journal publications and, remarkably, even in the apixaban label is based on the full data set including these 24 sites [25].

24.4.3 Overlooked Questionable Research Practices

Even when the FDA finds issues, they are simply disregarding the data from those sites internally instead of making it known that these issues had occurred [25]. This results in the persistence of questionable data in the peer-reviewed literature. In an assessment of FDA data, they found that 57 clinical trials had research sites that were cited with the most serious finding the FDA can give after an inspection, "Official Action Indicated." Protocol violations, inadequate or inaccurate recordkeeping, and failure to protect the safety of patients and/or issues with oversight or informed consent were each cited in over half of these sites. The most serious findings of submission of false information occurred in 39% of sites, problems with adverse events reporting occurred in 25% of sites, and miscellaneous violations were cited in 35% of sites. However, only 4% of the 78 resulting publications resulting from these trials mentioned the objectionable conditions or practices in the sites that provided the data. This makes malfeasance a lower risk endeavor because, even if caught, clinicians can still be fooled even though the FDA was not [25].

24.4.4 Better Outcomes

In an assessment of 139 meta-analyses with mortality outcomes, large differences in survival beyond those due to chance occurred in 8% of trials and in all cases showed more favorable treatment effects in the less developed countries (relative risk, 1.12 [95% confidence interval 1.06–1.18] [26]. This shows some research sites are likely intentionally manipulating survival data to make the companies' drug look better than it is. With the published data, clinicians may believe they are prescribing this expensive drug for their patients; however they will have an acceptable payback in terms of improvements in quality adjusted life years, while the reported difference is fictional.

While someone might contend that looking for sites that are clear outliers from the norm is an effective strategy to detect malfeasance, it does not protect the public from data that is being manipulated in a more subtle manner in some sites or in a systematic way across most sites [26]. It also does not mean that the rest of the studies in this meta-analysis had provided accurate information, just that the difference was not marked enough to make it obvious that the data was inaccurate.

24.5 Future Perspectives

In this chapter, we have seen how the FDA drug approval process is designed and that the ascribed steps, when done in sequence as intended by the designers, are rigorous and thorough. It assures that preclinical studies are sufficient before exposing normal volunteers to experimental drugs, it assures that studies that are proposed are not buried by the companies, and it has both prospective and retrospective inspections of study sites. It is designed to live up to the adage of "trust but verify."

Unfortunately, the confluence of several factors has riddled the FDA approval process full of loopholes and inconsistencies that place US clinical investigators and sites at a competitive disadvantage and make it virtually impossible for the FDA to truly ensure that the data it is reviewing is not tainted by malfeasance or sloppiness.

The FDA should know that citizens of developing countries are more susceptible to coercion and deserve more protections, from financial incentives to recruit and enroll subjects who might not be best served by participating, to collect data in a way or manner that speeds up recruitment of ineligible people, or to provide data to companies that those companies want rather than the real data that was generated.

The FDA should not be in the business of forcing US investigators or sites to uphold a separate set of rules from those in developing countries or drawing arbitrary thresholds. Differing rules for completing investigational new drug requests (INDs), differing inspection rates, and differing inspection intensities clearly and profoundly advantage foreign sites. If the FDA wants to instill confidence that clinical trials conducted in developing countries are of similar quality than those conducted in the US, the FDA needs a larger federal appropriation to conduct inspections of foreign clinical sites or needs to more rigorously negotiate with the companies so the funding is built into the Prescription Drug User Fee Amendments (PDUFA) in 2022. The FDA began piloting joint inspections and inspection data sharing with the EMA, becoming regular practice in 2011 [27]. Partnerships with other drug regulatory bodies should be encouraged as long as the level of transparency and trust is sufficiently high.

24.5.1 Regular, Systematic Inspections

We cannot just look for data outliers in clinical trial sites from developing countries and limit our inspections or scrutiny to them alone. Like in the US, all sites should anticipate the possibility that inspectors will come to assess their policies and procedures at any time. Identifying egregiously high thresholds for triggering inspections allows extensive lower-level or more balanced data manipulation to go undetected. Finally, when data manipulation is strongly suspected or proven, the FDA has a responsibility to alert the journals that published the information so the articles can be flagged, retracted, or updated and should never allow disregarded data to appear in the product labeling.

References

1. FDA Milestones (2020) Milestones in U.S. food and drug law history. fda.gov/about-fda/fdas-evolving-regulatory-powers/milestones-us-food-and-drug-law-history. Accessed 6 Jan 2021
2. FDA Infographic (2020) FDA drug approval infographic. fda.gov/media/82381/download. Accessed 6 Jan 2021
3. US OIG (2010) Challenges to FDA's ability to monitor and inspect foreign drug trials (OEI-01-08-00510). oig.hhs.gov/oei/reports/oei-01-08-00510.pdf. Accessed 12 Jan 2021
4. Ayalew K (2020) FDA perspective on international clinical trials. fda.gov/media/91849/download. Accessed 12 Jan 2020
5. Schuster BM (2015) For the love of drugs: using pharmaceutical clinical trials abroad to profit off the poor. Seattle J Soc Justice 13. Article 16
6. Thomas C (2011) Why Big Pharma now outsources its clinical trials overseas. The Ethical Nag. ethicalnag.org/2011/07/10/clinical-trials-outsourced-oversea/. Accessed 14 Jan 2021
7. White CM (2020) Current system of overseeing drug trials in developing countries by the FDA is dangerous. Ann Pharmacother 54:928–932
8. Hull D (2015) Reining in the commercialized foreign clinical trial. J Legal Med 36:367–401
9. FDA IND (2020) Investigational new drug IND applications. fda.gov/drugs/types-applications/investigational-new-drug-ind-application. Accessed 12 Jan 2021
10. Barlett DL, Steele JB (2011) Deadly medicine. Vanity Fair. vanityfair.com/news/2011/01/deadly-medicine-201101. Accessed 4 Jan 2021
11. Buchneva N (2018) To sign or not to sign FDA Form 1572? Applied Clinical Trials. appliedclinicaltrialsonline.com/sign-or-not-sign-fda-form-1572. Accessed 8 Jan 2021
12. Jonsen AR, Stryker J (eds) (1993) National Research Council (US) panel on monitoring the social impact of the AIDS epidemic. National Academies Press, Washington, DC
13. FDA FACT Sheet (2019) fda.gov/about-fda/fda-basics/fact-sheet-fda-glance. Accessed 8 Jan 2021
14. Hinzenrath DS (2016) FDA depends on industry funding; money comes with "strings attached". POGO. pogo.org/investigation/2016/12/fda-depends-on-industry-funding-money-comes-with-strings-attached/. Accessed 8 Jan 2021
15. Darrow JJ, Avorn J, Kesselheim AS (2020) FDA approval and regulation of pharmaceuticals, 1983–2018. JAMA 323:164–176
16. White CM (2020) Generic drugs not as safe as FDA wants you to believe. Ann Pharmacother 54:283–286
17. White CM (2020) Understanding and preventing (N-Nitrosodimethylamine) NDMA contamination of medications. Ann Pharmacother 54:611–614
18. White CM (2020) Dietary supplements pose real dangers to patients. Ann Pharmacother 54:815–819
19. Morrison N (2012) Indian CROs caught in Dateline sting insist report was "out of context". outsourcing-pharma.com/Article/2012/03/12/Indian-CROs-caught-in-Dateline-sting-insist-report-was-out-of-context. Accessed 3 Jan 2021
20. Woodhead M (2016) 80% of China's clinical trial data are fraudulent, investigation finds. BMJ 355:i5396
21. Venkatasubramanian KV (2017) EU recommends withdrawing drugs approved on unreliable data. chemistryworld.com/news/eu-recommends-withdrawing-drugs-approved-on-unreliable-data/3007019.article. Accessed 13 Jan 2021
22. de Denua S, Desai AS, Claggert B et al (2017) Spironalactone metabolites in TOPCAT – new insights into regional variation. N Engl J Med 376:1960–1962
23. Packer M, O'Connor CO, McMurray JJ et al (2017) Effect of ularitide on cardiovascular mortality in acute heart failure. N Engl J Med 376:1956–1964
24. Husten L (2017) As another trial goes bust investigators call for major changes in international trials. cardiobrief.org/2017/04/29/as-another-trial-goes-bust-investigators-call-for-major-changes-in-international-trials/. Accessed 13 Jan 2021
25. Seife C (2015) Research misconduct identified by the US Food and Drug Administration: out of sight, out of mind, out of the peer-reviewed literature. JAMA Intern Med 175:567–577
26. Panagiotou OA, Contopoulos-Ioannidis DG, Ioannidis JP (2013) Comparative effect sizes in randomised trials from less developed and more developed countries: meta-epidemiological assessment. BMJ 346:f707
27. Moscicki R (2014) FDA presentation–society for clinical research sites. sitesolutionssummit.com/wp-content/uploads/2014/11/FDA_Moscicki.pdf. Accessed 8 Jan 2021
28. Terry M (2015) Indian CRO quality questioned, as quest life sciences warned by WHO for HIV trial fraud. biospace.com/article/indian-cro-quality-questioned-as-quest-life-sciences-warned-by-who-for-hiv-trial-fraud-/. Accessed 12 Jan 2021

25 Current Status of Research with Brain Organoids

Sorin Hostiuc, Ioana Diaconescu, and Oana-Maria Isailă

Abstract

Cerebral organoids are three-dimensional biological structures trying to mimic some cerebral functions. This very hot topic in scientific research has attracted numerous researchers, but however has also generated a great number of controversies, due to its ethical and social implications. Differently from the heart, liver, or kidney organoids, these tissues are endowed with the potential for commanding or transferring thoughts, actions, and moral judgments, if eventually transplanted to human or animal organisms. The purpose of this chapter is to briefly summarize a few ethical issues that should be taken into account by researchers trying to publish research in this field, encompassing journal editors and ethical committees. Research should not try to circumvent but rather work within the accepted norms of ethics, and researchers should not try to present their findings in a thrilling manner, as they can easily lead to misunderstandings in the general population which could generate resistance or ridicule, not only praise.

S. Hostiuc (✉) · I. Diaconescu · O.-M. Isailă
Faculty of Dental Medicine, Department of Legal Medicine and Bioethics, Carol Davila University of Medicine and Pharmacy, Bucharest, Romania
e-mail: sorin.hostiuc@umfcd.ro

25.1 Introduction

Organoids are three-dimensional structures build in vitro from either pluripotent or induced pluripotent stem cells, whose purpose is to mimic the function of real organs. One of the types associated with the highest number of ethical issues and controversies is represented by brain organoids [1, 2]. The first studies were published by Lancaster et al., who showed that human pluripotent stem cells could recapitulate particular characteristics of cortical human development [3]. Publications have grown exponentially in numbers and importance. For example, just in early 2021, it was already shown the usefulness of advanced cerebral organoids for the study of later stages of neural development [4], drug screening for Creutzfeldt-Jakob disease [5], interactions with the SARS-Cov2 virus [6], and for analyzing the role of mitochondria in the development of human diseases [7].

25.2 Should Brain Organoids Be Regarded as Small Human Brains?

We should comparatively analyze cerebral organoids and human brains from a morphological and functional perspective. Morphologically, human brains are complex structures, comprised of different regions, with different cellular structures, encompassing neurons, supporting

J. Faintuch, S. Faintuch (eds.), *Integrity of Scientific Research*,
https://doi.org/10.1007/978-3-030-99680-2_25

cells, vessels, etc. Cerebral organoids are much simpler, yet still able to mimic the organization of various brain structures, including the cerebellum [8], midbrain [9], or hippocampus [10]. Vascularized cerebral organoids [11, 12] or cerebral organoid models to develop the microglial structure by using mesodermal progenitor cells inserted in cerebral organoids [13, 14] are relevant advances.

From a functional perspective, cerebral organoids were shown to have electric activity similar to neonatal EEG [15]. By using a reductionist approach, we may consider cerebral organoids as small human brain-like structures that, in the next decades, could be significantly closer to a human brain from both a morphological and functional point of view. Implications appertaining to the moral status were discussed elsewhere [1, 16, 17]. DeGrazia and DeGrazia defined moral status as an instance in which moral agents have specific duties concerning a certain agent. These obligations are based (at least partially) on the specific interest of that agent [18].

25.2.1 Are Organoids Moral Structures?

There are two distinct perspectives regarding the ethical use of this concept: (1) as a minimal level of moral standards that has to be fulfilled concerning a particular moral agent or (2) as an optimal/ideal level or moral standard, which becomes a theoretical, maximal desideratum, never achieved in practice but always seen as the threshold that has to be reached by moral agents [19]. From a descriptive point of view, moral status was evaluated concerning various elements like the presence of life (of all types) [20], the presence of life associated with other valid criteria like sensibility, capacity to have moral obligations, the personhood status [21], dignity [22], or the presence of morally relevant obligations [23–25]. Conscience, self-awareness, and intelligence are all attributes associated with moral agents; therefore, if we were to imagine that cerebral organoids were to develop such traits, it might be enough to consider them moral agents.

25.2.2 Are Artificially Created Organoids Sufficiently "Human"?

The human nature, the specific landmark to be used to establish the presence of moral status, and the presence of a potential to generate human beings have to be addressed [1]. The human nature is usually established through the presence of a unique characteristic that makes us humans, such as the presence of human DNA. Based on this characteristic, the human nature of the cerebral organoids is clear, as they consist of human-derived cells containing human DNA; the issue is more complicated in cases in which human/animal chimeras are created or if the genetic code is altered in a significant way. Le Bars developed a chimeric, murine-human model to study human brain development and disease by transplanting human cells into mouse brains [26]. Some authors have argued that, from an ethical point of view, the creation of these chimeras is deemed acceptable provided that the researchers have used a minimal number of human brain stem cells, the animals are significantly different from humans (to decrease the risk of developing neurological networks able to lead to human-like creatures, a criterion that, e.g., would render the creation of human-primate chimeras unacceptable), and as long as only dissociated human cells are used (otherwise breaching the concept of dignity, as human tissues or organs have stronger attachments to the human being compared to isolated cells) [27].

25.2.3 Stretching of the Limits of Ethical Acceptability

Small steps each stretching a little the limits might soon lead to the creation of de facto human-animal chimeras, in which the human neurons might alter the behavior or even intelligence of these creatures. This is currently seen in the development of human-primate chimeras. Tao Tan et al. used human pluripotent stem cells in cynomolgus monkey (*Macaca fascicularis*) embryos that were cultured ex vivo [28]. Due to these technological controversies, researchers performing them had to go to "friendlier" countries such as China [29]. This approach is a clear sign that the authors tried to circumvent ethical barriers. Scientific journals and the scientific community should sanction this approach. Only stating that the article complies with general norms of research ethics, or that it has obtained ethical approval, should not be enough. The journal editor should refuse to present controversial, immoral, or even illegal studies and should not seek to publish sensational results only as a publicity stunt or as a way to increase the impact factor of the journal.

25.2.4 Gene Engineering Procedures

The modification of the human genome using CRISPR-Cas9 or similar technologies was used to develop cerebral organoids with distinctive features, such as the inclusion of ancient genes in modern human brain cells [30]. These organoids are closer to human nature compared to chimeras but also have a higher potential to generate "augmented" organoids, as it is easier to create organoids with superior features (including appertaining to intellect or consciousness) by optimizing the human genetic material compared to mixing it with those of various animal species.

25.2.5 Moral Status, the Critical Threshold

Organoids should be as primitive as possible to minimize the risk of harming a moral entity, even if this approach would somehow limit the applicability in clinical research and practice, nevertheless still able to maximize the scientific output by using that specific landmark and not an inferior/lower one.

25.2.6 Can Organoids Become Active Parts of Human Beings?

Naturally the constructed biological structure consists of a single organ. The absence of a strongly interlinked net of organs and systems yields a practical impossibility to develop a full person; thus monsters or evil creatures will never directly emerge from cerebral organoids. However, we must be cautious as practical applications for cerebral organoids could flow in the direction of creating human-machine interfaces or brain transplantation, which would increase the complexity of this analysis significantly from a practical point of view.

25.3 Ethical Oversight in Human-Animal Chimeric Research

Attempts to implant human cerebral organoids in animal models have shown potentially significant implications for medical research. De facto they create human chimeras, a very controversial topic often overlooked in ethics guidelines primarily due to its novelty. As these issues are often unregulated, researchers have used mainly three approaches: (1) compliance with ethical regulations for similar animal studies, without also showing compliance with ethical requirements for studies using human biological materials; (2) ignoring the ethical regulations as the issue is mostly unregulated (often citing documents/guidelines that have nothing to do with research on human-animal chimeras); or (3) compliance with current regulations, usually working closely with independent ethical committees.

25.3.1 Conventional Ethical Restraints Only

Regarding the first possibility, human cerebral organoids were implanted in a mouse brain; soon after these structures interconnected with the mouse neuronal network creating complex synaptic structures, which were shown to alter the behavior of the chimeric mice, "compared to control mice, the mice transplanted with cerebral organoids showed an increase in freezing time in response to conditioned auditory stimuli, suggesting the potentiation of the startle fear response" [31]. The materials and method section posits that standard protocols for animal experiments were followed, whereas in the Ethics Declaration of the article, there is no statement. Another article, published by Cell Press, evaluated whether human cerebral organoids extend axons in the brains of non-human primates [32]. In the materials and methods section, it is indicated that university regulations on animal experimentation were adopted. Therefore, the study received approval from an ethical committee dealing with animal research for a study using human materials and, more importantly, neural cells to create human-primate chimeras, without any mention regarding compliance with ethical regulations on humans or the usage of human biological materials.

25.3.2 Specific Ethical Rules Minimized or Overlooked

Wang et al. demonstrated that human cerebral organoids transplanted into rat brains following traumatic brain injury increased cell survival rate and neurogenesis, improved neurological motor function, and reduced overall brain damage [33]. We believe that to comply with general research ethics requirements, animal cerebral organoids should be initially employed. Only after a clear benefit, with care, should human cerebral organoids be selected.

Incidentally the Materials and Methods section states that the use of human embryonic stem cells (hESCs) obeys the Ethical Guiding Principles for the Research of hESCs, without any further explanation or reference. There is no mention regarding an ethical committee approving the study as it created, de facto, human chimeras [33].

25.3.3 Delicate Questions Concerning Human-Animal Chimeras

Potential ethical issues involve the usage of human materials, the employment of animals in experimental research, and creating the actual chimeras. Most studies received approval or showed compliance with ethical regulations regarding animal experimentation only.

Additionally we must take into account that experimental beings containing human and animal genetic code (even if not within the same cell or nucleus) might have human-like behaviors, feelings, or actions. Here two issues must be discussed: if they are acceptable and in which conditions. Regarding the first question, our firm belief is that they are not acceptable yet. They should be acceptable provided their safety will be appropriately evaluated. Why proceed immediately to human-animal chimeras when the first logical step would be to create animal-animal chimeras? This approach is compliant with the Helsinki Declaration and would be easily performed with a single ethical oversight from an Animal Ethics Committee [34].

Regarding the conditions in which this type of research would be acceptable, Hermeren separated in vitro models using early embryos, from in vivo studies involving sentient animals. However the focus was supposed to be on animal welfare, mainly based on the principle of proportionality [35]. According to Hyun et al., "Research with the known, intended, or well-grounded significant potential to create humanized cognition, awareness, or other mental attributes, while not prohibited, should be subject to close scrutiny, taking the most careful steps to collect data pertinent to ethical protection of animal subjects, and an extraordinary degree of justification." It was added that committee composition should allow sufficient diversity to properly analyze the matters at hand [36]. Savulescu argued that if uncertainty arises regarding the moral status of a chimeric being, the genetically modified animal should be awarded the highest moral status consistent with its nature

[37]. This approach, with which we agree, would render impossible an exclusively animal ethics oversight on studies leading to human-animal chimeras, as they would be, de facto, considered to have a moral status equivalent to a human being.

Similarly journal editors should know that these studies should comply not only with animal but also human ethical requirements and should pay closer attention to the ethical section/statement of the manuscript. Whenever uncertainty arises, they could remit the research to an outside ethical committee.

25.4 Fake News and Sensationalization

One of the main misconceptions regarding cerebral organoids, which is often pushed by the mass and social media, is the fact that cerebral organoids are "small" or "tiny" or "mini" brains. For example, an article published in Stat News, discussing the implantation of human cerebral organoids in mice [38] is entitled "Miniature human brains grow for months when implanted in mice skulls," referring to an experimental article [39]. Nevertheless the implanted structures were cerebral organoids, not mini-brains, and of course the implantation occurred in the mouse brain, not the skull. Why are these differences relevant? First, cerebral organoids do not reproduce the complexity of the brain as a whole. Second, there is a big difference between implanting something in a mouse skull (which suggests, to a first-time reader, an empty skull without any other type of brain structure) and grafting simplified biological structures on a working brain.

In February 2021, Trujillo et al. published in *Science* an article entitled "Reintroduction of the archaic variant of NOVVA1 in cortical organoids alters neurodevelopment" [30]. The article's focus is a gene called NOVA1 (neuro-oncological ventral antigen-1), which is involved in synapse formation. Mutations in this gene may lead to various neurological disorders such as gait apraxia, low-grade astrocytoma, paraneoplastic opsoclonus-ataxia, etc. The researchers have replaced the modern human allele of this gene, using the CRISPR-Cas9 technology, with an ancestral gene common to Neanderthals and Denisovans in induced pluripotent stem cells from which brain organoids were built. These structures were shown to have different excitatory synaptic changes, affecting mainly glutaminergic signaling. Additionally, they had a somehow distinct morphology, with decreased number of rosettes, an increased number of apoptotic bodies, decreased cellular proliferation through a decreased number of Ki67+ and SOX2+ cells, and the presence of a progenitor cell layer with an aberrant structure [30]. In summary, the replacement of the contemporary with the ancient gene leads to some minor morphological changes of brain organoids.

25.4.1 Neanderthal Mini-brains

This very novel and exciting study from a scientific point of view was immediately "augmented" by various mass media outlets, seeking sensationalism. For example, National Post announced: "Scientists create Neanderthal 'mini-brains' using CRISPR gene-editing" [40], with the following subtitle: "While the brain tissue of modern humans are typically smooth and spherical, the tissue created with the ancient genes were smaller and had rough, complex surfaces" [40]. The brain tissue of modern humans is neither smooth nor spherical; on the contrary, it has a very complex structure that is one of the main particularities of the human species. This approach, apparently done to popularize scientific breakthroughs, may lead to significant misunderstanding by non-specialists. A step further was taken by cnet (www.cnet.com), in an article entitled "Scientists are now growing Neanderthal mini-brains in the lab," which had as a subtitle "They're being dubbed 'Neanderoids' and the next step is wiring them to robots that resemble crabs... because of course" [41]. The body of the article contains a solution for those who want to study a Neanderthal brain: "make your own Neanderthal in a petri dish" [41].

It is further stated: "In a both fascinating and terrifying twist, Muotri's team have already created human brain organoids and can detect electrical signals coming from the tissue. The next step is to wire those mini brains to crab-like robots, develop the Neanderoids to the point where they can do the same and pit them against each other" [41].

25.4.2 Brain-Machine Hybrids and Plane Thinking

It should be mentioned that this is not the first time when brain cells were associated/used in conjunction with non-biological structures. In 2002 it was created, at the University of Georgia, USA, a hybrid robot consisting of a cup-sized robot that was controlled through silicone electrodes containing an array of rat neurons [42]. In 2005, DeMarse and Dockendorf used an array of rat neurons to control the flight of a simulated aircraft, by using high-frequency stimulation inputs that produced a system in which the neuronal rat network was able to learn basic flighting tasks [43]. This has led to sensationalist titles of journal articles like "Piloting from a petri dish" [44] or "Plane thinking" [45].

25.4.3 Resurrection

In 2016, the NIH funded a project entitled "Non-randomized, Open-labeled, Interventional, Single Group, Proof of Concept Study With Multi-modality Approach in Cases of Brain Death Due to Traumatic Brain Injury Having Diffuse Axonal Injury," which aimed to use intra-thecal peptides, brain stem cells, laser, and transcranial IV laser or median nerve stimulations to potentially reverse brain death in subjects with traumatic brain injury and diffuse axonal injury [46]. The primary outcome of this study was reported with the title "Reversal of brain death as noted in clinical examination or EEG" [46]. Even though the clinical trial was not externally peer-reviewed and had no scientific foundation (brain death meaning the irreversible cessation of brain functions), media outlets presented the study using titles such as "Resurrected: A controversial trial to bring the dead back to life plans a restart" [47].

25.4.4 Man and Non-human Primates

An article published in early 2021, entitled "An early cell shape transition drives evolutionary expansion of the human forebrain," evaluated the differences between human, gorilla, and chimpanzee cerebral organoids. The authors have shown that humans', compared to gorillas' and chimpanzees', neuronal cells, have shown a delayed neuroepithelial transition leading to shorter cell cycles and subsequently a higher expansion of human cerebral organoids, which is controlled through an epithelial-mesenchymal transition regulator called ZEB2 [48]. CTV (www. ctv.ca) presents this finding as "Why are our brains so big? Researchers discover what separates us from other great apes." This was not, however, the purpose of the study, nor its result. The study has shown that human cerebral organoids are larger than those constructed using genetic material from gorillas and chimpanzees.

Brain size is not the fundamental element that makes us humans. An elephant brain is about three times larger than the human brain and contains about three times more neurons [49]. Is the elephant three times more intelligent than humans? The average male human brain is about 10% larger than a woman's brain [50]. Does this translate into a differential of intelligence? Of course not. The Neanderthal brain was larger than the current human brain. Were the Neanderthals more intelligent?

25.4.5 There Is Method in Misinformation Madness

The scientific distortions usually have a three-step approach: primary source (scientific article), secondary source (article aimed to popularize or sensationalize the information), and tertiary source (usually social media outlets).

The first step (primary source) is represented by a novel article, presenting a potentially significant advance in their field, usually published in a significant scientific journal (such as *Cell*, *Science*, *Nature*, etc.). Usually, the scientific article has a title that truly reflects its content and is not aimed to produce uproar. The next step (secondary source) is the publishing of a follow-up article, either in scientific journals (sometimes as a commentary) or the mass media aimed to popularize science. Typically there is shift in its main ideas, apparently preserving the scientific truth. However, more often than not, these articles do not objectively present the information that is either truncated or easily misinterpreted. The third step (tertiary source) occurs when the misinterpreted information is taken up by the social media mill. For example, a Facebook post regarding the Neanderthal brain states: "The Neanderthal brain was 15% larger than ours. Evidently, they were highly intelligent and comparisons made to them are actually compliments." Another post states: "They are figuring more and more every day, but as of right now no one can say they were smarter or dumber than modern humans. They were smart enough to survive the ice age so I'm betting they weren't all that dumb. Now liberals are a different story" [51]. We see how scientific information, which was trivialized by a secondary source (see above), becomes linked even with political views. This is a relatively benign example. Often results from scientific articles become misinterpreted in ways that hurt public health, as is the case with anti-vaccine theories.

References

1. Hostiuc S, Rusu MC, Negoi I et al (2019) The moral status of cerebral organoids. Regen Ther 10:118–122. https://doi.org/10.1016/j.reth.2019.02.003
2. Shepherd J (2018) Ethical (and epistemological) issues regarding consciousness in cerebral organoids. J Med Ethics 13–15. https://doi.org/10.1136/medethics-2018-104778
3. Lancaster MA, Renner M, Martin C-A et al (2013) Cerebral organoids model human brain development and microcephaly. Nature 501:373
4. Giandomenico SL, Sutcliffe M, Lancaster MA (2021) Generation and long-term culture of advanced cerebral organoids for studying later stages of neural development. Nat Protoc 16:579–602. https://doi.org/10.1038/s41596-020-00433-w
5. Groveman BR, Ferreira NC, Foliaki ST et al (2021) Human cerebral organoids as a therapeutic drug screening model for Creutzfeldt–Jakob disease. Sci Rep 11:1–9. https://doi.org/10.1038/s41598-021-84689-6
6. Tiwari SK, Wang S, Smith D et al (2021) Revealing tissue-specific SARS-CoV-2 infection and host responses using human stem cell-derived lung and cerebral organoids. Stem Cell Rep 16:437–445. https://doi.org/10.1016/j.stemcr.2021.02.005
7. Liput M, Magliaro C, Kuczyńska Z et al (2021) Tools and approaches for analysing the role of mitochondria in health, development and disease using human cerebral organoids. Dev Neurobiol. https://doi.org/10.1002/dneu.22818
8. Muguruma K (2018) Self-organized cerebellar tissue from human pluripotent stem cells and disease modeling with patient-derived iPSCs. Cerebellum 17:37–41. https://doi.org/10.1007/s12311-017-0905-2
9. Monzel AS, Smits LM, Hemmer K, et al (2016) A novel approach to derive human midbrain-specific organoids from neuroepithelial stem cells. bioRxiv
10. Todd GK, Boosalis CA, Burzycki AA et al (2013) Towards neuronal organoids: a method for long-term culturing of high-density hippocampal neurons. PLoS One 8:e58996. https://doi.org/10.1371/journal.pone.0058996
11. Shi Y, Sun L, Wang M et al (2020) Vascularized human cortical organoids (vOrganoids) model cortical development in vivo. PLoS Biol 18:e3000705. https://doi.org/10.1371/journal.pbio.3000705
12. Zhang S, Wan Z, Kamm RD (2021) Vascularized organoids on a chip: strategies for engineering organoids with functional vasculature. Lab Chip 21:473–488. https://doi.org/10.1039/d0lc01186j
13. Ormel PR, Vieira de Sá R, van Bodegraven EJ et al (2018) Microglia innately develop within cerebral organoids. Nat Commun 9:4167. https://doi.org/10.1038/s41467-018-06684-2
14. Wörsdörfer P, Dalda N, Kern A et al (2019) Generation of complex human organoid models including vascular networks by incorporation of mesodermal progenitor cells. Sci Rep 9:15663. https://doi.org/10.1038/s41598-019-52204-7
15. Trujillo CA, Gao R, Negraes PD et al (2019) Complex oscillatory waves emerging from cortical organoids model early human brain network development. Cell Stem Cell 25:558–569.e7. https://doi.org/10.1016/j.stem.2019.08.002
16. Boers SN, Clevers H, Bredenoord AL (2016) Organoid biobanking: identifying Organoids revive old and raise new ethical challenges for basic research and therapeutic use. EMBO Rep 17:1–4. https://doi.org/10.15252/embr.201642613

17. Lavazza A, Massimini M (2018) Cerebral organoids: ethical issues and consciousness assessment. J Med Ethics 1–5. https://doi.org/10.1136/medethics-2017-104555
18. DeGrazia D, DeGrazia PD (1996) Taking animals seriously: mental life and moral status. Cambridge University Press, Cambridge
19. Warren MA (1997) Moral status. Clarendon Press/ Oxford University Press, Oxford
20. Schweitzer A (1936) The ethics of reverence for life. Christendom 1:225–239
21. Mărculescu-Cojocea M (2005) Critica metafizicii la Kant şi Heidegger: problema subiectivităţii: raţiunea între autonomie şi deconstrucţie. Editura Trei, Bucureşti
22. Aquinas T (1991) Summa Theologiae: a concise translation. Christian Classics
23. Baird Callicott J (1989) In defense of the land ethic: essays in environmental philosophy. SUNY Press, New York
24. Hume D (2003) A treatise of human nature. Courier Corporation, North Chelmsford
25. Noddings N (2013) Caring: a feminine approach to ethics and moral education. University of California Press, Berkeley
26. Le Bras A (2020) A new chimeric model to study human brain development and disease. Lab Anim 49: 46–46. https://doi.org/10.1038/s41684-020-0477-5
27. Karpowicz P, Cohen CB, Van der Kooy D (2005) Developing human-nonhuman chimeras in human stem cell research: ethical issues and boundaries. CHIMBRIDS – Chimeras and Hybrids in Comparative European and International Research – Scientific, Ethical, Philosophical and Legal Aspects, pp 535–555
28. Tan T, Wu J, Si C et al (2021) Chimeric contribution of human extended pluripotent stem cells to monkey embryos ex vivo. Cell 184:2020–2032.e14. https://doi.org/10.1016/j.cell.2021.03.020
29. Ansede M (2019) Spanish scientists create human-monkey chimera in China. El Pais. https://english.elpais.com/elpais/2019/07/31/inenglish/1564561365_256842.html. Accessed 25 Apr 2021
30. Trujillo CA, Rice ES, Schaefer NK et al (2021) Reintroduction of the archaic variant of NOVA1 in cortical organoids alters neurodevelopment. Science 371. https://doi.org/10.1126/science.aax2537
31. Dong X, Xu S-B, Chen X et al (2020) Human cerebral organoids establish subcortical projections in the mouse brain after transplantation. Mol Psychiatry. https://doi.org/10.1038/s41380-020-00910-4
32. Kitahara T, Sakaguchi H, Morizane A et al (2020) Axonal extensions along corticospinal tracts from transplanted human cerebral organoids. Stem Cell Rep 15:467–481. https://doi.org/10.1016/j.stemcr.2020.06.016
33. Wang Z, Wang S-N, Xu T-Y et al (2020) Cerebral organoids transplantation improves neurological motor function in rat brain injury. CNS Neurosci Ther 26:682–697. https://doi.org/10.1111/cns.13286
34. World Medical Association (2013) Declaration of Helsinki. Ethical principles for medical research involving human subjects. World Medical Association
35. Hermerén G (2015) Ethical considerations in chimera research. Development 142:3–5. https://doi.org/10.1242/dev.119024
36. Hyun I, Taylor P, Testa G et al (2007) Ethical standards for human-to-animal chimera experiments in stem cell research. Cell Stem Cell 1:159–163. https://doi.org/10.1016/j.stem.2007.07.015
37. Savulescu J (2011) Genetically modified animals. In: Beauchamp TL, Frey RG (eds) The Oxford handbook of animal ethics. Oxford University Press
38. Begley S, McFarling UL, Sohn R, et al (2018) Miniature human brains grow for months when implanted in mice skulls. https://www.statnews.com/2018/04/16/human-brain-organoid-mice-neurons/. Accessed 16 Apr 2021
39. Mansour AA, Gonçalves JT, Bloyd CW et al (2018) An in vivo model of functional and vascularized human brain organoids. Nat Biotechnol 36:432–441. https://doi.org/10.1038/nbt.4127
40. Desai D (2021) Scientists create Neanderthal "mini-brains" using CRISPR gene-editing. National Post
41. Ryan J (2018) Scientists are now growing Neanderthal mini brains in the lab. https://www.cnet.com/news/best-dog-food-delivery/. Accessed 14 Apr 2021
42. Potter NN (2002) How can I be trusted?: a virtue theory of trustworthiness. Rowman & Littlefield, Lanham
43. DeMarse TB, Dockendorf KP (2005) Adaptive flight control with living neuronal networks on microelectrode arrays. In: Proceedings of IEEE international joint conference on neural networks, vol 3, pp 1548–1551
44. Grose TK (2005) Piloting from a petri dish. Washingtonian 14:18
45. Wood H (2004) Plane thinking. Nat Rev Neurosci 5: 898–898. https://doi.org/10.1038/nrn1572
46. Bioquark (2016) NCT02742857
47. Sheridan K, Diaz A, Liz Kowalczyk, et al (2017) A controversial trial to bring the dead back to life plans a restart. https://www.statnews.com/2017/06/01/brain-death-trial-stem-cells/. Accessed 15 Apr 2021
48. Benito-Kwiecinski S, Giandomenico SL, Sutcliffe M et al (2021) An early cell shape transition drives evolutionary expansion of the human forebrain. Cell. https://doi.org/10.1016/j.cell.2021.02.050
49. Herculano-Houzel S, Avelino-de-Souza K, Neves K et al (2014) The elephant brain in numbers. Front Neuroanat 8:46. https://doi.org/10.3389/fnana.2014.00046
50. Harrison PJ, Freemantle N, Geddes JR (2003) Meta-analysis of brain weight in schizophrenia. Schizophr Res 64:25–34. https://doi.org/10.1016/s0920-9964(02)00502-9
51. Kroneman O (2021) Olafkroneman.com–posts. https://www.facebook.com/Olafkronemanwriter/posts/3876225105734047. Accessed 15 Apr 2021

26 Ending Clinical Trials Prematurely

David L. Streiner

Abstract

Clinical trials can be terminated early for one of three reasons: there is a preponderance of adverse events in one arm, or both arms have more than the expected number of them; an interim analysis shows that no significant effect will be found with the projected number of participants (futility); or an interim analysis shows the clear superiority in the experimental condition. There is little controversy regarding truncating studies because of adverse events, but ending trials early because of futility, and especially for apparent superiority, is more controversial. Estimates of treatment effect early in the trial are based on smaller sample sizes, and hence the confidence intervals around those estimates are broad. There is much evidence that early estimates of treatment effect are inflated compared to trials of the same intervention that are completed. This likely does not constitute fraud or misconduct, but does reflect poor practice unless stringent criteria are followed.

D. L. Streiner (✉)
Department of Psychiatry and Behavioural Neurosciences, McMaster University, Hamilton, ON, Canada
e-mail: streiner@mcmaster.ca

26.1 Introduction

In a typical randomized controlled trial (RCT), participants are randomly divided into two groups, one of which receives a novel intervention and the other one either a placebo or treatment as usual (TAU). They are followed for a set length of time, which can be as short as a few days to as long as a number of years, and the results are then examined to determine which group had a better outcome. Despite being regarded as the gold standard design for determining the efficacy or effectiveness of an intervention, RCTs have been criticized on a number of grounds. These include their cost; the length of time between asking a question and getting an answer; as well as ethical issues regarding exposure to risks associated with a new treatment and the use of placebos.

26.2 Problems with Randomized Controlled Trials

Sertkaya et al. [1] estimated that the average cost a Phase 3 drug trial ranges between $11.9 million (for immunomodulation drugs) and $52.9 million (for pain and anesthesia trials); and the average time between enrollment to publication of results is approximately 5½ years [2]. The Women's Health Initiative, admittedly at the high end of the spectrum, cost $260 million in 2012 US dollars [3] and had a planned duration of 8½

J. Faintuch, S. Faintuch (eds.), *Integrity of Scientific Research*,
https://doi.org/10.1007/978-3-030-99680-2_26

years. One implication of this is that by the time the data have been published, new therapies may have been developed and the results of the trial are out of date. In addition to these problems, there are ethical considerations that arise from the design of RCTs. For these reasons, a number of techniques have been proposed to shorten RCTs, such as Bayesian adaptive designs [4]. Another, the focus of this chapter is to perform interim analyses to determine if it is possible to end a trial early because of evidence of effectiveness or because it appears that the study will not be able to achieve its aim with the proposed sample size. However, the decision regarding whether or not to terminate a trial prematurely itself may lead to difficulties, and these will be discussed.

26.3 Ethical Issues in Randomized Controlled Trials

There are various ethical issues that arise within the context of RCTs, for both the treatment and comparator groups, and depending on whether the treatment is effective or not. On the one hand, if the new treatment results in more adverse events (AEs) or serious adverse events (SAEs), then those in the experimental group are placed at increased risk from them. For example, a meta-analysis by Gøtzsche [5] found a higher risk of diarrhea and abdominal pain for patients receiving misoprostol compared to placebo and a tendency for more gross hemorrhage and ulcers in the non-steroidal anti-inflammatory drug groups versus placebo. Even more disturbingly, an RCT of clofibrate showed that while it did reduce plasma cholesterol and lowered blood pressure as was intended, it also led to a significant increase in non-cardiovascular mortality [6]. Moreover, the more potent the intervention, the more prevalent are AE and SAEs. Among 1160 patients who were receiving antiretroviral treatment for HIV, 545 (47%) had clinical AEs and 27% laboratory AEs probably or definitively attributed to the intervention, and of these, 16% were graded as severe and 9% as serious [7].

26.3.1 Demographic and Nocebo-Related Events

The risks from RCTs are not equally distributed. Those under the age of 9 are about 50% more likely to experience AEs than people in their 20s; and the risk increases for each decade over 30. The oldest age group (70–100) is 3.4 times more likely to have an adverse event [8]. Moreover, the risk is not limited to those in the active treatment condition. Just as the placebo effect can lead to positive changes in those taking an inactive agent, the nocebo effect means that those who receive one may have a negative outcome. A meta-analysis of ten double-blind, placebo-controlled phase 3 studies of modern cancer drugs found an incidence of grade 3 (severe or undesirable) and 4 (life-threatening or disabling) SAEs in the placebo groups of 18%. The most frequent AEs were hypertension, fatigue, and diarrhea [9].

26.3.2 Ethical and Moral Controversies of the Comparison Condition

On the other hand, if the treatment is effective, then those in the control arm of the study are denied its benefits, at least for the duration of the trial. This may not be an issue if the trial is short and the targeted condition is reversible, but can be a serious problem in longer trials. This is especially true if the deterioration while on placebo or TAU cannot be reversed, as is the case with dementing disorders, bipolar disorders [10], and schizophrenia [11]. In fact, some [12] have argued that it is unethical to use placebos at all if an effective treatment exists, as this would violate Article II.3 of the Declaration of Helsinki, which states that "In any medical study, every patient—including those of a control group, if any—should be assured of the best proven diagnostic and therapeutic methods." [13].

This is an issue of ongoing debate, as others have argued that, strictly interpreted, Article II.3 of the Declaration of Helsinki would ban all clinical trials, even those without a placebo-control arm [14, 15]. There is also an argument that it is more ethical to compare a new treatment against placebo rather than another active treatment because fewer participants are needed in the trial to demonstrate a difference, and hence fewer people are put at risk of AEs and SAEs [16, 17].

Thus, stopping a trial early can have a number of advantages. It would result in fewer participants being exposed to the harmful effects of new interventions or, on the other hand, being denied a useful treatment. Second, the results of the study, either positive or negative, would be disseminated more quickly to the clinical community, so that patients could benefit from the treatment (if the results are positive) or be spared being prescribed an ineffective or potentially harmful treatment (if negative). Finally, it would decrease the cost of the trial, in both dollars and time.

26.4 The Role of Data and Safety Monitoring Boards

One mechanism that has been put into place to shorten the duration of studies and minimize the risks is to examine the data part-way through the trial in order to determine if the trial could or should be stopped early. Most often, this is done by a Data and Safety Monitoring Board (DSMB). This is a group of experts, independent from the research team, that periodically reviews and evaluates the data with regard to safety and efficacy and can make recommendations regarding continuation of the study, modifications to its designs, or termination of the trial [18]. The committee usually consists of an expert in the clinical aspects of the disorder, an ethicist, one or more statisticians, and others with a background in conducting clinical trials. They generally meet on a regular basis, the frequency of which is determined by the flow of participants, and may have additional ad hoc meetings, usually triggered by reports of SAEs. In large trials, and especially those of long duration, there may be a predetermined number of interim analyses, whether or not there have been any AEs or SAEs.

26.4.1 Current Relevance of Data and Safety Monitoring Boards

Although the first RCT in medicine was conducted in 1948, examining the use of streptomycin for the treatment of pulmonary tuberculosis [19], DSMBs did not exist until the 1970s and became standard practice only in the late 1980s [20]. Since then, most regulatory agencies, funding organizations, and ethics boards require them in all studies where there is the possibility of AEs, even in social intervention trials where the study does not involve drugs or other medical treatments [21].

26.4.2 Adverse Events

There are three situations that may trigger a DSMB to recommend that a trial be terminated prematurely: the presence of AEs and SAEs, futility, and superiority of the experimental condition [22].

Adverse events, and even SAEs, are, unfortunately, a fact of life and prevalent in clinical care, whether it is delivered routinely or within the context of a research trial. In one review of Canadian hospitals, Baker et al. [23] found that the rate of AEs was 7.5 per 100 hospital admissions, while Ohta et al. [24] reported 16.3 AEs per 100 patients in a cardiac unit of a Japanese teaching hospital. Consequently, the mere presence of AEs or SAEs would not jeopardize a study. Indeed, the Food and Drug Administration in the United States says that "Risks to subjects are reasonable in relation to anticipated benefits, if any, to subjects and the importance of the knowledge that may reasonably be expected to result." [25] Concerns would be raised, though, if the prevalence of AEs and SAEs was substantially higher in one group relative to the other or exceeded what would be expected in routine clinical care.

26.4.3 Futility Concerns

Futility means that an interim analysis of the data shows that, given the planned sample size of the study, it would be impossible to prove that the intervention is superior to the comparator, either because there is little difference between the groups or because the control condition is superior. Consequently, it would be unethical to continue to enroll participants in the trial or at least into that arm of it if it were a multi-group study. The first, and perhaps most controversial, example of this was the University Group Diabetes Program, which looked at various treatments for type 2 diabetes. It discontinued the tolbutamide arm of the study after determining that it was ineffective, a decision so controversial that it ultimately ended up in the Supreme Court [26].

26.4.4 Unquestionable Advantage of the Intervention

The third situation where a trial may be discontinued is if an interim analysis shows clear superiority of the experimental treatment relative to the comparator. Briel et al. [27] suggest that this is occurring with increasing frequency, especially in the five most cited general medical journals. These studies were typically industry-funded and were in the areas of cardiology, hematology-oncology, and HIV/AIDS.

26.5 Statistical Issues with Interim Analyses

However, conducting one or more interim analyses to look at either benefit or futility raises a statistical problem: inflation of the p level. This is the value, usually 0.05, that determines whether a difference is declared to be statistically significant or not. The issue is that if the null hypothesis is true—there is actually no difference between the groups—then any one analysis has a 5% chance of committing a type I error, that is, declaring that there is a difference when in fact there is none. If two tests are performed, then the probability of a type I error rises to 9.75% and to over 14% if three tests are performed (i.e., two interim analyses plus the final one). More generally, the probability of a type I error is given by the formula $[1 - (1 - \alpha)^k]$, where α is the chosen critical value for declaring significance (as mentioned, usually 0.05) and k is the number of tests performed [28, 29].

26.5.1 Alpha-Spending Rules

A number of "α-spending" schemes have been developed to maintain a 0.05 level, irrespective of the number of analyses that are run. These divide 0.05 among the analyses in different ways. The first, and most simple, method was developed by Pocock [30, 31], which, much like the Bonferroni correction, simply divides the nominal α level equally among all of the analyses. For example, if there are three interim analyses, then a critical value of 0.0125 (i.e., 0.05/4) would be used for each as well as for the final analyses. However, this has been criticized because it is seen as too lenient, meaning that it can result in stopping trials before a sufficient number of participants have been enrolled to draw a firm conclusion. Thus, even if the results are statistically valid, the study would be perceived as lacking external validity [32]. Furthermore, this technique presupposes a fixed and pre-determined number of interim analyses and so is infeasible if ad hoc "peeks" at the data are necessitated by SAEs.

The Haybittle-Peto boundary [33] uses a fixed p level of 0.001 for all of the interim analyses and a value of 0.05 for the final one. Because these interim values are so conservative, they do not materially affect the overall p level. This makes it easy to implement, understand, and use if there needs to be additional analyses but has been criticized for the opposite reason: that it is too difficult to terminate a trial. Perhaps the most widely used stopping rules are based on the O'Brien-Fleming [34] and the Kim-DeMets [35, 36] methods. These "spend" α very conservatively in the beginning, by using extremely stringent critical values of p (e.g., 0.001), and slowly increasing it in subsequent analyses. The result is that the final analysis uses a value close to 0.05, and the sum of all the critical values remains at 0.05. Although these approaches are predicated on a pre-specified number of interim analyses, they can be modified while the trial is in progress if additional analyses are required because of SAEs.

Deciding which spending rule to use is, in one sense, a secondary question. The primary one is whether interim analyses should be done at all. There is no issue with looking at the data early and stopping a trial if there is a preponderance of AEs and SAEs in one group or the other or if they occur in both groups at a rate higher than what would be expected normally, which is why a clinician who has expertise in the disorder being treated is a key member of the DSMB. Rather, the question is whether it is wise to stop any trial prematurely, either for futility or superiority.

26.6 Stopping Trials Because of Futility

There are two general reasons to end trials early because of futility: operational and statistical. The major operational issue concerns an inability to recruit patients at a sufficient rate to meet the sample size requirements by the time the study is scheduled to end. For example, a review of all RCTs published in a 6-month period in six major journals found that 21% failed their recruitment goals at randomization and 48% at outcome assessment [37]. Similarly, of 122 trials in the UK funded by various agencies, fewer than a third achieved their targeted sample size [38]. In fact, a majority of studies that ended early did so because of an insufficient rate of accrual [39]. A second operational issue arises when a sufficient number of participants have been recruited, but the outcome occurs at a lower rate than anticipated, thus reducing the power of the analysis (i.e., the probability of detecting an effect when it is present). Here, though, we will focus solely on studies that terminate early because interim analyses suggest that the trial is unlikely to show the superiority of the experimental treatment.

26.6.1 Problems with Early Termination

As mentioned previously, there are a number of reasons that studies should be ended prematurely if interim analyses indicate that it would be futile to continue. The primary one is ethical; if patients will not improve with the intervention relative to placebo or TAU, then they are exposed to any risks from the treatment without any counterbalancing benefit. Also, if other treatments are available, patients in both groups are prevented or delayed receiving them. There are also economic considerations, because maintaining a trial center, paying staff, and enrolling patients are a very expensive undertaking.

On the other side of the equation, if a potentially successful trial is ended prematurely, there are costs associated with that decision. The primary one is that a possibly useful intervention is not pursued. Second, the time, inconvenience, and exposure to AEs and SAEs endured by the patients would have been in vain. And finally, all of the costs incurred in mounting and running the trial would have been for naught. The question, then, is whether the decision to terminate the trial early was the correct one. This issue was brought to the fore in recent drug company-sponsored trials of aducanumab (ADU), a monoclonal antibody developed by Biogen to slow memory loss in patients with Alzheimer's disease. Two phase III trials were terminated prematurely because interim analyses showed it was ineffective. However, the company said that a subsequent reanalysis demonstrated benefit for a higher dose, a conclusion that has generated considerable controversy [40, 41]. Without going into the details of the debate, which center around methodological aspects of the studies, this illustrates that decisions about futility are fraught, perhaps to a greater degree than decisions to continue trials.

A trial of chemotherapy for advanced non-squamous non-small cell lung cancer was ended for futility after approximately half of the 1300 patients were recruited. However, the patients were continued to be followed, and at the end, there was a statistically significant improvement in the primary endpoint, progression-free survival. This prompted the authors [42] to reanalyze the futility analyses, and they concluded that had any other time point been used for the interim analysis, the outcome would have been different, and the study would have continued.

Jitlal et al. [42] retrospectively analyzed ten RCT cancer trials, all of which reached their predetermined sample size. They found that after observing 50% of their target number of events, three of five trials with no benefit could have been stopped early, and the duration of two trials could have been reduced at a substantial cost saving (a third had already ended, and hence there was no saving). Disturbingly, though, of the four trials with moderate treatment effects, two would have been ended mid-way although they subsequently showed that the intervention was effective. They concluded that, "Unless there is very good evidence for futility, it is often best to continue to the planned end" (p. 910).

26.7 Stopping Trials Because of Benefit

The other side of the coin is whether trials should be ended early because an interim analysis demonstrates the superiority of the experimental intervention. Briefly recapitulating the benefits of doing so, it would allow the treatment to enter into clinical practice earlier, minimizes risks in the comparator group from not being treated with the preferred therapy, and saves costs to the funding agency or sponsoring company. The major issue, though, is the degree to which results from a trial that has been terminated mid-way are an accurate reflection of the true effect.

There is little question that early termination for benefit is becoming more prevalent. Montori et al. [43] estimated that none of the 620 RCTs published in high-impact medical journals between 1975 and 1979 were ended early for benefit; this increased to 0.5% (9 of 1938 trials) between 1985 and 1989 and to 1.2% (47 of 3859 trials) by 2000–2004. Of the 107 studies that were examined, the greatest number was in the area of cardiology, followed by cancer, HIV/AIDS, and critical care. On average, the trials ended after about 64% of the participants were recruited. Slightly over one-third of the studies were funded by for-profit organizations (e.g., drug companies). However, accompanying this increase in early termination is an ongoing debate among researchers and statisticians about whether doing so is a wise strategy.

On one side of the debate, Korn et al. [44] reviewed trials in oncology that were stopped early for positive results. Of the 18 studies that had sufficient follow-up information, in 17 of them, the treatment effect was similar or only slightly smaller at the last evaluation time compared with the effect when the trial ended. They concluded that "for trials with well-designed interim monitoring plans, the ability to stop early for positive results is an important component of the trial design, allowing the public to benefit as soon as possible from the study conclusions" (p. 1712).

26.7.1 Controversies Regarding Clear-Cut Benefit

However, concerns were raised about premature stopping for benefit over three decades ago by Pocock and Hughes [45]. They conducted a post hoc analysis of a completed trial and simulated the results from interim analyses using a number of stopping rules. They concluded that "clinical trials that stop early are prone to exaggerate the magnitude of treatment effect" (p. 209S). This was empirically demonstrated in a study comparing four versus five courses of consolidation therapy in patients with acute myeloid leukemia [46]. Six interim analyses were specified a priori, because of the large sample size ($N = 1078$) and the study's long duration (8 years). The first analysis, performed after 202 patients had completed the trial, found a statistically significant hazard ratio (HR) of 0.57 favoring five courses of treatment. The second, third, and fourth analyses were also significant and favored five courses, albeit with increasingly smaller HRs (0.53, 0.45, and 0.33). The last two interim analyses, while in the same direction, were not statistically significant; and the final analysis showed a non-significant HR in the opposite direction (−0.09). The reason that the trial was not terminated early based on the interim analysis was based on the judgment of the DSMB that the initial HRs showed an implausibly large treatment effect for such a relatively small change in the therapy. This was subsequently borne out by the final analysis, showing no difference between four and five courses of treatment, thus reinforcing Pocock and Hughes's contention.

A similar effect was found in the Candesartan in Heart Failure Assessment of Reduction in Mortality (CHARM) study [47]. Although six interim analyses showed highly significant HRs in favor of candesartan as compared to placebo, the trial was not stopped early. The HR at the time of the first interim analysis was 0.63, and this increased slowly at each subsequent time point, until by the final analysis the HR was 0.91, which was not significant. They attributed this effect to a number of factors, including that data on the primary endpoint (cardiovascular death and hospitalization for coronary heart failure) were not available early on and had to await adjudication by the endpoint committee and that the length of follow-up was relatively short. They concluded that this "illustrates the crucial importance of continuing a trial to its scheduled termination unless there emerges evidence that provided proof beyond a reasonable doubt that would influence clinical practice. Indeed, early termination of CHARM based solely on a statistical guideline would have been misleading" (p. 943).

More conclusive evidence that prematurely ended trials overestimate treatment effects was provided by Montori et al. [43]. They did a systematic review of 143 trials that were ended early because interim analyses showed the superiority of the new treatment. Their conclusion was that "Trials stopped early for benefit, particularly those with few events, often report treatment effects that are larger than typical of interventions that have been definitively studied" (p. 2209). A later study by the same team [48] compared 91 RCTs that were truncated early for benefit against 424 matching RCTs that were not ended prematurely. The ratio of the relative risks (RRs) was equal to or less than 1.0 in 87% of them, indicating larger effects in the truncated trials. The magnitude of the difference was related to the number of outcome events; the fewer of them (less than 500), the more likely the trials that were stopped early showed a large difference.

26.8 Why Prematurely Stopped Trials May Overestimate the True Effect

The evidence thus suggests that trials that are ended early based on interim analyses often overestimate the true effect of the intervention. The question then arises why this should be so. Imagine that we do a study where the true difference between the groups has an effect size of 0.5; that is, the mean of the experimental group is one-half a standard deviation higher than the mean of the comparison group on whatever scale is used for the outcome. Or, if the outcome is binary (e.g., improved/not improved; alive/dead), the RR is 2.0 in favor of the experimental group. Now imagine that we measure the treatment effect after every ten patients have been entered into the trial. The first interim analysis will likely yield an estimate that differs from the true value to some degree. Thereafter, as more subjects are entered into the study, the estimated effect will more closely approximate the true value—what has been called "regression to the truth"—but will still have results that vary randomly around the last computed value, sometimes higher and sometimes lower.

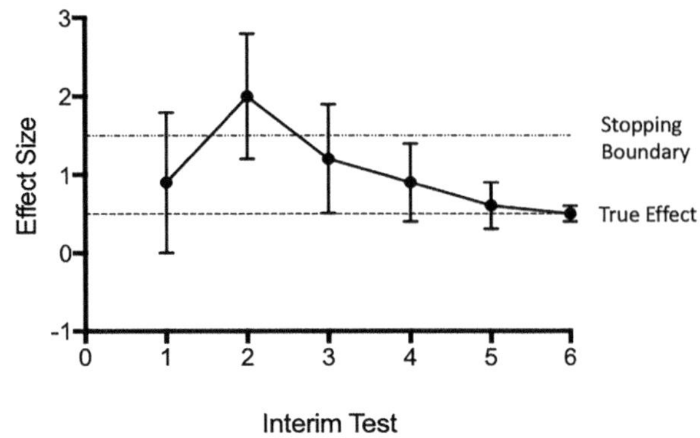

Fig. 26.1 Hypothetical results of five interim and one final analysis showing regression to the truth and a random high

How much higher or lower is determined by the standard error of the mean (SE_M), which is defined as $s/\sqrt{n}$, where s is the standard deviation of the scores and n is the sample size.[1] Toward the beginning of the trial, when n is quite small, the SE_M is therefore large, meaning that the values can deviate quite markedly around the true value. By chance, one of them may exceed the threshold and be statistically significant, what Pocock and White [49] refer to as a "random high." This is shown in Fig. 26.1. By definition, any difference that is statistically significant based on a small sample size must be large [50]. This is exacerbated by the alpha-spending rules discussed previously, because for the first few interim analyses, the critical value for significance is more stringent than $p < .05$, meaning that the difference needs to be quite large.

As the trial enrolls more patients, the SE_M becomes smaller, and the estimates of the treatment effect more closely approximate the true effect. As a consequence, trials that are terminated closer to the planned endpoint, or that have accumulated more outcome events, will be less likely to overestimate the actual effect size. This was demonstrated in the analysis by Bassler et al. [48], which found larger overestimates in studies that had fewer than 500 events as compared with trials that had more.

26.9 Other Problems Due to Terminating Prematurely for Benefit

There are additional difficulties that result from early termination: imprecision of the estimated effect and insufficient safety data. As mentioned in the previous section, the sample size determines the magnitude of the SE_M. In turn, the SE_M is used to calculate the confidence interval (CI) around the estimated effect. When there are two groups, the 95% CI is the mean difference $\pm 1.96 \times SE_M$.[2] The implication is that the CIs are wide early in the trial, when there are relatively few participants, and thus the estimates of the magnitude of the effect are imprecise. A number of authors have emphasized the fact that stopping early should require very strong evidence, i.e., proof beyond a reasonable doubt [51], and trials that have wide CIs do not provide that evidence. There have been many trials in which DSMBs have determined that the studies should continue despite significant interim analyses because of concerns that the results may have been spurious, concerns that were later borne out [47, 48]. Furthermore, clinicians may not believe the results of a trial that had small sample sizes [51], meaning that the time, effort, money, and the participants' possible exposure to risk were all for naught. There is also the danger that truncated trials with overly optimistic estimates of benefit may "freeze" subsequent research [27]; that is, there

[1] For two groups, the SE_M of the difference is calculated as $\sqrt{s_1^2/n_1 + s_2^2/n_2}$.

[2] The SEs and Cis are calculated differently for ORs and RRs and are described in many statistics texts.

is less motivation to perform follow-up studies that may reveal the overestimate of the original one.

26.9.1 Late or Incompletely Reported Events

Often, AEs and SAEs may not appear until weeks or months after the patients have been started on a new medication or undergone some intervention. For example, agranulocytosis does not manifest itself until months after patients have been started on the antipsychotic drug, clozapine [52]; and there are long-term complications associated with bariatric surgery [53]. These may come to the attention of the Food and Drug Administration through the Adverse Event Reporting System, but this is voluntary after a drug or device has been approved. Because of this, a systematic review of 37 studies across 12 countries found that the median rate of under-reporting was 94% [54]. More complete information about AEs and SAEs comes from trials, especially when participants are followed for a length of time. When studies are truncated, this follow-up period is similarly shortened, resulting in less information about potential risks of the intervention.

26.10 Guidelines for Stopping Trials Prematurely

Although not required for all trials, DSMBs have become much more prevalent over the past few decades, especially in multi-center trials and where there is the possibility of AEs or SAEs. And while they can only make recommendations, their opinions often carry great weight regarding the discontinuation of a trial. Thus, the procedures they follow must insure, to the greatest extent possible, that their decisions lead to the least possible harm. The following recommendations are taken primarily from Wheatley and Clayton [46] and Zannad et al. [51].

1. Alpha-spending rules should be very stringent toward the beginning of the trial (e.g., $p < .001$) and gradually become more lenient.
2. These stopping rules should be regarded as suggestions only and can be overridden in light of additional information.
3. This additional information can include the results of similar trials, clinical experience, and whether the results will be accepted by the clinical community.
4. The totality of evidence should be considered, such as endpoints other than the primary one and differences between subgroups (e.g., males and females).
5. The DSMB should specify a minimum follow-up period or number of outcome events.
6. The DSMB should be extra cautious with novel agents because of the possible freezing effect which can stifle future research.

26.11 Does Early Termination Constitute Misconduct or Fraud?

It is not always easy or straightforward to determine if an adverse event is related to the intervention. Patients rarely have just one disorder. Especially if they were recruited from a hospital or clinic, Berkson's bias [55] tells us they are more likely to have diseases other than the target one [56]. Thus, it may be difficult to ascribe an event, such as a heart attack or stroke, to the intervention, as opposed to a comorbid condition. However, in the long run, if these events are unrelated to the treatment, then they should balance out between groups.

Stopping for futility is a more difficult situation. In theory, it should be based on interim analyses with sufficient power to determine that the addition of more subjects will not suffice to make the results significant. In practice, though many interim analyses were underpowered, the decision relied on only one—and in some cases no—interim analysis or did not provide a basis for the stopping rule [57]. And, as has been seen [58], a "random low" can result in erroneously ending a trial for futility.

The most problematic situation is terminating a trial prematurely because of superiority. The rationale for doing so is clear; the results would reach clinicians more quickly and be incorporated into routine practice, fewer participants would be placed at risk of receiving inadequate treatment, and there could be significant cost savings. However, the risks are high. Many theoretical analyses, systematic reviews, and meta-analyses have shown that truncated studies overestimate the true effect of the intervention, often dramatically so [43, 45–48]. Moreover, as we have seen, prematurely terminated studies can yield imprecise estimates of the effect of treatment, provide insufficient evidence regarding AEs and SAEs, and may stifle other research that may provide better estimates of the effectiveness of a treatment (or its lack thereof).

26.11.1 Questionable Research Practices

The question is whether early termination is an example of research misconduct or fraud that should be sanctioned. I will not discuss trials that were prematurely ended by the sponsoring company for financial reasons and not for futility, although this raises ethical issues of its own, as discussed by Iltis [59] and others. In a review of high-profile cases of fraud in medical studies, George and Buyse [60] documented many cases in which data were fabricated or manipulated so that *more* participants could be enrolled in the studies, but none of which resulted in early termination. It could probably be argued that truncation because of apparent superiority is poor practice, unless there is very strong evidence, but not malpractice or fraud. Very strong evidence would consist of ideally two consecutive significant interim analyses, sufficient power, and results that are consistent with previous research or clinical findings. Even given these, it would likely be prudent to complete the trial, so that the treatment effect can be determined within a narrow CI, and there is sufficient evidence regarding adverse events.

References

1. Sertkaya A, Birkenbach A, Berlind A, Eyraud J, Eastern Research Group (2014) Examination of clinical trial costs and barriers for drug development: final. U.-S. Department of Health and Human Services. https://aspe.hhs.gov/ system/files/pdf/77166/rpt_erg.pdf. Accessed 14 Jan 2021
2. Ioannidis JP (1998) Effect of the statistical significance of results on the time to completion and publication of randomized efficacy trials. JAMA 279(4):281–286. https://doi.org/10.1001/jama.279.4.281
3. Roth JA, Etzioni R, Waters TM, Pettinger M, Rossouw JE, Anderson GL et al (2014) Economic return from the women's health Initiative estrogen plus progestin clinical trial: a modeling study. Ann Intern Med 160(9):594–602. https://doi.org/10.7326/M13-2348
4. Ryan EG, Bruce J, Metcalfe AJ, Stallard N, Lamb SE, Viele K et al (2019) Using Bayesian adaptive designs to improve phase III trials: a respiratory care example. BMC Med Res Methodol 19:99. https://doi.org/10.1186/s12874-019-0739-3
5. Gøtzsche PC (2000) Non-steroidal anti-inflammatory drugs. BMJ 320(7241):1058–1061. https://doi.org/10.1136/bmj.320.7241.1058
6. Oliver M (2012) The clofibrate saga: a retrospective commentary. Br J Clin Pharmacol 74(6):907–910. https://doi.org/10.1111/j.1365-2125.2012.04282.x
7. Fellay J, Boubaker K, Ledergerber B, Bernasconi E, Furrer H, Battegay M et al (2001) Prevalence of adverse events associated with potent antiretroviral treatment: Swiss HIV Cohort Study. Lancet 358(9290):1322–1327. https://doi.org/10.1016/s0140-6736(01)06413-3
8. Luo J, Eldredge C, Cho CC, Cisler RA (2016) Population analysis of adverse events in different age groups using big clinical trials data. JMIR Med Inform 4(4):e30. https://doi.org/10.2196/medinform.6437
9. Chacón MR, Enrico DH, Burton J, Waisberg D, Videla V (2018) Incidence of placebo adverse events in randomized clinical trials of targeted and immunotherapy cancer drugs in the adjuvant setting: a systematic review and meta-analysis. JAMA Netw Open 1(8):e185617. https://doi.org/10.1001/jamanetworkopen.2018.5617
10. Post RM, Leverich GS, Kupka RW, Keck PE Jr, McElroy SL, Altshuler LL et al (2010) Early-onset bipolar disorder and treatment delay are risk factors for poor outcome in adulthood. J Clin Psychiat 71(7):864–872. https://doi.org/10.4088/JCP.08m04994yel
11. Goff DC, Zeng B, Ardekani BA, Diminich ED, Tang Y, Fan X, Galatzer-Levy I et al (2018) Association of hippocampal atrophy with duration of untreated psychosis and molecular biomarkers during initial

antipsychotic treatment of first-episode psychosis. JAMA Psychiat 75(4):370–378. https://doi.org/10.1001/jamapsychiatry.2017.4595
12. Rothman KJ, Michels KB (1994) The continuing unethical use of placebo controls. N Engl J Med 331(6):394–398. https://doi.org/10.1056/NEJM199408113310611
13. World Medical Association Declaration of Helsinki (1997) Recommendations guiding physicians in biomedical research involving human subjects. JAMA 277(11):925–926. https://doi.org/10.1001/jama.1997.03540350075038
14. Ellenberg SS, Temple RT (2000) Placebo-controlled trials and active-control trials in the evaluation of new treatments. Part 2: Practical issues and specific cases. Ann Intern Med 133(6):464–470. https://doi.org/10.7326/0003-4819-133-6-200009190-00015
15. Temple RT, Ellenberg SS (2000) Placebo-controlled trials and active-control trials in the evaluation of new treatments. Part 1: Ethical and scientific issues. Ann Intern Med 133(6):455–463. https://doi.org/10.7326/0003-4819-133-6-200009190-00014
16. Streiner DL (1995) The ethics of placebo-controlled trials. Can J Psychiat 40(4):165–166
17. Streiner DL (1999) Placebo-controlled trials: when are they needed? Schizophr Res 35(3):201–210. https://doi.org/10.1016/s0920-9964(98)00126-1
18. National Institute of Dental and Craniofacial Research (n.d.) Data and safety monitoring board (DSMB) guidelines. https://www.nidcr.nih.gov/research/human-subjects-research/toolkit-and-education-materials/interventional-studies/data-and-safety-monitoring-board-guidelines. Accessed 4 Jan 2021
19. Streptomycin in Tuberculosis Trials Committee (1948) Streptomycin treatment of pulmonary tuberculosis. A Medical Research Council investigation. Br Med J 2(4582):769–782. https://doi.org/10.1136/bmj.2.4582.769
20. Jennison C, Turnbull BW (2000) Group sequential methods with applications to clinical trials. Chapman and Hall/CRC, Boca Raton
21. Silva DS, Goering PN, Jacobson N, Streiner DL (2011) Off the beaten path: conducting ethical pragmatic trials with marginalized populations. IRB 33(3):6–11
22. Deichman RE, Krousel-Wood M, Breault J (2016) Considerations for stopping a clinical trial early. Ochsner J 16:197–198
23. Baker GR, Norton PG, Flintoft V, Blais R, Brown A, Cox J et al (2004) The Canadian Adverse Events Study: the incidence of adverse events among hospital patients in Canada. Can Med Assoc J 170(11):1678–1686. https://doi.org/10.1503/cmaj.1040498
24. Ohta Y, Miki I, Kimura T, Abe M, Sakuma M, Koike K et al (2019) Epidemiology of adverse events and medical errors in the care of cardiology patients. J Patient Saf 15(3):251–256. https://doi.org/10.1097/PTS.0000000000000291
25. US Food and Drug Administration (n.d.) Institutional review boards. Publication no. 21 CFR §56.111; April 1, 2015. https://www.accessdata.fda.gov/scripts/cdrh/cfdocs/cfcfr/CFRSearch.cfm?fr1/456.111. Accessed 5 Jan 2021
26. Blackburn H, Jacobs DR Jr (2017) The University Group Diabetes Program 1961–1978: pioneering randomized controlled trial. Int J Epidemiol 46(5):1354–1364. https://doi.org/10.1093/ije/dyw168
27. Briel M, Bassler D, Wang AT, Guyatt GH, Montori VM (2012) The dangers of stopping a trial too early. J Bone Joint Surg Am 94(Suppl 1(E)):56–60. https://doi.org/10.2106/JBJS.K.01412
28. Armitage P, McPherson CK, Rowe BC (1969) Repeated significance tests on accumulating data. J Roy Statist Soc Ser A 132:235–244. https://doi.org/10.2307/2343787
29. Streiner DL (2018) Commentary no. 27: P-hacking. J Clin Psychopharmacol 38(4):286–288. https://doi.org/10.1097/JCP.0000000000000901
30. Pocock SJ (1977) Group sequential methods in the design and analysis of clinical trials. Biometrika 64(2):191–199. https://doi.org/10.1093/biomet/64.2.191
31. Pocock SJ (1992) When to stop a clinical trial. Br Med J 305:235–240. https://doi.org/10.1136/bmj.305.6847.235
32. Campbell D, Stanley J (1963) Experimental and quasi-experimental designs for research. Rand-McNally, Chicago
33. Haybittle JL (1971) Repeated assessments of results in clinical trials of cancer treatment. Br J Radiol 44(526):793–797. https://doi.org/10.1259/0007-1285-44-526-793
34. O'Brien PC, Fleming TR (1979) A multiple testing procedure for clinical trials. Biometrics 35(3):549–556. https://doi.org/10.2307/2530245
35. Kim K, DeMets DL (1987) Design and analysis of group sequential tests based on the type I error spending rate function. Biometrika 74(1):149–154. https://doi.org/10.1093/biomet/74.1.149
36. Lan KKG, DeMets DL (1983) Discrete sequential boundaries for clinical trials. Biometrika 70(3):659–663. https://doi.org/10.2307/2336502
37. Toerien M, Brookes ST, Metcalf C, de Salis I, Tomlin Z, Peters TJ et al (2009) A review of reporting of participant recruitment and retention in RCTs in six major journals. Trials 10:52. https://doi.org/10.1186/1745-6215-10-52
38. Walters SJ, dos Anjos B, Henriques-Cadby I, Bortolami O, Flight L, Hind D, Jacques RM et al (2017) Recruitment and retention of participants in randomised controlled trials: a review of trials funded and published by the United Kingdom Health Technology Assessment Programme. BMJ Open 7:e015276. https://doi.org/10.1136/bmjopen-2016-015276
39. Williams RJ, Tse T, DiPiazza K, Zarin D (2015) Terminated trials in the ClinicalTrials.gov Results

database: evaluation of availability of primary outcome data and reasons for termination. PLoS One 10(5):e0127242. https://doi.org/10.1371/journal.pone.0127242
40. Knopman DS, Jones DT, Greicius MD (2020) Failure to demonstrate efficacy of aducanumab: an analysis of the EMERGE and ENGAGE trials as reported by Biogen, December 2019. Alzheimers Dement. Online ahead of print. https://doi.org/10.1002/alz.12213
41. Sabbagh MN, Cummings J (2020) Open peer commentary to "failure to demonstrate efficacy of aducanumab: an analysis of the EMERGE and ENGAGE trials as reported by biogen December 2019". Alzheimers Dement. Online ahead of print. https://doi.org/10.1002/alz.12235
42. Jitlal M, Khan I, Lee SM, Hacksaw A (2012) Stopping clinical trials early for futility: retrospective analysis of several randomised clinical studies. Br J Cancer 107(6):910–917. https://doi.org/10.1038/bjc.2012.344
43. Montori VM, Devereaux PJ, Adhikari NKJ, Burns KEA, Eggert CH, Briel M et al (2005) Randomized trials stopped early for benefit: a systematic review. JAMA 294(17):2203–2209. https://doi.org/10.1001/jama.294.17.2203
44. Korn EL, Freidlin B, Mooney M (2009) Stopping or reporting early for positive results in randomized clinical trials: the National Cancer Institute Cooperative Group experience from 1990 to 2005. J Clin Oncol 27(10):1712–1721. https://doi.org/10.1200/JCO.2008.19.5339
45. Pocock SJ, Hughes MD (1989) Practical problems in interim analyses, with particular regard to estimation. Control Clin Trials 10(Suppl 4):209S–221S. https://doi.org/10.1016/0197-2456(89)90059-7
46. Wheatley K, Clayton D (2003) Be skeptical about unexpected large apparent treatment effects: the case of an MRC AML12 randomization. Control Clin Trials 24:66–70. https://doi.org/10.1016/S0197-2456(02)00273-8
47. Pocock SJ, Wang D, Wilhelmsen L, Hennekens CH (2005) The data monitoring experience in the Candesartan in Heart Failure Assessment of Reduction in Mortality and morbidity (CHARM) program. Am Heart J 149(5):939–943. https://doi.org/10.1016/j.ahj.2004.10.038
48. Bassler D, Briel M, Montori VM, Lane M, Glasziou P, Zhou Q et al (2010) Stopping randomized trials early for benefit and estimation of treatment effects: systematic review and meta-regression analysis. JAMA 303(12):1180–1187. https://doi.org/10.1001/jama.2010.310
49. Pocock SJ, White I (1999) Trials stopped early: too good to be true? Lancet 353(9157):943–944. https://doi.org/10.1016/S0140-6736(98)00379-1
50. Streiner DL (2006) Sample size in clinical research: when is enough enough? J Pers Assess 87(3):259–260. https://doi.org/10.1207/s15327752jpa8703_06
51. Zannad F, Stough WG, McMurray JJV, Remme WJ, Pitt B, Borer JS et al (2012) When to stop a clinical trial early for benefit: lessons learned and future approaches. Circ Heart Fail 5(2):294–302. https://doi.org/10.1161/CIRCHEARTFAILURE.111.965707
52. Alvir JM, Lieberman JA, Safferman AZ, Schwimmer JL, Schaaf JA (1993) Clozapine-induced agranulocytosis – incidence and risk factors in the United States. N Engl J Med 329(3):162–167. https://doi.org/10.1056/NEJM199307153290303
53. Ma IT, Madura JA (2015) Gastrointestinal complications after bariatric surgery. Gastroenterol Hepatol 11(8):526–535
54. Hazell L, Shakir SAW (2006) Under-reporting of adverse drug reactions: a systematic review. Drug Saf 29(5):385–396. https://doi.org/10.2165/00002018-200629050-00003
55. Berkson J (1946) Limitations of the application of fourfold table analysis to hospital data. Biom Bull 2(3):47–53. https://doi.org/10.1093/ije/dyu022
56. Roberts RS, Spitzer WO, Delmore T, Sackett DL (1978) An empirical demonstration of Berkson's bias. J Chronic Dis 31(2):119–128. https://doi.org/10.1016/0021-9681(78)90097-8
57. Walter SD, Han H, Guyatt GH, Bassler D, Bhatnagar N, Gloy V et al (2020) A systematic survey of randomised trials that stopped early for reasons of futility. BMJ Med Res Methodol 20:10. https://doi.org/10.1186/s12874-020-0899-1
58. Lesaffre E, Edelman MJ, Hanna NH, Park K, Thatcher N, Willemsen S et al (2017) Statistical controversies in clinical research: futility analyses in oncology – lessons on potential pitfalls from a randomized controlled trial. Ann Oncol 28(7):1419–1426. https://doi.org/10.1093/annonc/mdx042
59. Iltis AS (2005) Stopping trials early for commercial reasons: the risk-benefit relationship as a moral compass. J Med Ethics 31(7):410–414. https://doi.org/10.1136/jme.2004.008771
60. George SL, Buyse M (2015) Data fraud in clinical trials. Clin Investig (Lond) 5(2):161–173. https://doi.org/10.4155/cli.14.116

Part VIII

Corruption and Impropriety in the Healthcare Environment

27 Curtailing Corruption in the Recruitment of Public Sector Health Workers in Uganda

Henry Zakumumpa, Everd Maniple, and Monica Twesiime Kirya

Abstract

The chapter analyses how corruption affects public sector health worker recruitment in Uganda. It is based on a desk review of policies, literature, media reports and interviews with key informants. We found that health worker recruitment in Uganda is often marred by bribery, extortion, nepotism and favouritism, especially during short-listing and upon being entered on the payroll. While patronage and nepotism play a role, inadequate health financing has led to a ceiling on public service recruitment, which has created a perception of job scarcity among potential recruits. This in turn drives bribery and rent-seeking by politicians and technocrats. Measures such as independent oversight and digitisation of some recruitment procedures should be strengthened and scaled up country-wide. Furthermore, increasing health financing to remove the recruitment ceiling and fill available vacancies would help to reduce the perceptions of job scarcity driving corruption.

H. Zakumumpa
Makerere University, Kampala, Uganda

E. Maniple
Kabale University, Kabale, Uganda

M. T. Kirya (✉)
U4 Anti-Corruption Resource Centre, Chr. Michelsen Institute, Bergen, Norway
e-mail: monica.kirya@cmi.no

27.1 Historical Background

Uganda's public health system was inherited from the colonial government at independence in 1962. At the time, it was billed as one of the best in sub-Saharan Africa, with functional physical infrastructure and systems, yielding comparatively good health outcomes. However due to general mismanagement of the country in subsequent years, the health system along with other sectors also deteriorated to virtual collapse. In the immediate post-independence period, the public health system boasted of many qualified medical staff. However, most of the senior positions were occupied by the British and physicians of Asian origin. This was understandable because the local training capacity was still limited. In 1971 under former President Idi Amin, most of the British physicians left Uganda after a diplomatic fall-out. Before the end of 1972, the Asians were expelled from Uganda. The few qualified Ugandans could not fill all the gaps. Moreover, many of them escaped to exile for their safety or were killed over the next 14 politically turbulent years. By 1986 when the National Resistance Movement (NRM) government came to power, the public health system was non-functional. Most people only trusted the private sector, especially the faith-based sub-sector.

Since 1986 several reforms have led to progressive improvement and strengthening of the health system and other sectors of the government [1–3]. By 2021 the country had a health worker-

J. Faintuch, S. Faintuch (eds.), *Integrity of Scientific Research*,
https://doi.org/10.1007/978-3-030-99680-2_27

to-population ratio of 1.92/1000, which is still below the World Health Organization/WHO-recommended target of 2.28, however significantly better than in other times [2].

Health workers are poorly paid, with doctors' earnings ranging an equivalent of US$6000–12,000 per annum, depending on seniority and location. Nevertheless such values are compatible with those of several other African nations. Junior doctors working in rural areas receive a financial incentive, unlike those in urban areas [3]. However, the definition of 'rural' is contentious, with some 'non-rural' districts having very hard-to-reach and hard-to-stay areas, yet they are excluded from the incentive scheme. Other cadres earn much less. Salary payment for health workers in the public sector is regular, prompt and largely guaranteed by the central government.

27.2 Transparent Recruitment

Qualified and competent health workers are a fundamental pillar of strong health systems. Article 7 of the United Nations Convention Against Corruption (UNCAC) calls for 'the recruitment, hiring, retention, promotion and retirement of civil servants [and] other non-elected public officials...based on principles of efficiency, transparency and objective criteria such as merit, equity and aptitude' [4]. Likewise the 2010 WHO Global Code of Practice on the International Recruitment of Health Personnel [5] and the International Labour Organization/ILO's 2016 General Principles and Operational Guidelines for Fair Recruitment [6] advocate for transparency, fairness and protection of workers in recruitment processes. The 2008 WHO Kampala declaration also commits nations to provide all people with access to skilled health workers, as well as offering a safe working environment for effective retention and equitable distribution of the health workforce [7]. Corruption in the recruitment of health care professionals threatens countries' progress on all these commitments.

27.2.1 Common Corrupt Practices

These are patronage and clientelism, nepotism, cronyism, bribery, extortion and sextortion [8]. All of these practices have devastating consequences for health systems: they lead to the employment of poorly qualified health workers, setting the tone for other forms of corruption to proliferate, such as absenteeism, embezzlement, theft and bid-rigging. They also contribute to medical brain drain [9] and the deterioration of public health outcomes.

A 2012 study from Uganda found instances where health service jobs were provided in exchange of political support. Informal power networks influenced decision-making in the health sector, especially at subnational levels [10]. The appointment of relatives, friends and associated in positions of authority, without clear merit, is quite common in many countries, as are bribery and extortion, which are also known as 'job purchasing' [11]. In addition, sextortion, where sex (not money) is the currency of a corrupt transaction, affects many female health workers [12]. One study in Uganda mentioned that health workers were required to pay or 'offer personal services' to have their papers or pay cheques processed [13]. A 2019 review found that sexual harassment is prevalent in healthcare settings [14].

27.3 Political, Policy and Governance Context

Health is a key concern for the government and the general public, and the Public Service Commission (PSC) puts in place mechanisms that should standardise recruitment procedures [15–24]. Yet official procedures are often side-stepped by district officials. For example, due to the frequent and prolonged ban on recruitment by the central government, district health services often run short of staff in the inter-recruitment periods lasting up to 5 years.

27.3.1 Corruption and Misappropriation of Funds

Documented misdeeds involved the Global Alliance for Vaccines and Immunisations (GAVI) meant for immunisation in 2001 [15]; the misappropriation of funds from the Global Fund meant for treatment for HIV, TB and malaria in 2005 [16]; and the recruitment of the Director General of Health Services in 2017 [17]. More recently, the key reports rotate around lack of transparency over the contract to construct a major national specialised hospital [18] and the questionable contracts to install medical oxygen plants and intensive care units (ICUs) in regional referral hospitals during the COVID-19 pandemic [19]. Most of these high-profile cases involve people with political and kinship connections to people in power and involve very large sums of money.

27.3.2 Nepotism and Political Patronage

Nepotism is mentioned at all levels of the health system. Lower levels of the health system under the district local governments have been accused of corrupt recruitments favouring candidates from their areas or relatives. Higher-level appointments at the Ministry of Health headquarters are done by the President who has been accused of recruiting less qualified members of his family or adherents of his party. In addition presidential appointments to key positions may be based on tribal or religious affiliation [25]. Media reports also show that there is corruption in the deployment of health workers. For one to be deployed in the capital city Kampala, one has to part with a lot of money to appease the recruiting authorities at the District Service Commission or the Health Service Commission [20].

27.3.3 'Rent Seeking' Practices

Interventions to collect 'rent' from recruited healthcare workers by the political and technical district leadership was perceived to be common from the perspective of the 87 health workers we interviewed. Solicitation of bribes for advertised vacancies in the district local government services was frequently cited by participants.

It was discernible that prolonged delays by districts in conducting regular workforce recruitments (owing to a shortage of dedicated operational funds for the purpose) contributed to the phenomenon of increasingly dysfunctional district service commissions. It heightened the value of district jobs in the perception of potential recruits and also, it appears, fertile ground for 'rent seeking' by the recruitment authorities.

> Jobs are not easy to come by in district local government service. They are very difficult to get and there are many people out there looking for an opportunity to join. Some of these people are desperate after being 'on the street' (unemployed) for a long time looking for opportunity and are prone to schemes involving solicitation of bribes. (District Official, The President's Emergency Plan for AIDS Relief/PEPFAR intervention district)
>
> Of course, everybody would be happy to be employed by the government of Uganda, to be permanent and pensionable. It is not every day that that opportunity comes by. (Health worker, District local government service)

27.3.4 Wage Bill Constraints

> There was some kind of a ban on recruitment of health workers. For human resources for health we were constrained by a wage-bill ceiling. We would not think of recruiting more staff. Budget ceiling is the major challenge. Like this year, we do not have any possibility to absorb additional health workers. We have asked PEPFAR to renew their contracts in the mean time for at least one more year as we reorganize ourselves. (District Official, PEPFAR intervention district)

27.3.5 Delays in Accessing the Local Government Payroll

Some health workers reported that they were often compelled to bribe officials to reduce the long wait for accessing the salaries from the public sector payroll system. They reported experiencing delays in receiving their government salaries that lasted up to 5 months or even longer long after PEPFAR had weaned them from their support program. From our interviews with health workers, it emerged that some district technocrats took advantage of the cumbersome, but frequently systemic inefficiencies, in the administrative procedures for accessing the public sector payroll in case-study districts to extract 'rents'.

In some of the case-study districts, health workers indicated that they had received less than the expected total amount of the salary arrears. For instance, if a health worker had not been paid for 5 months, it was not uncommon to receive 3 months instead and to wait indefinitely without explanation. For a handful of the health workers we interviewed, we computed the wait before receipt of the balance on salary arrears to have lasted for up to 2 years from the date of data collection. Health workers had lost hope of recovery of their salary arrears.

27.4 Discussion

The findings show that there are two stages in the recruitment and deployment process at which corruption is most likely to occur—at short-listing stage and at entry onto the payroll. Accordingly, establishing appropriate safeguards against corruption in these two processes is likely to curtail corruption and its attendant impact on health worker motivation [16, 18, 20, 26, 27].

Short-listing emerged as a key moment fraught with corrupt activities. Only short-listed candidates can be interviewed by the District Service Commission; therefore, people pay bribes to appear on the short-list.

Corruption also often occurs at the point of being enrolled on the government payroll. Our study showed that accessing the payroll is long and cumbersome and that the responsible officials take advantage of this for rent-seeking. It is unfortunate that after jumping the hurdle of being recruited, accessing the payroll can take from 3 months up to 2 years in some instances. Health workers kept checking with district authorities but were told to bring money to 'facilitate the follow up' of their enrollment on the payroll. Often, this meant agreeing to surrender all the salary arrears for the period between recruitment and accessing the payroll. In the meantime, they continued to work without receiving any salary. A previous study had found that female health workers sometimes have to give 'personal favours' a euphemism for sexual favours, in order to be included on the payroll [16, 18, 20, 26, 27].

27.4.1 Mitigating Measures

Merit-based recruitment can be enhanced. For example, the NGO IntraHealth under its USAID-funded Uganda *CapacityPlus* programme introduced short-listing software to quicken the recruitment process and minimise instances of corruption. *E-shortlisting*, as the software is called, helped to reduce complaints regarding corruption in the recruitment process [21, 22]. The Uganda Ministry of Public Service has also been using e-recruitment to lessen corruption, streamline the hiring process and increase transparency in recruitment. However, the e-recruitment system is hampered by several challenges such as old, obsolete computers, unreliable electricity supply, intermittent Internet access and frequent system failures [22]. These challenges point to the importance of investing in digital infra-structure to realise the anti-corruption gains that may accrue from e-governance systems.

27.4.2 External Monitors

Putting in place independent third-party monitors to observe recruitment processes, as happened with PEPFAR's programme officers monitoring district-level recruitments, can also help to curb corruption and promote recruitment based on merit.

Our findings revealed that insufficient financing for health services and the concomitant ceiling on recruitment are a major driver of corruption in health worker recruitment in Uganda. Since the central government and districts cannot recruit all the health workers they need because they do not have enough funds, there is excess demand for the few positions available and corresponding opportunities for rent-seeking and influence-peddling [21–24].

Acknowledgements The findings were from a wider US Agency for International Development/USAID-funded study. The authors thank Daniela Cepeda Cuadrado for her assistance with compiling some of the results.

References

1. Tashobya CK, Ssengooba F, Cruz VO (eds) (2006) Health systems reforms in Uganda: processes and outputs. Health Systems Development Programme London School of Hygiene and Tropical Medicine (LSHTM), London. libraryhealthgoug/publications/health-information-systems/health-systems-reforms-uganda-processes-and-outputs. Accessed 5 Sept 2021
2. The Republic of Uganda (2020) Annual health sector performance report 2019/2020. The Republic of Uganda, Kampala. health.go.ug/cause/annual-health-sector-performance-report-financial-year-2019-20/. Accessed 5 Sept 2021
3. Lwamafa J (2010) Circular standing instruction no. 2 of 2010: payment of the hardship allowance in the ministry of public service. publicservice.go.ug/download/circular-standing-instruction-no-2-of-2010-payment-of-the-hardship-allowance-in-the-public-service/. Accessed 5 Sept 2021
4. United Nations General Assembly (2003) United Nations convention against corruption. unodc.org/documents/brussels/UN_Convention_Against_Corruption.pdf. Accessed 5 Sept 2021
5. World Health Organisation (2010) Global code of practice on the international recruitment of health personnel. who.int/hrh/migration/code/WHO_global_code_of_practice_EN.pdf. Accessed 5 Sept 2021
6. International Labor Organization (2016) General principles and operational guidelines for fair recruitment. ilo.primo.exlibrisgroup.com/discovery/fulldisplay/alma994942693502676/41ILO_INST:41ILO_V2. Accessed 5 Sept 2021
7. World Health Organisation (2008) Kampala declaration and Agenda for global action. who.int/workforcealliance/knowledge/resources/kampala_declaration/en/. Accessed 5 Sept 2021
8. Kirya M (2019) Promoting anti-corruption, transparency and accountability in the recruitment and promotion of health workers to safeguard health outcomes. Glob Health Action 13:1
9. Karan A, De Ugarte D, Barry M (2016) Medical "brain drain" and health care worker shortages: how should international training programs respond? AMA J Ethics 18(7):665–675
10. Baez-Camargo C (2012) Using power and influence analysis to address corruption risks: the case of the Ugandan drug supply chain. U4 Anti-Corruption Resource Centre Brief, Bergen. assets.publishing.service.gov.uk/media/57a08a81e5274a31e0000632/B2012-06.pdf. Accessed 5 Sept 2021
11. Lewis M, Pettersson G (2009) Governance in health care delivery: raising performance. elibrary.worldbank.org/doi/pdf/10.1596/1813-9450-5074. Accessed 5 Sept 2021
12. Jackson D, Clare J, Mannix J (2002) Who would want to be a nurse? Violence in the workplace–a factor in recruitment and retention. J Nurs Manag 10:13–20
13. Hagopian A, Zuyderduin A, Kyobutungi N et al (2009) Job satisfaction and morale in the Ugandan health workforce. Health Aff 28:863–875
14. Kabat-Farr D, Crumley ET (2019) Sexual harassment in healthcare: a psychological perspective. *OJIN: Online J Iss Nurs* 24(1)
15. Human Rights Watch (2013) "Letting the big fish swim": failures to prosecute high-level corruption in Uganda. reliefweb.int/report/uganda/"letting-big-fish-swim"-failures-prosecute-high-level-corruption-uganda. Accessed 5 Sept 2021
16. Kapiriri L, Martin DK The Global Fund Secretariat's suspension of funding to Uganda: how could this have been avoided? Bull World Health Organ 84:576–580
17. Mbonye AK (2018) Uganda's health sector through turbulent politics (1958–2018). worldcat.org/title/ugandas-health-sector-through-turbulent-politics-1958-2018/oclc/1108816935. Accessed 5 Sept 2021
18. Canevet J (2019) Lubowa hospital: a case study in corruption risk – and how open ccontracting could have helped. Transparency International. ti-health.org/content/lubowa-hospital-news-uganda-open-contracting/. Accessed 31 May 2021
19. The Republic of Uganda (2020) COVID-19 interventions report. 2019/2020. covid19.gou.go.ug/uploads/document_repository/authors/minstry_of_finance,planning_and_economic_development/document/COVID-19_Interventions_Report.pdf. Accessed 5 Sept 2021

20. Nassaka F (2021) Medics attribute low staffing levels in referral hospitals to corruption. ugandaradionetwork.net/story/medics-attribute-low-staffing-levels-in-referral-hospitals-to-corruption. Accessed 31 May 2021
21. Agaba A (2011) Computerized shortlisting saves time and money, gets health workers where they are needed. intrahealth.org/vital/computerized-shortlisting-saves-time-and-money-gets-health-workers-where-they-are-needed. Accessed 31 May 2021
22. Wadero AA (2021) Govt e-recruitment system below standard, says report. The Daily Monitor. monitor.co.ug/uganda/news/national/govt-e-recruitment-system-below-standard-says-report-3325346. Accessed 31 May 2021
23. Republic of Uganda, Ministry of Health. Annual health sector performance report. library.health.go.ug/sites/default/files/resources/Annual%20Health%20Sector%20Performance%20Report%202019-2020%20FY-1.pdf. Accessed 31 May 2021
24. Nabyonga-Orem J (2014) Uptake of evidence in policy development: the case of user fees for health care in public health facilities in Uganda. BMC Health Serv Res 14(639)
25. Mbonye AK (2018) Uganda's health sector through turbulent politics (1958–2018). worldcat.org/title/ugandas-health-sector-through-turbulent-politics-1958-2018/oclc/1108816935. Accessed 31 May 2021
26. Human Rights Watch (2013) "Letting the big fish swim": failures to prosecute high-level corruption in Uganda. reliefweb.int/report/uganda/%E2%80%9Cletting-big-fish-swim%E2%80%9D-failures-prosecute-high-level-corruption-uganda. Accessed 31 May 2021
27. The Republic of Uganda (2020) COVID-19 interventions report. covid19.gou.go.ug/uploads/document_repository/authors/minstry_of_finance, planning_and_economic_development/document/COVID-19_Interventions_Report.pdf. Accessed 31 May 2021

Legal and Regulatory Issues in Selfie Telemedicine

28

Maurice Mars and Richard E. Scott

Abstract

Telemedicine use has grown exponentially during the COVID-19 pandemic. Associated with this has been increased use of patient-generated health data. In this chapter, the key ethical and legal issues related to telemedicine will be discussed, and these will then be explored within the contexts of both doctor-initiated and patient-initiated selfie telemedicine. Regulation of medical apps, which are a source of patient-generated health data, will be summarised, and potential areas of fraud related to telemedicine will be explored.

28.1 Introduction

The COVID-19 pandemic triggered a global increase in telemedicine uptake and use that is anticipated to last beyond the resolution of the pandemic [1, 2]. Telemedicine, "the remote diagnosis and treatment of patients by means of telecommunications technology" [3], reduces the risk of infection by distancing patients and their physicians and reducing the number of people inadvertently contacted during transit and in waiting rooms or outpatient departments.

Telemedicine use during the pandemic has been facilitated by widespread relaxation of regulatory restrictions on telemedicine practice, related to the requirement of a prior doctor-patient relationship, data security, and remuneration. Not only have doctors been rapidly establishing and providing telemedicine services; patients too have seen the benefits. They are generating their own health data and initiating consultations with their own doctors, doctors with whom they have no previous relationship, or seeking clinical advice through direct to consumer online services.

28.2 Patient- or Person-Generated Health Data (PGHD)

PGHD has been defined as "clinically relevant data captured by patients outside of the traditional care setting" [4] and "health-related data created, gathered or inferred by or from patients and for which the patient controls data collection and data sharing" [5]. Originally termed patient-generated health data, the alternative of person-generated health data has been introduced to allow for the inclusion of patient-related data generated or submitted by patients or their family members, caregivers, and friends.

M. Mars (✉)
University of KwaZulu-Natal, KwaZulu-Natal, South Africa
e-mail: mars@ukzn.ac.za

R. E. Scott
University of Calgary, Calgary, AB, Canada

J. Faintuch, S. Faintuch (eds.), *Integrity of Scientific Research*,
https://doi.org/10.1007/978-3-030-99680-2_28

28.2.1 Internet of Things Healthcare Systems

The type of data varies and might include information from wearable wellness devices (step counters, exercise logs, heart rate monitors, nutritional diaries, sleep records) and other digital sources of health information. Such data have been informally incorporated into patients' written medical records by clinicians for some time, but with the growth in available devices and associated apps, and with the advent of electronic records of various forms, more automated, streamlined, and formal approaches are required [6].

The concept of patient-generated health data is not new, starting with people giving a verbal history to their physician or healer describing how they feel, the cause of their injury, or the site of their pain. Patients have been asked to record many things such as their weight, food intake, feelings of mood, physical activity, etc., in logbooks for review at a later date. Digital information and communication technologies (ICT) have both simplified and extended the range of "things" that can be recorded, rapidly transmitted, and acted upon. Fraud related to PGHD is also not new. People have been giving false information for their own gain, be it to obtain a cheaper health insurance premium, to gain employment, or to be given a sick note for time-off work. Healthcare professionals have also used PGHD to justify over-servicing.

28.3 Selfie

"Selfie," the Oxford Dictionaries' word of the year in 2013, is defined as "a photograph that one has taken of oneself, *esp.* one taken with a smartphone or webcam and shared via social media" [7]. It traces its origin to an inebriated Australian student who, on 13 September 2002, fell and cut his lip. Not knowing whether his wound required surgical repair, he posted a message on the chat forum of the Australian Broadcasting Company website, accompanied by a link to the photograph of his cut lip uploaded to his University's server. "I had a hole about 1 cm long right through my bottom lip. And sorry about the focus, it was a selfie" [8]. Thus, the first known use of selfie was a clinical photograph used for store and forward telemedicine and an unsolicited plea for medical advice. He was advised to have it sutured.

28.4 Telemedicine

The World Health Organization defines telemedicine as "the delivery of health care services, where distance is a critical factor, by all health care professionals using information and communication technologies for the exchange of valid information for diagnosis, treatment and prevention of disease and injuries, research and evaluation, and for the continuing education of health care providers, all in the interests of advancing the health of individuals and their communities" [9].

Telehealth has evolved to encapsulate a broader array of digital healthcare activities and can include non-clinical services. Both are subsets of eHealth, the use of information and communication technologies for health [10] or the more recent version digital health, "...a term encompassing eHealth (which includes mHealth—mobile health), as well as emerging areas, such as the use of advanced computing sciences in 'big data', genomics and artificial intelligence" [11].

Synchronous telemedicine refers to real-time communication between patient and physician or physician and physician. It is usually associated with some form of audio and visual conferencing but also includes telephone calls and the modern variants of video calls such as Facetime or WhatsApp. Asynchronous or store and forward telemedicine refers to care at a distance based on information, data, text, photographs, video recordings and sound recordings captured electronically and transmitted to a clinician for later interpretation. An example is images of a skin lesion attached to an email containing the patient's relevant history and clinical findings.

Hybrid telemedicine can, for example, involve photographs sent to a dermatologist, with the dermatologist then discussing the case and the photographs with the patient or the referring doctor over the phone.

28.4.1 Remote Monitoring

Monitoring is an ever-growing field, having its origins in monitoring astronauts' physiological responses to space flight. Home monitoring was the original concept, with patients or family members recording data, such as pulse rate, weight, mood, electrocardiogram, etc., and transmitting the information to a service provider or physician electronically. This has grown with the advent of both wearable devices and mobile phone health applications. Smartwatches can record activity—the number of steps taken, exercise duration, and heart rate—and other devices can monitor variables as diverse as blood glucose concentration, electrocardiogram/ECG, and foetal ultrasonography. In addition, there are over 400,000 mobile phone health and wellness applications [12].

28.4.2 Old and New Phone Applications

The COVID-19 pandemic has highlighted the use of the telephone for medical consultation and reinstated its place in telemedicine [13, 14]. Telephonic consultation is not new, with the first telemedicine report in the Lancet appearing in 1879, 3 years after the telephone was patented [15]. This too was a form of selfie with an agitated young mother seeking medical assistance for her infant in the middle of the night. The legend that the first conversation over the phone was a request by Bell to assistant Watson for help as he had spilt acid on himself, and thus a telemedicine request is incorrect and is from the script of the 1939 film *The Story of Alexander Graham Bell* [16]. By 1905, Einthoven was using the telephone to transmit electrocardiograms [17]. As technology evolves, telephone consultations are now moving to phone-based video consultations [18, 19].

28.5 Selfie: Person-Generated Health Data—Telemedicine

Selfie telemedicine can be doctor or patient initiated. For example, a doctor may request a patient to take a photograph of a postoperative wound to avoid having to have a formal face-to-face consultation if the photograph shows that all is well. Patient-initiated selfie telemedicine takes at least six different forms: seeking advice from their regular doctor for (1) a known condition; (2) a new condition; (3) seeking consultation and advice from a doctor with whom they have no prior relationship; (4) going online and purchasing a service from a direct to consumer service provider, from people who they do not know and who may not be duly licensed health professionals; (5) seeking services from their insurer who may refer them to an on-call doctor provided by the insurer who, although not having ever seen the patient, will have access to the patient's full medical record; and (6) posting and sharing personal information and/or images online with social media groups and seeking advice [20].

28.6 Telemedicine: The Ethical Issues

Medical ethics, a term first used in 1803 by English physician Thomas Percival [21], is based on four principles, not unlike ethics in other fields: autonomy (the right of a competent person to make informed decisions about their own medical care), beneficence (do good), non-maleficence (do no harm), and justice (fairness and equality). Aspects of these principles are found in the original Hippocratic Oath (400 BC), with modern medical ethics based on the mores of the developed world. The ethics of medical practice are the same for conventional face-to-face

Table 28.1 The main ethical and legal issues related to telemedicine

Ethical		
Autonomy	Confidentiality	Patient-generated health data
Authentication	Quality of information	Unsolicited requests for care
Dr-patient relationship	Quality of care	Dehumanisation of care
Consent	Continuity of care	Ethical judgement of machines
Legal		
Laws and regulations	Jurisdiction	Data storage
Licensure	Privacy	Record-keeping
Liability	Data security	Prescription

consultation and telemedicine consultation. Issues of concern for telemedicine include consent, quality of care, continuity of care, quality of information, licensure, jurisdiction, liability, data security, privacy, data storage, authentication, prescription, remuneration, and phone stewardship (Table 28.1).

While seemingly novel the ethical issues related to telemedicine have been debated for over a century, from the use of the telegraph to the telephone and radio and, more recently, videoconferencing and the internet. The digital migration of medical consultation has refocussed attention on seemingly mundane and routine tasks such as record-keeping and data storage. For example, what are the requirements for keeping a record of a telephone conversation, a WhatsApp chat message, an email, or a videoconference call? How are electronic data received by a doctor to be stored in a secure manner? How is this information to be incorporated into the patient's notes, which may be electronic or still paper-based?

28.7 Telemedicine Legislation

Malaysia was the first country to introduce legislation, the Telemedicine Act of 1997 [22]. Although there are telemedicine laws in each state in the United States, there are still few countries with specific telemedicine law [23]. In many countries relevant aspects of legislation on health, data protection and privacy, data security, electronic signature, and telecommunication are used to cover telemedicine.

28.7.1 Jurisdiction

Telemedicine offers the opportunity for inter-jurisdictional, cross-border practice, much needed in the developing world with its significant shortage of doctors and greater burden of disease. It is generally held that jurisdiction in a telemedicine consultation is that of the patient's locale or residence. The World Medical Association/WMA 2018 statement advises that "Cross-jurisdictional consultations should only be allowed between two physicians" [24], and Malaysian law makes it an offence for a doctor in another country to provide telemedicine services in Malaysia unless practised in association with a doctor licensed in Malaysia [22]. In the European Union/EU, medical licensure is valid across the union and thus national borders, and in the United States, the requirement to be licensed in the state of the patient has been relaxed in over half of the states.

28.7.2 Liability

Liability is linked to continuity of care. It has been generally held since the 1990s that the referring doctor is responsible for the ongoing care of their patient [25]. The WMA's 2009 guiding principles offered a more realistic approach, "The physician needs to give clear and explicit direction to the patient during the telehealth encounter as to who has ongoing responsibility for any required follow-up and ongoing health care" [26]. However, in 2018 the WMA reverted to the previous interpretation "The physician asking for another physician's advice or second opinion remains responsible for treatment and other decisions and recommendations given to the patient" [24]. These approach and interpretation do not allow for the situation in which the patient's ongoing care is taken over by the doctor to whom they were referred, albeit at a distance.

28.7.3 Licensure

Requirements for licensure to practice medicine in a given jurisdiction vary, with no current requirement for specific telemedicine licensure. In India, new regulations require practitioners to have undergone some form of telemedicine training [27].

28.8 Doctor-Patient Relationship

The doctor-patient relationship is considered a fundamental component to the delivery of high-quality medical care [28]. It is built on mutual trust and the knowledge the doctor has of the patient. At issue is whether a telemedicine consultation can be ethically undertaken if there is no prior doctor-patient relationship, i.e. the doctor does not know the patient and has not or cannot personally conduct a physical examination of the patient. Regulators see the lack of a prior relationship as a major shortcoming of some aspects of telemedicine. As a result in many jurisdictions, a prior relationship is required before there can be a telemedicine consultation. This can be overcome by the referring doctor who has examined the patient participating in the consultation and passing on this information [24, 29].

The 2018 statement on the Ethics of Telemedicine (WMA) states that "The patient-physician relationship should be based on a personal examination and sufficient knowledge of the patient's history" [24]. The American Medical Association's (AMA) policy requires a valid doctor-patient relationship before a telemedicine service is provided, defined as "A patient-physician relationship exists when a physician serves a patient's medical needs. Generally, the relationship is entered into by mutual consent between physician and patient (or surrogate)" [30]. The relationship is established "when a physician affirmatively acts in a patient's case by examining, diagnosing, treating, or agreeing to do so" [31]. The AMA has taken a pragmatic approach to this for telemedicine by proposing three ways in which the doctor-patient relationship can be developed: (1) a face-to-face consultation must occur, either in-person or through a videoconference "face-to-face" consultation; (2) a consultation with another physician who has an ongoing relationship with the patient; and (3) meeting evidence-based guidelines for establishing a patient-physician relationship developed by major speciality societies [32]. The second and third options may provide a means of establishing a doctor-patient relationship for store-and-forward telemedicine, where the doctor to whom the case is referred may never see or speak to the patient. Store-and-forward telemedicine is the most cost-effective and efficient form of telemedicine for much of the developing world, and its use has increased during the pandemic.

Interestingly, the requirement for audio- and videoconferencing appears to exclude the use of the telephone to develop a relationship. Is a prior doctor-patient relationship a prerequisite for telemedicine in those specialties in which the patient does not always see the specialist, for example, in aspects of routine radiology and pathology?

28.8.1 Selfie Telemedicine and the Doctor-Patient Relationship

In doctor-initiated selfie telemedicine, a relationship exists, as it does when patients initiate a telemedicine consultation with their current doctor. There is no relationship when the patient requests advice or a telemedicine consultation from a doctor with whom they have had no prior professional contact. Unresolved questions are whether the doctor is required to respond to the patient's unsolicited request? Is it ethical not to respond, and what are the obligations of the doctor with respect to confidential information received and legal requirements to keep records? When dealing with an online direct to consumer service, there is no existing relationship. Does the financial transaction involved constitute the establishment of a relationship? What are the obligations of the service providers with respect to informing the patient's regular doctor of the advice/treatment provided and recordkeeping [33]? Some insurance providers require patients to initiate contact about certain conditions and then refer the patient to an on-call physician with whom the patient may not have had previous contact. The physician has access to the patient's record, and in effect, this mirrors the situation of a *locum tenens* dealing with a patient for the first time [20].

28.9 Consent

Autonomy is "the right of competent adults to make informed decisions about their own medical care" and is one of the four principles of medical ethics [34]. Informed consent is the expression of autonomy and is required before any investigation or treatment. Consent has a long history in medical practice [35], but informed consent is a more recent construct based on Western medical, legal, and ethical concepts [36, 37]. These do not necessarily take socio-cultural factors into account, such as consent being a community, family, or patriarchal decision in some parts of the world. Care must be taken to ensure that realistic regulations are set in the developing world, rather than merely aping the developed world [38, 39].

Some regulators have viewed telemedicine as new, unproven, and potentially dangerous to both patients and practitioners, requiring a more onerous written, witnessed, informed consent process than a regular face-to-face consultation [29]. This approach, while well-intentioned, fails to differentiate between the extremes of a doctor seeking a second opinion from a colleague about a patient over the telephone and intercontinental robotic surgery.

28.9.1 Conventional Precautions

The WMA's 2009 Guiding Principles for the Use of Telehealth took the pragmatic approach that relevant legislation and regulations regarding consent should be followed: "Consent for telehealth should follow similar principles and processes as those used for other health services" and that, "to the extent possible informed consent shall be obtained by the physician before starting any service or intervention" [26]. The more recent 2018 WMA Statement notes that "Proper informed consent requires that all necessary information regarding the distinctive features of telemedicine visit be explained fully to patients". Consent should include a description of the technology to be used, the information that will be shared and with whom, and how privacy and data security will be maintained [24].

Instant messaging and chat groups are a form of telemedicine widely used in everyday practice, even if prohibited by national and institutional IT Governance Rules [40]. Consent for their use is seldom obtained from patients [41]. Data protection laws such as the General Data Protection Regulation of the European Union add a further dimension to the issue of consent as patients require an explanation of how their information will be stored, for what purpose it will be used, and who will access it.

28.9.2 Consent and Selfie Telemedicine

For doctor-initiated selfie telemedicine, consent is required for the transmission of data, its use, and storage, but not for its acquisition. In patient-initiated telemedicine, the patients in effect waive the need for consent as they are making an autonomous decision to share their information by whatever means and with whoever they choose.

28.10 Authentication

The issue of unqualified people practising as doctors dates back centuries and was one of the reasons for the development of licensure of medical practitioners in Europe, starting in Sicily in 1140 [42]. Authentication is important, but before the digital age, how did patients authenticate that their doctor is who they claim to be and that they are and qualified and licensed to provide specific services? Largely by word of mouth and the degree certificate, valid or not, hanging on the wall. How did doctors authenticate their patients? By identity/social security number or insurance number and details. How is authentication, if any, currently conducted in overworked services in the developing world?

A principle of the WMA 2018 statement on telemedicine is that the doctor-patient relationship is predicated on both parties being able to identify each other "when telemedicine is employed" [24]. A patient should therefore be able to authenticate that they are who they say they are, as should the doctor. To date, this has been poorly addressed in the telemedicine literature. With the growing use of telemedicine during the COVID-19 pandemic, and to avoid the possibility of identity and medical identity theft, a recent focus of telemedicine activity has been around patient authentication—know your patient [43].

28.10.1 Recent Recommendations

Solutions implemented during the COVID-19 pandemic require a patient to submit both a photograph of a government identity document containing a photograph and a selfie. Using an artificial intelligence system, the information provided by the patient is compared with information from the identity document, and biometric authentication is conducted through facial verification software that creates a 3D face map of the person. An addition to this process has the patient submit a selfie before a video consultation for assessment by facial recognition software against the original selfie and photograph [43]. Biometrics will ultimately be the solution, but this is still some way off, especially for impoverished patients in rural areas of the developing world.

28.10.2 Authentication and Selfie Telemedicine

For doctor-initiated telemedicine, the patient and doctor should be able to identify one another when using videoconferencing. For store-and-forward telemedicine, the patients may submit their data to a password-protected website or dedicated email address. It is assumed that the persons submitting information to the site are thus who they claim to be as is the doctor responding. For patient-initiated telemedicine to their regular doctor, a similar assumption is made, either on the basis of email address or access to the doctor's email or site. No guidance exists for authentication of doctors or patients when patients seek advice from doctors with whom they have no prior relationship or from direct-to-consumer sites. For direct-to-consumer sites, there are added legal concerns of practitioners' qualifications and licensure [33].

28.11 Privacy, Data Security, Data Storage, and Record-Keeping

In most countries, medical practitioners have legal and ethical obligations to keep private all confidential information about a patient, maintain their clinical record, and store it securely. Telemedicine by definition uses electronic information and communications technology/ICT, and while electronic medical records are used in much of the developed world, paper records persist in the developing world. Patient-generated electronic data is not necessarily in a format that can be imported into an electronic medical record or readily printed for inclusion in a paper record.

28.11.1 Selfie Telemedicine

Record-keeping becomes more complex in selfie telemedicine. In most reported doctor-initiated selfie telemedicine services, a secure web- or phone-based application or encrypted email account has been used for data transmission. The information is at least at one site from which it can be moved to a record. In patient-initiated telemedicine, the patient takes responsibility for the security and storage of their information on whatever devices have been used to acquire and send the data. The obligation of the healthcare professionals to keep a record of unsolicited information sent to them has not been addressed, unless they take on the patient, thus initiating a doctor-patient relationship. Information sent from and received on cell phones raises other issues of data storage and record-keeping, and the concept of cell phone stewardship has been introduced, to provide guidance on maintaining the privacy and security of health-related information stored on mobile devices [44]. Any selfie/PGHD telemedicine should form part of the patient's record, but few direct to consumer services seek information about the patient's regular practitioner [33].

28.12 Quality of Care and Standards of Practice

An ongoing concern about telemedicine consultation and, more particularly, store-and-forward consultation is the quality and quantity of information provided and the adequacy of photographs or other attachments. On the one hand, it has been argued that only relevant information should be shared electronically, with some advising that there should be no patient identifiers in any messages or emails. This can affect the quality of information available to the doctor providing the service. The absence of patient identification precludes patient record-keeping, as noted in the literature on the use of instant messaging used for telemedicine [45, 46].

Quality of care in telemedicine is related to the presence of clinical guidelines for telemedicine within a speciality and standard operating procedures for a telemedicine consultation. The WMA (2009) statement on telehealth noted that "The physician should use existing clinical practice guidelines, whenever possible, to guide the delivery of care in the telehealth setting, recognising that certain modifications may need to be made to accommodate specific circumstances" and that any such "modifications to clinical practice guidelines for the telehealth setting are approved by the discipline's clinical governing body or association" [26]. The problem, however, is that few countries and few specialist disciplines within a country have developed telemedicine guidelines. Developing world countries most in need of telemedicine are the least likely to have telemedicine guidelines, with the danger that national regulatory bodies propose one size fits all ethical solutions [27, 29, 38].

28.12.1 Selfie Telemedicine

In both doctor- and patient-initiated telemedicine, the quality of care will be a function of the quality of data provided and the prevailing guidelines. The doctor has the responsibility to determine the adequacy of the data before providing advice. In direct-to-consumer services, the quality of care will also be determined by the qualifications and scope of practice of the service providers [33].

28.13 Continuity and Duty of Care

It has been generally held that the referring doctor is responsible for the ongoing care of the patient in a telemedicine consultation. The WMA 2018 statement emphasises "the primary physician remains responsible for the care and coordination of the patient with the distant medical team" [24]. Their 2009 statement on the Guiding Principles for the Use of Telehealth for the

Provision of Heath Care noted that "The physician needs to clarify ongoing responsibility for the patient with any other health professionals who are involved in the patient's care" [26].

28.13.1 Selfie Telemedicine

In doctor-initiated selfie telemedicine, the doctor is responsible for the continuity of care. The same holds for patient-initiated telemedicine when the doctor enters a doctor-patient relationship and provides advice and in insurance-related care. The situation in direct-to-consumer telemedicine is less well described. In many examples, the service is linked to a single financial transaction with no further obligations or expectations [33].

28.14 Justice

The dilemma of justice and telemedicine is summarised by "Individuals should be treated the same, unless they differ in ways that are relevant to the situation in which they are involved" [47]. This encapsulates the debate between equality, in which everyone is given the same resources or opportunities, and equity, in which people are treated differently based on their need. Within the context of telemedicine, justice can be linked on the one hand to aspects of the digital divide and, on the other, on the disparate healthcare needs of people.

28.14.1 Disadvantaged Populations

The rural and particularly the rural poor with limited access to quality specialist healthcare are least likely to benefit from telemedicine. They may not have equal access to technology that meets, for example, the requirements of existing guidelines of high bandwidth connectivity for high-resolution cameras encompassing pan, tilt, and zoom features for telepsychiatry [48]. They may not be able to afford the technology or the connectivity required. Using the argument that something is better than nothing, is it then justifiable to offer equity and practice telepsychiatry at lower bandwidth and resolution and without camera control? Or should it not be offered because of the possibility of lesser quality and standard of care and greater risk of malpractice claims? Others have argued that telemedicine should not be seen as a means of saving costs if it deprives people of the gold standard face-to-face consultation [49].

The digital divide continues because of the constant evolution of technology, poverty, and the cost of providing connectivity to sparsely populated rural areas. The urban digital divide continues based on economics, education, language, literacy, and culture.

28.15 Mobile Phone Apps: Regulation

Globally it has been estimated that between Google Play and the Apple App Store, there are over 5.7 million apps available [50], and of these approximately 400,000 are health, healthcare, fitness, or wellness-related [12]. Health apps have been categorised into six domains: wellness management, disease management, self-diagnosis, medication reminder, physical medicine, and rehabilitation [51]. Very few apps have undergone any form of quality assessment of their content, usability, or efficacy [52, 53]. Reports on the impact and use of apps vary. Some indicate they can and do influence patient behaviour [54], even concluding their use with mobile phones provides a significant opportunity to impact health behaviours globally [55]. But, actual and continued use is often reported as being low [56, 57], as is app use by the desired populations [58]. Reports indicate many healthcare professionals are using apps themselves; however few recommend them to their patients [59].

Regulatory control of mobile apps is currently dependent on whether they are classified as medical devices and or if they provide treatment. The Food and Drug Administration (USA) differentiates between mobile applications, software applications that can be run on a mobile platform, and mobile medical applications, which are used as an accessory to a regulated medical device or transform a mobile platform into a regulated medical device. The FDA regulates "those software functions that are medical devices and whose functionality could pose a risk to patient safety if the device were not to function as intended" [60].

In the United Kingdom, the National Institute for Health and Care Excellence has categorised digital health technologies (DHTs) of which mobile apps are part into three functional groups: Tier 1, DHTs with potential system benefits but no direct user benefits; Tier 2, DHTs which help users to understand healthy living and illnesses but are unlikely to have measurable user outcomes; Tier 3a, DHTs for preventing and managing diseases. They may be used alongside treatment and will likely have measurable user benefits. Tier 3b, DHTs with measurable user benefits, including tools used for treatment and diagnosis, as well as those influencing clinical management through active monitoring or calculation. It is possible DHTs in this tier will qualify as medical devices. Tiers 2 and 3 meet the definition of selfie telemedicine [61].

Germany introduced an act to improve healthcare through digitisation and innovation (the Digital Healthcare Act) in December of 2019. It defines healthcare apps as "medical products with a low risk class whose essential function is mainly based on digital technologies, and which are intended to support the detection, treatment, alleviation, or compensation of injuries or disabilities". Apps registered with the German Federal Institute for Drugs and Medical Devices can be prescribed by doctors or psychotherapists, and their use is covered by health insurance that is compulsory for all Germans [62].

28.15.1 International Standards

The Global Harmonisation Task Force (1993), succeeded by the International Medical Device Regulators Forum, is made up of ten member countries. Although dealing with all medical devices, their current activities on software and a medical device and medical device evidence evaluation are relevant to medical apps [63].

In addition the International Organisation for Standardization/ISO 13485:2016 specifies requirements for a quality management system where an organisation needs to demonstrate its ability to provide medical devices and related services that consistently meet customer and applicable regulatory requirements. ISO 13485 is designed to be used by organisations involved in the design, production, installation, and servicing of medical devices and related services. It can also be used by internal and external parties, such as certification bodies, to help them with their auditing processes [64].

Of the approximately 400,000 health and wellness apps available, few are considered medical devices. In 2012, to address the need for certification of health apps, a set of guidelines was developed by the Association of American Medical Colleges, Mobile Marketing Association, Healthcare Information and Management Systems Society (HIMSS), the US Food and Drug Administration (FDA), and other USA federal agencies. These guidelines, now known as the Xcertia Guidelines, are based on operability, privacy, security, and usability, and the contents are regularly updated [65].

Alternative methods of evaluation and regulation have emerged, with hospitals developing guidelines for selecting products, insurers deciding which products will be reimbursed, and third-party app review websites which seek the evidence supporting the product [66].

28.16 A Model for Fraud

It is proposed that most fraudulent electronic communication activity can be described by three types of activity shown in Fig. 28.1: falsification or fabrication of information, creating misinformation or disinformation, and doing so through ignorance or complicity. Most casual fraudulent activity will arise as a result of individuals either falsifying or fabricating PGHD. More structured fraudulent activity will

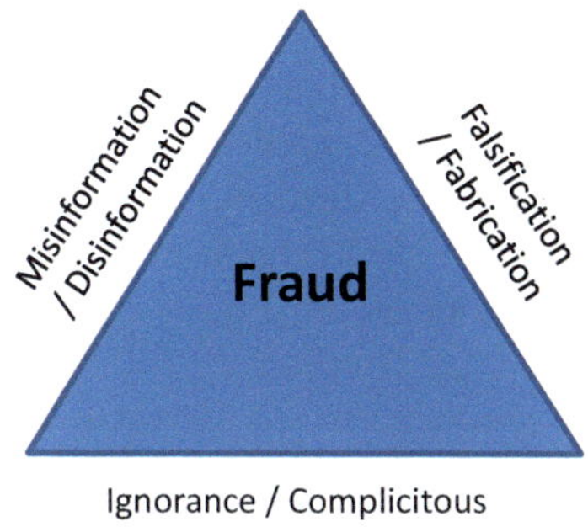

Fig. 28.1 A model for fraud

arise as a result of creating misinformation (false information that is spread regardless of intent to mislead) or disinformation (false or manipulated information deliberately intended to bias or mislead). Creating either of these types of incorrect or inaccurate information can be caused by individual ignorance or deliberate—often duplicitous—action of more than one individual.

28.16.1 Selfie Relevant Examples

Falsify: Not providing a clinician with salient insight (not providing images of all skin lesions). Fabricate: Sharing a video with a dental hygienist of someone else flossing but claiming it was you. Misinformation: Sharing an image of a forged COVID-19 vaccine record card to gain entry to a travel restricted location. Disinformation: Using deep fake techniques to create an image or video that falsely appears to show a famous individual criticising COVID-19 vaccine production by a manufacturer.

Other examples of potential fraud associated with selfies might include threats to "share" naked selfies originally intended for health-related purposes leading to bribery, creating fake reviews of a medical treatment or a specific clinician using "doctored" selfies, creating deep fakes by placing someone else's/your face on another person's body image in order to deceive, taking fake mirror selfies [67], distributing selfie-editing apps that once uploaded cannot be deleted and call up fraudulent ads to claim advertising funds [68], using posted selfies to create fake social media profiles that impersonate the person and promise explicit images in exchange for money and personal information [69], and experiencing identity theft through gathering of information from images shared online (full name, birthdate, tattoo, disability, social insurance number, even recreating fingerprints from "peace sign selfies") [70]. The possibilities seem endless and challenge the imagination of most people. But they happen.

28.17 Apps and Fraud

App use constitutes selfie telemedicine, and the most common fraud relates to both false advertising and lack of scientific verification of claims made. People are duped into purchasing and using the apps, some with associated risk [71]. This is due in part to regulatory lacunae.

Examples are as follows:

- Aura Labs marketed an app that enabled the user to "measure blood pressure with just your phone" "No cuff required". Unfortunately there were significant differences in the standard blood pressure measurements and those reported by the app; two of the endorsers of the product were relatives of the company's co-founder, and the five-star review was made by the company's CEO and President.
- Matis Ltd. marketed an app that converted a smartphone into a foetal heart monitor but never conducted any studies comparing their app with a foetal heart rate monitor or Doppler, for example.
- Runtastic GmbH and Cardio claimed to measure heart rate and performance under stress but never measured the apps' accuracy, while also claiming it was endorsed by MIT.
- Acne: At least two companies have claimed to treat acne with the power of the smartphone's light. Users are instructed to prop up the phone against their skin in acne-prone areas for a few minutes each day. These apps are no longer marketed.
- Breast augmentation: Cow Know claims that by listening to their app of the sounds of crying babies 20 times a day, a woman's breasts will grow because of the subliminal persuasion that they are breastfeeding.

28.18 (Selfie) Telemedicine and Fraud

The Centers for Medicare and Medicaid Services (USA) reported an 11,000% increase in telehealth services in the first month and a half of the public health emergency in the United States. With increased telemedicine remuneration during the pandemic, there is little doubt that the "routine" forms of medical fraud will now spill over into telemedicine. These include over-servicing, up-billing, charging for treatment not given, misrepresenting non-covered treatments as medically necessary, or prescription fraud [72].

Only one published fraud on patient-generated health data or selfie has been encountered, concerning patients not reporting their weight honestly [73]. The veracity of PGHD has always been an issue. Any withheld or misleading information could prove harmful. Yet, studies have shown that patients are not forthcoming in this regard. Up to 47% of patients have withheld information about experiencing at least one of four imminent social/psychological threats (depression, suicidality, abuse, and sexual assault) [74]. Furthermore, up to 81% of patients have avoided disclosing at least one type of relevant information related to understanding instructions, disagreeing with recommendations, describing their exercise regime, diet, medication consumption, medication adherence, or taking unprescribed medications of others. Ironically sicker patients are more likely to do so [75]. Even privacy and security concerns about the transfer of electronic personal health information (PHI) between healthcare professionals have led to withholding of information [76].

28.18.1 Patient Transparency and Honesty

Given the evidence above, which shows patients routinely withhold or embellish information (i.e. fraudulent behaviour), there is no reason to believe this will not occur with selfies or other forms of PGHD sent to healthcare professionals. Compounding these concerns are others related to the electronic capture of PGHD, such as completeness of a record, accuracy of the recording, collection period for the data, user experience in manipulating the data capture, sensor validation, and data standardisation [77].

28.19 The Future of Selfie Telemedicine

In terms of Rogers diffusion of innovation curve, the telemedicine world has moved or is in the process of moving rapidly from the "early adopters" to "the early majority". Laws and regulations take time to catch up with technological progress. With an estimated global mHealth market of US$312 billion by 2027 [78] and projected global cell phone market of $1351.8 billion by 2026 [79], the proliferation of apps and cell phones and their use to capture patient data and transmit selfies can be assured.

Healthcare costs are constantly rising as technology and knowledge advance. As the world's population ages and grows and birth rates decline, there will be an ever-growing shortage of healthcare professionals. Pragmatic solutions are needed to provide healthcare in both the developed and developing world. One of the few positive aspects of the COVID-19 pandemic is that it has brought telemedicine to the people and their healthcare providers. Telemedicine is here to stay.

References

1. Thomas EE, Haydon HM, Mehrotra A, Caffery LJ, Snoswell CL, Banbury A et al (2020) Building on the momentum: sustaining telehealth beyond COVID-19. J Telemed Telecare. https://doi.org/10.1177/1357633x20960638
2. Bestsennyy O, Gilbert G, Harris A, Rost J. Telehealth: a quarter-trillion-dollar post-COVID-19 reality? mckinsey.com/industries/healthcare-systems-and-services/our-insights/telehealth-a-quarter-trillion-dollar-post-covid-19-reality. Accessed 27 Apr 2020
3. Oxford English Dictionary (2021) Telemedicine. en.oxforddictionaries.com/definition/telemedicine. Accessed 27 Apr 2021
4. Tiase VL, Hull W, McFarland MM, Sward KA, Del Fiol G, Staes C et al (2020) Patient-generated health

data and electronic health record integration: a scoping review. JAMIA Open 3(4):619–627
5. Demiris G, Iribarren SJ, Sward K, Lee S, Yang R (2019) Patient generated health data use in clinical practice: a systematic review. Nurs Outlook 67(4): 311–330
6. Cresswell K, McKinstry B, Wolters M, Shah A, Sheikh A (2019) Five key strategic priorities of integrating patient generated health data into United Kingdom electronic health records. J Innov Health Inform 25(4):254–259
7. Oxford English Dictionary (2021) Selfie. en.oxforddictionaries.com/definition/selfie. Accessed 27 Apr 2021
8. Pearlman J (2013) Australian man 'invented the selfie after drunken night out'. telegraph.co.uk/news/worldnews/australiaandthepacific/australia/10459115/Australian-man-invented-the-selfie-after-drunken-night-out.html. Accessed 28 Apr 2018
9. WHO Director General. mHealth: use of appropriate digital technologies for public health: report by the Director-General. apps.who.int/iris/handle/10665/274134. Accessed 28 Apr 2018
10. World Health Organization. Global diffusion of eHealth: making universal health coverage achievable: report of the third global survey on eHealth. ghdonline.org/uploads/WHO_Report_-_Global_diffusion_of_eHealth-_Making_universal_health_coverage_achievable.pdf. Accessed 28 Apr 2018
11. World Health Organization (2019) WHO guideline: recommendations on digital interventions for health system strengthening. who.int/reproductivehealth/publications/digital-interventions-health-system-strengthening/en/. Accessed 27 Apr 2021
12. Georgiou M (2021) Developing a healthcare app in 2021: what do patients really want? Imaginovation. imaginovation.net/blog/developing-a-mobile-health-app-what-patients-really-want/. Accessed 27 Apr 2021
13. Hall Dykgraaf S, Desborough J, de Toca L, Davis S, Roberts L, Munindradasa A et al (2021) "A decade's worth of work in a matter of days": the journey to telehealth for the whole population in Australia. In J Med Inform 151:104483
14. Hong Z, Li N, Li D, Li J, Li B, Xiong W et al (2020) Telemedicine during the COVID-19 pandemic: experiences from Western China. J Med Internet Res 22(5):e19577
15. Aronson SH (1977) The Lancet on the telephone 1876–1975. Med Hist 21(1):69–87
16. Hjelm M (2005) Bell's first telephone call–not the first telemedicine consultation. J Telemed Telecare 11(4): 216
17. Einthoven W (1906) Het telecardiogram. Nederl Tijdschr v Geneesk 2:1517–1547
18. Nair PP, Aghoram R, Thomas B, Bharadwaj B, Chinnakali P (2021) Video teleconsultation services for persons with epilepsy during COVID-19 pandemic: an exploratory study from public tertiary care hospital in Southern India on feasibility, satisfaction, and effectiveness. Epilepsy Behav 117:107863
19. Uscher-Pines L, Sousa J, Raja P, Mehrotra A, Barnett ML, Huskamp HA (2020) Suddenly becoming a "virtual doctor": experiences of psychiatrists transitioning to telemedicine during the COVID-19 pandemic. Psychiatr Serv 71(11):1143–1150
20. Mars M, Morris C, Scott RE (2018) Selfie telemedicine – what are the legal and regulatory issues? Stud Health Technol Inform 254:53–62
21. Percival T (2014) Medical ethics. Cambridge University Press. cambridge.org/core/books/medical-ethics/362CABA86F05C455D49E9238F76F3110. Accessed 27 Apr 2021
22. Laws of Malaysia, Telemedicine Act 1997, Act 564 (1997) htapah.moh.gov.my. Accessed 27 Apr 2021
23. Intan Sabrina M, Defi IR (2021) Telemedicine guidelines in South East Asia—a scoping review. Front Neurol 11:1760
24. World Medical Association (2018) WMA statement on the ethics of telemedicine. wma.net/policies-post/wma-statement-on-the-ethics-of-telemedicine/. Accessed 27 Apr 2021
25. Stanberry B (2001) Legal ethical and risk issues in telemedicine. Comput Methods Prog Biomed 64(3): 225–233
26. World Medical Association (2009) WMA Statement on guiding principles for the use of telehealth for the provision of health care. In: Adopted by the 60th WMA General Assembly, New Delhi, India. wma.net/policies-post/wma-statement-on-guiding-principles-for-the-use-of-telehealth-for-the-provision-of-health-care/. Accessed 27 Apr 2021
27. Board of Governors in Supersession of the Medical Council of India 2020 Telemedicine practice guidelines. pp 1–48. mohfw.gov.in/pdf/Telemedicine.pdf. Accessed 27 Apr 2021
28. Kaba R, Sooriakumaran P (2007) The evolution of the doctor-patient relationship. Int J Surg 5(1):57–65
29. Health Professions Council of South Africa (2014) General ethical guidelines for good practice in telemedicine. hpcsa.co.za/Uploads/editor/UserFiles/downloads/conduct_ethics/Booklet%2010.pdf. Accessed 27 Apr 2021
30. American Medical Association (2021) Patient-physician relationships. American Medical Association. ama-assn.org/delivering-care/ethics/patient-physician-relationships. Accessed 27 Apr 2021
31. Blake V (2012) When is a patient-physician relationship established? Virtual Mentor 14(5):403–406
32. Farouk A (2016) Telemedicine prompts new ethical ground rules for physicians. American Medical Association. ama-assn.org/practice-management/digital. Accessed 27 Apr 2021
33. Resneck JS Jr, Abrouk M, Steuer M, Tam A, Yen A, Lee I et al (2016) Choice, transparency, coordination, and quality among direct-to-consumer telemedicine

websites and apps treating skin disease. JAMA Derm 152(7):768–775
34. British Medical Association (2020) Ethics toolkit for medical students. British Medical Association. bma.org.uk/advice-and-support/ethics/medical-students/ethics-toolkit-for-medical-students/autonomy-or-self-determination. Accessed 27 Apr 2021
35. Dalla-Vorgia P, Lascaratos J, Skiadas P, Garanis-Papadatos T (2001) Is consent in medicine a concept only of modern times? J Med Ethics 27(1):59–61
36. Beauchamp TL, Faden RR (1995) Informed consent; I. History of informed consent. In: Reich WT (ed) Encyclopedia of bioethics. Simon and Schuster Macmillan, New York, pp 1233–1238
37. Berg JW, Appelbaum PS, Lidz CW, Parker LS (2001) The legal requirements for disclosure and consent: history and current status. Informed consent. Oxford University Press
38. Townsend B, Scott R, Mars M (2019) The development of ethical guidelines for telemedicine in South Africa. S Afr J Bioethics Law 12(1):19–26
39. Kuziemsky CE, Hunter I, Gogia SB, Lyenger S, Kulatunga G, Rajput V et al (2020) Ethics in telehealth: comparison between guidelines and practice-based experience – the case for learning health systems. Yearb Med Inform 29(1):44–50
40. El Hadidy TS, Alshafei AE, Mortell AE, Doherty EM (2018) Smartphones in clinical practice: doctors' experience at two Dublin paediatric teaching hospitals. Irish J Med Sci 187(3):565–573
41. Morris C, Scott RE, Mars M (2021) Is consent not a consideration for instant messaging? In: Maeder AJ, Higa C, van den Berg MEL, Gough C (eds) 10th Global Telehealth Conference 2020. IOS Press BV, pp 88–95
42. Encyclopedia of Bioethics (2021) Medical ethics, history of Europe: I. Ancient and Medieval. C. Medieval Christian Europe. encyclopedia.com/science/encyclopedias-almanacs-transcripts-and-maps/medical-ethics-history-europe-i-ancient-and-medieval-c-medieval-christian-europe. Accessed 27 Apr 2021
43. Health IT Security (2020) Telehealth is the new normal, but so is online fraud. xtelligent Healthcare media. healthitsecurity.com/news/telehealth-is-the-new-normal-but-so-is-online-fraud. Accessed 27 Apr 2021
44. Scott RE, Morris C, Mars M (2021) Development of a "cellphone stewardship framework": legal, regulatory, and ethical issues. Telemedicine J eHealth 27(3): 316–322
45. Mars M, Morris C, Scott RE (2019) WhatsApp guidelines – what guidelines? A literature review. J Telemed Telecare 25(9):524–529
46. Mars M, Scott RE (2016) WhatsApp in clinical practice: a literature review. Stud Health Technol Inform 231:82–90
47. Andre C, Valasquez M (1990) Justice and fairness. Issues Ethics 3(2, Spring)
48. Keenan AJ, Tsourtos G, Tieman J (2021) The value of applying ethical principles in telehealth practices: systematic review. J Med Internet Res 23(3):e25698
49. Demiris G, Oliver DP, Courtney KL (2006) Ethical considerations for the utilization of tele-health technologies in home and hospice care by the nursing profession. Nurs Adm Q 30(1):56–66
50. Statista (2021) Number of apps available in leading app stores as of 1st quarter 202. statista.com/statistics/276623/number-of-apps-available-in-leading-app-stores/. Accessed 29 May 2021
51. Kao CK, Liebovitz DM (2017) Consumer mobile health apps: current state, barriers, and future directions. PM R 9(5):S106–S115
52. Jake-Schoffman DE, Silfee VJ, Waring ME, Boudreaux ED, Sadasivam RS, Mullen SP et al (2017) Methods for evaluating the content, usability, and efficacy of commercial mobile health apps. JMIR Mhealth Uhealth 5(12):e190
53. Paglialonga A, Lugo A, Santoro E (2018) An overview on the emerging area of identification, characterization, and assessment of health apps. J Biomed Inform 83:97–102
54. Scott RE, Mars M (2020) Behaviour change and e-health – looking broadly: a scoping narrative review. In: Maeder AJ, Champion S, Moores C, Golley R (eds) Stud Health Technol Inform. IOS Press, pp 123–138
55. Zhao J, Freeman B, Li M (2016) Can mobile phone apps influence people's health behavior change? An evidence review. J Med Internet Res 18(11):e287
56. Kayyali R, Peletidi A, Ismail M, Hashim Z, Bandeira P, Bonnah J (2017) Awareness and use of mhealth apps: a study from England. Pharmacy (Basel) 5(2)
57. Vaghefi I, Tulu B (2019) The continued use of mobile health apps: insights from a longitudinal study. JMIR Mhealth Uhealth 7(8):e12983
58. Robbins R, Krebs P, Jagannathan R, Jean-Louis G, Duncan DT (2017) Health app use among US mobile phone users: analysis of trends by chronic disease status. JMIR Mhealth Uhealth 5(12):e197
59. Byambasuren O, Beller E, Glasziou P (2019) Current knowledge and adoption of mobile health apps among Australian general practitioners: survey study. JMIR Mhealth Uhealth 7(6):e13199
60. U.S. Food and Drug Administration (2019) Policy for device software functions and mobile medical applications. In: U.S. Department of Health and Human Services (ed), pp 1–42
61. National Institute for Health and Care Excellence (2019) Evidence standards framework for digital health technologies. nice.org.uk/corporate/ecd7. Accessed 27 Apr 2021
62. Monitir GL (2020) Germany: new law allows health apps by prescription. Library of Congress Law. loc.gov/law/foreign-news/article/germany-new-law-allows-health-apps-by-prescription/. Accessed 27 Apr 2021

63. 2020 Forum IMDR. About IMDRF. International medical device regulators forum. imdrf.org/about/about.asp. Accessed 27 Apr 2021
64. International Organization for Standardization (2020) ISO 13485:2016(en) medical devices — quality management systems — requirements for regulatory purposes. iso.org/obp/ui#iso:std:iso:13485:ed-3:v1:en. Accessed 27 Apr 2021
65. Xcertia (2019) Xcertia. google.com/url?sa=t&rct=j&q=&esrc=s&source=web&cd=&ved=2ahUKEwi90ZLO2u_wAhWKgVwKHbhWDkUQFjAAegQIBRAD&url=https%3A%2F%2Fwww.himss.org%2Fsites%2Fhde%2Ffiles%2Fmedia%2Ffile%2F2020%2F04%2F17%2Fxcertia-guidelines-2019-final.pdf&usg=AOvVaw1HWSNoivwvaJInL6s5Y-j0. Accessed 27 Apr 2021
66. Cortez N (2018) The evolving law and ethics of digital health. In: Rivas H, Wac K (eds) Digital health. Springer, pp 249–269
67. Midkiff S (2021) Turns out, influencer mirror selfies are all a lie. Refinery 29. refinery29.com/en-ca/2021/03/10369627/fake-mirror-selfie-photos-influencers-tiktok. Accessed 27 Apr 2021
68. Goodfellow J (2020) Brands inadvertently funding fraudulent selfie apps. In: Influencer 360. campaignlive.co.uk/article/brands-inadvertently-funding-fraudulent-selfie-apps/1685855. Accessed 27 Apr 2021
69. Selbie T, Peters J (2021) Fake accounts used my pictures to sell sex. BBC. bbc.com/news/uk-scotland-56182060. Accessed 27 Apr 2021
70. McGoogan C, Demetriou, D (2017) Peace sign selfies could let hackers copy your fingerprints. The Telegraph, Tokyo. telegraph.co.uk/technology/2017/01/12/peace-sign-selfies-could-let-hackers-copy-fingerprints/. Accessed 27 Apr 2021
71. Akbar S, Coiera E, Magrabi F (2020) Safety concerns with consumer-facing mobile health applications and their consequences: a scoping review. J Am Med Inform Assoc 27(2):330–340
72. Giancola P, Stedman, C (2021) Telehealth fraud triggered by COVID-19 Pandemic. Snell & Wilmer. jdsupra.com/legalnews/telehealth-fraud-triggered-by-covid-19-8657616/. Accessed 27 Apr 2021
73. Steventon A, Chaudhry SI, Lin Z, Mattera JA, Krumholz HM (2017) Assessing the reliability of self-reported weight for the management of heart failure: application of fraud detection methods to a randomised trial of telemonitoring. BMC Med Inform Decis Mak 17(1):43
74. Levy AG, Scherer AM, Zikmund-Fisher BJ, Larkin K, Barnes GD, Fagerlin A (2019) Assessment of patient nondisclosures to clinicians of experiencing imminent threats. JAMA Netw Open 2(8):e199277
75. Levy AG, Scherer AM, Zikmund-Fisher BJ, Larkin K, Barnes GD, Fagerlin A (2018) Prevalence of and factors associated with patient nondisclosure of medically relevant information to clinicians. JAMA Netw Open 1(7):e185293
76. Agaku IT, Adisa AO, Ayo-Yusuf OA, Connolly GN (2014) Concern about security and privacy, and perceived control over collection and use of health information are related to withholding of health information from healthcare providers. J Am Med Inform Assoc 21(2):374–378
77. Codella J, Partovian C, Chang H-Y, Chen C-H (2018) Data quality challenges for person-generated health and wellness data. IBM J Res Dev 62(1):1–3
78. Reports and Data (2020) Mobile health (mHealth) market to reach USD 311.98 billion by 2027. globenewswire.com/news-release/2020/04/28/2023512/0/en/Mobile-Health-mHealth-Market-To-Reach-USD-311-98-Billion-By-2027-Reports-and-Data.html. Accessed 27 Apr 2021
79. Intelligence M (2020) Smartphones market – growth, trends, COVID-19 impact, and forecasts (2021–2026). mordorintelligence.com/industry-reports/smartphones-market. Accessed 27 Apr 2021

29 Healing the Healers: Addressing Moral Injury in Healthcare Workers During COVID-19

Aliza Naiman, Doron Amsalem, Maja Bergman, and Yuval Neria

Abstract

Moral injury (MI) describes the intense feelings of shame, guilt, anger, and betrayal that individuals may experience after violating their own moral codes or witnessing the moral transgressions of others. While MI is not a diagnosable mental health disorder, it has been associated with elevated symptoms of depression and anxiety, suicidal ideation, and substance misuse. MI overlaps with some of the diagnostic properties of posttraumatic stress disorder (PTSD) but has been proposed as a distinct phenomenon. Originally identified in military contexts, moral injuries have also been observed in civilian populations. Healthcare workers (HCWs) are at risk of encountering potentially morally injurious events (PMIEs) in the workplace. During the COVID-19 crisis, HCWs forced to provide care with limited resources reported self-blame after losing patients and while fearful of infecting loved ones. As vaccine roll-out continues, it is essential that we encourage healing among the very population that serviced the ill in their time of need.

A. Naiman · D. Amsalem · M. Bergman
Department of Psychiatry and New York State Psychiatric Institute, Columbia University Vagelos College of Physicians and Surgeons, New York, NY, USA
e-mail: aliza.naiman@rutgers.edu; Doron.amsalem@nyspi.columbia.edu; Maja.bergman@nyspi.columbia.edu

Y. Neria (✉)
Department of Psychiatry and New York State Psychiatric Institute, Columbia University Vagelos College of Physicians and Surgeons, New York, NY, USA

Department of Epidemiology, Columbia University Irving Medical Center, New York, NY, USA
e-mail: Yuval.neria@nyspi.columbia.edu

29.1 What Is Moral Injury?

Moral injury (MI), coined first by Jonathon Shay in 1995 and further elucidated by Litz and colleagues in 2009, is a promising phenotype pending within psychology and psychiatry. Litz et al. [1] classified moral injury as "perpetrating, failing to prevent, bearing witness to, or learning about acts that transgress deeply held moral beliefs and expectations (p. 697)."

Several highly stressful events constitute potentially morally injurious events (PMIEs), including both perpetrating transgressive, harmful acts (acts of commission and omission), witnessing as another inflicts them, and high-stakes betrayal [2]. The word "potentially" is central, as experiencing a morally dubious event is not directly indicative of moral injury as a clinical outcome, just as a trauma is not of post-traumatic stress disorder (PTSD). When balanced and finite, moral pain and guilt are evolutionarily adaptive responses motivating individuals to maintain group cohesion and reaffirm their

J. Faintuch, S. Faintuch (eds.), *Integrity of Scientific Research*,
https://doi.org/10.1007/978-3-030-99680-2_29

commitment to the moral code initially violated. To sustain a moral injury is to be symptomatic beyond the trajectory of normal trauma recovery [3]. Examples of self-perpetrated PMIEs are unintentionally killing a person in war, failing to save a child from running into an oncoming car, or a healthcare worker selecting who should receive care during a peak of the COVID-19 pandemic. PMIEs perpetrated by others include witnessing crimes against humanity as a refugee, learning of child abuse within the foster care system as a child protective specialist designated to safeguard such youth, or a nurse ordered by an authority to ration treatments [4, 5].

29.1.1 The Moral Injury Events Scale (MIES)

The first to create a unified measurement of PMIEs, Nash et al. [6] developed the MIES. A nine-item, self-report instrument, the MIES assesses self-perpetrated and witnessed transgressions as well as betrayal-based PMIEs. It also screens for event-related symptoms, offering clinical utility. Items on the Likert scale include "I saw things that were morally wrong" and "I am troubled by having acted in ways that violated my own morals or values."

The definition of moral injury (MI) developed within the military context [7]. However, PMIEs occur widely within civilian populations as well [8, 9]. Since early 2020, healthcare workers around the globe were forced to triage resources and ration personal protective equipment (PPE) while making life-and-death decisions surrounding patient care in unprecedented and unsupervised circumstances.

29.2 Effects and Outcomes of Moral Injury

Moral injury (MI) manifests in intense feelings of guilt/shame, betrayal, hopelessness, isolation, loss of meaning, and spiritual unease [10]. Various studies found an increased risk for self-reported mental health disorders such as depression and PTSD and internalizing behaviors such as social isolation and self-critical beliefs among people who experienced PMIEs [11, 12]. Suicidal ideation and attempts, elevated substance and alcohol use, and hostile and aggressive acts are all externalizing symptoms associated with MI [13, 14]. Currier et al. [15] connected these internalizing and externalizing symptoms to specific PMIE types, tracing self-inflicted PMIEs to negative cognitions such as shame and guilt, self-blame, feelings of badness, and self-destructive behaviors (i.e., "I am unlovable and evil. What I did was unforgivable and I am incapable of making moral decisions"). Conversely, acts of high-stakes betrayal or witnessing moral transgressions is associated with disgust, anger, mistrust, and revenge fantasies [3, 8, 16].

29.3 Moral Injury and PTSD

Moral injury (MI) is strongly related to PTSD as both can follow exposure to trauma. By surveying deployed veterans over 6 months, Currier et al. [15] found that moral injury reported at baseline (study initiation) was correlated with PTSD at follow-up assessments. Self-perpetrated moral injuries have been associated with increased severity across symptoms of guilt, shame, and re-experiencing when compared to interpersonal or other life-threatening traumas in veterans [17]. Increased suicidality, depression, and PTSD symptoms are also predictive of comorbid PTSD and MI [15, 18]. Suicidal ideation and attempts are more associated with self-perpetrated MIs than other MI types [8, 19]. Neria and Pickover [8] posit that this is because the core features of shame and guilt in MI contribute to the individual feeling isolated and burdensome to others. Burdensomeness and thwarted feelings of belonging, in turn, are concomitant with suicide risk [20].

29.3.1 Nosological Classification

Unlike PTSD, MI is not a recognized disorder in the Diagnostic and Statistical Manual of Mental

Disorders (DSM-5) [21]. Whether it is clinically freestanding from PTSD is undetermined. Nevertheless numerous parallels between PTSD's and MI's core features exist. PTSD's four symptom clusters include re-experiencing the event (e.g., dreams and flashbacks); avoidance (of trauma reminders); negative alterations in mood and cognition; and increased arousal/fear responses (e.g., startle response). Within the veteran population, it is likely that service members suffering from MI would be symptomatic on all four accounts. They would arguably re-experience the morally violating incident, avoid thinking of the event and its subsequent relational and situational triggers, struggle with feelings of isolation, and suffer anhedonia and emotional numbing [1]. Additionally, like PTSD, moral injury is also associated with self-destructive behaviors, alcohol and drug abuse, suicidal ideation, and interpersonal difficulties [3]. Hallmark MI symptoms of anger, guilt, shame, and self-blame provide additional overlap with PTSD [15].

29.3.2 Frequent Overlap with PTSD

The first condition for PTSD is experiencing a traumatic event. In the DSM-5, Criterion A includes directly or indirectly experiencing, witnessing, learning about, or being extensively exposed to an event involving a threat to one's life, serious physical harm, or sexual violence [21]. Particularly in military contexts, Criterion A events are also frequently PMIEs [3, 22]. For example, consider a soldier who, in a life-and-death situation, mistakenly kills a civilian he thought was carrying a weapon. In addition, the DSM-5 definition of PTSD was adapted to include symptoms in Cluster D, negative cognitions about the world or oneself [3]. These would contain morally based items such as "I am bad" and "there is something seriously wrong with me." Cluster E, arousal, also includes the self-harming and reckless behaviors commonly present in perpetration-based moral injury alongside PTSD [3, 18]. Neria and Pickover [8] emphasize that it is still undetermined if PMIEs that are also inarguable Criterion A events are more likely to result in a diagnosis of PTSD or MI when compared to PMIEs that do not involve life-threat or injury, intense stress, and violence. Traumas that also function as morally injurious events mark a challenge when distinguishing moral injury from PTSD as a distinct phenotype [23].

29.3.3 Unique Contexts

However, not all PMIEs meet the exclusive standards of Criterion A traumas. PMIEs don't always include actual or threatened death, serious injury, or sexual violence. Bearing witness to or ignoring another's suffering, though emotionally defeating, is not life-threatening, nor is betrayal [24]. A moral injury is an internal wound; the individual's identity, self-understanding, and core beliefs about the world are compromised, regardless of whether the body was endangered [25]. Furthermore, Litz and Kerig [3] posit that excessive blame and self-accusation are symptoms of PTSD in the DSM-5, but are necessitated reactions to moral violations (both perpetrated and witnessed).

While co-occurring cognitions are shared with PTSD, the additional presence of moral injury exacerbates feelings of shame and guilt and intervenes with recovery and self-forgiveness [22, 26]. In addition, while both moral injury and PTSD prompt social withdrawal, the motivation behind the avoidance differs and must inform treatment. Habituated avoidance in PTSD allays fear and evades triggers. In moral injury, the self is shameful or untrustworthy and must be concealed [22].

29.4 Moral Injury in Service Members

Because the battlefield is rife with PMIEs, moral injury's relevance in the military is explicit. According to the National Veteran Suicide Prevention Annual Report, since 2014, US Military veteran deaths by suicide outnumbered combat losses during the operations in Afghanistan and Iraq combined (averaging 17–20 deaths per day), suggesting that service members are haunted by their experiences long after they return home [27]. Suicide and moral injury are highly interconnected, with MI's core features of shame, guilt, and isolation contributing to suicidal ideation, while habituation to combat violence increases the lethality of such attempts [28]. In both Iraq and Vietnam veterans, killing in war time and suicidal ideation are related [19, 29].

Exposure to PMIEs and Criterion A traumas as a combatant is inevitable. In a study examining the longitudinal mental health effects of combat in veterans deployed in Operation Iraqi Freedom and Operation Enduring Freedom in Afghanistan, Hoge et al. [30] found that 71–86% of soldiers and marines deployed in Iraq reported participating in firefights, with 65% (Marine) and 48% (Army) of these firefights including enemy deaths. Study findings from Levi-Belz et al. [12] revealed that 49% of participants from Israeli Defense Force veterans responded "yes" to having shot at an enemy. In this same study, over 29% reported witnessing morally wrong acts, and 25% felt betrayed by trusted leaders. Military PMIEs include not only the killing of combatants and non-combatants and betrayals but also turning a blind eye on the enemy and civilian distress, lack of social support, and a sense of despair at war's political limitations [2].

29.4.1 The Moral Injury Outcome Scale (MIOS)

Using data collected from the MIOS consortium, Yeterian and colleagues [31] classified the most frequently self-perpetrated PMIES: (a) seeing immoral acts and doing nothing to stop them, (b) killing within the rules of engagement, (c) making a mistake or failing in a task that resulted in harm to others, (d) seeing others harmed as a result of decisions made or not made by someone else, (e) seeing others treat the helpless with disrespect, and (f) harm done to the individual or others or placing the patient or others in harm's way. A US Marine Corp study reported that 24% of the participants endorsed perpetrating moral transgressions whereas 18% felt they had violated their moral codes [5].

29.4.2 Mental Health and Quality of Life

Outcomes for veterans with transgressed moral frameworks are bleak and particularly for those who have killed. Maguen and colleagues [32] found killing associated not only with PTSD symptoms but also violent outbursts, dissociation, and functional impairment in the form of employment issues, interpersonal/spousal problems, financial difficulties, poor physical health, and alcohol and drug use. Twenty-six US Military combat veterans who testified to killing during their service discussed its impact on their own lives. In these vulnerable sessions, specific themes prevailed. Although the question veterans are most frequently asked upon their return is "Did you kill anybody?" reluctance to share quickly grows as they are ostracized and fear being perceived as "dangerous" and "crazy" [33]. Cut off, they feel those who were not overseas cannot understand what killing is like. To cope, veterans compartmentalize themselves and their actions and numb using alcohol. They report feelings of moral conflict, shame, and guilt, with some disgusted at their actions and others retrospectively disturbed by their tolerance or enjoyment of killing. The sanctioned violence of wartime leads many to lose their faith.

29.5 Moral Injury in Civilians

Moral injury has been identified in a wide range of populations including refugees and journalists covering mass migrations, law enforcement officers, social workers, and educators. In war-torn spaces, refugees often fall victim to moral injuries while living in areas rampant with violence and loss. Many are forced to make painful life-and-death decisions for themselves, family members, and others. Threatened on political, racial, or religious bases, they observe human depravity while fleeing. Refugees from some parts of the world are particularly exposed to PMIEs, including child soldiers or witnesses/victims of torture, and when forced to betray fellow refugees for survival [34]. An Australian study with 222 refugees from diverse cultural backgrounds found that within this understudied population, self-perpetrated or other-oriented moral injuries were associated with anger, depression, and overall increased psychopathology [35].

29.5.1 Press Correspondents

Journalists, as eye-witness historians, are at increased risk of experiencing traumatic events and PMIEs. Journalists without previous experience of a war zone, who are working alone, and perceive a lack of support and reduced resources on the field are most susceptible to MI, even as journalists overall are less affected by PTSD and depression. In addition, journalists who aided migrants outside of their duties or worked on the ground close to home were more likely to report experiencing guilt, with guilt and moral injury proving strongly associated [36].

29.5.2 Forces of Order in the Civil Environment

A study conducted by Weiss et al. [37] found that some 25% of law enforcement officers reported killing or injuring an assailant during their lifetime careers. A survey of police officers on the impact of killing on their mental health revealed that, within the first 3 years of service, over two-thirds of police officers experienced a life-threatening situation [38]. Ten percent of these police officers reported killing or severely wounding a civilian in that time. However, it is worth noting that their early career statuses and the perpetuated "code of silence" within the law enforcement community could contribute to the low percentage reported. Killing or seriously injuring someone was strongly associated with PTSD symptoms in police officers and less significantly related to symptoms of depression. Attempting to correlate this data more directly with MI, results from a study conducted across several national Finish law enforcement agencies indicated that MI significantly predicts PTSD and its core symptom clusters of avoidance, hyperarousal, and re-experiencing [39]. In tandem with the wealth of research emerging from the military sector, law enforcement officers are also at risk of developing MI and PTSD after taking a life.

29.5.3 Social Workers and Child Protective Services (CPS)

CPS and social workers witnessing child abuse and mistreatment, family fallouts, systemic issues, and biases within welfare, including lack of resources and unfair legal outcomes, experience shame, guilt, anger, numbness, and a loss of trust in the system [40, 41]. One-third of professionals revealed personal intent to quit their jobs as a result or knew of colleagues ready to quit [42]. Parents in the system sustain self-perpetrated moral injuries when they turn their offspring over to the system and witness and experience MI outcomes at the hands of others when they are shamed and stigmatized for their choices or learn about their children's mistreatment [41].

29.5.4 Community Educators

Levinson [43] writes that teachers in the public school system regularly face ethically problematic conditions and are forced to make morally challenging decisions with limited support and resources. As they attempt to serve as beacons for good, many educators find themselves grossly at odds with the racial and economic disparities in a system they feel is broken [44]. Such moral injury-related sentiments often yield burnout and withdrawal from the field or eventual acquiescence of and desensitization to committing ethical transgressions. A study on MI among professionals in the K-12 (kindergarten to 12th grade) education system found that most educators purported witnessing violations of or directly violating their moral codes at their jobs. As the majority reported witnessing or learning about others engaging in immoral activities instead of perpetrating them themselves, guilt among those morally injured within this population was low [45].

29.6 Moral Injury in Healthcare Workers During COVID-19

The World Health Organization has reported 3,325,260 international COVID-19 deaths as of May 13, 2021 [46]. Even as vaccine dissemination flourishes, new worrisome variants emerge. Travel bans, restrictions, quarantines, stay-at-home orders, and mask mandates attempt to enforce public health and safety, but social isolation, misinformation, and excess social media use have left many feeling distressed and confused [47]. Early studies have found elevated rates of depression and anxiety [48].

Before widespread vaccine availability, healthcare workers were expected to battle the influx of cases, shortages of beds and supplies, and risk of contraction to save the lives of those affected. In late April of 2020, the New York Times reported that Manhattan Dr. Lorna Breen had died by suicide. Dr. Breen had no notable psychiatric history. In her father's words, "she tried to do her job, but it killed her." Reminiscent of soldiers on the battlefield, he stated that "she was truly in the trenches of the front line [49]." Early outcome reviews from the COVID-19 pandemic also included reports of increased rates of depression (22%), anxiety (23%), distress (71.5%), and insomnia (38.9%) [50, 51]. It would be remiss not to include in this discussion the moral and mental health crisis facing our soldiers in scrubs today.

29.6.1 Overwhelming Public Health Challenges

Previous research explored the psychological impact on healthcare workers (HCWs). Working under stressful conditions, HCWs are at risk of suicide and high rates of anxiety, depression, and PTSD [52–54]. A representative review of the literature on burnout rates among Chinese healthcare professionals conducted in 2018 found that symptoms of burnout were present in 66.5–87.8% of the population [55]. Research also points to increased symptoms of pathology during national or global crises. One study reported that 2 months into the SARS outbreak in 2003, 20% of physicians and nurses self-reported symptoms of PTSD [56]. A questionnaire distributed to hospital staff just after the H1N1 pandemic in 2009 revealed that many grappled with feelings of stress, exhaustion, anxiety about infection and were highly distressed [57].

29.6.2 Healthcare Worker "Burnout"

This concept, previously used to describe industry-related symptoms, is known to affect professionals' productivity and perceived career satisfaction. It often results in HCWs leaving the occupation altogether [58]. Nevertheless, Talbot and Dean write that using the term "burnout" to describe doctors is a mis-conceptualization. Referring to physicians as a resourceful and resilient group, Talbot and Dean state burnout is a symptom of a more nuanced malady to which doctors fall victim: moral injury as a result of the Hippocratic Oath ("do no harm") clashing with the realities of healthcare's framework [59]. Even in non-pandemic times, doctors struggled to provide sufficient and proper care to the ill under pressure and while bombarded by financial considerations and conflicts of interests.

29.6.3 Life and Death Decisions

As cases rose and workloads increased, doctors and nurses were forced to triage resources between patients and subscribe to a hierarchical, needs-based approach as opposed to an egalitarian model of care. Examples of self-perpetrated PMIEs include taking a patient off a ventilator to transfer the machine to a new patient with a higher chance of survival or accidentally exposing a ward to COVID-19 after failing to properly screen a positive case [60]. Without clear guidance or training and with limited beds and medications, HCWs cannot provide equal or focused care to their patients, transgressing a deeply upheld occupational value [61]. Witnessing the rationing of medications as mandated by institutions or others is another relevant PMIE [60], along with enforcing rules separating families from their sick loved ones or struggling to provide adequate one-on-one attention to isolated, dying patients [62].

29.6.4 Contaminated Professionals

HCWs who test positive for COVID-19 might feel extremely guilty leaving their coworkers behind, albeit temporarily. Frontline healthcare workers may experience their health as insufficiently valued when risking ongoing exposure to the deadly virus without proper personal protective equipment (PPE) or protocols, even as the public cheers and bang pots in their honor [9]. Maguen and Price [63] write that HCWs putting themselves at risk create a ripple effect of guilt and fear as they jeopardize exposing their own family members to the disease. In prioritizing their jobs over their personal lives, HCWs may feel they are neglecting their spouses and children or wrestle with survivor's guilt as they continue to live while surrounded by death.

29.6.5 Birth Time Conflicts

Maternity nurses asked to submit examples of potentially morally injurious events sustained during this time cited enforcing rules separating laboring women from their partners or family members and separating mothers from their newborns when tested positive for COVID-19. In some countries, women were forced to have their labor induced or to have a caesarian section, doing so alone and in hospitals riddled with COVID-19 and without proper staffing. Nurses enacting these regulations might find themselves violating their standards of humane treatment, which in turn can lead to emotional numbing for self-protection, compassion fatigue, or even avoidance of patients [64]. Such PMIEs sustained by HCWs can doubtless result in moral injury [65].

In response to PMIEs, healthcare workers may experience shame, guilt, disgust, anger, and contempt. They may lose their frameworks of self-efficacy for feelings of self-betrayal and experience overall spiritual and existential crises [60, 66]. HCWs who experienced PMIEs at work report feeling animosity toward their institution of employment and/or alienation from their team and its leaders [67]. Although studies reporting quantitative data are lacking, early findings suggest that not only are healthcare workers longitudinally affected by PTSD, depression, and anxiety, but half of participants validated items on MIES scale, with participants directly exposed to COVID-19 reporting increased severity [68].

29.6.6 Remedial Interventions

Prevention efforts may include increasing currently limited access to professional mental health and spiritual support for HCWs. Educating and empowering frontline healthcare workers on the definitions and risks of PMIEs and MI as a potential outcome may also aid in catching symptoms early on and increase informal help-seeking behaviors at work, before they grow chronic or impairing [69]. Enacting "triage committees" for PPE and treatments can help distribute the weight of tough decisions cross-team so that they do not land only on the individual, who may feel solely responsible for poor outcomes [60]. In addition, a randomized controlled study conducted by Amsalem et al. [70] addressed stigma as a key cause preventing healthcare workers from seeking help. HCWs were assigned to one of three conditions: no intervention, a video intervention in which a protagonist shared her mental health struggle as an ICU nurse during COVID-19 and how receiving treatment improved her wellbeing, or the same video intervention with a follow-up, "booster" viewing of the clip. Participants' clinical symptoms and treatment-seeking intentions were assessed prior to randomization and after receiving the intervention, at follow-ups.

29.6.7 External Help

A high percentage of participants reached threshold self-report measures for anxiety, depression, and/or PTSD. Results revealed that participants in both video groups showed higher help-seeking intentions, with the "booster" condition yielding slightly stronger and longer efficacy. This study demonstrates the importance of encouraging treatment-seeking intentions among healthcare workers struggling with anxiety, depression, and PTSD as a result of the pandemic. Short, contact-based, video interventions that can be mass disseminated should be widely considered among the healthcare profession as a meaningful solution to destigmatizing and encouraging HCWs to reach out when suffering with MI.

References

1. Litz BT, Stein N, Delaney E, Lebowitz L, Nash WP, Silva C, Maguen S (2009) Moral injury and moral repair in war veterans: a preliminary model and intervention strategy. Clin Psychol Rev 29:695–706
2. Currier JM, McCormick W, Drescher KD (2015) How do morally injurious events occur? A qualitative analysis of perspectives of veterans with PTSD. Traumatology 21:106–116
3. Litz BT, Kerig PK (2019) Introduction to the special issue on moral injury: conceptual challenges, methodological issues, and clinical applications. J Trauma Stress 32:341–349
4. Shay J (2014) Moral injury. Psychoanal Psychol 31: 182–191
5. Jordan AH, Eisen E, Bolton E, Nash WP, Litz BT (2017) Distinguishing war-related PTSD resulting from perpetration- and betrayal-based morally injurious events. Psychol Trauma Theory Res Pract Policy 9:627–634
6. Nash WP, Carper TLM, Mills MA, Au T, Goldsmith A, Litz BT (2013) Psychometric evaluation of the moral injury events scale. Mil Med 178:646–652
7. Nash WP, Litz BT (2013) Moral injury: a mechanism for war-related psychological trauma in military family members. Clin Child Fam Psychol Rev 16:365–375
8. Neria Y, Pickover A (2019) Commentary on the special issue on moral injury: advances, gaps in literature, and future directions. J Trauma Stress 32:459–464
9. Williamson V, Murphy D, Phelps A, Forbes D, Greenberg N (2021) Moral injury: the effect on mental health and implications for treatment. Lancet Psychiat 8(6):453–455
10. Farnsworth JK, Drescher KD, Evans W, Walser RD (2017) A functional approach to understanding and treating military-related moral injury. J Contextual Behav Sci 6:391–397
11. Wisco BE, Marx BP, May CL, Martini B, Krystal JH, Southwick SM, Pietrzak RH (2017) Moral injury in U.S. combat veterans: results from the national health and resilience in veterans study. Depress Anxiety. https://doi.org/10.1002/da.22614
12. Levi-Belz Y, Greene T, Zerach G (2020) Associations between moral injury, PTSD clusters, and depression among Israeli veterans: a network approach. Eur J Psychotraumatol 11(1):1736411
13. Killgore WDS, Cotting DI, Thomas JL, Cox AL, McGurk D, Vo AH, Castro CA, Hoge CW (2008) Post-combat invincibility: violent combat experiences are associated with increased risk-taking propensity following deployment. J Psychiatr Res 42:1112–1121
14. Griffin BJ, Purcell N, Burkman K, Litz BT, Bryan CJ, Schmitz M, Villierme C, Walsh J, Maguen S (2019) Moral injury: an integrative review. J Trauma Stress 32:350–362
15. Currier JM, McDermott RC, Farnsworth JK, Borges LM (2019) Temporal associations between moral injury and posttraumatic stress disorder symptom clusters in military veterans. J Trauma Stress 32:382–392
16. Currier JM, Foster JD, Isaak SL (2019) Moral injury and spiritual struggles in military veterans: a latent profile analysis. J Trauma Stress 32:393–404
17. Litz BT, Contractor AA, Rhodes C et al (2018) Distinct trauma types in military service members seeking treatment for posttraumatic stress disorder. J Trauma Stress 31:286–295
18. Bryan CJ, Bryan AO, Roberge E, Leifker FR, Rozek DC (2018) Moral injury, posttraumatic stress disorder, and suicidal behavior among National Guard personnel. Psychol Trauma Theory Res Pract Policy 10:36–45
19. Maguen S, Metzler TJ, Bosch J, Marmar CR, Knight SJ, Neylan TC (2012) Killing in combat may be independently associated with suicidal ideation. Depress Anxiety 29:918–923
20. Bryan CJ (2011) The clinical utility of a brief measure of perceived burdensomeness and thwarted belongingness for the detection of suicidal military personnel. J Clin Psychol 67:981–992
21. American Psychiatric Association (2013) Diagnostic and statistical manual of mental disorders, 5th edn. https://doi.org/10.1176/appi.books.9780890425596
22. Held P, Klassen BJ, Brennan MB, Zalta AK (2018) Using prolonged exposure and cognitive processing

therapy to treat veterans with moral injury-based-PTSD: two case examples. Cogn Behav Pract 25: 377–390
23. Stein NR, Mills MA, Arditte K, Mendoza C, Borah AM, Resick PA, Litz BT (2012) A scheme for categorizing traumatic military events. Behav Modif 36:787–807
24. Barnes HA, Hurley RA, Taber KH (2019) Moral injury and PTSD: often co-occurring yet mechanistically different. J Neuropsychiatry Clin Neurosci 31: A4-103
25. Lancaster SL (2018) Negative outcomes after morally injurious experiences: a replication and extension. Psychol Trauma Theory Res Pract Policy 10:456–462
26. Gray MJ, Schorr Y, Nash W, Lebowitz L, Amidon A, Lansing A, Maglione M, Lang AJ, Litz BT (2012) Adaptive disclosure: an open trial of a novel exposure-based intervention for service members with combat-related psychological stress injuries. Behav Ther 43:407–415
27. Affairs O of P and I VA releases 2020 National Veteran Suicide Prevention annual report. va.gov/opa/pressrel/pressrelease.cfm?id=5565. Accessed 25 May 2021
28. Bryan CJ, Cukrowicz KC (2011) Associations between types of combat violence and the acquired capability for suicide. Suicide Life Threat Behav 41: 126–136
29. Rice TR, Sher L (2013) Killing in combat and suicide risk. Eur Psychiatry 28:261–261
30. Hoge CW, Castro CA, Messer SC, Mcgurk D, Cotting DI, Koffman RL (2004) Combat duty in Iraq and Afghanistan, mental health problems, and barriers to care. N Engl J Med 351:13–22
31. Yeterian JD, Berke DS, Carney JR, McIntyre-Smith A, St Cyr K, King L, Kline NK, Phelps A, Litz BT, Members of the Moral Injury Outcomes Project Consortium (2019) Defining and measuring moral injury: rationale, design, and preliminary findings from the Moral Injury Outcome Scale Consortium. J Trauma Stress 32:363–372
32. Maguen S, Metzler TJ, Litz BT, Seal KH, Knight SJ, Marmar CR (2009) The impact of killing in war on mental health symptoms and related functioning. J Trauma Stress 22:435–443
33. Purcell N, Koenig CJ, Bosch J, Maguen S (2016) Veterans' perspectives on the psychosocial impact of killing in war. Couns Psychol 44:1062–1099
34. Stotz SJ, Elbert T, Müller V, Schauer M (2015) The relationship between trauma, shame, and guilt: findings from a community-based study of refugee minors in Germany. Eur J Psychotraumatol 6:25863
35. Hoffman J, Liddell B, Bryant RA, Nickerson A (2018) The relationship between moral injury appraisals, trauma exposure, and mental health in refugees. Depress Anxiety 35:1030–1039
36. Feinstein A, Pavisian B, Storm H (2018) Journalists covering the refugee and migration crisis are affected by moral injury not PTSD. JRSM Open 9: 2054270418759010
37. Weiss DS, Brunet A, Best SR, Metzler TJ, Liberman A, Pole N, Fagan JA, Marmar CR (2010) Frequency and severity approaches to indexing exposure to trauma: the critical incident history questionnaire for police officers. J Trauma Stress 23:734–743
38. Komarovskaya I, Maguen S, McCaslin SE, Metzler TJ, Madan A, Brown AD, Galatzer-Levy IR, Henn-Haase C, Marmar CR (2011) The impact of killing and injuring others on mental health symptoms among police officers. J Psychiatr Res 45:1332–1336
39. Papazoglou K, Blumberg DM, Chiongbian VB, Tuttle BM, Kamkar K, Chopko B, Milliard B, Aukhojee P, Koskelainen M (2020) The role of moral injury in PTSD among law enforcement officers: a brief report. Front Psychol 11:310
40. Haight W (2017) Moral injury among Child Protection Professionals – implications for the ethical treatment and retention of workers. Child Youth Serv Rev 82: 27–41
41. Haight W, Sugrue E, Calhoun M, Black J (2017) Everyday coping with moral injury: the perspectives of professionals and parents involved with child protection services. Child Youth Serv Rev 82:108–121
42. Haight W, Sugrue E, Calhoun M, Black J (2017) "Basically, I look at it like combat": reflections on moral injury by parents involved with child protection services. Child Youth Serv Rev 82:477–489
43. Levinson M (2015) Moral injury and the ethics of educational injustice. Harv Educ Rev 85:203–228
44. Keefe-Perry LC (2018) Called into crucible: vocation and moral injury in U.S. public school teachers. Relig Educ 113:489–500
45. Sugrue EP (2020) Moral injury among professionals in K–12 education. Am Educ Res J 57:43–68
46. WHO Coronavirus (COVID-19) dashboard. covid19.who.int. Accessed 13 May 2021
47. Amsalem D, Dixon LB, Neria Y (2021) The coronavirus disease 2019 (COVID-19) outbreak and mental health: current risks and recommended actions. JAMA Psychiat 78:9
48. Gao J, Zheng P, Jia Y, Chen H, Mao Y, Chen S, Wang Y, Fu H, Dai J (2020) Mental health problems and social media exposure during COVID-19 outbreak. PLoS One 15:e0231924
49. Watkins A, Rothfeld M, Rashbaum WK, Rosenthal BM (2020) Top E.R. doctor who treated virus patients dies by suicide. nytimes.com/2020/04/27/nyregion/new-york-city-doctor-suicide-coronavirus.html. Accessed 13 May 2021
50. Tan BYQ, Chew NWS, Lee GKH et al (2020) Psychological impact of the COVID-19 pandemic on health care workers in Singapore. Ann Intern Med 173(4): 317–320
51. Pappa S, Ntella V, Giannakas T, Giannakoulis VG, Papoutsi E, Katsaounou P (2020) Prevalence of depression, anxiety, and insomnia among healthcare workers during the COVID-19 pandemic: a systematic

review and meta-analysis. Brain Behav Immun 88: 901–907
52. Smith GB, Hukill E (1996) Nurses impaired by emotional and psychological dysfunction. J Am Psychiatr Nurses Assoc 2:192–200
53. Brooks SK, Chalder T, Gerada C (2011) Doctors vulnerable to psychological distress and addictions: treatment from the Practitioner Health Programme. J Ment Health Abingdon Engl 20:157–164
54. Sanchez M, Simon A, Ford D (2019) PTSD in Tx ICU nurses. J Heart Lung Transplant 38:S93–S94
55. Lo D, Wu F, Chan M, Chu R, Li D (2018) A systematic review of burnout among doctors in China: a cultural perspective. Asia Pac Fam Med 17:3
56. Chan AOM, Huak CY (2004) Psychological impact of the 2003 severe acute respiratory syndrome outbreak on health care workers in a medium size regional general hospital in Singapore. Occup Med 54:190–196
57. Matsuishi K, Kawazoe A, Imai H et al (2012) Psychological impact of the pandemic (H1N1) 2009 on general hospital workers in Kobe. Psychiatry Clin Neurosci 66:353–360
58. Maunder RG, Lancee WJ, Balderson KE et al (2006) Long-term psychological and occupational effects of providing hospital healthcare during SARS outbreak. Emerg Infect Dis 12:1924–1932
59. Dean W, Talbot S, Dean A (2019) Reframing clinician distress: moral injury not burnout. Fed Pract 36:400–402
60. Borges LM, Barnes SM, Farnsworth JK, Bahraini NH, Brenner LA (2020) A commentary on moral injury among health care providers during the COVID-19 pandemic. Psychol Trauma Theory Res Pract Policy 12:S138–S140
61. Shortland N, McGarry P, Merizalde J (2020) Moral medical decision-making: colliding sacred values in response to COVID-19 pandemic. Psychol Trauma Theory Res Pract Policy 12:S128–S130
62. Young KP, Kolcz DL, O'Sullivan DM, Ferrand J, Fried J, Robinson K (2020) Health care workers' mental health and quality of life during COVID-19: results from a mid-pandemic, national survey. Psychiatr Serv 72:122–128
63. Maguen S, Price MA (2020) Moral injury in the wake of coronavirus: attending to the psychological impact of the pandemic. Psychol Trauma Theory Res Pract Policy 12:S131–S132
64. Horsch A, Lalor J, Downe S (2020) Moral and mental health challenges faced by maternity staff during the COVID-19 pandemic. Psychol Trauma Theory Res Pract Policy 12:S141–S142
65. Greenberg N, Docherty M, Gnanapragasam S, Wessely S (2020) Managing mental health challenges faced by healthcare workers during covid-19 pandemic. BMJ 368:m1211
66. Stovall M, Hansen L, van Ryn M (2020) A critical review: moral injury in nurses in the aftermath of a patient safety incident. J Nurs Scholarsh Off Publ Sigma Theta Tau Int Honor Soc Nurs 52:320–328
67. Gibbons SW, Shafer M, Hickling EJ, Ramsey G (2013) How do deployed health care providers experience moral injury? Narrat Inq Bioeth 3:247–259
68. Amsalem D, Lazarov A, Markowitz JC, Naiman A, Thomas ES, Lisa BD, Neria Y (2021) Psychiatric symptoms and moral injury among US healthcare workers in the COVID-19 era. BMC Psychiatry:1–8
69. Williamson V, Murphy D, Greenberg N (2020) COVID-19 and experiences of moral injury in front-line key workers. Occup Med 70:317–319
70. Amsalem D, Lazarov A, Markowitz JC, Smith TE, Dixon LB, Neria Y (2021) Video intervention to increase treatment-seeking by healthcare workers during the COVID-19 pandemic: randomised controlled trial. Br J Psychiatry:1–7

Is Multi-level Marketing of Nutrition Supplements Questionable? 30

Diana Cardenas and Vanessa Fuchs-Tarlovsky

Abstract

Multi-level marketing (MLM) of dietary supplements and other nutrition products that have nutritional value or promoted as remedies may be unnecessary and intended for conditions that are unsuitable for self-prescription. The respect of the principles of beneficence, nonmaleficence and autonomy is at stake. The general public should avoid nutrition and health-related multi-level products, while government agencies should become involved as well. Clinical nutrition scientific societies should define policies and recommendations on how, by whom and when dietary supplements should be prescribed.

30.1 Introduction

Dietary supplement (DS) demand for both adults and children has increased in recent decades [1–4]. DS can encompass vitamins, minerals, herbs or other botanicals and amino acids [5]. Some of the micronutrients can be useful when properly prescribed; however most are consumed worldwide despite unproven efficacy [6–16]. Harm is also a possibility depending on dosage or intrinsic toxicity [17–20].

Advertising emphasizes health, wellness, sport endurance, fertility, slower aging, prevention of degenerative diseases like cancer and other desirable outcomes [21]. Personalized nutrition is another reason for tailor-made supplementation [3]. The COVID-19 pandemic and the interest in boosting immunity have given a non-evidence-based place to DS in the treatment and prevention of the disease [22]. The market was estimated in 2019 as USD180.5 billion worldwide with 119.9 million independent commercial representatives [23]. For the USA and Europe, a market size of USD48.22 billion and 14.95 billion was announced respectively in 2019 and projected to more than double by 2027 [3, 24].

30.2 Multi-level Marketing (MLM)

In this form of network marketing, independent distributors sell products either in their customers' home, by telephone or through the Internet [25]. In the 'Wellness' category (36% of sales) [23], dietary supplements, weight management and sports nutrition products are the most relevant. Distributors (or sellers) are rewarded not

D. Cardenas (✉)
Cancer Center Gustave Roussy, París, France

V. Fuchs-Tarlovsky
Department of Clinical Nutrition, Hospital General de Mexico, Mexico City, Mexico

J. Faintuch, S. Faintuch (eds.), *Integrity of Scientific Research*,
https://doi.org/10.1007/978-3-030-99680-2_30

just for the sales they generate but also by those of the people they recruit.

30.2.1 Health Advisors

Whoever signs an agreement becomes a 'health advisor', independently of a degree in the health or nutrition field. There are no further requirements.

30.2.2 Legal and Ethical Questions

MLM is not wrong from a legal point of view. However there could be doubts concerning beneficence, nonmaleficence and autonomy. Economic scam and potential harm are not the only risks. The sale of dietary and nutrition supplements by physicians, dieticians and other healthcare professionals presents a conflict of interest and is not endorsed by scientific associations.

30.3 Major Corporations

Amway (Ada Township, Michigan, USA), also known as Nutrilite, was the first MLM business (USA, 1934). In the USA it is regulated by the Federal Trade Commission (FTC) [26] and in Europe by the Directive 2005/29/e of the European Council [27].

30.4 The Danger of Pyramid Schemes

In pyramid scams, somewhat similar to Ponzi schemes, main profit doesn't come from the sale of the product but by the creation of new distributorships. When too many franchises are sold, no more potential investors remain, and the pyramid falls apart [28]. Vemma Nutrition Company (Tempe, AZ, USA) operated a pyramid scheme disguised as an MLM business. It was shut down by the FTC in 2105 and ordered to pay millions of dollars of compensations. Eventually the company reopened [27, 29] (Fig. 30.1).

Also Herbalife (Los Angeles, CA, USA) and USANA (West Valley City, UT, USA) have been challenged. In 2011 a Belgian court found Herbalife to be a pyramid scheme [30], whereas in the USA the FTC determined in 2016 that it is just a 'problematic' scheme. Herbalife needed to change their model and partially refund its nearly 350,000 Herbalife distributors [26].

30.4.1 No Company Employees

Distributors purchase an inventory of their products to be able to get started or even a mandatory monthly stock [25] ('inventory-loading' or 'front-loading'). Thus the company pays no salaries and has much less legal responsibilities, which are transferred to the representatives.

30.5 Multi-level vs. Pyramid Models

MLM compensates their salespeople for selling products or services to ultimate consumers and not simply for recruiting people. However, this difference is blurred and difficult to prove [31, 32].

30.6 Compulsory Purchasing

Some of the MLM nutrition companies like Herbalife and USANA have a clear 'front-loading' demand [33]. Distributors can make money by selling products or by recruiting other salesmen and getting a share of their revenue. The company will continuously increase its sales and overall profits, but the distributor can incur heavy losses. According to the FTC, 99% of distributors in an MLM business lose money [26, 32].

30.6.1 Health Claims

DS are officially defined as foods, not drugs. They are not intended to diagnose, treat, cure or prevent any disease and are not allowed to

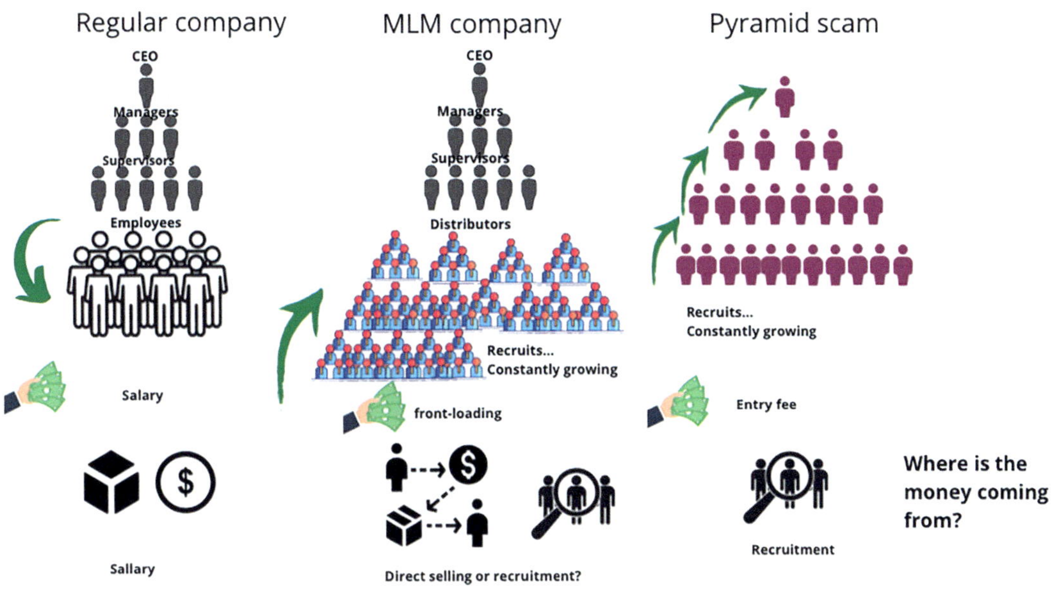

Fig. 30.1 Business scheme of a regular company, a multi-level marketing (MLM) company and a pyramid scam. Distributors can earn money by getting a salary, actual selling of products to customers and recruitment of distributors. In the MLM scheme, the revenue can be the result of the actual selling of products to customers or distributors buying products from the MLM company so they can remain distributors

publish related claims [34]. However distributors, who are not legal representatives of the company, are able to and even encouraged to make health claims, including interviews with consumers.

30.6.2 Company and Product Profiles

Among the top 20 MLM companies by global revenue in 2020, 9 are devoted to selling 'wellness' products (Table 30.1) [34–38]. Herbalife and USANA are specialized in dietary supplements and weight control items. On their websites, these companies claim a wide range of benefits to nutritional status and health.

30.7 Product Misrepresentation

In 2016, the organization Truth in Advertising (Madison, CT, USA) found that the majority of companies directly selling nutrition supplements (97%) had illegal claims, either directly or through their distributors [39].

30.8 Involved Healthcare Professionals

MLM organizations are rather traditional in the Western culture and are not commonly criticized by doctors, dietitians and other healthcare professionals or associations. Nevertheless the efficacy of dietary supplements to treat or control diseases other than those caused by nutrient deficiency is scarce and frequently lacks solid clinical trials [40, 41].

30.9 Beneficence and Professional Integrity

MLM companies instruct distributors to collect and publicize positive experiences from customers [42–44]. The main organization is legally not involved, and exaggerated or fake claims could thus emerge. Despite all risks and inconveniences, there is no lack of distributors, who dream with a profitable activity involving

Table 30.1 Top multi-level international marketing companies in 2020 in the dietary/wellness supplement domain

Company	2019 revenue	Supplements
Amway	$8.4B	DS Weight management
Herbalife Nutrition	$14.9B	DS Weight management
Nu Skin	$2.40B	DS
Pola	$1.24B	Health food
PM International	$1.11B	DS
Jeunesse	$1.1B	DS
USANA	$1.06B	DS Weight management
Medifast/OPTAVIA	$714M	DS Weight management
Arbonne	$672M	DS Weight management

DS dietary supplements, *B* billion US dollars, *M* Million US dollars

miraculous cures and occasionally endorsed by celebrities [32]. In Ghana, Western Africa, MLM distributors typically enjoy prestige and significant profit [45].

Autonomous purchase and consumption of supplements should entail truthful and complete information to the public [40], by knowledgeable professionals or under their guidance. This very seldom happens.

30.10 Nonmaleficence

Adulterated supplements appear from time to time in the market, and government agencies are rarely prepared to routinely monitor and check them in a timely and efficient form [36, 37]. Legitimate preparations may also trigger side effects, associated with as many as 23,000 emergency visits per year in the USA [46].

30.10.1 Healthcare Professionals Advertising Supplements

Professionals uncommonly sell products [47–49], yet they may recommend them, for direct or indirect profit [30]. Such suggestions do not necessarily follow evidence-based clinical data and could be ethically questionable [42].

References

1. Bright-See E (1983) Doctor, should I take vitamins? Can Fam Physician 29:2358–2367
2. Obbagy JE, English LK, Psota TL, Wong YP, Butte NF, Dewey KG et al (2019) Complementary feeding and micronutrient status: a systematic review. Am J Clin Nutr 109(Suppl_7):852s–871s
3. Fortune Business Insights (2021) North America dietary supplement market size 2016–2027. fortunebusinessinsights.com/dietary-supplements-market-102082. Accessed 10 Apr 2021
4. Dwyer JT, Coates PM, Smith MJ (2018) Dietary supplements: regulatory challenges and research resources. Nutrients 10(1)
5. Office of Dietary Supplements. Dietary Supplement Health and Education Act of 1994. ods.od.nih.gov/About/DSHEA_Wording.aspx. Accessed 10 Apr 2021
6. Cao J, Gao Z, Yan J, Li M, Su J, Xu J et al (2016) Evaluation of trace elements and their relationship with growth and development of young children. Biol Trace Elem Res 171(2):270–274
7. Maggini S, Pierre A, Calder PC (2018) Immune function and micronutrient requirements change over the life course. Nutrients 10(10)
8. García-Cortés M, Robles-Díaz M, Ortega-Alonso A, Medina-Caliz I, Andrade RJ (2016) Hepatotoxicity by dietary supplements: a tabular listing and clinical characteristics. Int J Mol Sci 17(4):537
9. Brown B, Wright C (2020) Safety and efficacy of supplements in pregnancy. Nutr Rev 78(10):813–826
10. Peeling P, Castell LM, Derave W, de Hon O, Burke LM (2019) Sports foods and dietary supplements for optimal function and performance enhancement in track-and-field athletes. Int J Sport Nutr Exerc Metab 29(2):198–209

11. Katta R, Huang S (2019) Skin, hair and nail supplements: an evidence-based approach. Skin Therapy Lett 24(5):7–13
12. Bach HV, Kim J, Myung SK, Cho YA (2016) Efficacy of ginseng supplements on fatigue and physical performance: a meta-analysis. J Korean Med Sci 31(12): 1879–1886
13. Dwyer JT, Coates PM (2018) Why Americans need information on dietary supplements. J Nutr 148 (Suppl_2):1401s–1405s
14. Garolla A, Petre GC, Francini-Pesenti F, De Toni L, Vitagliano A, Di Nisio A et al (2020) Dietary supplements for male infertility: a critical evaluation of their composition. Nutrients 12(5)
15. Bailey RL (2020) Current regulatory guidelines and resources to support research of dietary supplements in the United States. Crit Rev Food Sci Nutr 60(2): 298–309
16. Kilchoer B, Vils A, Minder B, Muka T, Glisic M, Bally L (2020) Efficacy of dietary supplements to reduce liver fat. Nutrients 12(8)
17. (2012) Dietary supplements: not for everyone. New findings indicate supplements may actually harm some people. Getting nutrients from real food is best. Duke Med Health News 18(1):4–5
18. (2012) Vitamins and supplements: 10 dangers that may surprise you. Consum Rep 77(9):18–23
19. Kmietowicz Z (2003) Food watchdog warns against high doses of vitamins and minerals. BMJ 326(7397): 1001
20. Incze M (2019) Vitamins and nutritional supplements: what do i need to know? JAMA Intern Med 179(3): 460
21. Marian MJ (2017) Dietary supplements commonly used by cancer survivors: are there any benefits? Nutr Clin Pract 32(5):607–627
22. Adams KK, Baker WL, Sobieraj DM (2020) Myth busters: dietary supplements and COVID-19. Ann Pharmacother 54(8):820–826
23. World Federation Statistical Database. wfdsa.org/wp-content/uploads/2020/07/Fact-Sheet-page-1-2019.pdf. Accessed 4 Apr 2021
24. Fortune Business Insights. Europe dietary supplement market size 2016–2027. fortunebusinessinsights.com/industry-reports/europe-dietary-supplements-market-101918. Accessed 4 Apr 2021
25. Muncy JA (2004) Ethical issues in multi-level marketing: is it a legitimate business or just another pyramid scheme? Mark Educ Rev 14:47–53
26. Federal Trade Commission. Multi-level marketing businesses and pyramid schemes. consumer.ftc.gov/articles/0065-multi-level-marketing-businesses-and-pyramid-schemes. Accessed 4 Apr 2021
27. Unfair Commercial Practices Directive. Directive 2005/29/EC of the European Parliament and of the Council of 11 May 2005. ec.europa.eu/info/law/law-topic/consumer-protection-law/unfair-commercial-practices-law/unfair-commercial-practices-directive_en. Accessed 4 Apr 2021
28. FBI. Common scams and crimes. Pyramid schemes. fbi.gov/scams-and-safety/common-scams-and-crimes. Accessed 4 May 2021
29. Vander Nat PJ, Keep WW (2002) Marketing fraud: an approach for differentiating multi-level marketing from pyramid schemes. J Pub Policy Market 21:139–151
30. Test-Aankoop V (2011) Herbalife International Belgium [Comm. Ct. Brussels], Nov. 23, AR 2004/7787, No. JC-DC/27. factsaboutherbalife.com/category/belgium-decision-2004-2013/. Accessed 4 Apr 2021
31. Cardenas D, Fuchs-Tarlovsky V (2018) Is multi-level marketing of nutrition supplements a legal and an ethical practice? Clin Nutr ESPEN 25:133–138
32. FitzPatrick RL, Scheibeler E (2011) What about this one? How to tell the difference between legitimate direct selling and pyramid recruitment schemes. PMA. pyramidschemealert.org/what-about-this-one-free-audio-report/. Accessed 4 Apr 2021
33. USANA watchdog. usanawatchdog.blogspot.com/. Accessed 4 Apr 2021
34. Food Dietary Supplements (2021) fda.gov/Food/DietarySupplements/UsingDietarySupplements/ucm480069.htm. Accessed 4 Apr 2021
35. Official Journal of the European Communities. Directive 2002/46/EC of the European Parliament and of the Council of 10 June 2002. eur-lex.europa.eu/legal-content/EN/TXT/PDF/?uri=CELEX:32002L0046. Accessed 4 Apr 2021
36. Sax JK (2015) Dietary supplements are not all safe and not all food: how the low cost of dietary supplements preys on the consumer. Am J Law Med 41(2–3): 374–394
37. Brown AC (2017) An overview of herb and dietary supplement efficacy, safety and government regulations in the United States with suggested improvements. Part 1 of 5 series. Food Chem Toxicol 107(Pt A):449–471
38. Direct Selling News (2020) Global list. directsellingnews.com/dsn-announces-2020-global-100-list/. Accessed 4 Apr 2021
39. Truth in advertising. truthinadvertising.org/mlm-health-claims-database/. Accessed 4 Apr 2021
40. Ricoeur P (2004) Éthique, Dictionnaire d'éthique et de philosophie morale (Ethics, dictionary of ethics and of moral philosophy). In: Canto-Sperber M (ed). Paris, France PUF, p 689
41. USANA (2017) usanahealth.net/en_US/essentials/VitaAntioxidant.html. Accessed 4 Apr 2021
42. Koehn D (2001) Ethical issues concerned with multi-level marketing schemes. J Bus Ethics 28:153–160
43. Groß C, Vriens D (2019) The role of the distributor network in the persistence of legal and ethical problems of multi-level marketing companies. J Bus Ethics 156:333–355
44. Biggart N (2008) Charismatic capitalism. Direct selling organizations in America. The University of Chicago Press, Chicago

45. Droney D (2016) Networking health: multi-level marketing of health products in Ghana. Anthropol Med 23(1):1–13
46. Geller AI, Shehab N, Weidle NJ, Lovegrove MC, Wolpert BJ, Timbo BB et al (2015) Emergency department visits for adverse events related to dietary supplements. N Engl J Med 373(16):1531–1540
47. Whitaker-Worth D, Shahriari M, Slade K, Grant-Kels JM (2012) The ethical controversies of office-based dispensing in academic health centers. Clin Dermatol 30(5):528–532
48. American Medical Association (1998) Should doctors sell non-health-related products to their patients? The AMA targets Amway. Health Care Business Digest 3: 22–24
49. Dumoff AJD (2000) Medical boards' prohibitions against physician supplement sales. Alt Compl Ther 6:226–232

31 Euthanasia Not On Trial: The Career of an Austrian Nazi Doctor After 1945

Sonja Schreiner

Abstract

Heinrich Gross (1915–2005), the Austrian Josef Mengele [1, 2], is notorious for his unethical behaviour and medical crimes in Central Europe. During WW II, he worked as a physician at *Am Spiegelgrund* in Vienna, a hospital comprising a *Kinderfachabteilung* (Children's Ward), where he and his colleagues systematically killed, tortured, and traumatized many disabled or handicapped children but also girls, boys, and teenagers stigmatized as hard to educate or "a(nti)-social". After 1945, a powerful network of politicians protected Gross, who published dozens of papers based on the brains surgically removed from his victims and conserved for later research and made career as a renowned, often consulted, and highly paid court expert determining the fate of countless victims until 1998. The chapter will highlight Gross's life-long and ongoing misconduct, and his clever adaptation to new times and circumstances, as well as the fraud committed when at old age he claimed to suffer from dementia and depression, allowing him to stay a free man. At the same time, he became a metaphor of evil, analysed in scientific publications and depicted in film, documentaries, and autobiographic literature, powerfully replacing court testimonies in a trial that never took place.

31.1 Introduction and Background

When I was in my 20s, Heinrich Gross had to face his second trial as defendant, after a long period, when judges had consulted him as a court expert, a decisive job that made him earn several millions of Austrian schillings (which equates some 100,000 euros). Gross's first charge for involvement in Nazi crimes dated back to 1950. His acquittal followed the sentence's revoke in 1951. In 1979, physician Werner Vogt (*1938) publicly accused him of having killed, tortured, and abused a tremendous number of children at *Am Spiegelgrund*, which ended up in Vogt as a defendant because of defamation. In the late 1990s and early 2000s, new evidence from German and Russian archives showed the ugly truth about Gross's activities during WW II. However as time had gone by, the Nazi doctor seemed to have become a weak man, a development that finished the trial, before it had begun.

S. Schreiner (✉)
University of Vienna, Wien, Austria

University of Veterinary Medicine Vienna, Wien, Austria
e-mail: sonja.schreiner@univie.ac.at; sonja.schreiner@vetmeduni.ac.at

J. Faintuch, S. Faintuch (eds.), *Integrity of Scientific Research*,
https://doi.org/10.1007/978-3-030-99680-2_31

31.1.1 A Chilling Personal Experience

A few years earlier, one of my fellow students and one of Gross's grandsons [3] were close friends, but my acquaintance did not know about the old man's past. Only years after that awkward encounter, when the trial was pending, she told me that one afternoon her friend's grandfather had entered the room and suddenly everything had become cold. She learnt to know what she had instinctively felt before. This was not a dramatic flashback of a young woman, but the experience described by one of Gross's surviving victims [4].

I do lively remember our collective disbelief, when a renowned court expert finally declared the Nazi villain unfit to stand trial because of senile dementia. On the same day, Gross gave a well-chosen interview broadcasted on Austrian television. As one of Derek Chauvin's prosecutors and bystanders of George Floyd's killing expressed it in April 2021 in remarkable words during the trial in Minneapolis, Heinrich Gross had "murdered in slow motion", but had escaped justice once more. Some of Gross's surviving victims (Friedrich Zawrel, Alois Kaufmann, and Johann Gross) published their memories, went into schools, spoke to the media, and were guests of honour, when documentaries on "euthanasia" [5, 6] or Elisabeth Scharang's movie *Mein Mörder* (2005) were presented to the public [7].

31.1.2 Academic Focus

Over the last two decades, a huge number of diploma theses were written on Gross, his colleagues (e.g. [8, 9]), and the trial [10]. Due to their status as so-called grey literature, a broader public does not know about these important works, although many of them are accessible online as *e-theses* (https://utheses.univie.ac.at/). Articles and books were published [11–15, 16], and even a novel on the children of *Am Spiegelgrund* appeared [17]. Analyses of newspaper articles and TV broadcasts followed [18]. Most of these scholars are familiar with the touching reports of the survivors [19, 20], but not with the poems [21]. Especially Kaufmann's modern and experimental lyrics demonstrate that contrary to Theodor Adorno's famous *dictum* that "to write poetry after Auschwitz is barbaric", that was quite timely and appropriate ([22]:10).

31.2 CV of a Villain

When Heinrich Gross joined the Nazi youth organization *HJ* (*Hitlerjugend*) in 1932, the Nazi party was forbidden in Austria. This would change in March 1938, after the so-called *Anschluss* (Annexation), when Austria became a part of the German Reich. In the years between, many Austrian citizens became illegal party members. So did Gross. In 1933, he entered the *SA* (*Sturmabteilung*), at this time an influential, elite Nazi group. In 1939, Gross graduated as doctor, and 1 year later, in a care unit in Ybbs, Lower Austria, he was already involved in a T4-transport, part of the Nazi Euthanasia Program to deport and kill mentally handicapped people.

He subsequently moved to *Am Spiegelgrund*, together with other physicians. In 1941 a pediatric ward had been opened where infants and children formally were treated, but informally "euthanized". The doctors betrayed the mourning parents by telling them that their sons and daughters had died of pneumonia—in most cases the final cause of death, but induced by overdoses of rectally administered phenobarbital. In advance, Gross (or one of his colleagues) informed Berlin about the predictable future of the little patients. When the prognosis was a negative one, they got the licence to kill. In addition to the use of barbiturates, nurses had to poison the children's food. If one of the victims did not die, he or she got a lethal injection causing a struggle for life that took many days. Gross even increased his deadly knowledge making visits to other institutions in the German Reich, thereby getting advanced training.

31.2.1 Illegal Neurological Biorepository

Brains and spines were taken from the little bodies and stored for further research. In 1942, Gross served in the German Army for less than 6 weeks, because his medical "duty" was classified as more important. By then he and his colleagues started extremely painful medical experiments, e.g. pneumoencephalography. Many children died, the brains were dissected, and Gross had a talk on a case study at the Vienna Biological Society.

31.2.2 Post-War Activities

In 1947, after being recruited again by the German Army and subsequently serving 2 years as a Russian prisoner of war (POW), he came back to Vienna and should testify against Anna Katschenka, who had worked with him as a nurse, and was accused of second-degree murder against the institutionalized children. Gross fled but was arrested in 1948 and charged of assisting in the killings in 1950. Even after the sentence, he had not gone to jail because he had already been in prison for 2 years under the Russians.

31.2.3 Neuropsychiatric Specialist and Prestigious Investigator

Rehabilitated through an influential political network of social democrats [23], in 1955 he became a neuropsychiatric specialist and returned to *Otto Wagner* Hospital, the former *Am Spiegelgrund*. He continued publishing on the children's brains, comprising nearly 600 specimens, and cooperated with renowned medical doctors. Nobody mentioned where the brains came from.

From 1960 on Gross became court expert and submitted 12,000 expertises until 1978. From 1958 to 1968, he also tested new pharmaceuticals on human beings, trained judges, and made profit from the reforms initiated by the minister of justice, Christian Broda, upgrading the role of psychiatrists in court. From 1968 on he was the director of the new and purpose-organized Ludwig Boltzmann Neurological Research Institute.

31.2.4 Journal Editor and Prize-Winning Scientist

Gross eventually moved his focus to questions of forensic psychiatry, founding a scientific society (*Arbeitsgemeinschaft für forensische Psychiatrie und Neurologie*) and a journal (*Forensia*). In 1975, he was awarded the first class medal *Ehrenkreuz für Wissenschaft und Kunst*, which was taken away from him in 2003.

31.2.5 Time for Reckoning

In 1975, as a court expert, Gross examined the former *Spiegelgrund* inmate Friedrich Zawrel, who with growing disbelief had followed his torturer's career in the newspapers. Gross did not recognize his victim. When Zawrel confronted him, Gross tried to shut him down forever using parts of a colleague's expertise from 1944. Zawrel contacted the press as well as Dr. Werner Vogt, the expert on the *Spiegelgrund* affair who in the most positive sense was known as rebellious intellectual in Austria. Vogt was eventually honoured by the Human Rights Prize of the Austrian League in 2019 [24].

31.2.6 Unpunished Malfeasance

Unfortunately enough Zawrel was hardly successful, and even Vogt had to face a trial in 1979 on different however related grounds and was sentenced. In a new trial (with a new judge), he was acquitted. Gross was not charged. The same happened in 1989, when new documents were found. Wolfgang Neugebauer, head of the *DÖW* (Archives of the Austrian Resistance), accused Gross of second-degree murder, yet prosecution could not proceed after so many decades.

In 1996, Matthias Dahl exposed in his doctoral thesis in Göttingen, published as a book [25] and accompanied by an article [26], all documents still in the morgue of the former *Spiegelgrund* Hospital. Neugebauer accused Gross again. As the prosecutors did not want to act, American TV reacted, followed by *The Sunday Times*. In 1997, the affair was discussed in the Austrian Parliament, and the pre-work for the murder trial was started. Viennese politician Sepp Rieder called Gross a murderer in public, and independent court experts declared him as able to go to the preparing interviews in 1998. The judge took hold of all materials including the brains that had been falsely declared as buried in 1989. The evidence helped to prove nine murders committed in 1944; however, the trial was delayed several times. In 2005, Heinrich Gross was accused for the very last time, based on new material from Moscow; nevertheless, he died in the same year as a free and, according to Austrian law, not guilty man.

31.3 The Survivors' Literary Testimonies

Autobiographies written by survivors and interviews are important and were presented on various platforms dedicated to lasting remembrance [27, 28]. The impressive texts mostly have co-authors, as can be seen from the comparison of the written style and the oral tradition. As described by Banerjee [29], such texts are relevant as literature, as autobiography, as historical, and especially as juridical testimonies, as they to some extent replace the appearance of witnesses before court, in a trial that would have been so necessary and healing, however never took place.

31.3.1 The Odyssey of Kaufmann

Alois Kaufmann made the start. He first published his memories in 1993. He begins in poetic language with a sleepless night, many years after the tortures ([20]:7). Writing is his therapy to overcome his traumata ([20]:9).

Alois begins to understand what really happens when one of his mates disappears. On July 20, 1944, his Jewish friend Herbert Reisenbaum throws a chair against Hitler's portrait after having listened to the radio informing about the "Führer's" failed assassination. Alois—the narrator is identical with his *persona*—does not understand why the other boys who are victims themselves hate and punish Herbert. When he wants to help, he is knocked out and gets unexpected treatment ([20]:25–26). One allegedly kind doctor distributing candies and asking fatal questions is Heinrich Gross. Kaufmann shows the ambivalent sides of a sadistic man who on a daily and sociopathic basis perfects his manipulating skills.

In the book chapter *Der Oberarzt und der "Judenbub"* (The Chief Physician and the Jewish Boy), we learn more about the fate of Herbert's parents deadly manipulated by Heinrich Gross, as we read their last letter to their beloved son, written immediately before their deportation ([20]:34–36). The passage is a telling example, successfully combining Alois's knowledge in the 1940s and information he got years or even decades after the events took place. As in a jigsaw puzzle, the psychological pieces fall into the right place and help to outline the horrendous psychogram of an evil however skilled psychiatrist, who is optimizing every single situation of his life for a positive outcome, not hesitating to literally walk over dead bodies.

It turns out that the Reisenbaum family had loaned money to Gross, when he was a medical student. Later on Herbert's mother asked him to protect her son. He promised to help because he would be able to boast the saving of a Jew. The boy does not touch him emotionally; he merely (ab)uses him ([20]:37). The chapter *Die Stunde Null* (The Zero Hour) describes Gross's sorrows ([20]:76). In despair he even tries to manipulate Alois and draw him on his side ([20]:77–78). As so many, Alois Kaufmann did not tell the world for a long time. He felt ashamed; he tried to forget.

31.3.2 The Pains of Zawrel

So did Friedrich Zawrel, whose memories [30] are perfectly well documented [31]. Zawrel's narration shows how Gross in the strong position of a court expert tries to silence the former victim and abuses him again. Manipulation of others and bettering his own conditions were his quite successful methods until this turning point of life. In the section *Epilog* (Epilogue), the most personal part of the book by Häupl [15], Friedrich Zawrel has finally found some peace and solace, expressed by the ravens and crows he had so often admired as mourning however free birds, by the time he was caught in the "cage" of *Spiegelgrund* ([32]: 645).

31.3.3 The "Education" of Johann Gross

Also Johann Gross published a book [19], with a foreword by the Austrian author Christine Nöstlinger and a postscript by Wolfgang Neugebauer. The book describes a difficult childhood ending up in the Nazi education system. One chapter, *Eis und Nadeln in den Schenkeln* (Ice and Needles in the Thighs), shows the horror of the daily routine the children were exposed to by sadistic doctors and nurses who did not care, but tortured and killed. Heinrich Gross was not a helper or a healer, but a soldier and warrior. He did not restore, but destroyed everything his Hippocratic oath should bind him to.

31.3.4 It Is Not Theoretical, It Is Human Beings

Friedrich Zawrel, Johann Gross, and Alois Kaufmann were healthy boys when they came to the *Spiegelgrund*. They had no handicap; they simply were registered as "difficult" children [33]. They had a certain age, which gave them a chance to survive, an opportunity for resilience. Little babies, born with disabilities—mental and/or physical—had not. One of these helpless little beings was a boy, whose fate was carefully reconstructed by Dahl, although he confused the dates ([34]:70), tacitly corrected by Czech ([11]:31).

31.3.5 Unworthy to Live

Adolf Guttmann was born on June 12, 1941, as a *prima vista* healthy child to a poor, but respectable family in a small village called Rastenfeld in Northern Lower Austria. My fiancé, the historian Franz Bauer, comes from this village and is only a few years younger than little Adolf was. He was familiar with the baby boy's elder brother Otto Guttmann, who died a few years ago, and with their mother, who died at a very high age as war widow, buried in a grave showing only the family name *Guttmann*, but no first names—an (in)-visible sign of the trauma the little family suffered from.

31.3.6 Anatomy of a Murderous Clinical Trial

Baby Adolf developed hydrocephalus and was taken away from his mother following the Nazi instructions concerning children showing evident disabilities. Adolf had no chance. Until today, the people of Rastenfeld say: "The Nazis did not give back the boy to his mum". They brought him to a hospital where he got no help or care but was further transferred to Vienna. His first housing was the children's clinic [11], a deadly cooperating partner of the *Spiegelgrund* [35], where he was admitted on March 30, 1943. Elmar Türk, with the knowledge and *placet* of the head of the clinic, Franz Hamburger, tried to develop a vaccine against tuberculosis and asked Ernst Illing, at this time head of the *Spiegelgrund*, for further examination and the results of the autopsy of the little boy on April 13, 1943 ([11]:31–32). Adolf belonged to the study controls. He on purpose was infected with tuberculosis, but got no vaccine. Developing a heavy infection, he finally expired (on June 18, 1943).

31.3.7 Unethical Immoral Clique

The resting place of his remains are unknown—except for his brain: It was buried, together with all the other body parts of children murdered by Heinrich Gross, Elmar Türk, Margarethe Hübsch, Marianne Türk, Helene Joc(k)l, Ernst Illing (the only one out of this group who was sentenced to death by hanging in 1946), Erwin Jekelius (who died in a Russian POW camp in 1952), and a number of nurses, among them Anna Katschenka [36, 37], who worked as children's nurse after her time in jail until her retirement. So did Margarethe Hübsch as chief physician in her own office and Elmar Türk as pediatrician (from 1949 to 1992; [11]:39).

This panorama of death shows that Adolf was not the victim of a single perpetrator, but of a whole group of (institutional) murderers, female and male, junior and senior, head of department and co-worker, doctor and nurse, people with medical skills or men and women bearing the mentality of conformist state servants. Every single person involved in this atrocious process was free from moral or ethics, not to say compassion for a helpless child and his family—and most of them continued to work in their jobs, earned a living, were respected members of the society, and had to face no (or only minor) consequences.

31.3.8 Inhumanity Is Everywhere

Adolf Guttmann's tragic fate was the doom of a little boy who was different from the norm whatever that may be—as different as hundreds of indigenous kids whose bodies were found throughout Canadian residential homes [38], calling back to memory the atrocities committed in Irish homes for young mothers, depicted in *The Magdalene Sisters* (2002), in the 1960s, and showing how important it is to never forget and continuously learn from history. Adolf Guttmann (1941–1943) is one of the many forgotten victims, because history and research focused on the perpetrators ([39]:7). What remains of many of them are the heartbreaking photographs taken by Heinrich Gross—some of them shown in the book by Häupl [15]—who took their life and dignity.

31.3.9 Final Remarks

When the famous Spanish writer Jorge Semprún was awarded the Peace Prize of the German Book Trade in 1994, he quoted in his acceptance speech Paul Celan's words from his probably best-known poem *Todesfuge* (Death tango) ([22]:199): "death is a gang boss from Germany", or actually from any totalitarian regime. Heinrich Gross's survivors would prefer "a gang boss from Vienna". A chorus line made up by the victims of Gross, Jekelius, Illing, Jo(c)kl, Hübsch, Katschenka, and Türk might well form a *crescendo* of consent.

References

1. Cymes M (2015) Hippocrate aux enfers. Les médecins des camps de la mort (Hippocrates in hell. The physicians in the death camps). Éditions Stock, Paris, France
2. Matalon Lagnado L, Cohn Dekel S (1994) Die Zwillinge des Dr. Mengele. Der Arzt von Auschwitz und seine Opfer (The twins of Dr. Mengele. The physician from Auschwitz and his victims). Rowohlt, Reinbek bei Hamburg, Germany
3. Mayr W (2000) Vom Hakenkreuz zum Ehrenkreuz (From the swastika to the medal). Der Spiegel. https://www.spiegel.de/politik/vom-hakenkreuz-zum-ehrenkreuz-a-f880556f-0002-0001-0000-000015986016. Accessed 26 May 2022
4. Thomas FP, Beres A, Shevell MI (2006) "A cold wind coming": Heinrich Gross and Child Euthanasia in Vienna. J Child Neurol 21:342–348. https://doi.org/10.1177/08830738060210040101 and https://doi.org/10.2310/7010.2006.00043
5. Nowak O (2019) Kindereuthanasie im Nationalsozialismus—Strukturen, Organisation und Methoden der Kinderfachabteilung Am Spiegelgrund (Euthanasia of children during the National Socialism—Structures, organization and methods in the pediatric service Am Spiegelgrund). Vienna, Austria (diploma thesis). https://doi.org/10.25365/thesis.60242. Accessed 26 May 2022
6. Vöros L (2010) Kinder- und Jugendlicheneuthanasie zur Zeit des Nationalsozialismus am Wiener Spiegelgrund (Euthanasia of children and adolescents in the time of National Socialism at the Spiegelgrund

in Vienna). Vienna, Austria (diploma thesis). https://doi.org/10.25365/thesis.9480. Accessed 26 May 2022
7. Koller B (2009) Die mediale Aufarbeitung der Opfer-Täter-Rolle in der Zweiten Republik dargestellt anhand des Spielfilms "Mein Mörder". (The media representation of the victim-perpetrator-relationship during the Second Republic as shown in the film "My Killer"). Vienna, Austria (diploma thesis). https://doi.org/10.25365/thesis.3895. Accessed 26 May 2022
8. Ertl KA (2012) NS-Euthanasie in Wien. Erwin Jekelius: Der Direktor vom "Spiegelgrund" und seine Beteiligung am NSVernichtungsprogramm (Erwin Jekelius: Director of "Spiegelgrund" and his participation in the Nazi annihilation program), Vienna, Austria (diploma thesis). https://doi.org/10.25365/thesis.18809. Accessed 26 May 2022
9. Pscheiden D (2015) Handlungsräume und Täterschaft von Medizinerinnen während der NS-Herrschaft am Beispiel der 'Spiegelgrund'-Ärztin Marianne Türk (Operating space and engagement of female physicians during the Nazi regime based on the example of Dr. Marianne Türk at Spiegelgrund). Vienna, Austria (diploma thesis). https://doi.org/10.25365/thesis.37560. Accessed 26 May 2022
10. Mayrhofer S (2020) Die Gerichtsverfahren gegen Dr. Heinrich Gross im Zusammenhang mit seiner Tätigkeit als Arzt in der Kindertötungsanstalt 'Am Spiegelgrund'. (The court proceedings against Dr. Heinrich Gross in connection with his activities as physician in the child killing institution "Am Spiegelgrund"). Vienna, Austria (diploma thesis). https://doi.org/10.25365/thesis.61430. Accessed 26 May 2022
11. Czech H (2001) Zuträger der Vernichtung? Die Wiener Universitäts-Kinderklinik und die NS-Kindereuthanasieanstalt "Am Spiegelgrund" (Contributors to the annihilation? The Pediatric Clinic of the University of Vienna and the Nazi euthanasia institution "Am Spiegelgrund"). In: Pollak A (ed) Festschrift 100 Jahre Universitätsklinik für Kinder- und Jugendheilkunde (Celebration edition 100 years of the Child and Adolescent University Clinic). Berger, Horn, Austria, pp 9–40
12. Czech J (2014) Der Spiegelgrund-Komplex. Kinderheilkunde, Heilpädagogik, Psychiatrie und Jugendfürsorge im Nationalsozialismus (The Spiegelgrund complex. Pediatrics, healing pedagogy, psychiatry and juvenile care during National Socialism). ÖZG 25(1+2):194–219
13. Czech H (2017) Beyond Spiegelgrund and Berkatit. Human experimentation and coerced research at the Vienna School of Medicine, 1939 to 1945. In: Weindling P (ed) From clinic to concentration camp. Reassessing Nazi Medical and racial research, 1933–1945. Routledge, London, pp 138–162. https://doi.org/10.4324/9781315583310
14. Czech H (2019) Der Spiegelgrund-Komplex. Kinderheilkunde, Heilpädagogik, Psychiatrie und Jugendfürsorge im Nationalsozialismus (The Spiegelgrund complex. Pediatrics, healing pedagogy, psychiatry and juvenile care during National Socialism). In: Biewer G, Proyer M (eds) Behinderung und Gesellschaft. Ein universitärer Beitrag zum Gedenkjahr 2018 (Disability and society. A university contribution to the celebration year 2018): Coralie Geier & Elvira Seitinger. University of Vienna, Vienna, Austria, pp 85–106
15. Häupl W (ed) (2006) Die ermordeten Kinder vom Spiegelgrund. Gedenkdokumentation für die Opfer der NS-Kindereuthanasie in Wien. (The killed children from Spiegelgrund. Remembrance documentation for the victims of the Nazi children euthanasia in Vienna). Böhlau, Vienna, Austria
16. Bird S, Fulbrook M, Wagner J, Wienand C (eds) (2016) Reverberations of Nazi violence in Germany and beyond. Disturbing pasts. Bloomsbury, London, UK
17. Sem-Sandberg S (2014) De utvalda (The chosen one). Albert Bönniers Förlag, Stockholm, Sweden
18. Michael I (2015) Die mediale Aufarbeitung der Verbrechen an Unmündigen in der Kinderfachabteilung *Am Spiegelgrund* 1940–1945 (The media representation of the crime against the underaged in the pediatric institution Am Spiegelgrund 1940–1945). Vienna, Austria (diploma thesis). https://doi.org/10.25365/thesis.38810. Accessed 26 May 2022
19. Gross J (2000) Spiegelgrund. Leben in NS-Erziehungsanstalten (Life in Nazi educational institutions). Ueberreuter, Vienna, Austria
20. Kaufmann A (2007) Totenwagen. Kindheit am Spiegelgrund (Hearses. Childhood at Spiegelgrund). Mandelbaum, Vienna, Austria
21. Kaufmann A, Podzeit-Lütjen M (eds) (2006) Dass ich dich finde. Kind am Spiegelgrund. Gedichte. (That I find you. Child at Spiegelgrund. Poems). Theodor Kramer Gesellschaft & Mandelbaum, Vienna, Austria
22. Sparr T (2020) Todesfuge. Biographie eines Gedichts (The death tango. Biography of a poem). Deutsche Verlagsanstalt, Munich, Germany
23. Neugebauer W, Schwarz P (2005) Der Wille zum aufrechten Gang—Offenlegung der Rolle des BSA bei der gesellschaftlichen Integration ehemaliger Nationalsozialisten (The will to have an upright gait—Exposure of the role of the association of socialist academics in the social integration of former Nazis). Czernin, Vienna, Austria
24. Ries E (2019) Dr. Werner Vogt: Menschenrechtspreis 2019 für sein Lebenswerk. http://www.liga.or.at/site/assets/files/2212/menschenrechtspreis2019_vogt_web.pdf. Accessed 26 May 2022
25. Dahl M (2004) Endstation Spiegelgrund. Die Tötung behinderter Kinder während des Nationalsozialismus am Beispiel einer Kinderfachabteilung in Wien 1940 bis 1945 (Final destination Spiegelgrund. The killing of disabled children during National Socialism based of the example of a pediatric service in Vienna 1940–1945). Erasmus, Vienna, Austria

26. Dahl M (2000) Die Tötung behinderter Kinder in der Anstalt "Am Spiegelgrund" 1940 bis 1945 (The killing of disabled children in the "Am Spiegelgrund" institution 1940–1945). In: Gabriel E, Neugebauer W (eds) NS-Euthanasie in Wien (Nazi euthanasia in Vienna). Böhlau, Vienna, Austria, pp 75–92
27. https://www.ueber-leben.at/home/alois-kaufmann. Accessed 26 May 2022
28. https://weitererzaehlen.at/interviews/alois-kaufmann-1. Accessed 26 May 2022
29. Banerjee M (2021) Biologische Geisteswissenschaften. Von den Medical Humanities zur Narrativen Medizin. Eine Einführung. (Biological humanities. From medical humanities to narrative medicine. An introduction). Winter, Heidelberg, Germany
30. https://friedrich-zawrel.zurerinnerung.at/. Accessed 26 May 2022
31. Lehmann O, Schmidt T (2001) In den Fängen des Dr. Gross. Das misshandelte Leben des Friedrich Zawrel (In the claws of Dr. Gross. The abused life of Friedrich Zawrel). Czernin, Vienna, Austria
32. Zawrel F (2006) Epilog. In: Häupl W (ed) Die ermordeten Kinder vom Spiegelgrund. Gedenkdokumentation für die Opfer der NS-Kindereuthanasie in Wien (The murdered children from Spiegelgrund. Remembrance documentation for the victims of the Nazi euthanasia of children in Vienna). Böhlau, Vienna, Austria, pp 645–646
33. Ischepp N (2012) Die "Asozialen" in der Anstalt Am Spiegelgrund—Folgen von Erziehung und Fürsorge im Nationalsozialismus (The "asocial" in the institution "Am Spiegelgrund". Consequences of care and education during National Socialism). Vienna, Austria (diploma thesis). https://doi.org/10.25365/thesis.22798. Accessed 26 May 2022
34. Dahl M (2002) "... deren Lebenserhaltung für die Nation keinen Vorteil bedeutet." Behinderte Kinder als Versuchsobjekte und die Entwicklung der Tuberkulose-Schutzimpfung / ("... to sustain their lives is of no advantage to the nation." Disabled and mentally retarded children as experimental subjects and the development of the tuberculosis vaccine). Med J 37(1):57–90
35. Sheffer E (2018) Asperger's children. The origins of Autism in Nazi Vienna. W.W. Norton & Company, New York, NY
36. Fürstler G, Malina P (2004) "Ich tat nur meinen Dienst." Zur Geschichte der Krankenpflege in Österreich in der NS-Zeit Facultas ("I was just doing my job". The history of healthcare in Austria during the Nazi regime). Facultas, Vienna, Austria
37. Primnig A (2013) Kindermorde in der Zeit des Nationalsozialismus am Wiener Spiegelgrund. Be- und Entschuldungsstrategien einer Krankenschwester vor Gericht (Killing of children at the time of National Socialism at the Spiegelgrund in Vienna. Accusation and defense strategies of a nurse in court). Vienna, Austria (diploma thesis). https://doi.org/10.25365/thesis.27116. Accessed 26 May 2022
38. https://www.bbc.com/news/world-us-canada-57674682. Accessed 26 May 2022
39. Enigl M (2006) Geleitwort. In: Häupl W (ed) Die ermordeten Kinder vom Spiegelgrund. Gedenkdokumentation für die Opfer der NS-Kindereuthanasie in Wien (The killed children from Spiegelgrund. Remembrance documentation for the victims of the Nazi children euthanasia in Vienna). Böhlau, Vienna, Austria, pp 7–9

Part IX

Science and the Internet

Curbside Consultations in Pediatric Dermatology: Risks and Benefits

32

Emily Duffy and Robert Sidbury

Abstract

Curbside consultations have always been a part of modern medical practice. Dermatology is given to this type of informal interaction in part due to the accessibility of physical findings, and this is amplified in pediatric dermatology due to a severe workforce shortage. The proliferation of email and especially cellphone cameras has dramatically increased the volume of such consultations many pediatric dermatologists encounter daily. In this chapter we review both the pros of curbside consultation including improved access and triage, convenience for patients, and decreased carbon footprint to name several, and the cons including potential inconvenience and volume overload for consultants, ethical and legal liability, and Health Insurance Portability and Accountability Act/HIPAA concerns, as well as potential loss of income. Brief actual or adapted "case reports" are transcribed, for a more realistic appraisal of the issues involved. Trends certainly suggest the number of curbside consultations is only increasing just as provider satisfaction among dermatologists is decreasing. Recognition and attention to both the risks and benefits of curbside consultation will be critical for a sustainable care model.

E. Duffy (✉)
Division of Dermatology, Department of Medicine, University of Washington School of Medicine, Seattle, WA, USA

Seattle Children's Hospital, Seattle, WA, USA
e-mail: Eduffy8@uw.edu

R. Sidbury
Division of Dermatology, Department of Pediatrics, University of Washington School of Medicine, Seattle, WA, USA
e-mail: robert.sidbury@seattlechildrens.org

32.1 Introduction

The scenario described below is familiar to primary provider and specialist alike, notably in the field of dermatology.

Dr. Z knew that today's clinic was going to be a bustling one, but she thought she was doing an admirable job thus far, being diligent but efficient with her patients and their families. During a quick break between patients, she skimmed her email inbox and noted a message from a primary provider, Dr. P, at her multispecialty clinic entitled, "Newborn with rash." She also noted that there were pictures attached but deferred opening the email at that time, opting to review the message at the end of morning clinic. Dr. Z stepped in to see her next patient, but a few minutes into the visit, her pager alerted her of a new message. She was caught off guard by the page since she was not on the consult service this week, and she stepped into the hallway to return the call. The same provider, Dr. P, answered the phone and stated that she was grateful that Dr. Z called back because she was seeing a young baby with a pustular rash that she was uncertain how to manage. As Dr. P then began to summarize the patient's course, Dr. Z could not

J. Faintuch, S. Faintuch (eds.), *Integrity of Scientific Research*,
https://doi.org/10.1007/978-3-030-99680-2_32

help but feel distracted by how far behind she was falling in her clinic.

Curbside consultations have been a bedrock of medical practice even before modern technologies became available and have evolved alongside medicine. In dermatology it has markedly changed over the past decade, with the explosion of digital smartphones and cameras; a "quick peek" at a rash is now just a text or email away. And for all the real benefits that informal consultations can bear, the risks are just as abundant. In pediatric dermatology, a relative paucity of trained and certified providers amplifies the push for easier and quicker access, at times generating strain for busy providers.

32.1.1 Informal Information and Care

1. Keating et al. define curbside consultation as "a process in which a physician seeks information or advice about patient care from another physician who has a particular expertise without obtaining a formal consultation between the patient and the consultant physician at that time."
2. Curbside consultations occur in all fields of medicine, from the "quick peek" in dermatology [1] to the "wet read" in radiology [2]. The curbside consult stands in contrast to traditional consultation, in which a physician receives a formal request from a treating provider to examine a patient and to document assessment and recommendations. Historically in dermatology, the curbside consult often entailed a quick aside in a hospital hallway or clinic corridor. The requesting provider may have shared a verbal description of a rash or lesion or requested a quick pop into a neighboring clinic. At times the curbside may not have even regarded a patient's skin but instead medical staff themselves. Guberman et al. reported, in an informal study of 50 "corridor consultations" in their hospital, that two thirds were from various hospital staff and one third were from the families of hospitalized patients [3]. Additionally, informal consults can arrive unsolicited from family and friends, sometimes in settings that are unexpected or, at times, even inappropriate [4].

32.1.2 The Many Faces of Informality

Curbside consults have taken all forms, from telegram to telephone and from faxes to emails to text messages. Groups on social media like Facebook allow members-only forums for medical providers to discuss difficult or interesting cases or to learn innovative treatments. Messaging platforms, such as WhatsApp, allow physicians from around the world to seek guidance from local colleagues and remote specialists [5]. Specialist societies offer listservs to members where challenging cases can be discussed [6]. As communications become faster and easier to send, however, the volume of direct digital curbside consultations has steadily increased and threatens to become intrusive to care. In 2019, Khorsand et al. reported on a survey of board-certified pediatric dermatologists. Half of the respondents received 0–5 curbside cases per week, but a third received 6–10 cases per week, amounting to an equivalent of one-half day worth of clinic time [1]. Similarly in an editorial response, Jon Dyer, also a pediatric dermatologist, noted that over one week he received a similar message load from colleagues, friends, and family, in addition to direct electronic medical record (EMR) messages, phone calls, and emails from providers and patients [7].

32.1.3 Hazy Limits

Does a review of photographs "count" as examining a patient? Though no one feature absolutely defines an informal consult, Olick et al. provide a list of indicators that suggest that a consult may be informal (Table 32.1) [8]. In sum, if a provider has not directly spoken with or examined a patient, has not independently reviewed their chart or data, or has not been in direct management of their care, a formal evaluation mutually

Table 32.1 Indicators of informal consultations

1.	No patient examination
2.	No communication with the patient
3.	No access to clinical records
4.	Formal consultation has not been arranged
5.	No payment involved
6.	Contact and exchanges just with the treating physician

Adapted from ref. [8]

entered upon by both parties has not occurred, and a therapeutic physician-patient relationship has not been established. These conditions can be an important consideration if a question of medical liability arises, as reviewed later. Curbside consults may be made to try to answer a "quick" or "easy" question, or they may aim to receive tailored advice for a unique or complex case [9]. Moon et al., looking at pediatric dermatology consultations from the emergency department, found that the majority of consultations dealt with infectious disease (39 percent), inflammatory skin conditions (16 percent, not including atopic dermatitis), and atopic dermatitis (14 percent) [10].

32.2 Benefits of Curbside Consultation

Access to specialist care can be particularly relevant in pediatric dermatology, where there is a relative paucity of certified providers [11, 12].

Dr. A sees a toddler in her office with what seems like severe atopic dermatitis. Given the child's significant symptoms, as well as the family's detailed questions regarding its etiology and management, she thinks that a dermatology referral is warranted. In the interim, however, she begins therapy with a "safe" (i.e., low potency) topical corticosteroid and schedules the patient to return in two weeks. When she sees the family back, the rash is only mildly improved and shows some worsened crusting on the face and arms, which she worries may represent infection. The family states that the earliest appointment that is available with pediatric dermatology is in three months; the mother is frustrated and tearful as neither the child nor the parents are getting much sleep. The provider wants to do more to help, but she feels out of her comfort zone in prescribing a stronger medication.

In a survey from 2016 to all society members, the average reported wait time to see a pediatric dermatologist was six weeks; more than 25 percent of respondents, however, reported a wait for a new-patient appointment of at least ten weeks [13]. For a nonurgent condition, such as warts or acne, a wait of two to three months would be undesirable but nevertheless manageable. For urgent concerns, however, such as severe atopic dermatitis or a concerning neoplasm, this sort of delay is untenable. For this very sort of concern, many practitioners set aside "urgent" appointment slots that can only be filled in the days before the appointment and require provider approval to schedule. Curbside consultations can fill the important role of triaging patients to go into these slots. Other methods to triage these slots may of course also exist, such as indicating urgency level for evaluation on a referral form. Sometimes, however, direct communication from provider to provider can relay timely updates or particular clinical concerns that may not be adequately conveyed on a boilerplate form.

32.2.1 Improving the Carbon Footprint

The geographic distribution of pediatric dermatologists can be a significant barrier to care for many families.

Dr. B is seeing an otherwise healthy 8-year-old boy in pediatric clinic. His dad noted the other day an unusual linear grouping of brown bumps along his child's leg and ankle. Neither patient nor parent is certain of how long this has been present, but the child is not bothered by it. Because the pediatrician is uncertain of the diagnosis, she considers referring the patient to see a pediatric dermatologist. From previous referrals for other patients, however, she knows that the closest pediatric dermatologist, Dr. Z, is approximately a three-hour drive away. She recalls, however, that she might have Dr. Z's email address from a previous exchange and, after obtaining consent from the family, snaps a picture of the skin findings to email to the consultant.

In a study performed in 2020, it was noted that 9 states in the USA have no pediatric dermatologist and that even 31 percent of counties with over 500,000 children lacked these providers. Ninety-eight percent of pediatric dermatologists work in a metropolitan area, versus two percent in a non-metropolitan area, and none in a rural area. While a void in a smaller state such as Delaware or Maine might be easier to manage (with a high density of providers in neighboring states), citizens of the Dakotas, Montana, and Wyoming do not have such proximate options [11]. Curbside consultation can serve as a valuable bridge to care, by either eliminating the need for in-person evaluation or de-escalating the perceived urgency by which a patient must be seen. For the vignette above, a pediatric dermatologist would likely feel very comfortable evaluating a photograph of the described lesion (likely an epidermal nevus, a common congenital skin finding with a typical clinical appearance) and reassuring of its benign nature. While the patient may eventually benefit from an in-person evaluation, consolidating a trip to this specialist along with other reasons to travel may serve to both convenience the family and to decrease environmental impact of this travel.

32.2.2 Convenience/Cost

Dr. B snaps a photo of the patient's skin lesion and emails it to Dr. Z, with plans to update the family once she has heard back from Dr. Z. Dr. Z responds at the end of the day with her presumed diagnosis and advises a nonurgent evaluation in the coming 6–12 months, if desired. After closing the loop with the family, she is satisfied that her informal consultation served the family, as well as the healthcare system, avoiding an unneeded in-person evaluation.

Curbside consults can provide possible diagnoses or treatment recommendations within minutes to hours, instead of days to weeks. With the use of text messages and emails, communication can be asynchronous; requestors don't need to sit by the phone to wait for a response, and consultants have some flexibility in when and how to respond [6, 8]. Instead of leaving a message with a third party, such as a nurse or secretary, the consultors can transmit precisely the information they want. Additionally, email or direct message through an EMR notebook has the benefit of being "self-documenting" [6], with records of history, photographs, and recommendations available for future reference. A study published in 2017 showed that smartphone photographs of pediatric skin conditions from external sources are relatively reliable. In an evaluation of 40 patients, in which parents submitted photographs of skin lesions or rashes and subsequently had in-person evaluation by a pediatric dermatologist of that same condition, there was overall diagnostic concordance of 83 percent [14]. Curbside consults can also decrease cost. Patients with informal consults often save money on co-pays, tests, and other studies, as well as transportation costs. So too patients who participate in a health management organization/HMO or other capitated reimbursement system can ultimately save the system money when seen on an informal (i.e., non-reimbursed) basis [4, 15].

32.2.3 Improved Patient Care and Provider Education

Curbside consultations have the potential to tailor treatment for a patient.

Dr. A decides to page Dr. Z to discuss her patient's worsening eczema and infection concerns. She mentions that after her own reading, she is concerned about a possible staphylococcal infection, but she is also considering empiric treatment for eczema herpeticum, given the severity of the patient's symptoms. There have been no known exposures to persons with herpes outbreaks. When Dr. Z calls back, they discuss the case, and Dr. Z requests photos of the child's skin. Dr. Z sees no areas of involvement that look concerning for eczema herpeticum, and so she advises that, based on her review, empiric antiviral therapy is likely not indicated. She recommends a bacterial culture, empiric antibiotics, and escalation of topical steroid therapy. She also sends an email to the dermatology clinic's schedulers to schedule this patient in an urgent clinic appointment.

In the example above, although Dr. A had done much research herself about how to care for this patient, Dr. Z was able to lend her expertise to tailor treatment recommendations and to avoid unnecessary therapies. Kuo et al. note that primary care physicians thought that the most important characteristics in choosing an informal consultant were the quality of their formal consultations as well as the consultant's superior skills or knowledge base [16]. Additionally, a study from Mayo Clinic in 2014 that surveyed both primary providers and consultants regarding the process of curbside consultation found that "all else being equal, humans are preferred" over referencing a book or a website. Survey participants identified three main reasons that curbside consultations were valuable: they received individualized answers that went beyond what could be learned in a book, they received answers quickly, and they "bolstered patients' confidence in and appreciation for their care" [9].

One participant noted: "It's the patient care, it's the education of staff, it's building that rapport to education and research" [9]. Golub additionally noted that informal consultations "can help physicians keep current with medical information" [17]. Consultants may also take the opportunity to share a recent informal consult with colleagues and/or trainees, which expands the knowledge base to others in the specialty field. Not only does the patient potentially benefit from individualized advice, but the requesting providers can build their knowledge. They can build on the repertoire of conditions that they are comfortable treating, as they can discuss caveats, concerns, and exceptions with the consultant in real time versus delayed review of an assessment and plan weeks or months after that patient presented initially.

32.2.4 Collegiality

Every physician, nurse practitioner, and physician's assistant has found themselves in a situation where they did not know the best approach in caring for a confusing or complicated patient. The advice of a colleague with greater experience is often clarifying and demystifying.

Dr. A ends her call with Dr. Z and conveys the treatment plan to the patient's family while they are still in clinic. The family is appreciative and makes an appointment to follow up with Dr. A in three days. Reflecting on this interaction at the end of a busy clinic day, Dr. A remarks to another co-worker that Dr. Z was very helpful in the care of a complex patient. They place a "star" next to Dr. Z's name on the lists of consultants that they keep in their workroom.

Many authors [1, 7] agree that collegiality and altruism are great motivators in their acceptance of curbside consultations and that they often feel "honored" [18] or "flattered" [9] to have been asked to provide their opinion. Moreover, providers involved with curbside consultation note that the interaction demonstrates transparency on the requestor's part as well as strength in the institution in being able to provide expert advice, both of which signal to the patient that they are receiving efficient, high-quality care [9]. On a more practical note, Kuo et al. observe that 77 percent of subspecialists indicate that curbside consultations are "essential" for maintaining good working relations with other providers [16], and this facile exchange of advice and access can also result in more formal consultations to this same specialist down the road [17].

32.3 Risks of Curbside Consultation

These include inefficient or inaccurate advice, practice inefficiency and provider burnout, infringement on patient ethics (privacy and autonomy), concerns for provider liability, and disparity in access to care.

32.3.1 Fear of Incomplete or Inaccurate Advice

If a consultant receives inaccurate or incomplete information, the subsequent recommendations can be affected. Data may not be obtained by the requestor and is therefore unavailable for evaluation by the consultant. Alternatively, a requestor may obtain all the needed information but may misremember or misinterpret the data, or the receiver may misunderstand or misinterpret what they hear.

Dr. Z ends her call with Dr. A and completes the rest of her work. As she prepares to leave the office, she replays the advice she gave to Dr. A in her head and starts to recognize some gaps in information. She realized she did not know if the child had history of previous skin infection or a previous exposure to methicillin-resistant Staphylococcus aureus /MRSA, which could change antibiotic therapy. She did not review antibiotic dosing with the primary provider, assuming that they would choose an appropriate dose, but she now wonders if she should have been more explicit. She made recommendations on which topical steroids to use but regrets that she did not specify the volume, worrying that they may receive too small a volume to treat the skin effectively. She considers calling Dr. A's office back but suspects that she may not reach her after hours.

Dr. Z is about to end her call with Dr. B, after describing the benign nature of an epidermal nevus, and is absentmindedly clicking through the photos one last time. In the final photo, she realizes there is an out-of-focus blurry dark spot in the corner of the shot. Dr. Z asks Dr. B if he knows anything about this other lesion, since, from what she can make out, there may be some irregularity in pigment and shape. Dr. B states that he did not ask the family about this lesion, but they did not express concern about it, and he suspects it has been there for "a while." They end the call, but later Dr. Z ponders if she should have recommended an earlier in-person evaluation to examine this potentially atypical lesion.

A family may state that they apply a topical steroid three times a week, and the requestor considers this frequency "often"; when the receiver hears that a medicine is used "often," though, their use of that word may align with "once a day" or "every few hours." Moreover, the way the requesting provider frames a problem may shape the recommendations that a consultant gives [17]. As in the case of Dr. B, the concern at hand was focused on the likely epidermal nevus, but if Dr. Z had performed a full skin examination, she likely would have spent more time evaluating the possible atypical melanocytic lesion.

A 2010 article by Chen et al. found that diagnoses and treatment plans were often discordant between primary providers and pediatric dermatologists, when using a store-and-forward consultation service, with discordance occurring in 42 percent of cases [19]. A prospective study by Burden et al. likewise showed that management advice from a consultant changed 60 percent of the time when a previous curbside consultation subsequently underwent formal, independent consultation [20]. Keating et al. found that specialists expressed more concern than generalists that recommendations may be based on incomplete or inaccurate information [15]. Their survey, however, also showed that concerns about the quality of advice given are not a one-way street; there was concern both from generalists and specialists.

On the tail end of the consult, as well, consultants expressed concern that their recommendations may be misunderstood or not implemented as intended. The consultant typically has no direct contact with the patient and cannot provide direct guidance about possible difficulties or complications with a suggested treatment. Moreover, the consultant often has no way to formally document their recommendations (though, the increased use of email now can serve as some form of record). Conversely, the requesting provider may not have an opportunity to evaluate the quality of the recommendations they receive; they may feel disinclined to question advice that was provided by a consultant through good will. So, while curbside consultations can be valuable in the right setting, understanding the inherent limitations of making advice based on secondhand information is critical, and a consultant may consider advising that formal consultation should be considered to provide a more complete evaluation.

32.3.2 Inefficiency/Burnout/Loss of Income

Though curbside consultation is meant to be convenient, inefficiencies abound. These inefficiencies often translate into increased workload, which has the potential to lead to dissatisfaction and burnout.

Dr. Z is wrapping up the end of her clinic and recognizes that she feels exhausted. What started out as a busy but manageable day quickly transitioned into a day where she felt like she had trouble keeping her head above water. After returning pages from Dr. P in the morning and Drs. A and B in the afternoon, she barely had time to grab a small bite between her two half-day clinics, and even then, she was running behind for most of her appointments. She decides to head home but anticipates spending at least two hours that evening completing clinic notes from today and previous days.

Even the initial question of who to consult may not be straightforward. Though many healthcare systems have an on-call provider available for each specialty, requesting providers may not contact that provider, either due to lack of familiarity or to the desire to confer with a friend in the specialty or a more widely acknowledged "expert" in the field. This ambiguity can create inefficiencies both for the requestor, who may try to contact more than one provider before they have success in receiving a response or must leave emails, pages, and messages before speaking with a provider, as well as for the specialist, who may either be unavailable that day or have their own clinical or surgical schedule that cannot accommodate much delay. Respondents in the survey by Cook et al. endorsed that they have a "go-to person in each specialty" or that they will "usually phone a friend" [9].

32.3.3 The Shadow Clinic

According to Khorsand, the equivalent of a full clinic's work is sometimes performed in the margins, in addition to a full, "official" workload [1, 7]. Shanafelt et al. published in 2019 that respondents in dermatology went from reporting a below-average rate of burnout, around 30 percent in 2011, to an above-average rate, at approximately 60 percent, just 3 years later, with a correlating drop in satisfaction in work-life integration during the same time [21]. They found that, in general, rates of physician burnout are higher than rates in the general US working population, and these rates of work dissatisfaction may correlate with increasing regulations and administrative burden. Adding additional workload can likely only intensify the problem, no matter how rewarding the actual interactions may be.

32.3.4 Lack of Financial Compensation

In a very concrete way, particularly in physicians who function in fee-based compensation models, increased curbside consultations may translate to decreased formal consultations and an overall decrease in income. While this effect may not translate to a real dollar value for many or may not be a primary concern to the consulting physician, it has the potential to contribute to the perception that the consultant's time, opinion, and expertise are not valuable; to some, the question at hand is "just about the skin."

32.4 Ethical and Legal Concerns: Privacy and Liability

Privacy concerns can create ambiguity and confusion when performing informal consultations.

Days after she was contacted, Dr. Z is discussing the details of a curbside consultation with a colleague, Dr. Y. They are reviewing photographs of the infant with a pustular rash on Dr. Z's phone and are discussing her recommendations. Dr. Y remarks that she is "impressed" that Dr. Z provided medical advice without seeing the patient in person. "I would have been too worried about getting sued if the patient had a bad outcome," remarked Dr. Y. Though Dr. Z was confident in her assessment of erythema toxicum neonatorum at the time of her evaluation, she now is concerned that she might not have been as thorough as possible in conveying other possible, more serious diagnoses in her differential diagnosis to the primary provider. She wonders how the patient is doing now.

In the case of direct email communication between a patient and provider, there exists a possible risk of misrepresentation by either party, with either the patient or provider not being, in actuality, who they claim to be. Additionally email servers may not be secure, and private health information can be hacked through insecure connections [6]. More commonly concerns arise regarding the security and confidentiality of patient images that are shared informally. Under the USA HIPAA Act of 1996 and a 2013 addendum, which could have corresponding legislation in other countries, if digital images are identifiable to a particular patient, they are considered protected health information (PHI). As such, there must be limited access to these images, and informal consultations run the risk of falling short on such legal safeguards [22].

32.4.1 Informed Consent

According to HIPAA regulations, when patient photos are taken consent must be obtained and reported, and images must be shared securely and only with providers who are involved in patient care. While cellphone images expand more specific and informed care to patients, a patient's privacy and autonomy may be threatened. Moreover few professionals have time for collecting or checking patient consent, within a context of informal contacts with colleagues.

32.4.2 Specific Consent

In some countries the consent document should state the envisaged uses of the material (clinical, teaching, publication) and by whom it will be manipulated (just involved physicians, other professionals).

32.4.3 Transmission and Storage

Analogously electronic sending of identifiable clinical material is sometimes allowed only by encrypted or well-protected media, and the consent document should inform how patient privacy will be ensured during storage, sharing, and utilization of such material.

32.4.4 Medicolegal Liability

Keating et al. report that most specialists surveyed believe that consultants should never be held liable for informal advice, and consultants are generally more concerned about malpractice lawsuit risk than generalists [15]. Khorsand similarly relates that pediatric dermatologists have similar concerns, but the majority (80 percent) are uncertain if their malpractice insurance applies in this setting [1]. In general, medical professionals are considered excluded from possible malpractice liability in the setting of an informal consultation. Multiple lawsuits over time involving physicians providing informal consultation consider this form of advice to be more a service to colleagues versus directly to a patient [23].

32.4.5 Physician-Patient Relationship

The available case law generally considers that a physician-patient relationship is necessary to create a duty or obligation for care towards a patient. Though there is no hard and fast delineation between a formal and informal consultation, the factors previously listed in Table 32.1 are some that may suggest a limited role for the provider [8]. Yet some interpret *Warren v. Dinter*, a 2019 Minnesota Supreme Court case, as representing a challenge to the claim that a physician-patient relationship is necessary to establish a medical malpractice claim [24]. Warren, the patient, was evaluated by an outpatient nurse practitioner for fever and abdominal pain; the nurse practitioner discussed Warren's care with Dinter, a hospitalist at another healthcare system, about the possible concern for infection and need for admission. Dinter is reported to have advised that the patient's symptoms were attributable to diabetes and suggested that admission was not needed; Warren returned home and was found dead by family three days later. An autopsy revealed that the cause of death was sepsis secondary to *Staphylococcus aureus*.

The court argued that there was "foreseeable harm" in this doctor's recommendation, even though a physician-patient relationship had not been established [24]. The American Medical Association issued an *amicus* brief on the case, arguing that Dinter should not have been held liable as there was no formal relationship established and that concern for possible liability will limit beneficial discussion of future cases between providers [25].

Kreichelt et al. provide guidance on ways to limit possible exposure to liability in informal consultation, including (1) giving advice based on general hypothetical clinical scenarios (not on a particular patient), (2) keeping curbside consultations brief, (3) recommending formal consultations for more detailed discussions or for repeat curbsides on a patient, and (4) asking that the consultant's name not be registered in the medical record [23]. Giving general advice, however, may be challenging if a patient's photo is included. Limiting named involvement in a patient's care is also considerably more challenging in the era of email communication. Some authors recommend against providing written advice over email [23], though this recommendation may be increasingly hard to adhere to over time, particularly given its convenient, asynchronous nature. In that setting, some consultants use a disclaimer at the end of an email stating the inherent limitations of the advice provided, though it is not clear how protective this practice may be.

32.4.6 Disparity in Access

Dr. Z is preparing for another full clinic day the following week. In reviewing her schedule, she notes that three of the ten patients on her morning schedule are patients of Dr. A, Dr. B, and Dr. P, all of whom were scheduled after discussing their cases. She observes to herself that she has been seeing a higher percentage of patients from those three providers' practices recently.

As mentioned previously, curbside consultation can facilitate evaluation of patients who may need more urgent assessment than what might be offered through a standard referral. While this likely benefits the patients that are seen, there exists the potential for disparity in access to specialist care. The provider does not know the patients that they don't see. Digital technologies have become increasingly available throughout all socioeconomic classes, but there are barriers for patients to directly contact a provider. Persons with decreased literacy or limited English fluency will inherently have less ability to write directly to a provider [26]. So, too, physicians who are a "squeaky wheel" (smart and vocal) may have an outsized ability to gain access to specialty care for their patients, which would subsequently disadvantage those patients whose providers do not take that extra step [27]. Calling on friends in a specialty field is certainly understandable, but it is important to consider how one's own practices may impact those patients with fewer connections and less access to care.

32.5 Future Directions

Dr. Z just completed a morning of clinic visits and sits down in her office to catch up on some work. Her afternoon has been set aside to review store-and-forward teledermatology cases, and perusing through her patient list, she sees that a range of diagnoses and acuity levels are referred. She notes a few requests for informal curbsides in her email inbox, but notably fewer since the telemedicine consult service was established.

32.5.1 The Telemedicine Impact

Though reimbursement through both private and public insurance has been one roadblock to telemedicine over previous years [28], its recent expansion during the COVID-19 pandemic facilitated the spread. The use of telemedicine continues to expand each year, and it has the potential to provide medical care to those who are underserved [28]. Teledermatology has gained some traction with pediatric dermatologists; in 2015, it was shown that 32 percent had used formal telemedicine systems (compared with 89 percent who had received informal curbsides), up from 2006, where 9 percent had used formal and 54 percent had used informal consults [29]. The most common diagnoses from pediatric teledermatology referrals included atopic dermatitis, molluscum contagiosum, verruca vulgaris, benign melanocytic nevi, and acne [30], all of which are commonly high-volume diagnoses in the office setting as well.

32.5.2 Documented and In-Depth Assessment

Placing a formal teledermatology consult allows the specialist to access a patient's information more fully, perhaps allowing more complete or nuanced advice. It also allows for a dedicated time and space to consider a patient's condition, instead of squeezing in an email or a phone call as time permits. Photographs provided in a formal consult may have the advantage of improved quality, since persons tasked with telemedicine photos often have formal training in these skills. Telemedicine visits, such as store-and-forward visits, in which the patient's chief complaint, relevant data, and photographs are uploaded by the referring office, are often completed more quickly than live visits, and these encounters allow for triaging patients; those with straightforward diagnoses can be managed remotely, while those with more complex or procedural needs can be referred to the office. Finally, using teledermatology consults in lieu of curbside consults, when appropriate, allows for the specialist provider to be reimbursed for their services, and this reimbursement can more easily facilitate setting aside dedicated time for this type of patient care.

32.5.3 Not a *Panacea*

Of course, formal teledermatology consults can carry some of the inherent risks that are seen with informal consults. Photos may not accurately represent a rash or lesion, a history provided may be in error, and charts may contain incomplete data. And increasing the formality of the consult may in turn increase potential liability for the consultant. Accessing consults through an EMR can create a formal record of recommendations and treatments, but it may also reduce efficiency inasmuch as EMRs contain inefficiencies. As a whole, however, a more facile use of the teledermatology consult likely will improve access and quality of care.

32.5.4 Guidelines for Curbside Dos and Don'ts

The American Academy of Dermatology and the Society for Pediatric Dermatology do not yet have such guidelines. A group from the Mayo Clinic has generated a thorough set of guidelines for curbside consults for the stakeholders in this practice, which issue advice for both the initiator and the consultant. These detailed guidelines provide precise recommendations about how to prepare for a consult as well as best practices when giving and receiving advice [9].

References

1. Khorsand K, Sidbury R (2019) The shadow clinic: emails, "curbsides," and "quick peeks" in pediatric dermatology. Ped Dermatol 36(5):607–610
2. Mahalingam S, Bhalla NM, Mezrich JL (2019) Curbside consults: practices, pitfalls, and legal issues. Clin Imaging 57:83–86
3. Guberman D, Vardy DA, Klaus SN (1994) Corridor dermatology consultation en passant. Arch Dermatol 130:233–234
4. Grant-Kels JM, Kels BD (2012) The curbside consultation: legal, moral, and ethical considerations. J Am Acad Dermatol 66(5):827–829
5. Ashique KT (2017) Curbside (corridor to the E-corridor) consultations and the dermatologists. Indian Dermatol Online J 8(3):211–214

6. Luo J, Logan C, Long TP, Bercovitch L (2009) Cyberdermatoethics I: ethical, legal, technologic, and clinical aspects of patient-physician email. Clin Dermatol 27(4):359–366
7. Dyer JA (2019) Illuminating the shadows: recognizing and valuing curbside care in the margins. Pediatr Dermatol 36(5):611–612
8. Olick RS, Bergus GR (2003) Malpractice liability for informal consultations. Fam Med 35(7):476–481
9. Cook DA, Sorensen KJ, Wilkinson JM (2014) Value and process of curbside consultations in clinical practice: a grounded theory study. Mayo Clin Proc 89(5): 602–614
10. Moon AT, Castelo-Soccio L, Yan AC (2016) Emergency department utilization of pediatric dermatology (PD) consultations. J Am Acad Dermatol 74(6): 1173–1177
11. Ashrafzadeh S, Peters GA, Brandling-Bennett HA, Huang JT (2020) The geographic distribution of the US pediatric dermatologist workforce: a national cross-sectional study. Pediatr Dermatol 37(6): 1098–1105
12. Prindaville B, Antaya RJ, Siegfried EC (2015) Pediatric dermatology: past, present, and future. Pediatr Dermatol 32(1):1–12
13. Prindaville B, Horii KA, Siegfried EC, Brandling-Bennett H (2019) Pediatric dermatology workforce in the United States. Pediatr Dermatol 36(1):166–168
14. O'Connor DM, Jew OS, Perman MJ, Castelo-Soccio LA, Winston FK, McMahon PJ (2017) Diagnostic accuracy of pediatric teledermatology using parent-submitted photographs. JAMA Dermatol 153(12): 1243–1248
15. Keating NL, Zaslavsky AM, Ayanian JZ (1998) Physicians' experiences and beliefs regarding informal consultation. JAMA 280(10):900–904
16. Kuo D, Gifford DR, Stein MD (1998) Curbside consultation practices and attitudes among primary care physicians and medical subspecialists. JAMA 280(10): 905–909
17. Golub RM (1998) Curbside consultations and the viaduct effect. JAMA 280(10):929–930
18. Kelsey A, Grant-Kels JM, Payette M (2015) A high-tech ethical twist on the curbside consult. J Am Acad Dermatol 72(2):349–351
19. Chen TS, Goldyne ME, Mathes EFD, Frieden IJ, Gilliam AE (2010) Pediatric teledermatology: observations based on 429 consults. J Am Acad Dermatol 62(1):61–66
20. Burden M, Sarcone E, Keniston A et al (2013) Prospective comparison of curbside versus formal consultations. J Hosp Med 8(1):31–35
21. Shanafelt TD, West CP, Sinsky C et al (2019) Changes in burnout and satisfaction with work-life integration in physicians and the general US working population between 2011 and 2017. Mayo Clin Proc 94(9): 1681–1694
22. Scheinfeld N, Rothstein B (2013) HIPAA, Dermatology images, and the law. Semin Cutan Med Surg 32(4):199–204
23. Kreichelt R, Hilbert ML, Shinn D (2008) Minimizing the legal risk with 'curbside' consultation. J Healthcare Risk Management 28(1):27–29
24. Bal BS, Teo W, Brenner LH (2019) Medicolegal sidebar: getting sued by someone else's patient—when does a curbside consultation carry medicolegal jeopardy? Clin Orthop Relat Res 477(10):2204–2206
25. Brief for the American Medication Association as Amicus Curiae, Warren v. Dinter (2018). searchltf.ama-assn.org/case/documentDownload?uri=/unstructured/binary/casebriefs/warren-v-dinter.pdf. Accessed 4 May 2021
26. Bovi AM (2003) Ethical guidelines for use of electronic mail between patients and physicians. Am J Bioethics 3(3):W-IF2
27. Grenier N, Bercovitch L, Long TP (2009) Cyberdermatoethics II: a case-based approach to teledermatology ethics. Clin Dermatol 27(4):367–341
28. Chuchvara N, Patel R, Srivastava R, Reilly C, Rao BK (2020) The growth of teledermatology: expanding to reach the underserved. J Am Acad Derm 82(4): 1025–1033
29. Fogel AL, Teng JMC (2015) Pediatric teledermatology: a survey of usage, perspectives, and practice. Ped Dermatol 32(3):363–368
30. Naka F, Makkar H, Lu J (2017) Teledermatology: kids are not just little people. Clin Dermatol 35(6):594–600

33 Social Media and How to Overcome Fake News in Health

Cristina M. Pulido and Sandra Racionero-Plaza

Abstract

This chapter explores the last findings on social media and fake news in health, focusing on what are the successful strategies for overcoming disinformation. Specifically evidence sharing of social impact in the electronic media could help overcoming fake news, a strategy in line with international recommendations.

33.1 Introduction

Social media is being increasingly used to find and share information about health; nevertheless, people are flooded with a lot of misinformation in these sources [1]. In a global survey done by Statista in 2020, only 35% of people said they trusted social media—and with the pandemic, people lost even more confidence about which type of information is reliable.

C. M. Pulido (✉)
Department of Journalism and Communication, Autonomous University of Barcelona, Bellaterra, Spain
e-mail: cristina.pulido@uab.cat

S. Racionero-Plaza
Department of Sociology, University of Barcelona, Barcelona, Spain
e-mail: sandraracionero@ub.edu

33.2 Social Media and Fake News in Health

Fake news and misinformation on health issues are not new; however, they are on the rise. At the same time, evidence also indicates that social media helps disseminate trustworthy scientific information [2–4]. In a study conducted by Thelwall and Thelwall [5], 1.9 million retweets done by users with nonofficial messages encouraged the following of safety rules, creating a compliant community during the pandemic [5], even in those countries where politicians were protagonists of evil rumors.

Yang et al. [6] point out that there are many legal and ethical considerations around free speech and censorship that need to be taken into account when addressing this challenge. The negative impact of false health information cannot be avoided; however, the debate should improve the information ecosystem. Stanescu [7] highlights that more than 3.4 million suspicious accounts targeting coronavirus discussions have been suspended on Twitter, and Facebook removed hundreds of accounts due to implication on misinformation campaigns.

33.2.1 Conspiracy Theories

Petit et al. [8] examined how online comments on vaccination news affected readers' credibility assessments and sharing behaviors. Participants

J. Faintuch, S. Faintuch (eds.), *Integrity of Scientific Research*,
https://doi.org/10.1007/978-3-030-99680-2_33

perceived false news with less level of credibility than true news, and this fact diminished their sharing. However, the interactions among users regarding that news did have an impact in the perception of credibility of that news, so that when either false or true, those news that received negative comments from users were perceived as having less credibility. On the contrary, positive comments mitigated such effect. Misinformation shared can promote even racist attitudes, and this should imply, among others, that international institutions should promote education for social media users rather than just removing information.

33.2.2 Mixed Contents

Cato et al.'s [2] research focused on Japanese citizens' use of Twitter, Facebook, and Instagram for sharing information. While many maintained social distance and kept in mind preventive measures that were scientifically reliable, some were endorsing such unsubstantiated approaches as eating fermented soybeans to fight COVID-19. It is concluded that the use of social media to inform health behavior has "both bright and dark sides" [2]. According to the authors, policy interventions should minimize the dark side and maximize the bright side of those uses and not only limit social media interactions.

33.3 Strategies for Overcoming Fake News in Health

Also Hartley and Vu [9] posit that online platforms might increase the proliferation of fake news and misinformation but also recognize and dismiss fake news. Petit et al. [8] recommend that science and health educators should pay attention to the writing of the content that they share online and try to include concise and factual evidence in their argumentation. The dissemination of scientific information should actively engage diverse social platforms [2]. In addition, Myers [10] insists that involving citizens in public health policies is crucial to acquire community resilience and overcome conspiracy theories.

33.3.1 Infodemic Response Checklist

Mheidly and Fares [11] propose (1) direct and effective communication between scientists and the public through direct channels and social media to empower people with facts; (2) promote websites of public health organizations via search engines; (3) verify the accounts of public health personnel on popular social media platforms; (4) promote the posts of public health and medical professionals; (5) monitor engagements on social media platforms to control the messages delivered; (6) establish programs that help people cope with stress and address their mental health concerns; (7) adopt an empathic style of communication to grab public attention and address concerns; (8) promote dialogue to understand people's perceptions and the motives behind their practices; (9) share personal experiences on social media to combat misinformation; (10) direct health communication strategies towards minority populations and people of different classes, races, and ethnicities; (11) develop educational material and speed up the sharing of evidence-based science to address existing wrong perceptions, correct behaviors, and promote healthy practices; and (12) increase investment in the research and development of health communication to explore and understand strategic ways of targeting different populations.

Among the listed behaviors and actions, the 11th is particularly important to help overcome discourses based on disinformation. In fact Pulido et al. [4] found that during the COVID-19 pandemic, people retweeted more tweets with scientific evidence than with false information, indicating that in such health crisis citizens were more engaged online with scientific evidence than with false information. Another study [12] on effective actions to fight misinformation emphasizes that the communication strategies of

Twitter accounts that belong to public health agencies and decision-makers should examine what messaging best meets their specific Twitter audiences. Engaging celebrities for spreading key messages [13] is another approach to overcome misinformation, but only if those shared messages are based on scientific evidence [14].

33.3.2 Quality Media Coverage

Mauri-Ríos et al. [15] posit that news dissemination must be ethical and responsible, in accordance with the following principles: (1) act responsibly; (2) promote personal and community measures in coverage; (3) the journalist must explain public health concepts in a way that is understandable to the general public; (4) it is necessary to respect the truth and the right of the public to know it; (5) it is desirable to consult as many reliable, scientific and proven sources as possible, but they must be carefully selected to distinguish what is known with certainty from what is not proven; (6) know the basics of epidemiology to gauge the relevance of new medical findings and avoid misinformation and fake news; (7) it is not recommended to publish information whose origin is unknown; (8) stories or images posted on social media should be verified and edited before posting; (9) in emergency situations, it is necessary to honestly investigate and publish relevant information, comment fairly, and generate criticism; (10) you must be realistic when reporting and avoid being alarmist or sensationalist or dramatizing the information to attract an audience; (11) it is essential to obtain information, data, or images by legal, lawful, and ethical methods; (12) it is recommended not to amplify the stigmatization of people affected by the disease; (13) informational attention must not only focus on the highest risk groups, and it is recommended to report on stories of overcoming the disease; (14) talking about racial profiling related to Covid-19 should be avoided.

33.3.3 National Risk Communication Plan

Whenever community engagement is a requirement [16], as is often the case in acute events, two-way communication should be a priority to gather public feedback, engaging with community-based public healthcare networks, such as nongovernmental organizations, media platforms, the business sector, education providers, healthcare services, and others, and promoting healthcare and hygiene practices by using different communication platforms.

33.3.4 Interdisciplinary Research

Within the computer science field, one of the crucial elements to fight disinformation is the design of personalization and algorithm-driven content. Linguistics is directly concerned with the use of language at all levels (tones, lexical, style communication), and such analysis on the role of language on constructing fake or true information is central to overcome infodemic (or promote it). Cognitive psychology can contribute on how to overcome disinformation with research on two ways of thinking, one more deliberate and analytic and the other one more intuitive and heuristic. According to authors such as Leitner et al. [17], other disciplines can contribute on areas such as the design of public policies to cope with infodemic and cognitive tools to improve decision-making autonomy.

33.4 Sharing Evidence of Social Impact for Overcoming Fake News

This is a time in history when social impact of science [18, 19] is at the forefront, defined as the social improvements achieved because of the implementation of the results of a particular research project or study, in order to measure the social improvements achieved by scientific research and that contribute to reach the UN Sustainable Development Goals. It is now unquestionable that science in all fields must contribute to improve the life of individuals around the globe, particularly of those belonging to vulnerable groups [20]. In Europe, the expert document entitled *Monitoring the Impact of EU Framework Programmes* [21] established key impact pathways used to assess the social impact of research of the Horizon Europe research framework program.

Research employing a social impact in social media methodology (SISM)—a novel methodology in both social media analytics and the evaluation of the social impact of research [22]—has shown quantitative and qualitative evidence of social impact shared by citizens in social media, such as Twitter and Facebook, those evidences encompassing both real and potential social impact, that is, impact that had already occurred during the research project as well as foreseen impact in the ground of current results. Such finding can have implications for fake news control.

A treatment, a program, a technology, or an intervention that improves the health status, well-being, economy, education, and quality of life of individuals and groups are the ones that people need and, therefore, disseminate the most. Likewise, the more the dissemination of this evidence of social impact, the more the social impact of the research studies that provided such evidence. Quotations from research participants explaining first-person improvements derived from the research, and quantitative results (statistical data) showing effectiveness of treatments/interventions/programs obtained in the project, were elicited in that protocol [22].

Pulido et al. [3] deepened the search for what type of interactions in social media is produced depending on the type of health information shared, either false information or evidence of social impact. In this case, the analysis of social media included Reddit, Facebook, and Twitter. Messages that focused on fake health information were mostly aggressive, while those based on evidence of social impact were respectful and transformative. Among the messages in social media that shared fake information and were aggressive were those on vaccines. They were mostly close statements, not allowing other social media users to engage in dialogue. On the contrary, sharing evidence of social impact in social media created online interactions based on understanding and respect, and these contexts contributed to overcoming false information about health and disease. One key reason for this to happen is that such messages with evidence of social impact allow for deliberation and for open dialogue and are grounded in the force of arguments. These achievements can contribute both to health literacy and to improve public health.

References

1. Swire-Thompson B, Lazer D (2019) Public health and online misinformation: challenges and recommendations. Annu Rev Public Health 41:433–451
2. Cato S, Iida T, Ishida K, Ito A, Katsumata H, McElwain KM, Shoji M (2021) The bright and dark sides of social media usage during the covid-19 pandemic: survey evidence from Japan. Int J Disaster Risk Reduc 54:102034
3. Pulido CM, Ruiz-Eugenio L, Redondo-Sama G, Villarejo-Carballido B (2020) A new application of social impact in social media for overcoming fake news in health. Int J Environ Res Public Health 17(7):2430
4. Pulido CM, Villarejo-Carballido B, Redondo-Sama G, Gómez A (2020) COVID-19 infodemic: more retweets for science-based information on coronavirus than for false information. Int Sociol 35(4): 0268580920914755
5. Thelwall M, Thelwall S (2020) Retweeting for COVID-19: consensus building, information sharing, dissent, and lockdown life. ArXiv 2020:02793. https://doi.org/10.1108/AJIM-05-2020-0134
6. Yang KC, Pierri F, Hui PM, Axelrod D, Torres-Lugo C, Bryden J, Menczer F (2021) The COVID-19 infodemic: Twitter versus Facebook. Big Data Soc 8: 1–16
7. Stanescu GC (2020) Disinformation during the coronavirus pandemic. An analysis of the types of fake news in four states. Soc Sci Educ Res Rev 361(7): 349–361
8. Petit J, Li C, Millet B, Ali K (2021) Can we stop the spread of false information on vaccination? How

online comments on vaccination news affect readers' credibility assessments and sharing behaviors. Sci Commun 43(4):1–28
9. Hartley K, Vu MK (2020) Fighting fake news in the COVID-19 era: policy insights from an equilibrium model. Policy Sci 53(4):735–758
10. Myers N (2021) Information sharing and community resilience: toward a whole community approach to surveillance and combatting the "infodemic". World Med Health Pol 13:1–12
11. Mheidly N, Fares J (2020) Leveraging media and health communication strategies to overcome the COVID-19 infodemic. J Public Health Policy 41(4): 410–420
12. Slavik CE, Buttle C, Sturrock SL, Darlington JC, Yiannakoulias N (2021) Examining tweet content and engagement of Canadian public health agencies and decision makers during COVID-19: mixed methods analysis. J Med Internet Res 23(3):1–18
13. Kamiński M, Szymańska C, Nowak JK (2021) Whose Tweets on Covid-19 gain the most attention: celebrities, political, or scientific authorities? Cyberpsychol Behav Soc Netw 24(2):123–128
14. Wang Y, Gao S, Gao W (2021) Can predominant credible information suppress misinformation in crises? Empirical studies of Tweets related to prevention measures during COVID-19 Yan. ArXiv 2102:00976
15. Mauri-Ríos M, Ramon-Vegas X, Rodríguez-Martínez R (2020) Media coverage of the covid-19 crisis: recommendations and proposals for self-regulation. Prof Inf 29(6):1–17
16. Ali S (2020) Combatting against covid-19 & misinformation: a systematic review. Hu Arenas 5:1–16
17. Leitner S, Gula B, Jannach D, Krieg-Holz U, Wall F (2021) Understanding the dynamics emerging from infodemic: a call to action for interdisciplinary research. SN Bus Econ 1:23
18. Flecha R (2014–2017) IMPACT-EV. Evaluating the impact and outcomes of European SSH Research. cordis.europa.eu/project/id/613202/fr. Accessed 4 May 2021
19. Reale E, Avramov D, Canhial K, Donovan C, Flecha R, Holm P, Larkin C, Lepori B, Mosoni-Fried J, Oliver E, Primeri E, Puigvert L, Scharnhorst A, Schubert A, Soler M, Soòs S, Sordé T, Travis C, Van Horik R (2018) A review of literature on evaluating the scientific, social and political impact of social sciences and humanities research. Res Eval 27(4):298–308
20. Oliver E, Scharnhorst A, Cabré J, Ionescu V (2020) SIOR: an egalitarian scientific agora. Qual Inq 26(8–9):955–961
21. v d Besselaar P, Flecha R, Radauer A (2018) Monitoring the impact of EU framework programmes: expert report. Luxembourg: Publications Office of the European Union. doi.org/10.2777/518781. Accessed 4 May 2021
22. Pulido CM, Redondo-Sama G, Sordé-Martí T, Flecha R (2018) Social impact in social media: a new method to evaluate the social impact of research. PLoS One 13(8):e0203117

Cyberbullying in the University Setting 34

María Carmen Martínez-Monteagudo and Beatriz Delgado

Abstract

There is a significant difference in the number of research projects carried out on cyberbullying in childhood and adolescence when compared to the analysis of this problem in the university environment. However, more and more studies are giving visibility to the online bullying that occurs in higher education. This chapter presents a review of cyberbullying in the university context, focusing its attention not only on the cyberbullying that occurs among students but also on the cyberbullying from students to professors and among teaching staff. The negative consequences of this problem are outlined for those agents involved as well as for the quality of teaching and the institution in general. Finally, a reflection is made on how universities deal with cyberbullying and what preventive and intervention measures can be carried out in the specific context of higher education.

34.1 Types of Cyberbullying: Harassment, Threats, False Profiles, Insulting, Blackmailing, and Other Modalities

The wide use of information and communication technologies (ICTs) has brought multiple advantages for individuals, companies, institutions, and society in general. It has substantially transformed the way in which we communicate, access information, and acquire knowledge in the field of education. However, it has also meant that some of the problems produced in real environments have found their equivalent forms in virtual environments. For example, bullying in academic contexts has given rise to a new variant, cyberbullying. Cyberbullying is defined as "an aggressive, intentional act carried out by a group or individual, using electronic forms of contact, repeatedly and over time against a victim who cannot easily defend him or herself" ([1], p. 376). Tokunaga [2] defines cyberbullying as any behavior by an individual or group carried out with the intention of causing harm or discomfort to others through the repeated dissemination of hostile or aggressive messages via digital media, particularly through using cell phones or the Internet.

Cyberbullying shares three basic characteristics with face-to-face bullying: (1) aggressive behavior; (2) power imbalance between the victim and the aggressor; and

M. C. Martínez-Monteagudo (✉) · B. Delgado
University of Alicante, Alicante, Spain
e-mail: maricarmen.martinez@ua.es; beatriz.delgado@ua.es

J. Faintuch, S. Faintuch (eds.), *Integrity of Scientific Research*,
https://doi.org/10.1007/978-3-030-99680-2_34

(3) repetition of the behavior. However, it also has its own characteristics that, in some cases, carry more negative consequences for those involved than traditional bullying [3]. These include, firstly, how the very nature of cyberspace facilitates the anonymity of the aggressor. In traditional bullying, the victim knows the aggressor or aggressors, with there being a possibility of avoiding them. In cyberbullying, the aggressor can remain anonymous or create a false profile to hide the identity. This, in turn, means greater uncertainty for the victim, who does not know whether the aggressor is one or several people, whether they are acquaintances or strangers, etc.

Moreover, the accessibility and permanence of the material published on the web are significant. In traditional bullying, aggression is usually confined to a specific space and time (generally, the educational environment); however, the characteristics of cyberbullying mean that information is constantly available to the public. Thus, cyberaggressors, taking refuge in anonymity, have the opportunity to disseminate harmful information about their victims to a very wide audience and can humiliate or harass at any time and in any place, maintaining aggressive behavior permanently in the virtual environment [4, 5].

Cyberbullying encompasses a wide variety of behaviors and methods [6, 7].

34.1.1 Classification of Principal Modalities

- Cyber insults: Brief and heated exchange between two or more people, which takes place through new technologies. This may be in the form of private e-mail exchanges or conversations in public contexts (chats, forums, etc.).
- Posting information on the Internet: Posting a compromising image (real or created through photo manipulation), sensitive data, or information that can harm or embarrass the victim and spreading it to their environment.
- Spreading rumors: Circulating rumors in which the victim is supposed to have behaved immorally, offensively, or disloyally, in such a way that other people could, without questioning what they read, exercise their own forms of retaliation or harassment.
- Harassment: Constant offensive messages sent to the victim by e-mail, in public forums such as chat rooms and discussion forums; sending multiple text messages to the targeted person's cell phone. This differs from insulting in that harassment is long term and unilateral (one or more aggressors versus a single victim).
- Belittling: Derogatory and false information regarding the victim that is disseminated through online media (websites, e-mails, instant messages, etc.), e.g., digitally altered photos, especially in a way that reflects sexual attitudes or that may harm the person in question.
- Impersonation: The aggressor impersonates the victim on online networks, sometimes using the victim's personal passwords to subsequently disseminate aggressive messages to other people as if these messages had been sent by the victim himself or herself.
- Disclosure and elicitation: Involves revealing compromising information about the victims to others. Information is either sent privately by the victims according to their own free will and then shared by the aggressors or is elicited from the victims and then disseminated to others.
- Creation of false profiles: Registering the victim, photo included, on a website where the aim is to vote for the ugliest, least intelligent person, etc. Here, the victim is typically given so many points that they appear at the top of the ranking.
- Exclusion: Not allowing a person to participate in a specific social network.
- Taunting: Taunting the victim on web services that have a person responsible for monitoring or moderating what happens there (chats, online games, virtual communities, etc.) in order to achieve a violent reaction that, once reported or seen by others, results in exclusion.
- Cyber persecution: Sending of repeat harassing and threatening electronic

communications. This is related to harassment, although it involves threats to a greater extent.

- Happy slapping: Physical aggression recorded on video with a cell phone which is then subsequently disseminated to online networks.

34.1.2 Most Commonly Employed Technologies

The diversity of Internet communication technologies (ICTs) has expanded the ways in which a person can bully [6, 7]. The main modalities are briefly presented below:

- *Instant messaging*: This refers to communication between individuals using a contact list via *WhatsApp*, *Messenger*, *Skype*, etc. The attack is committed by impersonating the victim's identity or impersonating another person to arrange a meeting and carry out the aggression. This is one of the most frequent forms that has been used for cyberbullying; although its use has been reduced in the light of the rise of social networks.
- *E-mail*: The sending of e-mail messages to harass a person or a group. On many occasions, it is not known who is behind the e-mail account that sends the messages.
- *Cell phone calls*: These are usually made at inappropriate times and are insulting, threatening, etc.
- *Text messages or SMS*: These usually contain voice messages, images, or video recordings, which attack the victim. Today it is a form of aggression that is in disuse due to its economic cost and has been replaced by instant messaging and social networks, which are free and more widespread.
- *Social networks*: *Facebook*, *Instagram*, and *Twitter* are the social networks with the greatest acceptance and largest number of users today. They are free and allow you to post your activities, personal information, and photographs. An alternative to social networks would be social websites (YouTube, Google Video, etc.) where it is possible to upload, share, and comment on videos.
- *Chat*: Chat rooms allow different people to meet. They are similar to SMS in that the aggressor can deceive the victim by pretending to be a different person.
- *Blogs*: These are virtual spaces with diverse information, which can be used to carry out cyberbullying and damage the victim's image.
- *Internet games*: Online players can vocalize their dissatisfaction with other players, insulting or threatening them. Players can block the game for other players and gain unauthorized access to their online gaming accounts.
- *Websites*: This involves creating websites to post offensive information and/or images of the victim. These websites can also be created to conduct surveys in order to humiliate another person (ugliest girl, fattest guy, etc.).

34.2 Cyberbullying Targets: Students, Professors, and Researchers?

Generally, research on cyberbullying has analyzed this problem with a focus on childhood and adolescence [8–10]. However, studies have shown that the university environment is not exempt from this phenomenon, and in fact, cases are progressively increasing. In this regard, the prevalence of cybervictimization has been found to range between 10 and 40% among university students [11–15], making the increased presence of cyberbullying in this student body visible. However, research on this remains scarce compared to the existing empirical evidence regarding cyberbullying against children and adolescents.

Moreover, studies conducted in higher education institutions have generally focused on analyzing the experiences of cyberaggression and cybervictimization among students, neglecting other university groups involved, including faculty members and staff. In fact, increasing research is suggesting that cyberbullying can take place not only from student to student but also from student to professor, from professor to student, or even between professors and other professionals working in university environments. The basic characteristics of these dynamics and descriptions of the targets of the bullying are outlined below.

34.2.1 Cyberbullying Among University Students

In recent years, there has been a considerable increase in the amount of research that is revealing the presence of cyberbullying among university students [11–13, 15–25]. Faucher et al. [11], using a sample of 1925 Canadian university students, found that 24.1% reported having experienced cyberbullying, while 5.1% claimed to have been the perpetrator of cyberbullying. Zalaquett and Chatters [15] found that 19% of a sample of 613 university students reported having been a victim of cyberbullying, while 5% admitted to have been the perpetrators. Cénat et al. [17], using a sample of 4626 French university students, found that 19% had been victims of cyberbullying. Zhou et al. [25] found in a sample of 655 Chinese university students that 37.6% had cyberbullied, and 57.3% reported that they had suffered from this type of harassment.

34.2.2 Physical, Social, Emotional, and Academic Repercussions

Cybervictims present anxiety, low self-esteem, depression, stress, helplessness, irritability, anger, sleep disorders, difficulties in concentrating, absenteeism, poor academic performance, and suicidal thoughts [11, 14, 22, 26–29]. Cyberaggressors often exhibit not only lack of empathy for victims, aggressiveness, drug use, and absenteeism [30, 31] but also anxiety, depression, and, in some cases, suicide [14, 32–34].

The presence of cyberbullying in the university setting could be a continuity of the cyberbullying that occurs in lower educational stages, producing a progressive escalation through the different educational stages. Previous empirical evidence corroborates how the role of aggressor and victim is generally perpetuated from childhood to adulthood and from early educational stages to university [35, 36]. Bauman and Newman [37] found that 3.7% of 79 US college students reported having occasionally been cybervictims, 84.6% of whom had also been bullied in primary education, 80.8% in secondary education, and 73% in both educational stages. In the same vein, Pörhölä [36] found that 50% of university aggressors reported that they had been aggressors during their earlier school years and, similarly, 50% of university victims reported having been bullied at school as well.

Therefore, students who had bullied their peers for years during their childhood and adolescence can be said to normalize such behaviors, considering them even acceptable and, therefore, continuing them at the university level. Students who have been bullied by their classmates at earlier school stages, when moving to the next educational level, are more likely to perceive their peers as a threat rather than as a positive challenge. They may fear their new peers, feel insecure, show defensiveness, and not expect to be approved or valued by their peers. All this, coupled with probable deficits in their social skills, may decrease their chances of integrating adequately into a group, increasing the risk of victimization and abuse [38].

Cyberbullying among university students has certain characteristics which hinder identification and intervention in the face of bullying. On the one hand, the agents involved (victims and aggressors) are of legal age, with possible legal repercussions [39]. Socially and culturally it is expected that having reached this age of maturity, victims of cyberbullying have the necessary skills to both cope with problematic situations and effectively solve any conflict. However, the acquisition of appropriate coping strategies does not occur simply because someone has reached adulthood. Research has suggested that the likelihood of a victim reporting the situation is low, feeling ashamed for not being able to resolve their situation, which has been associated with a lack of maturity by others [40]. In parallel, the characteristics of the university environment (overcrowding of classes, the limited role of the professor as only the academic lead, families more distanced from academic life) hinder the identification of cyberbullying in university classrooms.

34.2.3 Cyberbullying from Student to University Professor

Generally, cyberbullying has been characterized by a power imbalance between the victim and the aggressor. The hierarchical nature of universities favors power imbalances between the agents involved: student to professor, junior to senior professor, tenured to untenured professors, staff to faculty members, department heads to faculty, etc. These are in addition to the power imbalances that occur as a result of gender, age, race, sexual orientation, and social status. However, it should be noted that the use of ICTs to carry out cyberbullying can alter the a priori established hierarchy of power. Contrapower harassment occurs when a person with lesser power intimidates a person with greater power [41, 42]. DeSouza [41] states that contrapower harassment is widespread in the university environment. The very characteristics of virtual interaction allow for, in part, inappropriate behaviors by students towards their professors.

Students have constant access to their professors via e-mail, and this means of communication also favors a decrease in the formality of the exchanges produced, which, together with the perceived anonymity, the absence of non-verbal communication (paralinguistic signals), and the asynchrony of communication, results in a lack of civility and in cyberbullying. Other characteristics of the current university system are also conducive to cyberbullying toward faculty members. As such, the sense of entitlement and the consumerist attitudes of students towards the education they receive are pointed out as key factors. Academic entitlement refers to "expectations of high rewards for modest effort, expectations of special consideration and accommodation by professors when it comes to grades, and impatience and anger when their expectations and perceived needs are not met" ([43], p.1194). Consumerist attitudes towards education may lead a student to perceive that he or she is entitled to high academic grades in exchange for paying tuition.

34.2.4 Impact on the Targeted Professor

A lack of civility and cyberbullying can lead to serious consequences for the victim (the professor) but also for the university as an institution. Research has found that faculty members who are victims of cyberbullying by their students may exhibit high anxiety, depression, disengagement from students and professors, job insecurity, lack of motivation, alcohol abuse, embarrassment, perceived injustice, suicidal thoughts, job absenteeism, low job satisfaction, or decreased teaching quality [3, 41, 42, 44–46].

The limited research that has examined cyberbullying against professors [44, 47–49] has generally confirmed that e-mailing is the most prevalent means, followed by the use of websites such as *Rate My Professor*, *Facebook*, *YouTube*, or *Instagram*, in order to slander, humiliate, or post compromising information about professors [50–53]. Vance [49] conducted a cyberbullying survey of 225 students and 56 university professors working in online learning environments. Thirty-nine percent of faculty members reported that they had experienced cyberbullying at least once, and sixteen percent reported they had experienced it more than once. The majority of faculty members did not report cyberbullying, citing reasons such as doubt that the authorities could help in such a situation, thinking that it was not a crime, not knowing where to report it, and fear of retaliation.

Giumetti et al. [47] confirmed that 26% of US university employees had experienced cybervictimization by their students. Minor et al. [48], using a sample of 68 US professors, found that 33.8% reported being cyberbullied by their students. Two-thirds of the professors attempted to handle and resolve the conflict themselves, while in one-third of the cases, it was handled by their direct supervisor. Professors also reported concerns about reporting harassment, including reasons such as fear that it would affect future teaching opportunities, decrease the number of students enrolled in their classes, cause embarrassment, or fear of not being supported by their supervisors. Cassidy et al. [54] found that 17% of a sample of Canadian university faculty members had experienced cybervictimization. More recently, Coyne et al. [45], using a sample of employees from three universities in the UK, found prevalence rates ranging from 13.6% to 20.8% regarding online victimization.

34.3 Are There Professionals or Organizations Behind Cyberbullying?

Research has neglected to consider the cyberbullying that can also happen among professionals of the university institution. Personal variables play a very relevant role in cyberbullying, and the contextual factors of where such behaviors occur can also significantly influence not only the development of cyberbullying but also its encouragement and perpetuation. The higher education setting itself, due to certain characteristics (hierarchical nature, power imbalances), can lead to cyberbullying also occurring among faculty members.

The most frequent type of bullying in the academic setting is threats about professional status and prevention of access to key resources for the development of teaching or research work (economic resources, space, time) which are critical aspects, given the relevance of teaching and research achievements, intellectual rigor, and reputation in the academic setting. Harassment is less likely to occur through insults, threats, or physical aggression since these acts would be very explicit and, as a consequence, would be reprimanded by other colleagues [55, 56].

The causes for cyberbullying among colleagues can be multiple and complex. For example, the failure to meet expectations, as well as perceptions of injustice or frustration, can all provoke anger, intimidation, and cyberbullying [55]. University faculty members (especially those of higher rank, status, and tenure in the institution) usually present a very defined perception of entitlement, as their academic or professorial freedom provides them with (and leads them to expect) rights and autonomy related to their work. If such expectations are not met, a breach of the psychological contract occurs [57], which may lead to an aggressive reaction, usually being indirect or passive in nature. Aggression may be more direct or active if the aggressor believes that he or she will not receive retaliation or if in a powerful position.

However, in the university environment, perceptions of injustice unique to this context are produced. For example, evaluations by students are taken as a reference for decision-making on issues relevant to a particular faculty member (salary increases, promotion, reappointment, etc.) [54, 55]. If the academic content of the course or the testing set out by the faculty member is very strict, it can have an impact on the evaluation that the students give their professor, which generates feelings of stress and frustration (especially among untenured faculty members). This can lead the professors to "inflate" their students' grades and reduce demands on the student body [54], which may also cause issues with the rest of the faculty members.

Sometimes it is the faculty members themselves who must evaluate the performance of their colleagues or favor their promotion. In an environment where there is competition for economic resources, materials, and space, this can lead to perceptions of injustice that result, in turn, in hostility, and this frustration has generally been related to aggression. The causes of frustration among faculty members are also diverse: strict requirements for promotion, fear of dismissal, inadequate salary increases, student apathy, decreased funding, increased workload, and inconvenient student/class ratio. These issues often have a greater impact on junior or untenured faculty members, which can result in indirect or passive forms of aggression [55].

The job dissatisfaction caused by these situations is particularly relevant due to its close relationship with labor productivity [58]. If labor productivity drops or the professor "backs off," this will irremediably translate into not only a decrease in the merit pay and in the possibility of permanence in the institution but also in the lesser commitment to quality in the teaching-learning process of the student body [55].

34.4 Can Colleagues or Competitors Be Involved?

Keashly and Neuman [59] concluded that colleagues were more likely to be identified as bullies by other faculty members (63.4%), while superiors were more likely to be identified as bullies by frontline staff (52.9%). Sallee and Diaz [60] corroborated that lesbian, gay, bisexual, and queer (LGBQ) faculty members and staff were more likely to experience workplace harassment by superiors, colleagues, and subordinate staff members, highlighting the influence of gender, race, ethnicity, and sexuality.

In addition, research is currently being focused on the relationship between cyberbullying in universities and the gender variable. Cassidy et al. [61] found that female university professors were more likely to be victims of cyberbullying from both students and colleagues. Women who participated in the study expressed a higher level of concern about cyberbullying, and those who had experienced cyberbullying reported a wider range of negative consequences.

Lampman [42] also confirmed that women, minorities, younger professors, and those with less experience and lower professional status were more likely to be bullied. According to Crookston [62], most bullies are of higher-ranking positions and use their rank and hierarchy to cover up their behavior. Cassidy et al. [54], using a sample of Canadian university staff, found that 17% of participants had experienced cyberbullying by students (12%) or colleagues (9%) in the preceding 12 months. Gender differences have been noted, and racial minority status also appeared to make faculty members more vulnerable to cyberbullying. Cassidy et al. [63] concluded that students were generally cyberbullied by other students, whereas faculty members tended to be cyberbullied by other students or colleagues.

34.5 Are Universities Prepared to Deal with This Problem?

Despite the amount of evidence underlining the need for early identification of cyberbullying to prevent its negative consequences [44, 64], the role of institutional monitoring in higher education institutions is often insufficient. Very few universities have policies directed at cyberbullying. This means that these maladaptive behaviors, if addressed on campus, are dealt with through various disjointed policies (student behavior, human rights) that are very difficult for managers to justify and interpret. Clear and uniform policies should be established to address cyberbullying in the university context [64]. This situation needs to be addressed urgently as it has been found that members of the university community, especially faculty members, are in increasing need of the resources and skills to handle cyberbullying effectively [44].

In addition, a lot of cyberbullying takes place without the knowledge of school administrators and managers [63], as many cases of cyberbullying are kept silent by those affected. Students often do not communicate their problem out of fear of not being understood, being made to feel guilty, or having their use of technological devices restricted [64]. Similarly among faculty members, their experiences of bullying are unlikely to be reported to their superiors unless they can be assured that appropriate action will be taken to help the victim and deal with the aggressor [11, 54]. Many victims feel alone, guilty, ignored, or seen as a problem in the workplace [63].

It is essential that universities act with determination and clarity in order to establish cyberbullying control policies based on values that ensure individual rights such as freedom of expression and security of the person, as well as the care and support of all members of the university community.

34.6 Recommended Preventive Measures

Blizard [44] highlighted an approach with different levels of action to improve the prevention of cyberbullying in the university context. As such, higher education institutions should establish clear policies and institutionalized protocols and involve the entire university community (professors, managers, and students) in the development of these measures. In addition, training for faculty members and students on the issue (e.g., what cyberbullying is, how to prevent it) and on the responsible use of the Internet should be prioritized. Likewise, to strengthen the social support of those affected, support and counseling services should be formalized for those who have participated in or have suffered from cyberbullying [64, 65]. By doing this, faculties could provide information on support resources as a response to the first indicators of dissatisfaction or decreased emotional well-being. They could also establish contact with faculty members to create social support networks or even an anonymous hotline for the reporting of cyberbullying cases.

Another strategy that could improve the detection and intervention in cases of cyberbullying would be to strengthen communication and dialogue between faculty members and students. As Blizard [44] found, knowing the concerns and needs of students can help to establish a closer relationship and collaborative work, in which students are more actively involved in the learning process and professors can develop a closer and more effective tutoring methodology in the face of problematic situations.

The development of these skills can reduce the risk of situations of power imbalance and dissatisfaction that could lead those involved to use social networks as a means to hurt others or vent their frustration. In addition, this strategy would help to develop a more respectful university culture where kind behavior, social awareness, respect, and peaceful conflict resolution would be encouraged by all [11, 14].

Incorporating cyberbullying into the academic curricula can lead to improving the dialogue between both parties, providing strategies for identification, recognition, and understanding of the characteristics of maladaptive behavior. It is essential that the heads of educational institutions focus their efforts on establishing tutorial action programs in a generalized manner to improve socio-emotional skills (self-esteem, empathy, emotional intelligence) and management of stressful situations in all groups [44, 54], both in the academic and work environment. University administrators and policymakers should design specific interventions to address the gender and ethnicity aspects of this behavior.

Finally, disciplinary measures can also contribute in the design of effective programs. Some studies have found a positive assessment of these disciplinary measures for the control of cyberbullying by the university community and parents [11, 54].

References

1. Smith P, Mahdavi J, Carvalho M et al (2008) Cyberbullying: its nature an impact in secondary school pupils. J Child Psychol Psychiatry 49(4): 376–385
2. Tokunaga RS (2010) Following you home from school: a critical review and synthesis of research on cyberbullying victimization. Comput Hum Behav 26(3):277–287
3. Cassidy W, Faucher C, Jackson M (2019) You need a thick skin...: impact of cyberbullying in Canadian universities. In: Cassidy W, Faucher C, Jackson M (eds) Cyberbullying at university in international contexts. Routledge, London, UK
4. Garaigordobil M (2011) Prevalencia y consecuencias del cyberbullying: una revisión. Int J Psychol Psychol Ther 11(2):233–254
5. Hinduja S, Patchin J (2010) Bullying, cyberbullying, and suicide. Arch Suicide Res 14:206–221
6. Cowie H, Myers CA (2016) Bullying among university students. Routledge, London, UK
7. Kowalski R, Limber S, Agatston P (2008) Cyber bullying. Bullying in the digital age. Wiley-Blackwell, Blackwell, OK
8. DeSmet A, Aelterman N, Bastiaensens S et al (2015) Secondary school educators' perceptions and practices in handling cyberbullying among adolescents: a cluster analysis. Comput Educ 88:192–201

9. Escortell R, Aparisi D, Martínez-Monteagudo MC et al (2020) Personality traits and aggression as explanatory variables of cyberbullying in Spanish preadolescents. Int J Environ Res Public Health 17: 5705
10. Ho SS, Chen L, Ng A (2017) Comparing cyberbullying perpetration on social media between primary and secondary school students. Comput Educ 109:74–84
11. Faucher C, Jackson M, Cassidy W (2014) Cyberbullying among university students: gendered experiences, impacts, and perspectives. Educ Res Int 6:1–10
12. Finn J (2004) A survey of online harassment at a university campus. J Interpers Violence 19(4): 468–481
13. Lindsay M, Krysik J (2012) Online harassment among college students: a replication incorporating new Internet trends. Inf Commun Soc 15(5):703–719
14. Martínez-Monteagudo MC, Delgado B, Díaz-Herrero A et al (2020) Relationship between suicidal thinking, anxiety, depression and stress in university students who are victims of cyberbullying. Psychiatry Res 286: 112856
15. Zalaquett CP, Chatters SJ (2014) Cyberbullying in college: frequency, characteristics, and practical implications. SAGE Open 4:1–8
16. Abaido GM (2020) Cyberbullying on social media platforms among university students in the United Arab Emirates. Int J Adolesc Youth 25(1):407–420
17. Cénat JM, Smith K, Hébert M et al (2019) Cybervictimization and suicidality among French undergraduate students: a mediation model. J Affect Disord 249:90–95
18. Dilmaç B (2009) Psychological needs as a predictor of cyber bullying: a preliminary report on college students. Educ Sci: Theory and Pract 9(3):1307–1325
19. Martínez-Monteagudo MC, Delgado B, Inglés CJ et al (2019) Cyberbullying in the university setting. Relationship with family environment and emotional intelligence. Comput Hum Behav 91:220–225
20. Mitchell SM, Seegan PL, Roush JF et al (2018) Retrospective cyberbullying and suicide ideation: the mediating roles of depressive symptoms, perceived burdensomeness, and thwarted belongingness. J Interpers Violence 33(16):2602–2620
21. Molluzzo JC, Lawler JP (2012) A study of the perceptions of college students on cyberbullying. Inf Syst Educ J 10(4):84–109
22. Schenk AM, Fremouw WJ (2012) Prevalence, psychological impact, and coping of cyberbully victims among college students. J Sch Violence 11(1):21–37
23. Turan N, Polat O, Karapirli M et al (2011) The new violence type of the era: cyber bullying among university students: violence among university students. Neurol Psychiatry Brain Res 17(1):21–26
24. Walker CM, Sockman BR, Koehn S (2011) An exploratory study of cyberbullying with undergraduate university students. TechTrends 55(2):31–38
25. Zhou Y, Zheng W, Gao X (2019) The relationship between the big five and cyberbullying among college students: the mediating effect of moral disengagement. Curr Psychol 38:1162–1173
26. Aricak OT (2016) The relationship between mental health and bullying. In: Cowie H, Myers CA (eds) Bullying among university students: cross-national perspectives. Routledge, London, UK
27. Giménez AM, Hunter SC, Durkin K et al (2015) The emotional impact of cyberbullying: differences in perceptions and experiences as function of role. Comput Educ 82:228–235
28. Jasso JL, López F, Gámez M (2018) Assessing the links of sexting, cybervictimization, depression, and suicidal ideation among university students. Arch Suicide Res 22(1):153–164
29. Na H, Dancy BL, Park C (2015) College student engaging in cyberbullying victimization: cognitive appraisals, coping strategies, and psychological adjustments. Arch Psychiatr Nurs 29:155–161
30. Aricak OT (2009) Psychiatric symptomatology as a predictor of cyberbullying among university students. Eurasian J Educ Res 34:167–184
31. Hinduja S, Patchin J (2007) Offline consequences of online victimization. J Sch Violence 6:89–112
32. Nansel T, Overpeck M, Haynie D et al (2003) Relationships between bullying and violence among US youth. Arch Pediatr Adolesc Med 157:348–353
33. Nansel T, Overpeck M, Pilla R et al (2001) Bullying behaviours among US youth: prevalence and association with poor adjustment. JAMA 285:2094–2100
34. Seals D, Young J (2003) Bullying and victimization: prevalence and relationship to gender, grade level, ethnicity, self-esteem and depression. Adolescence 38:735–747
35. Cowie H (2019) Cyberbullying and its impact on young people's emotional health and well-being. Psychiat Online 37(5):167–170
36. Pörhölä M (2016) Do the roles of bully and victim remain stable from school ton university? Theoretical considerations. In: Cowie H, Myers CA (eds) Bullying among university students: cross-national perspectives. Routledge, London, UK
37. Bauman S, Newman ML (2013) Testing assumptions about cyberbullying. Perceived distress associated with acts of conventional and cyberbullying. Psychol Violence 3:27–38
38. Chapell MS, Hasselman SL, Kitchin T et al (2006) Bullying in elementary school, high school, and college. Adolescence 41:633–648
39. Myers CA, Cowie H (2017) Bullying at university: the social and legal contexts of cyberbullying among university students. J Cross-Cult Psychol 48(8): 1172–1182
40. Huang CL, Zhang S, Yang SC (2020) How students react to different cyberbullying events: past experience, judgment, perceived seriousness, helping behavior and the effect of online disinhibition. Comput Hum Behav 110:106338

41. DeSouza ER (2011) Frequency rates and correlates of contrapower harassment in higher education. J Interpers Violence 26(1):158–188
42. Lampman C (2012) Women faculty at risk: U.-S. professors report on their experiences with student incivility, bullying, aggression, and sexual attention. NASPA J Women High Educ 5(2):184–208
43. Greenberger E, Lessard J, Chen C et al (2008) Self-entitled college students: contributions of personality, parenting, and motivational factors. J Youth Adolescence 37(10):1193–1204
44. Blizard LM (2016) Faculty members' experiences of cyberbullying by students at one Canadian university: impact and recommendations. Int Res High Educ 1: 107–124. https://doi.org/10.5430/irhe.v1n1p107
45. Coyne I, Farley S, Axtell C et al (2017) Understanding the relationship between experiencing workplace cyberbullying, employee mental straits and job satisfaction: a disempowerment approach. Int J Hum Resour Manag 28(7):945–972
46. Wildermuth S, Davis CB (2012) Flaming the faculty: exploring root causes, consequences, and potential remedies to the problem of instructor-focused uncivil online student discourse in higher education. In: Wankel LA, Wankel C (eds) Misbehavior online in higher education: cutting-edge technologies in higher education. Emerald, Bingley, UK
47. Giumetti GW, McKibben ES, Hatfield AL et al (2012) Cyber incivility @ work: the new age of interpersonal deviance. Cyberpsychol Behav Soc Netw 15(3): 148–154
48. Minor MA, Smith GS, Brashen H (2013) Cyberbullying in higher education. J Educ Res Pract 3:15–29
49. Vance JW (2010) Cyber-harassment in higher education: online learning environments (Doctoral dissertation). digitallibrary.usc.edu/cdm/ref/collection/p15799coll127/id/309077. Accessed 4 May 2021
50. Binns A (2007) Staff suffer bullying by students on the web. Times High Educ. timeshighereducation.co.uk/208037.article. Accessed 4 May 2021
51. Daniloff D (2009) Cyberbullying goes to college. Bostonia. bu.edu/bostonia/spring09. Accessed 4 May 2021
52. Martin Q, Olson D (2011) College cyberbullying: the virtual bathroom wall. The J Technol Student Affairs, Summer [online edition]. studentaffairs.com/ejournal/Summer_2011/CollegeCyberbullying.html. Accessed 4 May 2021
53. McKay R, Arnold D, Fratzl J et al (2008) Workplace bullying in academia: a Canadian study. Empl Responsib Rights J 20(2):77–100
54. Cassidy W, Faucher C, Jackson M (2014) The dark side of the ivory tower: cyberbullying of university faculty and teaching personnel. Alberta J Educ Res 60(2):279–299
55. Keashly L, Neuman JH (2010) Faculty expediencies with bullying in higher education. Adm Theory & Prax 32:48–70
56. Westhues K (2004) Workplace mobbing in academe: reports from twenty universities. Edwin Mellen Press, New York, NY
57. Kiewitz C, Restubog SLD, Zagenczyk T et al (2009) The interactive effects of psychological contract breach and organizational politics on perceived organizational support: evidence from two longitudinal studies. J Manag Stud 46:806–834
58. Ambrose S, Huston T, Norman M (2005) A qualitative method for assessing faculty satisfaction. Res High Educ 46:803–830
59. Keashly L, Neuman JH (2008) Workplace behavior project survey. Minnesota State University, Mankato, MN
60. Sallee M, Diaz C (2012) Sexual harassment, racist jokes, and homophobic slurs: when bullies target identity groups. In: Lester J (ed) Workplace bullying in higher education. New York, Routledge
61. Cassidy W, Jackson M, Faucher C (2016) Gender differences and cyberbullying towards faculty members in higher education. In: Navarro R, Yubero S, Larrañaga E (eds) Cyberbullying across the globe: gender, family, and mental health. Springer, Switzerland
62. Crookston RK (2012) Working with problem faculty: a 6-step guide for department chairs. Jossey-Bass, San Francisco, CA
63. Cassidy W, Faucher C, Jackson M (2017) Adversity in university: cyberbullying and its impacts on students, faculty and administrators. Int J Environ Res Public Health 14:888
64. Faucher C, Cassidy W, Jackson M (2015) From the sandbox to the inbox: comparing the acts, impacts, and solutions of bullying in k-12, higher education, and the workplace. J Educ Train Stud 3:111–125
65. Symons MM, Di Carlo H, Caboral-Stevens M (2021) Workplace cyberbullying exposed: a concept analysis. Nurs Forum 56:141–150

35 Categories of Fake News from the Perspective of Social Harmfulness

Kamil Mamak

Abstract

Fake news impacts different spheres of life among individuals as well as whole populations, and discussion about how to fight it has led to what has been called a regulatory furor. Some emphasize that regulation should be the work of Internet platforms, while others expect states to take an active role. This chapter will demonstrate how nuanced any state interventions must be in relation to potential criminalization. I will categorize five types of fake news in terms of social harmfulness. Each of these categories requires different legal treatment, complicating potential legal responses to fake news, which need to be extremely subtle.

35.1 Introduction

Fake news is not a completely new phenomenon. It was known long before the Internet appeared (cf. [1]); however, we can say that the use of social media as a main mode of news consumption has without doubt created favorable conditions for its increased dissemination [2]. The outbreak of the COVID-19 pandemic showed again how problematic fake news is (cf. [3–6]).

The current discussion on how to fight this phenomenon through legislative means (cf. [7–10]) has recently been extended to the context of COVID-19 [11]. I will look at this problem from the perspective of potential state intervention to eradicate fake news from the Internet by means of criminal law [12]. Social harmfulness demonstrates that the fight against fake news through legal instruments is complicated, and there cannot be a single legal response to all such news.

35.2 What Is Fake News?

Fake news, disinformation, and misinformation are terms that could be seen as parts of the same multilevel problem connected with a kind of information disorder. Definitions have focused on different aspects of these phenomena (e.g., [13–15]). Some point out differences between the three terms, such as the definition in *Science*, according to which:

> "Fake news" is fabricated information that mimics news media content in form but not in organizational process or intent. Fake-news outlets, in turn, lack the news media's editorial norms and

K. Mamak (✉)
RADAR: Robophilosophy, AI ethics and Datafication Research Group, University of Helsinki, Helsinki, Finland

Department of Criminal Law, Faculty of Law and Administration, Jagiellonian University, Kraków, Poland
e-mail: kamil.mamak@uj.edu.pl

J. Faintuch, S. Faintuch (eds.), *Integrity of Scientific Research*,
https://doi.org/10.1007/978-3-030-99680-2_35

> processes for ensuring the accuracy and credibility of information. Fake news overlaps with other information disorders, such as misinformation (false or misleading information) and disinformation (false information that is purposely spread to deceive people). [16]

Other terms seem to combine, at least partially, all the designates from the remaining terms. For example, Molina et al. state that fake news is an umbrella term that covers false news, polarized content, satire, misreporting, commentary, persuasive information, and citizen journalism [17].

There is no place for a definitive decision regarding the definitions of those concepts. At this point, I should state that I am using the term broadly to cover all sorts of information disorders. For example, some websites with fake news regarding COVID-19, to which I will be referring, classify as fake news individual messages sent to people that mimic the correspondence of official institutions. I am more focused on the real or potential effects of fake news on legally protected individual or collective values.

35.3 Social Harmfulness as an Element of Crime

Social harmfulness, which is used in the title of this chapter in the context of criminal law, is connected with a concept which, in Anglo-American literature, is called the harm principle. J.S. Mill famously said that "the only purpose for which power can be rightfully exercised over any member of a civilised community, against his will, is to prevent harm to others" ([18], 94). The Italian thinker Cesare Beccaria also considered harm to be an important element of criminal law. In *On Crimes and Punishment*, Beccaria deliberated on this point in the context of the right to punishment: "Here, then, is the foundation of the sovereign's right to punish crimes: the necessity of defending the repository of public well-being from the usurpations of individuals" ([19], 10). Acts against the "repository of public well-being" could be understood as public harm or harm done to society [20]. Later in his book, Beccaria claimed that "(w)e have seen what the true measure of crime is, namely, harm to society" ([19], 24).

The harm principle has approximations in the continental systems. In German criminal law, which has an impact on many continental legal systems, it is tied to the concept of *Rechtsgut* [21]. Simply put, the aim of these concepts is to set limits on state interventions as regards both criminalization and reaction to individual human conduct. Detailed understandings of the concepts differ (see, e.g., [22–25]), as does the way they are applied in different jurisdictions.

One of the specificities of criminal law, in comparison to other branches of law, is that it is more nation-specific. The process of harmonization is relatively slow, even in the European Union, which is attempting to unify the national laws. The national character of criminal law is partially connected with the fact that traditionally the national state has a right to punish its own citizens. Without going too deeply into the reasons for that state of affairs at this point, I should say that it is useful to illustrate concrete issues connected with criminality with reference to a concrete legal system for purposes of conceptual clarity. In some countries, the social harmfulness of individual conduct is regarded more seriously within the criminal procedure [26]. The authorities are not prosecuting cases of small matter. In others—including Poland—social harmfulness is an element that defines the crime. Whatever the approach, the results are relatively similar: Acts which are characterized by a relatively low level of seriousness will not be punishable.

In Polish criminal law literature, it is quite commonly considered that the Criminal Code distinguishes five elements within the structure of crime ([26]; see also [27]), which, simply put, are (human) acts, unlawfulness, criminality, social harmfulness, and fault. Hence, individual human activity could be a crime if it meets all other criteria of crime. It has to be contrary to the law (unlawfulness), and it must be directly punishable by the legal system (criminality). It must be above a negligible level of social harmfulness, and it must be possible to attribute the fault at the time of the action to the perpetrator. If one of these elements is absent, there is no crime. All elements of the construction of crime are relatively independent and are evaluated separately.

35.3.1 Excusable Violations

According to Article 1, paragraph 2, Polish Criminal Code, "a prohibited act of negligible social harmfulness does not constitute a crime" [28]. From that short provision, a few important issues arise, of which I will mention three. The first is that every crime in Poland has to be socially harmful. If it is not, the act could be treated by the state as a crime. The other issue is that social harmfulness is graduated [29], so the criminal code allows human acts to have different levels of social harmfulness. The third issue is that there could be a socially harmful act which does not meet this criterion because only serious acts could be the basis for crime. In practice, even if there is an act which met the criteria of criminality (e.g., someone destroys public property), the prosecutor or judge could decide that this is not a crime because the act was not serious enough (e.g., [30]). Plebanek, talking about social harmfulness, specifies that a crime has a twofold nature: normative and social. In this understanding, social harmfulness is the social aspect of crime [31] or, in Plebanek's words, the "material aspect of crime" which needs to be present alongside the "formal aspect of crime."

In theory the more socially harmful an act is, the more severe the possible penalties are. From the perspective of fake news, it will be important to bear in mind these two aspects of social harmfulness. In other words, social harmfulness is important at the stage of creating law in order to limit criminalization only to those behaviors that are socially harmful enough to justify the state's reaction. Tarapata points out that only such behaviors may be of interest to criminal law, which, in the opinion of society, are excessively risky for legally protected goods [27].

I would like to focus on two things. First, illegality (as well as criminality) is different from social harmfulness. There could be acts which formally fulfill the elements of a crime described in criminal law but are not socially harmful; thus, they do not constitute a crime. On the other hand, there could be acts which are socially harmful but are not forbidden by law. The fact that the legal system does not consider a socially harmful act to be a crime could result from a deliberate decision taken by legislators but could also be an omission or failure to recognize that the behavior in question should be criminalized. (The latter possibility is especially likely in the case of relatively new phenomena, such as those resulting from new technologies.)

35.4 Social Harmfulness in the Dissemination of Fake News

From the perspective of (social) harmfulness, fake news is not bad per se; what is important is what we do with it, that is, its potential or actual effect. In other words, fake news written in a personal notebook kept on a desk is not yet harmful. Its dissemination, however, could be dangerous for society as it is made publicly visible.

Social harmfulness is a scale rather than a binary concept, as mentioned above. Taking that into consideration, I identify five categories of fake news from the perspective of social harmfulness. (1) Fake news that has negative "social harmfulness"; (2) fake news that is socially harmful, but not enough for criminalization; (3) fake news that is criminalized but whose social harmfulness does not justify intervention by the state; (4) fake news that is socially harmful and criminalized; and (5) fake news that is socially harmful and should be criminalized.

1. Fake news that has negative "social harmfulness"

Could we say that the dissemination of fake news under some circumstances could be beneficial? The Kantian view on lies precludes lies being good (see, e.g., [32]). If we accept the notion that fake news is a variation of lies, then, using the Kantian view, we could not say that any dissemination of fake news could be good. However, using the utilitarian approach and looking at the outcome of an act, there is a possibility that the dissemination of fake news could be socially beneficial. It seems that the idea of social harmfulness is closer to utilitarianism than Kantianism.

In March 2020, a popular report appeared entitled "An Italian priest who contracted coronavirus has died after donating his respirator to a younger patient who was also battling the illness" [33]. It was later discovered that this was not true: "The story was aggregated by a variety of regional and global news sites [. . .] However, that story turned out to be false" [34]. Many people disseminated what turned out to be fake news. It seems that this fake news was rather beneficial. "Goodness" of the act of dissemination could also be interpreted on the grounds of virtue ethics [35, 36], where we focus not on the evaluation of the act but more on the character of a person. That news was about a tragic event, but it taught society about sacrifice and commitment; hence, it had the potential to strengthen relationships. Even if someone disseminated the story while knowing that it was not true, it would be hard to say doing so was not socially beneficial.

35.4.1 Lions on the Streets

One story given worldwide coverage was that Vladimir Putin released lions to keep people indoors because of the COVID-19 outbreak. It turned out, as seems obvious now, that this was not true. Reuters wrote about the case [37]. Applying the same mode of thinking as that used in the case of Italian priest, we could also say that spreading this news could be seen as positive. On the one hand, it could keep some people at home, which is desirable in the face of the spread of coronavirus. On the other hand, such news was humorous and could make society feel better in hard times, which is also positive.

I am not claiming here that we should encourage the creation of fake news; such a view is associated with the risk of overuse. For example, the government could create fake news "for our own good," which seems unacceptable. However, an ex post assessment of fake news could conclude that the outcome of some fake news could be socially beneficial.

2. Fake news that is socially harmful, but not enough for criminalization

I believe that this group could cover some conspiracy theories about COVID-19 and the coronavirus. An up-to-date list of such claims is available, for example, on the website NewsGuard [38]. Two examples are "5G cell phone technology is linked to the coronavirus outbreak" and "The Italian Government is preventing migrants from being tested for COVID-19." Both examples are, in their own way, negative, and we could say for sure that this news is not socially beneficial. It creates a feeling of a planned pandemic, undermining trust in authorities and states. However, such claims are not crimes, and I believe they should not be. Not all negative phenomena need to be criminalized because criminalization should be the ultima ratio. There has to be a strong rationale for criminalization that justifies the limitation of freedom of expression. However the platforms, experts, and organizations that aim to fight fake news should debunk it or flag it as fake news. Noncriminal measures should be taken to solve the issues raised by this kind of fake news.

3. Fake news that is criminalized but whose social harmfulness does not justify intervention by the state

If the act, assessed in concrete terms, is not serious enough, "[. . .] treating (it) as criminal would be an exaggeration" ([26], 178). This could be confusing, bearing in mind that other jurisdictions might solve that issue differently and do not initiate criminal procedures in cases that are characterized by a low level of social harmfulness. Besides constructive differences, the end effect is similar in both the approaches mentioned. To say that some acts which meet the formal criteria of crime are not crimes is only possible in concreto. To be sure of that, we have to know the effect of such acts. In abstracto, a typical act from this category is socially harmful enough—otherwise, it should not be a crime at all. If some fake news becomes popular enough that the press describes it, it usually means that it is harmful enough to be considered a crime.

35.4.2 Fraudulent Crowd Funding

A text message was sent to entrepreneurs which informed them that governmental help connected to the pandemic situation was suspended because there was a small debt on their account with the social security agency [39]. The amount was minimal (around $US1.00), and the text message contained a link to make an online payment. Such an act met the criteria of fraud, which is a crime. However, if it ended up having a relatively marginal impact, it did not meet the requirements associated with social harmfulness. This is also problematic from the perspective of the hopes of automatized crime detection related to fake news. The formal criteria of crime are easier to automate than the actual weight of the act, which is context-dependent.

4. Fake news that is socially harmful and is criminalized

Dissemination of some fake news is already criminalized. The law is to some extent flexible enough to cover new phenomena [40]. Some crimes could be committed by the use of fake news; I mentioned fraud in the previous category, but there are many more, such as defamation or identity theft (see, e.g., [41]). A Europol report about COVID-19 crimes pointed out that some were connected to the dissemination of disinformation:

> The volume of fake news disseminated online around the COVID-19 pandemic has been substantial. Conspiracy theories, harmful narratives undermining public institutions, bogus scientific information, and links to online scams about the coronavirus pandemic have been steadily distributed via Facebook and other social media platforms despite efforts to clamp down on misinformation. [42]

5. Fake news that is socially harmful and should be criminalized

The answer to the question of whether fake news should be criminalized depends on many different factors (see, e.g., [43, 44]). It seems that it is unacceptable to criminalize all fake news because such legislation could introduce censorship. However, systems already penalize lies [45]. In my paper on the criminalization of medical fake news, I proposed the introduction of a new crime if other—noncriminal—measures do not work, and fake news has an increasing impact on public health [46]. Whoever publicly disseminates information evidently discrepant with medical knowledge is subject to a penalty.

For example, information was publicly spread that coronavirus could be treated by the injection of disinfectants [47]. There was, and still is, popular fake news regarding vaccines against COVID-19, showing their alleged harmfulness [38]. For example, "23 elderly people died in Norway after receiving the Pfizer COVID-19 vaccine" [48].

This category is problematic from the perspective of constant change within society. The law can only partially keep up with change regarding technology [40, 49]. Some new phenomena need to be incorporated into law.

35.5 Conclusions

It is unrealistic to hope that the law can solve the problem of fake news. The apparatus of the law can only intervene to a limited extent. Thus, the process of fighting fake news should engage other entities, such as platforms, experts, and civil society. Different stakeholders can play a role. At the same time, we could say that there is a sphere for law, and if some acts are serious enough, the state could and should intervene, prosecuting crimes committed with the use of fake news and even criminalizing behaviors connected with fake news that are serious enough.

References

1. Noor FA (2017) How 'fake news' was a tool of nineteenth century colonialism and conquest. Media Asia 44(2):88–93
2. Shu K, Wang S, Liu H (2019) Beyond news contents: the role of social context for fake news detection. In Proceedings of the Twelfth ACM International Conference on web search and data mining, 312–320. Association for Computing Machinery, New York, NY
3. Apuke OD, Omar B (2021) Fake news and COVID-19: modelling the predictors of fake news sharing among social media users. Telemat Informat 56: 101475
4. van der Linden S, Roozenbeek J, Compton J (2020) Inoculating against fake news about COVID-19. Front Psychol 11:566790
5. Orso D, Federici N, Copetti R, Vetrugno L, Bove T (2020) Infodemic and the spread of fake news in the COVID-19-era. Eur J Emerg Med 27(5):327–328
6. Pennycook G, McPhetres J, Zhang Y, Lu JG, Rand DG (2020) Fighting COVID-19 misinformation on social media: experimental evidence for a scalable accuracy-nudge intervention. Psychol Sci 31(7): 770–780
7. Alemanno A (2018) How to counter fake news? A taxonomy of anti-fake news approaches. Europ J Risk Reg 9(1):1–5
8. Baade B (2018) Fake news and international law. Europ J Int Law 29(4):1357–1376
9. Klein D, Wueller J (2017) Fake news: a legal perspective. Social Science Research Network, Rochester, NY. papers.ssrn.com/abstract=2958790. Accessed 4 May 2021
10. Verstraete M, Bambauer DE, Bambauer JR (2021) Identifying and countering fake news. Social Science Research Network, Rochester, NY
11. Hartley K, Vu MK (2020) Fighting fake news in the covid-19 era: policy insights from an equilibrium model. Policy Sci 53(4):735–758
12. Mamak K (2020b) Walka z Nieprawdziwymi Informacjami o COVID-19 Pokazuje, Jak Trudno Wprowadzić Regulacje Fake Newsów. [Fighting false COVID-19 shows how difficult to regulate fake news]. Centrum Cyfrowe (blog), Warsaw, Poland
13. Egelhofer JL, Lecheler S (2019) Fake news as a two-dimensional phenomenon: a framework and research agenda. Ann Int Commun Assoc 43(2): 97–116
14. Quandt T, Frischlich L, Boberg S, Schatto-Eckrodt T (2019) Fake news. Int Encycl Journ Stud:1–6
15. Tandoc EC Jr, Lim ZW, Ling R (2018) Defining 'Fake News'. Digit Journal 6:137–153
16. Lazer DMJ, Baum MA, Benkler Y, Berinsky AJ, Greenhill KM, Menczer F, Metzger MJ et al (2018) The science of fake news. Science 359(6380): 1094–1096
17. Molina MD, Shyam Sundar S, Le T, Dongwon L (2021) 'Fake news' is not simply false information: a concept explication and taxonomy of online content. Am Behav Sci 65(2):180–212
18. Mill JS (2003) In: Warnock M (ed) Utilitarianism and on liberty: including Mill's "Essay on Bentham" and selections from the writings of Jeremy Bentham and John Austin, 2nd edn. Wiley-Blackwell, Hoboken, NJ
19. Beccaria C (1995) Beccaria: "on crimes and punishments" and other writings. Cambridge University Press, Cambridge, UK
20. Forti G, Visconti A (2007) Cesare Beccaria and White-Collar Crimes' public harm. In: Pontell HN, Geis G (eds) International handbook of white-collar and corporate crime. Springer US, Boston, MA, pp 490–510
21. Persak N (2007) Criminalising harmful conduct: the harm principle, its limits and continental counterparts. Springer-Verlag, New York, NY
22. Baker DJ (2008) Constitutionalizing the harm principle. Crim Just Ethics 27(2):3–28
23. Ripstein A (2006) Beyond the harm principle. Philos Public Aff 34(3):215–245
24. Sadurski W (1990) Joseph Raz on liberal neutrality and the harm principle. Oxf J Leg Stud 10:122
25. Stewart H (2010) The limits of the harm principle. Crim Law Philos 4(1):17–35
26. Wróbel W, Zoll A (2014) Polskie Prawo Karne: Część Ogólna, [Polish criminal law: general part] Wyd. 3. Społeczny Instytut Wydawniczy Znak, Kraków, Poland
27. Tarapata S (2016) Dobro Prawne w Strukturze Przestępstwa: Analiza Teoretyczna i Dogmatyczna, [Legally protected good in the structure of a crime: theoretical and dogmatic analysis]. Wolters Kluwer, Warsaw, Poland
28. Wróbel W, Zontek W, Wojtaszczyk A (eds) (2014) Kodeks karny: przepisy dwujęzyczne [Criminal code]. Lex a Wolters Kluwer Business, Warsaw, Poland
29. Wolter W (1970) O stopniowaniu społecznego niebezpieczeństwa czynu karalnego. [On the grading of the social danger of a criminal act]. Krakowskie Stud Prawnicze 3:109–141
30. Gardocki L (2013) Pojęcie przestępstwa i podziały przestępstw w polskim prawie karnym. [The concept of crime and divisions of crimes in Polish criminal law]. Ann Univ Mariae Curie-Sklodowska 60(2): 29–40
31. Plebanek E (2009) Materialne Określenie Przestępstwa, [Material aspect of crime] Warsaw. Oficyna a Wolters Kluwer business, Poland
32. Sandel MJ (2010) Justice: what's the right thing to do? 1st edn. Farrar, Straus and Giroux, New York, NY
33. Anonymous (2020a) Coronavirus: Italian priest dies after giving his respirator to younger patient he did not know. news.sky.com/story/coronavirus-italian-priest-dies-after-giving-his-respirator-to-younger-patient-he-did-not-know-11962806. Accessed 4 May 2021
34. Orecchio-Egresitz, H (2020) There were reports that a 72-year-old Italian priest who died from coronavirus

gave up his ventilator. That didn't happen. insider.com/italian-priest-who-gave-away-ventilator-died-2020-3. Accessed 4 May 2021
35. Hursthouse R, Pettigrove G (2018) Virtue ethics. In: Zalta EN (ed) The Stanford encyclopedia of philosophy. Stanford University, Stanford, CA. plato.stanford.edu/archives/win2018/entries/ethics-virtue/. Accessed 4 May 2021
36. Russell DC (ed) (2013) The Cambridge companion to virtue ethics. Cambridge University Press, Cambridge, UK
37. Staff R (2020) False claim: Putin released lions on the streets of Russia to Enforce Coronavirus Lockdown. reuters.com/article/uk-factcheck-putin-lions-coronavirus-loc-idUSKBN21C371. Accessed 4 May 2021
38. Anonymous (2021a) Coronavirus misinformation tracking center. newsguardtech.com/coronavirus-misinformation-tracking-center/. Accessed 4 May 2021
39. Anonymous (2021b) Krótki Przewodnik Po Fake Newsach o Koronawirusie [Aktualizacja 05.01.2021]—CyberDefence24. [A brief guide to fake news about coronavirus, Updated 01/05/2021]. cyberdefence24.pl/krotki-przewodnik-po-aktualnych-fake-newsach-o-koronawirusie. Accessed 4 May 2021
40. Fairfield JAT (2021) Runaway technology: can law keep up? Cambridge University Press, Cambridge, UK
41. Mamak K (2020a) Prawnokarne Sposoby Walki z Fake Newsami. [Criminal law methods of combating fake news]. Centrum Cyfrowe, Warsaw, Poland
42. Anonymous (2020b) How COVID-19-related crime infected Europe during 2020. europol.europa.eu/publications-documents/how-covid-19-related-crime-infected-europe-during-2020. Accessed 4 May 2021
43. Cornford A (2012) Criminalising anti-social behaviour. Crim Law Philos 6(1):1–19
44. Malsch M (2007) Stalking: do criminalization and punishment help? Punishment Soc 9(2):201–209
45. Green SP (2018) Lying and the law. In: The Oxford handbook of lying. Oxford University Press, Oxford, UK
46. Mamak K (2021) Do we need the criminalization of medical fake news? Med Health Care Philos 24(2): 235–245
47. Perez M (2020) Trump suggests injecting coronavirus patients with light or disinfectants, alarming experts. forbes.com/sites/mattperez/2020/04/23/trump-suggests-injecting-coronavirus-patients-with-light-or-disinfectants-contradicting-experts/. Accessed 4 May 2021
48. Staff R (2021) Fact check: discussion of deaths in elderly vaccine recipients in Norway lacks context. reuters.com/article/uk-factcheck-norway-idUSKBN29P2R1. Accessed 4 May 2021
49. Mamak K (2019) Rewolucja Cyfrowa a Prawo Karne [Digital Revolution and Criminal Law]. Krakowski Instytut Prawa Karnego Fundacja, Kraków, Poland

Part X

The Ethics of Publications

36 Preregistration of Studies with Existing Data

Gaëtan Mertens and Angelos-Miltiadis Krypotos

Abstract

Preregistration of research plans is becoming an increasingly popular and common tool to enhance the transparency of a study's methodology. In a preregistration, researchers document their research plans and register them to a public repository prior to conducting their research. In this chapter, we provide arguments for why preregistration can protect scientific findings against questionable research practices (QRPs), such as outcome swapping, selective reporting of conditions, unwarranted data exclusions, and post hoc changing of hypotheses. Furthermore, we place particular emphasis on preregistering research plans when using existing data, and we give an overview of preregistration templates and public repositories for different types of research designs. We conclude this chapter with highlighting some of the common criticisms of preregistration and our counterarguments and provide future reflections.

Gaëtan Mertens and Angelos-Miltiadis Krypotos are contributed equally

G. Mertens (✉)
Department of Medical and Clinical Psychology, Tilburg University, Tilburg, The Netherlands
e-mail: g.mertens@tilburguniversity.edu

A.-M. Krypotos
Leuven Research Group on Health Psychology, KU Leuven, Leuven, Belgium

Department of Clinical Psychology, Utrecht University, Utrecht, The Netherlands
e-mail: a.m.krypotos@uu.nl

List of Abbreviations

HARKing	Hypothesizing after the results are known
NHST	*Null hypothesis significance testing*
QRPs	Questionable research practices

36.1 Introduction

Science relies for a large part on the collection and analysis of empirical data. Within the dominant hypothetico-deductive model of science, empirical data may be used to generate novel theories or test existing theories [1]. The underlying idea is that less accurate theories to predict and explain empirical data are gradually replaced with new theories that are more accurate or simpler than the old theories.

Until recently, researchers nearly always had to collect new empirical data to test their theories. However, due to rapid advances in the capacity to easily store and share data on online servers, datasets are now easily available to researchers across the world. While this facilitates the work of scientists, it also brings about new challenges to ensure accurate inferences based on data. To illustrate, can hypotheses still be validly tested on a dataset that has been collected for different

J. Faintuch, S. Faintuch (eds.), *Integrity of Scientific Research*,
https://doi.org/10.1007/978-3-030-99680-2_36

purposes? And how can it be ensured that the hypothesis was really specified independently of the data? Alternatively, are existing data merely useful for exploratory research (i.e., finding patterns in the data by the reanalysis of data)? In this chapter, we aim to answer these questions regarding the reuse of existing data to test hypotheses. First, however, we outline how the validity of hypothesis testing can be threatened through the use of questionable research practices.

36.2 Threats to the Validity of Scientific Inferences: Questionable Research Practices (QRPs)

Typically, when scientists want to test a theory, they will propose a falsifiable hypothesis and test this with empirical data [2]. Because empirical data are nearly always influenced by random noise, statistical models are applied to the data to quantify the reliability of the observations. In addition, statistical inference frameworks are used to deduce the population distributions from the collected sample data. Such deductions, as well as the correction for random noise, are typically expressed within the null hypothesis significance testing (NHST) framework using "*p*-values." This refers to the probability of obtaining test results at least as extreme as the results observed, under the assumption that the null hypothesis is correct. This approach is widespread across the empirical sciences, and many philosophers of science have defended this practice [3, 4].

However, *p*-values and the NHST framework has also received much criticism by other scientists and philosophers [5, 6]. This has become apparent in the study by Bem [7], where empirical data published in an influential psychology journal suggested the presence of precognition (i.e., the ability to predict random future events). This has alarmed psychologists [8] regarding the limitations of the NHST framework. Specifically, it has become apparent that scientists can use questionable research practices (QRPs) to influence the observed results and *p*-values, thereby compromising the validity of scientific findings. Please note that we do not want to suggest that only the NHST framework is sensitive to QRPs (for arguments that it is not when applied properly, see [4]). Indeed, also alternative inferential approaches (e.g., Bayesian hypothesis testing) suffer from limitations due to their sensitivity to misuse and QRPs [9].

Here we provide a short overview of such QRPs and how they can influence the reliability of hypothesis testing and scientific findings. However, note that we do not aim to provide a complete list of such QRPs here. We merely intend to illustrate how different QRPs can result in unreliable findings and incorrect hypothesis tests. For a more exhaustive list of QRPs, see [10].

36.2.1 Not Reporting All Collected Variables

Often, different measurements can be used to test a certain hypothesis. For example, psychologists could test whether social pressure elicits arousal in anxious individuals by measuring skin conductance or heart rate or perhaps both. Indeed, collecting multiple outcome measures is typically seen as good practice to check the generalizability of findings to different measures, as well as help in addressing different research questions within a single study [11]. However, it may be tempting for researchers to selectively report those dependent variables that showed a significant result only and disregard the rest of them. This is especially important since the more tests someone performs within NHST, the higher the changes of a false positive (i.e., the result of random measurement error).

Only presenting the outcome variables that confirm the hypothesis is misleading, as the result can be due to random noise, and the fact that the result was not obtained for the other outcome variables remains obscured. It is therefore commonly recommended by established reporting guidelines to always transparently report on all the collected outcome variables [12].

36.2.2 Failing to Report All Conditions

Much like including different measures in their studies, researchers often also include different conditions in their studies to control for different factors (including both a placebo condition and a wait-list control condition next to the main experimental intervention). When including multiple conditions, results can vary across conditions due to systematic or error variance. For instance, a certain intervention may work well when compared to the wait-list group, but not show a significant effect compared to the placebo group. A researcher can therefore be tempted to only report the comparison between two conditions in which the hypothesis was confirmed, and not report the other conditions. Once again, this compromises the validity of the findings. For instance, it is already well known that interventions tend to have artificially inflated effect sizes when compared to a wait-list control group instead of a placebo-control group [13].

36.2.3 Interim Analyses and Selectively Stopping Data Collection

A third way to influence the results of a study is to collect data and run statistical analyses until a significant result is detected. Due to a fundamental property of the *p*-values, namely, that they tend to decrease with an increasing sample size, additional data collection and uncorrected interim analyses can inflate the chance of a false-positive result. Particularly, *p*-values will always turn out to be significant given a large enough sample of observations [14]. Furthermore, *p*-values tend to fluctuate substantially [15], and it has been argued that the common evidence threshold of $\alpha = .05$ is too liberal and easily results in spurious findings [16]. Given these properties of the *p*-value, *uncorrected* interim analyses and collecting data until the alpha level is crossed will guarantee that a researcher can find a false-positive statistically significant result, thereby greatly increasing the number of spurious results in the literature. Thankfully, there are principled ways to perform interim analyses [17, 18].

36.2.4 Selectively Excluding Data

A fourth way in which results of scientific studies can be compromised is by selectively removing data from the dataset [19, 20]. Due to random noise, there is typically variability in the data of most scientific fields. Without clear pre-specified rules, it is often up to the researchers themselves to decide if certain outliers in the data are due to random error or because of a systematic error that may distort the results. This once again provides an opportunity to capitalize on chance and select those data points that selective support the hypothesis of the researcher, again increasing the chances of finding spurious results.

36.2.5 Changing the Hypothesis After Observing the Results

Another potential way in which the results of a scientific study can be influenced is by adjusting the hypothesis to the observed results. This practice is sometimes referred as "hypothesizing after the results are known" (HARKing) [21]. For instance, a treatment may work in one condition (e.g., low dosage) and not in another (e.g., high dosage). Even though the researchers had initially predicted the opposite pattern, it may be tempting, or it may even happen unintentionally when the hypothesis was not articulated clearly enough beforehand, for the researchers to change their hypothesis. This, however, does not constitute a valid test of a theory because the hypothesis is based on the observed results, rather than specified a priori. Once again, due to random noise in much of the scientific data, this can result in spurious results that are presented as a priori predicted by a flawed theory.

36.2.6 Falsifying and Fabricating Data

Arguably one of the most unethical ways to influence the results of a study is by outright fabrication or falsification of the data to obtain statistically significant "findings" [22]. Though this practice is most likely rare, between 0.3% and 4.9% of researchers self-admit having fabricated or falsified data, and between 5.2% and 33.3% reported personally knowing a colleague who had fabricated or falsified data [23].

36.3 The Consequences of QRPs for Scientific Findings

QRPs can drastically inflate false positives and thus produce unreliable research findings. This has alarmed more and more researcher in recent years. Particularly, several survey studies have shown that scientists self-admit engaging in QRPs [10, 23], and some researchers have raised concerns that as much as 50% or more of the findings in scientific journals are actually false-positive results, in part due to the common (intentional and unintentional) use of QRPs [24, 25]. For example, within the field of psychology, a large-scale replication project of research was only able to replicate 39% of published research findings [26]. Furthermore, a recent study showed that the results of registered reports (a type of journal submission where an article gets accepted before data are collected, merely on the hypotheses and methods to be followed; see below) reported significantly lower percentages of positive results, compared to traditional submissions, casting doubts on how reliable the reported results in the literature are [27]. Results such as these have led several researcher to conclude that psychology and related disciplines are currently suffering from a "replicability crisis" [28, 29]. Nonetheless, similar concerns about the replicability of findings have been raised in other scientific fields, such as cancer research [30], nutrition research [31], and neuroscience [32, 33], indicating that QRPs likely undermine the reliability of scientific findings in many different research areas. This state of (some parts of) the scientific literature is problematic as it has the potential to undermine public trust in scientific research [34] and does not provide stable foundation for further research to be built on.

36.4 Unique Challenges when Using Existing Data

The availability of existing datasets to test hypotheses on can add to the abovementioned problems. Particularly, given that the datasets are already available and that researchers may have preexisting knowledge of their properties, they can use this knowledge to increase their chances to observe statistically significant results, although it is more likely that these results will be spurious or biased. Furthermore, available datasets can often be very large (more than millions of observations) and include many different variables, thereby further increasing the opportunities for using QRPs and finding false-positive results [35, 36].

Additionally, it is more difficult to show with an existing database that a hypothesis was posited prior to looking at the data. When collecting new data, a hypothesis can be publicly announced (for instance, through a preregistration; see the next section) prior to collecting the data. However, when the data are already available, the independence of the hypothesis from the data is more difficult to prove. One exception is when a dataset can only be accessed after approval. In this case, the date of obtaining access to the data can be used to show that a hypothesis was developed independently of the data.

Finally, the widespread availability of datasets to (re-)analyze is relatively new, and there are few guidelines on how to do this correctly. Therefore, it can be argued that establishing guidelines on how to do reliable research and prevent QRPs is particularly important for studies making use of existing data [35, 37].

36.5 Preregistration as a Tool to Protect the Reliability of Scientific Findings

In order to combat these problems with QRPs and unreliable scientific findings, study preregistration has been proposed as a possible solution [38, 39]. In a preregistration document, scientists specify the details regarding their hypotheses, the design of their study, the way in which data are collected, the statistical analysis plan, and the evaluation of the results prior to the execution of the study. This preregistration is typically archived in a (publicly accessible) registration repository prior to conducting the research. By time-stamping this document (i.e., archiving when the preregistration was uploaded to the registry), it can be checked whether the preregistration was available prior to the execution of the study. The preregistration is typically publicly shared prior to the study execution or once the study is accepted, but it could also remain private and only accessible to a selected audience (e.g., co-authors, reviewer, etc.). The idea is that by preregistering important choices in the execution of a study (e.g., sample size, outcome measures, etc.), the flexibility of researchers to (intentionally or unintentionally) influence the results through QRPs is reduced [40] because researchers now specified a plan to follow. If deviations of this plan occur (which can of course happen in a research project), this should then be transparently reported and not presented as an a priori choice.

One of the first areas where the use of preregistrations became widespread is in clinical trials. Many scientific journals within the medical sciences now require clinical trials to be registered in a public repository (e.g., https://www.clinicaltrials.gov/) in order to be considered for publication. More recently, preregistration templates and repositories have also been developed for meta-analyses and systematic reviews (e.g., https://www.crd.york.ac.uk/prospero/) and within the behavioral sciences [41, 42]. For behavioral sciences, the two most common websites for preregistration of a study are osf.io and aspredicted.org, both of which allow researchers to upload their preregistrations for free. An important difference between the two repositories is that although preregistrations in osf.io can stay private for up to 4 years (after which preregistrations become available publicly), in aspredicted.org preregistrations can stay private forever.

36.6 Is Preregistration Necessary for Analyses on Existing Data? Exploratory and Confirmatory Studies

As mentioned above, the typical preregistration document includes a study's hypotheses, methods, planned sample, outcome measures, data transformations, and statistical analyses [42, 43]. In case of the analyses of preexisting data, though, such a complete preregistration is not possible because the study has already been done and the data have already been collected. As such, preregistration of the analyses of preexisting data requires using a different format. However, a first question that needs to be answered is whether the study requires preregistration. It depends on the goal(s) of that study.

First, it may be that a researcher would want to use preexisting data for purely exploratory purposes. In this case, no preregistration is required, given that the researcher has no concrete hypotheses but just performs various tests, attempting to find any interesting data pattern in the data. Although such type of analyses could seem spurious, they are not if they are introduced as such, because the reader can evaluate the provided evidence as stemming from pure exploration. Indeed, exploration is an integral part of the empirical cycle and of scientific discovery [1]. Still, when observing a novel phenomenon in exploratory research, often a follow-up confirmation study is appropriate (i.e., using novel data collection) in order to confirm the novel findings.

For such a "confirmatory study," a researcher could test a clearly specified hypothesis on a dataset that already exists (the European Social

Survey database, the UK Biobank data, etc.). In this case, the hypothesis is specified beforehand, and it is "confirmed" (or disconfirmed) by analyzing independent data (i.e., data that was not used to initially come up with the hypothesis, such as in a prior exploratory study). In a confirmatory study, a preregistration is appropriate and helps ensure that the test of the hypothesis is not tainted by QRPs [44].

We have previously introduced such a template for the preregistration of confirmatory studies on preexisting data [36]. This template consists of ten simple questions, ranging from stating the hypotheses and planned statistical analyses to the provision of clear statements regarding what is already known about the data. This last statement is particularly important because prior knowledge about the data (when dealing with a dataset that has been used before by the same researcher) could limit a study being truly confirmatory (perhaps the researcher already knows that the hypothesized pattern is present in the data). Apart from our own template, also other templates for analyzing secondary data have been recently introduced [37].

36.7 What and How to Preregister: Preregistration Templates and Repositories

Whereas a preregistration is mandatory in many medical journals, other fields are only taking their first steps in applying preregistration for their studies [45]. Therefore, the elements that should be included in the preregistration still differ widely, and, for some types of research designs and in certain scientific fields, no generally accepted standards or templates for preregistration are available as of now [46]. In Table 36.1, we provide a non-exhaustive overview of currently available templates and repositories for some of the most common types of research designs and studies to help guide researchers in choosing an appropriate preregistration format, including preregistration templates for studies using existing data [35–37].

When researchers want to preregister their study, they should complete an appropriate preregistration template for their study and submit this to a relevant repository before conducting their study (or, in the case of analyses on existing data, before inspecting the data and performing the statistical analyses). Furthermore, researchers should refer to this preregistration in the scientific paper resulting from the study. In some journals, preregistration is already required [47], and in other journals, papers with preregistration are designated with open science badges [48].

Finally, a special and powerful version of preregistration are *Registered Reports* [49]. In this format, researchers specify the background of a study, the hypothesis, the design, the sample, the procedure, and the statistical analysis plan, and this is reviewed by a journal *prior* to conducting the study. Once the reviewers and the editor accept this study plan, the authors receive an "in principle acceptance" by the journal, and they can start collecting the data and thereafter submit their final paper. The final paper will again undergo peer review to check whether the study plan was followed. Crucially, the journal will publish the final version of the manuscript, regardless of the obtained results. With this format, QRPs are maximally controlled, and publication of the findings is not based on the direction of the results but merely on the idea and methodology of the study. An updated list of journals offering this publishing format can be found on the Open Science Framework (https://osf.io/rr/).

36.8 Limitations and Critiques of Preregistration

The most common criticism against preregistration is that it does not fit all types of studies, [50], for example, in longitudinal studies, where the final dataset is available only years after the beginning of a study. Given that nowadays statistical methodologies develop rapidly, it is probable that by the end of a study, a better method for addressing the research question will be available. However, since this analysis was not

Table 36.1 Non-exhaustive review of preregistration templates and repositories for different types of studies

Type of study	Preregistration templates	Possible public repositories
Qualitative studies	Qualitative Preregistration template [46]	https://osf.io/
Quantitative behavioral research	As Predicted template (https://osf.io/fnsb6/) Pre-Registration in Social Psychology [42]; https://osf.io/ce3hr/)	https://osf.io/ https://aspredicted.org/
Analyses on existing data	Mertens and Krypotos [36] template (https://osf.io/3tbwc/) Weston et al. [35] template (https://osf.io/x4gzt/)	https://osf.io/
Randomized controlled trials	Use the protocol registration and results system (see www.clinicaltrials.gov)	https://www.clinicaltrials.gov/ https://www.clinicaltrialsregister.eu/
Systematic reviews/meta-analyses	Preferred reporting items for systematic review and meta-analysis protocols (PRISMA-P)	https://www.crd.york.ac.uk/prospero/ https://srdr.ahrq.gov/

Note: A more exhaustive list of preregistration templates can be found on the Open Science Framework: https://osf.io/zab38/wiki/home/

preregistered, researchers may worry that it is not allowed to perform it.

This idea stems from the common misconception that after the submission of a study's preregistration, researchers' hands are tied, and any deviation from the plan should be interpreted as engaging in QRPs. In our view, however, this is an unfortunate misrepresentation and misinterpretation of the goals and implementation of preregistrations. The preregistration is a plan, and a plan can be changed or updated [51]. As long this is done in a transparent manner, by updating the preregistration template before the data are inspected, with the authors describing the changes in their preregistration or by listing the deviations from the preregistration in the final paper, there is no reason to hang on to the choices made in the preregistration and for the authors to be accused of QRPs.

Another common argument is that preregistration cannot really protect from QRPs, simply because someone could preregister a study after the data have been inspected, or preregister multiple studies, keep the preregistrations private, and then only release the preregistration that better fits the direction of the results. Alternatively, and less dramatically, preregistrations could simply not be clearly specified enough and/or not followed by researchers and thereby not really protect against QRPs [52, 53].

We agree that indeed a study's preregistration is not a tool that can guarantee a 100% safeguard against QRPs and a preregistered study is not necessarily a good study. Still, we caution against throwing the baby out with the bathwater. Given that QRPs are fairly common and often happen unintentionally [23, 54], preregistration is a valuable tool to encourage researchers to be transparent in their choices. Furthermore, even if an intervention is not 100% effective, it can nonetheless be a useful tool to reduce, though perhaps not eliminate, (unintentional) QRPs.

36.9 Closing Remarks and Future Perspectives

We focused mostly on preregistration as a tool to reduce (unintended) QRPs and increase transparency and did not evaluate preregistration in its ability to test the severeness of a test, a topic that is most relevant for the field of philosophy of science [44].

Although the idea of preregistering studies has been around since the 1960s, researchers have only relatively recently started using preregistration for research in different areas (e.g., psychology, philosophy, social sciences). As such, for these fields, preregistration is a relatively new tool, and it is still being further developed and evaluated [45, 46]. Given the relatively quick developments in the field of open science in psychology and beyond, we anticipate that a study's preregistration will become more common for different types of research designs and scientific disciplines. This will be a good step towards promoting more transparency in our research. Such transparency is important for science consumers in evaluating scientific research and for having trust in scientific research findings [34].

Preregistration is not a catchall solution for other problems that may exists in different scientific fields, such as vague theories or poor external validity [55]. Finally, it is important that preregistration be sufficiently specific and actually followed. Indeed, a number of recent studies found that preregistrations are not always followed carefully, and this is often not reported transparently [52, 53]. In these cases, the usefulness of preregistration to protect against QRPs is obviously diminished, and papers may undeservedly receive credit for good practices that were not actually adhered to. That said, a preregistration does not and should not prevent researchers from choosing the optimal statistical models and data points for their research aims, and researchers should always be allowed to explore their data to discover new patterns and findings (provided that this is reported as such).

The wider acceptability of preregistration in the community is likely going to take more effort as it calls for a wider change in the scientific culture and the current way of doing science. Typically most of the important decisions (e.g., the exact hypothesis, the statistical model, data exclusions, etc.) are now commonly taken during or at the end of the study, while this should preferably be done beforehand. Furthermore, most researchers are under significant pressure to publish articles and journals and often prefer publishing positive results. Flexibility in the specification of the details of a study can help researchers find such positive results and thus publish more easily [54, 56]. As such, many researchers are still disincentivized from adopting the practice of preregistering their study. Nonetheless, given that preregistration provides important advantages to the transparency and reliability of scientific research, we expect that more funders, universities, and journals will require researchers to preregister their studies in the future.

References

1. De Groot AD (2014) The meaning of "significance" for different types of research [translated and annotated by Eric-Jan Wagenmakers, Denny Borsboom, Josine Verhagen, Rogier Kievit, Marjan Bakker, Angelique Cramer, Dora Matzke, Don Mellenbergh, and Han L. J. van der Maas]. Acta Psychol 148:188–194
2. Popper K (1959) The logic of scientific discovery. Hutchinson, London
3. Mogie M (2004) In support of null hypothesis significance testing. Proc R Soc Lond Ser B Biol Sci 7:271. https://royalsocietypublishing.org/doi/10.1098/rsbl.2003.0105
4. Lakens D (2021) The practical alternative to the p value is the correctly used p value. Perspect Psychol Sci 16:639–648. http://journals.sagepub.com/doi/10.1177/1745691620958012
5. Carver R (1978) The case against statistical significance testing. Harv Educ Rev 48(3):378–399. https://meridian.allenpress.com/her/article/48/3/378/21418/The-Case-Against-Statistical-Significance-Testing
6. McShane BB, Gal D, Gelman A, Robert C, Tackett JL (2019) Abandon statistical significance. Am Stat 73:235–245. https://www.tandfonline.com/doi/full/10.1080/00031305.2018.1527253
7. Bem DJ (2011) Feeling the future: experimental evidence for anomalous retroactive influences on cognition and affect. J Pers Soc Psychol 100(3):407–425. http://doi.apa.org/getdoi.cfm?doi=10.1037/a0021524
8. Wagenmakers E-J, Wetzels R, Borsboom D, van der Maas HLJ (2011) Why psychologists must change the way they analyze their data: the case of psi: Comment on Bem (2011). J Pers Soc Psychol 100(3):426–432. http://doi.apa.org/getdoi.cfm?doi=10.1037/a0022790
9. Tendeiro JN, Kiers HAL (2019) A review of issues about null hypothesis Bayesian testing. Psychol Methods 24(6):774–795. http://doi.apa.org/getdoi.cfm?doi=10.1037/met0000221
10. John LK, Loewenstein G, Prelec D (2012) Measuring the prevalence of questionable research practices with incentives for truth telling. Psychol Sci 23(5):524–532. http://journals.sagepub.com/doi/10.1177/0956797611430953
11. LoBue V, Reider LB, Kim E, Burris JL, Oleas DS, Buss KA et al (2020) The importance of using multiple outcome measures in infant research. Infancy 25(4):420–437. https://onlinelibrary.wiley.com/doi/abs/10.1111/infa.12339

12. Schulz KF, Altman DG, Moher D (2010) CONSORT 2010 Statement: updated guidelines for reporting parallel group randomised trials. Trials 11(1):32. https://trialsjournal.biomedcentral.com/articles/10.1186/1745-6215-11-32
13. Cuijpers P, Cristea IA (2016) How to prove that your therapy is effective, even when it is not: a guideline. Epidemiol Psychiatr Sci 25(5):428–435. https://www.cambridge.org/core/product/identifier/S2045796015000864/type/journal_article
14. Wagenmakers E-J (2007) A practical solution to the pervasive problems of p values. Psychon Bull Rev 14(5):779–804. http://www.springerlink.com/index/10.3758/BF03194105
15. Cumming G (2014) The new statistics. Psychol Sci 25(1):7–29. http://journals.sagepub.com/doi/10.1177/0956797613504966
16. Benjamin DJ, Berger JO, Johannesson M, Nosek BA, Wagenmakers E-J, Berk R et al (2018) Redefine statistical significance. Nat Hum Behav 2(1):6–10. http://www.nature.com/articles/s41562-017-0189-z
17. Lakens D (2014) Performing high-powered studies efficiently with sequential analyses. Eur J Soc Psychol 44(7):701–710. http://doi.wiley.com/10.1002/ejsp.2023
18. Schönbrodt FD, Wagenmakers E-J, Zehetleitner M, Perugini M (2017) Sequential hypothesis testing with Bayes factors: efficiently testing mean differences. Psychol Methods 22(2):322–339. http://doi.apa.org/getdoi.cfm?doi=10.1037/met0000061
19. Lonsdorf TB, Klingelhöfer-Jens M, Andreatta M, Beckers T, Chalkia A, Gerlicher A et al (2019) Navigating the garden of forking paths for data exclusions in fear conditioning. elife 8:e52465
20. Morís Fernández L, Vadillo MA (2020) Flexibility in reaction time analysis: many roads to a false positive? R Soc Open Sci 7(2):190831. https://royalsocietypublishing.org/doi/10.1098/rsos.190831
21. Kerr NL (1998) HARKing: hypothesizing after the results are known. Personal Soc Psychol Rev 2(3):196–217
22. Neuroskeptic (2012) The nine circles of scientific hell. Perspect Psychol Sci 7(6):643–644. http://journals.sagepub.com/doi/10.1177/1745691612459519
23. Fanelli D (2009) How many scientists fabricate and falsify research? A systematic review and meta-analysis of survey data. PLoS One 4(5):e5738. https://dx.plos.org/10.1371/journal.pone.0005738
24. Simmons JP, Nelson LD, Simonsohn U (2011) False-positive psychology: undisclosed flexibility in data collection and analysis allows presenting anything as significant. Psychol Sci 22(11):1359–1366. http://journals.sagepub.com/doi/10.1177/0956797611417632
25. JPA I (2005) Why most published research findings are false. PLoS Med 2(8):e124. https://dx.plos.org/10.1371/journal.pmed.0020124
26. Open Science Collaboration (2015) Estimating the reproducibility of psychological science. Science 349(6251):aac4716. http://www.sciencemag.org/cgi/doi/10.1126/science.aac4716
27. Scheel AM, Schijen M, Lakens D (2021) An excess of positive results: comparing the standard psychology literature with registered reports. Adv Methods Pract Psychol Sci 4:25152459211007467
28. Pashler H, Wagenmakers E (2012) Editors' introduction to the special section on replicability in psychological science. Perspect Psychol Sci 7(6):528–530. http://journals.sagepub.com/doi/10.1177/1745691612465253
29. Tackett JL, Lilienfeld SO, Patrick CJ, Johnson SL, Krueger RF, Miller JD et al (2017) It's time to broaden the replicability conversation: thoughts for and from clinical psychological science. Perspect Psychol Sci 12(5):742–756. http://journals.sagepub.com/doi/10.1177/1745691617690042
30. Wen H, Wang H-Y, He X, Wu C-I (2018) On the low reproducibility of cancer studies. Natl Sci Rev 5(5):619–624. https://academic.oup.com/nsr/article/5/5/619/4835582
31. Sorkin BC, Kuszak AJ, Williamson JS, Hopp DC, Betz JM (2016) The challenge of reproducibility and accuracy in nutrition research: resources and pitfalls. Adv Nutr 7(2):383–389. https://academic.oup.com/advances/article/7/2/383/4558081
32. Button KS, Ioannidis JPA, Mokrysz C, Nosek BA, Flint J, Robinson ESJ et al (2013) Power failure: why small sample size undermines the reliability of neuroscience. Nat Rev Neurosci 14:365–376
33. Botvinik-Nezer R, Holzmeister F, Camerer CF, Dreber A, Huber J, Johannesson M et al (2020) Variability in the analysis of a single neuroimaging dataset by many teams. Nature 582(7810):84–88. http://www.nature.com/articles/s41586-020-2314-9
34. Wingen T, Berkessel JB, Englich B (2020) No replication, no trust? How low replicability influences trust in psychology. Soc Psychol Personal Sci 11(4):454–463. http://journals.sagepub.com/doi/10.1177/1948550619877412
35. Weston SJ, Ritchie SJ, Rohrer JM, Przybylski AK (2019) Recommendations for increasing the transparency of analysis of preexisting data sets. Adv Methods Pract Psychol Sci 2(3):214–227. http://journals.sagepub.com/doi/10.1177/2515245919848684
36. Mertens G, Krypotos A-M (2019) Preregistration of analyses of preexisting data. Psychol Belg 59(1):338–352. http://www.psychologicabelgica.com/articles/10.5334/pb.493/
37. van den Akker O, Weston SJ, Campbell L, Chopik WJ, Damian RI, Davis-Kean PE et al (2019) Preregistration of secondary data analysis: a template and tutorial. PsyArXiv. https://doi.org/10.31234/osf.io/hvfmr
38. Munafò MR, Nosek BA, Bishop DVM, Button KS, Chambers CD, Percie du Sert N et al (2017) A manifesto for reproducible science. Nat Hum Behav 1(1):0021. https://doi.org/10.1038/s41562-016-0021
39. Nosek BA, Alter G, Banks GC, Borsboom D, Bowman SD, Breckler SJ et al (2015) Promoting an open

research culture. Science 348(6242):1422–1425. http://www.sciencemag.org/cgi/doi/10.1126/science.aab2374

40. Nosek BA, Beck ED, Campbell L, Flake JK, Hardwicke TE, Mellor DT et al (2019) Preregistration is hard, and worthwhile. Trends Cogn Sci 23(10): 815–818. https://linkinghub.elsevier.com/retrieve/pii/S1364661319301846
41. Krypotos A-M, Klugkist I, Mertens G, Engelhard IM (2019) A step-by-step guide on preregistration and effective data sharing for psychopathology research. J Abnorm Psychol 128(6):517–527
42. van 't Veer AE, Giner-Sorolla R (2016) Pre-registration in social psychology—a discussion and suggested template. J Exp Soc Psychol 67:2–12. https://doi.org/10.1016/j.jesp.2016.03.004
43. Nosek BA, Ebersole CR, DeHaven AC, Mellor DT (2018) The preregistration revolution. Proc Natl Acad Sci 115(11):2600–2606. http://www.pnas.org/lookup/doi/10.1073/pnas.1708274114
44. Lakens D (2019) The value of preregistration for psychological science: a conceptual analysis. Japanese Psychol Rev 62(3):221–230
45. Polonioli A, Vega-Mendoza M, Blankinship B, Carmel D (2021) Reporting in experimental philosophy: current standards and recommendations for future practice. Rev Philos Psychol 12:49–73. http://link.springer.com/10.1007/s13164-018-0414-3
46. Haven TL, Errington TM, Gleditsch KS, van Grootel L, Jacobs AM, Kern FG et al (2020) Preregistering qualitative research: a Delphi Study. Int J Qual Methods 1(19):160940692097641. http://journals.sagepub.com/doi/10.1177/1609406920976417
47. DeAngelis CD, Drazen JM, Frizelle FA, Haug C, Hoey J, Horton R et al (2005) Clinical trial registration. Arch Dermatol 141(1):76–77. http://archderm.jamanetwork.com/article.aspx?doi=10.1001/archderm.141.1.76
48. Kidwell MC, Lazarević LB, Baranski E, Hardwicke TE, Piechowski S, Falkenberg L-S et al (2016) Badges to acknowledge open practices: a simple, low-cost, effective method for increasing transparency. PLOS Biol 14(5):e1002456. https://dx.plos.org/10.1371/journal.pbio.1002456
49. Chambers CD (2013) Registered reports: a new publishing initiative at Cortex. Cortex 49(3):609–610. https://doi.org/10.1016/j.cortex.2012.12.016
50. Pham MT, Oh TT (2021) Preregistration is neither sufficient nor necessary for good science. J Consum Psychol 31(1):163–176. https://onlinelibrary.wiley.com/doi/10.1002/jcpy.1209
51. DeHaven AC (2017) Preregistration: a plan, not a prison. COS blog [Internet]. https://www.cos.io/blog/preregistration-plan-not-prison
52. Claesen A, Gomes S, Tuerlinckx F, Vanpaemel W (2019) Preregistration: comparing dream to reality. PsyArXiv
53. Bakker M, Veldkamp CLS, van Assen MALM, Crompvoets EAV, Ong HH, Nosek BA et al (2020) Ensuring the quality and specificity of preregistrations. PLoS Biol 18(12):e3000937. https://dx.plos.org/10.1371/journal.pbio.3000937
54. Grant DB, Kovács G, Spens K (2018) Questionable research practices in academia: antecedents and consequences. Eur Bus Rev 30(2):101–127. https://www.emerald.com/insight/content/doi/10.1108/EBR-12-2016-0155/full/html
55. Szollosi A, Kellen D, Navarro DJ, Shiffrin R, van Rooij I, Van Zandt T et al (2020) Is preregistration worthwhile? Trends Cogn Sci 24(2):94–95. https://linkinghub.elsevier.com/retrieve/pii/S1364661319302852
56. Fanelli D (2010) Do pressures to publish increase scientists' bias? An empirical support from US States data. PLoS One 5(4):e10271. https://dx.plos.org/10.1371/journal.pone.0010271

Preregistration in Animal Research 37

Céline Heinl, Gilbert Schönfelder, and Bettina Bert

Abstract

Attempts to replicate published studies have too often failed, and promising preclinical results could not be translated into clinical breakthroughs. Major causes for this lack of reproducibility are practical reasons such as selective reporting, poor study design, or insufficient method description. However, a rigid incentive system shaped by publishers, institutions, and funders intensifies the problem by favoring exciting results over robust ones. Poor quality of biomedical research not only slows down the scientific progress, it also represents an ethical issue. When animal experiments do not contribute to any knowledge gain, the justification for these experiments is not given, and animal lives are wasted. Preregistration of animal research, i.e., registration of the study protocol before the experiment has started, supports scientists in designing an excellent research study, prevents questionable scientific practices, and improves the reporting of all outcomes. With the online platform animalstudyregistry.org, scientists worldwide can now easily register their animal study and show their commitment to open science. To enable a broad uptake of preregistration into biomedical research, funders, publishers, and institutions need to acknowledge preregistration as an engagement for transparent and robust science.

37.1 Causes for Questionable Quality of Biomedical Research

In 2005 a theoretical analysis of the statistics used in the current literature led John Ioannidis to the conclusion that most published research findings are false [1]. Most attempts to replicate selected studies in cancer research that were published in renowned scientific journals failed [2]. If even the pure replication of published findings fails, it seems not surprising that promising results from basic and preclinical research rarely translate into clinical successes. The pharmaceutical industry observed already years ago that most tested treatments in clinical trials fail in phase 3 because of missing effectivity [3, 4]. This lack in quality and validity found in published biomedical

The original version of this chapter was previously published non-open access. A Correction to this chapter is available at https://doi.org/10.1007/978-3-030-99680-2_61

C. Heinl · B. Bert (✉)
German Federal Institute for Risk Assessment, German Centre for the Protection of Laboratory Animals (Bf3R), Berlin, Germany
e-mail: celine.heinl@bfr.bund.de; bettina.bert@bfr.bund.de

G. Schönfelder
German Federal Institute for Risk Assessment, German Centre for the Protection of Laboratory Animals (Bf3R), Berlin, Germany

Charité-Universitätsmedizin Berlin, Corporate Member of Freie Universität Berlin, Humboldt-Universität zu Berlin, and Berlin Institute of Health, Berlin, Germany
e-mail: gilbert.schoenfelder@bfr.bund.de

J. Faintuch, S. Faintuch (eds.), *Integrity of Scientific Research*,
https://doi.org/10.1007/978-3-030-99680-2_37

literature can have multiple causes that need to be addressed in order to speed up medical progress, to prevent the loss of resources and time, and to stop the waste of animal lives.

37.1.1 Basic Methodological Principles

The causes for the so-called "reproducibility crisis" are numerous and occur along the entire research process, from the first idea for the experiment to the publication of the results. Common mistakes already appear during the planning of a study. For example, scientists still too rarely apply crucial methods to reduce bias, such as blinding or randomization, in preclinical experiments [5]. A low statistical power further undermines the validity of the result [1]. Fundamental misconceptions of the study design can also impair the validity and translatability of the outcome. Scientists still often aim for a high standardization, and control as many parameters as possible, although this approach can reduce the external validity of results. Searching for cures in young, healthy, male inbreed mice living in a standardized environment with a common diet might be the wrong approach when working on diseases that concern an older, mainly female population, living diverse lifestyles and dealing with additional health conditions [6]. Introducing a controlled heterogeneity could boost the external validity of results [7, 8].

37.1.2 Manipulation of Hypotheses and Statistics

Scientists often exploit the flexibility of statistical analysis to get a p-value smaller than 0.05 by changing statistical tests, or defining exclusion or inclusion criteria *a posteriori*, or continuing experiments until significance is reached without an *a priori* sample size calculation. With a text-mining approach, Head and colleagues extracted significant p-values from thousands of research articles, and observed an unexpected accumulation of p-values just below the level of 0.05, which suggests intentional manipulation [9]. These data show the broad distribution of so-called p-hacking across different scientific disciplines. Similarly, it is common practice to define the hypothesis after the results are known (HARKing), although defining a clear hypothesis *a priori* is crucial to calculate a sample size beforehand and to obtain robust and unbiased results [10]. Certainly, recording as many variables as possible and screening data afterwards in order to find a significant result will strongly impair its validity. Exploratory research, which is performed with a broad question and a vague idea about a probable outcome, has its value, especially to create new hypotheses. However, it needs to be labeled as such to enable readers to better classify the validity of the results.

37.1.3 Poor Reporting of Results and Methods

Journal checklists and reporting guidelines have been established over the last years to motivate scientists to report all details of the study that might be necessary for the interpretation or replication of results. The Animal Research Reporting of In Vivo Experiments (ARRIVE) guidelines were proclaimed to specifically improve reporting of in vivo animal research [11, 12]. Within the last ten years, over a thousand scientific journals endorsed the ARRIVE guidelines, which were updated in 2020 [12]. However, too often, authors still do not report all points of the guidelines, even after being asked by editors to fill out a checklist during the publication process of their manuscripts [13].

In a survey conducted by the journal *Nature* amongst 1500 scientists, participants identified "selective reporting" as a major cause for the "reproducibility crisis" [14]. Indeed, in the published biomedical literature, there is a strong tendency to report rather novel, exciting, and significant findings, whereas so-called "negative" or "null" results are missing. A meta-analysis including over 500 studies investigated the effect of different treatments on the infarct volume in animal models of stroke. The analysis showed that only 2% of the studies did not report a significant effect [15]. This publication bias leads to an overestimation of treatment effects in published literature that can explain later failures in the translation into clinics. Thus, in stroke research, most positive effects are lost along the way, from the experimental animal models of stroke to phase III clinical trials [4].

37.1.4 Publish or Perish

The rigid incentive system of funders, institutions, and publishers exacerbates all the causes listed above. Scientists are primarily evaluated on their publication records when they apply for their next position or a research grant. These publications are, however, not judged by the quality of the work but mainly by the journal impact factor. Publishers prefer novel, exciting, and significant findings that will guarantee a high number of citations and with this a high impact factor for the journal. To prevail in a competitive research system and thus, being able to pursue their research, scientists are pressured to follow these evaluation criteria.

Within this competitive environment, an attitude of secrecy has also developed that further intensifies the lacking transparency in biomedicine. Thus, sharing raw data or method protocols that could help to understand and to replicate results remains marginal [16].

Selective reporting and unsatisfying quality of published fundamental research will lead to a loss of public money and can put human lives at risk when clinical trials are based on loose scientific fundaments. It has to be further acknowledged that biomedical literature is mainly based on animal experiments which adds an additional ethical issue.

37.2 The Ethical Challenge for Animal Research

In our society, animal testing is still a controversial topic. The public only tolerates animal experiments under the assumption that they will contribute to a knowledge gain that will enable scientific progress or lead to medical breakthroughs one day [17]. However, recent studies have pointed out that scientists have not published a considerable part of conducted animal experiments, therefore failing to contribute to potential knowledge gain. In a survey from 2012, it was generally estimated that only 50% of animal experiments were published. Asked about their own research, this estimate increased to 80% [18]. Wieschowski and colleagues found that in two German university medical centers, only 67% of processed protocols led to at least one publication six years after approval [19]. This number dropped to 58%, if doctoral theses were excluded from the analysis. Similarly, van der Naald and colleagues found that at the University Medical Center Utrecht in the Netherlands only 60% of animal study protocols led to a publication, and in these manuscripts, only 26% of approved animal experiments were reported [20]. When scientists were asked about the reason for not publishing animal studies, they named negative results as the main explanation, closely followed by methodological problems [18, 20]. Both informations would actually be valuable for their peers and would prevent an unnecessary doubling of experiments.

37.2.1 Waste of Animal Lives

If published studies are of poor experimental quality or the reporting is so insufficient that an interpretation seems impossible, the use of animals was not justified. Strech and Dirnagl proposed to extend the classical 3R concept, replacement, reduction, refinement defined by Russel and Burch, with another 3R for scientific value: robustness, registration, and reporting [21, 22]. Adding these three new Rs might restore the quality and transparency of published biomedical literature.

37.3 Preregistration of Animal Research

Preregistration for preclinical research has been discussed in the last decade as a promising tool that could effectively boost the quality of animal research and restore translation of preclinical research into medical application [23, 24]. Preregistration means that scientists deposit their study plan in an open registry before starting any experiments. The original idea for the study design, chosen methods, and statistical analysis is stored and is retraceable for other researchers and reviewers and cannot be changed anymore. This will later allow the comparison between the original study plan and the final outcome. As a result, preregistration fosters the publication of all gained results or encourages researchers to explain the reasons why specific outcomes or data have not been published.

Preregistration also supports scientists in planning their studies thoroughly right from the beginning. Researchers are thereby trained to avoid common mistakes in study design such as missing blinding or randomization that can impair the robustness of a result. Preregistration could also prevent questionable research practices such as *p*-hacking or HARKing. Indeed, in confirmatory studies, researchers must first define a clear hypothesis. If this does not apply, they must label their study as exploratory and define all planned statistical analyses before starting the experiments. This makes it nearly impossible to intentionally manipulate the result of the study in order to achieve a more attractive outcome. Preregistration can also improve the reporting of data, as key details from journal checklists or reporting guidelines can be covered in the preregistration form.

37.3.1 Experience with Clinical Protocols

It has been demonstrated that registration of clinical trials led to an increase in the publication of "negative" results, which probably better reflects the reality of scientific outcomes [25]. It is important to notice that also in clinical research, implementation of preregistration did not happen overnight. Only with the involvement of funders, publishers, and also scientists themselves, preregistration received entry into the scientific routine (Fig. 37.1). The idea of preregistration in preclinical research is rather new and might thus also take time until it prevails.

37.4 Preregistration in animalstudyregistry.org

In January 2019, the German Center for the Protection of Laboratory Animals (Bf3R), which is part of the German Federal Institute for Risk Assessment (BfR), launched animalstudyregistry.org (www. animalstudyregistry.org) [26]. Animalstudyregistry.org permits the preregistration of all research studies involving animals conducted worldwide. To preregister a study, scientists have to enter general information, details on the study design, methods, and statistics, as well as on the animals used. After submission the authors have two weeks to add any changes or to retract the entry, otherwise, it will automatically be registered. After registration, the content of the study cannot be changed anymore, and the study receives a digital object identifier (DOI) which the author of the study can use as a reference and which marks the study idea as his/her intellectual property. Anytime the author can download a copy of its registered study, which can be sent together with a manuscript to be reviewed.

The whole content of the study is not immediately publicly accessible with registration. The authors can choose a flexible embargo period that lasts up to five years, during which the study is only visible with its title, the institution, where it is conducted, a summary and optionally the name of the author. Only after termination of the embargo period, all study contents become entirely visible. The embargo option was implemented to address researchers' concerns about thefts of ideas [27, 28]. The duration was

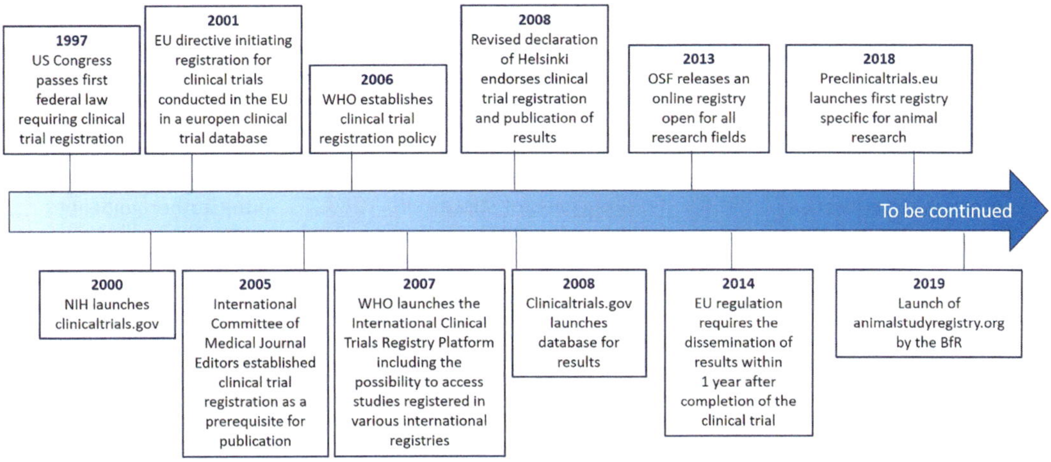

Fig. 37.1 Timeline of preregistration in clinical and preclinical research: The first federal law requiring clinical trial registration passed the US Congress in 1997. To enable clinical trial registration, the National Institutes of Health (NIH) launched the first online registry for clinical trials in 2000. In 2001, an European Union directive initiated the establishment for an EU database for the registration of clinical trials (Article 11 of the Clinical Trial Directive 2001/20/EC). In 2005, the International Committee of Medical Journal Editors established clinical trial registration as a prerequisite for publication, increasing the international pressure for registration of clinical trials, and thus taking a leading role in the initiation of a change of scientific practice. In 2006, the World Health Organization (WHO) established a clinical trial registration policy and launched the international clinical trials registry in 2007 that includes registrations from multiple international registries. In 2014, the EU releases Clinical Trial Regulation EU No. 536/2014 requiring the dissemination of results within 1 year after completion of the trial. In 2013, the open science framework (OSF) launched the first online registry for all research fields with a first focus on studies from psychology. In 2016, the NIH also required clinical trials funded by the NIH to disseminate their results. In 2018, scientists from Utrecht University launched the first registry conceived specifically for preclinical studies: preclinicaltrials.eu. In 2019, the German Federal Institute for Risk Assessment (BfR) launched the animalstudyregistry.org

based on an average research process from the idea for the project until publication.

To acknowledge the fact that studies can change during their conduction, authors have the possibility to add updates to their study anytime, where they can describe any kind of practical problem they were facing. This update function also allows authors to link the final outcome of their study to the preregistration by inserting links to data repositories, preprint servers, or journal articles or by directly uploading files. Within the last years, authors from all around the world joined the system, however, the uptake is still slow. To ensure that preregistration unfolds its full power, a broad enrollment will be crucial. Every single researcher can benefit. By asking distinct questions about the study plan, methods, statistical analysis, and specifics about the animals and housing conditions, preregistration assists researchers right from the start to plan their study thoroughly. These details are not only important to conduct robust experiments, they are also needed for the reporting according to the ARRIVE guidelines. Second, the DOI of a registered study together with the embargo period is the best proof for researchers that they had the idea for this study protocol already at an early time point. Third, with preregistration, researchers prove their commitment to transparency and quality. However, preregistration can only gather pace if the key players in the scientific landscape reward this commitment.

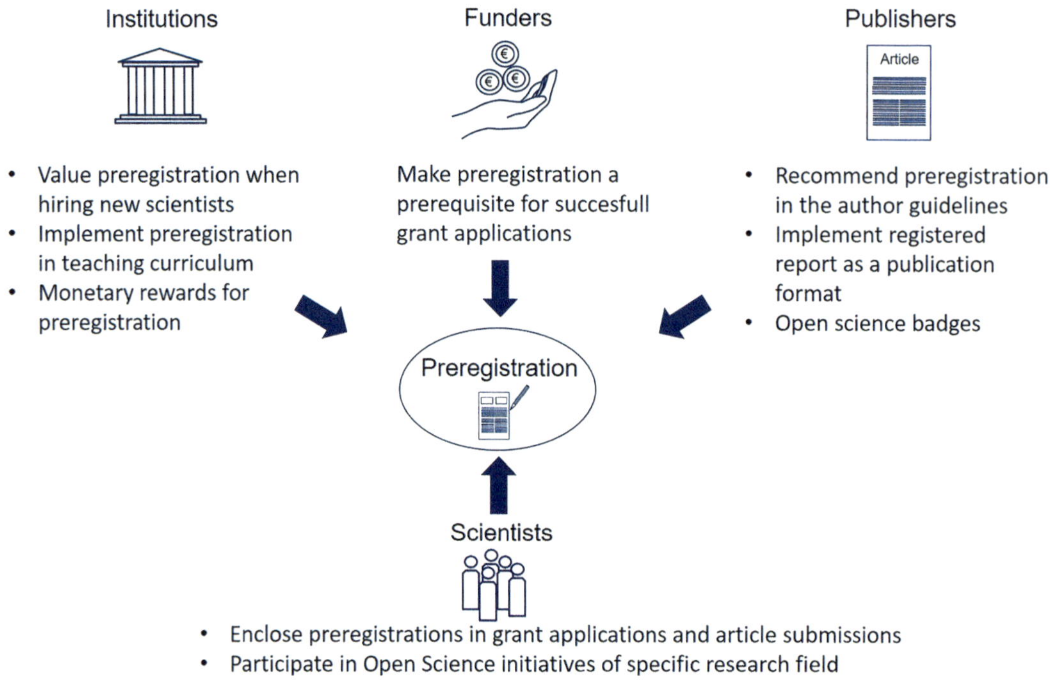

Fig. 37.2 Incentives that could promote a broad uptake of preregistration in preclinical research: Institutions have to value preregistrations when hiring new scientists and have to ensure that especially early career researchers receive a profound open science education, including information on preregistration. A monetary reward for preregistration could initiate first registrations in an institution and make it a topic of discussion. Funders have the greatest and most direct power to support preregistration by making it a prerequisite for successful grant applications. Publishers mostly come into play at the end of the research process and have thus an indirect role. They could recommend preregistration in author guidelines, offer the possibility to publish registered reports and promote preregistration by distributing open science badges. Scientists could demonstrate the quality and transparency of their research already now, by enclosing preregistration in grant applications and manuscript submissions. They could help to promote the change by participating in open science initiatives

37.5 The Future of Preregistration in Biomedical Research

Similar to other open science practices, the uptake of preregistration could still be improved in the field of biomedical research [16]. Several approaches could help to promote the implementation of preregistration. Stakeholders, as research institutions, funders, and publishers shaping the research incentive system, need to acknowledge their responsibility and value open science practices that increase the quality of science (Fig. 37.2).

Research institutions have to let go of the current evaluation measures, i.e., publication lists focusing on impact factors. They have to reward commitments to good scientific practice like preregistration by taking them into account when hiring new scientists. Institutions have to integrate a profound open science education into their curriculum to start a sustainable change in the research culture. Monetary rewards given for each preregistration to buy research material or for travel expenses can help to initiate the change [29].

Funders could make preregistration in a very straightforward way a prerequisite for successful applicants. The National Institutes of Health took this step already in clinical research and helped to establish preregistration in clinical research. The first third-party donors for preclinical research followed and made preregistration mandatory to support animal studies [30, 31]. The role of funders could play a pivotal role in the process, as funding of a study also stands at the beginning of a project.

Publishers have a powerful stand in the present research system. Their framework of action is, however, more complex as publication usually occurs at the end of a study, when it is already too late for preregistration. By attributing open science badges to preregistered studies, journals can raise awareness on the topic with a low-cost measure [32]. The American Association for Cancer Research and the American Heart Association have already implemented registration of animal research in animalstudyregistry.org within the editorial policies of their journals [33, 34]. Other publishers should follow this example. Multiple journals already offer the possibility to publish registered reports, which are characterized by two revision processes. In the first one, reviewers have to decide on the publication of a detailed study plan. If an in principle acceptance is given after the first revision, the second revision will mainly check whether the study was conducted as planned and will then accept the study for publication independent of whether or not the result was significant. Over 200 journals already offer registered reports, however, they are still very rare in biomedical research based on animals [35].

Bottom-up approaches could also initiate the change. Every scientist has already the possibility to preregister a study and to make this commitment to open science visible by enclosing it in grant application or in journal submissions. Initiatives for open science are pertinent for all research fields and can trigger a transformation in research culture.

References

1. Ioannidis JP (2005) Why most published research findings are false. PLoS Med 2(8):e124
2. Errington TM, Mathur M, Soderberg CK, Denis A, Perfito N, Iorns E et al (2021) Investigating the replicability of preclinical cancer biology. eLife 10:e71601
3. Arrowsmith J, Miller P (2013) Phase II and Phase III attrition rates 2011–2012. Nat Rev Drug Discov 12: 569
4. Schmidt-Pogoda A et al (2020) Why most acute stroke studies are positive in animals but not in patients: a systematic comparison of preclinical, early phase, and phase 3 clinical trials of neuroprotective agents. Ann Neurol 87(1):40–51
5. Crossley NA et al (2008) Empirical evidence of bias in the design of experimental stroke studies: a metaepidemiologic approach. Stroke 39(3):929–934
6. Dirnagl U (2016) Thomas Willis Lecture. Stroke 47(8):2148–2153
7. Richter SH, Garner JP, Würbel H (2009) Environmental standardization: cure or cause of poor reproducibility in animal experiments? Nature Method 6:257
8. Voelkl B et al (2018) Reproducibility of preclinical animal research improves with heterogeneity of study samples. PLoS Biol 16(2):e2003693
9. Head ML et al (2015) The extent and consequences of P-Hacking in science. PLoS Biol 13(3):e1002106
10. Kerr NL (1998) HARKing: hypothesizing after the results are known. Personal Soc Psychol Rev 2(3): 196–217
11. Percie du Sert N et al (2018) Revision of the ARRIVE guidelines: rationale and scope. BMJ Open Sci 2: e000002
12. Percie du Sert N et al (2020) The ARRIVE guidelines 2.0: updated guidelines for reporting animal research. PLoS Biol 18(7):e3000410
13. Hair K, Macleod MR, Sena ES (2019) A randomised controlled trial of an intervention to improve compliance with the ARRIVE guidelines (IICARus). Res Integr Peer Rev 4:12
14. Baker M (2016) 1,500 scientists lift the lid on reproducibility. Nature 533(7604):452–454
15. Sena ES et al (2010) Publication bias in reports of animal stroke studies leads to major overstatement of efficacy. PLoS Biol 8(3):e1000344
16. Serghiou S et al (2021) Assessment of transparency indicators across the biomedical literature: how open is open? PLoS Biol 19(3):e3001107
17. Clemence M, Leaman J (2016) Public attitudes to animal research in 2016. IPSOS MORI
18. ter Riet G et al (2012) Publication bias in laboratory animal research: a survey on magnitude, drivers, consequences and potential solutions. PLoS One 7(9):e43404
19. Wieschowski S et al (2019) Publication rates in animal research. Extent and characteristics of published and non-published animal studies followed up at two German university medical centres. PLoS One 14(11): e0223758
20. van der Naald M et al (2020) Publication rate in preclinical research: a plea for preregistration. BMJ Open Sci 4:e100051
21. Russell WMS, Burch RL (1959) The principles of humane experimental technique. Methuen, North Yorkshire, UK
22. Strech D, Dirnagl U (2019) 3Rs missing: animal research without scientific value is unethical. BMJ Open Sci 3(1):bmjos-2018-000048
23. Kimmelman J, Anderson JA (2012) Should preclinical studies be registered? Nat Biotechnol 30(6):488–489

24. Wieschowski S, Silva DS, Strech D (2016) Animal study registries: results from a stakeholder analysis on potential strengths, weaknesses, facilitators, and barriers. PLoS Biol 14(11):e2000391
25. Kaplan RM, Irvin VL (2015) Likelihood of null effects of large NHLBI clinical trials has increased over time. PLoS One 10(8):e0132382
26. Bert B et al (2019) Refining animal research: the animal study registry. PLoS Biol 17(10):e3000463
27. Deutsch S et al (2021) Web-based survey among animal researchers on publication practices and incentives for increasing publication rates. PLoS One 16(5): e0250362
28. Wieschowski S et al (2020) Attitudes towards animal study registries and their characteristics: an online survey of three cohorts of animal researchers. PLoS One 15(1):e0226443
29. Charité3R (2019) Charité 3R sponsors the first entries in animalstudyregistry.org. charite3r.charite.de/en/metas/news/artikel/detail/time_for_3r_how_to_improve_your_research_projects_by_3r_related_online_resources/. Accessed May 4, 2021
30. BMBF-DLR (2018) Confirmatory preclinical studies (Förderung von konfirmatorischen präklinischen studien). German Federal Ministry of Education and Research. gesundheitsforschung-bmbf.de/de/8344.php. Accessed 4 May 2021
31. BfR, G.F.I.f.R.A., Bf3R Research Funding—Replacement, Reduction and Refinement (2021) bfr.bund.de/en/bf3r_research_funding___replacement__reduction_and_refinement-62825.html. Accessed 4 May 2021
32. Kidwell MC et al (2016) Badges to acknowledge open practices: a simple, low-cost, effective method for increasing transparency. PLoS Biol 14(5):e1002456
33. AACR, American Association for Cancer R. Editorial Policies (2020) aacrjournals.org/content/authors/editorial-policies. Accessed 4 May 2021
34. AHA (2021) A.H.A. Transparency and Openness Promotion (TOP). Guidelines for authors publishing in an American Heart Association Journal. ahajournals.org/TOP-guidelines. Accessed 4 May 2021
35. Chambers C (2019) What's next for registered reports? Nature 573(7773):187–189

38 Plagiarism in Philosophy Research

M. V. Dougherty

Abstract

Academic philosophy faces a plagiarism crisis that is analogous to the replication crisis currently afflicting some scientific disciplines. The revelation of major instances of serial plagiarism in academic philosophy over the last decade suggests that the discipline is fundamentally under prepared to deal with plagiarism as a distinctive form of research misconduct. These cases have involved books issued by major publishers and articles published in established journals. Collectively they demonstrate not only a widespread inability to identify fraudulent manuscripts in the pre-publication process, but more seriously, a failure in post-publication responses to plagiarism. This chapter provides an overview of ongoing post-publication failures. Unretracted plagiarizing articles contaminate the body of published research in the years following publication. Without significant changes, the risk/reward ratio will continue to favor plagiarists in the field of philosophy. The enduring incentive to plagiarize will likely occasion new serial plagiarism cases that will further damage the quality of research in the field.

38.1 Introduction

Research conducted in humanities disciplines differs in fundamental ways from research conducted in the natural and biomedical sciences. Published books and articles in the humanities are typically not the culmination of a process involving the design of experiments, the collection of empirical data, or the subsequent analyses of data. For this reason, of the three basic classes of research misconduct—fabrication, falsification, and plagiarism—only the latter principally pertains to fraudulent research in humanities disciplines. Nearly every published retraction appearing for journal articles and book chapters in the humanities is due to plagiarism. These retractions are issued either for plagiarism in the primary sense, which is passing off the work of another as one's own, or plagiarism in the secondary sense (so-called self-plagiarism but more accurately *duplicate publication*), which is passing off one's previously published work in a new venue without disclosure of the prior publication. Across all academic disciplines, however, plagiarism in the primary sense only accounts for about 10% of all retractions, and duplicate publication accounts for only about 15% of them ([1]: 144).

The discipline of philosophy provides a useful case study for considering the unresolved problem of plagiarism—in the primary sense—in published research in humanities disciplines. To be sure, there have been some welcome advances in post-publication practices over the last decade.

M. V. Dougherty (✉)
Ohio Dominican University, Columbus, OH, USA
e-mail: dougherm@ohiodominican.edu

J. Faintuch, S. Faintuch (eds.), *Integrity of Scientific Research*,
https://doi.org/10.1007/978-3-030-99680-2_38

Retractions for plagiarism in philosophy were almost unheard of prior to 2009, even though occasionally accusations of plagiarism appeared in print [2–5]. Since 2009, a steady stream of high-profile cases of plagiarism in the discipline has surfaced [6–13]. Most of these cases have involved serial plagiarists in philosophy who have received multiple retractions, with some plagiarists reaching double-digit numbers. One of them received an ignominious listing on *The Retraction Watch Leaderboard*, which ranks individuals from all disciplines according to how many retractions they have earned ([14]; Retraction [15]).

38.1.1 Pseudonyms

Deserving of mention is a rarer form of research misconduct in philosophy that elicits lesser forms of correction than the "nuclear option" of retraction ([1]: 142): the use of undisclosed pseudonyms. Corrigenda and errata have appeared for articles that philosophers have published under unrelated names. The typical case is where an author publishes an article under a real name and afterwards publishes another article under a pseudonym that discusses, commends, critiques, or somehow engages the first paper. Doing so creates the illusion of a debate in the downstream published research literature. This practice of undisclosed pseudonyms is harmful insofar as it disrupts bibliographies and makes citations inconsistent and can be confusing to researchers who remain unaware of the multiplication of identities for a single researcher ([16]: 24–34).

38.2 Why Plagiarism Is a Form of Research Misconduct

An increase in studies that consider the problem of plagiarism in academic philosophy has brought renewed attention and increased theoretical clarity to the malevolent effects of plagiarism on the discipline [3, 16–19]. Plagiarism disrupts the allocation of credit for contributions to the discipline, falsifies the history of discovery, and duplicates findings in the body of published research literature. Every plagiarizing article is a Doppelgänger that draws attention away from the source texts and disrupts bibliometric data.

38.2.1 Repeat Offenders

Across all disciplines, plagiarism is typically serial plagiarism ([20]: 346). The standard situation involves a plagiarist who entangles many journals and publishers over the course of an academic career. Due to the serial tendency of academic plagiarism, every first discovery of a plagiarizing article should be treated as an index case, and the researcher's other published articles should be regarded as presumptively unreliable and carefully scrutinized.

Plagiarizing articles also create inefficiencies for the larger research enterprise in many ways, not the least of which is taking up finite space in journals and books that should have been reserved to genuine researchers. Furthermore, plagiarizing articles, as inauthentic products of research, waste the time and resources of researchers, publishers, editors, peer reviewers, and, eventually, whistleblowers. Through an illusion of research productivity, an undiscovered plagiarist is able to secure undeserved promotions, book contracts, grants, and offers of employment that should have gone to genuine authors. Despite the increased recognition of the manifold problems brought about by plagiarizing articles in recent years, several impediments remain to maintaining a reliable body of published research literature in philosophy.

38.3 Disguised Forms of Plagiarism

Academic plagiarism is commonly assumed to be the verbatim copying of another's words without adequate credit. The failure to recognize that copy-and-paste plagiarism—sometimes called "literal" plagiarism—is merely one of the many forms of plagiarism impedes efforts to maintain a reliable body of published research in philosophy.

In the last decade, expanded typologies of plagiarism have contributed to an analysis of the subtler forms of plagiarism, and these typologies include structural plagiarism, shake-and-paste plagiarism, and idea plagiarism ([21]: 6–13; [22]: 11–13; [23]: 9–10). Following this line of thought, Dougherty [17] argues that the body of published research in the humanities suffers from at least six disguised forms of plagiarism:

- *Translation plagiarism*: changing a source text into a different language
- *Compression plagiarism*: distilling a lengthy scholarly text into a short one
- *Dispersal plagiarism*: dividing a text and publishing it in several venues
- *Magisterial plagiarism*: concealing the authoritative status of ecclesiastical texts in theological writing
- *Exposition plagiarism*: conflating the authorial voices of canonical authors, of commentators, and of the plagiarist in historiographical writing
- *Template plagiarism*: replacing the key terms of a source text to produce the illusion of new research on a different topic

These various forms of disguised plagiarism exhibit some kind of concealment that creates a distance between the source text and the plagiarizing text not found in garden-variety copy-and-paste plagiarism. In translation plagiarism, for example, the manipulation of a source text into a different language changes the words and syntax, making the plagiarism undiscoverable through standard text-matching software programs. Specialized text-matching programs that focus on features such as citation patterns may have greater success in analyzing texts that are the products of translation plagiarism [22, 24, 25].

In compression plagiarism, a plagiarist distills a large text (e.g., a 450-page monograph) into a short one (e.g., a 5-page journal article). When whole paragraphs from the source text are elided into sentences in the plagiarizing version, and when long sentences are compacted into short ones, the resulting plagiarizing publication creates the appearance of a new and original work. Compression plagiarism may involve the appropriation of portions of the introduction, middle chapters, and conclusion of the larger source text to create a radically shorter whole that appears textually distant from its source.

Template plagiarism occurs when a plagiarist takes a published article on topic *X* and generates a new article on topic *Z* by simply substituting key terms. Template plagiarism can be found both in humanities articles and in scientific ones ([17]: 127–150; [26]: 584–585). Notable examples have been identified in a variety of very recent journal articles in philosophy ([8]: 261–264; [9]: 214–217).

When a plagiarizing text is the product of more than one kind of disguised plagiarism, involving, for example, both compression plagiarism and translation plagiarism, the discovery of the text's unoriginality is extraordinarily difficult. The discovery presupposes familiarity with more than one language and the ability to recognize the textually distant source. Furthermore, exhibiting the derivation of the short plagiarizing text from its lengthy source in a sufficiently persuasive and clear way to other researchers, editors, and publishers can be challenging. Nevertheless, there have been successful examples of demonstrating multi-leveled disguised plagiarism (see [27, 28]).

38.4 Inaction by Editors and Publishers

Correcting the body of published research in cases of research misconduct requires not only disclosure by whistleblowers but also the issuance of retractions by editors and publishers. The former does not guarantee the latter. A growing body of recent literature has examined the phenomenon of editorial inaction in cases of fraudulent publications in many fields. Even though uncontroversial research integrity guidelines state that the obligations of publishers include "Determin[ing] which publications warrant retraction, expression of concern, or correction" and "Issu[ing] freely available retraction/

correction notices," these obligations often go unfulfilled ([29]: 2, 3). Vorland et al. [30] observe that "editors are in the unique position to facilitate post-publication error correction" yet conclude nevertheless that "many journal editors do not fulfill this responsibility." Bülow et al. [31] address the same issue, explaining that, "for an editor to knowingly keep an unethical paper in one's journal, instead of retracting it, is a case of complicity." They argue that editors who fail to retract fraudulent or immoral research "are sharing a responsibility for this research by promoting rather than condemning it." The report of whistleblowers from a variety of fields reveals serious editorial inaction not only for fraudulent articles but even for those that exhibit clear invalidating errors [32].

With respect to the problem of editorial inaction in situations of demonstrated research misconduct, philosophy may have the dubious honor of being at the forefront. A 2017 study of a serial plagiarism situation (involving 43 previously documented articles) by one author of record concluded that there is "little uniformity in philosophy for correcting the scholarly record after the publication of plagiarized material" ([33]: 259). Of the 43 plagiarizing articles, only 23 were subject to published corrections by editors and publishers, consisting of 17 retractions and 6 expressions of concern. Three publishers quietly suspended sales of the plagiarizing material without ever publishing corrections, and eight took no public action for thirteen of the plagiarizing articles. The 2017 study presented itself as a "real-time snapshot" of the state of publishing in philosophy, and it concluded that the discipline of philosophy falls short of the standards of the natural and biomedical sciences for maintaining a reliable body of published research literature.

38.4.1 Editor Negligence

Some researchers have argued that the failure of editors to retract fraudulent publications is itself a distinctive and serious form of research misconduct ([34]: 225; [35]: 55, 51; see [36]: 162, 194). Inaction by an editor or publisher in a case of demonstrated plagiarism is arguably a graver failure than the precipitating act of plagiarism by the plagiarist. An act of plagiarism by a researcher is a singular instance of research misconduct, but editorial misconduct reflects a misuse of an office and institutional corruption. For an authority to omit to perform an official duty has far-reaching effects. All members of an academic community place significant trust in editors and publishers, granting them much power in the current economy of knowledge dissemination.

38.4.2 Long-Term Consequences

Left unchecked, fraudulent articles continue to exercise a corrosive effect on the downstream literature. An unretracted plagiarizing article can be likened to a toxic oil spill that will threaten a pristine environment unless containment measures are deployed quickly ([16]: 2). Some have likened fraudulent articles to viruses, drawing from public health models to conceptualize interventions that can mitigate the further contamination of the body of scientific publications [37, 38]. Editorial action impedes the implementation of containment measures and interventions necessary for maintaining a reliable research environment. Editorial inaction also discourages would-be future whistleblowers from performing the time-intensive and essential service of documenting plagiarism and reporting it, much as witnesses to crimes are less likely to come forward where there exists a corrupt judiciary that routinely ignores testimony.

38.5 Ambiguous Corrections

Published retractions for unreliable articles mitigate the damage and restore the trustworthiness of the body of published research. The mitigation is limited, however, if the retractions fail to explain precisely the reason for retraction. Retractions are commonly unclear and often use euphemisms that can leave a reader confused about the reason why an article is no longer to be judged as trustworthy.

The use of euphemisms and opaque language in published retractions stands in strict contrast with the precision found in scientific and academic publications. The two journalists who founded Retraction Watch have criticized "euphemistic language and notices that border on the preposterous" in many published retractions, and they have observed that "some of the best examples of this tortured wordplay involve retractions for plagiarism" ([39]: 120). Many publishers avoid the very use of the word *plagiarism*, substituting instead muted expressions such as *overlap*, *absent inverted commas*, *unattributed material*, *violation of an author agreement with the publisher*, or *erroneous citations*. Some retractions of plagiarizing articles provide no information beyond the fact that an article is being retracted. Such retractions consist of a single sentence such as *This article has been retracted* or *This article has been withdrawn at the request of the publisher*. A reader who assumes that the retraction is for some reason other than the reliability of the article (e.g., undisclosed duplicate publication) might be tempted to trust the retracted article if it is found to be useful.

38.6 Unfitting Corrections

More serious than the publication of opaque or euphemistic retractions for plagiarizing articles, however, is the publication of lighter forms of corrections—such as corrigenda—for demonstrated plagiarism. Bülow et al. critique publishers that offer corrections rather than retractions in cases of research misconduct; in arguing for the retraction of unethical papers, they defend the rule, "Corrections should only be used for honest errors" ([31]: 5). In severe plagiarism cases in philosophy, some publishers are willing to let plagiarists publish corrigenda or errata that give the appearance of saving the reliability of the article.

This practice of forgoing retractions in cases of demonstrated plagiarism and offering lighter corrections instead is pernicious. A plagiarist does not typically self-initiate a correction; in most cases a third-party whistleblower has already contacted an editor, requested a retraction, and provided the editor with meticulously documented evidence (usually in the form of a list passages taken without attribution from one or more uncredited sources). Only after the whistleblower's anonymized evidence has been shared with the plagiarist by the editor does the plagiarist begin to negotiate a light correction with an editor. Far too many editors give in to the plagiarist, so it is not uncommon to see extensive corrigenda published in philosophy journals that offer post facto credit to source texts that was missing in the original plagiarizing version of the article.

38.6.1 The Irreplaceable Role of Whistleblowers

The many hours of careful research, whereby a whistleblower painstakingly reduces a plagiarizing work to its uncredited sources, is published in the correction, but not as the discovery of the whistleblower. Rather, such corrections are typically published under the name of the plagiarist, who appears to the reader to be self-initiating the correction. In these cases, the plagiarist steals twice: first in the act of publishing the plagiarizing article and second in the act of publishing the whistleblower's discoveries as her or his own. The second act of theft includes the implicit approval of the editor or publisher. This common practice of allowing the whistleblower's discovery and demonstration of plagiarism to appear in print under the name of the plagiarist in a correction—thereby effacing the essential role of the whistleblower—remains a substantial impediment to research integrity in the discipline of philosophy ([16]: 120).

Whistleblowers need not be identified by name; but all published corrections should indicate when the evidence of research misconduct has brought forward by third parties. Retractions should be sufficiently detailed to allow readers to distinguish (1) "doing the right thing" or "good faith" author-initiated corrections that are issued when an author self-reports an error [40] from (2) ill-conceived corrections that efface whistleblowing activity and downplay research misconduct.

38.7 Institutional Conflicts

Some journals editors do not wish to issue retractions for fraudulent articles until a finding of research misconduct has been issued by the research integrity committee at the academic malefactor's home institution. Likewise, some institutional research integrity committees will not investigate suspicions of research misconduct until a retraction has been published by a journal. Whistleblowers are faced with conflicting advice urging them to report misconduct first to a journal editor [41] or first to an institutional research integrity committee [42]. A better option is multi-targeted whistleblowing that includes not only journal editors and institutional research integrity committees but also grant agencies and post-publication online peer review venues such as PubPeer, among others ([16]: 132–134).

38.7.1 Academic Research Integrity Committees

The posturing by institutional research integrity committees can lead to unusual situations. In one case, a research integrity officer lectured a journalist about "the need for strict confidentiality" and a "presumption of innocence" for a particular researcher in philosophy, even though at the time the researcher had already received 11 retractions for plagiarism [43]. In that case, editors at 2 journals had already published lengthy critical editorials that meticulously detailed the extensive plagiarism for 4 of the 11 retracted articles ([8, 9]; see also [44]). In an unrelated case involving two committees at different institutions that were investigating the work of a single researcher, the first found no plagiarism, but the second found "substantial plagiarism." In an apparent confirmation of the second investigation, journals and book publishers issued five retractions, three errata, and four corrigenda for plagiarism and the failure to acknowledge source texts adequately ([16]: 187).

38.7.2 Institutional Self-Preservation

Institutions that investigate their own researchers may have an incentive to downplay any finding of research misconduct for the sake of avoiding reputational damage. In worst-case scenarios, an institutional research integrity committee "functions surreptitiously as a *de facto* public relations firm or damage-control committee, issuing carefully-worded press releases meant to shield the reputation of the institution rather than support corrections of the scholarly record" ([16]: 129). The home institution of a plagiarist can impede or delay indefinitely the issuance of retractions by editors and publishers. The "inherent conflicts of interest" must be addressed when institutions investigate their own researchers if their findings are to be considered reliable by the larger research community ([45]: 1315).

38.8 Retractions that Fail to Retract

Even when a journal issues a clear and explanatory retraction, the correction of the body of research literature is not fully successful if the retraction is not promulgated adequately. Retracted articles continue to garner positive post-retraction citations in the downstream literature [46–48], and the many causes of this deleterious phenomenon have not been fully delineated.

In some cases, those whose articles have been retracted continue to cite the retracted articles in later publications, without referencing the retractions, thereby perpetuating the illusion that the articles are trustworthy [49]. These individuals may also continue to list their retracted articles on institutional CVs, in applications for grants, and in job applications, without any acknowledgment of their retracted status. Arguably these secondary actions are additional acts of research and professional misconduct, and they perpetuate the illusion that the version of record of the articles has not been changed.

38.8.1 Ineffective Retractions

Sometimes a published retraction remains unattached to the article it retracts, leaving readers uninformed of the article's change in status. A journal may fail to produce an electronic tether between the online version of an article and the published retraction statement, or a journal may fail to add the standard *Retracted* watermark to each page of the electronic version of a retracted article. Still further, a retraction may only be available on the publisher's website, but not on contracted secondary venues through which researchers access articles (e.g., JSTOR). The proliferation of academic social media websites and institutional repositories that maintain PDFs of articles without reference to the relevant published retractions exacerbates the problem. Sometimes an article that has been anthologized is retracted only in its earlier version, and the later version remains untouched. In all of these situations, the retraction fails to follow the retracted article through all the various ways in which researchers access the published research literature. These manifold failures mislead academic readers into believing that the articles are reliable, so the retracted articles continue to receive positive citations in the downstream literature ([16]: 221–230).

The retracted status of articles should also be reflected in the curated databases that serve each discipline. In this respect, the biomedical sciences fare well, as researchers accessing the MEDLINE database via PubMed will find retraction statements linked to retracted papers in all but 11% of cases ([50]: 327). In the two major research databases that serve philosophy—*The Philosopher's Index* and *PhilPapers*—there is no link at all, so the retracted status of articles in philosophy remains hidden to researchers who consult these databases. Again, the discipline of philosophy falls short.

38.9 Modelling the Sciences

Scientists typically lead institutional research integrity committees, and the procedural and evidentiary rules for these committees often have been written to apply primarily to research conducted in scientific disciplines rather than in humanities fields. These factors lead to strange situations where research integrity committees examine research misconduct cases in a humanities discipline using an incompatible perspective framed by the sciences. Articles in the humanities are not like scientific articles in major respects. A standard article from philosophy, for example, does not break down into the subsections of *introduction*, *literature review*, *method statement*, *presentation of data*, and *analysis*. Research integrity committees tend to focus on research integrity failures in the presentation of data section than those other parts of a scientific article.

Failures of attribution in other parts of a scientific article (e.g., introduction, literature review) are commonly treated as ineligible for findings of research misconduct. The US Office of Research Integrity (ORI) of the Department of Health and Human Services, for example, states that:

> ORI generally does not pursue the limited use of identical or nearly-identical phrases which describe a commonly-used methodology or previous research because ORI does not consider such use as substantially misleading to the reader or of great significance. [51]

The plagiarism of method statements or literature reviews will not trigger ORI investigations, which focus instead on data irregularities. Since nothing corresponds to a presentation of empirical data in the standard philosophy article, research integrity committees are often ill-equipped to issue a finding of plagiarism when approaching cases through a framework designed to assess articles in the sciences. Some research scientists state clearly that plagiarism not involving empirical data is never enough to justify a retraction [52, 53]. This view has been expressed elsewhere as:

> Borrowing sentences in the part of a paper that simply helps to better introduce the problem should not be seen as plagiarism. Even if our introductions are not entirely original, our results are—and these are the most important part of any scientific paper. ([54]: 658)

This perspective does not apply to philosophy articles, which typically lack a section for empirical results. Articles in philosophy do not divide into "essential" and "non-essential" or "core" and "non-core" sections ([16]: 63–67). For this reason, plagiarism in any section of a philosophy article should not be dismissed as innocuous.

38.10 Conclusion

Large sectors of academic philosophy appear to be substantially deficient in dealing with academic plagiarism. As long as researchers, editors, and publishers view the disclosure of plagiarism and the issuance of retractions as causes of embarrassment to be avoided, rather than as praiseworthy acts to be pursued, the risk/reward ratio will favor the plagiarist. A 2018 study based on rational crime theory demonstrated that the basic deterrents for plagiarism and other forms of research misconduct are lacking [55]. That conclusion is consistent with economic approaches to the "rationality" of plagiarism, where the enormous potential benefits can be seen to offset the comparatively minor dangers ([56]: 449).

Time is on the side of the academic plagiarists. Even if the plagiarism is discovered—and there is no guarantee that it ever will be—the period between publication and eventual retraction can take years and even decades in some cases [16, 57]. If plagiarists can perpetrate the fraud just long enough to establish themselves with some authority or standing in their field, then they will "have more tools to avoid or resist challenges to their abuses and prerogatives" ([58]: 925). Acts of plagiarism in distant publications can be overlooked on the pretext that they were merely minor offences committed long ago, even though those deficient publications were the very basis for the plagiarist's advancement in the field. In the highly competitive environment of academic philosophy, authentic researchers who eschew plagiarism are at a major disadvantage in competing for grants, job opportunities, promotions, and other professional advancements.

Seasoned plagiarists likely know that "whistleblowers often fare far worse than those they are accusing of misconduct," and further, they may know that "the most common outcome for those who commit fraud is: a long career" ([1]: 142). A starting point for a solution is the recognition of the existence of plagiarizing articles and a commitment to using post-publication tools to repair the body of published research.

Acknowledgments I am grateful to Lawrence Masek and Michelle Dougherty for comments on an earlier version of this chapter.

References

1. Oransky I (2020) Retraction Watch: what we've learned and how metrics play a role. In: Biagioli M, Lippman A (eds) Gaming the metrics: misconduct and manipulation in academic research. The MIT Press, Cambridge, MA, pp 141–148
2. Anonymous (1990) Bad manners? Quadrant 34(10): 65–69
3. Hansson SO (2008) Philosophical plagiarism. Theoria 74(2):97–101
4. Neale S (2001) No plagiarism here. Times Lit Suppl, Feb 8, 12–13. theotodman.com/Abstracts/Abstract_22/Abstract_22560.htm. Accessed 27 Mar 2021
5. Tudvad P (2007) SAK—an unscholarly biography of Kierkegaard (trans: Piety MG) The Torch, January. faklen.dk/english/eng-tudvad07-01.php. Accessed 27 Mar 2021
6. Anderson M (2011) Telling the same story of Nietzsche's life. J Nietzsche Stud 42(1):105–120
7. Dougherty MV, Harsting P, Friedman RL (2009) 40 cases of plagiarism. Bull Philos Mediev 51:350–391
8. The Editors (2020) The retraction of articles due to plagiarism. Vivarium 58(4):256–274
9. The Editorial Board (2020) A note from the Editorial Board. Rech Theol Philos Me 87(2):213–226
10. Paglieri F (2015) Reflections on plagiarism. Topoi 34(1):1–5
11. Quintanam C (2017) Here's what academics are saying. Chron High Educ, April 5. chronicle.com/article/Heres-What-Academics-Are/239699. Accessed 27 March 2021

12. Weinberg J (2018) Plagiarizes again—and is caught by philosophy prof.'s class (updated). Daily Nous: News for and About the Philosophy Profession, January 15. dailynous.com/2018/01/15. Accessed 27 Mar 2021
13. Wolters E (2014) I nonetheless deeply regret the incident. Critical Theory, July 12. http://www.critical-theory.com/i-nonetheless-deeply-regret-the-incident-zizek-responds-to-plagiarism-allegations/. Accessed 27 Mar 2021
14. Palus S (2016) Philosopher earns 14th retraction for plagiarism. Retraction Watch, June 8. retractionwatch.com/2016/06/08/philosopher-earns-13th-retraction-for-plagiarism. Accessed 27 Mar 2021
15. Retraction Watch (2016) The Retraction Watch Leaderboard. https://retractionwatch.com/the-retraction-watch-leaderboard. Accessed 27 March 2021
16. Dougherty MV (2018) Correcting the scholarly record for research integrity: in the aftermath of plagiarism. Springer, Cham
17. Dougherty MV (2020) Disguised academic plagiarism: a typology and case studies for researchers and editors. Springer, USA
18. Hansson SO (2017) The ethics of doing ethics. Sci Eng Ethics 23(1):105–120
19. Hansson SO (2018) Anonymous philosophical communication. Theoria 84(2):112–119
20. Fox M, Beall J (2014) Advice for plagiarism whistleblowers. Ethics Behav 24(5):341–349
21. Weber-Wulff D (2014) False feathers: a perspective on academic plagiarism. Springer, Heidelberg, Germany
22. Gipp B (2014) Citation-based plagiarism detection: detecting disguised and cross-language plagiarism using citation pattern analysis. Springer, Wiesbaden, Germany
23. Zhang YH (2016) Against plagiarism: a guide for editors and authors. Springer, USA
24. Kaur M, Gupta V, Kaur R (2021) Review of recent plagiarism detection techniques and their performance comparison. In: Gunjan VK, Zurada JM (eds) Proceedings of International Conference on Recent Trends in Machine Learning, IoT, Smart Cities and Applications: ICMISC 2020. Springer, Singapore, pp 157–170
25. Lekshmy V, Krishnan A, Aparnna S (2021) Role of non-textual contents and citations in plagiarism detection. In: Pandian AP, Palanisamy R, Ntalianis K (eds) Proceedings of international conference on intelligent computing, information and control systems: ICICCS 2020. Springer, Singapore, pp 601–611
26. Byrne JA, Christopher J (2020) Digital magic, or the dark arts of the 21st century—how can journals and peer reviewers detect manuscripts and publications from paper mills? FEBS Lett 594(4):583–589
27. Dougherty MV (2019) The corruption of philosophical communication by translation plagiarism. Theoria 85(3):219–246
28. Hansson SO (2019) Further developments after Professor Dougherty's recent article. Theoria 85(5):344
29. Collaborative Working Group (2018) RePAIR consensus guidelines: responsibilities of publishers, agencies, institutions, and researchers in protecting the integrity of the research record. Res Integr Peer Rev 3:15
30. Vorland CJ et al (2020) Toward fulfilling the aspirational goal of science as self-correcting. Eur J Clin Investig 50(2):e13190
31. Bülow W, Godskesen TE, Helgesson G, Eriksson S (2020) Why unethical papers should be retracted. J Med Ethics 0:1–6
32. Allison DB (2016) A tragedy of errors. Nature 530:27–29
33. Dougherty MV (2017) Correcting the scholarly record in the aftermath of plagiarism: a snapshot of current-day publishing practices in philosophy. Metaphilosophy 48(3):258–283
34. Teixeira da Silva JA (2016) An error is an error... is an erratum: the ethics of not correcting errors in the science literature. Publ Res Q 32(3):220–226
35. Shelomi M (2014) Editorial misconduct—definition, cases, and causes. Publica 2:51–60
36. Hames I (2007) Peer review and manuscript management in scientific journals: guidelines for good practice. Blackwell Publishing, Malden. MA
37. Montgomery K, Oliver AL (2017) Conceptualizing fraudulent studies as viruses: new models for handling retractions. Minerva 55(1):49–64
38. Nylenna M, Simonsen S (2006) Scientific misconduct: a new approach to prevention. Lancet 367(9526): 1882–1884
39. Marcus A, Oransky I (2017) Is there a retraction problem? And, if so, what can we do about it? In: Jamieson KH, Kahan DM, Scheufele DA (eds) The Oxford handbook of the science of science communication. Oxford University Press, New York, NY, pp 119–126
40. Vuong Q-H (2020) The limitations of retraction notices and the heroic acts of authors who correct the scholarly record: an analysis of retractions of papers published from 1975 to 2019. Learn Publ 33(2): 119–130
41. Doran M (2016) How to survive as a whistle-blower. Nature 532:405
42. Bouter LM, Hendrix S (2017) Both whistle blowers and the scientists they accuse are vulnerable and deserve protection. Account Res 24(6):359–366
43. Weinberg J (2020) Philosopher revealed as serial plagiarist (multiple updates). Daily Nous: News for and About the Philosophy Profession, November 22. dailynous.com/2020/11/11/philosopher-revealed-serial-plagiarist. Accessed 27 Mar 2021
44. La Rédactrice en chef et l'Éditeur (2020) Note aux lecteurs (Note to the readers). Arch Hist Doctrin Litt Moyen Age 78:i
45. Gunsalus CK, Marcus AR, Oransky I (2018) Institutional research misconduct reports need more credibility. J Am Med Assoc 319(13):1315–1316
46. Cosentino AM, Veríssimo D (2016) Ending the citation of retracted papers. Conserv Biol 30(3):676–678

47. Schneider J et al (2021) Reducing the inadvertent spread of retracted science: shaping a research and implementation agenda. f1000research.com/documents/10-211. Accessed 27 Mar 2021
48. Teixeira da Silva JA, Dobránszki J (2017) Compounding error: the afterlife of bad science. Acad Quest 30(1):65–72
49. Madlock-Brown CR, Eichmann D (2015) The (lack of) impact of retraction on citation networks. Sci Eng Ethics 21(1):127–137
50. Schmidt M (2018) An analysis of the validity of retraction annotation in PubMed and the Web of Science. J Assoc Inf Sci Technol 69(2):318–328
51. The Office of Research Integrity (1994) ORI policy on plagiarism. ORI Newsletter 3:1. ori.hhs.gov/ori-policy-plagiarism. Accessed 27 Mar 2021
52. Bouville M (2008) Plagiarism: words and ideas. Sci Eng Ethics 14(3):311–322
53. Chaddah P (2014) Not all plagiarism requires a retraction. Nature 511:127
54. Yilmaz I (2007) Plagiarism? No, we're just borrowing better English. Nature 449:658
55. Cox A, Craig R, Tourish D (2018) Retraction statements and research malpractice in economics. Res Policy 47(5):924–935
56. Hoover GA (2006) A game-theoretic model of plagiarism. Atl Econ J 34(4):449–454
57. Stern V (2017) The retraction countdown: how quickly do journals pull papers? retractionwatch.com/2017/07/07/retraction-countdown-quickly-journals-pull-papers. Accessed 27 Mar 2021
58. Martin B (2016) Plagiarism, misrepresentation, and exploitation by established professionals: power and tactics. In: Bretag T (ed) Handbook of academic integrity. Springer, Singapore, pp 913–927

Scientific Articles Retracted for Misconduct or Fraud in the Dental Literature

39

Antonio Rapani, Matteo Zotti, Federico Berton, Riccardo Pasquali, Teresa Lombardi, Roberto Di Lenarda, and Claudio Stacchi

Abstract

The number of retracted publications in dental literature has constantly grown from 2005 to 2016, with a tendency to stabilize in the last four-year period. Misconduct or fraud was by far the most common reason for retraction. A majority of these papers were authored in Asia, with many written in India. Most of the authors are present in only one publication. Nevertheless, some appear in multiple retracted articles: in particular, two research groups authored more than ten publications retracted for fabricated and/or unreliable data. The first of such groups was based in Norway, and the second is still active in Spain. Unfortunately, retracted papers continue to be very often cited after the publication of the retraction notice. Regrettably most of these citations were inappropriate, not considering the presence of the retraction notice and treating data from the affected articles as if they had been reliable.

39.1 Introduction

The publication of unreliable scientific articles causes serious negative effects both in medical research and clinical practice. Future investigations based on false assumptions may come to wrong conclusions, wasting time and money, and potentially increasing the risk of diagnostic errors and incorrect therapy. Recently, a study in the field of osteoporosis examined the effect on the current scientific evidence over a 15-year period of a group of 27 unreliable randomized clinical trials from a Japanese group. The authors concluded that 13 guidelines and systematic reviews would likely change their conclusions without the inclusion of the aforementioned retracted papers [1].

39.2 Reasons for Retraction

They include misconduct (fabricated or unreliable data, plagiarism, intentional duplicate publication) or honest scientific errors (in this last case, the retraction is often requested by the corresponding author himself). The US Department of Health and Human Services defines scientific misconduct as "fabrication, falsification, or plagiarism in proposing, performing, or reviewing research, or in reporting research results" ([2, 3].

What are the reasons pushing a researcher to ignore the basic rules of ethics and integrity in

A. Rapani · M. Zotti · F. Berton · R. Pasquali · R. Di Lenarda · C. Stacchi (✉)
Department of Medical, Surgical and Health Sciences, University of Trieste, Trieste, Italy
e-mail: fberton@units.it; rdilenarda@units.it; claudio@stacchi.it

T. Lombardi
Department of Health Sciences, University "Magna Græcia", Catanzaro, Italy

J. Faintuch, S. Faintuch (eds.), *Integrity of Scientific Research*,
https://doi.org/10.1007/978-3-030-99680-2_39

research? The current evaluation system which, in many countries, adopts the number of publications and citations as one of the principal indicators of scientific productivity could have its share of responsibility. Academic career progression and research funding are strictly related to scientific productivity, and authors are constantly under pressure to publish. Obviously, these considerations do not justify unethical behaviors. Perhaps a global reassessment of evaluation methods, privileging quality over quantity, could be helpful to contrast and limit intentional misconduct in scientific literature.

39.3 The Field of Dentistry

The first article was retracted in 2005 [4], and the number constantly increased during the following years [5, 6]. This tendency is likely to continue due to increasing attention to this phenomenon paid by the scientific community, leading to more accurate controls performed by journal editors during peer-review process (including the use of plagiarism detection software) and also after article publication. This growing trend is also common to other fields of health and life sciences [7, 8].

After a comprehensive electronic search on PubMed (ncbi.nlm.nih.gov/pubmed) and Retraction Watch website (retractiondatabase.org), our group encountered 211 retracted dentistry publications: 156 articles (73.9%) were retracted for misconduct or fraud from 2005 to 2020 (median year of publication of the retraction notice was 2015). The number of retractions has been constantly growing until 2016 but tended to stabilize over the subsequent four-year period (2017–2020) (Fig. 39.1). Misconduct or fraud were by far the most common reasons for retraction, in accordance with the general trend observed in health and life sciences publications [9, 10]. Among misconduct-related issues, plagiarism was the principal reason for retraction in journals with impact factor <2, while the presence of unreliable and/or fabricated data was the leading cause of retraction in journals with a higher impact factor (>2).

39.4 Implantology and Oral Pathology

These were the two dental disciplines showing the highest number of retractions for misconduct or fraud. However they are not the disciplines counting the highest number of published papers in the dental literature [11]: considerable economic interests moved by the aforementioned specialties and the presence of two specific research groups could underscore the result (Table 39.1).

A majority were authored in Asia ($n = 88$; 56.4%), with 47 papers written in India (30.1%). Complete distribution for continent and country is presented in Table 39.2. It should be noted that the main reason for retraction of papers written by Asian and, in particular, by Indian authors is plagiarism or duplicate publication. The great majority of these authors appears only in one publication and, very seldom, in two retracted papers.

Some researchers appear in multiple retracted articles (Table 39.3). One group in Norway (senior researcher Jon Sudbø, 12 retracted articles), and a second in Spain (senior researcher José Luis Calvo Guirado, 27 retracted articles) were the most remarkable.

39.5 Repeat Offenders

Jon Sudbø is a Norwegian dentist and physician who, in the beginning of the 2000s, published a series of articles with fabricated data in the field of oral oncology. These fraudulent papers were accepted by leading journals, including *The Lancet* and *The New England Journal of Medicine*: the first ([12], retracted in 2006) reported entirely fabricated data from 908 patients. Sudbø, who was consultant oncologist at the Radium Hospital in Oslo and associate professor at the University of Oslo, resigned from these positions after fraud was discovered. His medicine and dentistry authorizations and his doctorate in medicine were revoked in 2006 by the Norwegian Board of Health Supervision. In 2009 he was allowed to

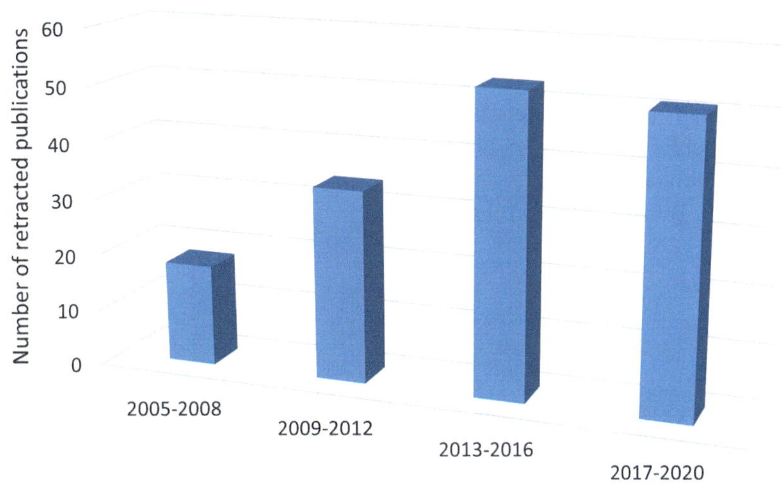

Fig. 39.1 Distribution of retracted publications in dentistry over time

practice again medicine and dentistry with restrictions, with no possibility to be involved into research.

José Luis Calvo Guirado is currently head professor at the Universidad Católica San Antonio de Murcia (Spain), and his main research topic is oral implantology. Calvo Guirado is the protagonist of the worst scandal ever in the field of dental research: in 2018 he had 18 papers retracted from one of the leading journals (*Clinical Oral Implants Research*) for duplication and manipulation of images and/or reporting unreliable data. Retractions continued in 2019 and 2020, for the same reasons however in different journals. Calvo Guirado's retracted papers were conducted within an extended research network involving many universities in Spain and abroad (Fig. 39.2). To the best of authors knowledge, at the moment it seems that no concrete measures have been taken by the competent authorities in response to these serious and repeated violations of ethics and integrity in research.

Sudbø wrote 31.6% of all retracted publications in oral pathology field, while Calvo Guirado authored 69.2% of all retracted papers on oral implantology topics. In December 31, 2020, Calvo Guirado's group was responsible of 96.4% of the retracted publications authored in the field of dentistry in Spain (which is the second country in the world for the number of retractions in dental science). Retractions of papers published by Calvo Guirado's group are still continuing after the present analysis: between January and May 2021, for duplication and/or manipulation of

Table 39.1 Retracted publications divided by discipline

Discipline	Number	Percentage (%)
Implantology	39	25.0
Oral Pathology	38	24.4
Periodontology	15	9.6
General Dentistry	13	8.3
Oral Surgery	13	8.3
Restorative Dentistry	13	8.3
Prosthodontics	10	6.4
Endodontics	7	4.5
Orthodontics	3	1.9
Dental Radiology	2	1.3
Pediatric Dentistry	2	1.3
Gnatology	1	0.6

Table 39.2 Retracted publications divided by country and continent

Nationality	Number	Percentage (%)	Continent	Number	Percentage (%)
India	47	30.1	Asia	88	56.4
Spain	28	17.9	Europe	51	32.7
Norway	12	7.7	America	14	9.0
Japan	10	6.4	Africa	2	1.3
China	9	5.8	Oceania	1	0.6
Iran	7	4.5			
Brazil	6	3.8			
USA	6	3.8			
Others	31	19.9			

images and/or reporting unreliable data ([13], retracted in 2021; [14], retracted in 2021; [15], retracted in 2021).

39.6 Background Problems

Bar-Ilan and Halevi [16] divided into three groups the reasons for retracting an article: (a) administrative errors (e.g., copyright violation, nonpayment of publication charge, accidental duplication by the journal), (b) ethical misconduct (e.g. plagiarism, duplicates, lack of ethical committee approval), and (c) scientific distortion (e.g., data falsification or fabrication). Citations of articles retracted for scientific distortion represent the most problematic situation, a sort of "hurdle for scientific advancement," as these papers mislead other scientists who trust in their reliability. The continued citation of retracted articles is not a novel issue and has been regarded as a major problem for research integrity for more than three decades both in medicine [17–19] and in dentistry [6, 20]. Post-retraction citations may be pertinent (considering the existence of a retraction and discussing it, together with the questionable reliability of the presented results and conclusions) or inappropriate (not considering the presence of a retraction and treating the cited findings as if they were reliable). In this last case, both authors of the citing article and the journal editor share a common responsibility for being inaccurate in checking the reliability of the cited source during manuscript preparation and peer-review process, respectively.

The 156 articles retracted for misconduct or fraud which were included in the present analysis received a grand total of 2262 citations (mean 14.5 ± 25.2). These articles were cited

Table 39.3 Authors present in more than five retracted publications (updated 31/12/2020)

Authors	Institution(s)	No. of retracted publications
José Luis Calvo Guirado	University of Murcia—Spain Murcia Catholic University—Spain	27
Rafael Arcesio Delgado Ruiz	University of Murcia—Spain Stony Brook University—USA	21
José Eduardo Maté Sánchez de Val	University of Murcia—Spain Murcia Catholic University—Spain	19
María Piedad Ramírez Fernández	University of Murcia—Spain Murcia Catholic University—Spain	15
Jon Sudbø	University of Oslo—Norway	12
Albrecht Reith	University of Oslo—Norway	10
Gerardo Gómez Moreno	University of Granada—Spain	8
Piedad Nieves de Aza Moya	University of Elche—Spain	8
Wanja Kildal	University of Oslo—Norway	7
Magne Bryne	University of Oslo—Norway	6

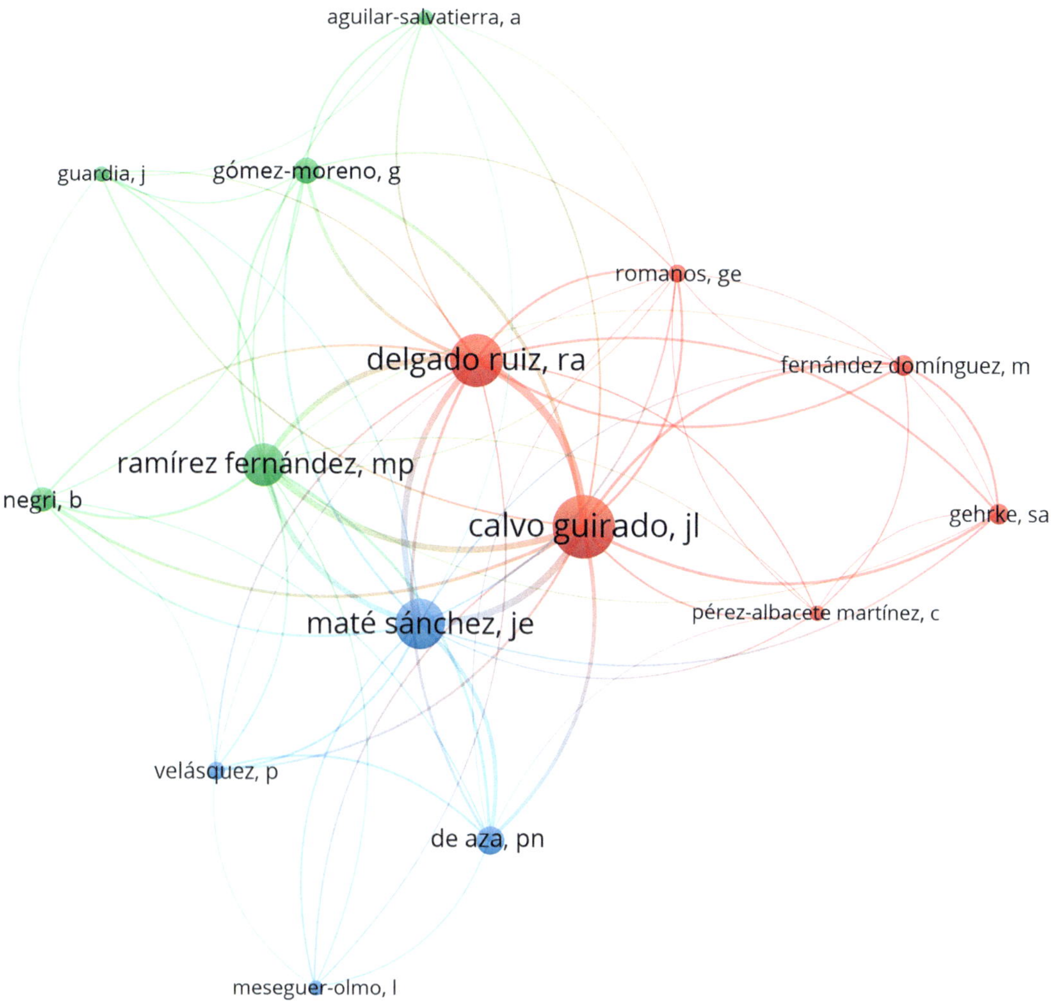

Fig. 39.2 Co-authors of José Luis Calvo Guirado with three or more retracted publications. Circle diameter and line thickness are proportional to the number of retracted papers (updated 31/12/2020)

712 times after retraction (mean 5.5 ± 7.3): these numbers do not include citations published in the same year of the retraction notice and in the subsequent one, in order to avoid questionable cases unaware of the retraction notice (Table 39.4).

Regrettably, 74% of the post-retraction citations included in the present analysis were inappropriate (Fig. 39.3), in accordance with others [5, 21].

Retraction notices are not always reported in public databases and in the journals they are often confined in secondary positions by the publisher, as if they were something shameful to hide. Not less than 78% of the retracted publications in the dental field are still available in full text on journal websites, and 11% of these articles present no watermark clearly indicating the presence of a retraction. Furthermore, retraction notices are often generic and unclear, with some of them (7%) giving no explanation at all, even if Committee on Publication Ethics (COPE) guidelines recommend that a detailed description of the reasons for retraction should be provided [6, 22]. Authors themselves should be more precise and careful when conducting literature

Table 39.4 Citations of articles retracted for misconduct or fraud. n: number; SD: standard deviation; PR: post-retraction

	n	%	mean	SD
Total citations	2262		14.5	25.2
PR citations	712	31.5%	5.5	7.3
PR pertinent citations	185	26.0%	1.4	2.5
PR inappropriate citations	527	74.0%	4.0	6.6

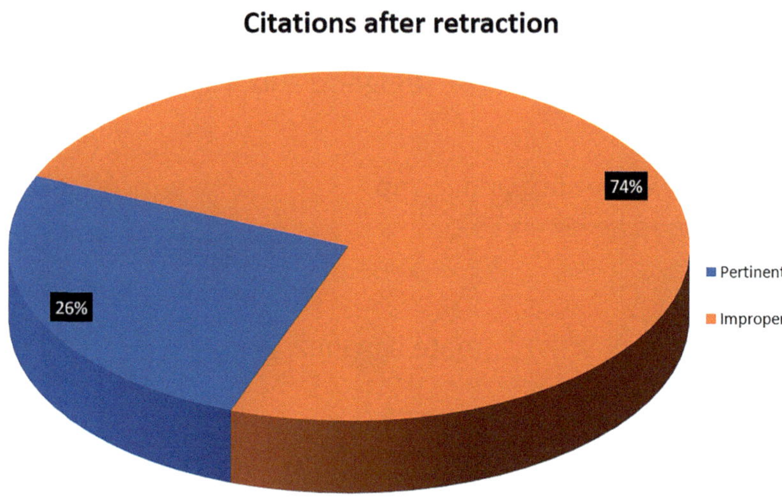

Fig. 39.3 Appropriateness of post-retraction citations

search, including a meticulous examination of the cited articles, not limited to a superficial reading of abstracts or bibliographic lists. Editorial offices and reviewers should as well carefully check the reference lists of newly submitted manuscripts, in order to immediately detect the presence of inappropriate citations and evaluate their weight in the general structure of the study.

These are a crucial point to ensure that unreliable data are not used as a basis for new research and that retracted publications are not cited in the future dental literature.

References

1. Avenell A, Stewart F, Grey A, Gamble G, Bolland M (2019) An investigation into the impact and implications of published papers from retracted research: systematic search of affected literature. BMJ Open 9:e031909
2. Federal Research Misconduct Policy (2021) https://ori.hhs.gov/federal-research-misconduct-policy. Accessed 2 Jun 2021
3. Khadilkar SS (2018) Scientific misconduct: a global concern. J Obstet Gynecol India 68:331–335
4. Nekora-Azak A (2004) Temporomandibular disorders in relation to female reproductive hormones: a literature review. J Prosthet Dent 91:491–493. Retraction: No authors listed. J Prosthet Dent 2005;94:305
5. Faggion CM Jr, Ware RS, Bakas N, Wasiak J (2018) An analysis of retractions of dental publications. J Dent 79:19–23
6. Rapani A, Lombardi T, Berton F, Del Lupo V, Di Lenarda R, Stacchi C (2020) Retracted publications and their citation in dental literature: A systematic review. Clin Exp Dent Res 6:383–390
7. Cokol M, Ozbay F, Rodriguez-Esteban R (2008) Retraction rates are on the rise. EMBO Rep 9:2
8. Steen RG (2011) Retractions in the scientific literature: Is the incidence of research fraud increasing? J Med Ethics 37:249–253
9. Campos-Varela I, Ruano-Raviña A (2019) Misconduct as the main cause for retraction. A descriptive study of retracted publications and their authors. Gac Sanit 33:356–360
10. Stavale R, Ferreira GI, Galvão JAM, Zicker F, Novaes MRCG, Oliveira CM, Guilhem D (2019) Research misconduct in health and life sciences research: A systematic review of retracted literature from Brazilian institutions. PLoS One 14:e0214272

11. Asiri FY, Kruger E, Tennant M (2020) Global dental publications in PubMed databases between 2009 and 2019. A bibliometric analysis. Molecules 25:4747
12. Sudbø J, Lee JJ, Lippman SM, Mork J, Sagen S, Flatner N, Ristimäki A, Sudbø A, Mao L, Zhou X, Kildal W, Evensen JF, Reith A, Dannenberg AJ (2005) Non-steroidal anti-inflammatory drugs and the risk of oral cancer: a nested case-control study. Lancet 366: 1359–1366. Retraction: Horton R. Lancet 2006;367: 382
13. Calvo-Guirado JL, Gómez-Moreno G, Barone A, Cutando A, Alcaraz-Baños M, Chiva F, López-Marí L, Guardia J (2009) Melatonin plus porcine bone on discrete calcium deposit implant surface stimulates osteointegration in dental implants. J Pineal Res 47:164–172. Retraction: No authors listed. J Pineal Res 2021;00:e12740
14. Calvo-Guirado JL, Gómez-Moreno G, López-Marí L, Guardia J, Marínez-González JM, Barone A, Tresguerres IF, Paredes SD, Fuentes-Breto L (2010a) Actions of melatonin mixed with collagenized porcine bone versus porcine bone only on osteointegration of dental implants. J Pineal Res 48:194–203. Retraction: No authors listed. J Pineal Res 2021;00:e12741
15. Calvo-Guirado JL, Ramírez-Fernández MP, Gómez-Moreno G, Maté-Sánchez JE, Delgado-Ruiz R, Guardia J, López-Marí L, Barone A, Ortiz-Ruiz AJ, Martínez-González JM, Bravo LA (2010b) Melatonin stimulates the growth of new bone around implants in the tibia of rabbits. J Pineal Res 49:356–363. Retraction: No authors listed. J Pineal Res 2021;00:e12731
16. Bar-Ilan J, Halevi G (2018) Temporal characteristics of retracted articles. Scientometrics 116:1771–1783
17. Budd JM, Sievert M, Schultz TR (1998) Phenomena of retraction: reasons for retraction and citations to the publications. JAMA 280:296–297
18. Budd J, Sievert M, Schultz TR, Scoville C (1999) Effects of article retraction on citation and practice in medicine. Bull Med Libr Assoc 87:437–443
19. Pfeifer MP, Snodgrass GL (1990) The continued use of retracted, invalid scientific literature. JAMA 263: 1420–1423
20. Theis-Mahon NR, Bakker CJ (2020) The continued citation of retracted publications in dentistry. J Med Libr Assoc 108:389–397
21. Hamilton DG (2019) Continued citation of retracted radiation oncology literature—do we have a problem? Int J Radiat Oncol Biol Phys 103:1036–1042
22. COPE Council (November 2019) COPE retraction guidelines. Version 2. https://doi.org/10.24318/cope.2019.1.4

Retraction of Scientific Papers: Types of Retraction, Consequences, and Impacts

40

Cristina Candal-Pedreira, Mónica Pérez-Ríos, and Alberto Ruano-Ravina

Abstract

The increase in the retraction of scientific papers is a worrying phenomenon that has been the subject of study in recent years. Of equal concern is the rise in scientific misconduct as the chief cause of retraction of publications, as well as the fact that retracted publications continue to be cited after their retraction. It is essential that retracted publications be correctly identified and monitored, in order to prevent the perpetuation of invalid results in the scientific literature, which can have consequences for both the scientific community and the general population.

C. Candal-Pedreira
Department of Preventive Medicine and Public Health, University of Santiago de Compostela, Santiago de Compostela, Spain

Health Research Institute of Santiago de Compostela (Instituto de Investigación Sanitaria de Santiago de Compostela - IDIS), Santiago de Compostela, Spain

M. Pérez-Ríos · A. Ruano-Ravina (✉)
Department of Preventive Medicine and Public Health, University of Santiago de Compostela, Santiago de Compostela, Spain

Consortium for Biomedical Research in Epidemiology and Public Health (CIBER en Epidemiología y Salud Pública/CIBERESP), Madrid, Spain

Health Research Institute of Santiago de Compostela (Instituto de Investigación Sanitaria de Santiago de Compostela - IDIS), Santiago de Compostela, Spain
e-mail: alberto.ruano@usc.es

40.1 Introduction

The retraction of scientific papers is an increasingly frequent phenomenon. Retraction of scientific papers is defined by the Committee on Publication Ethics (COPE) as "a mechanism for correcting the literature and alerting readers to articles that contain such seriously flawed or erroneous content or data that their findings and conclusions cannot be relied upon." Despite the stigma entailed by a paper's retraction, the main purpose of retraction is "to ensure the integrity of scientific literature rather than to punish the authors" [1].

The number and frequency of retracted publications, along with the quality of the peer review process, have been shown to be important indicators of the health of scientific literature. Even allowing for the increase in the number of scientific publications over recent decades, various authors have confirmed the notable increase in the proportion of scientific papers retracted, thus highlighting the fact that such retractions are not exclusively attributable to the mere rise in the number of papers published. In 1980, retractions accounted for 0.002% of total scientific publications, whereas in the years between 2005 and 2009 the proportion of retracted publications increased tenfold [2].

Other studies support these results. In the years between 1966 and 1996, a total of 235 papers were retracted, whereas in the period from 1997 through 2009 the number of retractions soared to

J. Faintuch, S. Faintuch (eds.), *Integrity of Scientific Research*,
https://doi.org/10.1007/978-3-030-99680-2_40

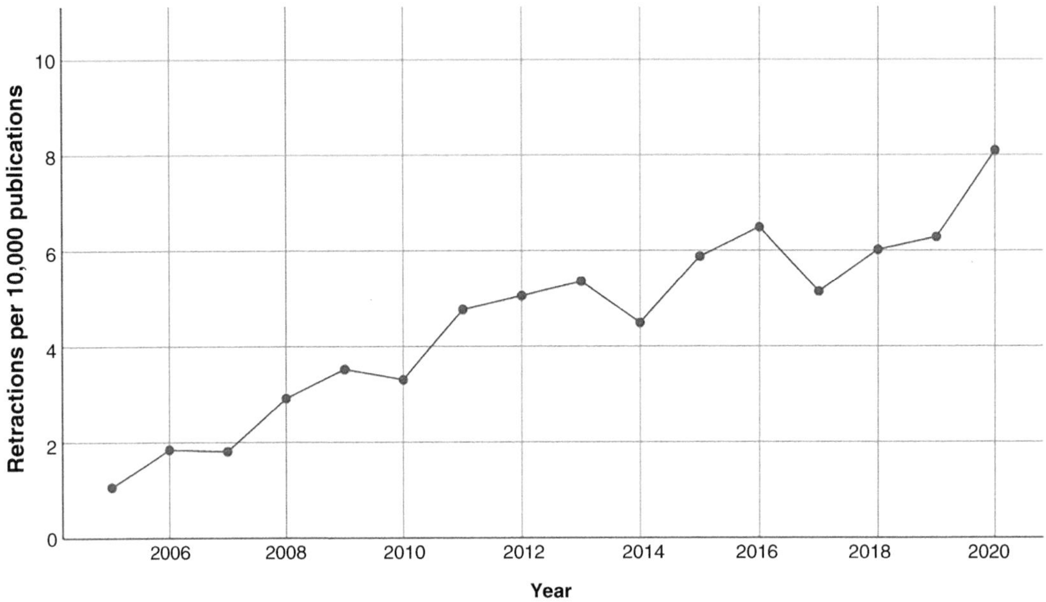

Fig. 40.1 Proportion of papers retracted with respect to papers published across the period 2005–2020

1164. Indeed, between 2000 and 2011, the number of retractions rose by 44% [3]. Furthermore, a more recent study has established that a total of 1082 papers were retracted in a period of only 3 years, from 2013 through 2016 [4].

Despite the slight decrease seen in 2014 and 2017 (Fig. 40.1), the proportion of papers retracted with respect to the total number of papers published displays a rising trend. In 2019, 6 of every 10,000 papers indexed in PubMed were retracted, whereas in 2020 this figure had risen to 8. These proportions were obtained by dividing the total number of retractions by the total number of publications in PubMed per year.

Added to this constant rise in retractions is the fact that only a small part of the papers liable to retraction is indeed retracted. This means that it is impossible to form an accurate idea of all the fraudulent or erroneous papers which may exist in the scientific literature: one can only ascertain the number of those that have already been retracted [5].

40.1.1 Underlying Causes

Different theories seek to account for the increase in retracted papers in the scientific literature. On the one hand, the pressure exerted on authors to publish (colloquially known in the scientific community as "publish or perish") may have caused an upsurge in flawed publications, which have then had to be retracted. The number of publications and citations that authors receive is a way of measuring their competence and academic impact and is vital for advancing their research careers [6].

On the other hand, there is also the possibility that manuscripts liable to being retracted are now being more quickly detected. Here, it should be noted that journals with a higher impact factor have a higher number of retractions. This may be due to the fact that publications in journals ranked in the upper quartiles enjoy greater visibility and are exposed to closer scrutiny by the scientific community [7]. In contrast the editorial committees of these journals are, in theory, made up of top-rated editors, who should reject papers at risk of being retracted, more frequently than do other journals having a lower impact factor.

40.2 The Retraction Process: Peer Review, Editorial Board Scrutiny, Notes of Concern, Publication Retraction

In the process of scientific publication, peer review is a fundamental step which seeks to ensure the quality of contributions. While the term "peer review" is tremendously unspecific and varies with the discipline, the most widely accepted definition is "...the assessment by experts of material submitted for publication" [8].

Reviewers are experts who voluntarily assess the strengths and weaknesses of a scientific manuscript prior to its publication in a scientific journal. The main aim is to "improve the reporting of scientific research and identify the most appropriate and highest quality" manuscripts, in order to publish these in the journal for which they are being reviewed [9].

There are different types of peer review, including: the *single blind review*, the most common type of peer review, where the reviewers remain anonymous to the authors; the *double blind review*, the model in which both reviewers and authors remain anonymous; the *triple blind review*, where additionally the author's identity is unknown to the editor; and the *open review*, where the reviewers and authors know each other's names [10].

Regardless of the type of peer review used, journals must describe the process transparently for authors and reviewers alike. Nonetheless, it has recently been observed that many scientific journals do not clearly describe such aspects [11, 12].

In recent years, peer review has been widely criticized for different reasons, ranging from delays in the publication of a manuscript to notable cases of fraud in the biomedical literature, coming to light after the peer review has been completed. Yet, despite the criticisms received regarding its integrity and the emergence of new and more modern review processes, peer review is still acknowledged by the greater part of the scientific community as being the most reliable way of assessing scientific literature, and has not been displaced by any other alternative mechanism [13].

40.2.1 Review Exempted Articles

New scientific-paper review processes have emerged, prompted in most cases by open access journals and preprint servers. A preprint is a complete manuscript uploaded by its authors to preprint servers before its peer review and publication in a journal. This enables the authors to receive feedback and comments from other researchers, and ensures swift dissemination of results [14].

Although the first preprint server, arXiv, dates from as far back as 1991, this form of publication has become popular in recent years, particularly as a result of the COVID-19 pandemic. In the first 4 months of the pandemic, more COVID-19-related preprints were published than those on Zika or Ebola throughout the duration of the latter two epidemics. Being an unknown disease of global proportions, there was an urgent need to generate and disseminate evidence that would help health professionals, as well as organizations tasked with implementing public health measures, to take evidence-based decisions. In this context, preprints are useful, inasmuch as they facilitate and expedite access to research results.

Although the benefits of preprints are evident, it should be borne in mind that one is dealing with unreviewed studies, so that as things stand, there is no filter that would guarantee the veracity, accuracy, or scientific quality of the preprint. Accordingly, preprints should not be used to take healthcare decisions. During the COVID-19 pandemic, politicians and physicians in some countries have used preprint-based information, enjoying a wide coverage in social networks, to promote certain treatments, a development that could lead to inappropriate use of medications, with the potential harm that this inevitably entails [14, 15].

Despite popular belief a peer review cannot guarantee that the results of published studies are correct, since reviewers do not always have access to the original data and, on occasions, do not devote enough time to conducting an exhaustive analysis of the document. When research integrity becomes compromised following the publication of a paper, the editors should, in each case, consider which corrective measures are best suited to the task [16, 17].

40.2.2 Non-fraudulent Problems

To remedy honest or unintentional errors in a paper that is already in print, correction will be the solution of choice. If the error invalidates the results of the study and its conclusions, however, it may be necessary to resort to retraction. In the case of honest errors, retraction with republication may be considered by the editor, allowing for a thorough correction of the paper. While it is also possible to republish a paper that has been retracted for containing intentional errors, this nonetheless poses a challenge for authors, since it is not common for such a manuscript to be reaccepted after retraction, given the authors' loss of credibility [1, 18].

40.2.3 Pre-retraction Interventions

Where there is suspicion that an error may have been committed knowingly but this has not yet been shown to be the case by an institutional investigation, the editor may publish a note or expression of concern. In contrast to "correction" or "retraction," "note (or expression) of concern" is a relatively novel term. It is a type of non-standardized editorial note which allows editors to act promptly, indicating that there may be doubts about the integrity of the paper, without actually going to the lengths of retracting or correcting it. If it is shown that the paper does indeed contain known errors, it will then be retracted [18, 19].

A scientific publication can be retracted in one of the two ways, i.e., totally or partially, depending upon whether the paper is wholly invalid or whether, in contrast, only a part is affected. Be that as it may however, the latest guideline published by COPE in 2019 advises against the use of partial retractions and instead recommends rectifying the part affected by means of correction or errata. This is because of the difficulty of determining which parts of a paper are and which parts of a paper are not reliable [1].

When a paper is retracted, a notice of retraction should be attached or linked to all versions, clearly identifying the retracted paper as such and stating who retracts it and the reason for doing so. In addition, the paper should be correctly identified as retracted, not only in the journal in which it was published but also in all scientific literature search engines, such as PubMed or EMBASE [1].

40.3 Reasons for Retraction

Some publications may be retracted as a result of honest errors. Coming within this category are errors in the study's approach, methodology, tabulation of data, statistical analysis, or conclusions [20]. In addition to the authors' errors, this category also includes errors incurred by editors and journals, such as publication of the same paper twice, repetition of tables, texts, etc.

Research or scientific misconduct is defined by the US Office of Research Integrity (ORI) as "fabrication, falsification, or plagiarism in proposing, performing, or reviewing research, or in reporting research results" [21].

The two most serious causes of scientific misconduct are *fabrication* and *falsification*. While fabrication is inventing the data used or the results, falsification is changing the data or results of the study as a consequence of the manipulation of materials, equipment, or processes. *Plagiarism* is regarded as being less serious because it tends to affect researchers to a far greater extent than it affects scientific literature per se. Plagiarism is defined as the "appropriation of another person's ideas, processes, results, or words without giving appropriate credit" [21, 22].

Although fabrication, falsification and plagiarism (known as FFP) are the most important forms of scientific misconduct, other questionable research practices have also been identified, ranging from problems of authorship (e.g., guest or ghost authorship), undeclared conflicts of interest to selection of nonindependent reviewers (friends, work colleagues, family members) and duplicate publication ("salami slicing") of text or images, among others [4, 23].

40.3.1 Immediate and Delayed Repercussions

Scientific misconduct can have serious consequences for the scientific community and general population alike. On the one hand, it fuels the population's lack of trust in scientific research and may well damage the reputation of institutions to which authors guilty of scientific misconduct belong or the very journals that publish such studies. It should also be borne in mind that every paper retracted for reasons of scientific misconduct implies a waste of both time and research resources. On the other hand, health professionals, among others, may base their clinical practice on fraudulent results, with the potential harm that this entails for patients [24].

40.4 Frequency of Retractions: Honest Error and Misconduct

A number of authors have analyzed the proportion of retractions found in the literature, whether because of scientific misconduct or honest error, and the trend in retractions over the years.

From 1982 through 2002, retractions due to honest error were more prevalent than those due to scientific misconduct, accounting for 61.8% and 27.1% of the total respectively, with the remaining 11% not being categorized. Between 1977 and 2012, however, another study concluded that only 21.3% of papers retracted during this period were due to honest error, as opposed to 67.4% retracted for reasons of scientific misconduct. Similarly, papers retracted due to scientific misconduct accounted for 65.3% of all retractions from 2013 through 2016 [4, 5, 25].

The results of studies conducted to date must be compared with caution, owing to the heterogeneity of design used to analyze these aspects. Moreover, the fact that there is no official classification of the causes of scientific misconduct means that authors have characterized the reasons for retraction differently. That said, the data suggest an increase in retractions due to scientific misconduct over the last two decades, while the proportion of papers retracted due to honest error has fallen [5, 26].

40.5 The Case of Andrew Wakefield (and Others)

In recent years, there have been notorious retractions in the sphere of biomedical literature, which have received widespread media coverage and enormous attention from the scientific community and the general population. This is for two reasons: the authors involved were internationally recognized in their respective fields; and the retractions were on very large scale, unveiling habitual or repeat offenders [27].

The Retraction Watch blog (www.retractionwatch.com) has drawn up a list of repeat offenders, ranked according to the number of retracted papers published by each. The field of anesthesia has been especially affected, since the top-ranked name is Yoshitaka Fujii, a Japanese anesthetist with 183 retractions, which account for 7% of all papers retracted across the period from 1980 through 2011. German anesthetist Joachim Boldt ranks second, with 135 retracted papers. In 2009 the results of an investigation of another anesthetist, Scott Reuben, were made public and ultimately led to the retraction of 25 of his papers for being fraudulent [28–31]. The impact of such results on clinical practice, i.e., potential harm to patients stemming from the unreliable results, by physicians who then use these studies as a guide to clinical practice is not known.

Notwithstanding the fact that the cases of scientific misconduct perpetrated by Fujii, Reuben and Boldt were all well known, in all likelihood the most important case is that of Andrew Wakefield, by virtue of the consequences it had on public health worldwide. In 1998, Wakefield and 12 co-authors published a case series in The Lancet, which suggested that the measles, mumps, and rubella (MMR) vaccine could cause autism in children. Despite being a study characterized by low-quality methodology, it began receiving a great deal of attention by the mass media and, in turn, by the general population. It was the latter that brought about an important reduction in MMR vaccine coverages in children around the world, and in the United Kingdom in particular, due to parents' concern that their children might develop autism.

The conclusions of Wakefield's study were used as the principal narrative of the so-called antivaccine groups. In 2008 and 2009, there were outbreaks of measles in the UK, the USA, and Canada, which were attributed to Wakefield's study [32]. In 2004, the paper received a partial retraction, and in 2010 The Lancet retracted it totally [33]. Paradoxically, to show that the paper's conclusions were false and obtain its subsequent retraction required far more time and effort than did its initial publication.

40.5.1 The COVID-19 Pandemic

A more contemporary example of a widely publicized retraction is the Surgisphere case, which refers to an observational study published in The Lancet in May 2020, during the COVID-19 pandemic. The authors tried to ascertain the effectiveness and safety of hydroxychloroquine or chloroquine for treatment of COVID-19, in the context of the global health crisis. The study included clinical data on 96,032 patients from 671 hospitals, which were situated on six different continents and were analyzed by a US company called Surgisphere Corporation. The results not only suggested that the use of hydroxychloroquine or chloroquine yielded no benefits in the treatment of patients diagnosed with COVID-19 but also indicated that there was evidence to show that the treatment increased the risk of ventricular tachycardia and intrahospital mortality.

As a consequence, the World Health Organization (WHO) halted the clinical trials that were being conducted with this medication in 17 countries. Furthermore, owing to the Surgisphere case, physicians around the world stopped treating patients diagnosed with COVID-19 with hydroxychloroquine, a drug which until then had been prescribed to most infected subjects requiring hospital admission. A number of researchers and health professionals expressed skepticism about the integrity and validity of the study. This led The Lancet to publish an expression of concern and launch an investigation, which concluded with the study being retracted due to the impossibility of gaining access to the original data of the results submitted by Surgisphere Corporation. This investigation led to the retraction of another observational study which had used the same data and had been published in the New England Journal of Medicine [34].

40.6 How Authors and Institutions Face Retractions?

Most researchers perceive retraction of their work as a stigma, and associate it with a negative impact on their reputation, even in cases where the retraction is due to honest error. Indeed, retraction has been defined as the "death penalty" for researchers. Ideally, it is the authors of the study who should be the sponsors of its retraction. Due however to the negative connotations of retraction, authors are reluctant to report their errors themselves. There is evidence to show that this only happens in 21.9% of cases of retraction, despite the fact that the scientific community has been shown not to penalize authors who self-report their errors by citing them any less often [35].

Furthermore in the case of retractions due to scientific misconduct, apart from being penalized by the scientific community when it comes to citations of the paper and the author's prior body of work, most authors with multiple retractions (i.e., repeat offenders) tend to become personally associated with misconduct, and their research activity dwindles. Indeed 4 years after a first retraction, fewer than 50% of the authors will publish as little as one paper per year [35, 36].

40.6.1 Academic Sanctions

When a researcher affiliated with an institution, such as a university or research center, receives a retraction because of scientific misconduct, the institution as well as the relevant research-integrity office/committee can impose different penalties. In many cases, the author is dismissed from the institution. An instance of this is the case of Dutch social psychologist Diederik Stapel who had 58 papers retracted and ranks sixth on the Retraction Watch list of repeat offenders: in addition to resigning from his post at the university, this researcher was forced to abandon his doctoral studies and do 120 h of community service. Moreover, he never managed to work in academic circles again [37].

In other cases, particularly in the USA, where the US Office of Research Integrity (ORI) has the authority to impose penalties in matters of scientific integrity, authors who have committed scientific misconduct are prohibited from receiving federal funds for investigating or working for any government agency, for a set period or, in some cases, forever [38–42]. Even so, it is likely that on many occasions, scientific misconduct results in no type of punishment or penalty for the author.

40.7 Are Papers Still Being Cited After Their Retraction?

Retracted papers continue to be cited after their retraction. Although some studies have reported a reduction of 35–69%, this nonetheless indicates that retracted papers are still being cited, albeit less frequently [24, 43, 44]. In contrast, other recent studies have suggested that, far from decreasing, citations received after retraction actually increase in the majority of cases [45–48].

A positive correlation has been reported between the impact factor of the journal in which a manuscript is published and the citations received after its retraction. A number of studies report that if a journal's impact factor is high, the paper will receive more citations after its retraction, as can be seen in Fig. 40.2 [25, 46]. Citations of papers published in journals in the first quartile decrease immediately after retraction, only to increase again in the long term [48].

Fig. 40.2 Post-retraction citations, by journal quartile

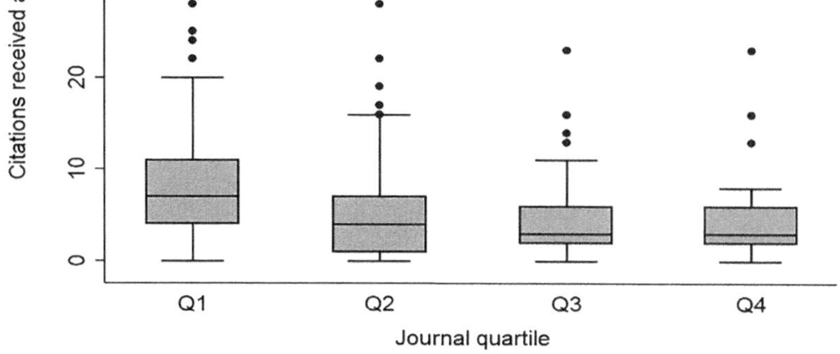

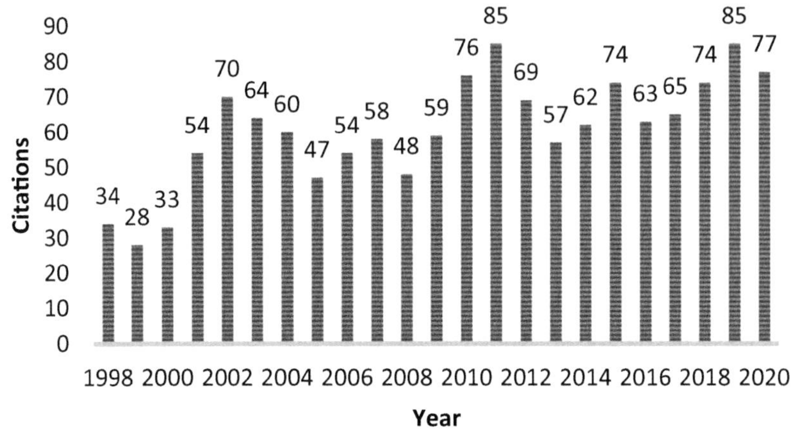

Fig. 40.3 Citations received by Wakefield et al.'s paper, 1998–2020

It has been observed that papers which were highly cited before retraction continue to be highly cited afterwards [49, 50]. This also occurs in the case of highly publicized retracted papers. A paradigmatic example is the above paper published by Wakefield in 1998. Despite being partially retracted in 2004, it received 881 citations from 2004 through 2019, more than those received before being retracted [33]. At present, this paper continues to be frequently cited, as can be seen from Fig. 40.3.

40.7.1 The Long Life of Retracted Publications

The appearance of open access journals and preprint databases has allowed for a wider dissemination of scientific results. Although the benefits for science are undeniable, this may also serve to perpetuate the existence of invalid papers in the biomedical literature. This is due to the fact that retracted papers can continue to exist publicly in their pre-retracted version, both on the Internet (e.g., repositories) and in personal libraries [51, 52].

One study concluded that 70–90% of citations are copied from the references used in other papers [53] and that 94% of post-retraction citations refer to the retracted paper as valid, without mentioning its retracted status. It is thus possible that a proportion of the citations received by retracted papers may be due to ignorance of the paper's retraction, whether because the references are sourced from another scientific paper or because the paper was not accessed by means of a traditional bibliographic search in MEDLINE or in the journal in which it was published [45, 46].

40.7.2 The Role of Self-Citations

On noting a reduction in the citations received, authors may cite the paper themselves without mentioning its retracted status, thereby increasing the number of citations that it receives. Yet there is not much evidence about who cites retracted papers [54]. It should be borne in mind that, though it may be deleterious as far as selfcorrection of scientific literature is concerned, it is nevertheless valid to cite a paper after its retraction, provided that its retracted status is clearly referred to. It is advisable to cite both the paper and the retraction notice, which should have different digital object identification numbers (DOIs). Similarly, if a paper has been cited and is subsequently retracted, it is advisable to ascertain the reason for retraction, in case this might affect the study's conclusions. If this occurs, a correction should then be published [55, 56].

40.8 Proposals to Track Retractions and Avoid Citations

To prevent the dissemination of invalid results, retractions should be correctly reported and monitored. The way to report a retraction is by publishing a retraction notice and identifying the paper as retracted in all its versions. Recent years have seen an improvement in the retraction notices published. In 2008, 9% of retraction notices made no mention of the reason for retraction, whereas in 2016 this percentage fell to 0.8% [57].

Yet retraction notices that do not comply with the recommendations continue to be published, possibly proving ambiguous or even deceptive. To avoid this, both the COPE and the International Committee of Medical Journal Editors (ICMJE) have issued guidelines and recommendations that seek to aid editors in drafting valid and useful retraction notices [1, 58]. In addition, the COPE proposed to draw up and adopt a checklist that would help follow these recommendations, with this being finally presented at the World Congress on Research Integrity in 2015 [59]. Along with the use of this checklist, it is thought that using templates with a standard format may contribute to the transparency and uniformity of retraction notices [23, 57].

40.8.1 Heterogeneous Standards

It has to be said that there is no stable, universal definition of what constitutes scientific misconduct. As a result, researchers, journals, universities, and other bodies tend to use a whole range of definitions, some more formalized than others [60]. To facilitate identification of research misconduct, a universally used definition and an outline of the causes of scientific misconduct must thus be drawn up.

As mentioned, the retraction of a paper entails stigma for authors and institutions alike, even where this is due to honest error. A good way of incentivizing authors to self-report errors in their papers, and thus correct the literature more efficiently, might be a simple change in terminology, since the word "retracted" has negative connotations in the scientific community. Some authors propose creating a new category for cases in which a paper has to be withdrawn from the literature because of honest error while retaining the label of "retracted" for cases of scientific misconduct; other authors advocate eliminating the term "retraction" altogether and replacing it with another one [61, 62].

At present, there are resources that can be used to monitor retractions and avoid citing papers after their retraction. A notable example is the Retraction Watch database (www.retractionwatch.com), which was set up in 2010 and has since cataloged more than 20,000 retractions. Although the database is not complete, it is available online and can be used by authors, to ascertain the status of a given paper or verify its citations.

Furthermore, there are some tools, such as CrossCheck (www.cross-check.com), iThenticate (www.ithenticate.com) or Turnitin (www.turnitin.com), which are accessible and capable of detecting plagiarism and multiple papers. Although there are already journals that routinely use these instruments when receiving manuscripts, their use should become standard practice in scientific publishing.

For their part, scientific journals could include a warning in their publication guidelines urging authors to review their citations. In addition to specifically indicating this, they could propose the immediate rejection of any paper that cited a retracted paper, should the case arise, something that would oblige authors to be far more cautious with the type of references they use.

At all events, it is evident that there is a need not only for improvement in the identification and traceability of citations of retracted papers but also for better training of researchers in the problems arising from the citation of this type of paper, something that only perpetuates the dissemination of bad science.

References

1. COPE Council (2019) COPE retraction guidelines. Committee on Publication Ethics. https://publicationethics.org/retraction-guidelines. Accessed 27 Jan 2021
2. Wager E, Barbour V, Yentis S, Kleinert S (2009) Retractions: guidance from the committee on publication ethics (COPE). Croat Med J 50:532–535
3. Van Noorden R (2011) Science publishing: the trouble with retractions. Nature 478:26–28
4. Campos-Varela I, Ruano-Raviña A (2019) Misconduct as the main cause for retraction. A descriptive study of retracted publications and their authors. Gac Sanit 33:356–360
5. Fang FC, Steen RG, Casadevall A (2012) Misconduct accounts for the majority of retracted scientific publications. Proc Natl Acad Sci U S A 109:17028–17033
6. Rawat S, Meena S (2014) Publish or perish: where are we heading? J Res Med Sci Off J Isfahan Univ Med Sci 19:87–89
7. Cokol M, Ozbay F, Rodriguez-Esteban R (2008) Retraction rates are on the rise. EMBO Rep 9:2–2
8. Olson CM (1990) Peer review of the biomedical literature. Am J Emerg Med 8:356–358
9. Callaham ML (2003) Journal policy on ethics in scientific publication. Ann Emerg Med 41:82–89
10. Ali PA, Watson R (2016) Peer review and the publication process. Nurs Open 3:193–202
11. COPE (2021) Peer review processes. COPE. https://publicationethics.org/peerreview. Accessed 28 Jan 2021
12. Klebel T, Reichmann S, Polka J, McDowell G, Penfold N, Hindle S et al (2020) Peer review and preprint policies are unclear at most major journals. PLoS One. https://www.ncbi.nlm.nih.gov/pmc/articles/PMC7577440/
13. Picciotto MR (2020) Peer review week 2020: trust in peer review. J Neurosci 40:7378–7378
14. Flanagin A, Fontanarosa PB, Bauchner H (2020) Preprints involving medical research—do the benefits outweigh the challenges? JAMA 324:1840–1843
15. Walker R, Rocha da Silva P (2015) Emerging trends in peer review—a survey. Front Neurosci. https://www.ncbi.nlm.nih.gov/pmc/articles/PMC4444765/
16. Kaufman KR, Malhi GS, Bhui KS (2019) When a corrigendum is not sufficient. BJPsych Open. https://www.ncbi.nlm.nih.gov/pmc/articles/PMC6611069/
17. Mulligan A (2005) Is peer review in crisis? Oral Oncol 41:135–141
18. Scott-Lichter D, The Editorial Policy Committee, Council of Science Editors (2012) CSE's white paper on promoting integrity in scientific journal publications, 2012 update, 3rd edn. CSE, Wheat Ridge. http://academy.rasep.ru/images/documents/rukovodstva/White%20Paper_DOI_Rus.pdf. Accessed 28 Jan 2012
19. Recommendations for the conduct, reporting, editing, and publication of scholarly work in medical journals. http://www.icmje.org/icmje-recommendations.pdf. Accessed 11 Feb 2021
20. Casadevall A, Fang FC (2012) Reforming science: methodological and cultural reforms. Infect Immun 80:891–896
21. ORI (2005) Public health service (PHS) policies on research misconduct – 42 CFR part 93. https://ori.hhs.gov/FR_Doc_05-9643. Accessed 4 Feb 2021
22. Fanelli D (2009) How many scientists fabricate and falsify research? A systematic review and meta-analysis of survey data. PLoS One 4:e5738
23. Moylan EC, Kowalczuk MK (2016) Why articles are retracted: a retrospective cross-sectional study of retraction notices at BioMed Central. BMJ Open 6:e012047
24. Furman JL, Jensen K, Murray F (2012) Governing knowledge in the scientific community: exploring the role of retractions in biomedicine. Res Policy 41:276–290
25. Nath SB, Marcus SC, Druss BG (2006) Retractions in the research literature: misconduct or mistakes? Med J Aust 185:152–154
26. Steen RG (2011) Retractions in the scientific literature: is the incidence of research fraud increasing? J Med Ethics 37:249–253
27. Bonnet F, Samama CM (2012) Les cas de fraude dans les publications: de Darsee à Poldermans. Presse Med 41:816–820
28. White PF, Rosow CE, Shafer SL (2011) The Scott Reuben Saga: one last retraction. Anesth Analg 112:512–515
29. Oransky AM& I (2015) How the biggest fabricator in science got caught. Nautilus. http://nautil.us/issue/24/error/how-the-biggest-fabricator-in-science-got-caught. Accessed 4 Feb 2021
30. Palus AS (2015) Scott Reuben notches 25th retraction, for a letter to the editor. Retraction Watch. https://retractionwatch.com/2015/11/23/scott-reuben-notches-25th-retraction-for-a-letter-to-the-editor/. Accessed 4 Feb 2021
31. Marcus AA (2020) A 'very cautious' process: journal retracts reviews by anesthesiologist found to have committed fraud a decade ago. Retraction Watch. https://retractionwatch.com/2020/05/19/a-very-cautious-process-journal-retracts-reviews-by-anesthesiologist-found-to-have-committed-fraud-a-decade-ago/. Accessed 4 Feb 2021
32. Rao TSS, Andrade C (2011) The MMR vaccine and autism: sensation, refutation, retraction, and fraud. Indian J Psychiatry 53:95–96
33. Suelzer EM, Deal J, Hanus KL, Ruggeri B, Sieracki R, Witkowski E (2019) Assessment of citations of the retracted article by Wakefield et al with fraudulent claims of an association between vaccination and autism. JAMA Netw Open 2(11):e1915552. https://www.ncbi.nlm.nih.gov/pmc/articles/PMC6902803/

34. Lipworth W, Gentgall M, Kerridge I, Stewart C (2020) Science at warp speed: medical research, publication, and translation during the COVID-19 pandemic. J Bioethical Inq 17(4):555–561
35. Lu SF, Jin GZ, Uzzi B, Jones B (2013) The retraction penalty: evidence from the web of science. Sci Rep 3: 3146
36. Mistry V, Grey A, Bolland MJ (2019) Publication rates after the first retraction for biomedical researchers with multiple retracted publications. Account Res 26: 277–287
37. McCook A (2016) No academic post for fraudster Diederik Stapel, after all. Retraction Watch. https://retractionwatch.com/2016/09/13/no-teaching-post-for-fraudster-diederik-stapel-after-all/. Accessed 18 Feb 2021
38. Oransky AI (2020) Former Maryland researcher banned from federal funding for misconduct. Retraction Watch. https://retractionwatch.com/2020/08/14/former-maryland-researcher-banned-from-federal-funding-for-misconduct/. Accessed 11 Feb 2021
39. Marcus AA (2011) Nursing researcher Scott weber draws penalties from ORI in plagiarism, fraud scandal. Retraction Watch. https://retractionwatch.com/2011/10/04/scott-weber-draws-penalties-from-ori-in-plagiarism-fraud-scandal/. Accessed 11 Feb 2021
40. Marcus AA (2021) Okinawa researcher suspended for faking data denies committing misconduct. Retraction Watch. https://retractionwatch.com/2021/02/03/okinawa-researcher-suspended-for-faking-data-denies-committing-misconduct/. Accessed 11 Feb 2021
41. Marcus AA (2021) Researcher to overtake Diederik Stapel on the retraction watch leaderboard, with 61. Retraction Watch. https://retractionwatch.com/2021/02/02/researcher-to-overtake-diederik-stapel-on-the-retraction-watch-leaderboard-with-61/. Accessed 11 Feb 2021
42. Marcus AA (2021) Former Texas postdoc earns 10-year federal funding ban for faking authors and papers to boost metrics. Retraction Watch. https://retractionwatch.com/2021/01/28/former-texas-postdoc-earns-10-year-federal-funding-ban-for-faking-authors-and-papers-to-boost-metrics/. Accessed 11 Feb 2021
43. Pfeifer MP, Snodgrass GL (1990) The continued use of retracted, invalid scientific literature. JAMA 263: 1420–1423
44. Mott A, Fairhurst C, Torgerson D (2019) Assessing the impact of retraction on the citation of randomized controlled trial reports: an interrupted time-series analysis. J Health Serv Res Policy 24:44–51
45. Budd JM, Coble ZC, Anderson KM (2011) Retracted publications in biomedicine: cause for concern. Association of College & Research Libraries National Conference, Philadelphia, PA, p 6
46. Bolboacă SD, Buhai D-V, Aluaş M, Bulboacă AE (2019) Post retraction citations among manuscripts reporting a radiology-imaging diagnostic method. PLoS One 14(6):e0217918
47. Kim SY, Yi HJ, Cho H-M, Huh S (2019) How many retracted articles indexed in KoreaMed were cited 1 year after retraction notification. Sci Ed 6:122–127
48. Candal-Pedreira C, Ruano-Ravina A, Fernández E, Ramos J, Campos-Varela I, Pérez-Ríos M (2020) Does retraction after misconduct have an impact on citations? A pre–post study. BMJ Glob Health 5(11): e003719
49. Redman BK, Yarandi HN, Merz JF (2008) Empirical developments in retraction. J Med Ethics 34:807–809
50. Snodgrass GL, Pfeifer MP (1992) The characteristics of medical retraction notices. Bull Med Libr Assoc 80: 328–334
51. Davis PM (2012) The persistence of error: a study of retracted articles on the internet and in personal libraries. J Med Libr Assoc 100:184–189
52. Budd JM, Sievert M, Schultz TR, Scoville C (1999) Effects of article retraction on citation and practice in medicine. Bull Med Libr Assoc 87:437–443
53. Simkin MV, Roychowdhury VP (2005) Stochastic modeling of citation slips. Scientometrics 62:367–384
54. Madlock-Brown CR, Eichmann D (2015) The (lack of) impact of retraction on citation networks. Sci Eng Ethics 21:127–137
55. Oransky AI (2018) Ask Retraction Watch: is it ok to cite a retracted paper? Retraction Watch. https://retractionwatch.com/2018/01/05/ask-retraction-watch-ok-cite-retracted-paper/. Accessed 11 Feb 2021
56. McCook AA (2016) What should you do if a paper you've cited is later retracted? Retraction Watch. https://retractionwatch.com/2016/11/01/what-should-you-do-if-a-paper-youve-cited-is-later-retracted/. Accessed 11 Feb 2021
57. Decullier E, Maisonneuve H (2018) Correcting the literature: improvement trends seen in contents of retraction notices. BMC Res Notes 11:490
58. ICMJE (2019) Recommendations for the conduct, reporting, editing, and publication of scholarly work in medical journals. ICMJE. http://www.icmje.org/recommendations/. Accessed 11 Mar 2021
59. COPE (2015) The standard retraction form proposed. https://publicationethics.org/file/7111. Accessed 13 Feb 2021
60. Aubert Bonn N, Godecharle S, Dierickx K (2017) European universities' guidance on research integrity and misconduct: accessibility, approaches, and content. J Empir Res Hum Res Ethics 12:33–44
61. Fanelli D, Ioannidis JPA, Goodman S (2018) Improving the integrity of published science: an expanded taxonomy of retractions and corrections. Eur J Clin Investig 48:e12898
62. Barbour V, Bloom T, Lin J, Moylan E (2017) Amending published articles: time to rethink retractions and corrections? bioRxiv. https://doi.org/10.1101/118356

The Role of Peer Review in the Scientific Process

41

Henriette De La Garza and Neelam A. Vashi

Abstract

Peer review has become an essential component of the scientific publication process. It ensures that papers published in scientific journals provide adequately analyzed and presented data, novel and meaningful information, answers to the unknown, and accurate conclusions based on professionally executed experimentation. This process aims to improve the quality of submitted manuscripts before they are published. Researchers believe that peer review is the central pillar of trust and that without it there would be no control in scientific communication. However, this process is not perfect, and it is entwined with advantages and disadvantages, which makes it prone to bias. The credibility of the scientific community and the perception of the public to accept new results strongly depends on the authenticity of the articles that are published. It is our duty to ensure conformance with ethical requirements and the publication of veritable research.

H. De La Garza · N. A. Vashi (✉)
Department of Dermatology, Boston University School of Medicine, Boston, MA, USA
e-mail: nvashi@bu.edu

41.1 Ethical and Scientific Importance of Peer Review

Peer review has been defined as a process of subjecting an author's scholarly work, research or ideas to the scrutiny of others who are experts in the same field [1]. It is intended to set higher scientific standards and improve the quality of suitable manuscripts. It acts as a filter by determining credibility, significance and originality. This filter also provides authors the opportunity to identify any error and improve the overall quality of their manuscript through the reviewer's suggestions, comments and corrections. This encourages authors to strive to produce faultless and meaningful manuscripts that will be of interest to the readers, and help to make advancements in their field.

Peer reviewed journals and their readers rely on the reviewers' expert and objective revision to provide a trusted form of scientific communication and prevent the dissemination of unverified research data. Without this screening system, there is a significant risk to advances in scientific knowledge. Currently, a scientific hypothesis or statement is generally not accepted by the academic community unless it has been published in a peer-reviewed journal [2]. A study done in 2013 revealed that 84% of researchers believe that peer review is the central pillar of trust and that without it there would be no control in scientific communication [3]. In another similar study done in 2016, 74% of authors stated that the

J. Faintuch, S. Faintuch (eds.), *Integrity of Scientific Research*,
https://doi.org/10.1007/978-3-030-99680-2_41

quality of their published paper was improved as a result of the peer review, with 82% stating that there is no control of scientific communication without it [4].

41.1.1 The Heavy Burden of Reviewing

It is a reviewer's duty to maintain the integrity and authenticity of the new developments in science. It has been concluded that the practice of peer review is based on faith in its effects, rather than on facts [5]. Despite its positive aspects, this process has been widely criticized. Among the most common complaints are the perceived bias by the editors and/or reviewers, a possible conflict of interest, as well as the lengthy and slow process. Critics also argue that the peer review process stifles innovation in experimentation, and acts as a poor screen against plagiarism [1].

41.2 Peer Review Process

The peer review system exists to validate academic work, provide a foundation for the credibility of scientific findings, and to improve the quality of research. Therefore, reviewers play a fundamental role in scholarly publishing. The process can vary slightly between journals but can be broadly summarized into ten steps.

1. Submission of Manuscript
   An author will submit the paper to the journal via an online system. Occasionally, journals may accept submissions by email.
2. Editorial Office Assessment
   An editor checks the manuscript's suitability and composition in accordance with the journal's guidelines to make sure it complies with the requirements and is within the scope of the journal. In this step, the manuscript also goes through a plagiarism check to see whether the paper has a large overlap with any published work. Commonly used platforms include iThenticate (www.ithenticate.com) and CrossCheck (https://crosscheck.ieee.org).
3. Appraisal by the Editor-in-Chief (EIC)
   The EIC checks that the paper is appropriate for the journal and assesses its originality and significance. If the manuscript does not comply with these criteria, the article will be rejected at this point.
4. EIC Assigns an Associate Editor (AE)
   Some journals have AEs who also assess the manuscript, assign reviewers, correspond with the author and coordinate with the reviewers as needed. At the end the AE evaluates the reviews and makes a recommendation for the acceptance or rejection of the paper.
5. Invitation to Reviewers
   Once the manuscript reaches this stage, the editor sends invitations to reviewers he or she believes would be appropriate for the manuscript. In some journals, the author may be required or given the option to suggest reviewers as well. Reviewers can either accept or decline to review a manuscript. As responses are received further invitations are issued, until the required number of acceptances is obtained.
6. Conduction of Review
   The reviewers read and revise the paper. They may make notes, comments, or a detailed point-by-point review. The review is then submitted to the journal, with a recommendation to accept or reject the manuscript. Reviewers can also request major or minor revisions before making any decision.
7. Journal Evaluation of Reviews
   The handling editor will consider all the reviews before making a decision. This is a complex process and even if a study is well-performed and the manuscript is also well-written, a manuscript may still not receive high enough scores for publication. Scores can be based on multiple attributes such as originality, relevance, proclivity to be highly cited, and specific journal appropriateness.
8. Communication of Decision
   The journal editor considers the feedback provided by the reviewers and makes a decision. The decision is sent by email to the

authors and includes any relevant reviewer comments.

9. Next Steps
   If the editor approves the manuscript for publication, it will be sent to production. If the article is rejected or sent back for either major or minor revisions, the editor will typically encourage resubmission and include the comments from the reviewers to help the authors improve their article. If the paper was sent back for revision, the reviewers should expect to receive a revised version within a specified deadline set to the authors.
10. Published
    After the manuscript is sent to production and is adequately edited, the article will be published.

When knowledgeable reviewers participate in the peer review process and evaluate submissions with care and sensitivity, their thoughtful comments often lead to improvements in the presentation of the work in several ways. The clarity of the manuscript is enhanced, ideas or parameters are clarified, relevant literature or important points are discussed more thoroughly, limitations of methodology are acknowledged, and broad or over-reaching conclusions are moderated [6]. The editors and reviewers should use stringent yet fair criteria to assess the manuscripts submitted to their journals.

41.2.1 Open Vs. Blinded Review

Single blind is the traditional method of reviewing and the most commonly used. In this type the names of the reviewers are hidden from the author. In double blind review, both the reviewer and the author are anonymous. In triple-blind review the author's identity is unknown to both the reviewers and the editor. Articles are anonymized at the submission stage to minimize any potential bias towards the authors. On the other end of the spectrum there is open review, with no blinding at all.

41.3 Reviewer Bias

The peer review process is far from perfect. Its lack of transparency and objectivity raise concerns that manuscripts might be judged based on aspects irrelevant to the content itself. Conflicts of interest (COI) can bias either in favor of or against a manuscript, can modify the communication of the findings and potentially introduce other unknown biases. Reviewer bias is the violation of impartiality in the evaluation of a manuscript [7].

41.3.1 The "Before" Stage

It includes the editor's evaluation of manuscripts before sending them to peer review and making judgments based on the characteristics such as the country of origin, institution, or a renowned author [7]. Social bias challenges the thesis of impartiality by suggesting that reviewers do not evaluate manuscripts, their content, or contributions to the literature independently of the author's identity. Similarly, in prestige bias, reviewers or editors give preferential evaluation of the submissions by prestigious authors or institutions. Many studies have found that journals favor authors located in the same country as the journal and there appears to be a strong preference for U.S authors [8]. In medical research a significant difference in the acceptance rates for manuscripts written in English compared to other languages is demonstrated [9]. Decisions by the editors regarding the assignment of specific peer reviewers can also introduce bias as well.

41.3.2 The "During" Stage

It includes the principles and preconceived notions by the peer reviewers [7], as a function of reviewer characteristics. Certain reviewers are more strict or flexible on identical submissions or even across multiple submissions [10]. Content-based bias can also ensue, in which reviewers can argue either in favor or against a manuscript based on its content. It is hypothesized that reviewers will evaluate more favorably the submissions of authors within the same field, interests, or way of thinking. Reviewers usually prefer research that is similar in cognitive orientation and content to their own, disfavoring those that fall outside the mainstream, including interdisciplinary and transformative research [11].

41.3.3 The "After" Stage

It encompasses delaying a publication or perhaps writing an editorial that might mitigate the findings of the study [7]. Publication bias can also occur in this stage and refers to the tendency of journals to publish research demonstrating positive outcomes, including results that have a positive direction, are statistically significant irrespective of the direction of result, or both [12].

41.4 Conflict of Interest (COI)

When a researcher is asked to review a paper submitted by someone he or she has a professional relationship with, such as a colleague, collaborator, former student or mentor, or someone with the same institutional affiliations, COI commonly arises.

Financial relationships also create obvious COI, the most apparent involving research related to a product or process that is owned or marketed by a for-profit entity. Alternatively, a potential reviewer may hold equity (stock or stock options) in a company, a researcher with stock in a company that manufactures a competing product is asked to review the same study, or an associate editor with stock in a company handles the review process. Receiving funding (grants, contracts, or gifts) for research projects or clinical trials from the company whose product is being examined could also give rise to COI [2, 13].

Most journals have policies that require that potential reviewers disqualify themselves from reviewing manuscripts if they have a COI that might compromise their objectivity.

Editors should avoid selecting peer reviewers with obvious COI. At the same time, reviewers must disclose to editors any COI that could bias their opinions of the manuscript.

41.4.1 Hasty or Faulty Assessment

Errors in identifying the true quality of submissions violate the ideal of impartiality in peer review by demonstrating that reviewers can fail to evaluate manuscripts in consistent ways [11]. Reviewer judgments should be reliable with respect to each other. However, in reality agreement between reviewers is not that common [14]. Another thing to note is that peer review is often conservative, that is, biased against groundbreaking or innovative research [15]. Many argue that conservativism threatens scientific progress by stifling the funding and public articulation of alternative and revolutionary scientific theories [16].

The threshold at which a conflict of interest is viewed as significant varies from journal to journal. Editors must strive to ensure that both appropriate expertise and objectivity are brought into the review process. Reviewers must be aware of potential biases that can influence their judgment and recommendations, and consider the manuscript from the perspective of the journal and the field of science.

41.5 Most Common Reasons for Manuscript Rejection

Most researchers face rejection of their manuscripts at some point in their career. Even top researchers and renowned authors have experienced rejection. The most common reasons for manuscript rejection are listed below.

1. A misleading title.
   The title is usually the first introduction readers have to the manuscript. A misleading title can lead to immediate rejection of a manuscript. Therefore, it is important to select a title that grabs attention, describes

the contents or main topics of the study, and highlights the importance of the research [17].

2. Lack of originality and novelty
   Reviewers may not find utility or relevance in a paper that presents unoriginal or predictable findings [18]. This can be seen as secondary analyses that extend or replicate published findings without contributing to the literature, or studies that report already known knowledge but positions it as novel by extending it to a new geography, population, or cultural setting. Lack of originality can also appear in the form of plagiarism. Too much overlap with other papers is a red flag to reviewers. Therefore, it is important to determine what has already been published and focus on creating research that proposes meaningful advancements in science or a change in practice.
3. Out of scope
   When a manuscript does not meet the journal's mission or would not be of interest to their readers, the manuscript can be easily rejected simply because it is not a good fit for that journal [19]. This also applies to manuscripts that do not follow the format specified by the journal's guidelines. Examples of common issues include unblinded manuscripts, omission of Internal Review Board/IRB or Ethical Committee approval no., omission of or insufficient COI statement, failure to register a trial, or when the type of article submitted is not accepted by that specific journal [20].
4. Flaws in study design
   Flaws in design or methodology significantly decrease chances for acceptance [18]. A poorly formulated research question, inappropriate statistical analysis, or unreliable/incomplete data are considered a fundamental problem.
5. Lack of significant results
   Conclusions that do not appear to be supported by the study data are not considered significant. If the results are not generalizable or have no clinical, theoretical, or practical implications, this may also lead to manuscript rejection.
6. Poor writing and organization
   Poorly written manuscripts tend to be rejected more frequently [21]. Concise articles are often preferred and accepted more often [22]. It is important to present the findings as efficiently as possible, using tables and graphs where and when appropriate.
7. Poor quality of English
   Manuscripts from authors for whom English is not a first language should be professionally edited before submitting to an English-language journal [20]. While minor errors are often overlooked, manuscripts that are not understandable or that contain a significant amount of typographical errors (typos), spelling or grammar mistakes will result in rejection. Some journals have copy editors that can go through a manuscript and make suitable corrections, but many journals do not and editors simply do not have the time to act as copy editors [23].
8. Citations/reference List
   Incorrect reference style or statements that are not accompanied by a citation can be problematic. It is also recommended to use the most recent literature as citations, as old references are generally not appropriate or preferred [20].
   Low quality or technical deficiencies of a manuscript are not the only reason for rejection [24].
9. Space constraints
   About 50% articles are rejected because of the limited amount of print space in most journals [25]. Each year, there is a growing number of articles submitted; however, editors have to choose between these papers, since they can only publish a limited number of articles.
10. Quality of peer reviews
    The quality of peer review varies widely according to the reviewers' interests, professional experience, educational background, or dedicated time. This can often be

subjective due to various interests and opinions. What a junior reviewer may consider relevant an experienced reviewer may not. While a reviewer can spend an entire day reviewing a manuscript, another may only spend a couple hours. This will ultimately impact the quality and direction of each review, eventually influencing the chances for publication.

11. Volume of submissions
    Journals that have a large number of submissions will also reject a large number of manuscripts, even those of high-quality.
12. Journal's decision-making policy
    Some journals follow a policy of rejecting any manuscript that requires major revisions, while other journals will complete another round of peer review if the authors agree to resubmit the manuscript after addressing these revisions.
13. The journal is looking for something specific at a particular time
    Sometimes editors may wish to publish a thematic issue or might be interested in a trending topic, in which case they might accept more manuscripts focusing on that particular topic. An example of this is the COVID-19 pandemic. Various researchers, authors, and journals sought to publish more about it.
14. The journal receives more than one submission on the same topic
    In such cases, the journal may choose to publish only one of the manuscripts, rejecting the others.

41.5.1 Streamlining Writing Practices

By understanding how manuscripts are assessed, authors can maximize the chances to get their work accepted for publication. Before writing the article, make sure to select a relevant research question. Assess that the study will add something new to the literature and will be of interest to the readers and the scientific community. Once the manuscript is written, ensure that the article is a good fit for the journal. Confirm that the manuscript is ready to submit and that it complies with the journal's guidelines. Submit a well-structured, perfectly polished, high-quality manuscript. Compose a good cover letter.

To increase the odds of acceptance authors should embrace the reviewer's comments. Rejection can be a stepping stone to success if researchers learn from and correct their mistakes. In fact, rejected manuscripts may ultimately lead to a superior end-product. Studies have shown that articles that were initially rejected and then got published elsewhere had a greater number of citations [26].

41.6 Author Fraud and Scientific Misconduct

The number of publications is often viewed as one's prestige, which can lead authors to behave wrongfully to achieve their goals [27]. Author misconduct includes a range of actions considered to constitute a breach of the conditions that an academic work must fulfill [28]. The extent of fraudulent research data is unknown, although it is believed that undiscovered fraud is much more common than it is supposed. Misconduct and fraud can be seen in many forms including:

- Fabrication or manipulation of data to draw fraudulent conclusions, including omission of conflicting findings, and/or distortion of data [29]. Adjusting data to suit the hypothesis is as unacceptable as making it up. Many authors are tempted to omit inexplicable outliers or exclude results that do not fit. When presenting results of randomized trials, authors should follow guidelines such as CONSORT and indicate clearly how many participants were screened, randomized, assessed and included in the analysis [30].
- Deliberately proving incomplete or poorly processed information. Authors should ensure to present complete findings and an

adequate quality control mechanism to minimize inaccuracy.
- Violation of patient's confidentiality. Collecting or publishing patient information for the purposes of research normally requires consent from the participants.
- Improperly attributing or denying authorship. The International committee of medical journal editors/ICMJE criteria for authorship state that all persons designated as authors should qualify and each should have participated sufficiently to take public responsibility for the contents [31]. An individual cannot be included as an author if he or she has not made substantial contributions to the design, analysis and interpretation of the data, or to drafting the manuscript.
- Ignoring to declare conflicts of interest.
- Plagiarism, in which the authors appropriate the ideas of another without crediting their true source and represent them as their own [32]. Citations are also important and they should be accurate. Wager and Middelton found that the median proportion of inaccurate citations across a wide range of medical journals was 39% and the median proportion of inaccurate quotations was 20% [33]. Journals can increase citation accuracy by including links to references or manually checking them [34].

Publishing about medicine carries considerable ethical responsibilities. The public expect that articles are a trustworthy source of information. Scientists, researchers, clinicians, and journals should therefore be professional and ethical. Failure to do so can result in stagnant advancements in science or even harm to patients. Fraudulent papers may also corrupt future research as others continue to trust and cite these papers [29]. The credibility of the scientific community and the perception of the public to accept new results strongly depends on the authenticity of the articles that are published. It is our duty to ensure conformance ethical requirements and the publication of veritable research.

References

1. Kelly J, Sadeghieh T, Adeli K (2014) Peer review in scientific publications: benefits, critiques, & a survival guide. EJIFCC 25(3):227–243
2. Rockwell S (2014) Ethics of peer review: a guide for manuscript reviewers case studies - guide for the discussion leader. Office, pp 1–19
3. Wiley RANO Survey (2013) Principles of peer review. Wiley Author Services. https://authorservices.wiley.com/asset/Principles_of_Peer_Review.pdf. Accessed 4 May 2021
4. Elsevier, The Publishing Research Consortium (PRC) Survey (2016) How researchers really feel about peer review. Elsevier, Amsterdam. elsevier.com/__data/assets/pdf_file/0004/205717/PeerReviewWeek2016_Infographic.pdf. Accessed 4 May 2021
5. Jefferson T, Alderson P, Wager E, Davidoff F (2002) Effects of editorial peer review: a systematic review. JAMA 287(21):2784–2786
6. Voight ML, Hoogenboom BJ (2012) Publishing your work in a journal: understanding the peer review process. Int J Sports Phys Ther 7(5):452–460
7. Haffar S, Bazerbachi F, Murad MH (2019) Peer review bias: a critical review. Mayo Clin proc 94(4):670–676
8. Ernst E, Kienbacher T (1991) Chauvinism. Nature 352:560
9. Ross JS, Gross CP, Desai MM, Hong Y, Grant AO, Daniels SR, Hachinski VC, Gibbons RJ, Gardner TJ, Krumholz HM (2006) Effect of blinded peer review on abstract acceptance. JAMA 295(14):1675–1680
10. Jayasinghe UW, Marsh HW, Bond N (2003) A multilevel cross classified modelling approach to peer review of grant proposals: the effects of assessor and researcher attributes on assessor ratings. J Royal Stat Soc Series A (Statistics in Society) 166(3):279–300
11. Lee CJ, Sugimoto CR, Zhang G, Cronin B (2012) Bias in peer review. J Am Soc Inform Sci Technol 64(1):2–17
12. Fanelli D (2010) Do pressures to publish increase scientists' bias? An empirical support from US states data. PLoS One 5(4):e10271
13. Resnik DB, Elmore SA (2018) Conflict of interest in journal peer review. Toxicol Pathol 46(2):112–114
14. Lindsey D (1991) Precision in the manuscript review process: Hargens and Herting revisited. Scientometrics 22:313–325
15. Braben DW (2004) Pioneering research: a risk worth taking. Wiley-Interscience, Hoboken, NJ
16. Stanford PK (2012) Getting what we pay for: unconceived alternatives and historical changes in scientific inquiry. Unpublished manuscript. University of California, Irvine, CA
17. The importance of titles (2021) https://springer.com/gp/authors-editors/journal-author/the-importance-of-titles/1410#:~:text=The title of your manuscript, people want to read further. Accessed 4 May 2021

18. Venkatesh S, Maymone MB, Vashi NA (2018) Peer reviews: the dreaded rejection. Dermatol Online J 24(3):13030/qt9h60v8r6
19. Elston DM (2019) Common reasons why manuscripts are rejected: advice for young investigators. J Am Acad Dermatol 80(6):1523
20. Chaitow S (2019) The life-cycle of your manuscript: from submission to publication. J Bodyw Mov Ther 23(4):683–689
21. Coates R, Sturgeon B, Bohannan J, Pasini E (2002) Language and publication in "Cardiovascular Research" articles. Cardiovasc Res 53(2):279–285
22. Patnayak R, Jena A (2014) Rejection of a manuscript: the other side of the story. Indian J Med Microbiol 32(3):350–351
23. Chernick V (2008) How to get your paper rejected. Ped Pulmonol 43(3):220–223
24. Pierson DJ (2004) The top 10 reasons why manuscripts are not accepted for publication. Resp Care 49(10):1246–1252
25. Reuters T (2012) Global Publishing: changes in submission trends and the impact on scholarly publishers. https://bibsonomy.org/bibtex/2ea40d7bf5b8995a511aba03008f89d7f/cram. Accessed 4 May 2021
26. Calcagno V, Demoinet E, Gollner K, Guidi L, Ruths D, de Mazancourt C (2012) Flows of research manuscripts among scientific journals reveal hidden submission patterns. Science 338(6110):1065–1069
27. Brice J, Bligh J (2005) Author misconduct: not just the editors' responsibility. Med Educ 39(1):83–89
28. Matías-Guiu J, García-Ramos R (2010) Fraude y conductas inapropiadas en las publicaciones científicas [Fraud and misconduct in scientific publications]. Neurologia 25(1):1–4
29. Marcovitch H (2007) Misconduct by researchers and authors. Gac Sanit 21(6):492–499
30. Moher D, Schulz KF, Altman DG (2001) The CONSORT statement: revised recommendations for improving the quality of reports of parallel-group randomised trials. Lancet 357(9263):1191–1194
31. Anonymous. (2010) Uniform requirements for manuscripts submitted to biomedical journals: writing and editing for biomedical publication. J Pharmacol Pharmacother 1(1):42–58
32. Groves T (2020) Scientific misconduct. https://authors.bmj.com/policies/scientific-misconduct/. Accessed 4 May 2021
33. Wager E, Middleton P (2008) Technical editing of research reports in biomedical journals. Cochrane Database Syst Rev 4:MR000002
34. Wager E (2007) Ethical publishing: the innocent author's guide to avoiding misconduct. Menop Int 13(3):98–102

Citation Misuses in the Biomedical Literature and Its Effects on Public Health

42

Estelle Dumas-Mallet and François Gonon

Abstract

To effectively contribute to scientific knowledge, biomedical observations have to be validated and debated by scientists in the field. Inside this debate, which mainly takes place in the scientific literature, citation of previous studies plays a major role. However, few academic studies have quantitatively evaluated the suitability and accuracy of scientific citations. Two types of misuse have been pointed out: citation bias and citation distortion. First, scientific citations usually favor positive results and those supporting the authors' conclusion. Second, many statements linked to a reference actually misrepresent the referenced findings. About 10% of all citations in biomedicine are strongly inaccurate and misleading for the reader. Finally, we illustrate with three examples how some citation misuses do affect public health: the opioid crisis in the US, the epidemic of unipolar depression and the unjustified fostering of hydroxychloroquine for COVID-19 treatment in France.

This chapter is partly based on the article "Dumas-Mallet E, Boraud T, Gonon F. Le mésusage des citations et ses conséquences en médecine. *Médecine/Science* 2021; in press." Reproduced with permission.

E. Dumas-Mallet · F. Gonon (✉)
Institute of Neurodegeneratives Diseases and CNRS UMR 5293, University of Bordeaux, Bordeaux, France
e-mail: francois.gonon@u-bordeaux.fr

42.1 Introduction

Biomedical research is a cumulative process. Although it starts with the collection of new observations, it becomes a source of new knowledge when these have been debated and validated by researchers in the field. This debate mainly takes place in scientific publications. In the introduction of each article, the authors present the state of knowledge and explain why they decided to collect new observations. In the discussion section they compare their results with what is already known on this topic and justify their conclusion in light of previous studies. The citation of previous studies therefore plays a role in each scientific article and, more generally, in the scientific debate leading to the advancement of science. It is thus essential that citation selection is not biased and that quotations are accurate.

In recent years, numerous articles have pointed out the weak replication validity of biomedical results (i.e., the fact that many early observations are disconfirmed by subsequent studies) [1–3]. In contrast, the use of citations in the scientific literature has only rarely been scrutinized. A handful of studies in the 1970s were focused on reference errors (e.g., publication date, journal pages) [4], but these errors have almost disappeared thanks to databases and dedicated software. However, other problems affecting citation use, i.e., quotation accuracy and citation biases, persist and impair the scientific debate.

J. Faintuch, S. Faintuch (eds.), *Integrity of Scientific Research*,
https://doi.org/10.1007/978-3-030-99680-2_42

42.1.1 Citation-Triggered Myths

Some case studies have shown that citations misuses can generate unfounded scientific dogmas that last for years. Among those studies, the most cited one is that of Steven Greenberg in 2009 [5]. He analyzed the citation network related to all studies addressing the belief that the amyloid β protein, which plays a role in Alzheimer's disease, is also produced by and injures skeletal muscles of patients with inclusion body myositis. The association can occur however pathogenetic links between them are obscure. Greenberg showed that studies that disconfirmed this hypothesis were much less cited than those that supported it, thus revealing a major citation bias. In addition, he observed several types of quotation distortions where the message delivered by the cited study was altered by the citing authors "in a manner that diverts its implications" [5].

Apart from Greenberg's publication, academic studies on citation misuses never addressed citation biases and quotation distortions together. We will therefore present them in two distinct sections, even if, as shown by Greenberg [5], these two main forms of misuse are often observed in the same publication.

This review is based on three systematic reviews that included meta-analyzes, one devoted to citation bias [6] and the other two to quotation distortion [7, 8] (Table 42.1).

42.2 Citation Biases

There are two types of citation biases (Table 42.1). The first is a general characteristic of the selected cited studies: those reporting a positive effect are more often cited than those favoring the null hypothesis. The second depends on the article citing the reference. It consists in preferentially citing previous works that support their conclusion. These two biases are often combined. For example Tatsioni et al. studied all the citations received by two observational studies published in 1993, claiming that a diet enriched in vitamin E reduces the risk of cardiovascular diseases [10]. At the end of 2006, they had been cited 1395 and 1234 times, respectively [10]. In 2000, however, a large randomized controlled trial (RCT) showed that vitamin E provided no benefit [10]. Other RCTs and a meta-analysis published in 2004 confirmed this lack of protective effect. However, the two initial studies were still widely cited in 2005 and half of the articles in which these citations appeared remained favorable to vitamin E, particularly those that failed to cite the RCTs [10]. Tatsioni et al. observed the same preferential use of citations favorable to initial studies, later contradicted by larger studies, in two other areas: the alleged protective effect of β carotene against cancer and of estrogens against Alzheimer's disease [10]. Similarly to Greenberg's study [5], that of Tatsioni et al. [10] illustrates the fact that citation biases generate unfounded dogmas and therefore slow down scientific advances.

42.2.1 Meta Analyses

Primary studies quantifying citation biases in the biomedical literature were only covered by one systematic review published in 2017 [6]. One meta-analysis of 21 primary data concluded that studies reporting a statistically significant effect were cited more often than those defending the null hypothesis (OR = 1.57; CI 1.34–1.83) [6]. Another meta-analysis of nine primary studies showed that publications favorable to the authors' conclusion were more often cited than those that questioned it (OR = 2.7; CI 2.0–3.7) [6]. Both meta-analyses only included primary studies investigating publications biases in the biomedical literature with one exception: a study in the field of sociology, which was included in both meta-analyses. Because the sample size of this latter study was the smallest one in both meta-analyses, its contribution to both summary effect sizes is negligible. Therefore, both meta-analyses and their summary effect sizes do reflect citations biases in the biomedical literature.

To identify recent studies published after this systematic review, we examined all articles citing it. We identified eight reviews and 19 primary

Table 42.1 Categorization of the different types of citation misuse

A: Citation biases according to Duyx et al. [6]
– Preferential citation of positive findings rather than those favoring the null hypothesis
– Preferential citation of previous work that supports authors' conclusion
B: Quotation distortions
(B.1) Appraising categorization according to Jergas & Baethge [7] *and Mogull* [8]
– Minor distortion: Without serious consequences for the reader
– Major distortion: Seriously misleading
(B.2) Descriptive categorization according to Pavlovic et al. [9]
– Citation of nonexistent finding
– Inaccurate interpretation of findings
– Inaccurately cited numerical data
– Wrong context
– Citation of quoted findings of another source
– Inaccurately cited method

studies in the biomedical domain. These eight review articles were related to issues other than citation biases. Therefore, the systematic review by Duyx et al. [6] seems the only review covering this topic. Among the 19 primary studies citing Duyx et al., only eight reported new observations about citation biases. Five studies quantified citation biases in epidemiological studies (e.g., association between diesel pollution and lung cancer). These five studies were carried out by the same group and were included in a sixth publication analyzing the corresponding citation networks [11]. This secondary analysis confirmed that citation biases favoring results in agreement with authors' hypothesis are frequent in epidemiology [11]. Finally, two studies exploring citation biases in the field of radiology have shown that results presented as positive by their authors are more often cited [12, 13]. In conclusion, citation biases are frequent in the biomedical literature. They can lead to overestimate positive results and generate information cascades resulting in unfounded dogma [5, 6, 10–13].

42.3 Quotation Distortions

The oldest quantitative study is probably that of De Lacey et al., published in 1986 in *BMJ*. According to these authors, no previous study had addressed the question [4]. The authors reviewed articles that appeared in the 1984 first issue of six biomedical journals (*BMJ, The Lancet, NEJM, Clinical Radiology, British Journal of Surgery,* and *British Journal of Hospital Medicine*). For each of these six issues, the authors randomly selected 50 citations and looked for discrepancies between the quotation statement and the cited source. De Lacey et al. detected between 10 and 20% of inaccurate quotations with an average of 15% for all 6 issues, with no obvious difference between the three most prestigious journals (*BMJ, The Lancet, NEJM*) and the other three [4]. They classified these distortions as trivial (2.7%), slightly misleading (6%) and seriously misleading (6.3%) [4]. For the latter, De Lacey et al. gave an example concerning the treatment of liver abscesses. The quotation statement claimed that patients treated with an antibiotic and surgical drainage had an 8% mortality. In fact, the cited study specified that the 8% mortality rate was observed in a subpopulation of patients with a single abscess, but that for patients with multiple abscesses the post-treatment mortality was 71% [4].

42.3.1 Systematic Reviews

Two systematic reviews have covered academic studies about quotation distortion. The first one identified 28 primary studies evaluating quotation accuracy in medical studies in the broad sense (i.e., including paramedical domains) and some of them were presentations at scientific meetings [7]. The second one, more recent however more restrictive, included only 15 peer-reviewed biomedical studies and excluded paramedical studies [8]. These 15 studies were already included in the first review. Following De Lacey et al. [4], these two systematic reviews classified quotation distortions either as minor (i.e., without serious consequences for the reader) or as major (i.e., seriously misleading) (Table 42.1) [7, 8].

The percentage of seriously misleading citations estimated by Jergas and Baethge was 11.9% (with a 95% confidence interval (CI) between 8.4 and 16.6%) [7]. According to Mogull, the overall percentage of distortions was 14.5% (with a 95% CI between 10.5 and 18.6%) [8]. Among them, two thirds represented major distortions (64.8% with a 95% CI between 56.1 and 73.5%) [8]. Therefore, according to both systematic reviews, the percentage of seriously misleading quotations is around 10% [7, 8].

To identify primary studies published later to these two systematic reviews, we examined all publications citing them. We found 26 separate articles, but only five of them provided new observations about quotation distortions in biomedicine [9, 14–17]. The first analyzed 300 quotations extracted from eight surgery journals and detected 6% of major distortions [14]. In the second article, this percentage reached 22% for 235 quotations referring to 51 studies published in psychiatric journals in German [15]. In the third, the analysis focused on 250 quotations taken at random from five prestigious journals (*Science, Nature, Science Advances, Nature Communication* and *PNAS*) [17]. A total of 21 quotations (i.e., 8.5%) were in fact not supported at all by the source study [17]. The fourth study tested the quotation accuracy in ten official UK government reports on mental health [16]. Among 236 statements, 45 (i.e., 19%) were strongly inaccurate [16]. Finally, the most recent study is especially interesting for its methodology and its large sample size [9]. It starts from 27 primary studies concerning different fields of biomedical research and selected because of the large number of citations they received. The authors reviewed the 4535 articles citing these 27 studies, which led them to examine the accuracy of 7438 quotations. They detected 688 inaccurate quotations, a rate of 9.2% [9]. They classified them descriptively (Table 42.1). The three main types of inaccurate quotations, accounting for 70% of the total, were citing non-existent data (38.4%), mentioning inaccurate quantitative data (16.6%) and biased interpretation of cited observations (15.4%) [9].

In conclusion, the quantitative studies are few, frequently ignored and leave unexplored several biomedical fields such as oncology [7, 8]. Yet all these studies concluded that major quotation distortions are far from rare in the biomedical literature: they affect 6–20% of all quotations.

42.4 Consequences for Public Health: The Opioid Crisis

In the United States, the rate of deaths from accidental overdose of opioids prescribed by a doctor increased fourfold between 1999 and 2010 [18]. This rate stabilized between 2010 and 2013 following measures that made the prescription of opioids more difficult [18]. At the same time, between 2010 and 2016 the death rates from accidental overdose of heroin and fentanyl (without medical prescription) were multiplied by 5 and 9, respectively [18]. Several studies suggest that the measures, taken in 2010 to reduce the use of prescription opioids, might have led opioid dependent patients to switch to these illicit drugs [18, 19]. Overall, the opioid crisis is believed to be responsible for 770,000 deaths in the US between 1999 and 2017 [19], and explains the drop in life expectancy since 2014 [20].

Since 1996, this crisis was precipitated by the aggressive promotion of analgesic opiates by the pharmaceutical industry [21]. This promotional discourse put forward three arguments. First, it pointed out that, beside cancer pain, other types of chronic pain are poorly treated even though they have a lasting effect on patients' quality of life. Second, analgesic opioids were said to be highly effective against all chronic pain. Third, contrary to previous fears, these pain relievers would very rarely lead to dependence or even drug addiction. In particular, after obtaining marketing authorization in 1996 for a delayed form of oxycodone, the company *Purdue Pharma (Stamford, CT, USA)* widely disseminated this discourse to both physicians and the general public [21]. The first argument still holds true. However, the effectiveness for chronic non-cancerous pain is more questionable and has only been studied for durations of <6 months [22, 23]. Compared to a placebo treatment, this efficacy is statistically significant, but clinically mild (−0.69 cm on 10-cm visual analog scale for pain) [22]. Moreover, the comparison with other treatments does not reveal superiority of opioids [22].

To support the claim that the risk of tolerance and addiction associated with opioids is negligible, the pharmaceutical industry as well as many authors cited a 1980 study published in *NEJM* with an unambiguous title: "Addiction rare in patients treated with narcotics" [24]. As of March 30, 2017, this study had been cited by 608 scientific articles, of which 439 supported its conclusion [25]. In fact, this study was a letter, five sentences long, mentioning that among 11,882 hospitalized patients who received at least once an opioid, only four cases of addiction were observed [24]. Among the 608 articles that cited this brief note, 491 failed to mention that patients were hospitalized and that its conclusion was therefore not relevant for the treatment of chronic non-cancerous pain in outpatient settings [25]. According to several observers, the wide dissemination of this inaccurate quotation contributed to the onset of the opioid crisis [21, 25–27]. We therefore see here a particular case of a more general observation: the misuse of citations is particularly frequent in promotional documents published by the pharmaceutical industry [28–30].

42.5 The *Epidemic* of Major Depressive Disorder: The Medicalization of Sadness

Between 1996 and 2005 the rate of antidepressant treatment almost doubled in the United States [31] and similar increases were also observed in most developed countries. Because the diagnostic boundary between major depressive disorder (MDD) and intense normal sadness is not easy to draw, the increase in antidepressant prescription might be, at least partly, due to overdiagnosis [32, 33]. Increase in medication rate and overdiagnosis might also be related to changes in the public understanding of MDD. Indeed, sociological survey showed that the percentage of Americans that attributed MDD to genetic problems, increased from 51% in 1996 to 64% in 2006, and that a neurobiological understanding of MDD enhanced the odds ratio of endorsing MDD medication from 2.02 in 1996 to 5.62 in 2006 [34].

A study published in *Science* in 2003 supported this biomedical conceptualization of MDD and was especially influential. Caspi and colleagues reported, for the first time, that the carriers of the short form of the gene coding for the serotonin transporter (5-HTT) were more vulnerable to MDD when exposed to stressful life events [35]. Many subsequent studies, particularly those with larger sample sizes [36], failed to confirm this 5-HTT gene by environment interaction [37]. Meta-analyses concluded that this interaction was either small (OR = 1.18 CI: 1.09–1.28) [38] or that it did not reach statistical significance [39–41]. This negative conclusion was further confirmed by a very large genome wide association study [42, 43]. Apart from this inconsistent gene-by-stress interaction, all these studies confirmed that exposure to stress and childhood maltreatment roughly doubles lifetime risk for MDD [41].

In 2016, de Vries et al. analyzed the citations received by Caspi's study and by the 72 subsequent studies that had investigated the same genetic susceptibility [37]. In 2015 these 73 studies had received a total of 9166 citations. Among them, 42% cited Caspi's study, 26% cited the 23 other studies in agreement with it, 9% cited the 11 unclear studies, and only 23% cited the 38 studies that disconfirmed it [37]. Therefore, among all primary studies and meta-analysis articles about the 5-HTT-gene susceptibility to

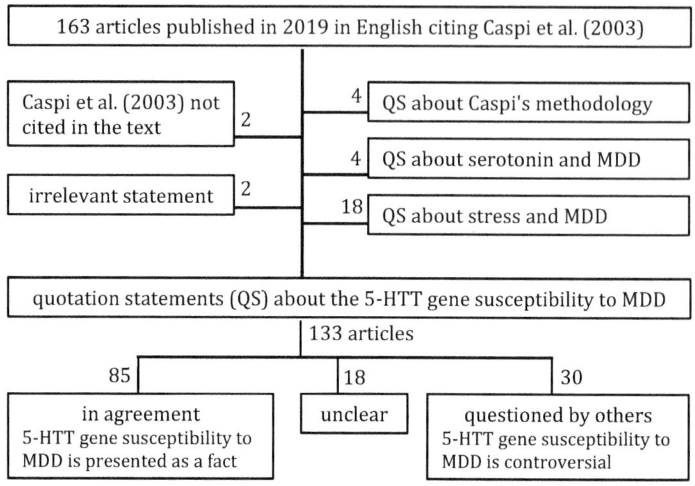

Fig. 42.1 Quotation statements in articles, review articles, editorials, and letters published in English in 2019 and citing Caspi et al. (2003) [35]. Books and conferences proceedings were excluded. *QS* Quotation statements, *5HTT* Serotonin transporter gene, *MDD* Major depression disease

MDD, the Caspi et al. (2003) [35] study was disproportionally cited (Fig. 42.1).

According to SCOPUS, since its publication in 2003, Caspi's study [35] had been cited 5820 times on April 30, 2020. In order to quantify the number of quotation statements that represented the 5-HTT-gene susceptibility to MDD as an obvious fact, we collected and read the 163 articles citing Caspi et al. (2003) [35] and published in English during 2019. As detailed in Fig. 42.1, two articles did not cite Caspi et al. in the text and 28 articles put quotation statements that were not related to the main conclusion raised by Caspi et al. (2003) [35]. Among the 133 other articles, 30 also mentioned that the 5-HTT-gene susceptibility to MDD was controversial and cited at least one meta-analysis article contradicting Caspi et al. (2003) [35]. Eighteen articles put unclear quotation statements. In particular, five articles represented Caspi et al. main conclusion as a fact at one point in their text, and then mentioned elsewhere that results about gene by environment interactions in psychiatry are often inconsistent, however failed to link both quotation statements. Finally, 85 articles only cited Caspi et al. (2003) [35] in support of a 5-HTT-gene susceptibility to MDD and did not cite any article contradicting it. Therefore, 85 out of 133 articles (64%) provided an unbalanced view of this topic and were seriously misleading.

This citation bias was exacerbated by mass media. Indeed, during the first month after its publication, Caspi's study was covered by at least 50 articles written in English in the general press [44]. Among the 13 subsequent studies analyzed by the meta-analysis published in 2009 [40], two smaller studies confirming Caspi's study were also covered (7 and 2 articles, respectively) [44]. Newspapers never covered the 11 subsequent studies that failed to replicate it [44]. Finally the 2009 meta-analysis, although published in a prestigious journal (*JAMA*), was covered by only four newspaper articles and all four mentioned that the initial finding "has not held up to scientific scrutiny" [44]. Even elite newspapers continued to refer to Caspi's study years after its invalidation. For example, on November 30, 2014, the *New York Times* wrote: "What distinguishes children who prove more versus less susceptible to developmental experiences? There is no single factor, but genetics seems to play a role. For instance, short alleles of the gene 5-HTTLPR, which transports serotonin, have been linked to depression."

42.5.1 Clinical Success of Antidepressant Drugs

Finally, the citation bias affecting clinical trials acted in synergy with a publication bias and both led to overestimate the efficacy of antidepressant drugs [45]. Publication bias consists in a preferential publication of positive results, while most negative ones remain unpublished. This bias is especially obvious concerning the clinical trials of antidepressants [46]. When all clinical trials registered by the Food and Drug Administration were considered, including the numerous unpublished ones, the difference between antidepressant and placebo treatments did not reach clinical significance, except for the more severely depressed patients [46, 47]. Moreover, positive antidepressant trials were cited three times as frequently as negative trials [45]. Therefore, this citation bias amplifies the effect of the publication bias.

In summary, citation biases and their exacerbation by mass media largely contributed to promote, in the general public, the view that the 5-HTT-gene susceptibility is a major risk factor for MDD. This neuro-essentialist conceptualization negatively affects several aspects of patients' stigmatization, focuses their expectations on antidepressant medication and overshadows psychological and social risk factors for MDD [48, 49]. As a result, social prevention that might address psychosocial risks for MDD, such as poverty or childhood maltreatment, is kept in the background and antidepressant medication is overvalued compared to psychotherapies [49].

42.6 The Promotion of Hydroxychloroquine as a Treatment for COVID-19

In the context of COVID-19, a French team published in April 2020 two studies without any control group [50, 51]. Both asserted the efficacy of a treatment combining hydroxychoroquine (HCQ) and azithromycin. The first one included 80 patients positive to the PCR test for SARS-CoV-2 and reported that 93% of them became negative 8 days after treatment initiation [50]. The authors compared this evolution with that reported by a Chinese study where the 191 patients had not received HCQ and most of whom were still positive on the PCR test 20 days after the first test [52]. Gautret et al. cited this Chinese study and, therefore, claimed that their treatment accelerated the disappearance of the virus [50]. This comparison was inappropriate, however, because 92% of the French patients presented only mild symptoms at treatment initiation and only one died [50], while all 191 patients included in the Chinese study were hospitalized, 63% of them suffered from a severe form of COVID-19 and 54 died from it [52]. The article by Gautret et al. [50] did not mention these characteristics of the cited Chinese study [52]. In addition, they cited another study showing that the severity of symptoms is proportional to the viral load [53] and concluded that their treatment prevented patients' deterioration because it accelerated the disappearance of the virus [50]. However, they failed to mention that "Mild cases were found to have an early viral clearance, with 90% of these patients repeatedly testing negative on RT-PCR by day 10 post-onset. By contrast, all severe cases still tested positive at or beyond day 10 post-onset [53]". Since 93% of the French patients had mild symptoms at treatment initiation, one can expect their viral load to drop spontaneously and there is no evidence that HCQ had anything to do with it.

The second French study involved 1061 PCR positive patients, 95% of whom had mild symptoms [51]. The authors followed the evolution of the response to the PCR test and, in the absence of a control group, put forward the same inappropriate comparison with the Chinese population of 191 hospitalized patients, to claim that their treatment was effective [52]. Altogether, both studies [50, 51] asserted the effectiveness of a treatment with HCQ by comparing with two previous studies [52, 53], although a fair quotation should have led to contradict this claim.

These two studies were published online on April 4 and May 1, 2020; however their results were made public several days earlier. They contributed to fuel the hope for a safe and effective treatment for COVID-19. Consequently, HCQ has been widely recommended in a large number of countries as a medication to cure COVID-19. In addition, it has been tested in numerous RCTs enrolling thousands of patients. These trials showed that HCQ was not an effective treatment for COVID-19 [54–56] and the HCQ controversy extinguished during the summer of 2020. However, during spring this controversy was tremendously amplified by traditional media and social networks and reached high intensity. Several researchers in France, Switzerland [57], and Brazil [58] have been singled out and even threatened with death for having criticized the HCQ treatment.

42.6.1 Misrepresentation Cascade

These major quotation distortions were in a way the consequence of previous violations of medical ethics. Indeed, the same French team had previously published a controlled trial [59], which was strongly criticized for its small sample size and its weak methodology [60]. However, Gautret et al. concluded from this controlled trial, that HCQ was an effective treatment for COVID-19 and that it was, therefore, no longer ethical to include a placebo arm in their subsequent studies. Accordingly in a newspaper article published in *Le Monde* on March 25, 2020, Didier Raoult, the leading investigator of these three HCQ studies, explicitly criticized the methods of the evidence-based medicine, namely the RCTs. As a result, the two studies discussed above [50, 51] did not include a control group and their authors had to compare their results with previously published data. To maintain that HCQ is efficient, the authors biased their comparison by omitting major information from the cited literature. This appalling story gives support to the recommendation made by Andre Kalil in May 2020: "The rapid and simultaneous combination of supportive care and RCTs is the only way to find effective and safe treatments for COVID-19" [61].

42.7 Conclusion

Several articles cited here offered recommendations to limit the misuse of citations and its consequences [7, 9]. Readers of the biomedical literature should be better informed of a possible citation misuse and should not hesitate to verify in the source articles the accuracy of the quotations. Second, reviewers in charge of reviewing a manuscript should carefully scrutinize the relevance and accuracy of all citations. They should also ask the authors to add references when the discussion of the previous literature is unbalanced. Third, editors of biomedical journals should ask their reviewers to be more watchful about the accurate use of citations. They might also warn, in the note to authors, that their reviewers will closely scrutinize their citation use. Finally, Penders [62], as well as Pavlovic et al. [9] offered numerous recommendations to the authors. They are wise, but insufficient to counteract the root cause of citation misuse according to Jergas et al. [7]: the confirmation bias, which makes us preferentially consider information in accordance with our opinions.

Ackowledgements Several views and arguments presented in this text were already expressed in our previous article published in French language in *Médecine/Science* [63]. We thank the editor of *Médecine/Science* for granting us the permission to use them in this chapter. Estelle Dumas-Mallet was supported by a grant from the Agence National de la Recherche (ANR-18-FRAL-0003-01).

References

1. Prinz F, Schlange T, Asadullah K (2011) Believe it or not: how much can we rely on published data on potential drug targets? Nat Rev Drug Discov 10:712
2. Ioannidis JP (2014) How to make more published research true. PLoS Med 11:e1001747
3. Dumas-Mallet E, Button K, Boraud T et al (2016) Replication validity of initial association studies: a comparison between psychiatry, neurology and four somatic diseases. PLoS One 11:e0158064

4. de Lacey G, Record C, Wade J (1985) How accurate are quotations and references in medical journals? Br Med J (Clin Res Ed) 291:884–886
5. Greenberg SA (2009) How citation distortions create unfounded authority: analysis of a citation network. BMJ 339:b2680
6. Duyx B, Urlings MJE, Swaen GMH et al (2017) Scientific citations favor positive results: a systematic review and meta-analysis. J Clin Epidemiol 88:92–101
7. Jergas H, Baethge C (2015) Quotation accuracy in medical journal articles-a systematic review and meta-analysis. PeerJ 3:e1364
8. Mogull SA (2017) Accuracy of cited "facts" in medical research articles: a review of study methodology and recalculation of quotation error rate. PLoS One 12: e0184727
9. Pavlovic V, Weissgerber T, Stanisavljevic D et al (2021) How accurate are citations of frequently cited papers in biomedical literature? Clin Sci (Lond) 135: 671–681
10. Tatsioni A, Bonitsis NG, Ioannidis JP (2007) Persistence of contradicted claims in the literature. JAMA 298:2517–2526
11. Urlings MJE, Duyx B, Swaen GMH et al (2021) Citation bias and other determinants of citation in biomedical research: findings from six citation networks. J Clin Epidemiol 132:71–78
12. Frank RA, Sharifabadi AD, Salameh J-P et al (2019) Citation bias in imaging research: are studies with higher diagnostic accuracy estimates cited more often? Europ Radiol 29:1657–1664
13. Treanor L, Frank RA, Salameh JP et al (2019) Selective citation practices in imaging research: are diagnostic accuracy studies with positive titles and conclusions cited more often? Am J Roentgenol 213: 397–403
14. Armstrong MF, Conduff JH 3rd, Fenton JE, Coelho DH (2018) Reference errors in otolaryngology-head and neck surgery literature. Otolaryngol Head Neck Surg 159:249–253
15. Baethge C (2020) Importance, errors, and patterns of quotations to psychiatric original articles. Pharmacopsychiatry 53:247–255
16. Hui A, Rains LS, Todd A et al (2020) The accuracy and accessibility of cited evidence: a study examining mental health policy documents. Soc Psychiatry Psychiatr Epidemiol 55:111–121
17. Smith N, Cumberledge A (2020) Quotation errors in general science journals: quotation errors in science journals. Proc Royal Soc A: Math Phys Eng Sci 476(2242):20200538
18. Jalal H, Buchanich JM, Roberts MS et al (2018) Changing dynamics of the drug overdose epidemic in the United States from 1979 through 2016. Science 361(6408):eaau1184
19. Healton C, Pack R, Galea S (2019) The opioid crisis, corporate responsibility, and lessons from the tobacco master settlement agreement. JAMA 322(21): 2071–2072
20. Woolf SH, Schoomaker H (2019) Life expectancy and mortality rates in the United States, 1959-2017. JAMA 322:1996–2016
21. Van Zee A (2009) The promotion and marketing of oxycontin: commercial triumph, public health tragedy. Am J Public Health 99:221–227
22. Busse JW, Wang L, Kamaleldin M et al (2018) Opioids for chronic noncancer pain: a systematic review and meta-analysis. JAMA 320:2448–2460
23. Chou R, Turner JA, Devine EB et al (2015) The effectiveness and risks of long-term opioid therapy for chronic pain: a systematic review for a National Institutes of Health Pathways to Prevention Workshop. Ann Intern Med 162:276–286
24. Porter J, Jick H (1980) Addiction rare in patients treated with narcotics. N Engl J Med 302:123
25. Leung PTM, Macdonald EM, Stanbrook MB et al (2017) A 1980 letter on the risk of opioid addiction. N Engl J Med 376:2194–2195
26. Becker WC, Fiellin DA (2017) Limited evidence, faulty reasoning, and potential for a global opioid crisis. BMJ 358:j3115
27. Rummans TA, Burton MC, Dawson NL (2018) How good intentions contributed to bad outcomes: the opioid crisis. Mayo Clin Proc 93:344–350
28. Keng A, Coley RM (1994) Evaluating the accuracy of citations in drug promotional brochures. Ann Pharmacother 28:1231–1235
29. Gonzalez-Santiago M, Bucher HC, Nordmann AJ (2008) Accuracy of drug advertisements in medical journals under new law regulating the marketing of pharmaceutical products in Switzerland. BMC Med Inform Decis Mak 8:61
30. Spielmans GI, Thielges SA, Dent AL, Greenberg RP (2008) The accuracy of psychiatric medication advertisements in medical journals. J Nerv Ment Dis 196:267–273
31. Olfson M, Marcus SC (2009) National patterns in antidepressant medication treatment. Arch Gen Psychiatry 66:848–856
32. Horwitz AV, Wakefield JC (2007) The loss of sadness. How psychiatry transformed normal sorrow into depressive disorder. Am J Psychiatry 164:1764–1765
33. Wakefield JC, Schmitz MF (2013) When does depression become a disorder? Using recurrence rates to evaluate the validity of proposed changes in major depression diagnostic thresholds. World Psychiatry 12:44–52
34. Pescosolido BA, Martin JK, Long JS et al (2010) "A disease like any other"? A decade of change in public reactions to schizophrenia, depression, and alcohol dependence. Am J Psychiatry 167:1321–1330
35. Caspi A, Sugden K, Moffitt TE et al (2003) Influence of life stress on depression: moderation by a polymorphism in the 5-HTT gene. Science 301:386–389
36. Duncan LE, Keller MC (2011) A critical review of the first 10 years of candidate gene-by-environment interaction research in psychiatry. Am J Psychiatry 168: 1041–1049

37. de Vries YA, Roest AM, Franzen M et al (2016) Citation bias and selective focus on positive findings in the literature on the serotonin transporter gene (5-HTTLPR), life stress and depression. Psychol Med 46:2971–2979
38. Bleys D, Luyten P, Soenens B, Claes S (2018) Gene-environment interactions between stress and 5-HTTLPR in depression: a meta-analytic update. J Affect Disord 226:339–345
39. Munafò MR, Durrant C, Lewis G, Flint J (2009) Gene X environment interactions at the serotonin transporter locus. Biol Psychiatry 65:211–219
40. Risch N, Herrell R, Lehner T et al (2009) Interaction between the serotonin transporter gene (5-HTTLPR), stressful life events, and risk of depression: a meta-analysis. JAMA 301:2462–2471
41. Culverhouse RC, Saccone NL, Horton AC et al (2017) Collaborative meta-analysis finds no evidence of a strong interaction between stress and 5-HTTLPR genotype contributing to the development of depression. Mol Psychiatry 23:133
42. Border R, Johnson EC, Evans LM et al (2019) No support for historical candidate gene or candidate gene-by-interaction hypotheses for major depression across multiple large samples. Am J Psychiatry 176: 376–387
43. Nierenberg AA (2019) A great gene environment depression story fades into the sunset. Psychiatric Ann 49:294
44. Dumas-Mallet E, Smith A, Boraud T, Gonon F (2017) Poor replication validity of biomedical association studies reported by newspapers. PLoS One 12: e0172650
45. de Vries YA, Roest AM, de Jonge P et al (2018) The cumulative effect of reporting and citation biases on the apparent efficacy of treatments: the case of depression. Psychol Med 48:2453–2455
46. Turner EH, Matthews AM, Linardatos E et al (2008) Selective publication of antidepressant trials and its influence on apparent efficacy. N Engl J Med 358: 252–260
47. Kirsch I, Deacon BJ, Huedo-Medina TB et al (2008) Initial severity and antidepressant benefits: a meta-analysis of data submitted to the Food and Drug Administration. PLoS Med 5:e45
48. Lebowitz MS, Appelbaum PS (2019) Biomedical explanations of psychopathology and their implications for attitudes and beliefs about mental disorders. Annu Rev Clin Psychol 15:555–577
49. Dumas-Mallet E, Gonon F (2020) Messaging in biological psychiatry: misrepresentations, their causes, and potential consequences. Harv Rev Psychiatry 28: 395–403
50. Gautret P, Lagier JC, Parola P et al (2020) Clinical and microbiological effect of a combination of hydroxychloroquine and azithromycin in 80 COVID-19 patients with at least a six-day follow up: a pilot observational study. Travel Med Infect Dis 34:101663
51. Million M, Lagier JC, Gautret P et al (2020) Early treatment of COVID-19 patients with hydroxychloroquine and azithromycin: a retrospective analysis of 1061 cases in Marseille, France. Travel Med Infect Dis 35:101738
52. Zhou F, Yu T, Du R et al (2020) Clinical course and risk factors for mortality of adult inpatients with COVID-19 in Wuhan, China: a retrospective cohort study. Lancet 395:1054–1062
53. Liu Y, Yan LM, Wan L et al (2020) Viral dynamics in mild and severe cases of COVID-19. Lancet Infect Dis 20(6):656–657
54. Fiolet T, Guihur A, Rebeaud ME et al (2020) Effect of hydroxychloroquine with or without azithromycin on the mortality of coronavirus disease 2019 (COVID-19) patients: a systematic review and meta-analysis. Clin Microbiol Infect 27:19–27
55. Juul S, Nielsen EE, Feinberg J et al (2020) Interventions for treatment of COVID-19: a living systematic review with meta-analyses and trial sequential analyses (The LIVING Project). PLoS Med 17: e1003293
56. Siemieniuk RA, Bartoszko JJ, Ge L et al (2020) Drug treatments for covid-19: living systematic review and network meta-analysis. BMJ 370:m2980
57. Peiffer-Smadja N, Rebeaud ME, Guihur A et al (2021) Hydroxychloroquine and COVID-19: a tale of populism and obscurantism. Lancet Infect Dis 21(5):e121
58. Ektorp E (2020) Death threats after a trial on chloroquine for COVID-19. Lancet Infect Dis 20:661
59. Gautret P, Lagier JC, Parola P et al (2020) Hydroxychloroquine and azithromycin as a treatment of COVID-19: results of an open-label non-randomized clinical trial. Int J Antimicrob Agents 56:105949
60. Kim AHJ, Sparks JA, Liew JW et al (2020) A rush to judgment? Rapid reporting and dissemination of results and its consequences regarding the use of hydroxychloroquine for COVID-19. Ann Intern Med 172(12):819–821
61. Kalil AC (2020) Treating COVID-19-off-label drug use, compassionate use, and randomized clinical trials during pandemics. JAMA 323:1897–1898
62. Penders PB (2018) Ten simple rules for responsible referencing. PLoS Comput Biol 14:e1006036
63. Dumas-Mallet E, Boraud T, Gonon F (2021) Le mésusage des citations et ses conséquences en médecine [The misuse of citations and their consequences in medicine]. Méd Sci 37(11): 1035–1041

Understanding Ghostwriting and Ghost Authorship As Problems of Research Integrity

43

Lisa DeTora

Abstract

Ghostwriting is a serious problem of research integrity; however, its true impact remains unclear, and little concrete advice is available on how to avoid it. Ongoing discussions and engagement are needed by all authors and contributors to research publications. Efforts made by professional groups to ensure authorship transparency with an aim of ending ghost authorship and ghostwriting are presented. Attention to publication planning and execution is necessary to ensure that ghostwriting is eliminated, and research integrity is protected, especially in the context of large studies sponsored by the pharmaceutical industry.

43.1 Introduction

The integrity, accuracy, and completeness of the published biomedical literature are critical for day-to-day clinical practice, public health policy, and treatment recommendations [1]. Recent articles suggest that ghostwriting in the medical literature represents an important failure of transparency that can affect the value of research studies [2–4]. Given the importance of the biomedical literature, any undue influences or biases that affect publications are a cause for grave concern [5]. Ghostwriting masks the process by which a publication is produced and may occur in connection with other bad practices. A consistent concern in this context has been the role of the pharmaceutical industry and professional writers in producing manuscripts.

Efforts made by professional groups to ensure authorship integrity and transparency also are presented [6, 7].

43.2 The International Committee of Medical Journal Editors (ICMJE)

The ICMJE [8] expressed concern about a lack of transparency in publications of clinical trials. Poor transparency obscured the role of corporate sponsors and contract research organizations (CROs) in the design, conduct and publication of clinical studies. Journal editors could not always determine the role of the funder or whether CROs had monitored clinical studies. Contracts between the pharmaceutical industry and academic authors might prevent researchers from publishing data independently or even seeing the results of the studies they had helped conduct, practices that limit the ability for authors to take accountability for their work. The ICMJE

L. DeTora (✉)
Writing Department, Hofstra University, Hempstead, NY, USA

J. Faintuch, S. Faintuch (eds.), *Integrity of Scientific Research*,
https://doi.org/10.1007/978-3-030-99680-2_43

also commented on failures to disclose or publish negative results, as well as redundant publication or salami slicing of favorable datasets which skewed the published literature and created confusion for researchers who needed to understand the extent of the clinical evidence base. Ghostwriting and guest authorship were described, however not explicitly named, as a usually paired practice in which named authors on published papers made little input while the actual writers of the text were not adequately mentioned in the final work [8].

Recommendations encompassed sharing complete study data with all invited authors of research papers, registering clinical trials in publicly accessible databases to enable researchers and editors to identify and differentiate unique datasets, and adhering to the ethical provisions in the ICMJE requirements [8, 9]. Authors, as Davidoff and colleagues suggested, had to be accountable for research quality and maintain intellectual independence when writing papers.

43.3 The World Association of Medical Editors (WAME)

WAME drew a distinction between ghost authorship (when a qualified author was left off the byline) and ghostwriting (when someone wrote a paper or large parts of it without their work being adequately disclosed). The lead author was responsible for avoiding ghost authorship. Ghostwriting, in contrast, was the responsibility of everyone involved in funding, drafting, or editing manuscripts for publication. WAME [10] also called on editors to be publicly transparent when they detected ghostwritten papers.

43.4 Government Concerns

Ghostwriting was one of several topics investigated by the United States Senate Committee on Finance when they examined transparency in the pharmaceutical industry, especially as regarded payments and gifts to prescribers. “Ghostwriting in Medical Literature” [11] detailed a general lack of clarity regarding the role of pharmaceutical companies in funding and authoring biomedical publications. Ghostwriting was defined as the hiring “of medical education, marketing or communications companies to draft articles that are presented to prominent physicians and scientists to sign on as authors to increase the likelihood that the article will be published in important medical journals” (p. 2).

Even when editorial assistance and writing support were acknowledged, the Committee found, statements about this support were vague. Acknowledgements and author bylines were not adequate to identify key stakeholders and contributors to studies and manuscripts, particularly the people responsible for the intellectual content of the work. Hence, the precise role and influence of writing and editorial support, including funding for the support, remained unclear. The existing peer review system, which assumed adherence to medical ethics, was not equipped to manage the realities of large-scale clinical research programs sponsored by the pharmaceutical industry. Existing medical school and journal policies against ghostwriting, also, were inadequate to prevent or even, in some cases, detect the problems associated with large-scale pharmaceutical studies.

The Committee also found that the NIH required stronger policies against ghostwriting. Most of these recommendations have been followed. As the ICMJE notes [9], trial registration and certain financial disclosures are now legally mandated in many locations.

43.5 Existing Authorship Criteria

Differing authorship criteria or even interpretations of those criteria, can make it difficult to create a definitive template for avoiding ghostwriting and ghost authorship. Certain journals may specifically include or exclude a professional medical writer as an author on the grounds of intellectual contributions. Decisions by editors, thus, may require authors and other contributors to manuscripts, to review author criteria each time a new publication is undertaken. Limitations of accepted authorship guidelines also have been identified. The ICMJE criteria have been criticized, as by Matheson [12, 13] and Moffat [14], for creating loopholes that can be exploited by the unscrupulous.

The most used authorship criteria are below.

- *International Committee of Medical Journal Editors (ICMJE)*: The ICMJE defined authorship criteria in 1988 to ensure that named authors had contributed to the intellectual content of a manuscript through study design and conception, study conduct and/or data analysis and/or interpretation. Authors also had to write or critically edit or revise the manuscript and approve the work to be published. Today, the ICMJE includes a fourth authorship criterion, supporting the work publicly. These criteria are considered a gold standard for many biomedical publications, especially the results of clinical trials and meta-analyses [9].
- *The World Association of Medical Editors (WAME)*: The WAME authorship guidelines [15] indicate that anyone who made in intellectual contribution to a work should be listed as an author. These criteria include writing as a potential, but not necessary, area of intellectual input. The WAME guidelines also suggest that one author serve as a guarantor for the integrity of the work as a whole.
- *The Council of Science Editors (CSE)*: The CSE suggests that authorship attribution establishes accountability for the work, and notes that journal editors are responsible for establishing, explaining, and enforcing authorship criteria to the best of their abilities. CSE policies also contain a listing of 12 areas that, in appropriate combination, can confer authorship: conception, design, supervision (substantive day-to-day oversight), resources, material, data collection and/or processing, analysis and/or interpretation, literature searching, writing, critical review, and other novel contributions (councilscienceeditors.org).
- *The Committee on Publication Ethics (COPE)*: COPE identifies accountability and intellectual contributions as minimal requirements for authorship attribution and provides many resources for considering authorship questions (publicationethics.org).

43.6 Ghostwriting and Related Problems

One problem in addressing the problem of ghostwriting is that definitions of this problem, like definitions of authorship, are slippery. The terms ghostwriting and ghost authorship are sometimes used interchangeably and may be considered to occur even if a vague acknowledgement of the work is made [11, 16, 17]. These concepts can be combined, even in discussions that examine both problems [17].

As noted in recent literature reviews [16, 17], ghostwriting can be defined as a form of plagiarism [18, 19], as any circumstance of failed acknowledgement in a published paper. Others classify it as research malpractice [20], as uncredited writing support, including in some cases any work by a hired writer [21]. The lack of a consistent definition of ghostwriting has complicated endeavors to identify the scope and prevalence of this problem. Some motivations for ghostwriting identified in the published literature include marketing and promotion pressures [5], as well as the logistics of running large-scale clinical trials and practices in corporations and medical communications companies that isolate professional writers from contributing authors [12, 16, 17, 22, 23].

Mass media use of the term ghostwriting is often nonspecific. News reports, for instance, might mention ghostwriting in association with failure to disclose financial relationships, using the term as a catch-all for any ethical issues in clinical research. Lack of clarity as to what constitutes ghostwriting can muddy public understanding of research ethics, limiting the ability of lay audiences to identify different problems of transparency [24].

Other problems of authorship attribution

- *Guest authorship*: Any paper listing coauthors who do not meet all pertinent authorship criteria is considered as guest authored, even if the named authors made some contribution to the work. Guest authorship may also be paid [25]. As described in a US Senate Committee report [11], guest authorship is often associated with ghostwriting, as when a paper is prepared and then submitted to authors with a request that the authors lend their name to the work.

- *Paid guest authorship*: Paying an expert to have their name attached to a publication of work they did not complete is unethical. These papers are necessarily ghostwritten. Paid guest authorship should not be confused with compensation for legitimate work or salaried employment in roles that lead to authorship [26].
- *Gift authorship*: Authors sometimes offer non-participating colleague a byline credit rather than an acknowledgement, often out of respect or to bolster the credibility of the work. Gift authorship differs from guest authorship in that it is usually a practice among current colleagues [27].
- *Relinquished authorship*: This practice is generally associated with corporate policies that limit the number of company employees as authors, leading some qualified authors to relinquish authorship. Some academic centers do not allow collaborations in which authorship is relinquished in this manner [28]. Relinquished authorship may also occur when pressure is placed on authors to represent data in ways they cannot support. Some journals may limit the number of byline authors, also leading to relinquished authorships.
- *Coerced authorship*: More hierarchically powerful people, such as grant holders or laboratory heads, may assert their right to be listed as the authors of work they did not directly participate in [28]. This may also be referred to as hostage authorship [29] In some cases, coerced authorship may be motivated by the intention to protect subordinates from public accountability for the published work.
- *Planted authorship*: When authors are named on a publication without their knowledge or consent [28].
- *Orphan papers*: The situation in which no one qualifies as an author because no one meets all the pertinent qualifications [14]. This situation is more common for restrictive authorship standards such as those of the ICMJE [17].
- *Ghost peer review*: Peer reviews may be completed by students under the supervision of the peer reviewer contacted by the journal. Most people surveyed found this practice acceptable even if the student was not credited with the peer review [30].
- *Ghost management*: This term was suggested to describe the bureaucratic apparatus at pharmaceutical companies and medical communications agencies that indirectly influences publications, often without adequate disclosure [31, 32].

The above terms have sometimes been augmented, embellished or combined, adding to a general confusion in ethical discussions and creating some problems for the study of poor authorship practices ([4, 33]; US 2001).

43.6.1 Difficulties in Identifying Ghostwriting Prevalence

Experts had trouble identifying the true extent and impact of ghostwriting. A systematic meta-analysis of studies of ghostwriting prevalence was attempted [17]. Among 848 papers of possible interest, 48 reported data on ghostwriting prevalence; only 16 were primary publications of research data. Published research studies varied in design making systematic meta-analysis impossible [17].

Of the 16 primary studies of ghostwriting prevalence, 12 were surveys of opinion and self-reported experiences from various expert groups, often retrospective. The reported prevalence of ghostwriting ranged from 0.2% to 75%, indicating a lack of consistency and alignment. Three additional papers reported on reviews of document repositories including early drafts of manuscripts; 2/3 were associated with a single marketed product. In these repositories, 57.3% to 75% of papers had possible evidence of ghostwriting or ghost authors, depending on the authorship criteria applied (which were not systematically linked to specific journal requirements). Ghostwriting tended to be associated with papers in which medical writing support was acknowledged.

Stretton [17] also observed that reported ghostwriting prevalence was strongly related to the definition of ghostwriting and authorship standards used, which could be manipulated to inflate or deflate study results. The secondary literature that reported ghostwriting prevalence consisted of 32 papers, many of which included erroneous or unverifiable data. A figure commonly reported in the secondary literature was that between 50% and 100% of papers had been ghostwritten; however, no source was identified for this figure [17]. Stretton concluded that, while ghostwriting and ghost authorship were to be avoided, manipulating definitions to estimate these practices was not an ethically advisable solution.

A subsequent review [16], assessed whether the indexed biomedical literature provided concrete advice to prevent ghostwriting. A PubMed database search (1900 to 2018), revealed 181 papers on ghostwriting. Manual review revealed eight primary studies of ghostwriting prevalence, none indexed after 2013. One additional study in the grey literature [34] presented survey data about ghostwriting from medical writing and publications professionals obtained in 2005, 2008, 2011, and 2014. Similar to Stretton, reports of prevalence ranged widely, in this case from $<1\%$ to over 90%.

Definitions of ghostwriting, ghost authorship, and authorship varied markedly across the 181 papers indexed in Medline [16]. Ghost authors were more likely to be identified when WAME author criteria were applied, as compared with the more restrictive ICMJE criteria. Ghost authors and ghostwriting were also more likely to be associated with studies listing many authors, with review papers rather than the primary or secondary results of clinical trials, and with papers published before 2005. The literature also contained multiple entries, most opinion pieces, by several authors including Matheson [12, 13] and Bosch [19, 20].

43.7 Recent Publishing about Ghostwriting

No new primary studies of ghostwriting prevalence appeared in PubMed or Medline between 2013 and 2021; however, ghostwriting remains a topic of concern, as evidenced by continued publishing. Zimba and Gasparyan [4] highlight ghost authorship as problem of industry-sponsored medical writers in a recent overview of scientific writing intended as a primer for research students. Additional ghostwriting problems may also occur, as when junior researchers ghost write peer reviews or grant applications for senior colleagues [30].

Kasenda et al. [35] reviewed 647 industry-sponsored randomized, controlled trial protocols that had undergone Internal review board/IRB or Ethical Committee/EC review between 2000 and 2003 at six research centers. Most (456/647) protocols mentioned publications policies, however 393/456 placed a condition, such as corporate review, on publication. Only 39/647 protocols specifically allowed researchers to publish their own data independently. Kasenda and colleagues phrased the right to review as the "right to disapprove or at least review" (p. 9); it was unclear whether this precise language was used in the protocols. The majority of publications reporting the results of these studies (260/388) included an industry author.

Yi et al. [36] conducted a survey among biomedical researchers working at universities in China (33 schools) or Europe (13 schools). All researchers had generally similar views on forms of plagiarism like copying text or data without attribution. However, researchers in China were more likely to view hiring a writer or obtaining free writing help without authorship attribution as plagiarism, than were researchers in Europe.

Nearly half of researchers overall did not view either of these practices as plagiarism.

A recent survey of health science students in Saudi Arabia [37] found that about half of respondents lacked knowledge of their institutional authorship and publication policies, and also about ghostwriting. Rasmussen et al. [2] reviewed publications policies at the 266 research 1 (very high activity) and 2 (high activity) institutions in the US and found that 93% had author criteria; all were clear in denouncing poor practice, including ghostwriting. However, fewer policies provided guidance for applying authorship criteria and handling authorship disputes (62%) or managing faculty-student authorship (49%).

43.8 Some Causes of Inadvertent Authorship Problems

Some common problems can cause ghostwriting or ghost authorship even among study teams and authoring groups who are trying to avoid poor practice.

- *Conflicting authorship standards*: The two most recent reviews of published literature about ghostwriting, including Stretton [17], which is the most robust analysis to date, link differences in authorship standards to problems with author bylines. This suggests that increased awareness of authorship guidelines is important [2, 16].
- *Insufficient training or knowledge*: Schroter and colleagues [38] noted that authorship ethics and publication ethics tend to be taught by osmosis from advisors, on a case-by-case basis rather than in a systematic way. The lack of formal training may be one reason for the findings reported by Badreldin et al. [37]. Rasmussen et al. [2] also noted that training and pragmatic advice were lacking in existing policies at academic centers.
- *Regional differences in understanding*: Yi et al. [36] noted that many researchers they surveyed believed that attributing authorship to a laboratory head was less important than plagiarism as a problem in research ethics. They also found differences in beliefs regarding the role of writers who did not conduct the research.
- *Working logistics*: Logdberg [23] links day-to-day practices that mechanize writing, similar to those cited in the Grassley report (2010), as isolating authors from professional writers. Logdberg [23] also comments that the scale of some trials creates logistical problems for authorship and accountability. Orphan papers may also result when no researcher has participated in enough roles to qualify as an author under ICMJE guidelines [14].
- *Varying levels of autonomy and accountability*: One of the tensions between academic researchers and employees of pharmaceutical companies is their level of individual autonomy and accountability. In organizations where accountability is associated with hierarchically higher positions, some professionals may want to incur the risks of authorship to protect their employees. Alfonso et al. [39] note that accountability is essential to truly confer authorship.

43.9 Preventive and Corrective Interventions

Since ghostwriting is a negative practice, a counterdefinition may be useful to promote good practice. Research publication and documentation is required to be transparent to maintain integrity and credibility. It must be clear who designed, conducted, and analyzed studies to ensure that qualified persons have done it. The identities of people who interpret and document research findings, including people who write much or all of manuscripts, must also be clear [8, 9]. Contributions should be clearly delineated in accordance with applicable authorship standards and as agreed by all the authors of a work. In other words, authorship should be attributed transparently and honestly, and writing support should be acknowledged accurately and completely.

43.9.1 Good Publication Practice (GPP)

Transparency and accurate attribution of authorship are considered as essential to good publications practice, or GPP (ismpp.org/gpp3), for the pharmaceutical industry [6, 7]. Designating a project lead intended to serve as the lead author can take place during the study design stage. A publication steering committee or authorship working group for large-scale trials can help teams remain on track and avoid miscommunication, especially if multiple publications are anticipated to result from a single study. Author contracts that detail expected responsibilities and qualifications for byline authorship are also helpful [6]. DeTora et al. [16] also recommend identifying an authorship champion for each publication, to ensure that everyone who contributed is properly credited.

43.9.2 Consult Available Guidelines

COPE and other groups provide educational materials to help authoring groups plan effectively and troubleshoot issues. The CRediT guideline, for instance, provides a taxonomy of publication contributions in a checklist format [40]. The Council of science editors/CSE also provides a listing of 12 categories of contributions that might be combined to confer authorship [41, 42]. Experts like Alonso et al. [39] provide specific advice for members of professional societies or researchers working in particular therapeutic areas.

43.9.3 Clarify the Role of the Medical Writer

Confusion about the role of medical writers may be a factor in the inconsistent use of terminology like ghost authorship or ghostwriting. Such confusion has persisted in part because of a tendency during the twentieth century to refer to all medical writing support as ghostwriting, even if properly disclosed [16, 17, 43].

A joint statement on the role of medical writers by medical writing and publications professional societies [44] outlines a process by which medical writers should be employed. The named authors of each publication should agree to the primary data to be presented as well as the literature to be cited and any discussion points before a professional writer begins any work on a manuscript. This process will help ensure that the authors can be accountable for the intellectual framing of the manuscript by requiring that they communicate proactively to the professional writer. Good practice might include a kick-off meeting as well as the circulation of multiple drafts for comment. Ideally, authors will have been apprised of their responsibilities in writing in advance of any manuscript preparation activities [6]. Asking authors to assess and interpret data and formulate and articulate their opinions before any writing work begins, ensures their active participation and avoids a situation in which a writer can introduce new ideas into a paper.

43.9.4 Consider Accountability

Alfonso et al. [39] consider accountability for the published work as a critical factor in assigning authorship. These authors note that the obligation to support any investigation associated with the published work is a factor over and above other types of intellectual contributions. In practical terms, contributors and potential authors should consider the source of data and their ability to participate in any investigation before interpreting information or agreeing to lend their name to any work.

43.9.5 Review Authorship Guidelines for each Publication

It is all too easy to assume an understanding of authorship criteria based on past publication experience. Algorithmic approaches to authorship may be tempting, but they are not necessarily aligned with the advice given by various expert groups such as ICMJE, COPE, CSE and WAME, each of which advise building consensus through team communication. Criteria and guidelines change over time and across scientific areas. Journals may also have specialized authorship guidelines. Therefore, a review of the authorship standards for each project will be helpful.

43.9.6 Promote Education and Training

Schroter and colleagues [38], Badreldin et al. [37], and Rasmussen et al. [2] each indicate a need for publications education and training among graduate students in the biomedical sciences and medical students. Corporate sponsors may also institute training programs or identify applicable training by professional societies.

43.10 A Shift to Positive Authorship Practices

A last consideration in discussions of ghostwriting is that maintaining a preoccupation about ghostwriting, or any other specific bad practice, might mask more serious ethical or logistical problems of authorship and review [24]. This is not to excuse ghostwriting or minimize the importance of instituting practices to prevent this serious problem, but to emphasize that true solutions to proper authorship attribution hinge on adopting positive practices. Therefore, a core message of the current chapter is that ongoing thinking and evaluation should be a usual practice in all authorship decisions.

References

1. Masic I, Jakovljevic M, Sinanovic O, Gajovic S, Spiroski M, Jusufovic R, Sokolovic S, Prnjavorac B, Zerem E, Djulbegovic B, Porovic S, Jankovic S, Hadzikadic M, Zunic L, Begic E, Nislic E, Begic N, Becirovic E, Cerovac A, Skrijelj V, Nuhanovic J (2018 Dec) The second mediterranean seminar on science writing, editing and publishing (SWEP – 2018). Acta Inform Med 24(4):284–299
2. Rasmussen LM, Williams CE, Hausfeld MM, Banks GC, Davis BC (2020) Authorship policies at U.-S. Doctoral Universities: a review and recommendations for future policies. Sci Eng Ethics 26(6):3393–3413
3. Yadav S, Rawal G (2018) Ghostwriters in the scientific world. Pan Afr Med J 30:217
4. Zimba O, Gasparyan AY (2020) Scientific authorship: a primer for researchers. Reumatologia 58(6):345–349
5. Sismondo S (2021) Epistemic corruption, the pharmaceutical industry, and the body of medical science. Front Res Metr Anal 6:614013
6. Battisti WP, Wager E, Baltzer L, Bridges D, Cairns A, Carswell CI, Citrome L, Gurr JA, Mooney LA, Moore BJ, Peña T, Sanes-Miller CH, Veitch K, Woolley KL, Yarker YE, International Society for Medical Publication Professionals (2015) Good publication practice for communicating company-sponsored medical research: GPP3. Ann Intern Med 163(6):461–464
7. Foster C, Wager E, Marchington J, Patel M, Banner S, Kennard NC, Panayi A, Stacey R, GPCAP Working Group (2019) Good practice for conference abstracts and presentations: GPCAP. Res Integr Peer Rev 4:11
8. Davidoff F, DeAngelis CD, Drazen JM, Hoey J, Højgaard L, Horton R, Kotzin S, Nylenna M, Overbeke AJ, Sox HC, Sox HC, Van Der Weyden MB, Wilkes MS (2001) Sponsorship, authorship, and accountability. Ann Intern Med 135(6):463–466
9. International Committee of Medical Journal Editors (2019) Recommendations for the conduct, reporting, editing, and publication of scholarly work in medical journals. https://icmje.org/recommendations/archives/2015_dec_urm.pdf. Accessed 30 May 2021
10. World Association of Medical Journal Editors (2005) Ghost writing initiated by commercial companies. https://wame.org/ghost-writing-initiated-commercial-companies. Accessed 30 May 2020
11. United States Senate (2010) Minority report. Ghostwriting in the medical literature. 111th Congress of the United States. 24 Jun 2010
12. Matheson A (2011) How industry uses the ICMJE guidelines to manipulate authorship--and how they should be revised. PLoS Med 8(8):e1001072
13. Matheson A (2016) The ICMJE recommendations and pharmaceutical marketing--strengths, weaknesses and the unsolved problem of attribution in publication ethics. BMC Med Ethics 17:20
14. Moffatt B (2013) Orphan papers and ghostwriting: the case against the ICMJE criterion of authorship. Account Res 20(2):59–71
15. World Association of Medical Journal Editors (2007) Authorship. https://wame.org/authorship. Accessed 30 May 2020
16. DeTora LM, Carey MA, Toroser D, Baum EZ (2019) Ghostwriting in biomedicine: a review of the published literature. Curr Med Res Opin 35(9): 1643–1651
17. Stretton S (2014) Systematic review on the primary and secondary reporting of the prevalence of

ghostwriting in the medical literature. BMJ Open 4: e004777
18. Anekwe T (2010) Profits and plagiarism: the case of medical ghostwriting. Bioethics 24:267–272
19. Bosch X, Ross JS (2012) Ghostwriting: research misconduct, plagiarism, or fool's gold? Am J Med 125(4): 324–326
20. Bosch X (2011) Exorcising ghostwriting. Ghostwriting could potentially have serious repercussions for science and should therefore be treated as research misconduct. EMBO Rep 12:489–494
21. Gotzsche PC, Kassirer JP, Woolley KL et al (2009) What should be done to tackle ghostwriting in the medical literature? PLoS Med 6:e23
22. Grossman L (1998) Ghostwriting. Lancet 351(9117): 1741
23. Logdberg L (2011) Being the ghost in the machine: a medical ghost-writer's personal view. PLoS Med 8: e1001071
24. DeTora LM (2016) The spectre of ghostwriting: eroding public trust in physicians, clinical trial integrity and biomedical authorship. Int J Clin Pract 70(7):630–633
25. Aliukonis V, Poškutė M, Gefenas E (2020) Perish or publish dilemma: challenges to responsible authorship. Medicina (Kaunas) 56(3):123
26. Puljak L, Sambunjak D (2020) Can authorship be denied for contract work? Sci Eng Ethics 26(2): 1031–1037
27. COPE/Committee on Publication Ethics (2021) Authorship and contributorship. https://publicationethics.org/files/authorship-cope-seminar-2021.pdf. Accessed 30 May 2021
28. Dartmouth University (2007) Authorship guidelines. https://dartmouth.edu/comply/docs/minutes-011007.pdf. Accessed 31 May 2021
29. Bülow W, Helgesson G (2018) Hostage authorship and the problem of dirty hands. Res Ethics 14(1):1–9
30. McDowell GS, Knutsen JD, Graham JM, Oelker SK, Lijek RS (2019 Oct) Co-reviewing and ghostwriting by early-career researchers in the peer review of manuscripts. elife 31(8):e48425
31. Sismondo S (2007) Ghost management: how much of the medical literature is shaped behind the scenes by the pharmaceutical industry? PLoS Med 4(9):e286
32. Sismondo S, Doucet M (2010) Publication ethics and the ghost management of medical publication. Bioethics 24(6):273–283
33. DeTora L, Foster C, Skobe C, Yarker YE, Crawley FP (2015) Publication planning: promoting an ethics of transparency and integrity in biomedical research. Int J Clin Pract 69(9):915–921
34. Hamilton CW, Jacobs A (2016) Ghostwriting prevalence among AMWA and EMWA members (2005 to 2014). Med Writ 25:6–14
35. Kasenda B, von Elm E, You JJ, Blümle A, Tomonaga Y, Saccilotto R, Amstutz A, Bengough T, Meerpohl JJ, Stegert M, Olu KK, Tikkinen KA, Neumann I, Carrasco-Labra A, Faulhaber M, Mulla SM, Mertz D, Akl EA, Bassler D, Busse JW, Ferreira-González I, Lamontagne F, Nordmann A, Gloy V, Raatz H, Moja L, Ebrahim S, Schandelmaier S, Sun X, Vandvik PO, Johnston BC, Walter MA, Burnand B, Schwenkglenks M, Hemkens LG, Bucher HC, Guyatt GH, Briel M (2016) Agreements between industry and academia on publication rights: a retrospective study of protocols and publications of randomized clinical trials. PLoS Med 13(6):e1002046
36. Yi N, Nemery B, Dierickx K (2020) Perceptions of plagiarism by biomedical researchers: an online survey in Europe and China. BMC Med Ethics 21(1):44
37. Badreldin H, Aloqayli S, Alqarni R, Alyahya H, Alshehri A, Alzahrani M, Al Tawalbeh A, Ismail WW (2021) Knowledge and awareness of authorship practices among health science students: a cross-sectional study. Adv Med Educ Pract 12:383–392
38. Schroter S, Montagni I, Loder E, Eikermann M, Schäffner E, Kurth T (2020) Awareness, usage and perceptions of authorship guidelines: an international survey of biomedical authors. BMJ Open 10(9): e036899
39. Alfonso F, Editors' Network, European Society of Cardiology (ESC) Task Force (2019) Authorship: from credit to accountability: reflections from the editors' network. Neth Heart J 27(6):289–296
40. Allen L, O'Connell A, Kiermer V (2019) How can we ensure visibility and diversity in research contributions? How the Contributor Role Taxonomy (CRediT) is helping the shift from authorship to contributorship. Learn Publ 32:71–74
41. Council of Science Editors. White paper on publication ethics. https://councilscienceeditors.org/resource-library/editorial-policies/publication-ethics. Accessed 30 May 2021
42. Friedman PJ. A new standard for authorship. Council of Science Editors Authorship Task Force. https://councilscienceeditors.org/resource-library/editorial-policies/cse-policies/retreat-and-task-force-papers/authorship-task-force/a-new-standard-for-authorship. Accessed 30 May 2021
43. Hamilton CW, Gertel A, Jacobs A, Marchington J, Weaver S, Woolley K (2016) Mythbusting medical writing: goodbye, ghosts! hello, help! Account Res 23(3):178–194
44. AMWA-EMWA-ISMPP (2017) Joint statement on the role of the professional medical writer. https://journal.emwa.org/writing-better/amwa-emwa-ismpp-joint-position-statement-on-the-role-of-professional-medical-writers. Accessed 31 May 2021

Part XI

Conflicts of Interest

Funding of Clinical Trials Through Non-Profit and Third-Party Organizations: A Case Study of the CREATE-X Trial in Japan

44

Akihiko Ozaki, Anju Murayama, and Tetsuya Tanimoto

Abstract

It is imperative to properly manage financial conflicts of interest (FCOI) in clinical trials. Non-profit or other third-party organizations obviously should be considered potential FCOI. This chapter discusses this under-recognized type of FCOI based on the Capecitabine for Residual Cancer as Adjuvant Therapy (CREATE-X) trial conducted targeting high-risk breast cancer patients in Japan and Korea. We also present issues of insufficient individual-level FCOI disclosure and health insurance fraud in the CREATE-X trial and the underlying structure of the profit-first doctrine of pharmaceutical companies.

Abbreviations

CPC	Chugai Pharmaceutical Company
CREATE-X	Capecitabine for residual cancer as adjuvant therapy
FCOI	Financial conflicts of interest
GCP	Good clinical practice
HCO	Healthcare organization
HCP	Healthcare professional
JPY	Japanese yen
NEJM	New England Journal of Medicine
pCR	Pathological complete response
US	United States

A. Ozaki (✉)
Department of Breast and Throid Surgery, Jyoban Hospital of Tokiwa Foundation, Iwaki City, Fukushima, Japan

Medical Governance Research Institute, Minato City, Tokyo, Japan
e-mail: akihiko-ozaki@tokiwa.or.jp

A. Murayama · T. Tanimoto
Medical Governance Research Institute, Minato City, Tokyo, Japan

44.1 Introduction

Cooperation between clinical investigators and pharmaceutical companies is imperative for the development of novel medical products. However such interactions which are often commercial, represent potential financial conflicts of interest (FCOI) [1]. This is particularly so when healthcare professionals prioritize the interests of companies over patient care, potentially harming patients [2]. This led to the development of regulatory measures for FCOI between pharmaceutical companies and healthcare sectors, chiefly in high-income countries, including the United States (US) [3, 4], Europe [5–8], Australia [9, 10], and Japan [11, 12]. A central tenet of such regulations has been transparency [13, 14], whereby pharmaceutical and other healthcare industries are increasingly required to disclose payments made to healthcare professionals (HCPs) and healthcare organizations (HCOs) through legal or self-regulation frameworks. Upon this premise transparency initiatives

J. Faintuch, S. Faintuch (eds.), *Integrity of Scientific Research*,
https://doi.org/10.1007/978-3-030-99680-2_44

developed mainly in western countries, combined with academic studies and investigative journalism utilizing payment databases, have clarified the prevalence and extent of financial entanglements between pharmaceutical companies and healthcare sectors [11, 15–17]. Importantly transparency has exposed the biases and potentials for harm in clinical practice associated with these FCOI [18–23].

44.1.1 Downsides of FCOI

FCOI with pharmaceutical companies can be problematic in any situation, including patient care [24–26], the operation of medical institutions or professional medical associations [27–30], the development of clinical practice guidelines [31–33], making professional recommendations for drug approval or regulatory processes [34–37], and various levels of medical research [38–40]. FCOI in medical research particularly those relating to clinical trials can be concerning, given that occasionally financial interactions with pharmaceutical companies have demonstrably biased both their execution and reporting [41]. Since the findings of clinical trials inform the decision-making behind drug regulation and clinical practice guidelines, any biases could significantly impact clinical practice. Recent studies revealed the high prevalence of FCOI among clinical trial authors [38, 39] and identified problems with the under-reporting of FCOI [42–44]. Since society rely on the integrity of clinical trial authors for accountability and FCOI disclosure, the development of a public FCOI disclosure registration would be warranted to address current limitations [45, 46].

44.1.2 Disclosure Loopholes using Funnel Funding

An emerging concern is the use of non-profit organizations and other third parties to funnel funding from related stakeholders into clinical trials, as a means to avoid disclosure obligations and other related operating restrictions [47–50]. By reviewing the Capecitabine for Residual Cancer as Adjuvant Therapy (CREATE-X) clinical trial, published in the New England Journal of Medicine (NEJM) [51], we will cover the use of non-profit organizations to secure funding, including the under-reporting of individual FCOI disclosure and violations of Japanese universal health coverage policies.

44.2 Overview of CREATE-X Trial published in the Prestigious Journal

CREATE-X was a clinical trial targeting patients with early but high-risk human epidermal growth factor receptor 2 (HER2, also known as ERBB2)-negative breast cancer. These were a subset of patients consisting of triple-negative and hormonal-receptor-positive breast cancer patients, who failed to achieve pathological complete response (pCR) following standard neoadjuvant chemotherapy, jointly conducted in Korea and Japan [51]. It is generally accepted that some breast cancer patients may develop distant metastasis and subsequently die following standard treatment [52]. Therefore exploration of improved treatment algorithms continues in seeking better patient outcomes. In this context, neoadjuvant chemotherapy increasingly became the standard of care, achieving downstaging of the disease with less extensive surgery, while also enabling the evaluation of the effectiveness of systematic treatment [53, 54]. As such, abundant evidence suggests that those failing to achieve pCR likely have worse outcomes than those who do and represent a specific subset of patients, such as those with triple-negative or ERBB2-positive breast cancer [54, 55].

44.2.1 Successful Results by Adding A Chemotherapeutic Agent

Compared with ERBB2-positive breast cancers which have a solid postoperative regime after neoadjuvant chemotherapy [54, 56], no effective add-on treatments for those without pCR were developed among triple-negative breast cancer patients before the CREATE-X study. While triple-negative breast cancer patients may benefit from evolving management efforts, including anti-programmed death-ligand 1 (PDL-1) treatment [54], they are generally regarded as a high-risk population for recurrence and subsequent death. Thus, it can be said that there was a reasonable rationale behind the CREATE-X trial. In the trial, ERBB2-negative breast cancer patients who did not achieve pCR after standard neoadjuvant chemotherapy were additionally given capecitabine/Xeloda, Chugai Pharmaceutical Company [CPC]). Those patients exhibited improved disease-free survival and overall survival. The effect was particularly pronounced among patients with triple-negative breast cancer.

The trial results were initially presented orally at the San Antonio Breast Cancer Symposium (SABCS) in December 2015, followed by formal publication in NEJM in June 2017. It is rare for a Japanese clinical trial on breast cancer to be presented orally at the SABCS or, for that matter, published in the NEJM [51]. As of June 2, 2022, the CREATE-X trial report published in the NEJM has been cited 942 times, according to Google Scholar. Naturally the trial received very favorable attention from Japanese breast cancer specialists. This was not surprising considering that many prominent Japanese specialists contributed to the study as authors or collaborating investigators.

44.3 Inappropriate Funding Disclosure

As both a breast cancer surgeon and attendee of the 2015 SABCS, the well-received attention the CREATE-X trial presentation garnered was not lost on the first author of this chapter. However as published, the article included no mention of CPC as either a funder or sponsor of the trial. Instead, they acknowledged that the trial was supported by a grant from the Japan Breast Cancer Research Group (JBCRC), a non-profit organization that supports the conduct of clinical trials, and the Advanced Clinical Research Organization (ACRO), a commercial clinical research organization (CRO).

44.3.1 Expensive Clinical Trial Budget Without Disclosure

In the trial, only from Japan 606 breast cancer patients were enrolled [51]. Given that a clinical trial of this scale requires a considerable amount of money for its operation, it may not be reasonable for entities other than CPC, the only one likely to benefit from positive results of this clinical trial, to support it. This means that CPC likely funded JBCRG and ACRO. In the International Committee of Medical Journal Editors (ICMJE) COI version used in the article, before the February 2021 update [57], authors were still required to disclose service or payments for any aspect of the submitted work. Nevertheless, none of the authors disclosed payments from CPC concerning the submitted work.

44.3.2 Was the Pharmaceutical Company Linked to Trial Stakeholders?

CPC is a Japanese pharmaceutical company and subsidiary of F. Hoffmann-La Roche, Ltd. (Basel, Switzerland). The mother company boasted the largest pharmaceutical sales globally in 2020, at roughly USD $49.5 billion [58]. CPC is a member of the Japan Pharmaceutical Manufacturers Association (JPMA) [59], which demands disclosure of payments made to HCPs and HCOs relating to promotion and research on company websites since 2013 [11]. Our investigation revealed that a total of 100 million Japanese yen (JPY) (~USD $910,000) and 236 million JPY (~USD $2.1 million) was paid to JBCRG and ACRO, respectively, from 2013 to 2016 [49]. At this point, it was unclear that any of this funding had been used for the CREATE-X trial. Important aspect however was the fact that there was a delivery of funding from CPC to the two organizations.

Some of the CREATE-X trial authors were in executive roles at this organization, including its auditor [60]. This means that they were in a position that afforded them access to all accounting information of the organization, including any funding from CPC during the study period, and should have disclosed these payments irrespective of how these funds were used. The first author of this chapter previously summarized this argument in a letter to the editor of NEJM in July 2017. While the letter itself was rejected for publication, it triggered communication between the journal editorial board and the authors [49]. Ultimately, a modified version of the manuscript was published online in September 2017 by Science and Engineering Ethics [49].

44.4 Connecting the Dots to Uncover FCOI Using A Third Party

With the publication of the manuscript came numerous conflicts for the first author of this chapter [61]. He belonged to the Cancer Institute Hospital in Tokyo. This is one of the premier cancer centers in Japan and treats the largest number of breast cancer patients in the country. Coincidentally, the head of the breast cancer center at the time also led the JBCRG and was an author of the published CREATE-X trial. Consequently, there was personal criticism for publishing the letter without making any consultation with him beforehand, including a lack of evidence that funding from CPC was used for the trial [61]. Although his criticism did not question the validity of the claims behind the failure to follow the ICMJE COI disclosure policy [57], it did emphasize the need to conclusively ascertain whether there was evidence linking the funding from CPC with the CREATE-X trial.

44.4.1 Hard Evidence of Financial Transfers Through the Non-Profit Organization from the Pharmaceutical Company

Unfortunately at the time it was not possible to address this point directly with open-access data alone. Instead, the investigation required internal data from ACRO. Subsequently, in November 2017, the authors received direct communication from a former ACRO official [61]. He was aware of our campaign about the CREATE-X trial and voluntarily reached out to provide internal data about ACRO funding. The information offered included hard copies of the ACRO bankbook containing all money transfers with pharmaceutical companies, HCPs, and other stakeholders [61]. It also had an excel file split out into separate sheets summarizing the receipts and disbursements for each clinical trial ACRO was involved with during the interval from April 1, 2006, to March 31, 2009 [61]. In one of these excel sheets, we found records of all financial transfers relating to the CREATE-X trial from CPC during this period. We confirmed the accuracy of the financial data presented in the excel worksheets by cross-checking the amounts and dates of transfers using the hard copies of the ACRO bankbook [61]. Additionally, memos left on the relevant transfers of the bankbook listing "The CREATE-X trial" clearly indicated that the transfer was made for the CREATE-X trial [61].

44.4.2 Potential Biases due to the Non-Disclosure of FCOI

Clinical trials funded by pharmaceutical companies often produce results favorable to the company's bottom line [39, 62]. Also, this was an open-label study without a placebo-controlled arm [51]. Open-label studies are susceptible to bias as the investigators know which patients take study medications [63, 64]. For example, in the CREATE-X trial, if there was a tendency to delay the timing of or reluctance in conducting imaging or other diagnostic testing in the capecitabine group, this could prolong the disease-free survival of this group. Further, if there were a penchant for using a more aggressive treatment and careful management in the capecitabine group versus the control group following disease progression, the overall survival of the capecitabine group would similarly be enhanced relative to the control group. It remains unclear whether there were any deliberate intentions concerning the failure of the CREATE-X trial authors to disclose their financial relationships with CPC. Nevertheless, it was inevitable that the non-disclosure of FCOI and subsequent revelations could cause doubt about the study findings.

44.4.3 Open and Accessible Deals to Improve Transparency

Funding using third-party organizations obscures the true financial relationships between parties without this information being disclosed by the organization originally providing the funding. Further, it is difficult to corroborate and clarify such relationships using only publicly available information. Therefore, it remains critical to create a system that ensures greater financial transparency. To this point, the Clinical Trial Act, enacted in Japan in 2018, requires that funding related to clinical trials should be appropriately disclosed [65]. While it is a valid criticism that the Clinical Trial Act has increased financial and administrative burdens [66], as far as FCOI are concerned, ongoing efforts in Japan to regulate FCOI have improved thanks to this law.

44.5 Inappropriate Individual-Level FCOI Disclosure

According to the latest version of the ICMJE COI form [57], individual-level financial relationships include fees for services such as speaking engagements, writing manuscripts, consulting, reimbursement of travel expenses, and profits gained from stock holdings and patents. As clarified to this point, the CREATE-X trial turned out to be a clinical trial fully supported by CPC, which also manufactured the study drug capecitabine.

Under the transparency guidelines published by the JPMA, only fees for speaking engagements, writing, and consulting are disclosed on an individual basis, while payments for travel, lodging, and meals are disclosed in an aggregated form [12]. Stocks and patent-related compensation are not questioned [12]. Nevertheless, even with the reporting of only disclosed fees on an individual basis, roughly a hundred doctors in Japan annually receive combined personal payments exceeding ten million JPN (~ USD$91,000) [67]. Individual FCOI in clinical trials should be accurately disclosed so that everyone can recognize them.

44.5.1 Identifying Underreported Individual Payments from the Pharmaceutical Company

To clarify the extent of existing individual FCOI among the Japanese authors of the CREATE-X trial and the accuracy of disclosure, we considered data on all individual payments for the fiscal year of 2016 disclosed by pharmaceutical companies belonging to the JPMA and their affiliated companies. The extracted data included fees for speaking engagements, writing, and consulting, and based on this, we calculated the total and author-based values. We categorized the companies as CPC, companies manufacturing or developing drugs used in breast cancer care other than CPC, and all remaining companies. We also reviewed the ICMJE COI forms submitted by the Japanese authors. All financial relationships among study authors and pharmaceutical companies, even if they are not directly related to the individual studies [57], should have been disclosed in those forms. The details of the methodology were published elsewhere [47].

Based on our calculations, the total payment amount received by all Japanese authors was 30,479,691JPY (USD $279,630) [47]. This amounted to 7,573,369 JPY (USD $69,480, 24.8%) from CPC, 22,535,304 JPY (USD $206,746, 73.9%) from companies manufacturing or developing breast cancer products other than the CPC, and only 371,018 JPY (USD $3404, 1.2%) from the other companies [47]. Of the 12 authors, nine (75.0%) received payments from CPC. Only three (33.3%) initially disclosed such payments, with three more disclosing the CPC payments in the subsequently revised COI forms [47]. Similarly, 11 (91.7%) authors received at least one payment from companies manufacturing or developing breast cancer products other than the CPC. Still, none of this was disclosed.

44.5.2 Insufficient Personal Awareness of FCOI Reporting

Personal awareness of FCOI reporting is generally low among Japanese HCPs. The most likely reason for this is that the standards for FCOI disclosure in Japan are much weaker than existing global standards [68]. Most international academic medical journals currently use the ICMJE or similar COI forms and require reporting any financial interactions during the past 3 years [69]. However, in domestic Japanese academic professional medical associations, FCOI below specific thresholds, typically 500,000 or one million JPY (USD 4550-9100) are not required to be disclosed [68, 70, 71].

44.5.3 Need for Transparent Database for FCOI Disclosure

Another concern with FCOI disclosure is that it relies too heavily on an individual's integrity. Assessment of disclosure inaccuracies in the CREATE-X study was possible only by extracting the relevant data and organizing it into a centralized database in collaboration with investigative journalists in Japan [11, 72]. Given the efforts made to create this database, future centralized databases should be prepared by pharmaceutical companies or national governments as in the US [15], the United Kingdom [17], France [25], and Australia [9].

44.5.4 Bolstering the Role of Societies and Institutions Against Inadequate COI Reporting

At present, Japan and its medical societies lack a well-developed system for judging HCPs or HCOs committing inadequate COI reporting or research misconduct more broadly. As a result, the situation is such that those denying accountability failures or misconduct tend to avoid social sanctions and continue to stay at influential roles [73].

44.6 Financial Damage in the Universal Health Care System

The trial was conducted within the framework of a universal health care system [51]. In clinical trials, researchers typically use research funds to purchase study drugs when not provided by pharmaceutical companies. However, the investigators conducted the clinical trial as if it was daily clinical practice [51]. Consequently, health insurers and patients unjustly covered the cost of study drugs [51]. This was a clear violation of existing law, as capecitabine was already approved for patients with advanced breast cancer with metastases or those with recurrent diseases but not adjuvant treatment of early breast cancer patients. Drugs not covered by public health insurance can be reimbursed if doctors specifically request them as necessary for the patient [48]. Traditionally Japanese insurers don't refuse reimbursement. In this light, the CREATE-X trial identified an exploitable loophole in the Japanese public insurance program amounting to 100 million JPY (~$910,000), which actually represented revenue for the CPC [48].

44.6.1 Company Profits Vs. Community Benefit

Despite the overwhelmingly positive and highly publicized effects of the trial, daily clinical practice with Japanese breast cancer patients remains unchanged. As it was conducted within the framework of clinical practice and not in accordance with Good Clinical Practice (GCP) [74], something not allowed in the health insurance framework, new indications cannot be adopted and become out-of-pocket costs for patients.

It appears puzzling why the CREATE-X trial was conducted not in compliance with GCP. Sadly, the most likely explanation is that the primary purpose of the trial was not to demonstrate the efficacy of the drug but rather to increase sales for CPC. More than 100 million JPY (~$910,000) worth of capecitabine used in the trial went directly to CPC sales. Surprisingly, in the past, such practice in clinical trials was pervasive among pharmaceutical companies and medical societies in Japan [48]. Among the doctors, these activities increased their status in medical communities. Companies recording the cost of the test drug used in the study as sales improved their bottom line. As a result, both companies and participating doctors benefited financially, while the benefits of the patients they enrolled and those that the enrolled patients represented were marginal [75].

44.6.2 Diovan Scandal: A Large-Scale Clinical Trial Misconduct

The Diovan Scandal was part of the routine custom of incorporating clinical practice into large-scale clinical trials. Five prominent Japanese university hospitals conducted post-marketing studies on valsartan/Diovan, with Novartis' financial and administrative support, seeking to demonstrate additional benefits of this hypertensive drug. The added element in this particular case was the falsification of study data led by the drug manufacturer [73]. These studies were published in prestigious medical journals such as the European Heart Journal and the Lancet and later retracted [76, 77]. The resulting controversy in Japan laid the groundwork for the eventual passage of the Clinical Trial Act, enacted in 2018 [66, 73].

44.7 Suboptimal Protection of Whistleblowers

In this particular case, the information provided by whistleblowers played a significant role in revealing the hidden financial ties between the CPC and CREATE-X trial authors [61]. Indeed, whistleblowers are crucial to worldwide efforts in exposing various compliance violations. In the US, awareness of such issues has been exceptionally high, and laws have been passed to protect the rights of whistleblowers, including the Qui Tam Act [50, 78]. This law allows whistleblowers to receive a certain proportion of the funds obtained through any government-led legal action against violators while also preventing retribution by enabling whistleblowers to maintain anonymity [50, 78]. Based on an analysis evaluating all press releases for state and federal settlements between 2003 and 2016 obtained from the US Department of Justice, the US Securities and Exchange Commission, and the US Environmental Protection Agency, Arnold and colleagues [78] revealed that whistleblowers led to the identification of misconduct in 28.6% (56) of 196 settlements. Among the 39 settlements that disclosed payments to the whistleblowers, the average whistleblower payout exceeded $39 million.

However, similar laws do not currently exist in Japan. Therefore, whistleblowers often find themselves in difficult positions, both personally and financially. The first author of this chapter found himself in such a situation following the revelations surrounding the CREATE-X Trial [61]. The same problem likely exists in other countries as well. We believe that a framework to protect whistleblowers better will facilitate greater accountability and transparency, along with better behavior by clinical trial sponsors and participating clinicians.

44.8 Responsibility of Individuals in FCOI Disclosure

It remains indisputable that in the conduct of clinical trials, the disclosure of FCOI of individuals and entities falls under the principal investigator's responsibility. However, there still exists controversy as to whether not reporting or under-reporting of individual-level FCOI should solely rest on the responsibility of the individual, the research entity, or the principal investigator. We believe that principal investigators have a responsibility to encourage individuals to disclose relevant FCOI appropriately. Still, ultimately it is difficult for them to ensure accurate disclosure by individuals. The primary reason for this remains the reliance on the integrity of individuals for these disclosures. Because some individuals continue to provide incomplete or inaccurate declarations even with amended declarations as shown in the CREATE-X trial [47], we believe that it is unreasonable to put all of the responsibility on the principal investigator. To improve this situation, efforts to publish the financial relationships of researchers in databases should be pursued [45, 46]. However, considering the feasibility of this, the current situation still requires that individual researchers understand and follow the rules accurately and as carefully as possible. Further, as we noted above, we call for a more rigorous framework to evaluate these FCOI disclosures embedded with penalties as necessary.

44.9 Responsibility of Medical Journals in FCOI Disclosure

Generally speaking, it would be difficult to lay the blame solely on the NEJM for the series of inappropriate FCOI disclosures associated with the CREATE-X trial. In principle, editorial offices establish guidelines for scientific publications covering both the editorial content and the conduct of clinical research. Further, the peer review process remains paramount for ensuring the quality and integrity of such research. As such, both journal publishers and the scientific community must rely on the integrity of those conducting and reporting on clinical studies.

Even so, despite clear abuse on the part of CREATE-X trial study authors, we still believe that the series of responses taken by the NEJM editorial office may not have been the best possible course of action given the shortcomings of existing FCOI management strategies. For example, we informed the NEJM editorial office of the inadequacies of CREATE-X trial authors in reporting or correcting FCOI, based on cross-checking pharmaceutical companies' reported payment disclosures. However, to the best of our knowledge, the NEJM editorial office appeared to have made no further effort on their part to clarify this [47]. As a past editor-in-chief of the Journal of the American Medical Association, Bauchner and colleagues [79] expressed similar concerns. They offered that reaching out to both funding agencies and the affiliated institutions of study authors should be pursued in instances of problematic or inadequate FCOI disclosure to determine appropriate further actions as needed.

The inadequacies of FCOI disclosure in CREATE-X was not an isolated case. Such disclosure issues have been problematic many times in the past [42–44, 80]. Consequently, it may well be necessary to consider more robust future measures to correct inappropriate FCOI disclosure practices. Such measures could include manuscript retraction, just as in situations of scientific misconduct that involve data falsification [81]. In particular, prominent journals belonging to the ICMJE, such as the highly influential NEJM and the Lancet, could incorporate more severe penalties, which would serve as a deterrent to encourage appropriate FCOI disclosure by HCPs and investigators.

44.10 Conclusion

This chapter constitutes a comprehensive analysis of the financial aspects behind the CREATE-X trial while also providing some contextual background relating to clinical trials in Japan. While the focus of the chapter was the funneling of money through a non-profit organization, we also evaluated the under-reporting of individual-level FCOI disclosure and abuse of the Japanese health insurance system. The context underlying these malpractices remains the profit-first doctrine of pharmaceutical companies and inadequate awareness of FCOI in the Japanese medical society. Therefore, it is imperative to develop more rigorous policing of FCOI between pharmaceutical companies and medical societies in Japan. Furthermore, a discussion should be initiated on whistleblower protections and accountability in FCOI disclosure at both the levels of authors and medical journals. We believe that this chapter will help readers learn more about FCOI in clinical trials beyond this specific case.

Conflicts of Interest All authors have several papers concerning conflicts of interest among Japanese clinical practice guideline authors. Drs. Ozaki and Tanimoto report receiving personal fees from Medical Network Systems, MNES Inc., outside the scope of the submitted work. Dr. Tanimoto reports personal fees from Bionics Co., Ltd., outside the scope of the submitted work. We report no other disclosures. The Medical Governance Research Institute is a non-profit organization that receives donations from a dispensing pharmacy company, Ain Pharmacies.

References

1. Institute of Medicine (US) Committee on Conflict of Interest in Medical Research, Education, and Practice (2009) Conflict of interest in medical research, education, and practice. In: Lo B, Field MJ (eds) The national academies collection: reports funded by National Institutes of Health. National Academies Press (US), Washington (DC). https://doi.org/10.17226/12598
2. Institute of Medicine (US) Committee on Quality of Health Care in America (2001) Crossing the quality chasm: a new health system for the 21st century. National Academies Press (US), Washington (DC). https://doi.org/10.17226/10027
3. Agrawal S, Brennan N, Budetti P (2013) The sunshine act--effects on physicians. N Engl J Med 368(22): 2054–2057. https://doi.org/10.1056/NEJMp1303523
4. Centers for Medicare & Medicaid Services (2014) OpenPaymentData.CMS.gov. https://openpaymentsdata.cms.gov/. Accessed 20 Jun 2020
5. Santos A (2017) The sun shines on Europe: transparency of financial relationships in the healthcare sector. https://haiweb.org/wp-content/uploads/2017/03/Sunshine-Act.pdf. Accessed 20 Jun 2020
6. Fabbri A, Santos A, Mezinska S, Mulinari S, Mintzes B (2018) Sunshine policies and murky shadows in Europe: disclosure of pharmaceutical industry payments to health professionals in nine European countries. Int J Health Policy Manag 7(6):504–509. https://doi.org/10.15171/ijhpm.2018.20
7. European Federation of Pharmaceutical Industries and Associations (2021) The EFPIA code. https://www.efpia.eu/relationships-code/the-efpia-code/. Accessed 6 Jul 2021
8. Mulinari S, Martinon L, Jachiet PA, Ozieranski P (2021) Pharmaceutical industry self-regulation and non-transparency: country and company level analysis of payments to healthcare professionals in seven European countries. Health Policy 125(7):915–922. https://doi.org/10.1016/j.healthpol.2021.04.015
9. Parker L, Karanges EA, Bero L (2019) Changes in the type and amount of spending disclosed by Australian pharmaceutical companies: an observational study. BMJ Open 9(2):e024928. https://doi.org/10.1136/bmjopen-2018-024928
10. Medicines Australia (2021) About the code. https://www.medicinesaustralia.com.au/code-of-conduct/about-the-code/. Accessed 6 Jul 2021
11. Ozaki A, Saito H, Senoo Y, Sawano T, Shimada Y, Kobashi Y, Yamamoto K, Suzuki Y, Tanimoto T (2020) Overview and transparency of non-research payments to healthcare organizations and healthcare professionals from pharmaceutical companies in Japan: analysis of payment data in 2016. Health Policy 124(7):727–735
12. Japan Pharmaceutical Manufactures Association (2018) Transparency guidelines for the relations between corporate activities and medical institutions. http://www.jpma.or.jp/english/policies_guidelines/pdf/transparency_gl_intro_2018.pdf. Accessed 12 Jul 2021
13. Grundy Q, Habibi R, Shnier A, Mayes C, Lipworth W (2018) Decoding disclosure: comparing conflict of interest policy among the United States, France, and Australia. Health Policy 122(5):509–518. https://doi.org/10.1016/j.healthpol.2018.03.015
14. Pham-Kanter G (2014) Act II of the sunshine act. PLoS Med 11(11):e1001754. https://doi.org/10.1371/journal.pmed.1001754
15. Marshall DC, Jackson ME, Hattangadi-Gluth JA (2016) Disclosure of industry payments to physicians: an epidemiologic analysis of early data from the open payments program. Mayo Clin Proc 91(1):84–96. https://doi.org/10.1016/j.mayocp.2015.10.016
16. Stoll M, Hubenschmid L, Koch C, Lieb K (2020) Voluntary disclosures of payments from pharmaceutical companies to healthcare professionals in Germany: a descriptive study of disclosures in 2015 and 2016. BMJ Open 10(9):e037395. https://doi.org/10.1136/bmjopen-2020-037395

17. Mulinari S, Ozieranski P (2018) Disclosure of payments by pharmaceutical companies to healthcare professionals in the UK: analysis of the Association of the British Pharmaceutical Industry's disclosure UK database, 2015 and 2016 cohorts. BMJ Open 8(10): e023094. https://doi.org/10.1136/bmjopen-2018-023094
18. Mitchell AP, Trivedi NU, Gennarelli RL, Chimonas S, Tabatabai SM, Goldberg J, Diaz LA Jr, Korenstein D (2021) Are financial payments from the pharmaceutical industry associated with physician prescribing? A systematic review. Ann Intern Med 174(3):353–361. https://doi.org/10.7326/M20-5665
19. Hwong AR, Sah S, Lehmann LS (2017) The effects of public disclosure of industry payments to physicians on patient trust: a randomized experiment. J Gen Intern Med 32(11):1186–1192. https://doi.org/10.1007/s11606-017-4122-y
20. Kanter GP, Carpenter D, Lehmann LS, Mello MM (2019) US nationwide disclosure of industry payments and public trust in physicians. JAMA Netw Open 2(4): e191947. https://doi.org/10.1001/jamanetworkopen.2019.1947
21. Murayama A, Senoo Y, Harada K, Kotera Y, Saito H, Sawano T, Suzuki Y, Tanimoto T, Ozaki A (2021) Awareness and perceptions among members of a Japanese cancer patient advocacy group concerning the financial relationships between the pharmaceutical industry and physicians: a mixed-methods analysis of survey data. medRxiv. https://doi.org/10.1101/2021.06.26.21259442
22. Nissen SE (2019) Can we trust the independence of cardiovascular guidelines? Circ Cardiovasc Qual Outcomes 12(12):e006225. https://doi.org/10.1161/CIRCOUTCOMES.119.006225
23. Fudman D, Feuerstein JD (2018) The call for greater transparency in conflicts of interest. JAMA Netw Open 1(8):e186342. https://doi.org/10.1001/jamanetworkopen.2018.6342
24. DeJong C, Aguilar T, Tseng CW, Lin GA, Boscardin WJ, Dudley RA (2016) Pharmaceutical industry-sponsored meals and physician prescribing patterns for Medicare beneficiaries. JAMA Intern Med 176(8):1114–1122. https://doi.org/10.1001/jamainternmed.2016.2765
25. Goupil B, Balusson F, Naudet F, Esvan M, Bastian B, Chapron A, Frouard P (2019) Association between gifts from pharmaceutical companies to French general practitioners and their drug prescribing patterns in 2016: retrospective study using the French Transparency in Healthcare and National Health Data System databases. BMJ 367:l6015. https://doi.org/10.1136/bmj.l6015
26. Khazzaka M (2019) Pharmaceutical marketing strategies' influence on physicians' prescribing pattern in Lebanon: ethics, gifts, and samples. BMC Health Serv Res 19(1):80. https://doi.org/10.1186/s12913-019-3887-6
27. Moynihan R, Albarqouni L, Nangla C, Dunn AG, Lexchin J, Bero L (2020) Financial ties between leaders of influential US professional medical associations and industry: cross sectional study. BMJ 369:m1505. https://doi.org/10.1136/bmj.m1505
28. Saito H, Ozaki A, Kobayashi Y, Sawano T, Tanimoto T (2019) Pharmaceutical company payments to executive board members of professional medical associations in Japan. JAMA Intern Med. https://doi.org/10.1001/jamainternmed.2018.7283
29. Ozieranski P, Csanadi M, Rickard E, Tchilingirian J, Mulinari S (2019) Analysis of pharmaceutical industry payments to UK health care organizations in 2015. JAMA Netw Open 2(6):e196253. https://doi.org/10.1001/jamanetworkopen.2019.6253
30. Fabbri A, Gregoraci G, Tedesco D, Ferretti F, Gilardi F, Iemmi D, Lisi C, Lorusso A, Natali F, Shahi E, Rinaldi A (2016) Conflict of interest between professional medical societies and industry: a cross-sectional study of Italian medical societies' websites. BMJ Open 6(6):e011124. https://doi.org/10.1136/bmjopen-2016-011124
31. Moynihan R, Lai A, Jarvis H, Duggan G, Goodrick S, Beller E, Bero L (2019) Undisclosed financial ties between guideline writers and pharmaceutical companies: a cross-sectional study across 10 disease categories. BMJ Open 9(2):e025864. https://doi.org/10.1136/bmjopen-2018-025864
32. Elder K, Turner KA, Cosgrove L, Lexchin J, Shnier A, Moore A, Straus S, Thombs BD (2020) Reporting of financial conflicts of interest by Canadian clinical practice guideline producers: a descriptive study. CMAJ 192(23):E617–E625. https://doi.org/10.1503/cmaj.191737
33. Saleh RR, Majeed H, Tibau A, Booth CM, Amir E (2019) Undisclosed financial conflicts of interest among authors of American Society of Clinical Oncology clinical practice guidelines. Cancer 125(22): 4069–4075. https://doi.org/10.1002/cncr.32408
34. Sawano T, Ozaki A, Saito H, Shimada Y, Tanimoto T (2020) Pharmaceutical company payments to Japanese government drug regulation committee members. Clin Pharmacol Ther. https://doi.org/10.1002/cpt.1892
35. Nejstgaard CH, Bero L, Hrobjartsson A, Jorgensen AW, Jorgensen KJ, Le M, Lundh A (2020) Association between conflicts of interest and favourable recommendations in clinical guidelines, advisory committee reports, opinion pieces, and narrative reviews: systematic review. BMJ 371:m4234. https://doi.org/10.1136/bmj.m4234
36. Abola MV, Prasad V (2016) Characteristics and conflicts of public speakers at meetings of the oncologic drugs advisory committee to the US Food and Drug Administration. JAMA Intern Med 176(3): 389–391. https://doi.org/10.1001/jamainternmed.2015.7805
37. Arthur W, Austin J, Wayant C, Vassar M (2019) Association of conflicts of interest for public speakers for the peripheral and central nervous system drugs

advisory committee of the US Food and Drug Administration with their statements. JAMA Neurol 76(3): 368–369. https://doi.org/10.1001/jamaneurol.2018.3997
38. Riechelmann RP, Wang L, O'Carroll A, Krzyzanowska MK (2007) Disclosure of conflicts of interest by authors of clinical trials and editorials in oncology. J Clin Oncol 25(29):4642–4647. https://doi.org/10.1200/JCO.2007.11.2482
39. Ahn R, Woodbridge A, Abraham A, Saba S, Korenstein D, Madden E, Boscardin WJ, Keyhani S (2017) Financial ties of principal investigators and randomized controlled trial outcomes: cross sectional study. BMJ 356:i6770. https://doi.org/10.1136/bmj.i6770
40. Kirkham JJ, Altman DG, Chan AW, Gamble C, Dwan KM, Williamson PR (2018) Outcome reporting bias in trials: a methodological approach for assessment and adjustment in systematic reviews. BMJ 362:k3802. https://doi.org/10.1136/bmj.k3802
41. Lundh A, Lexchin J, Mintzes B, Schroll JB, Bero L (2017) Industry sponsorship and research outcome. Cochrane Database Syst Rev 2:MR000033. https://doi.org/10.1002/14651858.MR000033.pub3
42. Rasmussen K, Schroll J, Gotzsche PC, Lundh A (2015) Under-reporting of conflicts of interest among trialists: a cross-sectional study. J R Soc Med 108(3): 101–107. https://doi.org/10.1177/0141076814557878
43. Bellomo TR, Hwang C, Spector-Bagdady K, Stanley JC, Corriere MA (2020) Industry compensation and self-reported financial conflicts of interest among authors of highly cited peripheral artery disease studies. J Vasc Surg 72(2):673–684. https://doi.org/10.1016/j.jvs.2019.09.053
44. Taheri C, Kirubarajan A, Li X, Lam ACL, Taheri S, Olivieri NF (2021) Discrepancies in self-reported financial conflicts of interest disclosures by physicians: a systematic review. BMJ Open 11(4):e045306. https://doi.org/10.1136/bmjopen-2020-045306
45. Rimmer A (2021) Briefing: why do we need a mandatory register of doctors' interests? BMJ 373:n1280. https://doi.org/10.1136/bmj.n1280
46. Coombes R (2021) Doctors' duty to declare their interests should be enforced, says GMC. BMJ 373: n1329. https://doi.org/10.1136/bmj.n1329
47. Ozaki A, Saito H, Sawano T, Shimada Y, Tanimoto T (2021) Accuracy of post-publication financial conflict of interest corrections in medical research: a secondary analysis of pharmaceutical company payments to the authors of the CREATE-X trial report in the New England Journal of Medicine. Bioethics. https://doi.org/10.1111/bioe.12854
48. Ozaki A, Takita M, Tanimoto T (2018) A call for improved transparency in financial aspects of clinical trials: a case study of the CREATE-X trial in the New England Journal of Medicine. Investig New Drugs 36(3):517–522. https://doi.org/10.1007/s10637-018-0577-x
49. Ozaki A (2018) Conflict of interest and the CREATE-X trial in the New England Journal of Medicine. Sci Eng Ethics 24(6):1809–1811. https://doi.org/10.1007/s11948-017-9966-3
50. Hashimoto T, Ozaki A, Tanimoto T (2021) Whistleblowing and misconduct at large pharmaceutical firms. JAMA 325(13):1331. https://doi.org/10.1001/jama.2021.0573
51. Masuda N, Lee SJ, Ohtani S, Im YH, Lee ES, Yokota I, Kuroi K, Im SA, Park BW, Kim SB, Yanagita Y, Ohno S, Takao S, Aogi K, Iwata H, Jeong J, Kim A, Park KH, Sasano H, Ohashi Y, Toi M (2017) Adjuvant capecitabine for breast cancer after preoperative chemotherapy. N Engl J Med 376(22): 2147–2159. https://doi.org/10.1056/NEJMoa1612645
52. DeSantis CE, Ma J, Gaudet MM, Newman LA, Miller KD, Goding Sauer A, Jemal A, Siegel RL (2019) Breast cancer statistics, 2019. CA Cancer J Clin 69(6):438–451. https://doi.org/10.3322/caac.21583
53. Thompson AM, Moulder-Thompson SL (2012) Neoadjuvant treatment of breast cancer. Ann Oncol 23(Suppl 10):x231–x236. https://doi.org/10.1093/annonc/mds324
54. Loibl S, Poortmans P, Morrow M, Denkert C, Curigliano G (2021) Breast cancer. Lancet 397(10286):1750–1769. https://doi.org/10.1016/S0140-6736(20)32381-3
55. Spring LM, Fell G, Arfe A, Sharma C, Greenup R, Reynolds KL, Smith BL, Alexander B, Moy B, Isakoff SJ, Parmigiani G, Trippa L, Bardia A (2020) Pathologic complete response after neoadjuvant chemotherapy and impact on breast cancer recurrence and survival: a comprehensive meta-analysis. Clin Cancer Res 26(12):2838–2848. https://doi.org/10.1158/1078-0432.CCR-19-3492
56. National Comprehensive Cancer Network (2021) NCCN guidelines version 5.2021 invasive breast cancer. https://www.nccn.org/professionals/physician_gls/pdf/breast.pdf. Accessed 12 Jul 2021
57. International Committee of Medical Journal Editors (2021) Disclosure of interest (updated February 2021). http://www.icmje.org/disclosure-of-interest/. Accessed 12 Jul 2021
58. Proclinical (2021) Who are the top 10 pharmaceutical companies in the world? (2021): top revenue from pharmaceutical segment. https://www.proclinical.com/blogs/2021-7/who-are-the-top-10-pharma-companies-in-2021. Accessed 12 Jul 2021
59. Japan Pharmaceutical Manufactures Association (2021) Member companies. http://www.jpma.or.jp/english/about_us/member.html. Accessed 12 Jul 2021
60. Japan Breast Cancer Research Group (2021) Our vision. https://jbcrg.jp/en/about/vision/. Accessed 12 Jul 2021
61. Tansa (2020) A cancer study's secret funder. https://en.tansajp.org/investigativejournal_category/cancerstudy/. Accessed 12 Jul 2021
62. Sun X, Briel M, Busse JW, You JJ, Akl EA, Mejza F, Bala MM, Bassler D, Mertz D, Diaz-Granados N,

Vandvik PO, Malaga G, Srinathan SK, Dahm P, Johnston BC, Alonso-Coello P, Hassouneh B, Truong J, Dattani ND, Walter SD, Heels-Ansdell D, Bhatnagar N, Altman DG, Guyatt GH (2011) The influence of study characteristics on reporting of subgroup analyses in randomised controlled trials: systematic review. BMJ 342:d1569. https://doi.org/10.1136/bmj.d1569
63. Hrobjartsson A, Thomsen AS, Emanuelsson F, Tendal B, Hilden J, Boutron I, Ravaud P, Brorson S (2012) Observer bias in randomised clinical trials with binary outcomes: systematic review of trials with both blinded and non-blinded outcome assessors. BMJ 344: e1119. https://doi.org/10.1136/bmj.e1119
64. Wood L, Egger M, Gluud LL, Schulz KF, Juni P, Altman DG, Gluud C, Martin RM, Wood AJ, Sterne JA (2008) Empirical evidence of bias in treatment effect estimates in controlled trials with different interventions and outcomes: meta-epidemiological study. BMJ 336(7644):601–605. https://doi.org/10.1136/bmj.39465.451748.AD
65. Ministry of Health, Labour and Welfare (2017) Clinical Trials Act (Act No. 16 of April 14, 2017). https://www.mhlw.go.jp/file/06-Seisakujouhou-10800000-Iseikyoku/0000213334.pdf. Accessed 12 Jul 2021
66. Nakamura K, Shibata T (2020) Regulatory changes after the enforcement of the new Clinical Trials Act in Japan. Jpn J Clin Oncol 50(4):399–404. https://doi.org/10.1093/jjco/hyaa028
67. Tansa (2018) Big pharma paid doctors 26.6 billion yen in a single year. Accessed 12 Jul 2021
68. Murayama A, Ozaki A, Saito H, Sawano T, Shimada Y, Yamamoto K, Suzuki Y, Tanimoto T (2020) Pharmaceutical company payments to dermatology clinical practice guideline authors in Japan. PLoS One 15(10):e0239610. https://doi.org/10.1371/journal.pone.0239610
69. Ngo-Metzger Q, Moyer V, Grossman D, Ebell M, Woo M, Miller T, Brummer T, Chowdhury J, Kato E, Siu A, Phillips W, Davidson K, Phipps M, Bibbins-Domingo K (2018) Conflicts of interest in clinical guidelines: update of U.S. preventive services task force policies and procedures. Am J Prev Med 54(1S1):S70–S80. https://doi.org/10.1016/j.amepre.2017.06.034
70. Yamamoto K, Murayama A, Ozaki A, Saito H, Sawano T, Tanimoto T (2021) Financial conflicts of interest between pharmaceutical companies and the authors of urology clinical practice guidelines in Japan. Int Urogynecol J 32(2):443–451. https://doi.org/10.1007/s00192-020-04547-3
71. Harada K, Ozaki A, Saito H, Sawano T, Yamamoto K, Murayama A, Senoo Y, Tanimoto T (2021) Financial payments made by pharmaceutical companies to the authors of Japanese hematology clinical practice guidelines between 2016 and 2017. Health Policy 125(3):320–326. https://doi.org/10.1016/j.healthpol.2020.12.005
72. Kobashi Y, Watanabe M, Kimura H, Higuchi A, Ozaki A (2019) Are pharmaceutical company payments incentivising malpractice in Japanese physicians? Int J Health Policy Manag 8(10):627–628
73. Sawano T, Ozaki A, Saito H, Shimada Y, Tanimoto T (2019) Payments from pharmaceutical companies to authors involved in the valsartan scandal in Japan. JAMA Netw Open 2(5):e193817. https://doi.org/10.1001/jamanetworkopen.2019.3817
74. Good Clinical Practice Network (Data unknown) ICH harmonised guideline integrated addendum to ICH E6 (R1): guideline for good clinical practice ICH E6 (R2) ICH consensus guideline. https://ichgcp.net/ja. Accessed 12 Jul 2021
75. Pharmaceuticals and Medical Devices Agency (2021) Outline of reviews and related services. https://www.pmda.go.jp/english/review-services/outline/0001.html?fbclid=IwAR3ssgRCKVytsqwQP4deaZLEx4AEu0qxFA8H_ico7fH2h1uMIMDdMJvOBOw. Accessed 12 Jul 2021
76. Sawada T, Yamada H, Dahlof B, Matsubara H, Group KHS (2009) Effects of valsartan on morbidity and mortality in uncontrolled hypertensive patients with high cardiovascular risks: KYOTO HEART study. Eur Heart J 30(20):2461–2469. https://doi.org/10.1093/eurheartj/ehp363
77. The Lancet Editors (2013) Retraction--valsartan in a Japanese population with hypertension and other cardiovascular disease (Jikei Heart Study): a randomised, open-label, blinded endpoint morbidity-mortality study. Lancet 382(9895):843. https://doi.org/10.1016/S0140-6736(13)61847-4
78. Arnold DG, Beck T, Stewart OJ (2021) Whistleblowing and misconduct at large pharmaceutical firms-reply. JAMA 325(13):1331–1332. https://doi.org/10.1001/jama.2021.0576
79. Bauchner H, Fontanarosa PB, Flanagin A (2018) Conflicts of interests, authors, and journals: new challenges for a persistent problem. JAMA 320(22):2315–2318. https://doi.org/10.1001/jama.2018.17593
80. Baselga J (2018) Failure to disclose conflict of interest in article published in JAMA on detection of cancer-related genes. JAMA 320(22):2380. https://doi.org/10.1001/jama.2018.16961
81. Botkin JR (2018) Should failure to disclose significant financial conflicts of interest be considered research misconduct? JAMA 320(22):2307–2308. https://doi.org/10.1001/jama.2018.17525

Pharmaceutical Company Payments to Clinical Practice Guideline Authors 45

Anju Murayama, Akihiko Ozaki, and Tetsuya Tanimoto

Abstract

Internationally, efforts promoting greater transparency and improved management strategies for conflicts of interest (COI) have gained traction in healthcare settings. This particularly pertains to the development and use of clinical practice guidelines (CPG). Mounting evidence indicates that pharmaceutical industry payments to GPG authors and developers influence clinical recommendations, including drug selection, often to benefit commercial interests and at the expense of patients. To prevent undue influence of COI and develop trustworthy CPG, authors and developing organizations should establish strict COI management policies, including full disclosure. Such policies should include details about the monetary values and funding sources of all payments and gifts from pharmaceutical companies. Authors and developers should refuse any payments or gifts while drafting CPG. CPG developers should establish clear and comprehensive COI definitions and create monitoring committees that implement COI policies, promote external review, and track COI declared by CPG authors using existing payment databases.

A. Murayama (✉)
Medical Governance Research Institute, Tokyo, Japan

Tohoku University School of Medicine, Sendai, Miyagi, Japan

A. Ozaki
Medical Governance Research Institute, Tokyo, Japan

Department of Breast and Throid Surgery, Jyoban Hospital of Tokiwa Foundation, Iwaki, Fukushima, Japan
e-mail: akihiko-ozaki@tokiwa.or.jp

T. Tanimoto
Medical Governance Research Institute, Tokyo, Japan

Department of Internal Medicine, Navitas Clinic, Kawasaki, Kanagawa, Japan

Abbreviations

COI	Conflicts of interest
CPG	Clinical practice guidelines
GIN	Guidelines International Network
NAM	National Academy of Medicine
rhAPC	Recombinant human activated protein C
USA	United States of America

45.1 Introduction

Over the last few decades, the role and use of clinical practice guidelines (CPG) have dramatically increased. The National Academy of Medicine (NAM) defines CPG as "statements that include recommendations intended to optimize patient care that is informed by a systematic review of evidence and an assessment of the

J. Faintuch, S. Faintuch (eds.), *Integrity of Scientific Research*,
https://doi.org/10.1007/978-3-030-99680-2_45

benefits and harms of alternative care options." [1] Before the introduction of CPG, clinical practice and the decisions of health care professionals were based mainly on experience and skill, i.e., the "art" of medicine [1]. Consequently, within a country, a given disease often had multiple treatments. To overcome these differences and provide all patients with the best possible treatments professional societies, specialty hospitals, patient advocacy groups, local governments, and national and international organizations pursued the development of CPG based on the most rigorous available scientific evidence. Nevertheless, risk bias exists in CPG due to several factors. Arguably the most relevant is the presence and inadequate management of conflicts of interest (COI) among CPG authors [2].

45.2 Conflicts of Interest Among Clinical Practice Guideline Authors

NAM also defines COI as "circumstances that create a risk that professional judgments or actions regarding a primary interest will be unduly influenced by a secondary interest." [2] Among healthcare professionals, COI can occur when they receive meals, gifts, or payments as speaking or consulting fees from commercial entities such as pharmaceutical companies and medical device companies. Unquestionably, healthcare professionals exercise their best professional judgment in prioritizing a patient's well-being over commercial interests. Nevertheless, any COI can unduly influence clinical practice. For example, clinicians could prematurely recommend certain drugs despite better alternative treatments. Even worse remains the possibility that a particular recommended drug should actually not be prescribed to certain patients.

Another avenue for COI among CPG authors includes potential bias in referencing scientific literature or the interpretation of clinical studies. Although the primary aim of CPG should be guiding healthcare professionals based on the best available evidence, this has not always been so [3–5].

45.2.1 Selective Rewards

COI remain pervasive among healthcare professionals, organizations, researchers, and other stakeholders. In particular, a minority of influential healthcare professionals with extensive experience and specialized knowledge, including university professors, medical society directors, journal editors, and CPG authors, received the vast majority of payments and gifts from pharmaceutical companies [6]. Campbell et al. [7] reported that physicians with experience writing CPG were 1.41 times more likely to receive payments from pharmaceutical companies.

45.2.2 Widespread Practice

Although the prevalence of COI may differ by specialty, country, or developing body, COI between CPG authors and pharmaceutical companies remains a universal problem. In 2002, Choudhry et al. [8] published the first study highlighting the financial relationships between CPG authors and pharmaceutical companies, with 100 authors representing more than 30 CPG published by medical professional societies in the United States (USA), Canada, and Europe between 1991 and 1999. Overall 87% of authors had relationships with pharmaceutical companies, of which 64% received speaking honoraria, and 38% accepted consulting fees [8]. Additionally, Cosgrove et al. [9] elucidated that 56% of authors for the Diagnostic and Statistical Manual of Mental Disorders had financial ties to pharmaceutical companies and that all authors for "Mood Disorders" and "Schizophrenia and Other Psychotic Disorders" also received industry-sponsored payments.

Bindslev et al. [10] found that 53% of Danish authors publishing CPG between 2010 and 2012 had COI and that 96% of CPG had at least one author with COI. In Canada, Shnier et al. [11] observed that 47.0% of CPG authors between 2012 and 2013 reported COI with pharmaceutical companies. Recently, Saito et al. [12] detected that 78.2% of authors for six prominent Japanese oncology CPG had financial relationships with pharmaceutical companies. Although not as widespread, Moynihan et al. [13] reported that among 33 Australian CPG, 35.1% of the authors had COI with pharmaceutical companies.

45.3 Transparency of COI Among CPG Authors

More often than not, transparency policy and legal frameworks shift followed significant scandals, including a participant's death in a failed USA gene therapy trial for ornithine transcarbamylase deficiency [14], the French Mediator Scandal [15], or the Diovan Scandal in Japan [16]. Since 2013, the USA Open Payments Program represents a leading example. Established under the legislative authority of the USA Physician Payment Sunshine Act, it mandates that all pharmaceutical and medical device companies disclose essentially all payments and gifts made to physicians and teaching hospitals [17]. This Open Payments Database and transparency initiatives like ProPublica allow the general public to scrutinize most financial transfers from industry to physicians and researchers [18, 19]. Similarly, France and a few other developed nations require companies to disclose payments to healthcare professionals in payment databases [20].

45.3.1 Self-Regulation and Professional Initiatives

Trade organizations such as the European Federation of Pharmaceutical Industries and Associations and the Japan Pharmaceutical Manufacturers Association required pharmaceutical companies to disclose their payments to healthcare organizations and healthcare professionals on a company's webpage only. The Association of the British Pharmaceutical Industry and Medicine Australia collected payment data from pharmaceutical companies and disclosed them on their comprehensive payment database, respectively namely Disclosure UK [21] and Disclosure Australia, however beyond any legal mandate [22].

45.3.2 Significant Budgets

In some countries independent non-profit journalist organizations coordinate structured payment databases including the Money Database in Japan [12, 23, 24], the Dollars for Docs in the USA [19, 25] and the Euros for Docs in Europe, to increase accessibility and usability for the general public [26, 27].

In 2014, 52.2% of USA physicians received a combined $1.94 billion in non-research payments for speaking honoraria, consulting fees, gifts, food, and beverages [18]. Another study by Goupil et al. [20] found that 87.8% of French general practitioners received gifts from 2013 to 2016.

45.4 Comparison Between CPG Authors and Others

Unsurprisingly CPG authors were more likely to receive payments from pharmaceutical companies, and more substantial than those awarded to other physicians. Using the Open Payments Database, Mitchell et al. [28] reported that 86.4% of National Comprehensive Cancer Network CPG authors received an average of $236,066 in research payments and another $10,011 in annual non-research payments, respectively. On average, non-research payments given to CPG authors were 1.29, 2.58, or 5.82 times higher than those received by medical, surgical, or radiation oncologists, respectively [29]. Similarly, Saleh et al. [30] found that the average American Society of Clinical Oncology CPG author received $1,168,956 in declared payments from pharmaceutical companies between 2013 and 2017, and another $187,503 in undeclared payments.

In Japan 255 CPG authors from six prominent Japanese oncology CPG received 2.12 times larger payments from pharmaceutical companies in 2016, than board-certified oncologists

($10,565 vs. $4982) [12, 31]. Concerning other medical specialties, USA CPG authors in dermatology, urology, and otorhinolaryngology received 6.7 times [19, 32], 3.4 times [25, 33], and 3.2 times [34, 35] higher payments, respectively, compared to their corresponding colleagues not involved in preparing CPG.

45.5 Undeclared and Under-Declared Conflicts of Interest

Papanikolaou et al. [36] determined that among 191 CPG published in six prestigious journals between 1979 and 1999, only 3.7% (7/191) reported author COI. Subsequently, a Nature survey from 2004 found that 58.1% of CPG (125/215) published by the National Guideline Clearinghouse failed to include detailed information about author COI [37]. Another study assessing the prevalence of COI among hyperlipidemia and diabetes CPG authors between 2000 and 2010 found that among 14 CPG, five failed to include author COI [38]. Additionally, while 73 CPG authors declared no COI, eight (11.0%) had unreported COI. This study reported that among 77 CPG authors not declaring COI, Medline and Google screening of previous publications indicated four (5.2%) had undisclosed COI [38].(Table 45.1).

45.5.1 More than a Decade of Recommendations

Dating back to 2011, published NAM guidelines recommend the declaration of all author COI in published CPG [1], analogously to the World Health Organization Handbook for Guideline Development [39]. The Guidelines International Network (GIN), comprising more than 100 CPG developing organizations, published nine principles for COI disclosure and management by CPG authors in 2015 [40]. In 2017 the Japanese Association of Medical Sciences published a COI Management Guidance on Eligibility Criteria for Clinical Practice Guideline Formulation [41].

Nevertheless unreported and underreported COI among CPG authors remain [12, 13, 19, 23, 24, 30, 42–44]. Saleh et al. [30] observed that among USA oncology CPG authors declaring no COI, 93.2% (55/59) received gifts and payments from pharmaceutical companies. These undeclared payments averaged $187,503. Checkett et al. [19] found that 55.0% of USA dermatology CPG authors (22/40) incorrectly reported their COI, including six with undeclared and 16 with underdeclared COI. In an Australian study by Moynihan et al. [13], 24.1% (83/344) of CPG authors had undeclared financial relationships with industry, mainly for research funding.

Senoo et al. [42] reported that 36.5% of Japanese hypertension CPG authors failed to disclose their financial ties to pharmaceutical companies, despite these interests exceeding declaration thresholds. There similarly were large numbers of undeclared and underdeclared payments among Japanese CPG authors in dermatology, hematology, urology, oncology, and infectious diseases [12, 23, 24, 43, 44].

45.6 Influence of Financial COI on CPG

There persists a view among many professionals that industry payments exert no undue influence. They remain steadfast in that only their professional knowledge, experience, and judgment guide their clinical practice, unaffected by COI. This contributes to the misperception that healthcare professionals with industry ties represent ideal experts to receive such funding for research [8, 37, 45]. However, incremental evidence of pharmaceutical company payments inappropriately influencing physician behavior dates back to the early 1990s [46]. Nejstgaard et al. [47] found that the relative risks for associations between financial COI and favorable recommendations for companies in CPG, advisory committee reports, opinion pieces, and narrative reviews were 1.26 (95% CI: 0.93–1.69),

Table 45.1 Conflicts of interest among clinical practice guideline authors

No. CPG No. authors	No. of authors or CPG with COI	Underreported or undeclared COI	Payments from pharmaceutical companies	Specialty or disease
More than 30 CPG 100 authors [8]	87% of authors with COI 64% of authors with speaking honoraria 58% of authors with research support	NR	NR	Multiple
18 CPG 170 authors [9]	56% of authors with declared COI	NR	NR	Psychiatry
14 CPG 288 authors [38]	48% of authors with declared COI 52% of authors with COI	11% of authors declaring no COI had COI	NR	Diabetes and hyperlipidemia
17 CPG 498 authors [80]	55.6% of authors with COI	NR	NR	Cardiology
54 CPG 135 authors [10]	53.1% of authors with COI	52.0% of authors did not disclose COI	NR	Multiple
13 CPG 192 authors [66]	Range: 0–94% of authors with declared COI	NR	NR	Type 2 diabetes mellitus
91 CPG and 51 CS [70]	67% of CPG or CS with COI	NR	NR	Oncology
4 CPG 125 authors [28]	86% of authors with COI	NR	Average: $10,011 for non-research $236,066 for research 2014	Oncology
28 CPG 350 authors [11]	47.0% of authors with declared COI 75.0% of CPG with COI	NR	NR	Multiple
3 CPG 49 authors [19]	81.6% of authors with COI	32.7% of authors with underreported COI 12.2% of authors with undeclared COI	Average: $157,177 for total $83,703 for non-research $11,250 for research $302,128 for associated research	Dermatology
2 CPG 45 authors [53]	72% of authors with any relationship in hepatitis C virus CPG 44% of authors with any relationship in cholesterol management CPG	Yes	NR	Hepatitis C virus Cholesterol management

(continued)

Table 45.1 (continued)

No. CPG No. authors	No. of authors or CPG with COI	Underreported or undeclared COI	Payments from pharmaceutical companies	Specialty or disease
15 CPG 83 authors [81]	53% of authors with COI (range: 0–100%)	62% of authors with not verified COI by open payment database	Median: $1000 of total payments	Gastroenterology
6 CPG 54 authors [25]	59.3% of authors with COI	37.0% of authors with undeclared COI	Median: $578 for total payments $51 for general payments	Urology
26 CPG 169 authors [54]	46% of authors with self-reported COI 74% of US authors with screened COI	Yes	Average: $65,476 for non-research payments $367,857 for research payments	Endocrinology
67 CPG NR for authors [82]	55% of CPG with COI	NR	NR	Multiple
5 CPG 49 authors [34]	80% of authors with COI	6% of authors did not disclose COI	Average: $18,431 for total $11,910 for general, $50,282 for research payments	Otolaryngology
26 CPG 314 authors [30]	58.6% of authors with declared COI	Among authors who declared no COI, 93.2% actually had COI	Average: $1,168,956 for disclosed $187,503 for undisclosed total payments	Oncology
7 CPG 43 authors [83]	32.6% of authors with COI	18.6% of authors with undeclared COI	$0 in median payments Interquartile range: 0–84) for total payments	Routine screening mammography
33 CPG 402 authors [13]	14% of authors with self-reported COI	24% of authors with undisclosed COI	NR	Multiple
6 CPG 326 authors [12]	78.2% of authors with COI	Unable to review due to anonymous COI disclosure or lack of COI statements.	Average: $10,565 for non-research payments	Oncology
145 CPG NR for authors [84]	25.5% of CPG disclosed COI 24.1% of CPG disclosed funding sources for CPG development	NR	NR	Multiple
45 CPG 296 authors [43]	90.6% of authors with COI	Among 17 authors who declared 2(17%) underreported their COI	Average: $26,600 for non-research payments	Dermatology
13 CPG 193 authors [23]	88.6% of authors with COI	92.3% of CPG failed to disclosed authors' COI	Average: $31,965 for non-research payments	Urology

(continued)

Table 45.1 (continued)

No. CPG No. authors	No. of authors or CPG with COI	Underreported or undeclared COI	Payments from pharmaceutical companies	Specialty or disease
21 CPG [85]	0% of authors by government-sponsored organizations 57% of authors by disease or condition interest groups 53% of authors by medical professional societies	NR	NR	Multiple
5 CPG 89 authors [55]	62.9% of authors with COI Range: 55.6–68.4%	25% of authors did not disclose COI worth of a total of $4,189,090)	Average: $109,311 for total $3020 for research $7582 for associated research payments	Rheumatology
1 CPG 198 CPG authors [42]	86.4% of authors with COI	36.5% of authors with undeclared COI	Average: $21,447 for non-research payments	Hypertension

CS clinical statement, *COI* conflicts of interest, *CPG* clinical practice guideline, *NR* not reported, *No.* number

1.14 (95% CI: 1.07–1.21), 2.62 (95% CI: 0.91–7.55), and 1.20 (95% CI: 0.97–1.49), respectively. The combined analysis of all documents reported a relative risk of 1.26 (95%: 1.09–1.44).

45.6.1 The "Xigris" Affair

As a former deputy editor and national correspondent for the New England Journal of Medicine, Steinbrook [48] discussed several COI cases, encompassing the Surviving Sepsis Campaign CPG for the management of severe sepsis and septic shock. Along with significant financial ties to the authors, Eli Lilly also funded the CPG development process [48]. Consequently, the published CPG were based on a single clinical trial conducted by Eli Lilly and developed to promote the sale of Xigris (recombinant human activated protein C: rhAPC) [49]. Eichacker et al. [49] pointed out that the CPG failed to disclose a significant bleeding risk with rhAPC and that two other prior trials with Xigris were terminated early for failing to reach their primary endpoints and an increased risk of bleeding among trial participants.

45.6.2 Strategies against Diminishing Revenue

Occasionally pharmaceutical companies may be disadvantaged by certain guideline recommendations and will therefore take steps to mitigate them. In 2005, when the National Institute for Clinical Excellence in the United Kingdom published a CPG for Alzheimer's disease restricting the use of Pfizer's and Eisai's acetylcholinesterase inhibitor Aricept (donepezil) along with drugs by other manufacturers, the involved companies petitioned the British courts to invalidate the CPG [48]. It took the submission of an economic model for the companies to scrutinize, demonstrating the reasoning behind these restrictions to resolve the dispute [50].

45.6.3 Lyme Disease and the Infectious Disease Society of America

Published in 2000 by the Society, this CPG rather narrowly defined Lyme disease. Consequently, many patients met roadblocks in accessing treatment, while physicians similarly faced restrictions in treating patients. The Society further aggravated the situation by restricting authorship of the revised 2006 CPG to a group of researchers with multiple COI that included highly biased perspectives on the diagnosis and treatment of Lyme disease, and strong financial ties to interested companies. This occurred despite repeated overtures from patients and their physicians seeking to participate in the development process.

The Connecticut Attorney General initiated an antitrust investigation against the Infectious Diseases Society of America and the Lyme disease CPG authors. As noted by Johnson and Stricker [4], the authors could have deliberately manipulated the Lyme disease definition in the CPG so that vaccine companies could "increase the effective rate for the vaccines so that fewer people taking the vaccine 'get' the disease." Moreover, a CPG that mandates testing for disease diagnosis directly promote the interests of diagnostic test companies. Similarly, a CPG that denies treatment to patients could be favorable to insurance companies. Indeed CPG authors received financial compensation from multiple interested parties and deliberately excluded dissenting opinions and voices of patients and practitioners, and the chairperson inappropriately selected the authors by forgoing a formal development process [4].

45.6.4 Steroids for Acute Spinal Cord Injury

In the 1990s, many neurosurgeons expressed concerns over the potential for infection, prolonged hospital stays, and excessive deaths with the CPG advocating high dose steroids, estimated to have exceeded 5000 [51]. Another neurological issue involved the recommendation of the thrombolytic drug alteplase (Activase, Genentech) for use in patients with acute stroke as recently as 2013. This occurred despite a lack of support by many emergency physicians, a lack of robust clinical evidence, and a high risk of bias in clinical trials with strong financial ties to the pharmaceutical company sponsors [51].

45.6.5 The French Connection

In 2011 the highest administrative court in France ordered the retraction of two CPG, one for type 2 diabetes and the other for Alzheimer's. The *Haute Autorité de Santé* (High Authority on Health) had drafted both. The justification for this decision included the lack of proper disclosure of the CPG authors' COI [52]. Following this ruling, another six CPG issued by the same Authority were voluntarily withdrawn for similar reasons. Ultimately, the French courts sustained this decision based on recent legislation requiring comprehensive declaration and disclosure of COI by all CPG authors. The legislation further prohibited authors from participating in any financial activities with pharmaceutical companies related to CPG.

45.7 COI Management Among CPG Authors and Developing Organizations

Given that poor COI management during the CPG development process could result in misconduct and direct harm or death of patients, NAM published a set of standards in 2011(Table 45.2) [1]. Overlapping suggestions by this and other organizations include the full disclosure of COI, minimizing CPG authors COI, eliminating COI for chairpersons, and including patients and multiple experts with broad backgrounds in the development process [3, 40].

Yet these remain persistent issues [23, 43, 53–56]. Of the 37 CPG developing organizations examined by Norris et al. [57], only 17 (45.9%) had established COI policies for CPG development in 2012. Furthermore, none of the organization's existing COI policies met NAM standards. In yet another study, Morciano et al. [58] surveyed the COI policies of 29 CPG developing organizations in 20 countries in 2016. Surprisingly, eight organizations (27.6%) meeting the study inclusion criteria did not have established COI policies for CPG development [58].

Table 45.2 Summary of recommendations on the CPG development process and COI policy by three major international CPG developing organizations

Organizations	Content
US National Academy of Medicine criteria [1]	
1.1	The CPG development process and funds for CPG development should be explicitly and publicly accessible.
2.1	All current and planned financial and non-financial COI should be disclosed.
2.2a	All COI among CPG authors should be reported and discussed prior to CPG development.
2.2b	All CPG authors should explain how their COI influence CPG development in a detailed manner.
2.3	All CPG authors should divest their and their family's COI and not have COI that could influence or be affected by CPG recommendations.
2.4a	Whenever possible, CPG authors should not have COI.
2.4c	CPG authors with COI should be in the minority.
2.4d	A chair or co-chair of CPG should not have COI.
2.4e	Financial backers should have no role in CPG development.
3.1	The CPG developing organizations should be multidisciplinary and balanced, comprising various subject matter experts, clinicians, and populations expected to be affected by the CPG.
3.2	Patient and public involvement should be facilitated by including patients and patient advocate organizational representatives in CPG authorship.
3.3	Strategies to increase the effective participation of patients and consumer representatives, including training in the appraisal of evidence, should be adopted by CPG developing organizations.
7.1	External reviewers should comprise a full spectrum of relevant stakeholders, including scientific and clinical experts, organizations, agencies, patients, and public representatives.
Guidelines International Network in 2015 [40]	
1	CPG developers should make all possible efforts not to include CPG authors with direct financial COI or relevant indirect COI. When including a CPG author with COI, CPG developers should guarantee a transparent development process, and a majority of CPG authors should be free of COI.
2	COI definition and management should apply to all CPG authors.
3	CPG development group should use a standardized form for COI disclosure.
4	CPG development group should publicly disclose all direct and indirect COI with specific monetary values, and these COI should be easily accessible for CPG users.
5	All CPG authors should declare and continually update their COI at regular intervals (e.g., annually).
6	CPG chairs should have no direct or indirect COI. When a chair unavoidably has direct or indirect COI, a co-chair with no COI should be appointed and lead the CPG development.
7	An expert with relevant COI and specific knowledge or expertise (namely expert advisor) may participate in CPG development, but there should be an appropriate balance of opinions among the experts.
8	CPG authors who decide about the direction or strength of a recommendation should not have any direct financial COI. CPG authors with direct financial COI should be physically absent from meetings involving recommendation decision-making.
9	A monitoring committee should be responsible for the formulation and implementation of rules on COI.
Guideline Panel Review in 2013 [3]	
1	The sponsor(s) is a professional society that receives substantial industry funding.
2	The sponsor is a proprietary company or is undeclared or hidden.
3	CPG chair(s) have any financial COI.
4	Multiple CPG authors have any financial COI.
5	There is any suggestion of CPG stacking that would pre-ordain a recommendation regarding a controversial topic.
6	There is no or limited involvement of an expert in methodology in the evaluation of evidence.
7	There is no external review.
8	There are non-physician experts, patient representation, or community stakeholders in CPG development.

There are two levels of COI management, respectively concerning the CPG authors (e.g., accurate and complete COI declarations when drafting guidance documents) and the CPG developing organizations (e.g., establishing and implementing strict COI policies for CPG authors) (Table 45.3).

45.8 Complete Declaration and Full Disclosure

Regarding the different types of authors COI, financial ones should include all payments from pharmaceutical companies (e.g., lecture honoraria, consulting fees, advisory services for pharmaceutical companies, travel, education reimbursement, patents or royalties, stock or shares, and research expenses). Additionally, CPG authors should declare the receipt of gifts or meals. Why? Evidence suggests that perks valued as little as $20 can influence a physician's prescription patterns [59]. A few professional societies including the American College of Physicians, the Canadian Task Force on Preventive Health and Care, and the National Clinical Effectiveness Committee in Ireland, all require CPG authors to declare all gifts received from pharmaceutical companies [58].

45.8.1 Duration of COI Declarations

The American College of Physicians asks for COI declarations covering the last 3 years. Similarly, the Japanese Association of Medical Sciences, representing the largest academic, professional society of physicians in Japan, also requires CPG developers and authors to disclose COI for the prior 3-year period. The French *Haute Autorité de Santé* and the American Pain Society both demand a 5 year span. In contrast the American College of Cardiology, the American College of Rheumatology, and the National Institute for Health and Care Excellence in the United Kingdom, require COI declarations dating back only 1 year before joining CPG development [55, 57, 58].

Imposing greater timeframes for COI disclosure, past and future, improves the ability to discern potential relationships between CPG authors and pharmaceutical companies. It may also benefit reviewers and readers of CPG by enabling users to understand the authors' background and attitudes towards CPG development, and ultimately the conclusions and recommendations presented. Beyond any defined period of COI disclosure such declarations should be continuously updated, following the recommendations of the GIN (2015) and the International Committee of Medical Journal Editors (2019) [40, 60].

45.8.2 Commercial Utilization of CPG

Pharmaceutical companies will utilize CPG to promote their products, and CPG authors could participate in related financial activities sponsored by pharmaceutical companies [61]. Miyata et al. [62] reported that 86.2% of physicians in Japan receive free CPG provided by pharmaceutical companies. The American Urological Association prohibits authors from giving lectures on behalf of pharmaceutical companies for 1 year after guideline publication [25]. Alternatively, if not prohibited from financial interests with pharmaceutical companies after publishing CPG, at a minimum, authors should inform speaking audiences or readers of their COI and continuously update any changes.

45.8.3 Explicit Impact

CPG developers clearly recognize the influence pharmaceutical companies have on CPG authors. This influence relates directly to the magnitude of shared financial interests, as studies reveal a greater impact on physician behavior with the increasing size of pharmaceutical company payments [40, 41, 59, 63]. The Japanese Association of Medical Sciences requires CPG authors to declare industry payments based on three tiers: one million or more Japanese yen (approximately $10,000 USA dollars), five million or more

Table 45.3 Recommendations for CPG developing organizations and CPG authors in order to produce rigorous and trustworthy CPG

Level	Recommendation
CPG author	
1	All CPG authors should declare and disclose all financial and some types of non-financial COI. These include payments or gifts from pharmaceutical companies with specific monetary value within the last 3 (or 5 years). Authors should continuously update their COI regularly after CPG publication.
2	CPG authors should cease and refuse to join financial arrangements with pharmaceutical companies nor receive any payments or gifts from pharmaceutical companies, especially 1 year before and during the development process and in the year following CPG publication.
3	CPG authors should describe how their financial and non-financial COI would influence recommendations and statements in the CPG.
4	A CPG chair should have no COI.
CPG developing organization	
1	CPG developing organizations should establish a strict and comprehensive COI policy in line with the recent movement towards greater transparency in COI disclosure.
2	Clear and comprehensive definitions of financial and non-financial COI should be established and implemented for all CPG authors.
3	CPG development organizations should declare the funding source for both the organization and for the development of CPG.
4	CPG developing organizations should disclose all financial and some types of non-financial COI declared by CPG authors with specific monetary values. The best practice would be to disclose the original COI declaration written by CPG authors.
5	CPG developing organizations should recruit predominantly experts with no COI.
6	CPG developing organizations should ensure that CPG authors include subject matter and methodological experts, statisticians, and patients with the disease concerned.
7	CPG developing organizations should include detailed descriptions of the CPG development process and COI management strategies, including the author selection process and the organizations' COI policies.
8	CPG developing organizations should establish a CPG monitoring or review committee to implement the organization's COI policies and to assess the accuracy of COI declared by CPG authors.
9	CPG developing organizations and monitoring committees should use the payment databases disclosed by pharmaceutical companies to assess the accuracy of COI declared by CPG authors in several countries with payment databases.
10	CPG developing organizations should collect public comments during CPG development and explain how those comments were incorporated into the CPG.
11	CPG developing organizations should establish penalties for CPG authors who fail to comply with the organization's COI policies.

Japanese yen (~ \$50,000), and ten million or more Japanese yen (~ \$100,000) or more. Such declarations include receipt of consulting fees, patents, royalties, and research payments.

45.8.4 Small Change

Even this policy warrants criticism. Requiring COI declarations for any gift represents a greater form of transparency than basing disclosure on specific monetary thresholds [23, 43]. The American Heart Association roughly defines a modest payment as <\$10,000 and a significant one as more than \$10,000 [56]. The American Thoracic Society analogously requires CPG authors to report COI to a specific dollar range, yet it lacks transparency in that this disclosure is accessible only by the society chair and other reviewers [64]. The GIN proposed a more rigorous method of COI management, namely explicit COI declarations and the disclosure of specific monetary amounts [40]. Indeed, CPG published by the American College of Physicians publicly disclosed the original COI declaration form written by CPG authors with specific monetary amounts on their official webpage.

45.9 Non-financial COI

By NAM standards non-financial COI include intellectual interests such as "authoring a publication or acting as an investigator on a peer-reviewed grant directly related to recommendations under consideration." [1] Taken further, the American Thoracic Society considers primary intellectual COI as authorship of original studies and peer-reviewed grant funding directly bearing on a recommendation, and secondary intellectual COI as participation in previous guideline panels or authorship on systematic reviews that provided recommendations [65]. Generally, non-financial COI also warrant disclosure in CPG, given their potential to influence CPG recommendations as much as financial ones. Some healthcare professionals and CPG authors acknowledge that non-financial COI likely bias CPG recommendations [8, 66, 67], yet few policies are in place requiring their declaration [58] or disclosure [65].

In 2019 the Canadian Medical Association announced that all CPG authors must adhere to the GIN principles, including non-financial COI declaration. On the other hand they also noted that caution should be exercised on this matter, as an overly broad definition of non-financial COI might inappropriately burden or infringe on the privacy of, or even create a misperception of non-existent pervasive COI among CPG authors [68]. In any case, the adequacy of the current definition of non-financial COI remains under debate, and there remains room for improving both the understanding and reporting of COI.

45.10 Refusal of Payments from Pharmaceutical Companies

CPG authors must strive to avoid the undue influence of secondary interests [1, 40]. One possible approach could include CPG authors forgoing any financial interactions with pharmaceutical companies up to 1 year before, during, and the year following CPG publication. Interestingly, one study found that expanding the reporting period beyond the previous 5 years to include the 3 years after publication resulted in an increase in undisclosed COI from 24% to 28% [13]. Unquestionably, discerning whether undisclosed COI between authors and pharmaceutical companies directly impacts CPG content or recommendations remains challenging. Nevertheless, one cannot exclude that authors may receive compensation from pharmaceutical companies for speaking or writing based on published guidelines. Internal company documents reveal that they consider CPG to be a marketing tool for drug sales [69]. Thus all CPG authors should, at a minimum, seek to avoid financial relationships with pharmaceutical companies for the period that includes the year before, during, and the year following CPG publication.

45.11 COI Management Strategies for CPG Developing Organizations

The criteria for establishing strict COI policies should incorporate NAM standards and the GIN's nine principles (Table 45.2) [1, 40]. Norris et al. [57] reported that 54.1% (20/36) of CPG developing organizations, whose guidelines were published and listed in the National Guideline Clearinghouse between January 2009 and November 2010, lacked any formal COI policies for authors or development staff. Still, inadequate policies may be worse than the absence of any formal policies. CPG developed under inadequate COI policies may have biased recommendations.

45.12 Chair Appointment and the Author Selection Process

The appointment of a chairperson overseeing CPG development represents a critical decision for ensuring the integrity and quality of published guidelines. Essentially, the chair often has the authority to select the authors and lead guideline development. Therefore, CPG developing organizations should appoint experts with no COI. Where a CPG chair has significant COI, there exists the potential for malfeasance. The previously discussed Lyme disease case exemplified this reality [4]. Additionally, Tibau et al. [70] found that 61% of first, second, or corresponding authors of oncology CPG had financial COI with pharmaceutical companies. Similarly, Horn et al. [34] reported that 88% of US otolaryngology CPG authors received payments from pharmaceutical companies. Further, Desai et al. [71] established that CPG chairs and co-chairs received on average more payments than other CPG authors. Adding to this, Saito et al. [12] noted that all Japanese oncology CPG chairs had significant financial ties to pharmaceutical companies. The lung carcinoma CPG and liver carcinoma CPG chairs received the largest single-year payments at USA dollars $158,217 and $152,156, respectively.

45.12.1 Authorship Diversification

Often, CPG author selection arises by nominations from the chair or prior authors. However, this selection method increases the risk of selecting authors with similar views, potentially contributing to biased and less rigorous CPG. To prevent this, many CPG developing organizations suggest that multiple stakeholders, not just subject matter experts, but also methodologists, statisticians, and patients be included as CPG authors. They further advocate for peer review by external reviewers and that CPG developers describe, in detail, the method and criteria used in author selection. Ironically, the CPG issued by the Infectious Disease Society of America on Lyme disease included a statement on author selection, serving as one of the key lessons learned from this case [4]. Lenzer et al. [3] recommended that CPG authors should receive comments from experts with scientifically based dissenting positions.

45.13 Funding Sources for the Organization and CPG Development

Although establishing what funding sources are appropriate for CPG development remains debatable, at a minimum both direct and indirect funding sources should be disclosed. Campsall et al. [72] found that 63.2% (60/95) of CPG developing organizations were funded by pharmaceutical companies, and some combination of membership dues (86.8%), donations from for-profit companies (52.6%), charitable donations (39.4%), or government funding (23.7%). Notably, 35.2% (102/290) of CPG lacked statements on funding and support for development [72]. While 57% of CPG disclosed direct funding from industry, many failed to mention it. CPG produced by organizations with more comprehensive COI policies generally included fewer positive recommendations (rate ratio: 0.91, 95% CI: 0.86–0.95), more negative recommendations (rate ratio: 1.32, 95% CI: 1.09–1.60), more COI disclosure statements (rate ratio: 1.36, 95% CI: 1.09–1.79), and greater disclosure of funding sources for development (rate ratio: 1.31, 95% CI: 1.10–1.56) [72]. Neuman et al. [38] reported that CPG developed by government-sponsored organizations were likely to include fewer authors with COI than those developed by other organizations (16% vs. 69%). Lenzer et al. [3] pointed out that industry-funding of developing organizations remains one of the indicators of untrustworthiness in published CPG. They concluded that government-sponsored CPG developing organizations might employ stricter policies and select qualified authors with fewer COI.

45.13.1 Weaknesses Are Always Possible

Government funding has not always been ideal in establishing independent CPG. The US dietary guidelines are affected by political calculations, and the National Guideline Clearinghouse was discontinued from government funding [73, 74]. The coronavirus diseases 2019 (COVID-19) CPG developed by the Japanese Ministry of Health, Labor, and Welfare did not require CPG authors to declare any COI and did recommend favipiravir based on an ongoing Japanese government-sponsored trial, but its efficacy was insufficient for COVID-19 and was not recommended by any agencies in the USA and European countries. Given these findings, the development of CPG by independent groups of authors could represent an option to avoid undue influence from both industry and politics [76, 77]. If a CPG developing organization receives funding from other entities or individuals, these organizations should state the role of such sponsors in a detailed manner [72].

45.14 External Review

Peer review of CPG by external experts would be beneficial to reduce bias [40, 73]. Many current CPG without appropriate management or adherence to COI policy rely instead on readership to judge the validity of included recommendations and content. However, CPG readers such as healthcare providers are often overwhelmed by day-to-day clinical practice, and generally have little time and relevant experience to review content and recommendations critically [3]. Just as with CPG authors, COI should also be declared by external reviewers. As Sonis and Chen [73] conveyed, there exists a potential risk that reviewers with COI might give biased comments on published CPG. To reduce the burden on readers, all CPG should be reviewed by external experts not involved in their development [3].

45.15 Establishing a Monitoring Committee

Campsall et al. [72] found that three CPG developing organizations produced guidelines with funding from pharmaceutical companies despite organizational policies prohibiting such contributions. Current COI management policies rely on individual integrity and ethics, including self-disclosure. Despite organizational requirements mandating full disclosure, CPG authors occasionally fail to declare large payments from pharmaceutical companies [13, 24, 55, 56, 70].

GIN recommends that CPG developing organizations establish a monitoring committee with the dual responsibilities of formulating and implementing COI policies [40]. Additionally, CPG developing organizations should use established payment databases to track financial COI reported by authors and industry-reported payments for consistency [19, 25, 43, 56, 70]. Currently however, only a few CPG developing bodies, such as the American Society of Clinical Oncology and American Gastroenterological Association, utilize these payment databases.

45.15.1 Sanctions

Monitoring committees should consider penalties for CPG authors failing to adhere to established policy [24]. A study of 19 CPG developing organizations found that just two had penalties for CPG authors who underreported their financial COI [58]. In the USA, the American Urological Association has a COI policy that includes imposing penalties on CPG authors that fail to comply. Initially, deviations result in a letter of warning from the CPG committee, followed by removing the author from the CPG, and ultimately expulsion from the Society [25, 58].

45.16 Future Perspectives

Ongoing efforts are utilizing artificial intelligence to examine and evaluate COI in healthcare. Graham et al. [78] reviewed 128,781 articles published in 159 journals using machine learning to evaluate the association between commercial publishing practices of academic journals and author declared COI. They found that journals with reprint fees were more likely to publish articles with author declared COI [78]. Machine learning-based artificial intelligence could identify and review the COI of manuscript authors [78] and facilitate the process of CPG author COI management [79]. Within several years, machine learning may assist monitoring committees in extracting COI information from previous articles and payment databases while reviewing ongoing COI among CPG authors. Nevertheless, even with the advent of a machine learning COI management process, all CPG developing organizations and authors must still strive to manage COI appropriately to ensure the integrity and validity of existing CPG.

Acknowledgement We thank the Tansa (formerly known as the Waseda Chronicle) for collaborating on this issue in Japan, and Dr. Derek Hagman for his constructive opinions and professional language editing.

Conflicts of Interest Anju Murayama, Drs Ozaki, and Tanimoto have several papers concerning conflicts of interest among Japanese and USA healthcare professionals. Drs Ozaki and Tanimoto report receiving personal fees from Medical Network Systems, MNES Inc., a dispensing pharmacy, outside the scope of the submitted work. Dr. Tanimoto reports personal fees from Bionics Co., Ltd., outside the scope of the submitted work. We report no other disclosures. The Medical Governance Research Institute is a non-profit organization creating the payment database (Money for Docs Database) in Japan that received donations from a dispensing pharmacy company, Ain Pharmacies and other private individuals.

References

1. Graham R, Mancher M, Wolman DM, Greenfield S, Steinberg E (eds) (2011) Institute of medicine. Clinical practice guidelines we can trust. The National Academies Press, Washington, DC, 290 p
2. Lo B, Field MJ (eds) (2009) Institute of Medicine. Conflict of interest in medical research, education, and practice. The National Academies Press, Washington, DC, 436 p
3. Lenzer J, Hoffman JR, Furberg CD, Ioannidis JP, Guideline Panel Review Working Group (2013) Ensuring the integrity of clinical practice guidelines: a tool for protecting patients. BMJ 347:f5535
4. Johnson L, Stricker RB (2009) Attorney general forces Infectious Diseases Society of America to redo Lyme guidelines due to flawed development process. J Med Ethics 35(5):283–288
5. Medicine AFABoI, Medicine A-AFACoP-ASoI, European Federation of Internal M (2002) Medical professionalism in the new millennium: a physician charter. Ann Intern Med 136(3):243–246
6. Moynihan R (2008) Key opinion leaders: independent experts or drug representatives in disguise? BMJ 336(7658):1402–1403
7. Campbell EG, Gruen RL, Mountford J, Miller LG, Cleary PD, Blumenthal D (2007) A national survey of physician-industry relationships. N Engl J Med 356(17):1742–1750
8. Choudhry NK, Stelfox HT, Detsky AS (2002) Relationships between authors of clinical practice guidelines and the pharmaceutical industry. JAMA 287(5):612–617
9. Cosgrove L, Krimsky S, Vijayaraghavan M, Schneider L (2006) Financial ties between DSM-IV panel members and the pharmaceutical industry. Psychother Psychosom 75(3):154–160
10. Bindslev JB, Schroll J, Gotzsche PC, Lundh A (2013) Underreporting of conflicts of interest in clinical practice guidelines: cross sectional study. BMC Med Ethics 14:19
11. Shnier A, Lexchin J, Romero M, Brown K (2016) Reporting of financial conflicts of interest in clinical practice guidelines: a case study analysis of guidelines from the Canadian Medical Association Infobase. BMC Health Serv Res 16(a):383
12. Saito H, Ozaki A, Sawano T, Shimada Y, Tanimoto T (2019) Evaluation of pharmaceutical company payments and conflict of interest disclosures among oncology clinical practice guideline authors in Japan. JAMA Netw Open 2(4):e192834
13. Moynihan R, Lai A, Jarvis H, Duggan G, Goodrick S, Beller E et al (2019) Undisclosed financial ties between guideline writers and pharmaceutical companies: a cross-sectional study across 10 disease categories. BMJ Open 9(2):e025864
14. Wilson JM (2009) Lessons learned from the gene therapy trial for ornithine transcarbamylase deficiency. Mol Genet Metab 96(4):151–157
15. Mullard A (2011) Mediator scandal rocks French medical community. Lancet 377(9769):890–892
16. Sawano T, Ozaki A, Saito H, Shimada Y, Tanimoto T (2019) Payments from pharmaceutical companies to authors involved in the valsartan scandal in Japan. JAMA Netw Open 2(5):e193817

17. Pham-Kanter G (2014) Act II of the sunshine act. PLoS Med 11(11):e1001754
18. Marshall DC, Tarras ES, Rosenzweig K, Korenstein D, Chimonas S (2020) Trends in industry payments to physicians in the United States from 2014 to 2018. JAMA 324(17):1785–1788
19. Checketts JX, Sims MT, Vassar M (2017) Evaluating industry payments among dermatology clinical practice guidelines authors. JAMA Dermatol 153(12): 1229–1235
20. Goupil B, Balusson F, Naudet F, Esvan M, Bastian B, Chapron A et al (2019) Association between gifts from pharmaceutical companies to French general practitioners and their drug prescribing patterns in 2016: retrospective study using the French Transparency in Healthcare and National Health Data System databases. BMJ 367:l6015
21. Rickard E, Ozieranski P, Mulinari S (2019) Evaluating the transparency of pharmaceutical company disclosure of payments to patient organisations in the UK. Health Policy 123(12):1244–1250
22. Pokorny AMJ, Bero LA, Moynihan R, Mintzes BJ (2021) Industry payments to Australian medical oncologists and clinical haematologists: a cross-sectional analysis of publicly-available disclosures. Intern Med J 51(11):1816–1824
23. Yamamoto K, Murayama A, Ozaki A, Saito H, Sawano T, Tanimoto T (2021) Financial conflicts of interest between pharmaceutical companies and the authors of urology clinical practice guidelines in Japan. Int Urogynecol J 32(2):443–451
24. Harada K, Ozaki A, Saito H, Sawano T, Yamamoto K, Murayama A et al (2021) Financial payments made by pharmaceutical companies to the authors of Japanese hematology clinical practice guidelines between 2016 and 2017. Health Policy 125(3):320–326
25. Carlisle A, Bowers A, Wayant C, Meyer C, Vassar M (2018) Financial conflicts of interest among authors of urology clinical practice guidelines. Eur Urol 74(3): 348–354
26. Stoll M, Hubenschmid L, Koch C, Lieb K (2020) Voluntary disclosures of payments from pharmaceutical companies to healthcare professionals in Germany: a descriptive study of disclosures in 2015 and 2016. BMJ Open 10(9):e037395
27. Mulinari S, Martinon L, Jachiet P-A, Ozieranski P (2021) Pharmaceutical industry self-regulation and non-transparency: country and company level analysis of payments to healthcare professionals in seven European countries. Health Policy 125(7):915–922
28. Mitchell AP, Basch EM, Dusetzina SB (2016) Financial relationships with industry among National Comprehensive Cancer Network guideline authors. JAMA Oncol 2(12):1628–1631
29. Marshall DC, Moy B, Jackson ME, Mackey TK, Hattangadi-Gluth JA (2016) Distribution and patterns of industry-related payments to oncologists in 2014. J Nat Cancer Inst 108(12):djw163
30. Saleh RR, Majeed H, Tibau A, Booth CM, Amir E (2019) Undisclosed financial conflicts of interest among authors of American Society of Clinical Oncology clinical practice guidelines. Cancer 125(22): 4069–4075
31. Ozaki A, Saito H, Onoue Y, Sawano T, Shimada Y, Somekawa Y et al (2019) Pharmaceutical payments to certified oncology specialists in Japan in 2016: a retrospective observational cross-sectional analysis. BMJ Open 9(9):e028805
32. Feng H, Wu P, Leger M (2016) Exploring the industry-dermatologist financial relationship: insight from the open payment data. JAMA Dermatol 152(12):1307–1313
33. Maruf M, Sidana A, Fleischman W, Brancato SJ, Purnell S, Agrawal S et al (2018) Financial relationships between urologists and industry: an analysis of open payments data. Urol Pract 5(3):180–186
34. Horn J, Checketts JX, Jawhar O, Vassar M (2018) Evaluation of industry relationships among authors of otolaryngology clinical practice guidelines. JAMA Otolaryngol Head Neck Surg 144(3):194–201
35. Pathak N, Fujiwara RJT, Mehra S (2018) Assessment of nonresearch industry payments to otolaryngologists in 2014 and 2015. Otolaryngol Head Neck Surg 158(6):1028–1034
36. Papanikolaou GN, Baltogianni MS, Contopoulos-Ioannidis DG, Haidich A-B, Giannakakis IA, Ioannidis JPA (2001) Reporting of conflicts of interest in guidelines of preventive and therapeutic interventions. BMC Med Res Methodol 1(1):3
37. Taylor R, Giles J (2005) Cash interests taint drug advice. Nature 437(7062):1070–1071
38. Neuman J, Korenstein D, Ross JS, Keyhani S (2011) Prevalence of financial conflicts of interest among panel members producing clinical practice guidelines in Canada and United States: cross sectional study. BMJ 343:d5621
39. World Health Organization (2014) WHO handbook for guideline development. WHO. https://apps.who.int/iris/handle/10665/145714. Accessed 4 May 2021
40. Schunemann HJ, Al-Ansary LA, Forland F, Kersten S, Komulainen J, Kopp IB et al (2015) Guidelines international network: principles for disclosure of interests and management of conflicts in guidelines. Ann Intern Med 163(7):548–553
41. Japanese Association of Medical Sciences (2017) Japanese Association of Medical Sciences COI management guidance on eligibility criteria for clinical practice guideline formulation. p 15. https://jams.med.or.jp/guideline/clinical_guidance_e.pdf. Accessed 4 May 2021
42. Senoo Y, Saito H, Ozaki A, Sawano T, Shimada Y, Yamamoto K et al (2021) Pharmaceutical company payments to authors of the Japanese guidelines for the management of hypertension. Medicine (Baltimore) 100(12):e24816
43. Murayama A, Ozaki A, Saito H, Sawano T, Shimada Y, Yamamoto K et al (2020) Pharmaceutical

company payments to dermatology clinical practice guideline authors in Japan. PLoS One 15(10): e0239610
44. Saito H, Tani Y, Ozaki A, Sawano T, Shimada Y, Yamamoto K et al (2019) Financial ties between authors of the clinical practice guidelines and pharmaceutical companies: an example from Japan. Clin Microbiol Infect 25(11):1304–1306
45. Nissen Steven E (2019) Can we trust the independence of cardiovascular guidelines? Circ Cardiovasc Qual Outcomes 12(12):e006225
46. Orlowski JP, Wateska L (1992) The effects of pharmaceutical firm enticements on physician prescribing patterns. There's no such thing as a free lunch. Chest 102(1):270–273
47. Nejstgaard CH, Bero L, Hrobjartsson A, Jorgensen AW, Jorgensen KJ, Le M et al (2020) Association between conflicts of interest and favourable recommendations in clinical guidelines, advisory committee reports, opinion pieces, and narrative reviews: systematic review. BMJ 371:m4234
48. Steinbrook R (2007) Guidance for guidelines. N Engl J Med 356(4):331–333
49. Eichacker PQ, Natanson C, Danner RL (2006) Surviving sepsis — practice guidelines, marketing campaigns, and Eli Lilly. N Engl J Med 355(16): 1640–1642
50. McKee S (2009) Eisai/Pfizer's Alzheimer's battle with NICE comes to an end Online: Pharma Times Online [updated July 2]. pharmatimes.com/news/eisaipfizers_alzheimers_battle_with_nice_comes_to_an_end_984556. Accessed 4 May 2021
51. Lenzer J (2013) Why we can't trust clinical guidelines. BMJ 346:f3830
52. Lenzer J (2011) French guidelines are withdrawn after court finds potential bias among authors. BMJ 342: d4007
53. Jefferson AA, Pearson SD (2017) Conflict of interest in seminal hepatitis C virus and cholesterol management guidelines. JAMA Intern Med 177(3):352–357
54. Irwig MS, Kyinn M, Shefa MC (2018) Financial conflicts of interest among authors of Endocrine Society clinical practice guidelines. J Clin Endocrinol Metab 103(12):4333–4338
55. Wayant C, Walters C, Zaaza Z, Gilstrap C, Combs T, Crow H et al (2020) Evaluation of financial conflicts of interest among physician-authors of American College of Rheumatology clinical practice guidelines. Arthritis Rheumatol 72(9):1427–1434
56. Dudum R, Sajja A, Amdur RL, Choi BG (2019) Analysis of American College of Cardiology/American Heart Association guideline author self-disclosure compared with open payments industry disclosure. Circ Cardiovasc Qual Outcomes 12(12):e005613
57. Norris SL, Holmer HK, Burda BU, Ogden LA, Fu R (2012) Conflict of interest policies for organizations producing a large number of clinical practice guidelines. PLoS One 7(5):e37413
58. Morciano C, Basevi V, Faralli C, Hilton Boon M, Tonon S, Taruscio D (2016) Policies on conflicts of interest in health care guideline development: a cross-sectional analysis. PLoS One 11(11):e0166485
59. DeJong C, Aguilar T, Tseng C-W, Lin GA, Boscardin WJ, Dudley RA (2016) Pharmaceutical industry–sponsored meals and physician prescribing patterns for medicare beneficiaries. JAMA Int Med 176(8): 1114–1122
60. International Committee of Medical Journal Editors (2019) Recommendations for the conduct, reporting, editing, and publication of scholarly work in medical journals [Updated December 2019]. icmje.org/icmje-recommendations.pdf. Accessed 4 May 2021
61. Booth CM, Detsky AS (2019) From the $80 hamburger to managing conflicts of interest with the pharmaceutical industry. BMJ 365:l1939
62. Miyata Y (2009) An internet survey of physicians' attitude towards gifts from drug companies. Med Educ 40(2):95–104
63. Yeh JS, Franklin JM, Avorn J, Landon J, Kesselheim AS (2016) Association of Industry Payments to physicians with the prescribing of brand-name statins in Massachusetts. JAMA Int Med 176(6):763–768
64. Schünemann HJ, Osborne M, Moss J, Manthous C, Wagner G, Sicilian L et al (2009) An official American Thoracic Society policy statement: managing conflict of interest in professional societies. Am J Respir Crit Care Med 180(6):564–580
65. Akl EA, El-Hachem P, Abou-Haidar H, Neumann I, Schunemann HJ, Guyatt GH (2014) Considering intellectual, in addition to financial, conflicts of interest proved important in a clinical practice guideline: a descriptive study. J Clin Epidemiol 67(11):1222–1228
66. Norris SL, Holmer HK, Ogden LA, Burda BU, Fu R (2013) Conflicts of interest among authors of clinical practice guidelines for glycemic control in type 2 diabetes mellitus. PLoS One 8(10):e75284
67. Abdoul H, Perrey C, Tubach F, Amiel P, Durand-Zaleski I, Alberti C (2012) Non-financial conflicts of interest in academic grant evaluation: a qualitative study of multiple stakeholders in France. PLoS One 7(4):e35247
68. Traversy G, Barnieh L, Akl EA, Allan GM, Brouwers M, Ganache I et al (2021) Managing conflicts of interest in the development of health guidelines. CMAJ 193(2):E49–E54
69. Spielmans GI, Parry PI (2010) From evidence-based medicine to marketing-based medicine: evidence from internal industry documents. J Bioethic Inq 7(1):13–29
70. Tibau A, Bedard PL, Srikanthan A, Ethier JL, Vera-Badillo FE, Templeton AJ et al (2015) Author financial conflicts of interest, industry funding, and clinical practice guidelines for anticancer drugs. J Clin Oncol 33(1):100–106
71. Desai AP, Chengappa M, Go RS, Poonacha TK (2020) Financial conflicts of interest among National Comprehensive Cancer Network clinical practice guideline panelists in 2019. Cancer 126(16):3742–3749

72. Campsall P, Colizza K, Straus S, Stelfox HT (2016) Financial relationships between organizations that produce clinical practice guidelines and the biomedical industry: a cross-sectional study. PLoS Med 13(5): e1002029
73. Sonis J, Chen OM (2020) Approval processes in evidence-based clinical practice guidelines sponsored by medical specialty societies. PLoS One 15(2): e0229004
74. Bero L (2017) Developing reliable dietary guidelines. BMJ 359:j4845
75. Hashimoto T, Murayama A, Mamada H, Saito H, Tanimoto T, Ozaki A (2021) Evaluation of financial conflicts of interest and drug statements in the coronavirus disease 2019 clinical practice guideline in Japan. Clin Microbiol Infect 28(3):460–462. https://doi.org/10.1016/j.cmi.2021.11.019
76. Keyhani S, Kim A, Mann M, Korenstein D (2011) A new independent authority is needed to issue National Health Care guidelines. Health Aff 30(2):256–265
77. Andreatos N, Zacharioudakis IM, Zervou FN, Muhammed M, Mylonakis E (2017) Discrepancy between financial disclosures of authors of clinical practice guidelines and reports by industry. Medicine (Baltimore) 96(2):e5711
78. Graham SS, Majdik ZP, Clark D, Kessler MM, Hooker TB (2020) Relationships among commercial practices and author conflicts of interest in biomedical publishing. PLoS One 15(7):e0236166
79. Shekelle PG (2018) Clinical practice guidelines: what's next? JAMA 320(8):757–758
80. Mendelson TB, Meltzer M, Campbell EG, Caplan AL, Kirkpatrick JN (2011) Conflicts of interest in cardiovascular clinical practice guidelines. Arch Intern Med 171(6):577–584
81. Combs TR, Scott J, Jorski A, Heavener T, Vassar M (2018) Evaluation of industry relationships among authors of clinical practice guidelines in gastroenterology. JAMA Intern Med 178(12):1711–1712
82. Napierala H, Schäfer L, Schott G, Schurig N, Lempert T (2018) Management of financial conflicts of interests in clinical practice guidelines in Germany: results from the public database GuidelineWatch. BMC Med Ethics 19(1):65
83. Combs T, Tritz D, Ivy H, von Borstel D, Horn J, Vassar M (2019) Financial conflicts of interest among authors of clinical practice guidelines for routine screening mammography. J Am Coll Radiol 16(11):1598–1603
84. Wang Y-B, Wang Q, Zhao M-J, Da-Lin HE, Wang Y-Y (2020) Conflict of interest and economic evidence in Chinese clinical practice guidelines pub-lished in 2018: a survey. Yixue Xinzhi Zazhi 30(2):139–58 (in Chinese)
85. Elder K, Turner KA, Cosgrove L, Lexchin J, Shnier A, Moore A et al (2020) Reporting of financial conflicts of interest by Canadian clinical practice guideline producers: a descriptive study. CMAJ 192(23):E617–EE25

How to Mitigate Unintentional Misconduct with Samples and Data in Biorepositories

46

Daniel Simeon-Dubach and Zisis Kozlakidis

Abstract

Unintentional misconduct with samples and data can have a huge impact on the quality of the research, on the reproducibility of the data and the trust in research in general and in researchers specifically. It is also a waste of human and financial resources and may threaten the sustainability of research teams and infrastructure. Bad management may cause unintentional misconduct. A sound quality management system could be a relevant tool to address these issues. Many guidelines, standards, and recommendations are available to prevent unintentional misconduct or at least to mitigate its effect. This will build trust in biomedical research and will ensure that the financial and work force investments are justifiable.

D. Simeon-Dubach (✉)
Medservice, Walchwil, Switzerland
e-mail: biobanking@medservice.ch

Z. Kozlakidis
International Agency for Research on Cancer, World Health Organization, Lyon CEDEX 08, France
e-mail: kozlakidisz@iarc.fr

46.1 Introduction

The goal of medical research is to generate and apply new knowledge for the benefit of individual patients and society as a whole [1]. Achieving this goal requires transparent scientific methodology, honesty in data collection and interpretation, and a realistic assessment of the implications of the results. Underlying all steps toward gaining new knowledge is the need to apply high ethical standards in research. Misconduct can be prevented if research ethics principles are understood and consistently applied.

Fraud and misconduct are the two terms that are often used interchangeably [2]. However, there is a significant difference between the two. The legal definition of fraud is "the knowing violation of the standard of good faith and fair dealing, as understood in the community, with deception or breach of trust, for money." [3]. Thus, fraud is an intentional deception for personal gain or to the detriment of another person, such as the intentional falsification and/or fabrication of research data and the misleading reporting of results. Misconduct does not necessarily involve money however it can also be intentional and in this case can also be prosecuted.

Several studies have found that more than 40% of researchers surveyed were aware of misconduct but did not report it [2]. In one study, one in six of the authors of clinical drug trials surveyed reported that they personally knew of falsification

J. Faintuch, S. Faintuch (eds.), *Integrity of Scientific Research*,
https://doi.org/10.1007/978-3-030-99680-2_46

in research that had occurred in the past 10 years [3].

46.2 Scientific and Social Implications

Fraud can cause a study to lose all credibility. These intentional violations of research ethics principles can have profound effects on the scientific community and lay people. It can also result in ineffective or harmful treatments becoming available or in patients being deprived of effective treatment. Therefore, such cases may also have criminal consequences.

In this chapter we have restricted ourselves to non-intentional misconduct, as an act of bad management during the course of biomedical research. It also includes failure to follow established protocols when that failure results in unreasonable risk or harm to people.

46.3 Existing Recommendations and Guidelines

Valuable information can be found in the EQUATOR Network (Enhancing the QUAlity and Transparency Of health Research) (equator-network.org). The EQUATOR network, initiated earlier this century, promotes transparent and accurate reporting and wider use of robust reporting guidelines. This is intended to address underreporting systematically and on a global scale.

46.4 Integrity in Publication

A reporting guideline is a simple, structured tool for health researchers to use when writing manuscripts with a number of examples already implemented [4–7]. The following four points need to be ensured: (1) Readers can understand the content. (2) Other researchers are able to independently reproduce the results. (3) For clinical data, a physician is sufficiently informed to make clinical decisions, and (4) The work has enough quality to be included in a systematic review.

A reporting guideline can be viewed as a checklist, flowchart, or structured text that guides authors in reporting a particular type of research and has been developed using an explicit methodology. Additionally, it can be updated as new requirements and/or technologies might emerge, to ensure its consistent applicability [8, 9].

46.5 Non-Reproducibility

Much has been written about the alarming number of preclinical studies that later turned out to be non-reproducible [10, 11]. Success rates for new development projects in phase II trials are very low, with insufficient efficacy being the most common reason for failure [12]. This points to the limitations of the predictability of disease models and also to the fact that the validity of the targets studied is often questionable.

Estimates of irreproducibility of preclinical studies range from 51% to 89% [13–17], with available series of reports (between 53 and 257 studies). One must admit that even in the "best" case, not every second published study would be reproducible [15–17]. Nevertheless this non-reproducibility leads to approximately US$ 28 billion per year being spent on non-reproducible preclinical research—in the United States alone [18].

46.6 Poor Stewardship

Cumulative errors in the categories of (1) biological reagents and reference materials, (2) study design, (3) data analysis and reporting, and (4) laboratory protocols were the main causes of irreproducibility. Biological reagents and reference materials were the source of error in about one third of the cases, study design and data analysis and reporting, respectively, in about one quarter each, and laboratory protocols in about 10% [18].

The level of adherence to the guidelines and their impact on improving reproducibility has been disappointing [19]. Given the size, scope, and complexity of the challenge of reproducibility in medical research, there is no panacea to address the problem. However, one issue that has had a demonstrable impact on similar challenges in other fields is the expanded development and adoption of standards and best practices [20, 21].

46.7 Quality Issues for Biological Samples and Reagents

A survey among recipients of U.S. National Cancer Institute/NCI funds found that about half of the respondents reported difficulties in obtaining biospecimens of adequate quality [22]. Low-quality biospecimens led more than half of the respondents to question their own results and four of five researchers to limit the scope of their research [22]. Thus, there is no question that the quality of samples and data is a major problem.

Biological reagents and reference materials are the main source of error in about one third of the cases [18]. On many occasions, they are provided by biobanks, as in the case of the European Virus Archive, providing well-characterized SARS-CoV-2 reference material throughout the COVID-19 pandemic [23]. It is more important that exact statements about the quality of samples can be made rather than inform that all samples and data are of "maximum quality". For certain studies, samples and data of good but clearly defined quality are sufficient to provide an answer to the scientific questions.

46.8 Institutional Biobanks

They see themselves as a research infrastructure [24–26], usually providing samples and data of clearly defined quality. Quality challenges tend to be more difficult in the case of so-called "corridor biobanks". These are collections of samples and data initiated and operated by a researcher or research group, typically represented by a short line of freezers along a hospital corridor.

In some circumstances these collections got started because there were no other sources for the samples and data needed [27, 28]. However in very many cases samples and data would be encountered in established biobanks as well. Another excuse for them is immediate access to samples and data, without waiting for acquisition of fresh materials.

46.8.1 Lack of Careful Planning

Sadly researchers often do not know which biobanks/sample collections are active in their own institution and organization [29, 30], thus reducing the push action in terms of samples utilization. This is also a call for existing biobanks to do more to let researchers know that they exist and what products and services they offer. In doing so, these biobanks should start affirming their visibility and impact first within their own institution(s) and subsequently, once operations have a long-term support, looking further away [31].

From the perspective of the researchers there is often no particular incentive or requirement imposed by their institutions or funding agencies to even investigate if existing samples are qualitatively suitable, thus fulfilling their research needs.

46.8.2 Low Level of Professionalism

Often, biobanking staff are trained "on the job," in particularly in corridor biobanks. This may result in lack of knowledge and experience regarding the regulatory, ethical, and social requirements of modern biobanking [32–34]. Ultimately all of the challenges listed above contribute towards unintentional misconduct in biomedical research with samples and data.

46.8.3 Quality Management System

There is a clearly demonstrated need for established quality processes in biobanks.

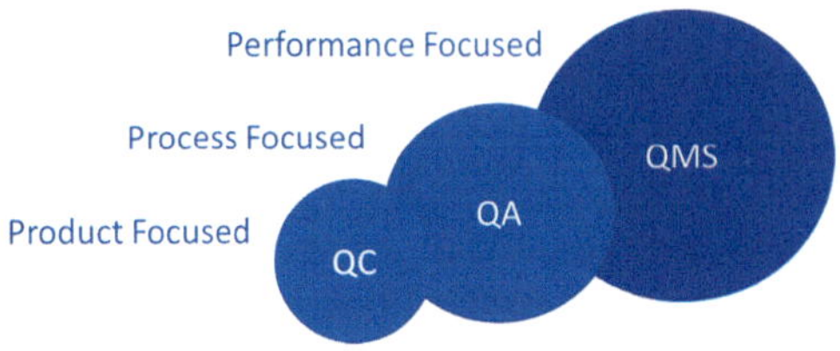

Fig. 46.1 A Quality Management System (QMS) should include Quality Assurance (QA) and Quality Control (QC) programs that cover all repository operations

Therefore, every biobank should have a quality management system (QMS). The goal is a systematic quality management. The QMS should include quality assurance (QA) and quality control (QC) programs that cover all repository operations. QMSs ensure that system, process, and product quality in an organization is reviewed and improved. QA focuses on processes, QC on products, and QMS on performance (Fig. 46.1).

In terms of unintentional failures, QA, QC and QMS are equally relevant. It must be possible to ensure that sufficient information is available to substantiate the quality of samples and data. Some data must necessarily be documented currently, while others can be collected retrospectively. Examples of experimental data that can hardly be collected retrospectively are warm/cold ischemia time [35], and thaw/freeze cycles.

46.8.4 Biobank Heterogeneity

Both in terms of the biological samples and data they collect, biobanks are as different as the multiple domains of science, yet QMS is recommended for all of them. Customization is the key word. Tissues from the central nervous system (CNS) are relatively inert to the effects of warm ischemia [36, 37]. Therefore, for a CNS biobank, these data are not absolutely central for quality assessment. This is then accompanied by a lower risk of unintentional failure based on unknown or unclear quality. This process is akin to a risk assessment approach, where the risk in focus is the unintentional misconduct.

46.8.5 Risk Analysis

The first step is the identification of the individual risks [38]. Each process specific to the biobank concerned must be defined, from identification (anonymization) of patients/donors, sample collection and data collection, processing and storage, to provision to researchers.

The second step is to evaluate the risk in terms of its relevance to quality. The probability of occurrence and the possible amount of the risk must be considered. Once the risks have been identified and their significance assessed, the third and last step is that of risk management. The aim of this latter step is to deal with the risks, to identify measures to limit or avoid them, and finally to monitor their success.

Risk analysis is a time-consuming activity, especially for research infrastructures that are within resource-restricted settings. One could theoretically fully document all possible steps all of the time, but doing so would not always be necessary and certainly not economical. Thus, a QMS should ensure that a biobank can provide samples and data that are fit for each purpose, including those that do not have to meet the maximum quality requirements. The primary focus must be on what the researchers require from the biobank. Where necessary, the quality of systems, processes, and products in a biobank should be reviewed and improved. It is highly likely that there will then be a continuous increase in quality with respect to all aspects over time.

46.9 BRISQ, SPREC and Other Toolkits (EQUATOR Database)

BRISQ (Biospecimen Reporting for Improved Study Quality) is a set of recommendations that should apply to any study in which human samples are used [39, 40]. Biological samples are subject to a number of several factors during collection, processing, and storage that can significantly alter their molecular composition and consistency. These pre-analytical factors in turn affect the experimental results and the reproducibility of the scientific results [41, 42].

The purpose of BRISQ is to provide readers of scientific publications with more consistent and standardized information to better evaluate, interpret, compare, and reproduce experimental results. Several journals have implemented BRISQ as an official standard for article submission [43–46]. BRISQ is not the first set of recommendations within such context. Several others have been compiled in the EQUATOR network. Annotating biospecimens to comply with BRISQ will require increased effort, as will other changes and improvements to biobanking, as additional actions/elements are incorporated into routine. On the other hand, we cannot afford to invest in research infrastructures such as biobanks and be surprised by irreproducible or unpublishable outcomes.

46.9.1 Pitfalls and Challenges

Despite the increasing recognition of the potential impact of variability in biobank processes on the interpretation of biospecimen analyses, adequate documentation of important pre-analytical factors is not common in research publications. Actually the development of biospecimen reporting guidelines cannot be separated from the realities of the research laboratory, the funding environment, or the article review process.

What can be expected from the implementation of BRISQ, based on the experience with the implementation of other, similar guidelines? One analysis found that even within journals recommending the use of BRISQ, manuscript-level acceptance and reporting of relevant biospecimen information is not widespread or consistent [40].

46.10 The SPREC "Barcode"

The Sample PRE-analytical Code (SPREC) [47–50] addresses a well-annotated process chain history of pre-analytical variables, from biomaterial collection to sample storage. The SPREC allows quick and easy comparability of the quality of biological samples and improves the interoperability of collaborating institutions by establishing a standard.

The SPREC is a "sample barcode" or "preanalytical barcode" that provides details on preanalytical sample processing in a standardized format. Specifically, it is a seven-element long code, with each element corresponding to a critical preanalytical variable.

These guidelines (BRISQ, SPREC) have been implemented very recently and in a limited form due to their voluntary nature [51]. As such, the following gaps remain: (a) difficulty to compare across research infrastructures and within projects—i.e., we are missing the common denominator; (b) difficulty to establish the relative contribution of different quality elements into unintentional misconduct; and (c) uncertainty by biobanks as well as funding agencies to determine the best course of action.

46.10.1 Best Practices and Standards

Best practices are used to maintain quality as a supplement to mandated or regulated standards. A best practice is a generally accepted technique, process, or method that leads to better results than those achieved by other means. They have been developed through experience and research and are proven to lead to a desired and evidence-based outcome [52]. Repository management and other staff should decide which practices best fit their circumstances to adopt and include them in their standard operation procedures (SOPs).

Currently, biobanks are entering a new management phase, characterized by professionalism in all areas of biobanking, driven by and whenever possible based on newly developed standards and best practices [53]. It is anticipated that all of these activities will eventually have a significant impact on the future of modern biobanking, as has been observed in other areas of clinical research [54].

46.10.2 Quality and Efficiency for Less Misconduct

Sample quality was already intensively discussed [55]. In later iterations sustainability of biobanks has been focused. The financial, social, and operational dimensions of sustainability were discussed intensively [56–59] in the last decade. It is hoped that as the professionalism level increases, the instances of unintentional misconduct will be identified much easier, addressed and hence will decrease overall.

The fourth edition (2018) of the ISBER Best Practices (International Society for Biological and Environmental Repositories) is a real success story and will continue to influence biobank collection and discovery even in the future [60] (isber.org/bestpractices). They are the first step in preparing biobanks for the application of international standards, such as the International Standardization Organization/ISO.

46.10.3 ISO Standard 20,387 Biotechnology: Biobanking

As analyzed in General requirements for biobanking [61], this respected standard can already be applied for. The fifth edition of ISBER Best Practices is already in the hands of a task force. The wide distribution and adoption of the ISBER best practices globally is expected to raise the professional standards and conduct of biobanks and thus reduce unintentional misconduct; however the latter can only be measurable as a mid- to long-term impact.

46.11 Outlook

Solving the issue of non-individual misconduct, within the context of biobanking, requires engagement of all parties at the individual, collective/community, and institutional levels. Each individual party must also recognize the issue as a genuine concern. Only in this way is it possible for the appropriate measures to be initiated. A number of basic guidelines, standards, and recommendations have been developed to prevent unintentional misconduct or at least to mitigate its effect. This is the only way to ensure that the investments continuously generate an optimal output.

Disclaimer Where authors are identified as personnel of the International Agency for Research on Cancer/WHO, the authors alone are responsible for the views expressed in this article and they do not necessarily represent the decisions, policy or views of the International Agency for Research on Cancer/WHO.

References

1. Biros MH, Fish SS, Taggart P (1999) Research fundamentals VI: misconduct in biomedical research. Acad Emerg Med 6(8):840–848
2. Gupta A (2013) Fraud and misconduct in clinical research: a concern. Perspect Clin Res 4(2):144–147
3. Sheehan JG (2007) Fraud, conflict of interest, and other enforcement issues in clinical research. Cleve Clin J Med 74(Suppl 2):S63–S67
4. De Geest S, Zullig LL, Dunbar-Jacob J, Helmy R, Hughes DA, Wilson IB et al (2018) ESPACOMP medication adherence reporting guideline (EMERGE). Ann Intern Med 169(1):30–35
5. Campbell M, Katikireddi SV, Hoffmann T, Armstrong R, Waters E, Craig P (2018) TIDieR-PHP: a reporting guideline for population health and policy interventions. BMJ 361:k1079
6. Campbell M, McKenzie JE, Sowden A, Katikireddi SV, Brennan SE, Ellis S et al (2020) Synthesis without meta-analysis (SWiM) in systematic reviews: reporting guideline. BMJ 368:l6890
7. Tong A, Synnot A, Crowe S, Hill S, Matus A, Scholes-Robertson N et al (2019) Reporting guideline for priority setting of health research (REPRISE). BMC Med Res Methodol 19(1):243
8. Agha R, Abdall-Razak A, Crossley E, Dowlut N, Iosifidis C, Mathew G (2019) STROCSS 2019 guideline: strengthening the reporting of cohort studies in surgery. Int J Surg 72:156–165
9. Page MJ, McKenzie JE, Bossuyt PM, Boutron I, Hoffmann TC, Mulrow CD et al (2021) Updating guidance for reporting systematic reviews: development of the PRISMA 2020 statement. J Clin Epidemiol 134:103–112
10. Collins FS, Tabak LA (2014) Policy: NIH plans to enhance reproducibility. Nature 505(7485):612–613
11. Begley CG, Ioannidis JP (2015) Reproducibility in science: improving the standard for basic and preclinical research. Circ Res 116(1):116–126
12. Arrowsmith J (2011) Trial watch: phase II failures: 2008-2010. Nat Rev Drug Discov 10:328–329

13. Begley CG, Ellis LM (2012) Drug development: raise standards for preclinical cancer research. Nature 483(7391):531–533
14. Prinz F, Schlange T, Asadullah K (2011) Believe it or not: how much can we rely on published data on potential drug targets? Nat Rev Drug Discov 10(9):712
15. Hartshorne JK, Schachner A (2012) Tracking replicability as a method of post-publication open evaluation. Front Comput Neurosci 6:8
16. Vasilevsky NA, Brush MH, Paddock H, Ponting L, Tripathy SJ, Larocca GM et al (2013) On the reproducibility of science: unique identification of research resources in the biomedical literature. Peer J 1:e148
17. Glasziou P, Meats E, Heneghan C, Shepperd S (2008) What is missing from descriptions of treatment in trials and reviews? BMJ 336(7659):1472–1474
18. Freedman LP, Cockburn IM, Simcoe TS (2015) The economics of reproducibility in preclinical research. PLoS Biol 13(6):e1002165
19. Baker D, Lidster K, Sottomayor A, Amor S (2014) Two years later: journals are not yet enforcing the ARRIVE guidelines on reporting standards for pre-clinical animal studies. PLoS Biol 12(1):e1001756
20. Pohle A, Blind K, Neustroev D (2018) The impact of international management standards on academic research. Swiss Med Wkly 10(12):4656
21. Buts C, Van Droogenbroeck E, ooms M, Willems K. (2020) The economic impact of standards in Belgium. Int J Stand Res 18(1):44–64
22. Massett HA, Atkinson NL, Weber D, Myles R, Ryan C, Grady M et al (2011) Assessing the need for a standardized cancer HUman Biobank (caHUB): findings from a national survey with cancer researchers. J Natl Cancer Inst Monogr 42:8–15
23. Coutard B, Romette JL, Miyauchi K, Charrel R, Prat CMA (2020) The importance of biobanking for response to pandemics caused by emerging viruses: the European virus archive as an observatory of the global response to the zika virus and COVID-19 crisis. Biopreserv Biobank 18(6):561–569
24. Parodi B (2019) Biobanking and biomolecular resource research infrastructure: the Italian node. Biomed Sci Eng 3(s2):100
25. Glasa J, Kollar D, Cvapek P, Glasova H, Antosova M, Pella D et al (2020) Establishing a national biobank. Biobanking infrastructure initiative in Slovakia - public policy, legal and ethical issues. Health Pol Tech 9(1):53–55
26. Permuth JB, Dezsi KB, Vyas S, Ali KN, Basinski TL, Utuama OA et al (2021) The Florida pancreas collaborative next-generation biobank: infrastructure to reduce disparities and improve survival for a diverse cohort of patients with pancreatic cancer. Cancers (Basel) 13(4):809
27. Naveen R, Anuja AK, Rai MK, Agarwal V, Gupta L (2020) Development of the myocite biobank: cost-efficient model of public sector investigator-driven biobank for idiopathic inflammatory myositis. Indian J Rheumatol 15(6):194–199
28. Blutke A, Renner S, Flenkenthaler F, Backman M, Haesner S, Kemter E et al (2017) The Munich MIDY pig biobank - a unique resource for studying organ crosstalk in diabetes. Mol Metab 6(8):931–940
29. Linsen L, Vanhees K, Vanoppen E, Ulenaers K, Driessens S, Penders J et al (2019) Raising to the challenge: building a federated biobank to accelerate translational research-the university biobank Limburg. Front Med (Lausanne) 6:224
30. Mendy M, Caboux E, Wild CP (2019) Centralization of the IARC biobank: combining multiple sample collections into a common platform. Biopreserv Biobank 17(5):433–443
31. Makkonen E (2019) Investigation of the needs for national and local biobank services and improvement of biobank awareness. Master thesis, Tampere University, Finland. https://trepo.tuni.fi/handle/10024/117693. Accessed 4 May 2021
32. Hartung ML, Baber R, Herpel E, Specht C, Brucker DP, Schoneberg A et al (2021) Harmonization of biobank education for biobank technicians: identification of learning objectives. Biotech 10(2):7
33. Rush A, Christiansen JH, Farrell JP, Goode SM, Scott RJ, Spring KJ et al (2015) Biobank classification in an Australian setting. Biopreserv Biobank 13(3):212–218
34. Williams RR, Gupta D, Yong WH (2019) Orientation and training of new biobank personnel. Methods Mol Biol 1897:51–63
35. Guo D, Wang A, Xie T, Zhang S, Cao D, Sun J (2020) Effects of ex vivo ischemia time and delayed processing on quality of specimens in tissue biobank. Mol Med Rep 22(5):4278–4288
36. Jensen PSH, Johansen M, Bak LK, Jensen LJ, Kjaer C. (2021) Yield and integrity of RNA from brain samples are largely unaffected by pre-analytical procedures. Neurochem Res 46(3):447–454
37. Mock A, Rapp C, Warta R, Abdollahi A, Jäger D, Sakowitz O et al (2019) Impact of post-surgical freezing delay on brain tumor metabolomics. Metabolomics 15(5):78
38. Janssens PM (2014) Practical, transparent prospective risk analysis for the clinical laboratory. Ann Clin Biochem 51(Pt 6):695–704
39. Moore HM, Kelly AB, Jewell SD, McShane LM, Clark DP, Greenspan R et al (2011) Biospecimen reporting for improved study quality (BRISQ). Cancer Cytopathol 119(2):92–101
40. Meredith AJ, Simeon-Dubach D, Matzke LA, Cheah S, Watson PH (2019) Biospecimen data reporting in the research literature. Biopreserv Biobank 17(4):326–333
41. Ellervik C, Vaught J (2015) Preanalytical variables affecting the integrity of human biospecimens in biobanking. Clin Chem 61(7):914–934
42. Mateos J, Carneiro I, Corrales F, Elortza F, Paradela A, Del Pino MS et al (2017) Multicentric study of the effect of pre-analytical variables in the quality of plasma samples stored in biobanks using different

complementary proteomic methods. J Proteome 150: 109–120
43. Simeon-Dubach D, Moore HM (2014) BIO comes into the cold to adopt BRISQ. Biopreserv Biobank 12(4): 223–224
44. Simeon-Dubach D, Burt AD, Hall PA (2012) Quality really matters: the need to improve specimen quality in biomedical research. J Pathol 228:431–433
45. Simeon-Dubach D, Burt AD, Hall PA (2012) Quality really matters: the need to improve specimen quality in biomedical research. Histopathology 61(6): 1003–1005
46. Nature Editorial (2013) Announcement: reducing our irreproducibility. Nature 496:398
47. Betsou F, Lehmann S, Ashton G, Barnes M, Benson EE, Coppola D et al (2010) Standard preanalytical coding for biospecimens: defining the sample PREanalytical code. Cancer Epidemiol Biomark Prev 19(4):1004–1011
48. Betsou F, Bilbao R, Case J, Chuaqui R, Clements JA, De SY et al (2018) Standard PREanalytical code version 3.0. Biopreserv Biobank 16(1):9–12
49. Benson EE, Betsou F, Amaral R, Santos LM, Harding K (2011) Standard PREanalytical codes: a new paradigm for environmental biobanking sectors explored in algal culture collections. Biopreserv Biobank 9(4): 399–410
50. Lehmann S, Guadagni F, Moore H, Ashton G, Barnes M, Benson E et al (2012) Standard preanalytical coding for biospecimens: review and implementation of the sample PREanalytical code (SPREC). Biopreserv Biobank 10(4):366–374
51. T'Joen V, Vaneeckhaute L, Priem S, Van WS, Bekaert S, Berneel E et al (2019) Rationalized development of a campus-wide cell line dataset for implementation in the biobank LIMS system at Bioresource Center Ghent. Front Med (Lausanne) 6:137
52. Hallmans G, Vaught JB (2011) Best practices for establishing a biobank. Methods Mol Biol 675:241–260
53. Simeon-Dubach D, Kozlakidis Z (2018) New standards and updated best practices will give modern biobanking a boost in professionalism. Biopreserv Biobank 16(1):1–2
54. Ohmann C, Canham S, Demotes J, ChÃªne G, Lauritsen J, Martins H et al (2017) Raising standards in clinical research - the impact of the ECRIN data Centre certification programme, 2011-2016. Contemp Clin Trials Commun 5:153–159
55. Engel KB, Vaught J, Moore HM (2014) National Cancer Institute biospecimen evidence-based practices: a novel approach to pre-analytical standardization. Biopreserv Biobank 12(2):148–150
56. Simeon-Dubach D, Watson P (2014) Biobanking 3.0: evidence based and customer focused biobanking. Clin Biochem 47(4–5):300–308
57. Henderson MK, Simeon-Dubach D, Albert M (2015) Finding the path to biobank sustainability through sound business planning. Biopreserv Biobank 13(6): 385–386
58. Henderson MK, Goldring K, Simeon-Dubach D (2017) Achieving and maintaining sustainability in biobanking through business planning, marketing, and access. Biopreserv Biobank 15(1):1–2
59. Henderson MK, Goldring K, Simeon-Dubach D (2019) Advancing professionalization of biobank business operations: performance and utilization. Biopreserv Biobank 17(3):213–218
60. Kozlakidis Z, Seiler C, Simeon-Dubach D (2018) ISBER best practices fourth edition: a success story. Biopreserv Biobank 16(3):242–243
61. Furuta K, Allocca CM, Schacter B, Bledsoe MJ, Ramirez NC (2018) Standardization and innovation in paving a path to a better future: an update of activities in ISO/TC276/WG2 biobanks and bioresources. Biopreserv Biobank 16(1):23–27

Combatting Corruption in the Pharmaceutical Sector

47

Marc-Andre Gagnon

Abstract

The pharmaceutical sector is a complex and highly regulated sector in which minor technical or regulatory decisions about products' safety, effectiveness, intellectual property or reimbursement can add or reduce billions in sales for drug companies. Because a lot of money is at stake, many corporate strategies and corrupt practices are at play to influence regulatory decisions or physicians' prescribing habits. This chapter identifies types of corruption in the pharmaceutical sector, the main fields in which these types of corruption are applied, and explores what should be the main priorities to combat corruption. Well-being of patients and the improvement of health outcomes should remain the ultimate goals in the ways that the pharmaceutical sector is structured by public policy.

47.1 Introduction

The pharmaceutical sector is a complex and highly regulated sector in which minor technical or regulatory decisions about products' safety, effectiveness, intellectual property, or reimbursement conditions can add or reduce billions in sales for drug companies. Because a lot of money can be at stake, many corporate strategies are at play to influence regulatory decisions or physicians' prescribing habits. While the sector is known for different corporate scandals due to fraud and misconduct, most of unethical corporate strategies to increase sales volume or prices can be completely legal. In this case, corruption must be understood beyond illegal activities and include all legal strategies that increase earnings while undermining the purposes of medical practice and public health.

An illustration of these strategies and their impact is the opioid crisis that is observed especially in the United States and Canada, with a death toll of around 4000 Canadians and around 50,000 Americans every year [1, 2]. A series of out-of-court settlements and court decisions have revealed that drug companies such as Purdue Pharma, Abbott, and Johnson and Johnson prevented the disclosure of adverse drug reactions related to their opioid products, organized "false, misleading and dangerous marketing campaigns," and have engaged with "front groups" of patients and medical specialists who insisted the drugs were effective for daily pain and minimized the risk of addiction [3]. In Canada, Purdue Pharma sponsored medical education that promoted expanded use of their product over a seven-year period at the University of

M.-A. Gagnon (✉)
School of Public Policy and Administration, Carleton University, Ottawa, ON, Canada
e-mail: Marcandregagnon@cunet.carleton.ca

J. Faintuch, S. Faintuch (eds.), *Integrity of Scientific Research*,
https://doi.org/10.1007/978-3-030-99680-2_47

Toronto, where "key opinion leaders" on pain management were paid by the drug company and material distributed to students was produced by the drug company [4].

Purdue Pharma directly paid more than $2 million to healthcare professionals in Canada in 2017 to influence their prescribing habits [5]. Similar strategies have been observed for many non-opioid classes of drugs such as antipsychotics, antidepressants, anti-hypertensive drugs, cholesterol reducers, antidiabetic drugs or anti-cancer drugs [6–9]. Another example is the campaign of misinformation to hide away adverse effects of the blockbuster drug Vioxx (a non-steroidal anti-inflammatory drug withdrawn in 2004), which led to the massive prescribing of the drug by physicians. It is estimated that the use of this product instead of existing safer alternatives killed between 33,000 and 45,000 people in the United States alone [10, 11].

47.2 Types of Corruption

A standard definition relates to the principal vs. agent problem at the root of many types of economic and political corruption. Transparency International [12] defines corruption as "the abuse of an entrusted power for private gain." As Rose-Ackerman and Palifka [13] note, the key term is "entrusted power," which refers to the tasks one (the agent) is expected to perform according to certain rules set by another (the principal) to achieve specific purposes. In this case, corruption is understood as the agent "breaking the rules" for private gains, undermining the capacity to achieve set purposes. The examples of corruption according to such definition are bribery, influence peddling, extortion, fraud or misconduct. These practices are often illegal and can lead to significant financial penalties or jail sentences.

In the pharmaceutical sector, these types of behavior do exist in different ways [14]. In the United States, financial penalties due to corrupt practices normally relate to illegal promotion, defrauding health programs, poor manufacturing practices, kickbacks or concealment of data [6]. In many countries procurement can also be a crucial point for such corruption, for example by buying drugs at higher prices than necessary, by the selling of counterfeit drugs, or by favoring some suppliers over others because of gifts or kickbacks [15]. Because of the lack of price transparency on the international drug market, and because of the systematic use of confidential rebates between buyers and sellers at the international level [16], corruption in drug procurement can be difficult to assess.

47.2.1 The Tip of the Iceberg

Standard corruption is often considered to be simply the tip of the iceberg. What distinguishes the pharmaceutical sector is the importance of different sets of perfectly legal practices and strategies and influences that can increase earnings for some to the detriment of public health. An important literature exists analyzing the regulatory capture of public institutions by private interests, which is a form of legal corruption. Regulatory capture can be defined as the "result or process by which regulation, in law or application, is consistently or repeatedly directed away from the public interest and toward the interests of the regulated industry, by the intent and action of the industry itself" [17].

Beyond the influence over regulatory institutions, scholars like Sismondo [9] and Gagnon [18] uses the notion of "ghost management" to refer to broader systematic strategies to shape science, media, culture and public narratives, to influence experts and policymakers, and to influence regulations. Often, ghost management entails not only the production of knowledge, but also the selective production of ignorance or uncertainty through the dissemination of false, misleading, or contradictory information.

Other scholars have developed the concept of institutional corruption to discuss these realities [19–21]. Lessig [19] offers the following definition:

> Institutional corruption is manifest when there is a systemic and strategic influence which is legal, or even currently ethical, that undermines the institution's effectiveness by diverting it from its purpose or weakening its ability to achieve its purpose, including, to the extent relevant to its purpose, weakening either the public's trust in that institution or the institution's inherent trustworthiness.

This definition emphasizes a "political economy influence" at play in the pharmaceutical sector, weakening public institutions and public health in ways that serve private interests. In their own way, the notions of regulatory capture and of ghost management also emphasize such political economy of influence shaping institutions and knowledge to the benefit of private interests. Each approach offers some depth into different dimensions of how commercial interests can shape knowledge and practices in specific industries.

47.3 Corporate Influence and Unethical Practices in the Pharmaceutical Sector

The extent of institutional corruption and unethical practices in the pharmaceutical sector has been well documented [7, 9, 22–25]. While one could find a political economy of influence at work in every interstice of pharmaceutical and medical activities, one can identify five main fields: (1) research and clinical trials; (2) dissemination of medical knowledge; (3) the interaction with the medical profession; (4) the interaction with regulatory agencies and authorities; and (5) the interaction with civil society.

47.3.1 Research and Clinical Trials

The basic and translational research for potential new drugs is normally done in academic settings through public funding. Once a promising molecule is found (a "silver bullet"), a start-up company is created to develop the product and begin clinical trials. Most start-ups exist only to be acquired by larger drug companies that will finish developing the product (through large phase III clinical trials) and market it worldwide. The global pharmaceutical sector is dominated by a handful of drug companies. According to the website Value.today, in January 2022, 19 of the global 200 largest corporations in terms of market value were drug companies, which can serve as a good proxy for what is often called "Big Pharma."

47.3.1.1 Contract Research Organizations and Commercial Sponsorship

Most phase III clinical trials are outsourced to contract research organizations (CROs), which capacity to obtain contracts often depends on their ability in getting the results preferred by their contractor. Because the results of clinical trials determine whether or how drugs will be approved, used and reimbursed, there is strong commercial pressure to make sure that results are favorable, which occasionally leads to biases overstating the benefits of the treatment while hiding away potential harm [26]. Commercial sponsorship of clinical trials has been suspected to create systematic biases in the results [27–30]. They also could shape the research agenda in the health sector in favor of products maximizing commercial gains instead of health outcomes [31, 32].

47.3.1.2 Unethical Behavior and Non-accountability

Fisher [33] analyzed the exploitation of marginalized and racialized populations in clinical trials bearing little resemblance to real word conditions. In the case of India, Rajan [34] relates the case of two clinical trials on human papilloma virus (HPV) vaccines on tribal girls in Andhra Pradesh in 2009. The HPV vaccines were administered through "vaccination camps" held at schools and hostels, and consent for vaccination was sometimes given by the teacher or by the hostel warden while parents were not told about the vaccinations. The consent forms stated that the vaccine would be administered to adolescent girls but many of the girls who received the vaccine were prepubescent.

47.3.2 Dissemination of Medical Knowledge

In most circumstances, clinical trial results are not directly available to physicians. The prescribing physician obtains relevant information about drug product from the medical literature, which is something slightly different. A new drug can gain financial success only if it is possible to convince prescribers about the products' significant benefits and low risks. Ghostwriting has become a usual strategy to influence the medical literature [35, 36]. The notion of "ghost management" was developed to show the extent of the use of ghostwriting and refers to a whole system of management behind closed doors used to influence scientific results in favor of corporate interests [9, 24, 37]. Sismondo [9] considers that ghost management must be understood through carefully thought-out publication planning strategies that drug companies develop in collaboration with medical communications companies to shape narratives about their products in a way that will maximize sales.

47.3.2.1 A Plethora of Encouraging Literature

Internal documents from Pfizer revealed that, between 1998 and 2000, the company directly initiated the writing of at least 85 scientific articles on the antidepressant drug *sertraline* (Zoloft). During this period, the entire scientific literature on this active substance consisted of only 211 articles [37]. In this way, Pfizer produced a raft of articles showing the drug in a positive light, lessening the impact of the critical studies. In the same way, Wyeth (now owned by Pfizer) generated about 50 articles in favour of hormone replacement therapy [38]. Merck developed a ghostwriting campaign to promote *rofecoxib* (Vioxx): 96 articles were published, some of which omitted to mention patient deaths in the drug's clinical trials [39].

Gagnon [18] explains that corporate influence over medical literature is not only about inflating the number of favorable scientific publications. It is also about suppressing scientific results that could harm sales and neutralizing independent academics and whistle-blowers. Pharmaceutical companies consider private-sector clinical research outcomes private, confidential results as part of their intellectual property. They assume the right not to publish certain results, in the name of trade secrecy. Even if it is normally mandatory to register all clinical trials and report all results, this mandate is not being enforced and drug companies can therefore select what data they want to see published [40] and the reporting of results remain incomplete [41].

47.3.2.2 Sweep the Dirt Under the Carpet

Major pharmaceutical companies have systematically failed to publish unfavourable studies on the "new generation" of antidepressants, known as selective serotonin reuptake inhibitors (SSRIs). Of the 74 clinical trials that were conducted on these antidepressants, 38 produced positive results, while the other 36 showed the drug to have questionable or no efficacy. However, while 94% of the positive studies were published, only 8% of the unfavourable studies were published as negative results, and 15% of the negative studies were published in terms that suggested that the results were positive [42]. Doctors reading the scientific literature got a biased view of the "risks and benefits" of SSRIs, which explains why they so readily prescribed these antidepressants to their patients.

47.3.2.3 Obstacle Clearance Policies

A strategy reportedly employed by certain drug companies is to intimidate and neutralise independent researchers who produce studies that show the product in an unfavourable light. In Merck's internal e-mails, which came out during lawsuits over the harm caused by its drug *rofecoxib* (Vioxx), a hit list of "rogue" researchers emerged who had criticised Vioxx and who had to be "discredited" and "neutralized." This intimidation was the result of the work of an entire team that monitored everything that was said about the product [43]. Similarly, in the case of the antidiabetic drug *rosiglitazone* (Avandia), which was withdrawn from the market in 2010 for safety reasons, a report by the US Senate explained that the main strategy of GlaxoSmithKline executives when confronted with the publication of negative clinical results, was to downplay the importance of these results and to intimidate independent researchers [35].

47.3.3 Interaction with the Medical Profession

In the United States, while the pharmaceutical industry spent $24 billion in research and development in 2004, it spent $58 billion in promotional campaigns [44], of which $54 billion was spent targeting healthcare professionals ($43 billion for physicians). It represents average promotional spending of $61,000 per physician annually to influence their prescribing habits. Scientific literature confirms that when physicians receive their information directly from the pharmaceutical industry, prescribing habits—as measured by cost, frequency, and appropriateness—tend to deteriorate [45].

47.3.3.1 Generous Disbursements

The Centers for Medicare and Medicaid Services/CMS Open Payment Data shows that, in the United States, drug manufacturers paid $9.35 billion to 627,000 physicians every year (directly or through an institutional affiliation) which represents a yearly average of $15,000 per physician. The use of "key opinion leaders" is also a central way in which drug companies influence prescribing habits. These paid physicians hired to talk to other physicians at educational meetings are selected based not on the appropriateness of their presentations, but on their impact on sales [46]. Finally, the influence over clinical guidelines and diagnostics textbooks, for example when authors have conflicts of interests (COIs) with specific companies, can also lead to inappropriate practices for commercial gains [21, 47–51].

47.3.4 Interaction with Regulatory Agencies and Authorities

47.3.4.1 Regulatory Capture

Legislators—yielding to pressure—have on occasion enacted laws or regulations that benefit private interests to the detriment of public interest, or weaken the government's ability to regulate and advance public interest [52]. In the United States Medicare Part D, providing prescription drug coverage to seniors, can certainly be considered as a poster child for regulatory capture. The act that established Medicare Part D specifically forbade the government from negotiating prices [53] even if the price of patented drugs in the United States is, on average, more than four times the median price of OECD countries [54]. The fact that such legislated impediment could have been instituted to reduce market competition and allow the artificial increase of prices can be explained through the idea of regulatory capture, in which the regulation is consistently directed away from the public interest in favor of private interests. While the dynamics of regulatory capture are mostly observed at a national level, a growing literature accounts on how private interests manage to shape public law at the international level, especially in the case of international trade agreements [55–57].

47.3.4.2 Heavyweight Trade Associations

According to the Center for Responsive Politics based on data from the Senate Office of Public Records, total declared spending on lobbying was $3.53 billion in 2020. The pharmaceutical sector ranked by far as the top lobbying industry with declared spending of $308 million. In the United States, corporations are also allowed to spend as much as they want to convince people to vote for or against a political candidate through Political Action Committees. According to the 5-4 Supreme Court decision, if the funds are not being spent in coordination with a political campaign, they "do not give rise to corruption or the appearance of corruption." Following the decision, independent expenditure-only committees (Super PACs) were created to raise unlimited sums of money to overtly advocate for or against political candidates. For the 2019–2020 electoral cycle, Super PACs amassed $3.4 billion (Open secrets database). For many, such political funding can be compared to open forms of corruption [58].

47.3.4.3 Revolving Door Lobbyists

In addition to direct lobbying and electoral contributions, revolving doors and ubiquitous conflicts of interests in Government are also considered important means of regulatory capture [59]. According to the Center for Responsive Politics, the pharmaceutical sector is the top-ranking sector for hiring revolving door lobbyists (with around 1000 industry lobbyists who previously worked with government).

47.3.4.4 Financing the Regulatory Agency Through User Fees

Another way that contributes to regulatory capture is to create a dependence of the regulator on the regulated, such as financing regulatory agencies through user fees. Such user fees can favor Big Pharma by making it more difficult for smaller players to enter the market, as is normally the case with most regulatory burdens and costs. User fees can also increase the ease with which new, potentially harmful products are approved for use, for example by pressuring scientists working with the regulatory agency to approve drugs despite safety concerns [60].

47.3.5 Interaction with Civil Society

Civil society refers here to charities, non-governmental organizations, trade unions, social movements and other groups that can influence policymakers, culture and habits of thought. The technique of creating front groups (sometimes called astro-turf organizations) to create the illusion of support from civil society has a long history in the era of corporate capitalism [61]. Some companies, such as Pfizer, are known to be very proactive in developing means to capture civil society [55]. Pfizer has prioritized the funding and creation of think tanks all over the world. Former Reagan administration official, Catherine Windels, was in charge of "Worldwide Policy Mobilization at Pfizer" and is sometimes described as the "godmother of think tanks" [62]. During her 22 years career at Pfizer, she helped create new think tanks and networks of think tanks in Europe, Canada, Africa and Asia, and worked closely with many leading institutes in the United States. Among others, she is the former Secretary Treasurer of the very influential Fraser Institute in Canada.

47.3.5.1 Organized Patient Groups

Most patient groups rely on corporate donations to fund their activity. Not surprisingly, they often end up defending the interests of drug companies (which is to get drugs approved and reimbursed at high prices) despite claims that their funding does not influence their discourse [63]. It is indeed likely that these groups did not change their discourse because of their funding. However, one must consider that drug companies will fund only groups who already embraced the "right" discourse to make sure that this "patient voice" has more impact than others in the regulatory debates. While drug companies might fund specific groups to support specific products, drug companies can also band together to fund specific groups to support specific regulation in favor of their sectors, just like when they hire marketing firms to create non-profit groups to push for higher drug prices in the name of patients [64]. In many ways, patient organizations have become a central part of the communication strategies used by Big Pharma [65]. Some companies such as GlaxoSmithKline, Novartis, Roche or Genentech detail their financial grants to different charities and patient groups in the United States. If their grants and contributions are proportional to their market size, one can estimate that drug companies spend almost a billion in grants to patient groups and education in the United States [24].

47.3.5.2 Direct to Consumer Advertising (DTCA)

DTCA for prescription drugs is legal in the United States and New Zealand, despite evidence that DTCA has a negative impact for patients by increasing inappropriate prescribing [66, 67]. According to Statista website, DTCA for prescription in the United States has reached $6.6 billion in 2020.

47.4 How to Reduce Corruption in the Pharmaceutical Sector

There is a significant literature exploring the different ways and means to tackle corruption in the pharmaceutical sector [14, 52, 68–70]. The British Medical Journal (BMJ) ran a series of articles to reduce commercial influence in health (bmj.com/commercial-influence), from achieving greater independence in research and clinical trials [71], to improving conflict of interests management [72], to the role that can be played by medical students [73].

47.4.1 Top Priorities in Combatting Pharmaceutical Corruption

A first building block to counter corruption is a strong, transparent, and accountable universal health care system, independent from commercial interests [74]. Corruption has many faces, but policymakers must start with tackling practices that are the most harmful to the population and that most undermine trustworthiness of affected institutions [75]. Standard corruption in health care remains a priority but, as explained above, it is not the dominant form of corruption in the pharmaceutical sector, which is often undermined by a political economy of influence.

1. A first priority, as identified in Goldacre [40], is to allow for the full transparency of clinical trial data to restrain disinformation campaigns based on selective publications. Greater transparency of data can significantly reduce the ghost management of evidence and increase accountability in the pharmaceutical sector [76]. Note that important steps have been made in that direction in the last years in different countries, and Canada has somewhat taken leadership on this issue by publishing all clinical trial data submitted to its regulatory agency in charge of drug approval [18].
2. A second priority would be to delink profits from drug sales. In the current system, the more drugs that a company is selling, the more benefits the company will make. Financial incentives are thus pressuring drug companies to sell always more drugs, even when it can be harmful or inappropriate to do so, making unethical and corrupt practices profitable. A potential alternative is to stop paying for the drugs in and of themselves but for the desired therapeutic effect they bring about. This type of "value-based pricing" makes payments to drug companies as a proportional counterpart to the health benefits they bring to the population. It reduces incentives to oversell drugs or oversell the benefits of drugs through disinformation campaigns [77]. Such approach would also provide financial incentives for medical research that is focused less on me-too drugs of little therapeutic value, and more on breakthrough drugs that could significantly improve health outcomes. While value-based pricing is attractive, it has several obstacles to overcome since it can induce higher prices [78]. Even in high-income countries, it may need to be accompanied by other strategies such as tendering and price-volume agreements.
3. The third priority would be to make the medical profession more independent from the pharmaceutical sector to ensure that medical practices are based on evidence instead of marketing claims. Commercial influence over prescribing physicians can be observed in almost every interstice of the medical profession: from bias in academic curriculum to continuing medical education, from influence over clinical guidelines to medical advertising in scientific journals; from corporate research funding to free lunches; from massive

promotional spending to paid key opinion leaders. Disclosure of financial conflict of interests (FCOI) is certainly a necessary step, but it is not sufficient. The ultimate goal should be to exclude people with FCOI from decision-making capacity to ensure greater independence in the shaping of medical practices. As much as possible, marketing towards medical professionals should be abolished and payment for medical research should come from public funds and go to researchers who do not have direct conflicts of interest. More public funding for clinical research could help to prevent researchers from engaging in trials that are being conducted for marketing purposes. A potential way to do this could be to tax promotional activities and use tax revenues to fund public research [77].

Mapping the different levels of corruption and exploring ways of combating corrupt practices remains necessary to ensure that the well-being of patients and the improvement of health outcomes remain the ultimate goals in the ways that the pharmaceutical sector is structured by public policy.

References

1. Katz J (2017) Short answers to hard questions about the opioid crisis. nytimes.com/interactive/2017/08/03/upshot/opioid-drug-overdose-epidemic.html. Accessed 4 May 2021
2. Special Advisory Committee on the Epidemic of Opioid Overdoses (2020). Opioid-related harms in Canada. health-infobase.canada.ca/substance-related-harms/opioids. Accessed 4 May 2021
3. Hoffman J (2019) Johnson & Johnson ordered to pay $572 million in landmark opioid trial. nytimes.com/2019/08/26/health/oklahoma-opioids-johnson-and-johnson.html. Accessed 4 May 2021
4. Persaud N (2014) Questionable content of an industry-supported medical school lecture series: a case study. J Med Ethics 40:414–418
5. Grant K (2018) Drug makers in Canada disclose doctor payments as transparency debate heats up. theglobeandmail.com/canada/article-drug-companies-paid-nearly-75-million-to-doctors-health-care/. Accessed 4 May 2021
6. Almashat S, Lang R, Wolfe S, Carome M (2018) Twenty-seven years of pharmaceutical industry criminal and civil penalties: 1991 through 2017. citizen.org/article/twenty-seven-years-of-pharmaceutical-industry-criminal-and-civil-penalties-1991-through-2017/#:~:text=Financial%20penalties%20continued%20to%20pale,years%20(2003%2D2012)May 2021
7. Gotzsche P (2013) Deadly medicines and organized crimes. CRC Press, Boca Raton
8. Prasad V (2020) Malignant; how bad policy and bad evidence harm people with cancer. John Hopkins University Press, Baltimore
9. Sismondo S (2018) Ghost-managed medicine: big pharma's invisible hands. Mattering Press, Manchester
10. Abraham C (2005) Vioxx took deadly toll: study. theglobeandmail.com/life/vioxx-took-deadly-toll-study/article1113848/. Accessed 4 May 2021
11. Graham DJ, Campen D, Hui R, Spence M, Cheetham C, Levy G, Shoor S, Ray WA (2005) Risk of acute myocardial infarction and sudden cardiac death in patients treated with cyclo-oxygenase 2 selective and non-selective non-steroidal anti-inflammatory drugs: nested case-control study. Lancet 365(9458): 475–481
12. Transparency International (2021) What is corruption. transparency.org/en/what-is-corruption. Accessed 4 May 2021
13. Rose-Ackerman S, Palifka B (2016) Corruption and government; causes, consequences, and reform. Cambridge University Press, Cambridge
14. Kohler JC, Martinez MG, Petkov M, Sale J (2016) Corruption in the pharmaceutical sector. Transparency International, London
15. Kohler JC, Dimancesco D (2020) The risk of corruption in public pharmaceutical procurement: how anti-corruption, transparency and accountability measures may reduce this risk. Glob Health Action 13(Suppl 1): 1694745
16. Berdud M, Chalkidou K, Dean E, Ferraro J, Garrison L, Nemzoff C, Towse A (2019) The future of global health procurement: issues around pricing transparency. cgdev.org/publication/the-future-global-health-procurement-issues-around-pricing-transparency. Accessed 4 May 2021
17. Carpenter D, Moss D (2014) Preventing regulatory capture. Cambridge University Press, Cambridge
18. Gagnon MA et al (2021) In: Fierlbeck, Graham, Herder (eds) Clinical trial data transparency in Canada; mapping the progress from laggard to leader. pp 114–128
19. Lessig L (2013) Institutional corruption. Defined J Law Med Ethics 41(3):553–555
20. Light DW, Lexchin J, Darrow JJ (2013) Institutional corruption of pharmaceuticals and the myth of safe and effective drugs. J Law Med Ethics 41(3):590–600
21. Rodwin M (2011) Conflicts of interests and the future of medicine. Oxford University Press, Oxford
22. Angell M (2004) The truth about the drug companies: how they deceive us and what to do about it. Random House, New York, p 2004

23. Dukes G, Braithwaite J, Moloney JP (2014) Pharmaceuticals, corporate crime and public health. Edward Elgar Press, Chetelham
24. Gagnon MA (2021) Ghost management as a central feature of accumulation in corporate capitalism: the case of the global pharmaceutical sector. In: Benquet M, Bourgeron T (eds) Accumulating capital today: contemporary strategies of profit and dispossessive policies. Routledge, New York, pp 163–177
25. Lexchin J (2016) Private profits versus public policy; the pharmaceutical industry and the Canadian state. University of Toronto Press, Toronto
26. Penston J (2010) Stats.Con; how we've been fooled by statistics-based research in medicine. London Press, London
27. Lexchin J, Bero LA, Djulbegovic B, Clark O (2003) Pharmaceutical industry sponsorship and research outcome and quality: systematic review. BMJ 326(7400): 1167–1170
28. Lexchin J (2012) Those who have the gold make the evidence: how the pharmaceutical industry biases the outcomes of clinical trials of medications. Sci Eng Ethics 18(2):247–261
29. Lundh A, Lexchin J, Mintzes B, Schroll JB, Bero L (2017) Industry sponsorship and research outcome. Cochrane Database Syst Rev 2(2):MR000033
30. Sismondo S (2008) Pharmaceutical company funding and its consequences: a qualitative systematic review. Contemp Clin Trials 29(2):109–113
31. Fabbri A, Lai A, Grundy Q, Bero LA (2018) The influence of industry sponsorship on the research agenda: a scoping review. Am J Pub Health 108:e9–e16
32. Testoni FE, García Carrillo M, Gagnon M-A, Rikap C, Blaustein M (2021) Whose shoulders is health research standing on? Determining the key actors and contents of the prevailing biomedical research agenda. PLoS One 16(4):e0249661
33. Fisher JA (2020) Adverse events; race, inequality and the testing of new pharmaceuticals. New York University Press, New York
34. Rajan KS (2017) Pharmocracy: value, politics, and knowledge in global biomedicine. Duke University Press, Durham
35. U.S. Senate Committee of Finance (2010). Grassley, Baucus release committee report on Avandia. finance.senate.gov/release/grassley-baucus-release-committee-report-on-avandia. Accessed 4 May 2021
36. Lacasse JR, Leo J (2010) Ghostwriting at elite academic medical centers in the United States. PLoS Med 7(2):e1000230. https://doi.org/10.1371/journal.pmed.1000230
37. Sismondo S (2007) Ghost management: how much of the medical literature is shaped behind the scenes by the pharmaceutical industry? PLoS Med 4 (9): 1429-1433
38. Fugh-Berman A (2010) The haunting of medical journals: how ghostwriting sold 'HRT'. PLoS Med 7(9)
39. Ross JS et al (2008) Guest authorship and ghostwriting in publications related to rofecoxib: a case study of industry documents from rofecoxib litigation. JAMA 299(15):1800–1812
40. Goldacre B (2013) Bad pharma; how drug companies mislead doctors and harm patients. Signal, Toronto
41. Liu X, Zhang Y, Li W et al (2021) Evaluation of oncology trial results reporting over a 10-year period. JAMA Netw Open 4(5):e2110438
42. Turner EH et al (2008) Selective publication of antidepressant trials and its influence on apparent efficacy. N Engl J Med 358(3):252–260
43. Rout M (2009) Vioxx maker Merck and Co drew up doctor hit list. healthleadersmedia.com/strategy/vioxx-maker-merck-and-co-drew-doctor-hit-list. Accessed 4 May 2021
44. Gagnon M-A, Lexchin J (2008) The cost of pushing pills: a new estimate of pharmaceutical promotion expenditures in the United States. PLoS Med 5(1):e1. https://doi.org/10.1371/journal.pmed.0050001
45. Spurling GK, Mansfield PR, Montgomery BD, Lexchin J, Doust J, Othman N, Vitry AI (2010) Information from pharmaceutical companies and the quality, quantity, and cost of physicians' prescribing: a systematic review. PLoS Med 7(10):e1000352
46. Moynihan R (2008) Key opinion leaders: independent experts or drug representatives in disguise? BMJ 336: 1402–1403
47. Brody H (2007) Hooked: ethics, the medical profession, and the pharmaceutical industry. Rowman & Littlefield, Lanham
48. Kassirer JP (2005) On the take. Oxford University Press, New York
49. Lexchin J (2017) Doctors in denial: why big pharma and the Canadian medical profession are too close for comfort. Lorimer, Toronto
50. Nejstgaard CH, Bero L, Hróbjartsson A, Jørgensen AW, Jørgensen KJ, Le M, Lundh A (2020) Conflicts of interest in clinical guidelines, advisory committee reports, opinion pieces, and narrative reviews: associations with recommendations. Cochrane Database Syst Rev 12(12):MR000040
51. Whitaker R, Cosgrove L (2015) Psychiatry under the influence: institutional corruption, social injury, and prescriptions for reform. Palgrave/Macmillan, New York
52. Lexchin J, Kohler JC, Gagnon MA, Crombie J, Thacker P, Shnier A (2018) Combating corruption in the pharmaceutical arena. Indian J Med Ethics 3(3): 234–239
53. National Academies of Science, Engineering, and Medicine (2017) Making medicines affordable: a national imperative. The National Academies Press, Washington

54. Patented Medicine Price Review Board (2020). Annual report 2019. pmprb-cepmb.gc.ca/home. Accessed 4 May 2021
55. Drahos P, Braithwaite J (2002) Information feudalism; who owns the knowledge economy? The New Press, New York
56. Gleeson D et al (2019) Analyzing the impact of trade and investment agreements on pharmaceutical policy: provisions, pathways and potential impacts. Glob Health 15(Suppl 1):78
57. Sell SK (2003) Private power, public law; the globalization of intellectual property rights. Cambridge University Press, Cambridge
58. Lessig L (2011) Republic, lost: how money corrupts congress and a plan to stop it. Twelve, New York
59. Public Citizen (2005) Congressional revolving doors: the journey from Congress to K Street. policyarchive.org/handle/10207/10858. Accessed 4 May 2021
60. Lurie P, Wolfe SM (1998) FDA medical officers report lower standards permit dangerous drug approvals. Public Citizen; [cited June 15, 2021]. citizen.org/our-work/health-and-safety/fda-medical-officers-report-lower-standards-permit-dangerous. Accessed 4 May 2021
61. Miller D, Dinan W (2008) A century of spin. Pluto, London
62. Powerbase (2010) Catherine Windels. Public Interest Investigations. powerbase.info/index.php?title=Catherine_Windels. Accessed 15 June 2021
63. Batt S (2017) Health Advocacy Inc. UBC Press, Vancouver
64. Elgin B (2019) Pharma's secret to lobbying for higher drug prices?. bloomberg.com/news/articles/2019-04-25/pharma-s-charity-allies-help-its-fight-to-keep-drug-prices-high. Accessed 4 May 2021
65. McCoy MS (2018) Industry support of patient advocacy organizations: the case for an extension of the sunshine act provisions of the affordable care act. Am J Pub Health 108(8):1026–1030
66. Franquiz MJ, McGuire AL (2021) Direct-to-consumer drug advertisement and prescribing practices: evidence review and practical guidance for clinicians. J Gen Intern Med 36:1390–1394
67. Mintzes B (2012) Advertising of prescription-only medicines to the public: does evidence of benefit counterbalance harm? Annu Rev Pub Health 33:259–277
68. Çaliyurt KT (ed) (2020) Integrity, transparency and corruption in healthcare & research on health. Springer, Singapore
69. Mialon M, Vandevijvere S, Carriedo-Lutzenkirchen A et al (2020) Mechanisms for addressing and managing the influence of corporations on public health policy, research and practice: a scoping review. BMJ Open 10:e034082
70. Moynihan R et al (2019) Pathways to independence: towards producing and using trustworthy evidence. BMJ 367:l6576
71. Lexchin J, Bero LA, Davis C, Gagnon M (2021) Achieving greater independence from commercial influence in research. BMJ 372:n370
72. Grundy Q, Dunn AG, Bero L (2020) Improving researchers' conflict of interest declarations. BMJ 368:m422
73. Fabbri A (2020) Conflict of interest in medicine: how medical students can become catalysts for change. blogs.bmj.com/bmj/2020/09/11/conflict-of-interest-in-medicine-how-medical-students-can-become-catalysts-for-change/. Accessed 4 May 2021
74. García PJ (2019) Corruption in global health: the open secret. Lancet 394(10214):2119–2124
75. Klitgaard R (1988) Controlling corruption. University of California Press, Berkeley
76. Fierlbeck K, Graham J, Herder M (2021) Transparency, power, and influence in the pharmaceutical industry. University of Toronto Press, Toronto
77. Gagnon MA (2013) Corruption of pharmaceutical markets: addressing the misalignment of financial incentives and public health. J Law Med Ethics 41(3):571–580
78. Roy V (to be published) Capitalizing a cure: how finance controls the price and value of health. Routledge, New York

48 Conflicts of Interest Between Neurologists and Pharmaceutical and Medical Device Industries

Nathaniel M. Robbins and Jade E. Smith

Abstract

Financial relationships between physicians and pharmaceutical and medical device companies (Industry) represent a conflict of interest (COI). A medical COI exists when any outside interest competes with the physician's primary professional responsibility to place the patient first. Fraud and manifest bias represent one end of the COI spectrum. Subtler COI are more widespread and often problematic. Many scholars have advocated for the disclosure of Industry relationships to mitigate bias and preserve the benefits of collaboration. The Physician Payment Sunshine Act established the "Open Payments" database (OPD), in which companies are legally required to report all payments to U.-S. physicians. Recent scholarly work using the OPD has identified that physicians routinely underdisclose Industry-related financial relationships. In this chapter we will discuss financial COI between neurologists and Industry. We review pertinent ethical issues and discuss potential solutions to enable neurologists to best work with Industry without compromising professional integrity.

N. M. Robbins (✉)
Dartmouth Geisel School of Medicine, Lebanon, NH, USA
e-mail: Nathaniel.m.robbins@dartmouth.edu

J. E. Smith
Dartmouth College, Hanover, NH, USA

Abbreviations

CME	Continuing medical education
COI	Conflict of interest
CPG	Clinical practice guidelines
IRB	Institutional review board
OPD	Open Payments database
RCT	Randomized control trial

48.1 Introduction

In a seminal editorial in the New England Journal of Medicine, Arnold Relman wrote that to represent patients' interests fairly, physicians should have no economic conflict of interest with the medical-industrial complex [1]. Published in 1980, this editorial set the stage for decades of discourse on physician's financial associations with pharmaceutical and medical device companies (Industry). In medicine, a conflict of interest (COI) exists when any outside interest competes with a physician's primary professional responsibility: to place the patient's interests first. Financial relationships between physicians and Industry can introduce such conflicts and have been scrutinized by the public, the media, and even the national government. Sensational stories of corrupt physicians get picked up by the popular and academic press. A congressional inquiry was carried out. And now with passage of the Physician Payment Sunshine Act, drug and device

J. Faintuch, S. Faintuch (eds.), *Integrity of Scientific Research*,
https://doi.org/10.1007/978-3-030-99680-2_48

companies are mandated to publicly disclose payments to physicians, allowing the public to dive into the web of physician-Industry relationships more than ever before.

Conflicts of interest among neurologists are of particular consequence due to their extensive financial ties to Industry and high prescription drug costs. From 2013 to 2016, over half of all U.S.-based neurologists received at least one payment from Industry each year, totaling over USD $354 million [2]. Industry funding has been associated with higher prescription rates of sponsored products and increased publication of articles that report positive effects. Rising prescription rates fueled by secondary interests can have national economic consequences [3], and biased medical advice can comprise clinical care, potentially placing patients at risk and undermining patient-doctor trust and credibility of the medical system as a whole [4].

48.2 Conflicts of Interest in Medicine

In medical practice the primary professional responsibility of each physician is to care for and protect the interests and well-being of their patients. This can be pursued through various avenues—patient care, research, education, guideline development, policy formulation, advocacy—each of which is subject to unique secondary interests.

Conflicts of interest can be conceptualized along a broad ethical spectrum. On one end lie situations where professional and personal interests actually align or competing obligations do not compromise professional judgement. Biases that result from COI, conscious or subconscious, compromise one's primary obligation to the patient and push you further down the spectrum towards intentional dishonesty and fraud [5]. Sensational cases of physician fraud make for compelling news stories and occupy most media coverage, but seemingly innocuous cases of competing interests are much more common and can greatly erode public trust in the medical field.

48.2.1 Pharmaceutical Financial Interfaces

When a physician holds stock in a drug company, prescribing that drug is not solely in the patient's best interest. When they are compensated for clinical trial referrals, they are in the conflicted position of determining whether their patients are appropriate for the study. When pharmaceutical companies support medical research, physicians face an implicit demand to interpret data in favor of the product to help secure further funding and publish the results, or they might receive remuneration for acting as a study investigator. Industry courts physicians with gifts, utilizing financial gains to promote their products and alter prescribing patterns.

48.3 Types of COI Between Neurologists and Industry

Many neurologists receive compensation from Industry, from occasional meals and travel payments, to large speaking and consulting fees. The Open Payments database divides Industry payments into four main categories—ownership (such as stock or patent rights), individual research payments, institutional research payments, or general payments.

The first two categories introduce obvious competing interests: physicians with direct ownership of Industry stocks or patents are incentivized to promote the company or product, which might not align with patients' interests. Direct payments to investigators for industry-funded research might similarly bias findings. Industry can also pay research grants to an institution, rather than a physician. While less direct, the benefits of winning grants are often passed along to the physician as salary support or incentives, reduced clinical responsibilities, esteem, and promotion. The last category, general payments, are direct payments to physicians not made in connection with a research agreement or protocol. Examples include meals, travel expenses, honoraria, consulting fees, and other speaking or faculty payments. General payments exemplify how pharmaceutical companies devote money to marketing, rather than research and development. The benefits of academic-industrial collaborations emerge through synergistic research efforts, not through product promotion, making direct general payments particularly controversial. Industry can also influence the medical literature through other means, such as sponsoring continuing medical education (CME) opportunities and paying authors of clinical practice guidelines (CPGs) [6, 7].

48.3.1 Nonfinancial COI

These might include personal beliefs, career advancement, or intellectual legacy. Social context affects how investigators conduct research, and presumably all investigators have a vested personal and intellectual interest in their research topic as part of their professional role [8]. This ubiquity is a major delineation between financial COI and nonfinancial interests. Personal beliefs and experience, academic drive, and theoretical approach are inherent to the individual and who they are as investigators, not secondary interests that compromise the primary interest of producing unbiased research [9]. They also differ from financial COI when considering methods of mitigation. Disclosure, a common strategy with financial COI, introduces unique ethical pitfalls in the context of nonfinancial COI. Should someone studying racial inequities in healthcare be required to disclose their own racial background, or is it fair to have authors list their career goals? Another method of tackling COI, eliminating the conflict, is not possible with nonfinancial interests, as personal beliefs and experiences cannot be eliminated.

48.4 Scope of COI Between Neurologists and Industry

Compared to other specialties, neurologists have among the most extensive financial relationships with Industry [10]. This may be due to the relatively high prevalence of neurologic disorders [11], and the fact that the market for neurologic medications is the third highest amongst any specialty [3]. Due to the exorbitant drug prices within the field of neurology, conflicts of interests among neurologists can have extensive ethical and economic implications.

From 2013 to 2016, Industry paid USD $354 million to U.S.-based neurologists, of which USD $303 million (95.6%) were general, non-research-related payments [2]. Within this 4-year window, consulting and speaking fees to neurologists increased, while research payments remained generally stable, potentially highlighting the increasing focus on Industry marketing. While Industry payments within neurology are widespread, a majority of payments go to a minority of physicians. Most neurologists receive less than $1000 per year, but over 200 neurologists each received more than $100,000 during some years. Several received over $1 million [2].

48.5 Why Industry Collaboration Should Not Be Eliminated

Some authors have argued that Industry-physician relationships must be eliminated since they inherently compromise research quality and patient care [12]. Accordingly, the American Medical Student Association has advanced a policy outlawing gifts, meals, pharmaceutical samples, and drug representatives at medical schools, and setting stringent restrictions on other activities related to Industry [13]. However, Industry now funds the majority of healthcare research in the United States [14] and has facilitated numerous public health advances by translating academic discoveries into beneficial therapies, encompassing statin therapies for cardiovascular disease, which have contributed to decreasing cardiovascular mortality throughout the developed world [15]. The first HIV protease inhibitor was borne of academic-Industrial collaboration, transforming HIV/AIDS from a death sentence to a chronic disease [16]. The benefits of Industry-academic collaboration were highlighted again in 2020–2021 with the unprecedented development and distribution of the mRNA SARS-CoV-2 vaccines.

48.5.1 The Relevance of Drug Manufacturing and Testing

Industry is uniquely built to translate academic discoveries into use, and research collaborations with Industry have been cited as the most efficient way to improve public health [16]. Academia and Industry each have an essential role in this process. Academia conducts years of basic research to understand disease mechanisms and identify potential drug targets. Industry then screens these drug targets and tests thousands of therapeutic molecules, using its substantial funding and infrastructure to support the practical steps of translation. The question, then, is how to best manage conflicts of interest while preserving the collaborations on which medical advances depend.

48.6 Bias Is Inherent to COI

Some argue that COIs are not the issue, and instead bias is the real enemy [17]. Bias arising out of accepting gifts from Industry is seen as a deliberate choice on the part of the physician, in which they decide to compromise their morals for personal gain. It follows that if bias is intentional, a more ethical physician could remain unswayable in the face of competing interests. Succumbing to bias is then an ethical shortcoming on the part of the physician, and our efforts need to be redirected from potential COI to corruptible physicians. However, from work in the social sciences, we know that bias is not a matter of deliberate choice. Instead, an unconscious and unintentional self-serving bias is inherent to COI [18].

48.6.1 Academic Integrity vs. Moral Weaknesses

When we stand to gain from reaching a certain conclusion, we unknowingly default to an interpretation that favors our own interests—a self-serving bias [19]. This bias is unintentional and unconscious—individuals are unable to remain objective even when highly motivated to be impartial and succumb to bias despite explicitly instructed about it. In one study, 20 physicians were asked if attending continuing medical education seminars sponsored by drug companies would influence their prescribing practices. All but one physician denied that the seminars would influence their behavior. Although primed to the potential of bias when entering the seminar and confident in their own objectivity, the physicians increased their usage of the drugs described at these seminars [20]. Alerting people to their biases does little to mitigate their influence. Instead, it tends make us more confident in our own judgement and quicker to recognize bias in others—a phenomenon known as the "bias blind spot" [18, 21].

In a survey study by Steinman and colleagues, most physicians (61%) thought that Industry marketing did not influence their own prescribing, but only 16% believed that other physicians were similarly unaffected [22]. The vast majority of this bias does not represent a moral shortcoming on the part of physicians. Bias is often subconscious, meaning that even well-meaning individuals can be influenced. That being said, numerous cases of blatant fraud and misconduct have been publicized, in which practitioners motivated by personal gain intentionally hide relationships or willfully distort the truth.

48.7 Industry-Bias Within Neurology Research and Literature

Results from clinical studies on drugs and medical devices are crucial for determining the safety and efficacy of new treatments. In this way, research greatly impacts how physicians practice medicine and which treatments they offer patients. However, clinical research is increasingly sponsored by the companies that manufacture these products. When industry companies directly perform or fund research studies, the conclusions are more likely to favor the sponsor's product [12, 23–25]. These results cannot be explained by methodological differences, and in many cases Industry-sponsored trials use better methodology [12, 25]. Instead, this positive Industry bias manifests within opinion pieces, positive study result framing, and altered outcome reporting.

48.7.1 Industry Relationships Influence Views Expressed in Scientific Reports

Editorials and commentaries provide the means for researchers to express personal opinions within the academic literature. Such pieces are afforded a level of subjectivity beyond that in peer-reviewed trials and allow experts to contribute to the discourse on controversial topics.

In the early 2000s, a meta-analysis showed a significant increase in the risk of myocardial infarction associated with the use of rosiglitazone (GalxoSmithKline) in patients with diabetes, sparking the "rosiglitazone controversy." What followed was an outpouring of commentaries and letters, reviews and meta-analyses, in which physician-scientists took strong and opposing stances. A subsequent study by Wang et al. found a clear link between financial conflicts of interest and views on rosiglitazone. Authors who had a favorable view of the risk of myocardial infarction with rosiglitazone were more likely to have financial conflicts of interest with manufacturers of antihyperglycemic drugs in general, and with rosiglitazone manufacturers in particular, than authors who had an unfavorable view, and were similarly more likely to recommend the use of rosiglitazone amid the controversy [26].

48.7.2 Industry Sponsorship Influences Rhetoric and Conclusions of Neurology RCTs

Although randomized control trials (RCTs) are considered the golden standard of medical research, researchers frame clinical trial findings through their choice of rhetoric. In this way, bias can be introduced into an otherwise objective study. Gilstad et al. found that data across 18 different donepezil trials (Aricept, Eisai/Pfizer) concerning Alzheimer's disease were highly consistent, however rhetoric varied based on study sponsorship. Uniformly positive rhetoric was present in all Industry-funded trials, which described donepezil as "well-tolerated and efficacious." Conversely, trials without vendor support referred to the donepezil's effect as small or absent, sometimes even emphasizing the need for better treatments [27]. While choices of data presentation do not indicate misconduct or invalidate such studies, rhetorical enthusiasm has obvious promotional value. Many readers skim an article's abstract and then skip to the discussion and conclusions, and therefore are heavily influenced by rhetoric encased in the author's interpretation of the results. Industry authors often co-write articles, helping relieve time-burdens for overworked physicians. While the Industry sponsorship of the trial and author Industry affiliations are often disclosed, this collaborative structure permits objective data to be subjectively interpreted in a biased fashion.

48.7.2.1 Enhanced Findings ("Spin")

These behaviors may be unconscious, highlighting the dangers posed by industry relationships even when the academics are well-meaning. Scientific literature is assumed to be objective and dispassionate, but secondary interests can greatly shape how data are presented by authors, and promotional rhetoric is often embedded even within this academic space. Favorable rhetoric and conclusions end up promoting the multibillion-dollar commercial success of drugs whose efficacy remains contested by non-Industry sponsored researchers.

48.7.3 Selective Reporting and Outcome Switching in Neurology RCTs

Bias can be introduced through selective and altered outcome reporting. Before beginning any clinical trial, researchers are required to specify the primary and secondary outcomes of their study. However, through the omission of negative findings and the modification of prespecified outcomes, researchers can make interventions appear effective when they are not.

In a 2009 report in the New England Journal of Medicine, Vedula et al. examined reporting practices for trials of gabapentin (Neurontin, Pfizer) by comparing the manufacturers' internal research documents with published reports of study findings. They exposed many instances of the prespecified primary outcomes, the most important measure of efficacy, being reported inaccurately in the publications. These outcome changes were often made when the protocol-defined primary outcome did not reach statistical significance. The findings were either not published in full or were published with a changed primary outcome. Although there are legitimate reasons to amend a trial's outcomes, in this review, *all* changes between what was written in the internal company research report and what was reported to the public led to a more favorable presentation of gabapentin's efficacy. By defining the outcome once the data are known, chance findings can be presented as evidence of a drug's effectiveness, threatening the integrity of scientific knowledge [28]. Negative trials are harder to publish in well-respected journals, further compounding the issue of selective reporting even for well-meaning authors.

48.8 Manifestations of Industry-Bias Within Neurology Clinical Practice

Once a drug has been successfully researched and developed, the role of industry changes from that of scientist to that of corporate promoter. The pharmaceutical industry invests up to 40% of its revenue in promotional efforts, spending upwards of USD $12 billion each year on marketing [6, 29]. This spending proves to be a worthwhile investment for pharmaceutical manufacturers. Several studies have shown that financial relationships with Industry greatly alter the prescribing practices of physicians [9, 20, 30–32].

48.8.1 Pharmaceutical Sales Representatives Alter Prescribing Practices

These representatives use gifts, friendship, and flattery to develop relationships with healthcare prescribers. Most physicians remain skeptical of representatives, believing that they prioritize product promotion above patients' welfare. Nonetheless, physicians meet with pharmaceutical representatives about four times a month on average [6], and these interactions effect prescribing practices. Additionally, the initial skepticism towards sales representatives fades through repeated exposure and gifts—receiving a gift from Industry and the number of gifts received is correlated with the belief that pharmaceutical representatives have no impact on prescribing behaviors [6].

48.8.1.1 Profitable Friendship

Physician-Industry relationships are associated with increased prescribing of brand-name drugs. In single hospital studies, Industry-sponsored symposiums and meals are linked to increased use of the promoted drug over alternatives within the drug class, choice of the promoted drug when presented with a clinical scenario, and formulary addition requests for the sponsor's drug [33–35]. Many physicians self-report that they rely a great deal on information provided by sales representatives when making their prescribing decisions, often instead of consulting scientific data [6]. Even for providers who report weighing scientific materials over promotional ones, studies suggest that their beliefs align more strongly with the commercial sources of information, and they often cannot distinguish between correct and incorrect information presented to them by drug representatives [36]. The inability to accurately extract educational information from promotional materials, and the unconscious internalization of information from commercial sources, makes these marketing interactions so powerful. The free meals, pens, and notepads help get prescribers in the door and in front of a representative, but the gifts themselves also have a powerful effect on human relationships.

Gifts as small as an Industry-sponsored lunch may foster a subconscious obligation to reciprocate through changes to prescribing practices. DeJong et al. has shown that sponsored meals of less than $20 can increase prescribing of branded products, and larger gifts elicit larger impacts on prescribing behaviors [30]. Another study showed that gift recipients prescribed 2.3 more claims per patient, prescribed medications costing $50 more per claim, and prescribed 7.8% more branded drugs [34]. Despite this, prescribers typically report that they are not biased by financial relationships or interactions with Industry, highlighting the influence of COIs and the enduring "bias blind spot."

48.9 Targeting Prominent Thought Leaders

A small number of physicians receive the vast majority of payments from Industry, and most physicians receive little or no payments from industry [2, 37]. The champions are often prominent academics who have substantial influence on health policy through their roles as journal authors, journal editors, medical educators, specialty and subspecialty board members, and members of prominent committees that write guidelines and set policy. Their key positioning allows them to disseminate bias that results from these Industry relationships, whether conscious or subconscious, to the greater medical community.

48.9.1 Research Bias Through IRB Members and Journal Editors

Like individual physicians, academic institutions have financial ties to Industry and are subject to conflicts of interest. Universities and medical centers receive Industry research grants, arrange patent agreements, and own substantial equity in start-up businesses. Relationships with industry may analogously affect members of institutional review boards (IRBs). IRBs review and approve research protocols and are responsible for protecting the safety and well-being of research participants. Relationships between IRB members and Industry are common, and members sometimes participate in decisions about protocols sponsored by companies with which they have a financial relationship. One study found that 36% of IRB members in academic institutions had at least one relationship with Industry in the past year, and 15% had protocols come before their IRB that were sponsored by a company with which they had a relationship or by a competitor of that company. In these cases of conflicting interests, 35.5% reported that they voted on the protocol, nonetheless, rather than recuse themselves [38].

One study found that 78% of physicians on the editorial boards of high-impact orthopedic journals receive Industry payments [39], and similar rates have been found among oncology journal editors [40]. The process by which editors decide to accept manuscript for further review, and by which they assign reviewers, is often opaque, and therefore the authors and readers are deprived of the ability to assess for Industry influence. Furthermore, unlike journal authors, editors' COI are rarely published.

48.9.2 Clinical Practice Bias Through Thought Leaders and Guideline Authors

One study found that academic neurologists publishing in high-impact journals averaged $19,586 in annual general payments compared to the national average of around $6000 [37]. More payments in the form of speaking and consulting fees may reflect a concerted effort from Industry to shape medical opinion and practice; by paying key opinion leaders to speak at company events, Industry pays for leaders' widespread influence on the prescribing patterns of other physicians [41].

Clinical practice guidelines (CPGs) authors are often academic thoughts leaders, and therefore have disproportionate influence on medical practice and more extensive industry relationships than the average physician [7]. As with editorials, CPG authors are allowed a greater degree of subjectivity and can make recommendations based on their expertise, even when evidence is conflicting or lacking. CPGs have the potential to influence national or international practice and are often considered the gold standard within a field. Since CPG authors are often invited or have editorial prerogative, COI disclosure requirements can be less stringent than for journal authors.

48.10 Disclosure as a Method for Addressing COIs Among Neurologists

Physicians are now required to disclose COI for publication in major academic journals, and most organizations require financial disclosures from faculty and lecturers. Approaching the issue from the side of industry, the Physician Payment Sunshine Act was passed in 2010 and mandates that drug and device manufacturers report all payments to physicians and academic medical centers. While financial disclosure is a reasonable compromise that permits ongoing collaboration, it does not eliminate the conflict, nor protect against subconscious bias. Nevertheless, is does publicize conflicts of interest to those who need to know about it and can be followed by subsequent administrative review and action.

48.10.1 The Physician Payment Sunshine Act

The Physician Payment Sunshine Act was signed into law in 2010 as part of the Affordable Care Act. Industry companies are now legally required to submit annual data listing all payments over $10 made to physicians and teaching hospitals. Physicians and teaching hospitals are offered a 3-week period during which to review their data and challenge inaccurate reports, after which the data are published on the Centers for Medicare and Medicaid Services' Open Payments database (OPD) website for online browsing and downloading. This increased transparency has facilitated research into the prevalence and depth of financial ties between physicians and Industry.

The OPD has several limitations. First, only U.S. providers are included, excluding foreign physicians or non-clinical researchers. Second, overdisclosure of payments by industry is not punished while underdisclosure carries significant penalties, incentivizing Industry to report when in doubt, sometimes inaccurately. The process to dispute entries is difficult—the Center for Medicare Services does not arbitrate, and the companies themselves, who are given the responsibility of correcting inaccuracies, dedicate few resources to resolving disputes. Another challenge is how to incorporate this database into the academic literature and the practice of medicine. Journals and professional societies do not cross check voluntary disclosures against this database, nor do most institutions. Studies suggest patients are not aware of this database [42] and therefore do not check it to see if their own physicians have COI. Some authors have argued that OPD actually undermines public trust in medicine since the database does not distinguish between harmful Industry relationships and benign ones [43]. Despite these problems, the OPD database represents the best action to-date to try to mitigate problems that arise from physician-Industry COI by promoting disclosure.

48.10.2 What Is the Optimal Level of Disclosure?

Too many disclosures can be distracting and detract from the relevant ones; when an author has ties to the company manufacturing the study drug but discloses 30 additional relationships at the bottom of the manuscript, no reader can be expected to identify the relevant conflict hidden among the benign ones. This covering of relevant conflicts is not done maliciously—in fact, it could be a result of good-intentioned physicians doing their best to adhere to journal policies and avoid underdisclosure. The inconsistency among journal disclosure policies adds to the confusion over which relationships should be disclosed. While some journals require complete disclosure, others only require the disclosure of financial relationships relevant to the submitted manuscript, although relevance is not clearly defined. While payments from a company sponsoring a RCT are clearly relevant, what about payments from a company that makes a competing product? And if they are writing a review about the need for better drug therapies in epilepsy, are payments from an epilepsy device manufacturer relevant? Or if that author has royalties from a book on epilepsy or a patent on a blood test to stratify medication risks? There is a continuum of clearly relevant financial COI and unrelated payments which is not yet well-defined.

48.10.3 Widespread Failure to Disclose COI Among Neurologists

Incomplete or inaccurate voluntary disclosures have been identified within the fields of orthopedics, otolaryngology, cardiac surgery, hematology, pulmonology, sports medicine, plastic surgery, vascular surgery, and general surgery, as we previously reviewed [44].

Neurologists in particular have extensive financial relationships with pharmaceutical and medical device companies, and many fail to completely disclose these secondary interests. We recently quantified disclosure discordance among academic neurologists publishing in high-impact clinical neurology journals. Among 200 authors with Industry relationships according to the OPD, 54% had completely missing disclosures; these authors erroneously reported "no conflicts" in their published COI statements. Only 2% of these authors had self-disclosures that matched the payments reported by Industry [37].

This underdisclosure does not necessarily imply fraud or misconduct. Some authors may have been unaware of payments due to the sheer volume and value of payments—the most highly remunerated authors in this sample received over 650 general payments valued over $800,000. Although some journals did not mandate complete disclosure of all Industry relationships, underdisclosure of payments *directly relevant to the article in question* were found. Of 3013 relevant payments (i.e., the article directly tested a drug or device made by a company that paid the author), 50.9% were not self-disclosed by the authors, totaling $5,782,197 [37]. We are aware of one instance when a neurologist wrote an editorial in a premier journal that criticized a new device for epilepsy without disclosing financial relationships worth in excess of $1 million per year, paid by a company that manufactured a competing device [45]. This oversight was not uncovered for many years, after which time a correction was issued online, however the article was not withdrawn.

Of note, the Smith et al. study [37] did not evaluate payments which were indirectly related, such as those from companies that manufacture competing drugs or devices. When considering such payments, the true extent of undisclosed conflicts of interests is likely much higher. Table 48.1 details the rates of underdisclosed payments by payments type by authors of articles in leading clinical neurology journals.

Table 48.1 Nondisclosure rates by payments type for authors of articles in major clinical neurology journals [37]

	Nondisclosure rate for...	
Payment type	*...all payments*	*...directly relevant payments*
General	53.6%	43.3%
Research	75.0%	14.9%
Associated research	79.7%	76.7%

Directly relevant payments refer to payments from industry companies that manufacture products tested or discussed in the author's manuscript

48.10.4 Does Disclosure Foster or Inhibit Public Trust?

Disclosing Industry relationships, regardless of whether they manifest in bias, can decrease patient trust or trust in research results. A survey of people with multiple sclerosis found that patients would have less trust in their physician if the physician was compensated for subject participation in clinical trials. Similarly, industry consulting or giving industry-sponsored talks also adversely affects patient trust [46]. Disclosure of COI may also exacerbate bias; authors with more financial disclosure may be viewed by readers as having more expertise and therefore being more trustworthy [47]. Or, after ridding themselves of the weight of bias through self-disclosure, some authors may feel morally licensed to editorialize even more [48].

Even medical professionals experience skepticism of industry, which can bias the scientific community against effective treatments. Kesselheim et al. asked over 500 internists to evaluate abstracts describing clinical trials of hypothetical drugs. The abstracts were designed to have high, medium, or low methodologic rigor, and the funding source for each was attributed to industry, the NIH or none. When the funding was thought to come from Industry, readers thought the study was poor quality and were half as willing to prescribe the drug being studied, although the studies were identical [49]. In other words, vilification of Industry is so salient that commercial sponsorship alone reduces confidence in research results, independently of a trial's quality. This reticence to adopt effective treatments due to Industry distrust has been observed in neurology in the case of adoption of intravenous alteplase (Activase, Genentech) for treatment acute ischemic stroke.

48.11 Future Perspectives

In other to maintain public trust and preserve the integrity of the medical profession, it is vital to formulate and implement policies to navigate industry-physician financial COI. With this in mind, we offer the following recommendations:

1. Editors, CPG authors, journal reviewers, and all physicians should be required to disclose all industry COI. This allows collaborators, consumers of their intellectual property, and patients the opportunity to contextualize all scholarly work and medical advice.
2. Non-U.S.-based physicians and non-physician researchers should be held to the same standards as U.S. physicians, and companies should be required to report payments to these persons.
3. Given the inaccuracies of the Open Payments database, and the difficulty in reconciling disputes, a professional society (e.g., the American Medical Association) should maintain a database of Industry financial COI, which can be sourced from the OP database. All clinicians should then have the opportunity to register their objections to the relationships listed, and add relationships not listed.
4. The above database can then be used by journals, conferences, and other entities to reduce "disclosure fatigue" (inability to maintain diligence in accurate disclosure due to the sheer abundance of relationships, and due to time requirements necessary to thoroughly and accurately disclose in so many different

venues according to so many different sets of rules). Cross-checking self-disclosures should be mandatory, and failure to disclose should be considered a serious professional breach of conduct that harbors serious consequences.
5. Journals and other academic organizations should work to standardize disclosure policies, both in the United States and abroad, to minimize inadvertent disclosure oversight and disclosure fatigue.
6. Hospitals and other clinical organizations should have a link to each clinician's disclosures (through the above maintained database) to ensure that patients can contextualize their provider's advice in the presence of potential COI.

It is our belief that implementing the above policies will limits nefarious activity without prohibiting the important benefits that stem from outstanding academic-Industry collaboration.

References

1. Relman A (1980) The new medical-industrial complex. N Engl J Med 303:963–970
2. Robbins N, Meyer M, Bernat J (2019) Scope and nature of financial conflicts of interest between neurologists and industry. Neurology 93:438–449
3. De Lott L, Burke J, Kerber K et al (2016) Medicare part D payments for neurologist-prescribed drugs. Neurology 86:1491–1498
4. Grande D, Shea J, Armstrong K (2011) Pharmaceutical industry gifts to physicians: patient beliefs and trust in physicians and the health care system. J Gen Intern Med 27:274–279
5. Stead W (2017) The complex and multifaceted aspects of conflicts of interest. JAMA 317:1765
6. Wazana A (2000) Physicians and the pharmaceutical industry. JAMA 283:373
7. Choudhry N (2002) Relationships between authors of clinical practice guidelines and the pharmaceutical industry. JAMA 287:612
8. Bero L, Grundy Q (2016) Why having a (nonfinancial) interest is not a conflict of interest. PLoS Biol 14: e2001221
9. Bero L (2017) Addressing bias and conflict of interest among biomedical researchers. JAMA 317:1723
10. Tringale K, Marshall D, Mackey T et al (2017) Types and distribution of payments from industry to physicians in 2015. JAMA 317:1774
11. Feigin V, Vos T, Alahdab F et al (2021) Burden of neurological disorders across the US from 1990-2017. JAMA Neurol 78:165
12. Lexchin J (2003) Pharmaceutical industry sponsorship and research outcome and quality: systematic review. BMJ 326:1167–1170
13. AMSA PharmFree (2021) The campaign | best practice policies. pharmfree.org/campaign?id=0003. Accessed 25 May 2021
14. Dorsey E (2010) Funding of US biomedical research, 2003-2008. JAMA 303:137
15. Rosenblatt M (2013) How academia and the pharmaceutical industry can work together. Ann Am Thorac Soc 10:31–38
16. Johnston S, Hauser S, Desmond-Hellmann S (2011) Enhancing ties between academia and industry to improve health. Nat Med 17:434–436
17. Schwid S, Gross R (2005) Bias, not conflict of interest, is the enemy. Neurology 64:1830–1831
18. Dana J (2003) A social science perspective on gifts to physicians from industry. JAMA 290:252
19. Gilovich T (1993) How we know what isn't so. The Free Press of Glencoe, New York
20. Orlowski J, Wateska L (1992) The effects of pharmaceutical firm enticements on physician prescribing patterns. Chest 102:270–273
21. Rosenbaum L (2015) Understanding bias—the case for careful study. N Engl J Med 372:1959–1963
22. Steinman M, Shlipak M, McPhee S (2001) Of principles and pens: attitudes and practices of medicine housestaff toward pharmaceutical industry promotions. Am J Med 110:551–557
23. Bekelman J, Li Y, Gross C (2003) Scope and impact of financial conflicts of interest in biomedical research. JAMA 289:454
24. Jørgensen A, Hilden J, Gøtzsche P (2006) Cochrane reviews compared with industry supported meta-analyses and other meta-analyses of the same drugs: systematic review. BMJ 333:782
25. Lundh A, Lexchin J, Mintzes B et al (2017) Industry sponsorship and research outcome. Cochrane Database Syst Rev 2(2):MR000033
26. Wang A, McCoy C, Murad M, Montori V (2010) Association between industry affiliation and position on cardiovascular risk with rosiglitazone: cross sectional systematic review. BMJ 340:c1344–c1344
27. Gilstad J, Finucane T (2008) Results, rhetoric, and randomized trials: the case of donepezil. J Am Geriatr Soc 56:1556–1562
28. Vedula S, Bero L, Scherer R, Dickersin K (2009) Outcome reporting in industry-sponsored trials of gabapentin for off-label use. N Engl J Med 361: 1963–1971
29. Caudill T (1996) Physicians, pharmaceutical sales representatives, and the cost of prescribing. Arch Fam Med 5:201–206
30. DeJong C, Aguilar T, Tseng C et al (2016) Pharmaceutical industry–sponsored meals and physician

prescribing patterns for medicare beneficiaries. JAMA Int Med 176:1114
31. Fleischman W, Agrawal S, King M et al (2016) Association between payments from manufacturers of pharmaceuticals to physicians and regional prescribing: cross sectional ecological study. BMJ:i4189
32. Yeh J, Franklin J, Avorn J et al (2016) Association of industry payments to physicians with the prescribing of brand-name statins in Massachusetts. JAMA Intern Med 176:763
33. Fickweiler F, Fickweiler W, Urbach E (2017) Interactions between physicians and the pharmaceutical industry generally and sales representatives specifically and their association with physicians' attitudes and prescribing habits: a systematic review. BMJ Open 7:e016408
34. Wood S, Podrasky J, McMonagle M et al (2017) Influence of pharmaceutical marketing on Medicare prescriptions in the District of Columbia. PLoS One 12:e0186060
35. Spurling G, Mansfield P, Montgomery B et al (2010) Information from pharmaceutical companies and the quality, quantity, and cost of physicians' prescribing: a systematic review. PLoS Med 7:e1000352
36. Molloy W, Strang D, Guyatt G et al (2002) Assessing the quality of drug detailing. J Clin Epidemiol 55:825–832
37. Smith J, Wahle C, Bernat J, Robbins N (2021) Financial conflicts of interest of United States–based authors in Neurology Journals. Neurology 96:e1913–e1920
38. Campbell E, Weissman J, Vogeli C et al (2006) Financial relationships between institutional review board members and industry. N Engl J Med 355:2321–2329
39. Mehlman C, Okike K, Bhandari M, Kocher M (2017) Potential financial conflict of interest among physician editorial board members of Orthopaedic Surgery Journals. J Bone Joint Surg 99:e19
40. Verma V (2017) Financial relationships with industry of editorial board members of the Three Journals of the American Society for Radiation Oncology. Int J Rad Oncol Biol Phys 99:286–291
41. Sismondo S (2013) Key opinion leaders and the corruption of medical knowledge: what the Sunshine Act will and won't cast light on. J Law Med Ethics 41: 635–643
42. Stein G, Kamler J, Chang J (2018) Ophthalmology patient perceptions of open payments information. JAMA Ophthalmol 136:1375
43. Butcher L (2015) IN PRACTICE: what the open payments program means for neurologists. Neurol Today 15:32–33
44. Robbins N (2018) Ethical issues pertaining to conflicts of interest between neurologists and the pharmaceutical and medical device industries. Semin Neurol 38: 589–598
45. Osorio I (2014) The NeuroPace trial: missing knowledge and insights. Epilepsia 55:1469–1470
46. Klein E, Solomon A, Corboy J, Bernat J (2016) Physician compensation for industry-sponsored clinical trials in multiple sclerosis influences patient trust. Mult Sclerosis Rel Disord 8:4–8
47. Taha A, Petrocelli J (2015) Disclosures about disclosures: can conflict of interest warnings be made more effective? J Empir Legal Stud. ssrn.com/abstract=2284212. Accessed 25 May 2021
48. Cain D, Loewenstein G, Moore D (2005) The dirt on coming clean: perverse effects of disclosing conflicts of interest. J Leg Stud 34:1–25
49. Kesselheim A, Robertson C, Myers J et al (2012) A randomized study of how physicians interpret research funding disclosures. N Engl J Med 367:1119–1127

Part XII

Predatory Practices

Predatory Journals and Conferences 49

Andrea Cortegiani, Giulia Catalisano, and Andrea Manca

Abstract

The adjective *"predatory,"* referred to scientific publications, presentations, and conferences, is currently used to identify entities that prioritize self-interest at the expense of the ethical codes that rule scientific divulgation. The profit-driven behavior that predatory publishers adopt exploits the open access publishing system and is characterized by deviation from best editorial practices, lack of transparency, spread of false or misleading information and intrusive solicitation practices. The number of predatory journals and conferences is worryingly high as is the income that predatory publishers and conferences organizers are getting from these deceitful practices. Unfortunately, it is not uncommon to find scholars who try to take career and curricula advantages by joining predatory practices through the publication of articles or the acceptance of an editorial board membership. Choosing a trustable journal to publish or discerning a dependable conference organizing body is not always easy. This chapter analyzes the main characteristics of predatory publishing and examines the several distinctive features and tools that can help scholars in the process of recognizing and avoiding predatory organizations.

A. Cortegiani (✉)
Department of Surgical, Oncological and Oral Science (Di.Chir.On.S.), University of Palermo, Palermo, Italy

Department of Anesthesia, Intensive Care and Emergency, Azienda Ospedaliera Universitaria "Policlinico Paolo Giaccone", Palermo, Italy
e-mail: andrea.cortegiani@unipa.it

G. Catalisano
Department of Surgical, Oncological and Oral Science (Di.Chir.On.S.), University of Palermo, Palermo, Italy

A. Manca
Department of Biomedical Sciences, University of Sassari, Sassari, Italy

49.1 Introduction

The crucial objective of scientific publishing is to advance knowledge. Since its conventional start on March 6, 1665, with the publication of the journal "Philosophical Transactions of the Royal Society" (London, UK) [1], scientific publishing has been integrally encouraged and sustained by respectful entities such as publishers and scientific societies, pursuing well established rules and ethic codes, despite including business elements. Regrettably, during the last 20 years, a worrying phenomenon has developed, which threatens the integrity and credibility of scientific divulgation, in both the forms of scientific publications in journals and scientific presentations to conferences: *predatory publishing*. Generally, this phenomenon is profit-driven and ignores the ethical codes that rule scientific publications.

J. Faintuch, S. Faintuch (eds.), *Integrity of Scientific Research*,
https://doi.org/10.1007/978-3-030-99680-2_49

49.2 "Predatory" Publishing

The term was coined by Professor Jeffrey Beall, a librarian from the University of Colorado Denver, in a landmark article published in *Nature* in 2012. Predatory publishers would distinct themselves for publishing "*counterfeit journals to exploit the open-access model in which the author pays*" [2]. In the effort of identifying such kinds of publications Beall created the Beall's list, a popular blacklist of potential, possible, or probable predatory journals which was discontinued in 2017.

In 2019, an article published in *Nature* with the eloquent title of "Predatory journals: no definition, no defence" illustrated the process and the conclusions of the consensus held in Ottawa, Canada, among 43 experts and stakeholders from ten different countries, to agree on a shared definition for predatory publishers. The consensus definition was: "*Predatory journals and publishers are entities that prioritize self-interest at the expense of scholarship and are characterized by false or misleading information, deviation from best editorial and publication practices, a lack of transparency, and/or the use of aggressive and indiscriminate solicitation practices*" [3]. Predatory conferences are a more recent phenomenon. Flattering and tempting emails to take part as speakers or delegates to unconvincing conferences are crowding scholars' mail inboxes.

49.2.1 The Size of the Issue

Giving an idea of the extent of predatory publishing is not easy: they often have an unstable nature making it difficult to keep track of them. A robust estimation (2014) counted 420,000 articles from approximately 8000 presumed deceptive journals, demonstrating an eight-fold increase in 5 years with an estimated profit of about USD $75 million [4]. Predatory publishing is being noticed also in the juridical context. One group handling more than 700 journals and organizing predatory conferences, was sanctioned a total of USD $50.1 million for deceptive claims to academics and researchers about the nature of their conferences and publications, and for hiding steep publication fees [5].

49.3 The National Institutes for Health (NIH-HHS, USA)

Within the same context, the NIH were identified among the largest funders for that organization, potentially resulting in a waste of public money [6]. Biomedical and many other fields of sciences are affected [7–14]. Anesthesiology and related disciplines count 212 potentially predatory journals, with more than 12,000 articles published at a median charge of USD $634.5 per article [9]. Neurosciences [8], rehabilitation [13], critical care medicine [11], orthopedics [7, 14], palliative care [12], and pediatric urology [10] similarly deserve to be mentioned.

49.3.1 The Threat to Ethics and Quality

The business of predatory publishing produces low quality evidence, endangering patient care in the hands of physicians not trained to recognize questionable quality publications. Predatory journals lack, partially or totally, in editorial control and peer reviewing. As a consequence the quality, authenticity and ethical nature of the data may not be reliable or sufficiently checked [15, 16]. Journal editors have the key role of deciding whether a paper should be rejected, reviewed or accepted for publication. This prestigious position is usually assigned to experts to guarantee quality in the decision process. Many predatory journals indiscriminately recruit academics to build legitimate looking, however hardly functional editorial boards.

49.3.2 Fake or Misguided Journal Editors

In 2015, a group of Polish scientists conceived a sting operation by displaying an ad hoc created fictitious scientist profile, and making it apply as editorial board member for journals selected from Beall's list. Several not only accepted but they did so against the payment of a fee [17]. Unethical behavior around the formation of editorial boards has been widely reported, with predatory journals publicizing opinion leaders and important scientists among their editorial board members who never consented to such a position, or were not fully aware of the quality of the journal [9, 18]. Additionally, researchers with incongruent curricula compared to the scope of the journals can often emerge, as well as the frequent repetition of the same names in the editorial board of dozens of presumed predatory journals [9, 18]. As it was observed during COVID-19 pandemic, a loosening in the reviewing process can lead to the spread of low quality or misleading data and to the retraction of articles [19].

49.3.3 A Two-Way Mechanism

It appears that the interaction between predatory journals and scholars may perversely benefit both parties (see Fig. 49.1). Scholars, especially the younger ones, under the pressure of the "publish or perish" rush might be victims of the false promises and harmful practices of predatory journals, but they can also knowingly resort to such practices to get curricula advantages [20–23], including quick peer-review process and bogus impact factors and metrics [24–26]. Although young researchers from low-middle income countries appear to be the most interested, scholars from high-ranking universities in the United Stated, Germany and Italy are not strangers to these practices [16, 27–29], encompassing those applying to qualify for associate or full professor [29].

49.3.4 Editorial Board Memberships

Scholars can easily add editorial board memberships to their curricula by answering to frequently received emails, while journals can

Fig. 49.1 Who is the predator and who is the prey? The figure describes a scholar exploiting the predatory system of publishing to get career advantages. Authors' own figure

profit from international participants, which is one of the criteria for potential indexing in legitimate databases such as PubMed (for evidence-based literature search) and Scopus (for scholarly metrics evaluation), causing the infiltration of predatory journals into standard scientific databases [9, 21, 30–33]. Thus, the idea that naïve, inexperienced and unaware scholars are the victims of this mechanism does not depict appropriately the whole situation: both parts have tangible interests, financially and career-wise.

49.3.5 Grey Zones and Distinctive Features

Discerning between clearly deceitful journals and simply low-quality/young journals might not be always easy. Clearly deceitful journals act on a "pay-to-publish" scheme exploiting the open-access model only to make profit, while emerging journals need to publish a minimum number of articles to secure indexing in legitimate databases, although they might not have the full spectrum of core and ancillary competences to align with the international regulations of scientific editorial practice such as the Committee on Publication Ethics (COPE) or the International Committee of Medical Journal Editors (ICMJE) [34].

49.3.6 Dubious Profiles

Purely predatory journals should be suspected on indexing of quality and on the ability of scholars to analyze the main features reported on websites and emails [35, 36]. Editorial related distinctive features typically list unreliable location for editorial office; scarcely credible editorial board members and absence of an editor-in chief; advertisement of false metrics and improper use of the term "impact factor" [26]; repetitive and flattering invitations to submit new manuscripts, join the editorial board or contribute as a reviewer. Journal related features: journals' names echoing those of well-known publications; excessive broad field of interest extended to several scientific disciplines. Submission related features: ambiguous or absent description of article processing charges; inadequate information about manuscript handling policy and ethics regulations; poor English form [34].

49.3.7 Choosing a Trustable Journal

There is not one single tool that can guarantee the quality of a journal, and that is why scholars should consider several items. Lists of journals such as the anonymously updated Beall blacklist and the Cabells' Scholarly Analytics may contribute to recognize predatory journals, but their reliability and updating is not sufficient to let them be the only parameter considered [37]. The Cabell's Scholarly Analytics (www.cabells.com/) is a scholarly, for-profit service that hosts and maintains a whitelist and a blacklist of journals. The whitelist is intended to provide academics with accurate information and reputable outlets for publication [38]. To be indexed journals must fulfill a number of criteria, which cover aspects such as Business Practices, Policies, Peer Review or Editorial Services [37]. The aim of the blacklist is to warn authors against journals with questionable practices and entails journals that have exhibited behavior that can be indicative of deception [37, 39, 40].

Indexing in a scientific database such as PubMed, PubMed central (PMC) or Scopus may be an element of reassurance, although certain predatory journals may penetrate under open-access policies [9, 30–32]. The possibility to retrieve articles published in predatory journals in such databases, especially in the case of PMC, may also lead to the erroneous belief that the journal is formally "indexed" in such databases, generating trust in scholars who found them. In fact, open access articles may be retrieved on PMC any time the article processing charges (APCs) are paid by an institution or funder and not directly by the authors. This is very different from stating that the journal in which the article was published is officially indexed on Medline and therefore meeting the National Institutes of Health (NIH) criteria [6].

49.3.8 The Directory of Open Access Journals (DOAJ)

It represents a trustworthy reference to evaluate the quality and legitimacy of open-access journals and publishers [33, 41]. DOAJ is a "*community-cured online directory that index and provide access to high quality, open access, peer review journals*" listing, to date, 16.451 journals. It is funded through donations made by sponsors and members such as several recognized organizations, publishers, and universities. DOAJ's website (doaj.org) is easily surveyable to find information about articles, journals, or publishers. Getting indexed, as the other services, is free of charge. A multi-step assessment process is ensured with approximately 50 quality control checks and constant re-evaluation including structured information about the journal, the editorial process quality and transparency, the copyright license and the open-access statement [41].

49.3.9 Think. Check. Submit.

This is the expression of an international initiative endorsed by legitimate publishers and several institutions such as COPE, DOAJ, and Open Access Scholarly Publishers Association (OASPA) designed to educate researchers, promote integrity, and build trust in credible research and publications [42, 43].

49.3.10 Potential Solutions

Relieving pressure on the "publish or perish" mechanisms may be a solution at the institutional level to discourage publication on predatory journals: to date 19,737 individuals and 2226 organizations have signed the Declaration on Research Assessment (DORA). The DORA was first developed in San Francisco in 2012 and focuses on the improvement of the evaluation methods of scholars' outputs, suggesting innovative approaches. For example, journal-based metrics should not be used as a measure of the quality of individual research articles or in hiring, promotion or funding. The scientific content of a paper is much more important than publication metrics or the identity/standings of the journal, especially for early-stage researchers [16, 44].

49.4 Predatory Conferences

These can be defined as "fake/bogus conferences that fail to be conducted with the standards of quality, transparency and integrity that characterize legitimate and reputable conferences" [45]. A basic search in PubMed/MEDLINE to outline the extent of publications dealing with predatory journals/publishing and conferences, respectively, yielded 194 retrieved items for the former and only 19 for the latter (effective date: June 22, 2021). In fact, the issue of fake, low-quality, deceptive conferences is not only relatively less known, but also considerably less investigated. As a rule of thumb the excessive, gratuitous flattery and deference of the invitations ("*We would be honored by your Eminent presence*") should set off the first alarm bells [25, 46].

49.4.1 Is this Conference for Real?

A distinguishing feature of predatory conferences relates to the nature of the organizing bodies, often consisting of private business companies rather than scientific societies, institutions, committees, universities, or professional associations [45, 46]. The event names tend to overlap with those of established conferences. Unaware prominent scientists are displayed by means of their photographs and hijacked names, thus misleading those that pay to attend and fall prey to such exploitation [45]. Based on the works by Cobey et al. (2017) [45] and Cortegiani et al. (2020) [34], it should be possible avoid the threats posed by scam invitations to attend pseudoscientific, lucrative meetings:

1. Genuine conferences do not invite scholars to serve as speakers/chairmen upon payment of a registration fee.
2. Legitimate venues do not delegate conference organization to business companies. If private companies are involved/hired to manage the event organization and logistics, as is often the case of modern scientific conferences, the event patronage is evidently and unquestionably retained by an institutional organizing body (such as a university, a national or international scientific society).
3. In their communications, legitimate scientific organizers avoid any form of obsequiousness.

4. They are meticulous and as specific as possible about the details, which means that they never address "Dear Dr or Professor." The communication is professional, free from grammar/spelling errors, and avoids descriptors overplaying the global reach of the event ("World/Wordly," "International," "Global").
5. They do not lure scientists into purchasing special or promotional packages, e.g., by including tourist attractions and targeting tourist information about the conference venue.

49.4.2 Zeroing Attendance to Fraudulent Conferences

Lang and colleagues [47] found that the most important features considered by 150 faculty members and medical residents of the University of Calgary were: (1) the conference focus on their specific area of interest (92%), (2) the opportunity to present their work (60%) and receiving recommendations from mentors (48%). Of lowest importance to them was the chance to obtain credits for continuing professional development and get honoraria for presentation [48]. Based on their extensive survey, the authors suggested to:

- use conferences as an opportunity to find potential collaborators for research and to speak to experts with similar interests,
- favor professional or scientific society meetings, which usually offer high impact with high quality science, and typically have experts of the field,
- avoid any conference invitation from unsolicited emails or unknown sources; avoid conferences run by pro-profit organizations, which may be of little to no scientific or educational value,
- attend more general conferences early in your training; attend meetings that invite high quality speakers,
- always seek advice from mentors or colleagues prior to committing to or paying for a conference if you are suspicious, it may be "predatory",
- do not attend conferences that only provide broad guideline talks that can easily be read in textbooks or journals,
- only attend conferences that are sponsored by known and recognized organizations.

49.4.3 Potential Solutions

Think. Check. Attend. [49] is a valuable initiative, stemming from the *Think. Check. Submit.* campaign [42] that aims to guide and assist researchers and scholars when choosing trusted conferences to attend and to present their research at. It consists of a 19-item checklist enquiring the organizers and sponsors, the agenda/editorial committee, and the conference proceedings [48], as well as a conference checker tool [50] by which direct assistance can be obtained. Yet paying attention to avoid careless purchases – as we typically do when shopping on the Internet – remains the smartest possible move (Fig. 49.1).

Acknowledgements We acknowledge Dr. Antonio Corrado (experienced physician and illustrator) for providing the figure.

References

1. Butterworth JF, Vetter TR (2019) Predatory journals undermine peer review and cheapen scholarship. Anesth Analg 128:11–12
2. Beall J (2012) Predatory publishers are corrupting open access. Nature 489:179
3. Grudniewicz A, Moher D, Cobey KD (2019) Predatory journals: no definition, no defence. Nature 576(7786):210–212
4. Shen C, Björk BC (2015) "Predatory" open access: a longitudinal study of article volumes and market characteristics. BMC Med 13:1–15
5. Court rules in FTC's favor against predatory academic publisher OMICS Group; Imposes $50.1 million judgment against defendants that made false claims and hid publishing fees. Federal Trade Commission. ftc.gov/news-events/press-releases/2019/04/court-rules-ftcs-favor-against-predatory-academic-publisher-omics. Accessed 6 Jun 2021
6. Manca A, Cugusi L, Cortegiani A, Ingoglia G, Moher D, Deriu F (2020) Predatory journals enter biomedical databases through public funding. BMJ 371:m4265

7. Yan JR, Baldawi H, Lex JR, Simchovich G, Baisi LP, Bozzo A, Ghert M (2018) Predatory publishing in orthopaedic research. J Bone Jt Surg 100(21):e138
8. Manca A, Martinez G, Cugusi L, Dragone D, Dvir Z, Deriu F (2017) The surge of predatory open-access in neurosciences and neurology. Neuroscience 353:166–173
9. Cortegiani A, Longhini F, Sanfilippo F, Raineri SM, Gregoretti C, Giarratano A (2019) Predatory open-access publishing in anesthesiology. Anesth Analg 128:182–187
10. O'Kelly F, Fernandez N, Koyle MA (2019) Predatory publishing or a lack of peer review transparency?—a contemporary analysis of indexed open and non-open access articles in paediatric urology. J Pediatr Urol 15: 159.e1–159.e7
11. Cortegiani A, Sanfilippo F, Tramarin J, Giarratano A (2019) Predatory open-access publishing in critical care medicine. J Crit Care 50:247–249
12. Cortegiani A, Garofalo E, Bruni A, Sanfilippo F, Longhini F (2019) Predatory open-access publishing in palliative and supportive care. J Pain Symptom Manag 57:e1–e3
13. Manca A, Martinez G, Cugusi L, Dragone D, Mercuro G, Deriu F (2017) Predatory open access in rehabilitation. Arch Phys Med Rehabil 98:1051–1056
14. Rupp M, Anastasopoulou L, Wintermeyer E, Malhaan D, El Khassawna T, Heiss C (2019) Predatory journals: a major threat in orthopaedic research. Int Orthop 43:509–517
15. Forero DA, Oermann MH, Manca A, Deriu F, Mendieta-Zerón H, Dadkhah M, Bhad R, Deshpande SN, Wang W, Cifuentes MP (2018) Negative effects of "predatory" journals on global health research. Ann Glob Heal 84:584–589
16. Cortegiani A, Misseri G, Gregoretti C, Einav S, Giarratano A (2019) The challenge of the predatory open-access publishing outbreak. Eur J Anaesthesiol 36:810–813
17. Sorokowski P, Kulczycki E, Sorokowska A, Pisanski K (2017) Predatory journals recruit fake editor. Nature 543:481–483
18. Ruiter-Lopez L, Lopez-Leon S, Forero DA (2019) Predatory journals: do not judge journals by their editorial board members. Med Teach 41:691–696
19. Cortegiani A, Catalisano G, Ippolito M, Giarratano A, Absalom AR, Einav S (2021) Retracted papers on SARS-CoV-2 and COVID-19. Br J Anaesth 126: e155–e156
20. Cobey KD, Grudniewicz A, Lalu MM, Rice DB, Raffoul H, Moher D (2019) Knowledge and motivations of researchers publishing in presumed predatory journals: a survey. BMJ Open 9(3):e026516
21. Pond BB, Brown SD, Stewart DW, Roane DS, Harirforoosh S (2019) Faculty applicants' attempt to inflate CVs using predatory journals. Am J Pharm Educ 83(1):7210
22. Hedding DW (2019) Payouts push professors towards predatory journals. Nature 565:267
23. Frandsen TF (2019) Why do researchers decide to publish in questionable journals? A review of the literature. Learn Publ 32:57–62
24. Moher D, Srivastava A (2015) You are invited to submit. . . . BMC Med 13:1–4
25. Mercier E, Tardif PA, Moore L, Le Sage N, Cameron PA (2018) Invitations received from potential predatory publishers and fraudulent conferences: a 12-month early-career researcher experience. Postgrad Med J 94:104–108
26. Gutierrez FRS, Beall J, Forero DA (2015) Spurious alternative impact factors: the scale of the problem from an academic perspective. BioEssays 37:474–476
27. Moher D, Shamseer L, Cobey K et al (2017) Stop this waste of people, animals and money. Nature 549:23–25
28. More than 5,000 German scientists have published papers in pseudo-scientific journals. ndr.de/der_ndr/presse/More-than-5000-German-scientists-have-published-papers-in-pseudo-scientific-journals,fakescience178.html. Accessed 6 Jun 2021
29. Bagues M, Sylos-Labini M, Zinovyeva N (2019) A walk on the wild side: 'predatory' journals and information asymmetries in scientific evaluations. Res Policy 48:462–477
30. Severin A, Low N (2019) Readers beware! Predatory journals are infiltrating citation databases. Int J Public Health 64:1123–1124
31. Manca A, Moher D, Cugusi L, Dvir Z, Deriu F (2018) How predatory journals leak into PubMed. CMAJ 190: E1042–E1045
32. Manca A, Cugusi L, Dvir Z, Deriu F (2017) PubMed should raise the bar for journal inclusion. Lancet 390: 734–735
33. Cortegiani A, Ippolito M, Ingoglia G et al (2020) Citations and metrics of journals discontinued from Scopus for publication concerns: the GhoS(t)copus project. F1000Research 9:415
34. Cortegiani A, Manca A, Giarratano A (2020) Predatory journals and conferences: why fake counts. Curr Opin Anaesthesiol 33:192–197
35. Cobey KD, Lalu MM, Skidmore B, Ahmadzai N, Grudniewicz A, Moher D (2018) What is a predatory journal? A scoping review. F1000Research 7:1001
36. Shamseer L, Moher D, Maduekwe O, Turner L, Barbour V, Burch R, Clark J, Galipeau J, Roberts J, Shea BJ (2017) Potential predatory and legitimate biomedical journals: can you tell the difference? A cross-sectional comparison. BMC Med 15:1–14
37. Strinzel M, Severin A, Milzow K, Egger M (2019) Blacklists and whitelists to tackle predatory publishing: a cross-sectional comparison and thematic analysis. MBio 10(3):e00411-19
38. Cabell's International - About. cabells.com/about. Accessed 9 Jun 2021
39. Murphy JA (2019) Predatory publishing and the response from the scholarly community. Ser Rev 45: 73–78

40. Cabell's International - About predatory reports. cabells.com/about-predatory. Accessed 9 Jun 2021
41. Directory of Open Access Journals. doaj.org/. Accessed 7 Jun 2021
42. Think. Check. Submit. thinkchecksubmit.org/. Accessed 7 Jun 2021
43. Cortegiani A, Shafer SL (2018) "Think. Check. Submit." to avoid predatory publishing. Crit Care 22:1–2
44. Home. DORA. sfdora.org/. Accessed 7 Jun 2021
45. Cobey KD, De Costa E, Silva M, Mazzarello S, Stober C, Hutton B, Moher D, Clemons M (2017) Is this conference for real? Navigating presumed predatory conference invitations. J Oncol Pract 13:410–413
46. Asadi A (2019) Invitation to speak at a conference: the tempting technique adopted by predatory conferences' organizers. Sci Eng Ethics 25:975–979
47. Lang R, Mintz M, Krentz HB, Gill MJ (2019) An approach to conference selection and evaluation: advice to avoid "predatory" conferences. Scientometrics 118:687–698
48. Check - Think Check Attend. thinkcheckattend.org/check/. Accessed 24 Jun 2021
49. Think Check Attend. thinkcheckattend.org/. Accessed 24 Jun 2021
50. Conference checker - Think Check Attend. thinkcheckattend.org/conference-checker/. Accessed 24 Jun 2021

How Open Access Publishing Developed in the 2010s and the Potential for Publishing Misconduct

50

W. Angus Wallace and Ben Ollivere

Abstract

The drive to encourage medical journals to move in the direction of Open Access (OA) publishing started in the 2000s but it was not until the 2010s that OA publishing became a reality and has resulted in a transformation in the publishing scene across the globe. In the UK, this was initially encouraged by the publication of the Finch Report, commissioned by the UK Government, and further influenced by Plan S developed in Europe. The reasons for the drive to Open Access Publishing has been described but the unintended consequences were not predicted and are now coming back to bite us. Who expected that the business model for OA publishing would be so attractive, and who predicted how predatory journals would emerge out of the arena like a phoenix arising from the ashes? In relation to academic publishing, the most appropriate saying would now appear to be "We are living in interesting times."

50.1 Background to Open Access Journals

Open access (OA) is a set of principles and a range of practices through which research outputs are distributed online, free of cost or other access barriers [1]. With open access strictly defined (according to the 2001 definition), or "libre open access," barriers to copying or reuse are also reduced or removed by applying an open license for copyright in addition [1].

Conventional (non-open access) journals cover publishing costs through access tolls such as subscriptions, site licenses or pay-per-view charges. Open-access journals are characterized by funding models which do not require the reader to pay to read the journal's contents. The open access movement is mainly focused on "peer reviewed research literature" [2].The term "open access" itself was first formulated in three public statements in the 2000s: the Budapest Open Access Initiative in February 2002, the Bethesda Statement on Open Access Publishing in June 2003, and the Berlin Declaration on Open Access to Knowledge in the Sciences and Humanities in October 2003 [3]. The initial concept of open access refers to an unrestricted online access to scholarly research, primarily intended for scholarly journal articles.

Historically, this has centered mainly on print-based academic journals, but it can be applied to all forms of published research output, including peer-reviewed and non- peer-reviewed academic

W. A. Wallace (✉) · B. Ollivere
University of Nottingham, Nottingham, UK
e-mail: Angus.Wallace@RCSEd.ac.uk; Benjamin.Ollivere@nottingham.ac.uk

J. Faintuch, S. Faintuch (eds.), *Integrity of Scientific Research*,
https://doi.org/10.1007/978-3-030-99680-2_50

journal articles, conference papers, PowerPoint and video presentations, webinars, theses [4], book chapters [1], monographs [5], and images [6].

50.2 The Finch Report from the United Kingdom

In October 2011, the UK government set up a working group on expanding access to published research findings, known as the "Finch Group" after its chair, Dame Janet Finch [7]. It was tasked with "proposing a programme of action and also invited to make recommendations to government, research funders, publishers, and other interested parties, on how access to research findings and outcomes can be broadened for key audiences such as researchers, policy makers and the general public."

The Finch report, published in June 2012, was widely regarded as the most important government review in favor of open access to date. The report stated "Our view is that the UK should embrace the transition to open access, and accelerate the process in a measured way which promotes innovation but also what is most valuable in the research communications ecosystem." Subsequently a number of "hybrid" journals developed, defined as journals that carry both open-access and traditional-access articles: an extra fee, the Article Processing Charge (APC)—has to be paid to make an article in these journals freely available to all. Unfortunately, the established publishers saw this as solving a problem of having a stable income stream, and some charged an excessively large APC.

50.3 Plan S

Plan S was developed by a group of the largest research funders in Europe, who took the view that they wanted to maximize dissemination and value for the taxpayer, by facilitating the publication of their funded research in journals that have an entirely open-access model [8]. They believed that by using this approach they would destabilize the publishing world and would apply disproportionate pressure to the hybrid journals both financially and academically.

The aims of free access to scientific information deserves praise however the enormous effort, responsibility, ethics and costs behind carefully curated articles should not be minimized. Even Wikipedia, which enjoys multiple volunteers and a few professional editors for doing a serious job, is prone to inaccuracies.

The National Institutes of Health (USA) adopted a definition of open access (articles become available after 6 months), and respected academic publications like JAMA and Lancet carefully follow their guidelines. They have little in common with countless journals of questionable reputation, which similarly announce themselves as open access. If adequately scrutinized these journals of questionable reputation will be identified as nothing less than predatory publications. Editorial patterns, peer review and analysis of scientific rigor are entirely lacking or perfunctorily conducted. This doesn't exclude the possibility of well qualified articles appearing within their pages, because the authors were misled by the journal title and believed they were submitting their articles to a high-quality journal. Yet the risk of encountering junk material is not negligible.

50.4 Financial Thresholds Vs. Scientific Criteria

Trauma and orthopedics are areas in which case series, prognostic studies, registry articles, and retrospective reviews have a significant role to play, in certain circumstances even more than standard protocols. Moreover, not all who publish such reports are able to comply with the disbursements of the open access model. Maybe it's not time yet to discard the subscription model, which boasts quite a good performance track. Just as an example, The Bone & Joint Journal is typically rich with insightful information and even breakthrough contributions, whose authors do not currently require official financial help. Nevertheless, such printed volumes are sent to scores of subscribers in all continents, causing a significant impact in their activities. Probably a lot of them are "partial adopters" of new techniques and prefer the paper edition, as we do.

Plan S has unquestionable merits however it is not necessarily ideal for observational studies which, as alluded to, are still fundamental in many domains. The other point is that more than the publication model, it's the publisher and the editorial committee who really count. In an era not strange to controversy and even blatant fraud, all texts emerging from poorly known publications should be viewed with a grain of salt.

50.5 Open Access Journal Legislation in North America

Peter Suber is a University Professor, Director of the Harvard Office for Scholarly Communication, Director of the Harvard Open Access Project, and a Senior Researcher at the Berkman Klein Center for Internet & Society. He has authored an excellent book on "Open Access" which we can recommend [3].The US National Institutes of Health (NIH) Public Access Policy ensures that the public has access to the published results of all NIH funded research. It requires scientists to submit final peer-reviewed journal manuscripts that arise from NIH funding to the digital archive PubMed Central upon acceptance for publication. To help advance science and improve human health, the Policy requires that these papers are accessible to the public on PubMed Central no later than 12 months after publication [9].

In 2015, North America topped the world rankings with the highest regional Internet penetration rate (86.9%), and Europe follows closely with an Internet penetration rate of 70.4%. This has created an enabling environment for the development of OA digital repositories and e-journal interfaces in the regions [9].

50.6 What About the Downside?

Academics are employed by universities which require, in return, that these academics carry out and publish their research, freeing them to dive deeply into their research topics and publish specialized texts that may not have general market appeal. In the UK this is monitored through the evaluation of University Departments nationally, using the Research Assessment Exercise (RAE) up to 2008, which was then replaced by the Research Excellence Framework (REF) in 2014. Its stated aims are to provide accountability for public investment in research, establish "reputational yardsticks" [10], and thereby achieve an efficient allocation of resources. The REF has a significant emphasis on publication in high Impact Factor Journals. However, critics have argued that there is too much focus on the impact of research outside the university system, and that impact may have no real relevance to the quality of the research being reported.

50.7 Free Lunch Vs. Fair Pay

Balanced against this, creative people who live by royalties, such as novelists, musicians, and moviemakers, may consider this scholarly tradition a burden and sacrifice for these scholars [3]. Unfortunately, this Open Access environment has now moved in the direction of music and the theatre as well, with the general public expecting to be able to access music and shows inexpensively or free through channels like YouTube and streaming sites like Spotify and Amazon Music Unlimited. This is causing a real crisis in the music industry with the Musicians' Union in the United Kingdom "Fighting for Fair Pay for Musicians" [11]. As a result, the UK Government is investigating whether artists are paid fairly for streaming music. As well as the issue of pay, the UK Government is investigating how playlists and algorithms distort the music market, and whether new music is being strangled by the dominance of big names like Ed Sheeran, Ariana Grande and Drake [12].

50.8 Ongoing Controversies with and Without Open Access Publication

During 2019 and 2020, three of the most prestigious medical journals published research papers that have been considered controversial, two of which have been subsequently withdrawn because they were either fraudulent or the research on which they were based was found to be faulty and thus non-reproducible.

These retracted papers were originally published in

1. The New England Journal of Medicine
2. The Lancet
3. The Journal of the American Medical Association Network Open (JAMA-NO)

50.9 The New England Journal of Medicine (NEJM) Paper November 2019

Stone and the EXCEL Trial Investigators published the "Five-Year Outcomes after PCI or CABG for Left Main Coronary Disease" [13]. Concerns were raised about the study after additional unpublished data from the trial had been obtained by the BBC's Newsnight program in December 2019 [14]. It was also reported that the trial had been funded by the stent maker Abbott (Abbott Vascular, Temecula, CA, USA).

Elisabeth Mahase, Clinical news reporter at the BMJ, published a report on December 17, 2019 on "Surgical association withdraws support for stent advice after controversy over study" [15] in which she stated:

> The data reportedly showed that, when using the universal definition of heart attack, patients with left main coronary artery disease treated with stents are 35% more likely to die than those treated with conventional open heart surgery.
>
> In the paper, the researchers only published the findings using an alternative definition of heart attack. Using this definition, the paper claimed that stents and heart surgery are equally effective for people with left main coronary artery disease.
>
> After reviewing the unpublished data (using the universal definition), EACTS has now withdrawn its support for guidance it co-wrote with the European Society of Cardiology (ESC), and invited the ESC to work with it to develop a new joint section of the myocardial revascularisation guidelines 'as a matter of urgency'.
>
> Surgeon David Taggart, from Oxford University, was chair of the surgical committee for the trial, but resigned during the submission of the paper to NEJM over problems with how the findings were being presented.
>
> Taggart told The BMJ that he thought the trial was 'very well designed' and was happy with the protocol, which outlined that the study would present 'two different sets of data' for the two heart attack definitions, therefore allowing comparison of the findings with previous research. He explained that the alternative definition was untested and based 'purely on a biochemical definition' unlike the standard definition, which combined 'some biochemical and some other data'. But he added, 'I resigned over the failure to appropriately describe the significant difference in mortality that we saw. To see that excess mortality was the most important part of the study. NEJM initially insisted that it should be the central emphasis of the paper. But when the paper was resent to NEJM without that matter being resolved, that's when I decided I could not go along with it'.

The authors of this chapter feel that the commercial sponsorship of the trial may have been a significant factor that resulted in the problems alluded to.

50.10 The Lancet Paper in May 2020

Mehra et al. published a highly referenced article in the Lancet electronic edition in May 2020 [16]. Three of the four original authors submitted a retraction of their paper [17] for the reasons stated below:

> After publication of our Lancet Article, several concerns were raised with respect to the veracity of the data and analyses conducted by Surgisphere Corporation and its founder and our co-author, Sapan Desai, in our publication. We launched an independent third party peer review of Surgisphere with the consent of Sapan Desai to evaluate the origination of the database elements, to confirm

the completeness of the database, and to replicate the analyses presented in the paper.

Our independent peer reviewers informed us that Surgisphere would not transfer the full dataset, client contracts, and the full ISO audit report to their servers for analysis as such transfer would violate client agreements and confidentiality requirements. As such, our reviewers were not able to conduct an independent and private peer review and therefore notified us of their withdrawal from the peer-review process. We always aspire to perform our research in accordance with the highest ethical and professional guidelines. We can never forget the responsibility we have as researchers to scrupulously ensure that we rely on data sources that adhere to our high standards. Based on this development, we can no longer vouch for the veracity of the primary data sources. Due to this unfortunate development, the authors request that the paper be retracted. We all entered this collaboration to contribute in good faith and at a time of great need during the COVID-19 pandemic. We deeply apologise to you, the editors, and the journal readership for any embarrassment or inconvenience that this may have caused.

The authors of this chapter feel that, when large database analyses are carried out, it is important that all co-authors are confident about the completeness of the database and the professionalism of those carrying out the analyses.

50.11 The JAMA-Network Open (JAMA-NO) in December 2020

The authors of a 2019 meta-analysis in a JAMA journal by Hodkinson et al. [18] on "Accelerometer- and pedometer-based physical activity interventions among adults with cardiometabolic conditions: a systematic review and meta-analysis" have retracted the paper after discovering that a quarter of the studies they had used in the analysis did not belong [19].

The retraction is the first for the journal, which had published some 2800 articles before having to pull one, as Frederick P. Rivara, the editor in chief, told Retraction Watch. One in 2800, we should note, is quite close to the 4 in 10,000 rate of retraction in the overall literature.

The authors of this chapter feel that there have been too many publications of meta- analysis and that it is really difficult for reviewers to have a full and detailed knowledge of many of the papers they are invited to review.

50.12 Conclusions

The philosophy of Open Access publishing for medical and scientific research is laudable and has achieved a huge step-forwards over the past decade. However, the law of unintended consequences, often cited but rarely defined, is that the actions of people and especially those of governments, always have effects that are unanticipated or "unintended." Economists and other social scientists have heeded its power for centuries; for just as long, politicians and popular opinion have largely ignored it [20]. We shall all need to decide how we personally and professionally respond to the increasing push to provide and use Open Access and how we respond to its range of consequences.

References

1. Suber P (2004) Open Access Overview - Focusing on open access to peer- reviewed research articles and their preprints. http://bit.ly/oa-overview
2. Swan A (2012) Policy guidelines for the development and promotion of open access. United Nations Educational, Scientific and Cultural Organization, Paris
3. Suber P (2012) Open access. MIT Press, Cambridge, MA
4. Schöpfel J, Prost H (2013) Degrees of secrecy in an open environment. The case of electronic theses and dissertations. ESSACHESS – J Commun Stud 6(2 (12)):65–86
5. Schwartz M (2012) Directory of open access books goes live. Library Journal
6. Agency ES (2014) Terms and conditions for the use and distribution of sentinel data (PDF). European Space Agency (ESA)
7. Finch J, Finch_Group_Members (2012) Accessibility, sustainability, excellence: how to expand access to research publications. UK Government. https://www.researchinfonet.org/finch/
8. Marc S, cOAlition S (2018) Plan S - making full and immediate Open Access a reality. https://www.coalition-s.org/. Accessed 4 Sep 2018
9. UNESCO (2017) Global open access portal (GOAP) - Europe and North America. Copyrighted 2017. http://www.unesco.org/new/en/communication-and-

information/portals-and-platforms/goap/access-by-region/europe-and-north-america/. Accessed 4 May 2021
10. UK_Research_and_Innovation (UKRI) (2020) REF2021 - Research excellence framework - what is the REF? https://www.ref.ac.uk/about/whatref/. Accessed 4 May 2021
11. Union M (2021) #WorkNotPlayMU - fighting for fair pay for musicians. https://musiciansunion.org.uk/campaigns/fighting-for-fair-pay-for-musicians. Accessed 4 May 2021
12. BBC, Savage M (2020) MPs to investigate whether artists are paid fairly for streaming music. https://www.bbc.co.uk/news/entertainment-arts-54551342. Accessed 4 May 2021
13. Stone GW et al (2019) Five-year outcomes after PCI or CABG for left main coronary disease. Randomized Controlled Trial 381(19):1820–1830
14. Cohen D, Brow E (2019) Surgeons withdraw support for heart disease advice. BBC Newsnight
15. Elisabeth M (2019) Surgical association withdraws support for stent advice after controversy over study. BMJ 367:17006
16. Mehra MR, Ruschitzka F, Patel AN (2020) Hydroxychloroquine or chloroquine with or without a macrolide for treatment of COVID-19: a multinational registry analysis. Lancet. https://doi.org/10.1016/S0140-6736(20)31180-6
17. Mehra MR, Ruschitzk F, Patel AN (2020) Retraction— hydroxychloroquine or chloroquine with or without a macrolide for treatment of COVID-19: a multinational registry analysis. Lancet 395(10240):1820
18. Hodkinson A et al (2019) Accelerometer- and pedometer-based physical activity interventions among adults with cardiometabolic conditions: a systematic review and meta-analysis. JAMA Netw Open 2(10):e1912895
19. Hodkinson A et al (2020) Letter-notice of retraction. Hodkinson et al. Accelerometer- and pedometer-based physical activity interventions among adults with cardiometabolic conditions: a systematic review and meta-analysis. JAMA Netw Open 3(12):e2032700. https://doi.org/10.1001/jamanetworkopen.2020.32700
20. Norton R (2002) Unintended consequences. http://www.econlib.org/library/Enc/UnintendedConsequences.html; https://www.econlib.org/library/Enc1/UnintendedConsequences.html

51 The Challenges from Predatory Journals and Fake Medical News

W. Angus Wallace

Abstract

This chapter describes how and why predatory journals have flourished worldwide and the problems they have created, chiefly through making the publication of research easier but carried out with less scrutiny and poorer peer-review processes. In addition, the problem of multimedia publishing of lectures, lecture series, web publishing of video presentations and related but unreviewed educational material is described, with the consequences in relation to the negative effect this is having on patient information and education. The reader will also be introduced to the disadvantages of providing open access material to the Internet and how misinformation emanates from this and can be worsened by the activities of advertisers and influencers. These have resulted in an epidemic of fake medical news occurring in the 2020s. Efforts now being made to address this by providing reliable online medical information are précised.

W. A. Wallace (✉)
University of Nottingham, Nottingham, UK
e-mail: Angus.Wallace@RCSEd.ac.uk

51.1 Background

Publishing research can generate a substantial income, and in recent years, companies were set up around the world designed to capitalize on this business opportunity. Established journals have sophisticated peer-review processes with experienced peer reviewers and are proud to maintain high academic standards. However, this was not the main goal of many of the new start-ups, which were more focused on commercial success rather than maintaining standards. To be fair to all new journals, there is an ethical spectrum within journal publishing, but there is a real problem separating the truly ethical journals from the predatory ones, particularly as many well-established journals are now turning to open access, often as a hybrid model, but how is the 'trusting' researcher able to tell the difference?

Many of these new start-ups, loosely called predatory journals, also had to create their own peer-review process. They went out into the clinical and academic community and invited those who were interested to become editors, subeditors and reviewers, without properly vetting their credentials [1].

51.2 The Arrival of Citation Indexes

The main metric used for demonstrating the scientific value of journals established in the 2000s is the 'impact factor'. The impact factor (IF) or

J. Faintuch, S. Faintuch (eds.), *Integrity of Scientific Research*,
https://doi.org/10.1007/978-3-030-99680-2_51

'journal impact factor' (JIF) of an academic journal is a scientometric index calculated by Clarivate (previously called Thomson Reuters ISI (Institute for Scientific Information)) and based on the Journal Citation Reports (JCR). These reflect the yearly average number of citations that articles published in the last 2 years in each journal received. The pioneering Institute for Scientific Information (ISI) was founded by Eugene Garfield, PhD (1925–2017), in the USA in 1960. He and his institute should also be credited for the original concept of citation indexes in science, proposed in 1955 [2]. Although the IF is used worldwide, many journals in Asia, Latin America, Middle East, China and Far East felt they were disadvantaged by the Western-based impact factor. In the last years, certain publications in these countries have moved towards using other metrics that are 'kinder' to their journals.

51.3 Alternative Impact Factors

Since starting around 2012, the world of academic publishing has become flooded with alternative impact factors and misleading metrics, many of which have been launched by new companies set up to provide these alternative metrics for a fee. This has been highlighted by a paper on 'The story of fake impact factor companies and how we detected them', published in the *Electronic Physician* [2]. These reportedly fake impact factors can damage the prestige and reliability of scientific research and scholarly journals, such as the universal impact factor (UIF) [3], global impact factor (GIF) [4], CiteFactor (*African Journals OnLine*) [5] and International Scientific Indexing (ISI) impact factor [6].

Other alternative indexes with more solid background, yet lacking the track record of IF, are the SCImago Journal Rank (SJR), related to the Scopus database/Elsevier [7], the International Organization of Scientific Research (IOSR) journals that are indexed in the 'Indian Citation Index' as recommended by the Indian National Medical Commission (NMC) new guidelines [8], the Google-based Journal Impact Factor (based on Google Scholar) [9] and the CiteScore, a new metric also produced by Elsevier to help track journal performance and make decisions [10].

51.3.1 Impact Factor: Is It the Real One?

Often in predatory journals, when the impact factor is quoted, it is not the actual impact factor we understand it to be. One journal has stated 'The impact factor shown here is equivalent to Citescore and is, therefore, used as a replacement for the same. Citescore, produced by Elsevier, is part of the Scopus basket of journal metrics [10] and can be a little higher or different compared to the impact factor produced by Journal Citation Report' [11].

The use of fake impact metrics should be considered a 'red flag' to researchers who wish to publish their work in properly accredited and respected journals. In other words, if you are considering submitting an article to a journal that does not use the traditional journal impact factor [12], then beware that you may be submitting your research article to a predatory journal.

51.3.2 Why Does Peer Review Matter?

Peer review has been the established standard that has maintained the quality of scientific research and our confidence in what we read over the past 50 years. Perhaps the best description of the value of peer review was from an article on 'Five reasons why peer review matters' published online by Roganie Govender on the website of the publisher, Elsevier [13]. She explains that it is your peers in the field examining the work and raising questions about aspects that may need greater clarity. The peer reviewer is expected to offer an unbiased view, highlighting when certain claims cannot be drawn from your work, given the scope of the study design and methodology:

1. Time to reflect—The reviewer is given an opportunity to reflect on someone else's work and to provide thoughtful comment using his/her own knowledge and expertise of the subject. On receiving feedback, authors have time to reflect on how their research is viewed by experts in the field and gives them an opportunity to make constructive changes to their research work before publication.
2. Research quality—'Which' magazine, in the UK, has become the consumer's guide to buying new products? Peer review does the same for scientific publications—it provides some quality assurance to consumers of research.
3. Understanding our ethical responsibility as researchers—Research is undertaken in the hope that our findings will contribute to the betterment of care, treatment or patient satisfaction. The way in which we conduct research and the basis upon which we make claims should therefore be subject to scrutiny. Both authors and reviewers are responsible for sharing this ethical responsibility.
4. Training—Engaging in the process of peer review, either as a reviewer or as an author, results in both receiving feedbacks and is an opportunity for academic dialogue. Peer-reviewing the work of others helps you to think more critically about your own work, as well as the work you are evaluating.
5. Helping each other—Peer review brings both the author and the reviewer closer to the rest of the scientific community. There is a lot of satisfaction in knowing that you are both contributing to the improvement of a piece of work, and both the article author(s) and the reviewer should share this sense of collegiality.

51.4 What Is a Predatory Journal?

In 2019, *Nature* published an article entitled 'Predatory journals: no definition, no defense' that is recommended reading. Leading scholars and publishers from ten countries have now agreed a definition of predatory publishing that can protect scholarship. They identified 18 fundamental questions, each requiring 3 rounds of debate and a total of 12 h of discussion, to reach their conclusions [14].

> Predatory journals and publishers are entities that prioritize self-interest at the expense of scholarship and are characterized by false or misleading information, deviation from best editorial and publication practices, a lack of transparency, and/or the use of aggressive and indiscriminate solicitation practices

False impact factors are often used by predatory publishers. Consulting Journal Citation Reports' master journal list can confirm if a publication is indexed [15].

51.5 The History of 'Beall's List' of Predatory Journals

Jeffrey Beall, now an associate professor and librarian from the University of Colorado at Denver, recognized the potential damage being created by predatory journals, and in 2008, as a personal endeavour, he launched his online 'Beall's List' of predatory journals. It became a widely followed piece of work by the mid-2010s. Unfortunately, its influence led some publishers of journals on the list to threaten defamation lawsuits against Beall, as well as lodging official complaints against Beall's work to the University of Colorado. On Sunday, 15 January 2017, Beall took down his site—www. scholarlyoa.com. It vanished from the Internet without explanation. Beall's business partners now say he was forced to shut down the website for legal and harassment reasons [16].

Since then, a new website 'Stop Predatory Journals' [17] has taken over the role of 'Beall's List'. The new site owners have explained on their website, 'After Jeffrey Beall took down his list of predatory journals in January 2017 to avoid continued harassment and threats, a small group of scholars and information professionals decided to anonymously rebuild and resurrect that list. The anonymity of the maintainers allows for them not to be harassed, but the structure of the site and the platform allows for a community-based approach to curating and maintaining these lists. It is hoped that a community of people will develop around the site to vet and investigate journals already on the lists and contribute new submissions'.

On a personal basis, I now use 'Stop Predatory Journals' [18] which lists hundreds of journals. This is my guide to journals that might be predatory, recognizing that it is not completely accurate but it is a useful guide.

51.6 How Can Predatory Journals Be Controlled?

In 2018, Yan et al. published an important paper [19] in which they reported that the scientific community needed to increase its awareness of how to identify and avoid predatory journals. This is especially important for junior researchers because 65% of the publications in predatory journals are by first-time authors [20], with many recounting that they had been lured by the promise of quick publication, only to regret it after being hounded for additional publication fees [21]. It is now well recognized that poor publication practices of predatory journals are likely to have a negative effect on clinical care [22–24].

Predatory ''journals can often be spotted by including articles with spelling and grammatical errors on their websites, promises of rapid publication (for additional fees) and a lack of information about retraction policies, copyright, manuscript handling or digital preservation. Yan et al. have recommended the following strategy when searching for a journal in which to publish one's work in Think. Check. Submit [25].

The 'Think. Check. Submit.' website was [25] launched in 2015 and has been designed to help researchers identify trusted journals for their research [25]. Through a range of tools and practical resources, this international, cross-sector initiative aims to educate researchers, promote integrity and build trust in credible research and publications. Central to the initiative is a checklist based on best practices in scholarly communication. This widely used checklist is available in nearly 40 languages (from Albanian to Vietnamese).

51.7 Cowboys in the Nineteenth-Century Wild West

The era known as the Wild West or the American Frontier began after the Civil War in 1865 and ended around 1895. The myth-making cowboys and gunslingers lasted only 30 years. The frontier area west of the Mississippi River during the late 1800s included the territories of Dakota, Nevada, Oregon, Utah, Idaho, Montana, Arizona, California, New Mexico and Colorado. The untamed territories were noted for their lawlessness, which gave rise to wild, rowdy, unrestrained, disorderly and unruly behaviour—which is what made for great stories in print and on screen. But while much of the culture that is attributed to the Wild West was just normal colonial culture in many parts of America at different times, the 'Wild West' became ingrained in people's imaginations.

51.8 The Multimedia Medical Cowboys in the 2020s

The way multimedia has been developed and is being used in the 2020s is remarkably like the Wild West era in the USA in the late 1800s—lawless, unrestrained publication of PowerPoint presentations, podcasts, videos and some blatantly commercial presentations. This has happened as a result of the explosion of 'social media'. YouTube was the first on the scene, but in medical circles, it is VuMedi (Oakland, CA, USA—vumedi.com) that has dominated. These platforms are not properly peer-reviewed or refereed, so virtually anyone can upload his/her own material and publish it, even if the material is inaccurate or misleading. There is a perception by the general public that VuMedi, as a medical platform, must be refereed. However, material submitted to VuMedi is uploaded by the authors with minimal checks on the accuracy or validity of the information. Those running VuMedi simply facilitate the display of the presentations in a high-quality way, but they do not referee or monitor the material that is uploaded. Much funding is required to run organizations like VuMedi, and they attract funding through charging some of their clients (hospitals and commercial companies) and advertising. Consequently, there is a huge amount of information on the Internet that is not refereed, not censored in any way and frankly can be inaccurate and misleading.

To clarify the situation with VuMedi, I emailed them to identify how I could publish a video on their platform. They explained that, as a single surgeon, I could simply upload my presentation onto their platform and they would publish it. But if I were an organization or commercial company, then a fee would be payable, either on a one-off or subscription basis (currently quoted for pilot programmes at $35 k and annual brand channels at $72 k).

51.9 Peer-Reviewing Multimedia Content

Several educational organizations and established medical journals have realized that there is now a much-needed peer-review process required for multimedia publications, and they are beginning to create appropriate platforms. This has been pioneered effectively by the American Academy of Orthopaedic Surgeons (AAOS) with their Orthopaedic Video Theater [26, 27]. Both the *Journal of Bone and Joint Surgery* and the *American Journal of Sports Medicine* are now providing appropriate platforms. The new *Video Journal of Sports Medicine*, being promoted by the American Orthopaedic Society for Sports Medicine, should be able to provide viewers with reassurance that the content has been evaluated and scrutinized both for accuracy and for uncontrolled bias by peers [28].

51.10 The Role of Social Media Influencers

An influencer is someone who has the following:

(a) The power to affect the purchasing decisions of others because of his or her authority, knowledge, position or relationship with his or her audience.
(b) A following in a distinct niche, with whom he or she actively engages. The size of the following depends on the size of his/her topic of the niche [29].

It is important to note that these individuals are not merely marketing tools but rather social relationship assets, with which brands can collaborate to achieve their marketing objectives.

These influencers generate personal income by passing on advertising information to their followers many of whom will be patients with underlying medical problems. Although some influencers are altruistic, a significant number carry out their activities purely for commercial gain. This can be very disruptive to the patient-doctor relationship, when the information promulgated is misleading or inaccurate.

Although social media influencers have focused on the commercial world, there is now some evidence that they have become more active in the medical world, with both individuals and companies reacting to and stimulating medical information.

51.11 Patients Searching for Medical Information

Within the UK, the NHS (National Health Service) does provide patient information from its website [30]. During the COVID-19 pandemic, the UK Government like other national governments also provided web-based information [31]. However, search engines such as Google have become both a blessing and a curse because it is so easy for patients to access the other information on the Internet. Once patients have identified the medical information that they want to find for the medical condition they believe they have, they cling on to this information and bring it with them to the outpatient clinic. Often they believe the information they have found themselves is more accurate than the diagnosis and advice given to them by their medical adviser in the outpatient clinic.

51.12 Social Media and Fake News

The misuse of social media has been worsened by the example given by Donald Trump, the ex-president of the USA, and also because of the multiple announcements of inaccurate information promulgated during the COVID-19 pandemic. In his article on 'How liars create the 'illusion of truth'', Tom Stafford emphasizes an often quoted law of propaganda – 'Repeat a lie often enough and it becomes the truth' – which was actually attributed to the Nazi Minister of Propaganda, Joseph Goebbels [32].

51.13 The Way Forwards?

We are now in the midst of our 'Medical Information Wild West' period that started from about 2012. The original Wild West problem was eventually resolved by law enforcers—more powerful sheriffs. We need to design similar enforcers over the next 10 years to reduce medical misinformation in the future. This will be an even greater challenge than controlling the American cowboys.

There needs to be a reassessment of how fact-checking of Internet-published material can be carried out and what organizations should be encouraged to help with this exercise. OrthoEvidence™, based in Burlington, ON, Canada (myorthoevidence.com) and headed by Dr. Mohit Bhandari, is a good example of how this is now being achieved. This site is available as the global online source for high-quality and timely orthopaedic-only evidence-based summaries, pre-appraised by orthopaedic medical experts for the orthopaedic fraternity and other doctors [33]. We now need a similar innovation to help and support the general public. This is still being resolved. Those organizations which understand the problem realize how difficult it will be to address, so it is likely to continue as a challenge for the next decade.

References

1. Wallace WA (2019) Publish and be damned: the damage being created by predatory publishing. Bone Joint J 101-B(5):500–501
2. Jalalian M (2015) The story of fake impact factor companies and how we detected them. Electron Physician 7(2):1069–1072. www.ncbi.nlm.nih.gov/pmc/articles/PMC4477767/
3. Universal_Impact_Factor. http://universalimpactfactor.com
4. Global_Impact_Factor. http://globalimpactfactor.com
5. African_Journals_Online. https://www.ajol.info
6. International_Scientific_Indexing(ISI)_-Impact_Factor. http://www.isi-impactfactor.com/
7. SCImago_Journal_Rank(SJR). https://www.scimagojr.com
8. International_Organization_of_Scientific_Research (IOSR)_Journals. http://www.citationindex.org/
9. Xia J, Smith MP (2018) Alternative journal impact factors in open access publishing. Learned Publishing 31(4):403–411. https://onlinelibrary.wiley.com/doi/full/10.1002/leap.1200
10. Citescore-1. https://www.elsevier.com/editors-update/story/journal-metrics/citescore-a-new-metric-to-help-you-choose-the-right-journal
11. CiteScore-2. https://www.elsevier.com/connect/editors-update/citescore-a-new-metric-to-help-you-choose-the-right-journal
12. Journal_Impact_Factor. https://clarivate.com/webofsciencegroup/solutions/journal-citation-reports/
13. Govender R (2015) 5 reasons why peer review matters. www.elsevier.com/reviewers-update/story/career-tips-and-advice/5-reasons-why-peer-review-matters
14. Grudniewicz A et al (2019) Comment - predatory journals: no definition, no defence. Nature 576:210–212. https://www.nature.com/articles/d41586-019-03759-y
15. Clarivate. J Cit Rep. https://clarivate.com/webofsciencegroup/solutions/journal-citation-reports/
16. Straumsheim C (2017) No more 'Beall's List'. https://www.insidehighered.com/news/2017/01/18/librarians-list-predatory-journals-reportedly-removed-due-threats-and-politics
17. Anonymous (2017) Stop predatory journals. https://predatoryjournals.com/about/
18. Stop_Predatory_Journals. https://predatoryjournals.com/

19. Yan JR et al (2018) Predatory publishing in orthopaedic research. J Bone Joint Surg (Am Vol) 100(21): e138
20. Van Noorden R (2013) Open access: the true cost of science publishing. Nature 495(7442):426–429
21. Nicoll LH, Chinn PL (2015) Caught in the trap: the allure of deceptive publishers. NA&E 25(4):1–11. https://naepub.com/predatory-publishing/2015-25-4-4/
22. Franko OI, Okike K (2014) Open access and online journals in orthopaedics: what does the future hold? JBJS Rev 2(7):e7
23. Andrew S (2017) Pay-to-view blacklist of predatory journals set to launch. Nature. https://www.nature.com/news/pay-to-view-blacklist-of-predatory-journals-set-to-launch-1.22090
24. Sarewitz D (2016) The pressure to publish pushes down quality. Nature 533(7602):147
25. ThinkCheckSubmit (2015) https://thinkchecksubmit.org/. Accessed 25 May 2021
26. A.A.A.o.O. (2020) Orthopaedic Video Theater (OVT). https://www.aaos.org/education/orthopaedic-video-theater/. Accessed 27 May 2021
27. Stambough JB et al (2020) The past, present, and future of orthopedic education: lessons learned from the COVID-19 pandemic. J Arthroplasty 35(7S):S60–S64
28. A.A.O.S.f.S. (2021) Video J Sports Med. https://journals-sagepub-com.ezproxy.nottingham.ac.uk/home/vjs. Accessed 27 May 2021
29. Influencer_Marketing_Hub (2021) What is an influencer? – social media influencers defined. https://influencermarketinghub.com/what-is-an-influencer/
30. N.N.H. (2021) Health A to Z. https://www.nhs.uk/conditions/. Accessed 27 May 2021
31. UK_Government (2020) Coronavirus (COVID-19). https://www.gov.uk/coronavirus
32. Stafford T (2016) How liars create the 'illusion of truth' - repetition makes a fact seem more true, regardless of whether it is or not. Understanding this effect can help you avoid falling for propaganda. https://www.bbc.com/future/article/20161026-how-liars-create-the-illusion-of-truth. Accessed 26 Oct 2016
33. Bhandari M (2009) OrthoEvidence™. https://myorthoevidence.com/

52 Cheating in the Academic Environment

Thomas Lancaster

Abstract

The term cheating is an emotive one when used in the academic environment. It suggests that a student is obtaining an unfair advantage over others or is aiming to put the value of the educational system in jeopardy. In reality, student cheating is a difficult subject to address. Students are encouraged to act with academic integrity, but they are not always equipped with the core skills they need for success. Students do not always find faculty members modelling academic integrity values through their own behaviour. This chapter briefly introduces key concepts on educational cheating, including a discussion on the factors that draw students to cheat. Some methods of cheating that educational providers need to be aware of are explored. These methods include plagiarism, collusion, contract cheating, the use of file sharing sites, the misuse of text spinning software, exam cheating and opportunities for students to breach security systems. Student cheating opportunities are changing in relation to advances in technology. The academic environment needs to continually consider how teaching and assessment will remain current, valid and fit for purpose in an ever-changing world.

T. Lancaster (✉)
Imperial College London, London, UK
e-mail: thomas@thomaslancaster.co.uk

52.1 Introduction

The subject of cheating in the academic environment is one that is multifaceted, bringing together students, instructors, administrative staff, leadership teams and other members of the academic community. Often, the discussion is boiled down into two simple statements. Students should not cheat. Instructors should stop students from cheating. But the reality of academic cheating is set inside a much more complex environment, one that is full of grey areas.

52.1.1 Terminology

The term cheating is often used to describe the situation where students attempt to gain an unfair advantage over their peers or the wider educational system. The term is emotive. Within the university environment, the terms used tend to be more nuanced, like academic integrity, honesty, misconduct and dishonesty. For example, cheating is often referred to in the academic literature as a breach of academic integrity.

Academic misconduct describes the process where students attempt to get an advantage over their peers that is unfair or not deserved. Most educational establishments will have their own definition of academic misconduct described in their regulations, along with processes to follow when misconduct is suspected. The notion of an unfair advantage is important. It is perfectly

J. Faintuch, S. Faintuch (eds.), *Integrity of Scientific Research*,
https://doi.org/10.1007/978-3-030-99680-2_52

reasonable (and expected) for a student to work harder than others, to put extra time into revision or to try and predict which areas of a course syllabus are most likely to appear on an exam paper and to prepare accordingly. That would be working strategically and would be a fair advantage. As soon as cheating is involved, for instance, students copy text without acknowledgement to submit as their own, the advantage becomes unfair and a form of academic misconduct or dishonesty occurs.

52.1.2 Academic Integrity

This refers to the system of values that everyone involved in the educational process should subscribe to. The International Center for Academic Integrity uses six fundamental values: honesty, trust, fairness, respect, responsibility and courage [1]. The final value, courage, is often considered the most important, as it can be hard for individuals to do the right thing when they know that others are taking shortcuts or when they have to speak out against problems with the wider educational system.

For students to succeed, they need support to embrace the fundamental values of academic integrity. Alongside that, students need to be provided with the opportunity to develop the core academic skills needed for success in their own discipline, such as the ability to write, to value information and to reference. Although cheating is an essential part of the discussion, the onus of the system should not be in detecting students cheating or trying to set up ways to trap them. Suspect methods do not help anyone to gain confidence in the educational system. Some cheating is inevitable, but in so many cases, cheating can be avoided if the right encouragement and support for students are there. To ensure that academic integrity is at the forefront of the discussion also requires everyone involved with the educational system to develop an understanding of why students cheat.

52.1.3 Role Modelling

The examples of cheating methods presented are deliberately taken from the student perspective. If students cheat, they miss out on core learning, they lack essential skills to succeed in the future and they gain an unfair advantage over their peers. But academic integrity is something that has to be modelled by everyone who is part of the educational community. When faculty members do not acknowledge their sources on their lecture slides, students can feel that it is appropriate for them to also take shortcuts. If a researcher outsources the production of a paper, is that any different to a student paying for presentation slides to be produced for them?

52.2 The Reasons Student Cheating Happens

Eaton provides a useful summary of the factors that may encourage students to plagiarize [2]. These include the pressure on them to perform, the competition for grades and the perception that other students are cheating and getting away with it, whether this is true or not. The local response to cheating is also a consideration. If this is ignored or not taken seriously, students may believe that cheating is acceptable. Two other factors highlighted include when students are unfamiliar with institutional requirements, as may be the case when they arrive from a different educational system, and the situation where students lack self-control.

A survey of students at Australian universities investigated the reasons they elected to outsource their assessed work [3]. They found first that the opportunity to cheat existed and second that the students were dissatisfied with their teaching and learning environment.

Within an analysis of student requests to outsource work that had been placed on the Twitter social network [4], cheating happened where a student felt unable to persevere further. This could happen through frustration, when students found assignments tedious or when they simply did not understand the material.

A short-term problem such as illness can offer students a reason to cheat if they feel they would fail otherwise and they are not encouraged to mitigate against such illness. Even a situation which would delay a student from completing a course by a year can provide a cheating incentive. Not only does a delay increase the costs of studying; it also postpones the ability to receive a graduate level wage. Ultimately, this is a decision of risk versus reward. If the risk of being caught is low and the reward is high, this can often provide a reason that validates cheating in the mind of a student.

52.3 Methods of Student Cheating

Plagiarism can take place due to a lack of knowledge or support, but it is hard to argue that getting a friend to complete an assignment can ever be accidental, particularly if money changes hands. In many cases, educational establishments will put sanctions or penalties in place surrounding student cheating. These can offer a deterrent effect. Penalties should only be awarded after following due process, be considered in light of academic regulations, reflect the level of cheating and offer the student a right to respond.

52.3.1 Sanctions and Rehabilitation

Simply applying penalties alone is not a suitable outcome. Students also need to receive the opportunity to redeem themselves, correct their mistakes and develop any core skills and knowledge they have missed. Academic misconduct processes can be harrowing for students; however, they deserved the use of these processes might be. Educational establishments should ensure that support is available for students before, during and after academic misconduct hearings.

52.3.2 Plagiarism and Similarity Checking

It describes the situation where a student uses text or ideas of another person without acknowledgement. Often, this can take the form of words, sentences or paragraphs being copied and pasted from the Web or other sources, sometimes with minor changes being made to them. Plagiarism is not always intentional. It can represent poor academic practice.

Modern software tools make written plagiarism of this type fairly straightforward to detect. Text matching technology, often made available through commercial providers, enables student's work to be scanned against a large database of text [5]. This might include such sources as work from previous students, pages crawled from websites and content licensed from journal publishers. In this sense, plagiarism is often considered to be a solved problem. But text matching tools are not themselves infallible. If students suitably transform their text, or if they plagiarize from a source outside the database being used, the submission will not be indicated as unoriginal.

Text matching technology does also offer a benefit for students looking to improve how they write and reference. Although opinions about how best to use such technology in the teaching and learning process vary, commentators have indicated that such tools can be used to provide students with immediate formative feedback [6]. But if providing access to such software is to be considered, so too has to be the conditions under which students can be allowed to access it. Similarity reports are not always intuitive for students to understand. Sometimes students can end up being tempted to keep modifying their work just to reduce the level of similarity this has, rather than improving their writing style and learning how to paraphrase.

52.3.3 Self-Plagiarism

The idea that students cannot get academic credit by submitting the same work twice is often confusing to them. After all, it is their work. Therefore, teaching on self-plagiarism needs to be particularly emphasized.

52.3.4 Collusion

Most often collusion between students occurs when several hand in solutions that are very similar to one another. In essence, this is just one form of plagiarism.

One model for helping students to understand when unacceptable collusion begins suggests that students can be allowed to discuss the problems they are working on verbally but stresses they should not work on answers together nor ever share their materials with one another [7]. But studies have indicated that the boundary between collusion and cheating is thin and both staff and students are often unsure where to draw the line [8].

Similarly, within the research environment, collusion is often difficult to distinguish from usual professional exchange and cooperation. The misconduct becomes more clear when techniques or finances are questionable or openly fraudulent and several parties are involved in the cover-up or sharing of spoils.

52.3.5 Contract Cheating

Students can get a third party to complete custom assessed work for them, a real threat to the integrity of education [9]. Terms such as essay mills and ghostwriting are often used alongside contract cheating, but these terms can be limiting since assignment outsourcing can include reports, computer programmes, artwork and presentations other than essays. Assignment solutions can often be outsourced quickly and cheaply, with workers from countries such as Kenya competing for business [10, 11].

Clues and statements that are out of place within the student submissions should be noticed [12]. Solutions involving the automated analysis of student writing style are possible too. One student has shown that such automated tools offer promise, but training of assessors is needed before they become useful [13].

Investigations have shown that students run the risk of blackmail from disgruntled writers and sham companies who demand money from them, to avoid revealing their misconduct to their educational establishment [14].

52.3.6 Misuse of File Sharing Sites

File sharing sites have many uses some of which may be legitimate, but through such sites, students can exchange notes, share questions and access assignment answers [15]. Data collected during the Covid-19 pandemic saw a sharp increase in student use of such sites [16].

One model of file sharing site use is of interest. Here, students can post questions they are struggling with, expecting to receive an answer within minutes. The answer is available for all subscribers to the service, but not easily viewable otherwise. Nor can the questions be promptly searched, since they commonly take the form of images. Students have been observed using such file sharing sites to have homework completed for them or to even request and receive answers within the short time duration of examinations.

Even when file sharing sites are just used to house existing college and university intellectual property, such as lecture notes, commentators have noted the substantial revenue that providers stand to lose [17]. Each taught programme from which content finds its way into a major file sharing service can be worth USD $700,000 per year.

52.3.7 Text Spinning

The same techniques that are used to produce plagiarized content to fool online search engines such as Google into thinking that web pages are original can also be used to disguise plagiarized student work. The use of text spinning tools is one way in which work can be taken and then modified and disguised from its original form with very little effort needed on the part of the student [22].

These often involve online paraphrasing tools [18]. One simple hands-off method involves students uploading their work to such a tool. The tool will automatically go through the work and replace words and phrases with variant versions, such as synonyms. The end result is not always satisfying to read, but it is one that may fool both automated tools designed to identify plagiarism and human markers.

52.3.8 Doctored Single or Serial Translations

A further such method involves using the same technology which is used in document translation [19]. A student can start off with a document written in a foreign language, automatically translate it to English and then have text that will not be detected as plagiarized when simple text matching software is used. Since some students writing outside of their native language will use translation software anyway, working out how to support students in a legitimate manner offers another barrier to detection. A more complicated version of this process involves running text through multiple chains of translation. Each time the document is translated, the word forms used will change further. Such translations can be detected when the end result reads oddly, but like other forms of misconduct, this does require markers to be alert about and actively looking for spun text.

52.3.9 Exam Cheating

Exams appear to offer greater potential for security due to face-to-face supervision. But the value of exams is dependent on the quality of the questions, the processes used to ensure the content of exams papers are protected and the steps are taken to ensure robust proctoring.

Notes taken into the exam room, concealed out of sight, are an ancient trick. They have been updated by smartwatches, often looking like traditional watches to the untrained proctor. Cell phones can be hidden inside calculator shells. Even special pens can be purchased to which reveal crib notes when manipulated in a certain way.

In South-East Europe, it is apparently common for students to wear miniature earpieces so that they can have answers verbally communicated to them from outside an examination room [20, 21]. Such devices are said to be almost impossible to see and can be concealed by long hair. Hidden cameras can also be used to communicate the contents of exam papers, if the illicit helpers cannot access these through different means.

52.3.10 Onlinespiepr146 Tests

Where exams are taken online, remote methods of proctoring are possible, but the use of such methods can also be controversial. Students may find being monitored by camera distracting, or they may fear being falsely accused of cheating if they simply move their head or mouth answering to themselves. At the same time, if an exam is taken remotely and not monitored, it is hard to ever verify who is actually taking the exam. Other students can also find it frustrating if they believe their peers have the option of cheating available to them.

52.3.11 Security Breaches

One of the biggest risks to educational standards comes when students or their representatives receive unauthorized access to computer systems. Through such fraud, an individual can change the marks awarded, transfer a failing grade to a pass, generate fake certificates showing that an academic award has been obtained or receive access to questions or their solutions in advance of an examination.

Unauthorized access may not always involve computer systems. Physical security measures can be breached to print exam questions or update the answers on exam scripts. In some universities, students are taught how to breach security, often under the guise of an ethical hacking programme. Even where students themselves do not have the technical skills needed to access computer systems, social engineering tactics can be employed, or students can hire people to assist with this. It would be hard to argue that student involvement with a security breach could be anything other than intentional. But what happens in the situation which is engineered to put the blame on a student other than the person who really committed a security attack? This is an example of the difficult type of question academic establishments will need to consider if they ever wish to address student cheating.

52.4 Future Perspectives

Student cheating has to be addressed through programme design, through the engagement of students as academic integrity partners, by developing suitable academic regulations and by providing students with academic integrity role models to develop themselves after.

Helping students to cheat is a big business, with contract cheating providers alone turning over billions of dollars a year in business. A movement to online tuition and assessment brings with it many advantages in allowing more people the opportunity to study and improve their knowledge but also makes one of the main ways of detecting student cheating more difficult that of face-to-face contact and instructors getting to know students. As educational providers move in this direction, they also need to be thinking about how to address the movement away from face-to-face teaching. It should not be possible for a third party to log on to a system in place of a student, complete all of the assessments for that student and then have their client walk away with a qualification at the end of the process.

52.4.1 Artificial Intelligence

Perhaps the biggest future threat to academic integrity comes with the growth of artificial intelligence. It is not unfeasible to think that artificial intelligence tools will become easily available which both mimic the writing style of an individual and provide high-quality content that matches a topic that a student has been asked to write about.

Given that businesses may then become reliant on automated writing support, how would an educational environment ensure that students are ready for the workplace and proficient in the use of such tools? What adjustments to the sector's understanding of academic integrity and the ownership of intellectual materials are necessary? Such potentially contentious issues are ones that researchers and practitioners alike will need to strive to address.

References

1. International Center for Academic Integrity (ICAI) (2021) The fundamental values of academic integrity, 3rd edn. academicintegrity.org/the-fundamental-values-of-academic-integrity/. Accessed 4 May 2021
2. Eaton SE (2021) Plagiarism in higher education: tackling tough topics in academic integrity. ABC-CLIO, Santa Barbare, CA
3. Bretag T, Harper R, Burton M, Ellis C, Newton P, Rozenberg P, Saddiqui S, van Haeringen K (2019) Contract cheating: a survey of Australian university students. Stud High Educ 44(11):1837–1856
4. Amigud A, Lancaster T (2019) 246 reasons to cheat: an analysis of students' reasons for seeking to outsource academic work. Comp Educ 134:98–107

5. Lancaster T (2013) The use of text matching tools for the prevention and detection of student plagiarism. In: Dias P, Bastos A (eds) Plagiarism phenomenon in Europe: research contributes to prevention. Scientific and Cultural Association of the Faculty of Philosophy of the Catholic Portuguese University, Lisbon
6. Rolfe V (2011) Can Turnitin be used to provide instant formative feedback? Brit J Educ Technol 42(4): 701–710
7. Culwin F, Naylor J (1995) Pragmatic anti-plagiarism. Proceedings of the 3rd all Ireland conference on the teaching of computing
8. Barrett R, Cox AL (2005) ‘At least they’re learning something’: the hazy line between collaboration and collusion. Assess Eval High Educ 30(2):107–122
9. Clarke R, Lancaster T. Eliminating the successor to plagiarism? Identifying the usage of contract cheating sites. citeseerx.ist.psu.edu/viewdoc/summary?doi=10.1.1.120.5440. Accessed 4 May 2021
10. Lancaster T (2019) Profiling the international academic ghost writers who are providing low-cost essays and assignments for the contract cheating industry. J Inf Commun Ethics Soc 17:72–86
11. Lancaster T, Clarke R (2016) Contract cheating: the outsourcing of assessed student work. In: Handbook of academic integrity. Springer, pp 639–654. link.springer.com/referenceworkentry/10.1007%2F978-981-287-098-8_17. Accessed 4 May 2021
12. Rogerson AM (2017) Detecting contract cheating in essay and report submissions: process, patterns, clues and conversations. Int J Educ Integr 13:10
13. Dawson P, Sutherland-Smith W, Ricksen M (2020) Can software improve marker accuracy at detecting contract cheating? A pilot study of the Turnitin authorship investigate alpha. Assess Eval High Educ 45(4):473–482
14. Yorke J, Sefcik L, Veeran-Colton T (2022) Contract cheating and blackmail: a risky business? Stud Higher Educ 47:53–66
15. Rogerson AM, Basanta G (2016) Peer-to-peer file sharing and academic integrity in the internet age. In: Bretag T (ed) Handbook of academic integrity. Springer, Heidelberg, pp 273–285
16. Lancaster T, Cotarlan C (2021) Contract cheating by STEM students through a file sharing website: a Covid-19 pandemic perspective. Int J Educ Integr 17:3
17. Dixon Z, George K (2021) Monitoring uncharted communities of crowdsourced plagiarism. J Acad Ethics 19(2):291–301
18. Rogerson AM, McCarthy G (2017) Using internet based paraphrasing tools: original work, patchwriting or facilitated plagiarism? Int J Educ Integr 13:2
19. Akbari A (2021) Spinning-translation and the act of plagiarising: how to avoid and resist. J Furth High Educ 45(1):49–64
20. Lancaster, T, Clarke, R (2017) Rethinking assessment by examination in the age of contract cheating. pp 215–228. eprints.staffs.ac.uk/3608/. Accessed 4 May 2021
21. Lancaster T, Glendinning I, Foltýnek T, Dlabolová D, Linkeschová D (2019) The perceptions of higher education students on contract cheating and educational corruption in South East Europe. J Educ Thought/Revue de La Pensée Educative 52(3):209–227
22. Lancaster T, Clarke R (2009) Automated essay spinning–an initial investigation. academia.edu/2810694/Automated_Essay_Spinning_an_Initial_Investigation. Accessed 4 May 2021

Part XIII

Corrective Interventions

53 Relevance and Challenges of Whistleblowing in the UK's National Health System

Megan Reitz and John Higgins

Abstract

In healthcare and elsewhere, the frequent response to actual and potential malpractice is to increase the intensity and volume of formal processes of inspection—alongside an emphasis on individual courage to speak up, responsibility and accountability. What gets disappeared or ignored are the social norms and relational practices which make it more or less safe for people to challenge the day-to-day realities of workplace power. It is these practices that shape who gets heard and with what consequences.

53.1 Introduction

53.1.1 The Espoused Reality in Health in the UK

Speaking truth to power in the UK's National Health System (NHS) is greatly shaped by what is known as the Francis Report [1], published in 2013, in response to the discovery of gross failings in care leading to a number of excess deaths at an NHS Hospital Trust prior to 2009. The report identified 290 separate recommendations.

Earlier inquiries into similar failures of care, and the failure of people to speak up about what they knew to be going on, had already resulted in specific policies around whistleblowing [1] and complaints [2] procedures and policies, which have continued to be updated.

Of particular note was the establishment of a new role to be implemented across healthcare organisations—the Freedom to Speak Up Guardians [3]—which were to be further bolstered by a statutory requirement for there to be a duty of candour [4]. However, failures in care continue across many areas of health delivery. Reporting after the event tells of people failing to do what the above policies and procedures were meant to ensure, that is, speaking up when they see something going on that they know is detrimental to the well-being of patients.

53.2 Policy and Procedure Vs. People Speaking up

The Francis Report itself found, quoting the inquiry in the case of the hospital it was investigating, that there was a "a culture 'not conducive to providing good care for patients or a supportive working environment for staff'"

M. Reitz (✉)
Hult International Business School – Ashridge Executive Education Campus, Berkhamsted, UK
e-mail: Megan.reitz@ashridge.hult.edu

J. Higgins
The Right Conversation, London, UK

J. Faintuch, S. Faintuch (eds.), *Integrity of Scientific Research*,
https://doi.org/10.1007/978-3-030-99680-2_53

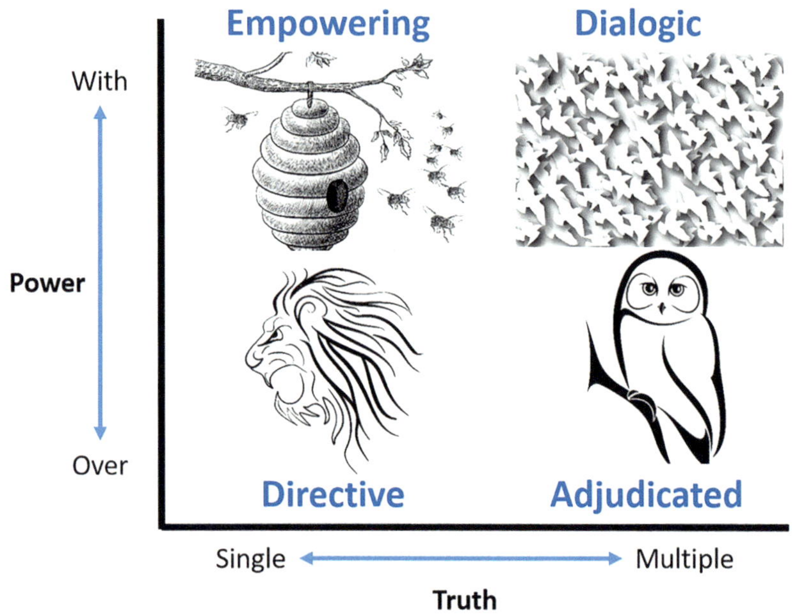

Fig. 53.1 Organisational cultures: the truth—power framework

[1]. This takes us to the heart of the matter: what makes an organisation's culture and how can it be engaged with and perhaps disrupted, in a way that makes a useful difference to employees' ability to speak up?

What enables and prevents people from speaking up is very rarely a technical issue but is part of the workplace culture—where culture can be understood as being the relational and conversational practices which shape who gets heard and who feels more or less able to speak up. In our original multi-method research into the subject of speaking up at work [5], we identified how an organisational culture can be seen to be shaped by the relationship between two dimensions, truth and power, and the way these are constructed. We identified four 'archetypal' cultures, as shown in Fig. 53.1.

53.2.1 Easy Answers to Difficult Questions

Truth and power are deeply contested terms—and what we offer here is a shorthand which raises as many questions as it answers. It also highlights for us one of the major challenges that exists within the wider managerial and leadership discourse, namely, a habit of rushing to superficial, fast and action-oriented conclusions in the face of what are known as 'wicked' problems, that is, challenges that are not readily solved unless addressed with considerable thought and attention, or may actually never be resolvable, and may therefore need to be engaged with as tensions to be held.

53.2.2 Dialogue Instead of Ready-Made Solutions

This is why in much of our work we are keen advocates of the dialogic turn, a belief in the potential that comes from supporting enriched conversational practice over and above a privileging of prescriptive process and policy on its own.

As shown in Fig. 53.1, we explored power in terms of 'power over' and 'power with', as described by Joyce Fletcher [6], and truth as having either a singular or multiple meaning. This framing draws on the work of Michel Foucault [7] and his argument that truth and power are entwined terms, with each informing the other. In health, for instance, there is often experienced to be an epistemological hierarchy in play, with those traditions most closely allied with the scientific method seen as being 'truthier' than those that come from more subjective traditions. Within the multidisciplinary practice of health, this immediately establishes a hierarchy as to who is more or less likely to get heard, whatever the espoused policy wishes to be the case and whatever the relative merits of particular epistemological traditions within the specifics of the moment.

53.2.3 Focus on the Patient Vs. Organisational Success

In the UK, there has been an attempt to create a dominant truth, what is called 'new public management', as epitomised by the Milton Friedman [8]-inspired school of shareholder value maximisation. In this world, a single dominant group is identified as having the most important perspective, and the organisation as a whole is expected to organize itself around this single, easily reported figure, that is, the share price. Often a 'directive' leadership style ensues. Within a publicly funded health system such as the NHS, such a single perspective is in practice impossible, with competing political narratives fighting it out alongside the competing internal ones, with a focus on the patient being much advocated and relatively little practised. This has resulted in the growth of regulatory agencies such as the Care Quality Commission who, in terms of our cultural typology, exercise leadership by adjudicating between the often conflicting priorities of clinicians and managers. Of note is that regulatory agencies themselves are also often in need of help, in adjudicating between the myriad of professions (each of which has its own cultural type) that make up a modern healthcare system.

53.2.4 How Much Priority to Clinical Benefits?

Given such a complex and contested context, where there is an endless game of lobbying and special interest pleading, getting heard becomes difficult and highly political (with a big and small 'p'), and new policy initiatives are interpreted within a multitude of competing agendas, with institutional units seeking to bolster the weight and validity of their perspective. Within the leadership of hospitals, this can result in medical directors being accused of 'shroud-waving', that is, claiming all non-clinically led changes will result in death and patient catastrophe.

53.2.5 Is Whistleblowing for Life-Saving or for Freedom of Speech in General?

In the case of the Freedom to Speak Up Guardians (FSUGs), some individual guardians have tried to raise the sights of the role in order to make a more systemic difference. Rather than being the conduit of 'what's gone wrong', they have tried to link their role to their healthcare units' ability to achieve their strategic goals, for example, emphasising 'speaking up' as part of the creation of a culture for learning, inclusion and innovation. FSUGs can be seen to have the potential to be both empowering and dialogic—empowering when they provide mechanisms for people to speak up within the context of the given strategic plan of their health unit—and dialogic when they provide a forum for involving a wider range of perspectives to be heard in the development of the plan in the first place.

Well-intentioned policy around reporting of malpractice and identification of speak-up opportunities is often undermined when it does not fit with the cultural context. A directive 'single truth' and 'power over' stance may often be assumed, especially when public pressure on politicians invites them to 'do something' in the light of a headline scandal.

53.3 Relational and Conversational Practices That Influence Speaking-up at Work

We wish to draw attention to the following four areas that, if not paid attention to, can confound policies and procedures meant to ensure ethical, 'speak-up cultures':

(a) The basic human need to belong to a group
(b) Conversation is a relational act
(c) How short-termism shapes the managerial agenda
(d) How senior people live in a 'bubble'

53.3.1 The Basic Human Need to Belong to a Group

Whistleblowing and speaking-up are often seen according to a long-standing Western tradition of individualism, usually heroic individualism, which can be seen in endless Hollywood tropes, where the courageous individual stands up against the power of the system—and wins.

What is missing in the movies as in much organisational study is why so many people decide to stay silent, going along with the established consensus, even when it is obviously not a good thing, that is, detrimental to patients in the case of healthcare. As Margaret Heffernan [9] spells out, people are incredibly skilled at learning to accept what would appear to an outsider to be unacceptable—we train ourselves not to notice (we become 'wilfully blind') in order to avoid what would be an intolerable burden—and that intolerable burden is becoming an outsider to the community to which you belong. In the scandal that led to the UK's Francis Report, what was striking was the longevity of the intolerable practices and the capacity for people to become encultured into silence.

53.3.1.1 Turning a Blind Eye

In our research, we found that the two main reasons people do not speak up at work in health or elsewhere, whether it is about malpractice or an opportunity, are because people do not want to be thought of negatively in the eyes of others and nor do they want to shame or upset others.

In the fairy story of the *Emperor's New Clothes* [10], we like to identify with the little boy who pointed out that the emperor was in fact wearing nothing at all. We would suggest however that for most of us, it would be more honest and insightful to realise we are more likely to be in the shoes of the crowd who went along with the emperor's conceit. Maybe as they looked at the eminence, they wondered whether it was just them that was seeing him in all his nakedness. No doubt they were also mindful of his power and authority and thought better of pointing out what seemed obvious to them for fear that they might have their heads chopped off.

53.3.1.2 Collectivism or Individualism?

We are also drawn to the work of feminist [11] and group psychologists [12] who propose that the focus on the individual as a standalone unit is an unhelpful starting point in understanding how people in work groups come together and get things done. As social mammals, we exist in groups who have distinctive languages and behavioural customs which act as significant markers of whether or not you belong or not. Whilst much of the work on whistleblowing acknowledges the danger experienced by those wanting to speak up and therefore often advocates anonymous phone lines, it can ignore the social reality, seeing 'belonging' as a barrier to be overcome rather than a defining social activity which needs to be worked with.

53.3.2 Conversation Is a Relational Act

Hearing is an active skill, connected not only to individual capacity but also to the context within which both speaker and hearer operate. There is therefore a challenge for whistleblowers and others who speak up when they feel they only have the opportunity to speak into a process, rather than a living relational context. As human beings, we are wired to read and respond to each other, adjusting how we are in a myriad of micro-gestures that encourage or discourage people from hearing us or feeling able to speak their truth. In our research, we have found that when people are unable to read the response to what they are saying, they can stay silent if they perceive the risks to be greater as a result.

53.3.2.1 Swallowed by Machinery

Too often, once someone has been identified or has self-identified, as a whistleblower, they become an object in a process—all relational subtlety can be lost if an overly de-personalised legal process kicks in. Additionally, in an age of constant organisational restructuring (the UK's NHS has on average experienced a major upheaval every 18 months or so since its inception in the 1940s), what gets lost are meaningful interpersonal connections. Constant upheaval removes the opportunities for people to get to know each other through the normal give and take of ordinary, 'small conversations'. Instead, people are expected to launch straight into the 'big conversation'. What policies can ignore is that small conversations are needed and provide the necessary social warp and weft, for the big ones to take place.

53.3.2.2 It Can Be Used Against You

If an overly legally minded process replaces relational connection, it can create a world of silence, where suddenly every move is judged and assessed through the prism of legal truth, which sets a high bar and can be intimidating to those not well versed in it. Whilst such a process looks good on a management consultancy's flow chart, with its neat flows and clearly delineated decision points, it can bear little relation to how people in groups create shared meaning together and the role that relationships play in facilitating productive conversations which encourage yet further, deeper, conversations to take place.

53.3.3 How Short-Termism Shapes the Managerial Agenda

One of the papers we are most drawn to [13] highlights the importance of managerial horizons when it comes to managers being open to hear disruptive voices, be they voices of activists, whistleblowers or innovators. Unfortunately, many managers are given tight, annual (or shorter), performance targets, and these lessen their desire to prioritise actions that may create short-term 'pain' in order to achieve long-term gain—they focus instead on 'quick win' policies.

Many managers we have worked with explain how difficult it is to prioritise any social conversation in their meetings (particularly the virtual meetings), because they are focused on getting to the tangible actions which are seen as the main purpose of meetings. In a world of constant restructuring and reshuffling of managers (who are also looking to move up the corporate hierarchy), there can also be little managerial incentive to engage with wicked/complex issues that might upset the smooth running of the unit or the individual manager's career plan.

53.3.3.1 Sweep the Dirt Under the Carpet

In the case of one hospital CEO we knew well ('Jeff'), he decided to end the hospital's playing of the 'budget game'. Given what this level of activity would cost, and what he knew the income was going to be, he published a deficit budget right from the start of the year. In other words, he was going to be honest—this was in part motivated by a desire to be transparent in dealings with the clinical heads of the various departments, who saw little point of being honest in their dealings with Jeff when they could see that Jeff was having to 'manage' the books every year in order to play the hospital budget game demanded by government and regulator.

This desire to be honest, to blow the whistle on the financial reporting game, resulted in Jeff being removed from post (he was close to retirement anyway and that is what gave him the courage to act). A turnaround interim CEO was brought in—who guaranteed to deliver the impossible budget expected of him. Cut to the end of the year and the hospital was bang on target with Jeff's original budget.

53.3.4 How Senior People Live in a 'Bubble'

As well as contextual bubbles, senior leaders also create and get trapped in their own truth bubbles, which get in the way of truth being spoken or heard. Amongst a subset of 115 respondents who identified as being at executive or board level, not one of them would admit to being influenced by someone's gender or ethnicity when it came to how they listened to people. One possibility is that senior leaders in the UK are amongst the most enlightened in the world. The more probable one is that they are blind to the privileges (or lack thereof) that come with being a particular ethnicity or gender [14].

Becoming aware of this blindness requires a serious commitment to self-reflection or metacognition. Any process put in place may not go far enough as it will be interpreted through the lens of people who do not see matters such as gender or ethnicity as having a material impact. For example, attending a course of unconscious bias training (which in itself is widely challenged as a concept) is unlikely to make a lasting difference.

The 'bubble' is reinforced by another issue our data points towards—senior people in all sectors overestimate the degree to which more junior employees speak up as well as their personal listening skills and approachability. They therefore underestimate the challenges their colleagues face [15].

There is the potential for formal processes to help burst the bubble; however, it is only if they stimulate the conversations and relational understanding and practice that forms the bedrock of an environment where speaking-up can feel psychologically safe. 'Pulse surveys', a recent popular addition to the leaders' dashboard, are a good example here—employee's experience is translated into a series of graphs presented to the board by human resources during a packed meeting agenda. Relational connection and care are difficult to nurture in these circumstances—they are another example of a non-relational approach to address what is a relational problem.

Senior leaders often need to empathise and then skilfully invite people to speak up, not simply instigating a helpline—but this requires them to commit to a personal change in how they engage with people. It requires a change in their habits.

53.4 Speaking-Up Through an Activist Lens

Speaking-up is an act of activism—it seeks to challenge the status quo by bringing news of difference to the attention of those who either do not know about something or are choosing to ignore it. In our recent work into activism in the workplace [16], of which whistleblowing is a particular form, we have identified a continuum of organisational responses from non-existent through to actively stimulating it.

53.4.1 The Right Policies Within a Receptive Environment

Formal whistleblowing policies and procedures can be perceived as being associated with the first three categories in our taxonomy: suppression, facadism and defensive engagement. Formal policies can suppress people's willingness to speak up if they put the burden of responsibility on the individual to encounter and engage with the full weight of the formally mandated hierarchy of the organisation. When leaders then attempt to identify whistleblowers, as was the case in the widely reported instance of the West Suffolk Hospital where doctors were asked to provide fingerprints [17] to uncover who sent a whistleblowing letter to the husband of a deceased patient, trust deteriorates to such a degree that employees are silenced regardless of any further policies or procedures.

53.4.2 Facadism

This occurs when leaders say the 'right' things, but no action is undertaken to show genuine commitment to the issue. It can be easier to craft a policy—or even simply *talk about* crafting one—than engage with the complicated and messy reality of actual organisational practice. Facadism is not necessarily signifying that leaders do not care at all about an issue. As with a hospital CEO we worked with recently, it can happen when a leader makes statements of intent but then persistently decides to prioritise shorter-term, operational issues.

53.4.3 Defensive Engagement

Institutional leaders often commit to actioning the bare minimum required for legal or reputational purposes. One healthcare trust who asked us to work with them was running a training programme, predominantly to provide something visible and tangible to show to the regulators when they came to assess how they were addressing concerns around people's willingness to speak up. They wanted to 'tick the boxes' and persistently refused to engage in deeper inquiry into the reasons for employee silence.

53.4.4 Dialogic Engagement

Dialogic engagement is an approach and a philosophy, which actively invites people into an ongoing process of individual and collective sense-making. It works from the assumption that people have to inquire into their experiences in their own time and in the company of others if meaningful action is to arise—otherwise people simply comply with, ignore or disobey instructions in which they have no investment, belief or understanding.

Formal policies may at their best provide a starting point for such an engagement, but they are absolutely not its end point. It requires a mental handbrake turn to whistleblowing, from a position where it is about policing what people do to inviting people to see speaking-up as a normal part of their day-to-day practice. Whistleblowing loses its special status and becomes business as usual. This attitude is what lies at the root of Amy Edmondson's work on error reporting and psychological safety in US hospitals [18].

53.5 What Makes a Difference (and Could Make a Difference) to People Speaking Up in Health

Formal policies and procedures can make a difference to people in the health workplace. They can provide legal protection, and social justification for taking a stand on something that the speaker or whistleblower believes needs paying attention to. They legitimise someone's actions. But the consequence of those actions, and the way policies and procedures are followed in practice, is an expression of organisational culture, which is wrapped up in conversational practices and the practice of power and authority.

Rather than starting from the position of saying 'We need a whistleblowing line or speak-up guardian round here', a more impactful starting point might be to ask 'What is it about our culture that makes it hard for people to get heard, when they know something that matters to the well-being of patients?' For those of an appreciative turn, they would then want to take that question and rather than start with the negative framing, begin by asking 'When are the times that people are really able to get heard around matters of patient well-being? And how can we build on that?'

53.5.1 About the Authors

Megan is Professor of leadership and dialogue at Hult International Business School. She has written *Dialogue in Organizations* (2015) and co-authored *Mind Time* (2018) and *Speak Up* (2019).

John is a researcher, author, tutor and coach specialising in how people use and abuse power at all levels of the workplace and in society. He is a research director at the Right Conversation and research partner at GameShift. His latest book is *Leadership Unravelled: The Faulty Thinking Behind Modern Management* (2022).

References

1. Powell T (2013) The Francis report (report of the mid Staffordshire NHS Foundation Trust public inquiry) and the government's response 2013. commonslibrary.parliament.uk/research-briefings/sn06690/. Accessed 4 May 2021
2. NHS England Customer Contact Centre. NHS England complaints policy 2017. england.nhs.uk/wp-content/uploads/2016/07/nhse-complaints-policy-june-2017.pdf. Accessed 4 May 2021
3. National Guardian's Office (2018) Freedom to speak up guardians. nhsproviders.org/media/4970/freedom-to-speak-up-guardians.pdf. Accessed 4 May 2021
4. NHS screening programmes: duty of candour. gov.uk/government/publications/nhs-screening-programmes-duty-of-candour/duty-of-candour. Accessed 4 May 2021
5. Reitz M, Higgins J. Being silenced and silencing others: developing the capacity to speak truth to power. Hult Research 2017. iedp.com/media/3469/being-silenced-and-silencing-others.pdf. Accessed 4 May 2021
6. Fletcher J (1999) Disappearing acts: gender, power, and relational practice at work. MIT Press Books, Cambridge, MA
7. Foucault M (2000) Power: the essential works of Michel Foucault 1954–1984. Penguin Modern Classics, London
8. Friedman M (1970) A Friedman doctrine: the social responsibility of business is to increase its profits. nytimes.com/1970/09/13/archives/a-friedman-doctrine-the-social-responsibility-of-business-is-to.html. Accessed 4 May 2021
9. Heffernan M (2011) Willful blindness: why we ignore the obvious at our peril, 6th edn. Doubleday Canada, Toronto
10. The Hans Christian Andersen Centre. andersen.sdu.dk/vaerk/hersholt/TheEmperorsNewClothes_e.html. Accessed 15 May 2021
11. Gilligan C (1982) In a different voice: psychological theory and women's development. Harvard University Press, Cambridge, MA
12. Dalal F (1998) Taking the group seriously: towards a post-Foulkesian group analytic theory. Jessica Kingsley Publishers, London
13. Sherf E, Tangirala S, Venkataramani V (2019) Why managers do not seek voice from employees; the importance of managers personal control and long-term orientation. Organ Sci 30:447–466
14. Fuchs B, Reitz M, Higgins J (2018) Do you have advantage blindness? Harv Bus Rev. hbr.org/2018/04/do-you-have-advantage-blindness. Accessed 15 May 2021
15. Reitz M, Higgins J (2020) Speaking truth to power: why leaders cannot hear what they need to hear. British Medical Journal Leader. bmjleader.bmj.com/content/early/2020/11/06/leader-2020-000394. Accessed 15 May 2021
16. Reitz M, Higgins J, Day-Duro E (2021) The do's and don'ts of employee activism: how organizations respond to voices of difference. meganreitz.com/s/Activism_REPORT_FINAL_PRINT_NoBleed.pdf. Accessed 15 May 2021
17. Murray J (7 September 2020) Errors at West Suffolk hospital contributed to woman's death. theguardian.com/society/2020/sep/07/errors-west-suffolk-hospital-contributed-womans-death-susan-warby. Accessed 1 May 2021
18. Edmondson A (2004) Learning from mistakes is easier said than done: group and organizational influences on the detection and correction of human error. J Appl Behav Sci 40(1):66–90

Professional Development Courses for Unprofessional Physician Behavior

54

Charles P. Samenow, Linda L. M. Worley, Charlene Dewey, and William Swiggart

Abstract

This chapter examines the lessons learned from 30 years of experience with a large cohort of medical professionals referred to continuing medical education (CME) courses for unprofessional behavior. The courses offered are aimed at (1) maintaining proper boundaries, (2) prescribing controlled drugs, and (3) addressing unprofessional behaviors that compromise safety in clinical practice (aka the distressed physician course). These courses provide skills and training to remediate lapses in professional behavior for physicians who engage in mid-level lapses of professionalism. While a small subset of physicians are unfit to practice medicine, most physicians are able to return to practice with enhanced knowledge and skills.

C. P. Samenow (✉)
George Washington University, Washington, DC, USA
e-mail: Csamenow@mfa.gwu.edu

L. L. M. Worley
University of Arkansas, Little Rock, AR, USA

C. Dewey · W. Swiggart
Vanderbilt Center for Professional Health, Nashville, TN, USA

54.1 Introduction

Founded in 1998, the Vanderbilt Center for Professional Health (CPH) has served as an international resource for physicians in need of continuing medical education and professional development and remediation in their practice of medicine. Physicians (as well as some physician assistants, nurse practitioners, and dentists) from all 50 states and Canada have been referred to CPH to participate in intensive transformative educational experiences aimed at addressing the root cause(s) of their difficulties with professional behaviors, in order to implement lasting remediation in the building of essential new skills. CPH offers three intensive professional development/continuing medical education (CME) courses that address (1) prescribing controlled drugs, (2) maintaining proper boundaries, and (3) distressed physicians [1–3]. Table 54.1 illustrates courses affiliated with CPH that utilize the same curriculum.

Healthcare practitioners are generally referred to the program(s) through state physician health programs (PHP), health professions/regulatory boards, medical staff organizations, group practices, and attorneys and on occasion by self-referral. CPH does not conduct or provide assessments or fitness for duty evaluations of physicians. Nor does CPH provide ongoing treatment. The CPH is academic in its mission and charges fees only to cover the administrative and personnel costs of programming. The programs

J. Faintuch, S. Faintuch (eds.), *Integrity of Scientific Research*,
https://doi.org/10.1007/978-3-030-99680-2_54

Table 54.1 Courses affiliated with the Vanderbilt Center for Professional Health

Prescribing controlled drugs course
University of Florida at Gainesville, FL, USA.
The Center for Personalized Education for Physicians (CPEP), Glendale, CO, USA
Maintaining proper boundaries course
Sante Center For Healing (no longer offered)
Distressed physician course
Professional Renewal Center – Kansas, Lawrence, KS, USA
University of Florida at Gainesville, FL, USA.
University of Virginia Effective Coping and Communication Skills Program, Charlottesville, VA, USA
Universidad de la Republica, Montevideo, Uruguay

are shared with other professional development groups in the spirit of collaboration, research, and improved practice. All research activities at CPH have been approved by the Vanderbilt Medical Center Institutional Review Board including a thorough consent to use/enter participant's de-identified data into the CPH database.

54.2 Types of Unprofessional Behaviors Addressed

Continuing medical education is an appropriate intervention for mid-level lapses in professional behavior. An appropriate referral is a physician who does not pose an imminent danger to the safety of patients or others, is not impaired with a mental health problem or substance-related disorder, and/or is not involved in egregious criminal behavior. Generally, these physicians' lapses in professional behavior are the result of a skills or knowledge deficit as opposed to pathological characterological traits. Some physicians require remediation through participation in multiple courses. Physicians identified with more serious mental health or addiction problems are referred for more intensive treatment.

54.2.1 The Proper Prescribing of Controlled Drugs Course

Misprescribing refers to a set of prescribing behaviors that depart from the standard of care including, but not limited to, issuing prescriptions without a legitimate medical purpose or in inappropriate dosage or quantities, given the characteristics of the patient to whom they are prescribed [4]. Little research has been conducted on providers who misprescribe and unintentionally contribute to this problem.

Physicians are referred to the prescribing controlled drugs course for the following reasons:

1. Complaint driven: A complaint has been made by patients or a family member who believes the provider has overprescribed and ignored their (patient/family member) concerns. Alternatively, a pharmacy or other providers may recognize concerning prescribing practices and file a complaint.
2. State regulatory agencies: The provider has been identified as a high prescriber through the state's prescription monitoring database. We are also seeing states begin to identify providers who prescribed opiates and benzodiazepines together.
3. Self-referrals: The provider feels he/she is poorly trained in screening, intervening, and referring patients in need of treatment for substance abuse including legal medications, illicit drugs, and alcohol and requires up-to-date education in safe prescribing of medications with addictive potential.
4. Victimized by office practices: The provider is a victim of patients and/or staff fraudulently obtaining prescriptions for controlled substances through vulnerable office practices.
5. "Duped" by patients: The provider has been overwhelmed by aggressive, drug-seeking

patients who render the provider incapable of effectively setting limits or making appropriate referrals.
6. Practicing out of scope: The provider is providing management in areas that are not commonly associated with their specialty of practice.

54.2.2 Maintaining Proper Boundaries Course

The Federation of State Medical Boards (FSMB) has defined two levels of sexual misconduct by physicians, representing incremental difference in the degree of sexual transgression. These two levels of misconduct are *sexual impropriety* and *sexual violation* [5]. The corresponding behaviors for each category are summarized in Table 54.2. The CPH course addresses individuals who demonstrate either type of behavior. The most common behaviors seen in physicians referred to the course are (1) a sexual relationship with a patient; (2) a sexual relationship with a nurse or staff member; (3) a complaint from a patient, family member, or nurse about inappropriate flirting, sexual advances, or behavior; and (4) cybersex instances such as accessing pornography.

Currently, the evidence is unclear about why physicians sexually offend [6–9]. In our cohorts, poor education around sexual boundaries remains the principal problem. Anecdotally, physicians attending the course are of the opinion that they are not properly educated or trained about their vulnerability, the seductive patient or the power differential. Physicians fail to recognize nonsexual boundary crossings including dual relationships, gifts and services from patients, and a loose unprofessional atmosphere in the office that forms a slippery slope toward problems and sanctions. Unresolved family of origin issues is also an important factor in boundary violating behavior. A classic example of a boundary-violating physician seen in the course is a lonely physician with poor self-esteem who has difficulty saying "no." Finally, a small subset of physicians have unresolved trauma histories or demonstrate compulsive sexual behaviors that they act out upon.

54.2.3 The Distressed Physician Course

Disruptive physician behavior represents a threat to public health by negatively impacting the safe and effective care of patients. Disruptive behavior also undermines effective staff retention, increases malpractice risk, and jeopardizes patient safety specifically related to the culture of avoidance that ensues [10–12]. Maladaptive role-modeling perpetuates this cycle to the detriment of medical students', residents', and allied health professionals' empathy, personal wellness, and overall professionalism [13].

Disruptive physician behavior may be verbal (foul language, intimidating language, belittling or demeaning words) or nonverbal (facial expressions or other body language, throwing things). The behavior also may be aggressive (see above), passive (chronically late, failure to write chart notes), or passive-aggressive (hostile or inappropriate notes or derogatory comments about colleagues or institutions) [14]. Participants in the course demonstrate a full range of these disruptive behaviors. Because the focus of the course is aimed at helping the physicians regulate their behavior (as opposed to the impact of their behavior on others), we choose to use the term distressed physician behavior for our courses.

54.3 The CME Courses

Each of the CPH courses consist of small groups of physicians (6–12) with a high faculty to participant ratio (3–4 faculty per group). Each course lasts three full days with homework assignments between days. The distressed physician course has an additional two follow-up sessions (1/2 day each) in the 6 months following the course. Faculty include the CPH program directors along with psychiatrists, psychologists, drug and alcohol counselors, and clinical social

Table 54.2 Summary of FSMB types of professional sexual misconduct

Sexual violation may include physician-patient sex, whether or not initiated by the patient, and any conduct with a patient that is sexual or may be reasonably interpreted as sexual: • Sexual intercourse • Kissing in a romantic or sexual manner • Touching breasts, genitals, or any sexualized body part for any purpose other than appropriate examination • Encouraging the patient to masturbate in the presence of the physician • Offering to provide practice-related services, such as drugs, in exchange for sexual favors
Sexual impropriety may comprise behavior, gestures, or expressions that are seductive, suggestive, or sexually demeaning to a patient: • Examining the patient intimately without consent • Disrobing or draping practices that reflect a lack of respect for the patient's privacy • Subjecting a patient to an intimate examination in the presence of medical students or other parties without the explicit consent of the patient • Examination or touching the genitals without the use of gloves • Using the physician-patient relationship to solicit a date • Initiation by the physician of conversation regarding the sexual problems, preferences, or fantasies of the physician

workers. The courses take place in person (except during COVID-19 when they were translated to an online format) and take advantage of a group process where participants are encouraged to share with each other and provide each other feedback. The courses are a mix of didactics, group exercises, self-assessments, homework, skill sessions, and role-plays/activities.

54.3.1 Transformative Learning

Transformative learning theory is emerging as an effective way of changing behavior in physicians and trainees. Elements of transformative learning have also been identified in narratives reflecting on the personal growth of physicians [15, 16]. The *process* of transformative learning includes the following:

1. A disorienting dilemma
   Example: Separating good intentions from poor behavior
2. Self-examination
   Examples: Reflection exercises, feelings chart, and emotional "check-ins"
3. A critical examination of assumptions
   Examples: Family of origin exercises (genograms, family roles, family lessons), review of standard of care, principles of addiction, and rules of medicine
4. Recognition of a connection between one's discontent and the process of transformation
   Examples: Facilitated group process and feedback and identifying triggers
5. Exploration of options for new roles, relationships, and action
   Examples: Group process and homework assignments
6. Planning a course of action
   Example: Intent to change exercise
7. Acquiring the knowledge and skills for implementing one's plan
   Examples: Communication skills, assertiveness training, mindfulness, best office practices, building emotional intelligence, role-play, clinical skills and knowledge, work/life balance, and grounding techniques
8. Provisional trying of new roles
   Examples: Role-plays and standardized patient exercises
9. Building competence and self-confidence in new roles and relationships
   Examples: Intent to change exercises, group feedback, and support
10. A reintegration into one's life on the basis of conditions dictated by one's new perspective.
    Examples: Follow-up, mentorship, referrals to therapy, and coaching.

54.4 Lessons Learned

54.4.1 Clinical Knowledge Deficits Are Only Part of the Problem

Rarely do our providers have major deficits in clinical knowledge. The most common causes of unprofessional behavior relate to poor boundaries, underdeveloped emotional intelligence, dual relationships, unresolved family of origin issues, poor office practice/management, communication deficits, and practicing out of scope.

54.4.2 Adverse Childhood Experiences (ACEs) [15] Matter

A critically important component of the course is understanding how each participant's personal upbringing influences their present-day practice and vulnerabilities. To best decipher these contributors, participants complete in-depth genogram exercises and identify their family of origin teachings and roles. Within all three CPH courses, participants acknowledge higher rates of exposure to trauma than average along with complex family dynamics and strongly engrained role expectations. We have found a pattern of childhood adverse events such as abuse, neglect, bullying, and high/unobtainable expectations. Some physicians avoid conflict, shy away from challenging/disappointing patients, and project their own fears/needs onto patients. Such dynamics contribute to physicians giving in to demands of patients even when they go against clinical judgment. We have seen these dynamics in physicians of all demographics who engage in unprofessional behaviors [16].

54.4.3 The Rules of Professional Behavior Are Not Formally Taught in Traditional Medical Education

Over and over again, we see physicians who had no formal training in the proper procedures and documentation for prescribing controlled substances, maintaining appropriate boundaries, conflict resolution skills, assertive communication, or self-care. Very few providers attending the course have ever read the *Drug Enforcement Administration* (DEA) handbook [17] containing guidelines for prescribing controlled substances. Most of our participants do not know how to check state prescription monitoring programs [18] and document that they have done so. Very few are aware of the requirements of the Controlled Substances Act in terms of informed consent, scope of practice, and substance use screening. This corresponds with the research literature [19].

The same is true of the physicians in the maintaining proper boundaries course. The most common explanation for the behavior is that the physician was "never taught the rules." These physicians are oblivious to the challenges inherent in dual relationships. They are often unaware of rules concerning dating and having sex with patients. The majority of participants in our courses indicate they had no formal training on these topics during medical school or in residency. Lack of a professional mentor is common.

54.4.4 Communication Is Key (and Is Also Not Taught)

It is well known that deficient communication skills are connected to bad outcomes and contribute to poor prescribing practices [20, 21]. It is extremely common for participants in the courses to struggle with "saying no" and to demonstrate a pattern of passive or conflict avoidant communication styles. Curriculum preparing providers with the communication skills necessary to effectively set limits and to navigate difficult conversations are practically nonexistent in medical school and residency programs across the country. Coupled with unresolved family of origin issues, these providers are simply not equipped to deal with the manipulative nature of drug seeking and strongly assertive patients, staff, and colleagues.

54.4.5 Practitioner Self-Care Is Important

Our cohort demonstrates the critical role that physician burnout [22] plays in prescribing practices, maintaining appropriate boundaries, and preventing disruptive behaviors. Many of our providers have office practices that are fairly chaotic and disorganized which leave them vulnerable, for example, not securing blank prescription pads, allowing access to pads to those who should not have access, and not properly using chaperones. Frenetic scheduling with feelings of overload leads to rushed environments where it becomes easier to provide a patient with a prescription, as opposed to holding a difficult conversation or administering proper screening. Loneliness and unmet emotional needs set physicians up for violating boundaries. Physicians who feel medicine has robbed them of their family and recreation life may act out in a disruptive manner. Finally, financial pressures also may contribute to some physician practices devolving into "pill mills" as opposed to comprehensive clinical care. While recent news articles have portrayed some physicians being rewarded with millions of dollars by pharmaceutical companies for prescribing opioids, especially name-brand versus generics [23], our courses have not seen many physicians engaged in this type of behavior.

54.4.6 All Specialty and Provider Types Involved; However, there Are Patterns

The most typical provider referred to our course is a mid-career family medicine/general practice internist Caucasian male physician who operates in solo practice. While no single one of these demographics can be seen as a risk factor, when combined, there does seem to be a perfect storm of variables that increase a provider's risk. Similar patterns have been seen in the research literature [24–26]. From our experience across all CPH courses, we have found that mid-career physicians seem more prone to burnout and represent an era of physician who received less training in addictions, communication skills, boundaries, and self-care. Physicians in solo practice have fewer checks and balances for correcting poor clinical and operational practices. Nevertheless there may be a selection bias in who has access to the course and/or which referring agencies are willing to send their physicians for remediation.

Similar to national cohorts, we see a large number of surgeons and intensivists in the distressed physician course. Psychiatry and family medicine are highly represented in the maintaining proper boundaries course. Finally, family medicine and internal medicine are common specialties in the proper prescribing course. More research is needed to determine the reason for these patterns.

54.5 Results

The majority of participants rank their satisfaction with the course as very high. This is an important finding given that most physicians enter the course angry and resentful about having to attend. Very few physicians are referred back to the courses for remediation which is an indication that many are able to return to practice safely. The high number of continued referrals from licensing boards, physician health programs, and practice groups indicates that these referring entities view remediation as being successful for previously referred participants. Analysis of intent to change exercises, letters written to future participants, and pre-/post-course assessments demonstrate that participants graduate from the courses with new knowledge and skills as well as viable action plans to be implemented upon returning to their respective workplace and at home. Many of our participants have sought further work through therapy and coaching, something they resisted to prior to the course(s). Finally, we have published follow-up results regarding our cohort's behavioral change [23, 27–30].

54.6 Conclusions

Continuing medical education is an effective way at remediating unprofessional behavior in physicians. The effectiveness of this modality is based on the high faculty to participant ratio, the group process, and an interactive curriculum that is based on transformative education theory. Essential to the process is teaching these physicians skills sets to remediate deficits that have not been taught elsewhere. Reinforcement through role-plays, standardized patients, and group exercises is also critical to the translation of these skills to clinical practice. Skills and exercises pull from best practices in cognitive behavioral strategies, mindfulness, family of origin work, and communication skills. Homework and follow-up sessions can provide additional remediation and accountability.

References

1. Swiggart W, Pichert J, Elasy T, Dixon G, Spickard A (1999) Continuing medical education course on proper prescribing of controlled substances in the United States. Fed Bulletin J State Med Boards US 86(1): 20–28
2. Spickard A, Swiggart W, Manley G, Dodd D (2002) A continuing education course for physicians who cross sexual boundaries. Sex Add Comp 9(1):33–42
3. Samenow CP, Swiggart W, Spickard A (2008) A CME course aimed at addressing disruptive behavior. Phys Exec 34(1):32–40
4. Dineen KK, DuBois JM (2016) Between a rock and a hard place: can physicians prescribe opioids to treat pain adequately while avoiding legal sanction? Am J Law Med 42(1):7–52
5. Boards FoSM (2006) Addressing sexual boundaries: guidelines for State Medical Boards. fsmb.org/grpol_policydocs.htm. Accessed 4 May 2021
6. Myers M, Gabbard G (2008) The physician as patient: a clinical handbook for mental health professionals. American Psychiatric Press, Washington, DC
7. Irons R, Schneider J (1999) The wounded healer: addiction-sensitive approach to the sexually exploitative professional. Jason Aronson Publishers, Northvale, NJ
8. Tschan W (2014) Professional sexual misconduct in institutions: causes, consequences, prevention and intervention. Hogrefe Publishing, Boston, MA
9. Roback H et al (2007) Problematic physicians: a comparison of personality profiles by offense type. Can J Psych 52(5):315–322
10. The Joint Commission (2011) National patient safety goals. jointcommission.org/assets/1/6/NPSG_EPs_Scoring_HAP_20110706.pdf. Accessed 6 Jan 2021
11. Leape LL, Fromson JA (2006) Problem doctors: is there a system-level solution? Ann Int Med 144(2): 107–115
12. Rosenstein AH (2002) Nurse-physician relationships: impact on nurse satisfaction and retention. Am J Nurs 102(6):26–34
13. Paice E, Smith D (2009) Bullying of trainee doctors is a patient safety issue. Clin Teach 6:13–17
14. American Medical Association (2009) Report of the council on ethical and judicial affairs. ama-assn.org/sites/ama-assn.org/files/corp/media-browser/public/about-ama/councils/Council%20Reports/council-on-ethics-and-judicial-affairs/i09-ceja-physicians-disruptive-behavior.pdf. Accessed 6 Jan 2021
15. Felitti VJ, Anda RF, Nordenberg D et al (1998) Relationship of childhood abuse and household dysfunction of many of the leading causes of death in adults: the adverse childhood experiences (ACE study). Am J Prev Med 14(4):245–258
16. Samenow C, Yabiku S, Ghulyan M, Swiggart W (2012) Do family of origin issues contribute to physician sexual boundary violation? Measuring family dynamics and boundary violating behavior among 613 physician referred to a CME course. HEC Forum 24(2):115–126
17. Rannazzisi J, Caverly M (2006) United States Department of Justice Drug Enforcement Administration Office of Diversion Control "practitioner's manual: an informational outline of the controlled substances act" 2006 edition. deadiversion.usdoj.gov/pubs/manuals/pract/. Accessed 1 Jun 2021
18. State prescription drug monitoring programs. US Department of Justice and Drug Enforcement Administration. deadiversion.usdoj.gov/faq/rx_monitor.htm. Accessed 1 Jun 2021
19. Loeser JD, Schatman ME (2017) Chronic pain management in medical education: a disastrous omission. Postgrad Med 129(3):332–335
20. Ha JF, Longnecker N (2010) Doctor-patient communication: a review. Oschner J Spring 10(1):38–43
21. Hauser K et al (2017) The prescription talk: an approach to teach patient-physician conversation about drug prescription to medical students. GMS J Med Educ 34(2):1–16
22. Spickard A, Gabbe SG, Christensen JF (2002) Mid-career burnout in generalist and specialist physicians: definitions, risk factors and prevention. JAMA 88(12):1447–1450
23. Zezima K (2017) Study: doctors received more than $45 million from drug companies marketing opioids. The Washington Post. washingtonpost.com/news/post-nation/wp/2017/08/09/study-doctors-received-more-than-46-million-from-drug-companies-marketing-opioids/?utm_term=.ca110e802fcf. Accessed 1 Jun 2021

24. DuBois JM, Chibnall JT, Anderson EE, Eggers M, Baldwin K, Vasher M (2016) A mixed-method analysis of reports on 100 cases of improper prescribing of controlled substances. J Drug Issues 46(4):457–472
25. Khaliq AA, Dimassi H, Huang CY, Narine L, Smego RA Jr (2005) Disciplinary action against physicians: who is likely to get disciplined? Am J Med 118(7):773–777
26. Kohatsu ND, Gould D, Ross LK, Fox PJ (2004) Characteristics associated with physician discipline: a case-control study. Arch Intern Med 164(6):653–658
27. Spickard A et al (2002) Changes made by physicians who misprescribed controlled substances. J Med Licens Disc 88(3):110–115
28. Dewey CM, Ghulyan MV, Swiggart WH (2016) Misprescribing controlled substances: an evaluation of a professional development program. Subst Abus 37(3):412–418
29. Swiggart WH, Ghulyan MV, Dewey CM (2012) Using standardized patients in continuing medical education courses on proper prescribing of controlled substances. Subst Abus 33(2):182–185
30. Swiggart WH, Bills JL, Penberthy JK, Dewey CM, Worley LLM (2020) A professional development course improves unprofessional physician behavior. Jt Comm J Qual Patient Saf 46(2):64–71

Is the Current Informed Consent Model Flawed? 55

Bert Heinrichs and Serap Ergin Aslan

Abstract

Informed consent is a widely acknowledged ethical principle that plays a crucial role, both in research ethics and medical ethics. However, empirical findings as well as theoretical considerations suggest that the current model of informed consent might be flawed. In particular, the understanding of information disclosed to patients and research participants during the consent process proves to be limited. This, in turn, casts serious doubt on the validity of the consent. In this paper, we will review the current state of research on informed consent and discuss the charge of inadequacy. Subsequently, we will examine some suggestions that have been presented in the literature to improve the current model of informed consent. At the end, we will summarize the current state of the discussion and give a brief outlook on future developments.

B. Heinrichs (✉)
Institute for Neuroscience and Medicine: Brain and Behaviour (INM-7), Forschungszentrum Jülich, Jülich, Germany

Institute of Science and Ethics (IWE), University of Bonn, Bonn, Germany
e-mail: b.heinrichs@fz-juelich.de

S. Ergin Aslan
University of Bonn, Bonn, Germany
e-mail: sergin@uni-bonn.de

55.1 Introduction

Today, informed consent is widely recognized as the most prominent principle of both medical ethics and research ethics. In what follows, we will focus on the area of research ethics. However, much of what we present can be applied to the field of medical ethics as well. It remains controversial exactly when informed consent became a recognized ethical principle. Some authors point to early twentieth-century court cases [Mohr v. Williams (1905), Pratt v. Davis (1906) and Schloendorff v. Society of New York Hospitals (1914)] that took place between 1905 and 1911 and the discussions in context as the origin of informed consent ([1]: 56–60). Actually, as early as 1894, a relevant case was heard before the German Imperial Court [Reichsgericht] ([2]: 21–24). However, some have contradicted the view that this is the origin of our modern understanding of informed consent [3, 4]. What is clear, in any case, is that informed consent became one of the most important ethical principles in the course of the second half of the twentieth century. Informed consent obligations and requirements, as we know them today, were developed from the Nuremberg trials to the mid-1970s. Nevertheless, it was not until the early 1970s that the term became the subject of intense debate ([1]: 91; [5]: 118).

J. Faintuch, S. Faintuch (eds.), *Integrity of Scientific Research*,
https://doi.org/10.1007/978-3-030-99680-2_55

55.1.1 Current Concepts and the Right to Autonomy

Informed consent was initially used to protect "the right to choose whether to permit others to invade one's physical integrity" ([1]: 28). Discussions were primarily focused on the doctor's or researcher's obligation to disclose information. Beauchamp [6] notes that only around the early 1970s serious discussion of the meaning and ethics of informed consent began in medicine, research, law and philosophy ([6]: 515). With an increasing emphasis on respect for the autonomy of the person, strong paternalism lost its support and the emphasis and attention to the importance of the right to self-determination and making an autonomous decision grew steadily. In the mid-1970s, the discussion had begun to focus on the understanding of information by patients or potential research subjects ([6]: 515). Today, literally all codes and institutional rules of medical ethics and research ethics contain provisions of informed consent and provide that they are essential under normal circumstances.

55.1.2 Basic Elements

Nowadays, most ethicists agree that five elements are constitutive of informed consent: (1) competence, (2) disclosure, (3) understanding, (4) voluntariness and (5) consent [1, 7–10]. Thus, for informed consent to be valid, (1) a person P must be competent, (2) he or she must have been adequately informed, (3) he or she must understand that information, (4) he or she must act voluntarily and (5) finally, he or she must actually consent.

55.2 Criticism of the Informed Consent Model

Although the model of informed consent just outlined is widely recognized, it has met with criticism for decades [11–14], particularly with regard to its practical implementation [15–18]. Doubts were expressed for a number of reasons. The standards of disclosure are especially controversial. Many codes provide for "adequate" information as standard [19–22]. However, the notion of "adequacy" is of little help when it comes to practical implementation. As a consequence, there are various attempts to determine the standard of disclosure more precisely, three of which are particularly important: (1) the professional standard, (2) the reasonable person standard and (3) the subjective standard [1, 5, 23–27].

55.2.1 The Focus on Disclosure and the Ability to Decide

In their influential book *Principles of Biomedical Ethics,* Beauchamp and Childress maintain that "disclosures usually should cover the aims and methods of the research, anticipated benefits and risks, any anticipated inconvenience or discomfort, and the subjects' right to withdraw, without penalty, from the research" ([5]: 123). Although this gives some orientation as to what potential research participants should be informed about, some critics argue that it is rather unrealistic to inform potential participants about all scientific details of a study [23, 26, 28]. Furthermore, critics claim that the current model of consent is excessively demanding. If taken seriously, few adults are actually capable of giving consent [29–31].

What is more, some suggest that information overload can even prevent adequate understanding ([5]: 133). Bester and colleagues, for example, focus on factors such as the complexity of the information or its scope which can overwhelm the person's ability to make decisions in the consent process [32]. They believe that informed consent has become so central and important to the way clinicians practice and that they may fail to recognize situations where the person's ability to provide informed consent is compromised. In this regard, they criticize the information overload in the consent process, which in their opinion can strain the person in the decision-making and damage his or her ability to make decisions.

Empirical data indicate that potential research participants often do not grasp the information they get [9, 17, 33–36]. The first problem that comes to mind in this context is that the subjects do not understand the risks of the experiments. Stanley et al. determined the understanding of patients undergoing femoral popliteal bypass or carotid surgery for the risks associated with surgery, after obtaining consent in a routine manner. According to the data, 25% of patients said they had a poor understanding of risks and complications ([37]: 788–791).

55.2.2 Patient-Oriented Attention vs. Non-therapeutic Investigation

The data of Paul Appelbaum and colleagues show that a fundamental misunderstanding is prevalent among research subjects as related to implications for their clinical care of participating in a clinical trial ([33]: 20–24; [34]: 319–329; [36]: 1689–1697). Participants thought that the research was designed for their direct benefit or, at least, that they would benefit from it. Appelbaum and colleagues referred to this misunderstanding as "therapeutic misconception" and showed that it is widespread among research subjects. Research subjects systematically misinterpret the risk-benefit ratio since they are unable to distinguish the consequences of participating in research from ordinary treatment, that is, they do not understand the underlying scientific methodology ([36]: 1689–1697; [38]: 288–290).

Many ethicists acknowledge that the consent process is not just about providing information material and having research participants sign a consent form. A document alone does not guarantee that participants understood the information provided. Among others, Meisel and Kuczewski emphasized this problem and spoke of "a myth of informed consent" ([39]: 2522).

55.3 Empirical Studies on the Implementation of Informed Consent

Many empirical studies support the prevailing view that research participants absorb the information poorly during the consent process [33, 34, 40–44]. Leeb, Bowers and Lynch [35] conducted a prospective study to find out how much information patients can remember. For this purpose, they interviewed 100 patients who were already informed in a standard format before an operation and who had given their consent to the intervention. The patients were able to maintain an average retention rate of only 35% ([35]: 280–282).

A few years later, Bergler et al. [40] investigated the understanding of hypertensive patients enrolled in a controlled study. The data showed that 95% of patients said they expected to receive the best possible treatment. They were less able to remember the possible side effects of the drugs tested as compared to their effects ([40]: 435–440).

Appelbaum et al. [33, 34] reported that after a detailed and comprehensive information process, "half the subjects failed to grasp that treatment would be assigned on a random basis, four out of twenty misunderstood how placebos would be used, five out of twenty were not aware of the use of a double-blind, and eight out of twenty believed that medications would be adjusted according to their individual needs" ([33]: 23). They referred to this kind of misunderstanding as "therapeutic misconception"—a term that is now widely used and generally acknowledged.

The situation described was confirmed by various other studies [36, 43, 45]. Joffe et al. [43] studied cancer patients who had already participated in multiple oncology studies. A significant percentage did not fully understand the clinical studies they were participating in, and 31% did not grasp that the main reason for conducting the clinical trials of cancer was to improve the treatment of future cancer patients ([43]: 1772–1777). Also 29% mistakenly believed that the treatment under investigation was found to be the best treatment for their cancer, and another 40% were unsure. As many as 38% believed that participating in a clinical trial posed no additional risks or discomfort compared to standard treatments for their cancer type, and another 26% were unsure about additional risks.

Lidz, Appelbaum et al. [36] interviewed 155 test subjects from 40 different studies. The authors remarked that "it appears that subjects often sign consents to participate in clinical trials with only the most modest appreciation of the risks and disadvantages of participation" ([36]: 1696).

55.4 Strategies to Improve the Informed Consent Process

Interventions related to enhanced consent forms focus on improving the consent process or the printed material that serves to convey the relevant information. The consent forms were modified using various strategies: changing the content and information materials, writing style, format, length, the use of simpler language, images or changed layout that highlights important points. In several studies, the influence of modifying or enhanced consent forms has been tested [16, 46–55].

Bjørn and colleagues, for example, studied the effect of changes in written fact leaflets on understanding the content of the form with two different scenarios, one for a new drug for hypertension in people over 75 and the other for the effect of a new local anaesthetic in connection with the sterilization of women between 25 and 45 ([48]: 263–267). For this purpose, they revised written patient information leaflets, which were used in various randomized controlled trials by a pharmaceutical company, in terms of language, style and layout, and then tested the original and the revised leaflets against each other. They reported that "the perception of the readability and comprehensibility of a standard information leaflet can be improved by professional linguistic revision of the leaflet" ([48]: 266).

The randomization of treatment was correctly understood by 26% of the research participants who received the original hypertension leaflet and by 42% who received the revised hypertension leaflet, while most of the research participants in the sterilization study got it right in both cases ([48]: 265). Bjørn and colleagues hypothesized that the problem of understanding the randomization method could be explained by an age effect or by a cohort effect.

In a prospective, randomized study, a standard consent form and a modified, shortened version were compared among 100 patients with asthma ([51]: 246–252). There was a significant increase in the information retained by patients who received the modified version.

In another protocol, there was no significant improvement in understanding ([50]: 668–674): "simplifying informed consent material alone makes the forms more appealing and easier to read but will not improve comprehension" ([50]: 674). This finding was supported by various other studies including Campbell et al. [56], Agre and Rapkin [46] and Coyne et al. [49].

55.4.1 Non-conventional Approaches

Multimedia interventions were also tested [46, 53, 56–62]. While some studies showed that media interventions can play an important role in improving understanding [58, 63], others indicated that interactive computer tools did not significantly improve understanding [46, 47, 56, 60].

Dunn et al. [58] examined the effect of an extended consent process, which consisted of a computerized slide show with more structure and review of key information, on the understanding of the information given to 80 patients with psychotic disorders and to a control group of 19 healthy persons, who were randomized to receive routine consent or extended consent. The understanding of the patient with psychotic disorders was significantly improved through the computerized consent process. Both patients and controls who received the extended consent process had a better understanding than those who received routine consent ([58]: 142–150).

This latter finding is supported by a study of Wirshing et al. [63]. They evaluated a short educational video designed to improve the process of informed consent for people with severe mental and medical illnesses. Knowledge about consent information and the consent process was significantly improved after viewing the experimental videotape.

55.4.2 Negative Findings

The effectiveness of three presentation methods in the consent process was evaluated, namely, videotape, computer and booklet format, vs. the standard written consent form ([46]: 1–7). The analysis did not reveal any differences between the use of computer and video-based information or between tools for booklets and standard consent forms. Moreover, the authors conducted a crosstab analysis, which examined the relationship between response patterns to the knowledge quiz and the various consent methods: "In the full sample of 441 subjects, those receiving booklet or standard version of consent information were more likely to demonstrate specific errors, such as misunderstanding privacy, compensation, or alternative interventions. Very similar findings were evident in analyses including patient subjects only" (cf. [46]: 1–7; [47]: 11–19).

When a computerized tool was tested against an information brochure for cancer patients who were considering participating in an early-stage clinical study, even though the computer-based tool helped to improve the understanding of participation, the problem of therapeutic misconception among the participants could not be completely solved ([64]: 1–10). The authors of the study emphasized that "regardless of the arm to which they were randomized, most of the participants believed that the early-phase clinical trial they had the option to participate in was designed to examine whether the investigational drug would treat their cancer" ([64]: 7).

55.4.3 Complex Projects and Unfamiliar Details

The studies presented above show that the use of the multimedia intervention methods and extended consent forms has some positive effect on the understanding of research participation. However, it is still questionable how good these approaches really are. It remains, for instance, unclear whether the research participants understand complex details of research projects. Furthermore, it is unclear whether such approaches also work for specific groups such as cognitively impaired adults or minors. Problems with these strategies are not only related to their practical limitations and adverse effects but rather to their problematic notions of autonomy, consent and information.

55.4.4 Not Just Understanding

Some have argued that they stick to the false idea that consent is *a solitary act* and assume that the quality of consent correlates directly with understanding ([12]: 33–49). According to this line of argument, they all follow "the simple but erroneous logic that deeper understanding leads to better consent and in turn to more autonomy" ([12]: 36).

55.5 Extended Discussions of the Research and the Participant's Involvement

The effect of extended discussions was examined in various studies [57, 65–69], in which a discussion was led by a member of the study team or a neutral educator.

In a randomized study by Aaronson et al. [65], a telephone nursing intervention was used in addition to the usual disclosure of information. The effectiveness was tested among 180 cancer patients. This approach was an effective means for increasing awareness and understanding of the problems associated with clinical trials ([65]: 984–996). This finding was supported by a study by Fitzgerald et al. [70]. In contrast, Benson and colleagues showed no significantly improved understanding by research participants ([57]: 455–475). Different methods, namely, by videotape, by oral disclosure and by a neutral educator as well as by videotape plus oral disclosure, were compared. This latter negative finding was supported by Sarkar et al. [67].

55.5.1 The Essence of the Consent Process

Neil Manson and Onora O'Neill claimed that there are "gaps between actual cognitive capacities and the decision-making capacities required" ([13]: 6) to consent to the proposed action. Apparently, these gaps cannot be closed by the strategies available so far, which attempt to make consent simpler and more user-friendly by improving the procedures for providing information and better information sheets and multimedia technologies or by using intermediaries like consultants ([13]: 6).

55.5.2 Two-Way Exchange vs. One-Way Road

The problem, according to Manson and O'Neill, arises from the view that communication is viewed as merely the transmission of content or the disclosure of information—they call this "the conduit metaphor" ([13]: 35–38). This way of thinking portrays the communication rather as a passive transfer of information. This means that information recipients are merely seen as "containers" in which information is stored and immediately they can process it deliberately.

They did not claim that the conduit/container model—and the metaphors that often obscure it—is completely wrong, but rather it hides too much ([13]: 65). They focused on informed consent as a communication transaction between agents ([13]: 69). This allowed them to highlight the importance of various background obligations and skills. Communication is "a species of action, which communicative acts presuppose the practical and cognitive commitments both of those who seek to communicate, and of those with whom they seek to communicate" ([13]: 55). This idea was picked up by Heinrichs [12]. He emphasized that the current standards of consent are too demanding and that many research participants are in fact unable to meet the high cognitive requirements for understanding.

55.5.3 An Agreement Between Persons vs. an Extensive Legal Document

According to him, understanding the complex details of research projects is not essential for a valid consent. Drawing on Searle's speech act theory, he interpreted informed consent as a special form of speech act between people. In fact, he suggested that to consent should be presented in the special meaning it has in the concept of informed consent as a declarative speech act, which serves to constitute institutional facts ([12]: 39). The main focus of this approach is to view the concept of informed consent as an activity that two people must participate in together ([12]: 46). Dealing with consent as a speech act between two agents does not necessarily require high cognitive skills. In particular, a detailed understanding of the specific scientific goals or the methods used is not essential for an informed consent to be valid. The main focus must lie on ensuring supportive conditions for a successful communication and not on an improved information disclosure.

55.6 Summary and Outlook

Is the current informed consent model flawed, and if so, in what ways?

1. The current disclosure standards written into many regulations are unrealistic and excessively demanding.
2. Approaches to improve the process of information disclosure are partly promising but are unlikely to change the fundamental problem of limited participant's understanding.
3. If one assumes that the validity of consent depends on full understanding of all information provided, then the current model must indeed be said to be flawed.

4. However, if one considers a changed basic understanding of informed consent, then one does not necessarily have to draw this conclusion.
5. Such an understanding views informed consent as a communicative act between persons and does not see the success of this act, or at least not exclusively, in information transfer and isolated decision-making.
6. Instead, it must be about the parties involved developing a common understanding of what both parties agree to do and what the consequences are or may be.
7. Therapeutic misconception remains a major challenge even with this revised model of informed consent. It must be clear to all concerned that it is a research project and not a medical therapy. This is the basic prerequisite for all further communication processes.
8. Whether a communicative understanding of informed consent can actually prevail in practice and, more importantly, whether it can meet the legal requirements remains to be seen and needs to be further explored.
9. If this is not the case, then other ways must be sought to address the problems that informed consent raises.
10. Empirical research and conceptual thinking on informed consent will continue. Since it is one of the central ethical principles of research ethics and medical ethics, an ongoing critical discussion is certainly appropriate.

References

1. Faden RR, Beauchamp TL (1986) A history and theory of informed consent. Oxford University Press, Oxford
2. Heinrichs B (2006) Forschung am Menschen. De Gruyter, Berlin
3. Katz J (2002) The silent world of doctor and patient. JHU Press, Baltimore, MD
4. Pernick MS (1982) The patient's role in medical decision making: a social history of informed consent in medical therapy. In: President's commission for the study of ethical problems in medicine and biomedical and behavioral research, vol 3, pp 1–35. PMID 12041401
5. Beauchamp TL, Childress JF (2019) Principles of biomedical ethics, 8th edn. Oxford University Press, Oxford
6. Beauchamp TL (2011) Informed consent: its history, meaning, and present challenges. Camb Q Healthc Ethics 20(4):515–523
7. Brock DW (2008) Philosophical justifications of informed consent in research. In: Emanuel EJ, Grady C, Crouch RA, Lie RK, Miller FG, Wendler D (eds) The Oxford textbook of clinical research ethics. Oxford University Press, Oxford, pp 606–612
8. Capron A (2008) Legal and regulatory standards of informed consent in research. In: Emanuel EJ, Grady C, Crouch RA, Lie RK, Miller FG, Wendler D (eds) The Oxford textbook of clinical research ethics. Oxford University Press, Oxford, pp 613–632
9. Meisel A, Roth LH (1981) What we do and do not know about informed consent. JAMA 246(21): 2473–2477
10. National Commission for the Protection of Human Subjects of Biomedical and Behavioral Research (1978) The Belmont report. DHEW Publication, Washington, DC. hhs.gov/ohrp/regulations-and-policy/belmont-report/index.html. Accessed 13 Mar 2021
11. Corrigan O, Liddell K, McMillan J (2009) The limits of consent. A socio-ethical approach to human subject research in medicine. Oxford University Press, Oxford
12. Heinrichs B (2019) Myth or magic? Towards a revised theory of informed consent in medical research. J Med Philos 44(1):33–49
13. Manson NC, O'Neill O (2007) Rethinking informed consent in bioethics. Cambridge University Press, Cambridge
14. Miller FG, Wertheimer A (2010) The ethics of consent. Theory and practice. Oxford University Press, Oxford
15. Fellner CH, Marshall JR (1970) Kidney donors–the myth of informed consent. Am J Psychiatry 126(9): 1245–1251
16. Flory J, Emanuel E (2004) Interventions to improve research participants' understanding in informed consent for research: a systematic review. JAMA 292(13): 1593–1601
17. Herz DA, Looman JE, Lewis SK (1992) Informed consent: is it a myth? Neurosurgery 30(3):453–458
18. Macklin R (1999) Understanding informed consent. Acta Oncol 38(1):83–87
19. Council for International Organizations of Medical Sciences (2002) International ethical guidelines for biomedical research involving human subjects. cioms.ch/wp-content/uploads/2021/03/International_Ethical_Guidelines_for_Biomedical_Research_Involving_Human_Subjects_2002.pdf. Accessed 13 Mar 2021
20. European Parliament and the Council of the European Union (2014) Regulation (EU) No 536/2014 of the European Parliament and of the Council of 16 April

2014 on clinical trials on medicinal products for human use, and repealing Directive 2001/20/EC. ec.europa.eu/health/sites/health/files/files/eudralex/vol-1/reg_2014_536/reg_2014_536_en.pdf. Accessed 13 Mar 2021
21. United Nations Educational, Scientific and Cultural Organization (2005) Universal declaration on bioethics and human rights. portal.unesco.org/en/ev.php-URL_ID=31058&URL_DO=DO_TOPIC&URL_SECTION=201.html. Accessed 13 Mar 2021
22. World Medical Association (2013) Declaration of Helsinki. Ethical principles for medical research involving human subjects. wma.net/policies-post/wma-declaration-of-helsinki-ethical-principles-for-medical-research-involving-human-subjects/. Accessed 13 Mar 2021
23. Beauchamp TL (2010) Standing on principles: collected essays. Oxford University Press, Oxford
24. Braddock CH III, Fihn SD, Levinson W, Jonsen AR, Pearlman RA (1997) How doctors and patients discuss routine clinical decisions. Informed decision making in the outpatient setting. J Gen Intern Med 12(6):339–345
25. Briguglio J, Cardella JF, Fox PS, Hopper KD, TenHave TR (1995) Development of a model angiography informed consent form based on a multiinstitutional survey of current forms. J Vasc Interv Radiol 6(6):971–978
26. Dranseika V, Piasecki J, Waligora M (2017) Relevant information and informed consent in research: in defense of the subjective standard of disclosure. Sci Eng Ethics 23(1):215–225
27. Morin K (1998) The standard of disclosure in human subject experimentation. J Leg Med 19(2):157–221
28. O'Neill O (2002) Autonomy and trust in bioethics. Cambridge University Press, Cambridge
29. Appelbaum PS, Roth LH (1982) Competency to consent to research: a psychiatric overview. Arch Gen Psychiatry 39(8):951–958
30. Appelbaum PS, Mirkin SA, Bateman AL (1981) Empirical assessment of competency to consent to psychiatric hospitalization. Am J Psychiatry 138(9):1170–1176
31. Roth LH, Meisel A, Lidz CW (1977) Tests of competency to consent to treatment. Am J Psychiatry 134(3):279–284
32. Bester J, Cole CM, Kodish E (2016) The limits of informed consent for an overwhelmed patient: clinicians' role in protecting patients and preventing overwhelm. AMA J Ethics 18(9):869–886
33. Appelbaum PS, Roth LH, Lidz CW, Benson P, Winslade W (1987) False hopes and best data: consent to research and the therapeutic misconception. Hast Cent Rep 17(2):20–24
34. Appelbaum PS, Roth LH, Lidz C (1982) The therapeutic misconception: informed consent in psychiatric research. Int J Law Psychiatry 5(3–4):319–329
35. Leeb D, Bowers DG Jr, Lynch JB (1976) Observations on the myth of "informed consent". Plast Reconstr Surg 58(3):280–282
36. Lidz CW, Appelbaum PS, Grisso T, Renaud M (2004) Therapeutic misconception and the appreciation of risks in clinical trials. Soc Sci Med 58(9):1689–1697
37. Stanley BM, Walters DJ, Maddern GJ (1998) Informed consent: how much information is enough? Aust N Z J Surg 68(11):788–791
38. Berg JW, Appelbaum PS, Lidz CW, Parker LS (2001) Informed consent: legal theory and clinical practice, 2nd edn. Oxford University Press, New York, NY
39. Meisel A, Kuczewski M (1996) Legal and ethical myths about informed consent. Arch Intern Med 156(22):2521–2526
40. Bergler JH, Pennington AC, Metcalfe M, Freis ED (1980) Informed consent: how much does the patient understand? Clin Pharmacol Ther 27(4):435–440
41. Daugherty CK, Banik DM, Janish L, Ratain MJ (2000) Quantitative analysis of ethical issues in phase I trials: a survey interview of 144 advanced cancer patients. IRB 22(3):6–14
42. Howard JM, DeMets D (1981) How informed is informed consent? The BHAT experience. Control Clin Trials 2(4):287–303
43. Joffe S, Cook EF, Cleary PD, Clark JW, Weeks JC (2001) Quality of informed consent in cancer clinical trials: a cross-sectional survey. Lancet 358(9295):1772–1777
44. Miller C, Searight HR, Grable D, Schwartz R, Sowell C, Barbarash RA (1994) Comprehension and recall of the informational content of the informed consent document: an evaluation of 168 patients in a controlled clinical trial. J Clin Res Drug Dev 4(8):237–248
45. Featherstone K, Donovan JL (2002) "Why don't they just tell me straight, why allocate it?" The struggle to make sense of participating in a randomised controlled trial. Soc Sci Med 55(5):709–719
46. Agre P, Rapkin B (2003) Improving informed consent: a comparison of four consent tools. IRB 25(6):1–7
47. Agre P, Campbell FA, Goldman BD et al (2003) Improving informed consent: the medium is not the message. IRB Suppl 25(5):11–19
48. Bjørn E, Rossel P, Holm S (1999) Can the written information to research subjects be improved?–An empirical study. J Med Ethics 25(3):263–267
49. Coyne CA, Xu R, Raich P et al (2003) Randomized, controlled trial of an easy-to-read informed consent statement for clinical trial participation: a study of the Eastern Cooperative Oncology Group. J Clin Oncol 21(5):836–842
50. Davis TC, Holcombe RF, Berkel HJ, Pramanik S, Divers SG (1998) Informed consent for clinical trials: a comparative study of standard versus simplified forms. J Natl Cancer Inst 90(9):668–674
51. Dresden GM, Levitt MA (2001) Modifying a standard industry clinical trial consent form improves patient

information retention as part of the informed consent process. Acad Emerg Med 8(3):246–252
52. Murphy DA, O'Keefe ZH, Kaufman AH (1999) Improving comprehension and recall of information for an HIV vaccine trial among women at risk for HIV: reading level simplification and inclusion of pictures to illustrate key concepts. AIDS Educ Prev 11(5):389–399
53. Nishimura A, Carey J, Erwin PJ, Tilburt JC, Murad MH, McCormick JB (2013) Improving understanding in the research informed consent process: a systematic review of 54 interventions tested in randomized control trials. BMC Med Ethics:14–28
54. Stiles PG, Poythress NG, Hall A, Falkenbach D, Williams R (2001) Improving understanding of research consent disclosures among persons with mental illness. Psychiatr Serv 52(6):780–785
55. Taub HA, Baker MT, Sturr JF (1986) Informed consent for research. Effects of readability, patient age, and education. J Am Geriatr Soc 34(8):601–606
56. Campbell FA, Goldman BD, Boccia ML, Skinner M (2004) The effect of format modifications and reading comprehension on recall of informed consent information by low-income parents: a comparison of print, video, and computer-based presentations. Patient Educ Couns 53(2):205–216
57. Benson PR, Roth LH, Appelbaum PS, Lidz CW, Winslade WJ (1988) Information disclosure, subject understanding, and informed consent in psychiatric research. Law Hum Behav 12(4):455–475
58. Dunn LB, Lindamer LA, Palmer BW, Golshan S, Schneiderman LJ, Jeste DV (2002) Improving understanding of research consent in middle-aged and elderly patients with psychotic disorders. Am J Geriatr Psychiatry 10(2):142–150
59. Fureman I, Meyers K, McLellan AT, Metzger D, Woody G (1997) Evaluation of a video-supplement to informed consent: injection drug users and preventive HIV vaccine efficacy trials. AIDS Educ Prev 9(4): 330–341
60. Llewellyn-Thomas HA, Thiel EC, Sem FW, Woermke DE (1995) Presenting clinical trial information: a comparison of methods. Patient Educ Couns 25(2):97–107
61. Strevel EL, Newman C, Pond GR, MacLean M, Siu LL (2007) The impact of an educational DVD on cancer patients considering participation in a phase I clinical trial. Support Care Cancer 15(7):829–840
62. Weston J, Hannah M, Downes J (1997) Evaluating the benefits of a patient information video during the informed consent process. Patient Educ Couns 30(3): 239–245
63. Wirshing DA, Sergi MJ, Mintz J (2005) A videotape intervention to enhance the informed consent process for medical and psychiatric treatment research. Am J Psychiatry 162(1):186–188
64. Kass NE, Sugarman J, Medley AM et al (2009) An intervention to improve cancer patients' understanding of early-phase clinical trials. IRB 31(3):1–10
65. Aaronson NK, Visser-Pol E, Leenhouts GH et al (1996) Telephone-based nursing intervention improves the effectiveness of the informed consent process in cancer clinical trials. J Clin Oncol 14(3): 984–996
66. Freer Y, McIntosh N, Teunisse S, Anand KJ, Boyle EM (2009) More information, less understanding: a randomized study on consent issues in neonatal research. Pediatrics 123(5):1301–1305
67. Sarkar R, Sowmyanarayanan TV, Samuel P et al (2010) Comparison of group counseling with individual counseling in the comprehension of informed consent: a randomized controlled trial. BMC Med Ethics
68. Simes RJ, Tattersall MH, Coates AS, Raghavan D, Solomon HJ, Smartt H (1986) Randomised comparison of procedures for obtaining informed consent in clinical trials of treatment for cancer. Br Med J 293(6554):1065–1068
69. Tindall B, Forde S, Ross MW, Goldstein D, Barker S, Cooper DA (1994) Effects of two formats of informed consent on knowledge amongst persons with advanced HIV disease in a clinical trial of didanosine. Patient Educ Couns 24(3):261–266
70. Fitzgerald DW, Marotte C, Verdier RI, Johnson WD Jr, Pape JW (2002) Comprehension during informed consent in a less-developed country. Lancet 360(9342):1301–1302

56 Perverse Incentives: A Psychoanalysis of Fraud

Hub Zwart

Abstract

Prevention of scientific misconduct (e.g. fabrication, falsification, plagiarism, also known as FFP) and other questionable research practices has become issues of concern. This is connected with broader issues within the research integrity debate. Should we primarily focus on individual responsibilities or on institutional responsibilities or both? In terms of diagnosis and therapy, the initial tendency in integrity discourse has been to focus on individualisation: on detecting and penalising individual deviance. Currently, however, more emphasis is given to environmental factors, for example, the quality and resilience of the research ecosystem. To highlight the relevance of a (Lacanian) psychoanalytical approach to fraudulent or questionable research, I will briefly analyse three case histories, as *Fallgeschichte*, in the psychoanalytic sense of the term. After discussing a first exploratory case history, presented by Lacan himself, I will outline my conceptual frame of reference, notably by zooming in on Lacan's theorem of the four discourses. Subsequently, two additional case histories will be addressed: the case of the Majorana fermion (reflecting current developments in quantum computing and elementary particle physics) and the case of Dutch psychologist Diederik Stapel (representing social psychology).

Part of this chapter was adapted from the open access book by the same author entitled *Tales of Research Misconduct*. A Lacanian Diagnostics of Integrity Challenges in Science Novels. Springer, 2017, available at https://link.springer.com/book/10.1007/978-3-319-65554-3

H. Zwart (✉)
Erasmus University Rotterdam, Rotterdam, The Netherlands
e-mail: zwart@esphil.eur.nl

56.1 Introduction

Research integrity is essential for fostering excellence in science and for maintaining societal trust in researchers and research institutions [1]. As indicated by the EU-funded project PRINTEGER (https://printeger.eu) however, researchers, research managers and policymakers may have different perspectives on how best to address the challenges involved. While researchers tend to endorse bottom-up conceptions of research integrity, seeing it as professional attitudes and best practices that should be facilitated and supported, managers and policymakers tend to opt for more regulatory modes of action, focussing on norms, regulations and financial damages [2, 3]. This results in a shift of focus towards institutional responsibilities and the maintenance of a resilient research culture [3].

J. Faintuch, S. Faintuch (eds.), *Integrity of Scientific Research*,
https://doi.org/10.1007/978-3-030-99680-2_56

56.2 Individual Deviance Versus Ecosystem Resilience

It is precisely here that psychoanalysis, notably the Lacanian strand, has something to offer, I will argue. Psychoanalysis zooms in on contemporary research cultures and their discontents [4]. Misconduct may reflect systemic challenges, turning scientific research into an "impossible profession" [5, 6]. Researchers are spurred on by relentlessly demanding but potentially conflicting expectations (e.g. "excellence" versus "relevance") and may easily become tormented subjects, driven by a frantic will to know, but challenged and frustrated, either by intractable research objects or by the increasingly compelling expectations of the knowledge production system (the scientific superego) or both. These challenges and contradictions may result in various symptoms, ranging from moral conflicts and workaholism via burn-out down to fraud [7], p. 73.

56.3 First Case History: The Ubiquity of Plagiarism

Both in his *Écrits* and in his Seminars, French psychoanalyst Jacques Lacan (1901–1981) discusses a case study concerning a plagiarist, originally published by Ernst Kris [8], and involving an academic patient whose career was seriously thwarted by an obsession with plagiarism ([9], 393 ff., [10], 119–120, [11]). An inexplicable compulsion to steal other peoples' ideas gave rise to a chronic inhibition: an inability to publish his research. At a certain point, when the patient finally manages to finish a manuscript, he discovers a book in the university library that allegedly already contains all his ideas. His therapist Ernst Kris kindly asks his patient for the book, reads it, ascertains that there is not much originality in it and kindly informs the patient that his self-accusation proves unjustified. The plagiarism is "self-fabricated" as it were. Moreover, it turns out that a close colleague has repeatedly stolen and published the patient's ideas without proper acknowledgement, so that, when it comes to plagiarism, he is a victim rather than a perpetrator. According to Kris, what is troubling the patient is the conviction that only ideas conceived by others can be truly interesting. In response to this interpretation, the patient makes an awkward confession: his favourite dish happens to be "fresh brains".

In his comments on this clinical vignette, Lacan argues that the patient's culinary confession actually shows that we should not too easily assure ourselves that there is no reason for our feelings of guilt. Although no accusation is raised against him, the patient is nonetheless tormented by the prospect that some misdemeanours may be discovered. Thus, Lacan argues that the paralysing experience of guilt is not directed towards illicit acts but rather points to secret, illicit intentions. According to Lacan, regardless of whether plagiarism has actually been committed, the paralysing sense of guilt stems for the unconscious *desire* to copy others, fuelled by the conviction that only the thoughts of others are worth publishing. This is also the meaning of the favourite menu. The desire (i.e. "brain-picking") is still there but has found a new target (a psychic mechanism known as displacement). An oral form of incorporation of brain content is now adopted to act out the secret wish.

From a Lacanian perspective, this points to a more fundamental problem, especially in the humanities, where words and phrasings are important, namely, our excessive dependence (as academic authors) on established discourse. Contrary to what outsiders tend to think, in academia, originality is not only exceptional but may even raise suspicion. This is why, as Lacan's friend Michel Foucault nicely phrased it in his inaugural lecture [12], we surreptitiously strive to slip into established discourse and try to envelop ourselves in the words and phrases coined by (known and unknown, acknowledged and unacknowledged) others. In academia, we speak a language that is already there and process ideas which have already been produced by previous contributors. Every sentence we (as academics) utter contains a plethora of references, and we inevitably use (or even parasitize upon) the work of countless others, whether we are conscious of it or not. And this is what causes the patient's suffocating sense of guilt. In academic publications, we acknowledge the extent to which we rely on and build on the work of others via references, but our references only partially acknowledge our pervasive dependence on others, on existing forms of discourse.

56.4 Why Is Plagiarism a Problem?

Plagiarism therefore remains a problematic concept. Academics are continuously surrounded by concepts, language, arguments and phrases. They participate in and are immersed in a discourse that is already ongoing, especially in the social sciences and the humanities, where language is decisive. Can we rightfully and meaningfully claim some of these phrases, ideas or arguments to be "ours"? Most if not all the words we use have already been used by others. Therefore, Lacan argues, we may question whether such a thing as intellectual property really exists [13], p. 14. We cannot "own" ideas, for they are always already there. Even if we conduct an experiment to procure new knowledge, we build on many things which are already in place (starting with the concept of an experimental design). We would not be able to think or write at all in the absence of a discourse already established. We dwell in a thriving world of signifiers, to which we can only marginally contribute. On closer inspection, originality may turn out to be an untenable egocentric prejudice, while awareness of our excessive and chronic dependence on established discourse may entail a painful narcissistic offence. Not *I* speak, but *it* speaks ("ça parle"). We are parasites, appropriating the discourse of others, while originality is something marginal at best, occurring in the folds and margins of a world of language that always already pervades our linguistic environment [14], p. 568.

Rather than denying that citations and references are important, Lacan's argument implies that citations play multiple functions besides acknowledgement [15], p. 40. Via citations, we seek support and credibility through the words of others, preferably established experts. Like the patient described above, we are highly uncertain about our own original, self-expressed ideas and therefore in dire need of a discursive scaffold. Citations indicate that we participate in a discourse that is already there, adding credibility to our own ideas. Lacan not only builds on the linguistic theories of Ferdinand de Saussure concerning the anonymity and chronic dependence of speaking and thinking subjects on language but also on Martin Heidegger, who relentlessly emphasises the thraldom and subjugation of humans vis-à-vis language—as indicated by one of his most famous phrases, put forward on several occasions: *Die Sprache spricht* ("language speaks"). To the extent that our work aligns with that of others, however, its added value may decrease, and academic discourse often proves highly repetitive.

Other authors have brought forward similar ideas. In scholarly discourse, Friedrich Nietzsche argues, "it" speaks [16], §547. Modern science replaces self-expression with anonymity [17]. For Nietzsche, the most important scientific virtue of all is self-denial. *Was liegt an mir!* It is not me that counts! This phrase summarises the core of the scientific ethos, Nietzsche argues, the quintessence of being "in science".

As indicated, this view was taken up many years later by Michel Foucault: "Qu'importe qui parle?" For Foucault, the most fundamental ethical principle of contemporary scientific discourse resides in a basic indifference towards authorship [18], 789, (cf. Zwart 2001). Science is, first and foremost, a discursive phenomenon where self-expression and originality are marginalised or even eliminated, and language use tends to be technical and impersonal, so that even author's names tend to have functional relevance, in the context of information retrieval and quality assessment of research teams. In normal science, academic authorship comes very close to anonymity, and there is a certain moral quality in the stoical acceptance of this fact. And yet, in contemporary discourse, plagiarism remains a serious concern, as one of the three basic forms of scientific misconduct.

In short, we end up with a puzzling result. Plagiarism is both reprehensible and inevitable. How to get out of this cul-de-sac? Additional case histories may help. Before introducing the other two cases, however, I will first outline my conceptual and methodological framework (already briefly introduced above), namely, Lacan's analysis of the vicissitudes of "university discourse", as one of the four genres of discourse he distinguishes [15].

56.5 University Discourse and Its Vicissitudes

According to Lacanian psychoanalysis, humans are first and foremost erring, "divided" subjects, driven by desire and tormented by questionable and enigmatic objects (objects of desire). Human existence revolves around a basic split or *Spaltung*, as reflected by human practices such as arts, religion or science: erring quests in search of an elusive (absent, irretrievable) object which, allegedly, will satisfy our desire. When, to mention just one example, paleoanthropologist Eugène Dubois travelled to the Dutch East Indies on his improbable quixotic quest to find the missing link (in the form of a *Homo erectus* skull), he was basically a tormented subject, dropping out of a more regular career in anatomy, in search of his object *a*—the entity which, after many trials and tribulations, both saved and ruined his life [19].

According to Lacan, this predicament can be clarified with the help of symbols, where the symbol $\$$ (pronounced as "barred S") represents the erring, craving subject (the subject pole) while *a* ("object *a*") represents the elusive object of desire (the object pole). These erring, craving subjects are searching for a lifeline, something which will make their lives more focussed and consistent, for instance, in the form of a scientific career, and which will justify their hardships.

The interactive relationship between barred subject and object *a* is represented in shorthand by the "matheme of desire": $\$ \lozenge a$. In this psychoanalytical equation, the lozenge or *poinçon* ($\lozenge$) points in two directions, indicating that, while subjects are focussed on (or even obsessed with) an inexorable object (which promises to satisfy their desire), this object (*a*) is not a purely passive entity (not an "object" in the traditional epistemological sense) but rather an *actant* (a non-human agent), actively drawing (fixating) the subject's attention and intentionality. The lozenge is reminiscent of an optic contrivance, enabling subjects to zoom in or out, in response to the object's irresistible appeal.

Subsequently, four types of discourse are distinguished by Lacan [15], all of them allegedly offering a way out of this impasse. They provide a scaffold for tormented subjects to (temporarily) function efficiently, within a relatively stable constellation, until anomalies accumulate again and inherent disruptive tensions rise to the surface. These four discourses constitute a sequence, in the sense that one particular type of discourse may evolve into another. Building on Hegelian dialectics, Lacan develops a scheme (in his terminology: a "matheme") for presenting and analysing these four genres of discourse. His scheme consists of four positions which may be occupied by four types of actors. These four positions are referred to (in clockwise order) as *agent*, *other*, *product* and *truth*:

Agent →	Other
Truth	Product

The *Agent* is the position from where a particular genre of discourse is initiated, while the *Other* is the one who is addressed by it. The *Product* refers to the (intended or unintended) results of the discursive process, while *Truth* must be understood in the Hegelian sense of the term as well: the truth of the constellation which is initially obfuscated but increasingly manifests itself (via accumulating symptoms).

These four positions can be occupied by four types of actors (four "variables"). Three of these four variables are subjects, namely, the Master (S_1), the servant or scholar (S_2) and the divided subject ($\$$), while the fourth variable is an object:

the object *a* (the object of desire, the target of our will to know).

A first solution to the impasse, represented by the matheme of desire, is provided by what Lacan refers to as the *discourse of the Master*. Divided, erring subjects ($\$$) need a Master, providing guidance, a guru who addresses them and calls upon them and whose instructions they may follow: an authoritative Master voice, on whose sayings erring subjects allegedly can rely. This primary subject, allegedly in control (reminiscent of the Master in Hegelian dialectics), is represented by the symbol S_1, which may refer to the Master as such, but also to the *word* or the imperative of the Master (S_1 as "Master signifier").

In academia, this means that erring subjects ($\$$) may become scholars (S_2), qualified stewards of an authoritative oeuvre, say Aristotle or Marx. The Master initiates the discursive process as the Agent, whose imposing textual legacy addresses the servant (S_2 as Other), putting the latter to work. Scholarly servants (S_2) may dedicate years of their life to reading and explaining the works and words of the Master (S_1), claiming that only the words of the Master have substance and value and that only the ideas of Master thinkers (Aristotle, Marx, etc.) are worthy of our consideration, sacrificing their own autonomy and originality. Their originality literally becomes something marginal, tugged away in inconspicuous footnotes and glosses.

The symbol S_2 stands for these labouring servants (scribes, scholars, interpreters, etc.) but also for the skills and knowledge (the know-how, the expertise) they develop in the course of the process. Although these labouring servants may address enigmatic objects (*a*), they do not address things "out there" directly. They do so *via* the word of the Master. They point out what Masters (Aristotle, Marx, etc.) have to say about such things, preferably about enigmatic entities such as *final causes*, *prime matter* or *surplus* value, that is, concepts or signifiers which only have meaning within the context of the Master's oeuvre. This may result in the production of knowledge concerning reality, but as a by-product as it were. For the basic objective of these scholars is to come to terms with the oeuvre of the Master:

S_1 (agent) →	S_2 (other)
$\$$ (truth)	*a* (product)

In the context of scholasticism, the name Aristotle functioned as a guarantee of truth, as indicator of validity and reliability. Aristotle's oeuvre entailed an imperative to scholastic scholars (S_2, the ones addressed): read me, interpret me and gloss me. Scholastic scholars devoted their life to this assignment, this textual practice. These scribes or scholars do not interact with objects of knowledge directly. Insofar as reference is made to real (mundane) objects, the Master acts as intermediary. When speaking about objects, the words of the Master are consulted. If scholastic scholars speak about animals, they preferably quote what Aristotle or other authoritative voices have to say about them. Knowledge concerning objects is a by-product of the reading process. Preferably, the focus is directed on questionable entities, for example, "final causes" or "prime matter", those things which later ("Enlightened", that is, anti-Aristotelian) authors tried to eliminate (e.g. David Hume: "Commit it to the flames").

S_1 (the agent, initiating the discourse, functioning as authority; for instance, Aristotle)	S_2 (the other, the one addressed, allowing divided subjects to become acknowledged experts of an authoritative oeuvre)
$\$$ (on closer inspection, the Master is not an authoritative source of guidance but a tormented subject himself, whose sayings remain fundamentally questionable: the obfuscated Truth)	Knowledge concerning enigmatic entities or objects (*a*) allegedly at work in the real, although they can only be known and addressed via the word of the Master (Product).

Although the servants (as scholastic scribes) initially occupy a subordinate position (bereft of the right to speak in their own voice), there is obviously a benefit for them involved in the constellation. Their discursive servitude allows them to become reliable and consistent ($\$ \rightarrow S_2$). They have an assignment to fulfil, and the Master's discourse provides structure and guidance to their life and thinking. The insights provided by the oeuvre make an apparently erratic reality understandable, which constitutes a source of enjoyment (*a* as by-product).

Thus, this genre of discourse allows divided subjects ($\$$) to become qualified experts (S_2). Integrity literally (etymologically) means "wholeness", and through their labour, the scholarly subjects acquire consistence. Although they gave up their autonomy (their freedom to err), they receive something in return: recognition as an acknowledged, qualified expert.

Eventually, however, the constellation proves untenable, for various reasons. First of all, to the extent that the expertise of the servants (as stewards of a legacy) increases, they become increasingly aware of the fact that the (word of the) Master (S_1) is not as reliable as was initially expected. Inconsistencies may pop up, and anomalies accumulate or even cases of fraud, such as misattributions, making it questionable whether a particular work was really written by, say, Aristotle or was actually the work of epigones. The alleged author may be a pseudo-Aristotle, abusing the name of the Master to enhance the credibility and prestige of the work. Or, to give a more recent example, it may become questionable whether the works of young Marx may already be considered as belonging to the oeuvre of Marx as Master thinker, etc. Whereas in the current era, plagiarism became a problem (appropriating the works and words of others), in the past, scholars faced the opposite problem: authors claiming that their work was written by someone else. In more recent versions of this type of discourse, scholars may dispute whether particular comments are genuinely and sufficiently "Marxist".

Finally, rather than consulting the (questionable) authority of the Master, the qualified experts may want to address the objects of their will to know (the target of their intentionality) more directly. Consulting the Master becomes a questionable and unnecessary detour. The qualified experts (S_2) may have developed more robust and reliable methods to question their objects. This results in the inevitable emancipation of the servants (S_2), who now occupy the position of the agent themselves, dethroning the Master, pushing the Master signifier (S_1) beneath the bar, to interact with the object directly. Historically, this is known as the scientific revolution (an anticlockwise quarter-turn of the quadruped scheme), resulting in what Lacan refers to as university discourse:

S_2	a
S_1	$\$$

Now, the qualified, acknowledged experts (S_2) take the floor as autonomous agents, initiating the discourse and addressing the enigmatic object *a* (as other): the target of their "cupido sciendi", their desire to know—questioning the object through experimental designs, forcing it to answer their questions. By playing the role of qualified experts, the discourse provides a scaffold which allows erring subjects ($\$$) to acquire consistency, by conducting their research in a methodological fashion (requiring self-discipline and training), and thereby become a reliable source of information. This is the objective of an academic education: transforming erring subjects (i.e. students, $\$$) into qualified, reliable experts (S_2), functioning as agents.

Although this constellation may seem convincing at first glance, various tensions may once again arise. For instance, notwithstanding the methodological skills and know-how of the qualified experts (S_2), the enigmatic object (*a*) may refuse to live up to expectations, so that anomalies accumulate. In addition, there is an obfuscated truth, in the form of an ideology (an ideological bias), addressing experts from beneath the bar (S_1), urging them to verify a particular worldview,—although S_1 may also refer to the imperative of contemporary university life: go on, produce more knowledge, more papers, more grants, more citations, never enough! Due to such tensions, the constellation may go awry. The expert (S_2) becomes a divided, tormented subject again ($\$$), susceptible to perverse incentives, torn between incompatible expectations, resulting in various possible symptoms, from burn-out down to misconduct and fraud (as unintended by-products of university life).

Again, an anticlockwise turn to the left is now called for, so that university discourse gives way to the "discourse of the analyst", zooming in (above the bar) on the problematic interaction between divided subjects and objects of desire:

a	$
S_2	S_1

The focus is now on the role of the object *a* as agent ("actant"), drawing the subjects' attention, triggering their desire. When studying a case of misconduct, for instance, the focus of attention is *not* on the particularities of the research field involved (S_2 is bracketed beneath the bar) but on misconduct as a syndrome which affects the interaction between subject pole and object pole. These subjects may try to capture their elusive object by fabricating or falsifying their results. And eventually, this type of discourse may give rise to new rules, codes and regulations, to a normative stance, cleansing the academic ecosystem of perverse incentives (S_1 as product or by-product)

56.6 The Majorana Fermion

Before diving into the case study as such, let me first provide some background information about the research field in question.

In elementary particle physics, every type of particle is associated with an antiparticle, having the same mass, but with opposite charge (e.g. electrons complemented by positrons). These particles and antiparticles annihilate each other, producing photons as result. Some particles are their own antiparticle, however, and this applies to Majorana fermions, whose existence was hypothesised theoretically by the Italian physicist Ettore Majorana in 1937 but still awaits empirical confirmation.

After predicting their existence, Ettore Majorana mysteriously disappeared in 1938. He left a note saying he made a decision that had become unavoidable. He may have committed suicide, but his body has never been found and it has been suggested that, being a devout catholic, he became a monk.

In recent years, Majorana particles became the focus of intense interest because they may possibly serve as building blocks (as stable, error-proof "qubits") for quantum computing: a new and cutting-edge research field which is expected to revolutionise computer technologies.

So far, Majorana particles remained enigmatic entities, but recently, research groups claimed to have observed traces ("signatures") of these exotic objects under hyper-extreme conditions [20, 21]. The publication by Zhang et al. in *Nature* was heralded as the third and final step in the Majorana trilogy: from the prediction by Ettore Majorana in 1937, via the group's disappointed first efforts at confirmation in the past, up to the recent publication, presented as "definite proof" of the particle's existence [22]. Allegedly, this experiment closed the first (exploratory) chapter in the quest for Majorana particles and opened the way to a new episode: working towards quantum information processing, based on their unique properties.

Recently, however, the research group involved, led by professor Leo Kouwenhoven at TU Delft, in response to critical comments and questions from peers, indicated growing doubts and concerns as to the validity of their previously published results. In fact, an integrity committee has already been installed to investigate whether the research was conducted and reported in accordance with appropriate standards of research integrity [23].

Obviously, besides scientific prestige, substantial financial interests are at stake, as quantum computing is heralded as the next era in global computing and ICT. Although no further details were provided, the authors apparently alerted the editors of *Nature* to potential problems in the manner in which the data had been processed, affecting the reliability of the conclusions [24]. Ongoing investigations remain confidential until completion.

This case history concurs with the "matheme" of university discourse. S_2 (a prominent and highly esteemed research group) becomes intrigued (or even obsessed) with an enigmatic entity (the Majorana fermion, as object *a* of elementary particle physics) whose ontological status remains highly questionable. Majorana research counts as high-risk research. It can lead to everlasting fame but also to disaster.

The discursive constellation can be summarised as follows. A prominent research group (S_2) acquires international fame by announcing a breakthrough concerning research which seems as hazardous as it is inexorable and enticing. The forbidding Majorana particle is an alluring actant (*a*), an allusive phantom object, which allegedly can only be spotted under hyper-exceptional circumstances, namely, in a cryo-state of 273 degrees below Celsius. The "object" is inevitably drawing prominent research groups onto hazardous terrain. Besides academic prestige, enormous financial prospects are beckoning. Thus, it seems plausible that these researchers fell victim to their *cupido sciendi*, their desperate will to capture the allusive object (*a*):

S_2	*a*
S_1	$

Allegedly, the qualified experts (S_2) are in control as agents, producing reliable and objective results. On closer inspection, however, this proves not to be the case. By giving in to their desire, they became divided subjects ($), caught in the tension between methodological prudence and scientific carefulness on the one hand, and the imperative (speaking from beneath the bar) to gain priority in research, as a precondition for world fame, substantial research funding and, eventually, the Nobel Prize. Initially, when everything seemed in order, the principal investigator (Leo Kouwenhoven) was the centre of attention, but now, integrity enquiries zoom in on the first author of the questionable paper (Hao Zhang), who is now under the suspicion of being a tormented subject who, triggered by perverse incentives, resorted to questionable practices ($). Misconduct would be the unintended by-product of an untenable situation (bottom-right position).

56.7 Crystalline Science and Its Sinister Traps

The truth of the constellation is that research, rather than being pure and disinterested, is driven (contaminated) by hidden imperatives, for example, global competition, a situation which easily generates perverse incentives. In accordance with neoliberal ideology of globalisation (S_1), addressing these researchers from beneath the bar, research teams strive to be the winners in the race for fame and funding. S_1 may refer here to Microsoft, the ICT powerhouse or Master agent. In response to promising initial results, Microsoft had indicated its willingness to invest substantial ("big") funding to facilitate the work.

Ettore Majorana himself was the first victim of the particle he predicted. His breakthrough resulted in an impossible situation, which either resulted in suicide or in a decision to revert to servitude. If the story of Majorana entering a monastery (to spend the remainder of his life in anonymity and seclusion) is true, it would entail a relapse into the discourse of the Master, so that the qualified expert (S_2) becomes a servant again, a monk, functioning in service of the Master signifier, the Word of God. In other words, the two versions of his disappearance are quite dissimilar. In the case of suicide, his despair (*$* in the lower-right position) would have been pushed to the extreme, but in the case of conversion, S_1 would have been elevated ("aufgehoben") into the position of the agent once again, and the quadruped scheme would have taken a clockwise quarter-turn to the right. Ettore would have remained a devoted scholar (S_2), but now in the upper-right position (as devoted servant), the one who is called, who is addressed, by the ultimate Master (God).

The objective of bringing the inherent antagonisms and inconsistencies of university discourse to the fore, by questioning its inherent ideology, for instance, requires a different discourse, another anticlockwise turn of the quadruped scheme, towards what Lacan refers to as discourse of the analyst, resulting in an oblique perspective on technoscience. The focus of attention is no longer on scientific niceties of elementary particle physics (S_2) but on the interactions between divided subjects ($) and phantom objects (*a*), now operating as actant, initiating the process, and easily becoming an obsession:

a	$
S_2	S_1

In this constellation, the enigmatic object (*a*) as "actant" (non-human agent) addresses the researchers (as Other), trapped in their predicaments and relapsing into the position of erring subjects ($). The focus of attention is not on the research field as such (e.g. quantum computing: S_2, bracketed beneath the bar) but on the dynamics of the interaction between object (*a*) and subject ($). A similar dynamic of fatal attraction may occur in other fields. In the end, there may be a "positive" (normative) result, in the form of integrity policies and procedures: more stringent regulations for conducting, reporting and publishing (S_1). Let this suffice as a Lacanian reading of the Majorana case. In the next section, I will turn attention to my final case, involving psychologist Diederik Stapel.

56.8 The Stapel Case

Diederik Stapel became professor of social psychology in *Groningen* (in 2000) and *Tilburg* (in 2006). Here, he was appointed as director of the prestigious *Tilburg Institute for Behavioural Economics Research* (TIBER) and as dean of the *Faculty of Social and Behavioural Sciences*. In 2011, he formally and publicly admitted to have fabricated and manipulated data for at least 55 publications, from 2003 onwards. Because of the prominence of the perpetrator and the scale of the fraud, it quickly became a widely publicised misconduct case, both in the Netherlands and abroad,—and something of an epistemic trauma: for academic research in general and for social psychology in particular. After his formal dismissal and public condemnation (by three committees, established by three universities), Stapel became marginalised. In 2012, he resurfaced with an autobiographical account in Dutch entitled *Ontsporing* ("derailment"), the first in a series of literary books written in the wake of his exposure [25].

We will analyse the case from two perspectives. Stapel's academic papers reflect the logic of university discourse. To understand what is happening, however, we have to assess this discourse from an oblique (tilted) perspective, where the floor is given to the tormented subject ($) more or less directly (the discourse of the analyst), using *Derailment* as our starting point.

Derailment takes us straight to the decisive moment. Stapel has just been informed by a close colleague and friend (the chair of the department) that three young colleagues accuse him (the powerful professor) of faking his data. At a conference, they had taken the department chair aside. They had been following Stapel for months and had developed well-founded suspicions regarding his fraudulent ways. Stapel denies the accusations flat out. What could they have found out? The fraud simply seems too weird to be credible. No one will believe them. Still, the department chair decides to inform the rector.

In despair, and realising that his version of the story from now on has to be as convincing as possible, Stapel jumps into his car and visits some of the places where the fabricated data allegedly had been collected. Eventually, he ends up at Utrecht central station where he claimed to have gathered data for one of his most impressive papers, published in the journal *Science* in April 2011 (just a few months before the exposure), a highlight in his career [26]. In this paper, it is nicely demonstrated how, in a disordered, littered environment, research subjects are significantly more prone to stereotypical (i.e. racial) prejudices than in clean and orderly places. A unique opportunity for conducting the field experiment presented itself when the cleaners at Utrecht station (where thousands of travellers pass through on a daily basis) went on strike, so that the area quickly changed into a dirty and disordered ambiance.

Research subjects were allegedly invited to fill in a questionnaire concerning prejudice and stereotypes and to do so on one of the six chairs lined up by the research team.[1] Respondents ("Caucasians") could choose any chair, except that the first chair in the row was already taken by either a black or a white ("Caucasian") team member, disguised as a normal passenger. And indeed, according to Stapel, the data convincingly and consistently showed that messy environments resulted in more prejudice (operationalised as the increase of the distance between the chair chosen by the respondents and the chair occupied by the black team member). Stereotyping, Stapel and his co-author concluded, is a "mental cleaning device" that helps people to cope with chaos (p. 253).

During his desperate visit to Utrecht station, however, Stapel discovers that the experiment could never have been conducted as reported. It is impossible to line up chairs and enrol volunteers in a crowded and hasty place like this. For anyone acquainted with the situation, the whole design seems extremely impractical and highly improbable. On his way back home, driving through the dreary night, he finally realises that his world is on the verge of collapse and that a full confession is the only realistic option left to him.

Social psychology represents "university discourse" as specified by Lacan:

S_2	a
S_1	$\$$

As a prolific author of scientific papers, in collaboration with numerous co-authors, Stapel occupies the upper-left position of agent (S_2). Interestingly, as Stapel points out in his autobiographical retrospect, his identification with social psychology, his decision to commit himself to university discourse, was actually the outcome of an existential struggle. Initially, he wanted to become an actor. Before his conversion to academic psychology, he had been a Harold Pinter adept. The theatre of the absurd, of which Harold Pinter is a prototypical protagonist, enacts an absurdist interpretation of reality, and Stapel performed Pinter's plays already in high school. As an adolescent, he acted in a soap series broadcasted on Dutch television.

Stapel's conversion from actor to scientist implies that an erring subject, experiencing the world as absurd ($\$$), is transformed (converted) into a subject of science (S_2), a reliable and qualified source of information, someone who is able to act methodically and systematically, in accordance with the scientific method. Stapel's conversion from theatre to psychology implies a renouncement of absurdism and a willingness to adopt a more rational and scientific worldview (S_1 beneath the bar). In principle, social psychology provides the methodological tools that allow us to understand human behaviour (including unacknowledged biases) in a rational manner.

A clash between colliding worldviews (between social scientific rationalism and existential absurdism) was concluded (pushed into the lower-left position) in favour of the rational scientific stance requested by university discourse. From now on, S_2 (the objective scientific agent) is supposedly an individual *without* a worldview, although in reality, science continues to be spurred on by the rational worldview of experimental social science (seeing humans as rational decision makers who are nonetheless unconsciously biased). Thus, by becoming a social scientist, Stapel repressed or suspended a (fundamental, philosophical) question which until then had occupied him, namely, whether, in the final instance, human reality is rational or absurd. Stapel considered both options, wavered between two vocations, two callings: the absurdist option (a career in absurdist theatre) and the rationalist option (a career in science). Eventually, he converts to social psychology, placing S_2 in the upper-left position, pushing the disconcerting *truth* of absurdism beneath the bar. But this struggle of scientific rationalism (fuelling his research practice as a knowledge production system) against the disconcerting, but nonetheless revelatory truth of absurdism, continues to haunt him at times and even resurges in times of crisis:

[1] The *Science* paper mentions "six" chairs [26], p. 251; in *Derailment*, the number is reduced to "five" [25], p. 29.

the return of the repressed. In fact, on closer inspection, some of his experimental designs, including the Utrecht experiment, bear striking similarities with theatre (people sitting on a row of chairs, with a train station as a backdrop, etc.).

As a qualified university expert, Stapel focusses on discovering a decisive but enigmatic causal factor determining human behaviour, the "x effect", as he phrases it: that which makes the difference between significant and insignificant and between publishable and unpublishable results, the object *a* of social psychology research (upper-right position). In the Utrecht experiment, the object *a* is the empty chair, situated between research subject and disguised collaborator. An empty chair is a theatre prop par excellence. According to Lacan, the object *a* is not an object in the traditional epistemological sense, tangible and visible with our natural senses. Rather, it is an "object" which denotes absence; it is something enigmatic and negative, a stain or a trace: something like a particle signature obtained under exceptional conditions or an empty chair in a psychological trial. The uncertainty concerning the number of chairs (was it five, as claimed in *Derailment*, or six, as reported in the paper?) seems symptomatic of this. The chair is symptomatic of the unconscious bias Stapel wants to bring to the fore. The object *a* is the chair which is *not* taken, due to the enigmatic x effect. It is a chair which, from the perspective of the research subject, is both available and not available (standing next to the collaborator of colour). For a biased subject, it is a chair which is not there. And indeed, the chair has *never been there*, because the experiment was never conducted.

In the case of Stapel, after a successful start of his career, university discourse increasingly refuses to operate. Increasingly, his research practices become entangled in frustrations. Reality seems too messy and recalcitrant, or too absurd, to comply with his theories. Despite all his hard work, conducted when Stapel was still a reliable expert, his experiments begin to falter, and at a certain point, he even fails to publish a single academic paper for a whole year. Increasingly, he faces a split between outward recognition (academic tokens of acknowledgement) and internal crisis. And then, one day, sitting alone in his office in Groningen, he makes some subtle changes in his data. *Derailment* describes the event in meticulous detail. He quickly pushes the *Undo Typing* button, but this ruins the significance of his results, so that he soon relapses into altering some figures again. *Redo typing*, *Undo typing*—eventually he gives up his resistance. Before long, falsification becomes an addiction. He forces reality into compliance. His epistemic crisis gives rise to despair and eventually to fraud: $\$$ as an (unintended) by-product (lower-right position). The end result is resurgence of despair and a collapse into his former position as an erring adept of absurdism.

Psychoanalytically speaking, the crisis reveals that, during all those years, his scientific vocation actually remained something of a compromise. The kind of research he preferably practised, constituted a middle position between two vocations, for social psychology, as Stapel recognises in retrospect, is both a science and an "art". His outdoors laboratories (e.g. Utrecht central station) are basically "theatres" where human behaviour can be primed and moulded in various directions [25], p. 102. An important ingredient of both theatre and social psychology is deception. Research subjects must be naïve. As soon as they are aware of the aim of the experiment, this will affect their responses and ruin the results. Ideally, research subjects *believe* that they know what the research is about, but this actually proves to be an illusion, a distraction and a lure. While research subjects erroneously presume to know in what kind of situation they are involved, they are suddenly exposed to an unexpected stimulus, while their responses are being meticulously monitored, so that these responses may be considered as indications of a disavowed flaw (e.g. racial prejudice).

The set-up of Stapel's fictitious experiment is a cunning exemplification of this. The respondents believe that, by filling in questionnaires, they participate in a survey, while the chairs are merely placed there for their convenience (filling in questionnaires while standing upright seems impractical). They are unaware that the row of chairs is actually a trap. By keeping their distance vis-à-vis the black male, the empty chair *betrays* their obfuscated prejudices. Stapel's experiment is a play within a play, with Utrecht station as the theatre of life, where a plethora of prejudices are steering our daily behaviour (often in subliminal, unconscious ways). To come to terms with them, social psychologists design devices that *capture* them. Stapel's "play" was never been performed and never conducted: it was fiction. Stapel is actually a playwright and his article a parody of fashionable science.

This means that the prejudices attributed to the research subjects actually reflect the prejudices of Stapel and his research community. It is the prejudice of the social psychologists that their research subjects will prove prejudiced (by keeping their distance to non-Caucasian males). In other words, the prejudices enacted in Stapel's paper are actually the (prejudiced, stereotypical) expectations of the researcher (Stapel) himself and his readers, projected onto research subjects. He *wants* and *expects* his research subjects to be biased, even if this proves not to be the case, for otherwise he could simply have *conducted* the experiment instead of fabricating it. The hidden prejudice *must* be there. The "x effect" proves so difficult to capture because it is the "object *a*" of social psychology. It is that which *must* be brought to the fore, with the help of a contrivance, a trap, functioning as lozenge or *poinçon*, allowing Stapel to zoom in on human prejudice. The objective of social psychology (revealing prejudice) is considered sufficiently relevant and important to legitimise the use of deception. Social psychology aims to expose the Mr. Hyde in us, something which (notably in "Caucasians") refuses to give itself away so easily. Thus, instead of capturing the "x effect" (racial prejudice), the paper captures (reflects) the way in which social psychologists frame and perceive the behaviour of their research subjects. To phrase it somewhat bluntly, social psychologists believe that Caucasians are racists, especially those who seem unaware of their unconscious flaws.

Stapel assumes an "oblique" perspective: while respondents are ticking boxes, feigning innocence, Stapel secretly exposes them to a stimulus in disguise. These naïve respondents erroneously believe that they are unprejudiced, but the empty chair is the trap which exposes them. This is in line with the vocation of experimental social psychology as such, whose founding fathers (Lewin, Festinger, etc.) were Jewish immigrants coming from Europe and personally interested in the question whether Americans might be as prejudiced and anti-Semitic as their fellow Caucasians in Europe.

Yet, to Stapel's frustration, psychology's traps no longer function. The prejudice described in *Science* is the prejudice that research subjects are prejudiced; it is the prejudice which social psychological theory *attributes* to research subjects, existing in the minds and theories of social psychologists themselves. After concocting the missing, elusive data (his object of desire), Stapel's readers (social psychologists) believe the reported results because the outcomes concur with their headstrong conviction that research subjects are secretly prejudiced.

56.9 Deception

To understand how this could happen, Lacan's analysis of university discourse provides important hints. Stapel temporarily resolved the struggle between scientific objectivity and persistent absurdism by formally endorsing the first and disavowing the latter. His susceptibility to absurdism (the calling of his youth) was pushed beneath the bar (into the lower-left position) in order to endorse social psychology as a professional scientific research practice (S_2 in the upper-left position). As an experimental researcher committed to social psychology (S_2), Stapel is desperately looking for what he refers to as the "x effect" [25], p. 103, that is, a decisive (but often intractable and allusive) factor which makes the difference between significant and non-significant results (*a* in the upper-right position). In the Utrecht experiment, the empty chair symbolises this "x effect". Yet, he discovers that social reality has become a messy and convoluted realm, where countless factors interact. The problem is not that research subject are no longer biased (probably they still are). The problem rather is that *one particular effect* ("x"), whose discovery will lead to a publication in a top journal, has to compete with multiple other (disturbing) factors. To some extent, this is acknowledged. Stapel consciously connects the *x* effect with another factor, namely, littered versus clean environments. In littered circumstances, the trap is expected to do the trick: racial prejudices will be unmasked.

What if the mousetrap fails to work? Stapel aims to conduct research under real-life circumstances. Psychological laboratory experiments in the traditional sense lack external validity and relevance. In Utrecht central station and comparable settings, however, too many factors seem at work. Instead of deceiving his *research subjects*, Stapel basically decides to deceive his *readership*. Deception (of research subjects) is a perfectly acceptable ingredient of social psychological research, but Stapel changes the perspective as it were. His *Science* paper becomes an experiment indeed, a trap of the second order, so that peers, reviewers, colleagues, etc. are unwittingly lured into the position of research subjects: will they fall into this trap, and will they really believe that this (absurd) experiment actually happened, basically because their second-order prejudice (viz. that research subjects are prejudiced) is confirmed by it? Stapel's experiment allegedly confirms what social psychologists expect to happen, namely, that the "x effect" (as coping mechanism) will give itself away in a disorderly environment. Stapel's parody, a piece of fiction which reads like a *Science* paper, confirms what purports to be "normal science" although it actually reflects—a worldview.

Notwithstanding the success of his *Science* paper, the failure to achieve significant results is unsettling for Stapel and eventually produces despair ($\$$ in the lower-right position). The absence, the intractability of the x effect (object *a* of social psychology research), reinforces his inner split, his inability to satisfy his *cupido sciendi*. This *Spaltung* reveals an insurmountable gap between the knowledge provided by social psychology and the disconcerting truth of human existence (e.g. that life is a chaotic mess).

In Stapel's case, the struggle between rationalistic leanings and absurdist experiences continues to rage beneath the bar. On the one hand, Stapel (as qualified expert) was fuelled by the imperative which, according to Lacan, drives all contemporary research practices: "go on, produce more knowledge! Never enough!" [15], p. 120. In contemporary research, scientists have never published enough papers, have never achieved enough impact and have never acquired enough grants. Therefore, they continue to publish. Even if researchers are never literally *told* to do so, the imperative (S_1) functions as an unconscious injunction, coming from beneath the bar (lower-left position), fuelling contemporary social psychology as well (p. 121). Although rational science works hard to keep the absurdist alternative at bay, its research practices are so relentlessly demanding that they may be experienced as absurd (a "rat race").

Psychoanalytically speaking, this confirms the role of Stapel as a tormented subject facing an epistemic crisis. His misconduct is symptomatic, pointing to frustration and despair, as "by-products" of university discourse research ($\$$ in the lower-right position). S_2 becomes a divided (*spaltet*) subject, caught in the force field between resurging absurdism and the recalcitrance of his research object. By taking his deceptions one step further, building on the congruence between experiments and plays, his research really becomes theatre, so that, eventually, social psychology (as a particular strand of university discourse) gives way to absurdism ($S_2 \rightarrow \$$), resulting in an anticlockwise quarter-turn of the scheme, towards the discourse of the analyst.

56.10 Perverse Incentives

The discourse of the analyst provides an oblique (sideways) perspective, focussing not on fermions or social prejudices (S_2 pushed beneath the bar) but seeing science as an existential theatre of interactions between desiring subjects and allusive objects (as represented by the matheme of desire). We now see Stapel's misconduct as symptomatic of the systemic challenges which research practices are experiencing. This does not imply whitewashing their misdemeanours. Although the focus is on fostering understanding and deepening self-knowledge (so that *Derailment*, as a self-analysis, reflects the discourse of the analyst in terms of discursive structure), the eventual outcome or product is a critical reconsideration of the normative framework that is guiding contemporary science (S_1 in the lower-right position). If perverse incentives have become an inherent component of the research infrastructure, misconduct can only be addressed through deliberative reflection and systemic change, fostering self-reflection and moral deliberation, resulting in normative components, including an updated and reaffirmed integrity code (S_1).

References

1. Hermerén G, Zwart H, Marušić A, Fanelli D (2019) Mutual learning exercise (MLE) on research integrity. ec.europa.eu/research-and-innovation/en/statistics/policy-support-facility/mle-research-integrity. Accessed 4 May 2021
2. Horbach S (2019) To spill, filter and clean: on problematic research articles, the peer review system, and organisational integrity procedures. Radboud University/Ridderprint, Nijmegen, The Netherlands/Alblasserdam
3. Zwart H, Ter Meulen R (2019) Editorial: addressing research integrity challenges: from penalising individual perpetrators to fostering research ecosystem quality care. Life Sci Soc Policy 15:5
4. Zwart H (2017) Tales of research misconduct: a Lacanian diagnostics of integrity challenges in science novels, Library of Ethics and Applied Philosophy, vol 36. Springer, Heidelberg
5. Freud S (1925/1948) Geleitwort zu *Verwahrloste Jugend* von August Aichhorn. (Introduction to wayward youth by August Achhorn) *Gesammelte Werke XIV*. Imago, London, pp 565–567
6. Freud S (1937/1950) Die endliche und die unendliche Analyse. (Terminable and interminable analysis) *Gesammelte Werke XVI*. Imago, London, pp 59–99
7. Lacan J (1960/1974/2005) Discours aux Catholiques/*Le triomphe de la religion*. (Discourse to the Catholics: the triumph of religion). Éditions du Seuil, Paris
8. Kris E (1951/1975) Ego psychology and interpretation in psychoanalytic therapy. In: Selected papers. Yale University Press, New Haven, CT, pp 237–251
9. Lacan J (1966) Écrits (writings). Éditions du Seuil, Paris
10. Lacan J (1965/1966) Le séminaire XIII: L'objet de la psychanalyse (The seminar XIII: the object of psychoanalysis). staferla.free.fr/. Accessed 4 May 2021
11. Krüger S (2012) Fresh brains: Jacques Lacan's critique of Ernst Kris's psychoanalytic method. Am Imago 69(4):507–542
12. Foucault M (1971) L'ordre du discours. (The order of the discourse). Gallimard, Paris
13. Borch-Jacobsen M (1990) *Lacan: le maître absolu*. (Lacan: the absolute master). Flammarion, Paris
14. Lacan J (1958–1959/2013) *Le Séminaire VI: Le désir et son interprétation*. (The seminar VI: the desire and its interpretation). Éditions de La Martinière et Le Champ Freudien Éditeur, Paris
15. Lacan J (1969–1970/1991) Le séminaire XVII: L'envers de la psychanalyse. (The seminar XVII: the opposite of psychoanalysis). Éditions du Seuil, Paris
16. Nietzsche F (1881/1980) Morgenröte (Daybreak). Sämtliche Werke. Deutscher Taschenbuch Verlag, München
17. Zwart H (2010) The Nobel Prize as a reward mechanism in the genomics era: anonymous researchers, visible managers and the ethics of excellence. J Bioethical Inq 7:299–312
18. Foucault M (1994) "Qu'est-ce qu'un auteur." Dits et Écrits (What is an author. Sayings and writings). Gallimard, Paris, pp 789–809
19. Zwart H (2019) Psychoanalysis of technoscience: symbolisation and imagination, Philosophy and Psychology in Dialogue. LIT Verlag, Berlin
20. Manna S, Wei P, Xie Y, Law K, Lee P, Moodera J (2020) Signature of a pair of Majorana zero modes in superconducting gold surface states. Proc Nat Acad Sci 117(16):8775–8782
21. Zhang H et al (2018) Quantized Majorana conductance. Nature 556:74–79
22. TU Delft (2018) Majorana trilogy completed. eurekalert.org/pub_releases/2018-03/duot-mtc032318.php. Accessed 4 May 2021
23. QuTech (2020) Expression of Concern about Quantized Majorana conductance publication. qutech.nl/2020/05/16/expression-of-concern-about-quantized-majorana-conductance-publication/. Accessed 4 May 2021
24. Zhang H, Liu CX, Gazibegovic S et al (2020) Expression of concern: quantized Majorana conductance. Nature 581(7870):E4
25. Stapel D (2012) Ontsporing (Derailed). Prometheus, Amsterdam
26. Stapel D, Lindenberg S (2011) Coping with chaos: how disordered contexts promote stereotyping and discrimination. Science 332:251–253

57 Development and Implementation of a National Research Integrity System: The Case of the Estonian Code of Conduct for Research Integrity

Mari-Liisa Parder, Marten Juurik, Kristi Lõuk, Katrin Velbaum, Kadri Simm, and Margit Sutrop

Abstract

Numerous codes of conduct for research integrity (RI) have been developed, but there are few reports on the process of developing such a code. The current chapter explores the process of development and ratification of the Estonian Code of Conduct for Research Integrity (2017). Reasons behind the code included the need to regulate RI on the national level, raise awareness, prevent misconduct and keep up with other countries. Drafting the code from scratch was preferred because it created a sense of responsibility and acknowledged the local cultural context. The most divisive issues were the length of the document, its declarative nature and its ambiguous role. The study confirmed that in order to promote the culture of RI, the code needs to be reinforced by the other elements of the RI system.

57.1 Introduction

From the 1990s on, there has been a boom in the creation of codes of ethics. Research codes of ethics have emerged partly as a reaction to public scandals, partly as an attempt to introduce more transparency and ethical rigour into research practices [1–4]. Numerous initiatives have been taken to achieve agreement on proper research standards [5]. However, there are few examples in the literature exploring the process of development and ratification of a code, stakeholder engagement and negotiations and collection of feedback. Examples include the Netherlands code [6] and codes related to specific topics or areas [7–11].

The Estonian Code of Conduct for Research Integrity (Estonian code) [12] has been created by different partners and representatives, underwent detailed discussion and was signed by almost all Estonian research and development institutions (RDIs). The Estonian code includes values and principles of action which are distributed among five main areas: (1) research planning; (2) research conduct; (3) authorship, publishing and application of research results; (4) the researcher in the research community; and (5) observance, promotion and application of RI.

57.1.1 The Code Roadmap

The code of conduct is a central element of the research integrity (RI) system. However, its implementation and effectiveness are dependent upon the presence of the other elements of the RI system (e.g. education, regulations and other governance mechanisms). As our results

M.-L. Parder · M. Juurik · K. Lõuk · K. Velbaum
Centre for Ethics, University of Tartu, Tartu, Estonia

K. Simm · M. Sutrop (✉)
Department of Philosophy and Centre for Ethics, University of Tartu, Tartu, Estonia
e-mail: margit.sutrop@ut.ee

J. Faintuch, S. Faintuch (eds.), *Integrity of Scientific Research*,
https://doi.org/10.1007/978-3-030-99680-2_57

demonstrate, an inclusive and stakeholder-engaged process might often be the key to successful acceptance and adoption of a code. The value of the current chapter is that it reflects this process, providing a close-up view of deliberation points and mapping the steps to be considered in the development and implementation of the code.

57.2 Ethics Codes and the RI System

Ethics codes (or codes of conduct) function between ethics and law [13]. Ethics highlights the values and principles which are the basis of our actions. However, values can come into conflict, which creates the need for interpretation and deliberation points [14]. Ethical principles and standards for research are based on their aims, and David Resnik [15] divides them into two aspects, epistemic (the theoretical drive towards knowledge) and practical (finding and implementing a solution to a practical problem). Codes usually have multiple objectives, ranging from the declaration of shared understanding by a professional community of their obligations to adhere to certain ethical standards to signalling dedication to a responsible activity (see more [16]).

57.2.1 Pressing Needs

Reasons behind the sudden boom for generating codes of ethics for researchers include reactions to major misconduct cases but also the realisation of cultural and societal differences in understanding what is right and wrong and, more broadly, in how values and norms are understood [16]. Major roles are played by different pressures from publishing and competition for funds [17] but also by various (and sometimes conflicting) roles and requirements for researchers [18]. Not less important are conflicts arising from the twin duties of sharing data and protecting confidentiality [19]. In addition, scientists are often expected to take on roles outside of academia in order to transfer their knowledge to society, which in turn raises potential conflicts of interests (e.g. professional, personal or financial) [20]. Codes are also important in order to overcome role ambiguity and role autonomy within the academy (e.g. president, dean, administrator, etc.) [21]. Different areas are highlighted as to where research misconduct may arise, and such challenges are illustrated with actual cases from the respective scientific field [22]. Various reasons for following codes have been emphasised, with the fear of losing colleagues' trust as the most essential reason indicated for adherence to a code. However, people may also follow a code for internal reasons [23].

57.2.2 International Priorities

Four arguments have been generally put forward: (1) international science needs standards for solving disagreements arising between scientists from different cultures; (2) where no local standards have been proposed, it is possible to rely on international ones; (3) good international standards encourage the emergence of local standards; and (4) ethics codes foster trust between scientists from different countries [24].

57.2.3 Pitfalls and Weaknesses

For successful implementation of a code, six main obstacles have been highlighted [25]: (1) insufficient awareness of the content of the code; (2) lack of motivation to follow the values, principles or rules agreed upon in a code; (3) absence of recognition that a code should be applied in a specific situation; (4) conflict in the values making up a code, making it hard to follow all of them; (5) contradiction between personal, organisational or societal values expressed in the code, making them difficult to follow; and (6) conflict between the declared values and those adhered to in everyday work. Several authors have indicated that the effectiveness of the codes is related to how well they are implemented and communicated [6, 26].

We argue that ethics codes cannot function properly without being incorporated into the RI system and being enforced by the latter. Ethics codes are an element of the integrity system, but the integrity system comprises more than just a code [27]. In broad terms, it consists of policy, instruction and oversight [28] and includes stakeholders (e.g. governance boards) and procedures [29].

57.2.4 Beacons of Integrity

This reflects the idea emphasised in the Singapore Statement [30] that research institutions have the responsibility to create and sustain environments that encourage integrity through education and to have clear policies. The Bonn PRINTEGER Statement [31] stresses that it is the responsibility of leaders of institutions to set the standards for acceptable conduct and to promote a strong integrity culture which involves sharing and giving substance to ethical norms, values and beliefs, emphasising their value for the organisation and attaching incentive structures to them. Codes are important, but they need a "resilient integrity culture" [32] to be effective, for example, a supportive research environment. In 2010, the European Science Foundation recommended that national integrity structures be set up and codes of conduct be adopted [33]. They brought out that successful approaches to promoting good research practices include establishing an adequate institutional framework, which is composed of ethics committees and RI offices, both at institutional and national levels [34].

57.2.5 Universal Yet Custom Built

Lex Bouter [35] argues that multiple stakeholders (research institutions, funding agencies, scientific journals) have a role to play. At the same time, the RI governance system should be adjusted to fit each country, considering its size, research infrastructure and available resources; governance should be incorporated at institutional and national levels [34]. If unintended consequences emerge, the integrity system should be revised, or policies should be adapted accordingly [36]. It was also recommended that the involved parties "create a national RI body that could help coordinate, monitor, educate, communicate and promote RI in a country" [36]. Thus, not only is it important to establish a national body, but there should also be national level RI officers [36]. For oversight of RI investigations, an appeal system is recommended, which is especially important for countries where there are no national RI bodies in place [36]. Therefore, for the RI system to function properly, a national code of conduct is needed, and inversely, for the national code to work, an RI system is needed.

For developing a code, the process is as important as the content of the code itself. The development of the code can happen top-down or bottom-up. In our view, the latter is more suitable for achieving broad consensus.

57.3 The Estonian Context

Estonia is a small country with a population of 1.3 million. Since the regaining of independence in 1991, research funding in Estonia has been competition-based, which was a major change compared to the Soviet times [37]. The Estonian Science Foundation (reorganised as the Estonian Research Council (ETAg[1]) in 2012) was established to create a system where funding decisions would be made independently (by the research community), not by politicians and the state. ETAg manages different competitive funding instruments and monitors how the grants are used, including adherence to ethical principles and good practices of research [37]. Research and development activities are carried out by public research institutions (mostly universities) and private research institutions. Most Estonian researchers are affiliated with universities, where the majority of the research is carried out [38]. External positive evaluation is required

[1] The abbreviation for Estonian Research Council in English should be ERC; however, since the abbreviation "ERC" is more commonly known as European Research Council, the Estonian abbreviation ETAg (*Eesti Teadusagentuur*) is used in this article.

from research institutions for applying for public funding.[2]

In terms of the RI system, Estonia has had a code of ethics for scientists since 2002. Research ethics committees had already existed from the 1990s on, as well as training programmes and guidance materials at an institutional level. The leading competence centre for research ethics and integrity issues in Estonia has been the interdisciplinary Centre for Ethics of the University of Tartu (CEUT) founded in 2001. From 2015 until 2018, CEUT has studied RI issues within an H2020 project PRINTEGER on the national level. It was identified as a problem that in the Estonian RI system, there is no national body for dealing with misconduct cases and that at the university level, an ad hoc approach is used [40]. The stakeholder panel, organised in the framework of the PRINTEGER project, involved representatives from universities, funding organisations and journal editors who focused on issues related to the understandings of RI and misconduct. The panel took place in November 2015 and turned out to be a catalyst for RI issues in Estonia. It was agreed at this stakeholder panel that besides a Code of Ethics for Estonian Scientists [41], there is a need for guidelines both for individual scientists and for institutions.

57.4 Research Questions and Methodology

1. How was the code created?
2. What were the motivations behind creating the code?
3. What are the views of partners on the final version of the code?

During the process of development of the code, which began in February 2016 and ended in November 2017, there were ten organisations[3] represented in the working group. Institutions participating in the process were distributed among four categories (one partner could belong to several categories): (1) partners facilitating the initiation of the process and members of the ETAg working group (16 representatives), (2) scientists who were not members of the working group but who contributed significantly during the consultation periods (28 persons), (3) representatives of the research organisations in Estonia who were invited to join the agreement of the Estonian code (26 institutions) and (4) partners who have implemented the Estonian code in practice (three institutions at that time) (see Table 57.1).[4]

Interviewees were selected from these four main categories, keeping in mind the inclusion of organisations from different groups as well as people with different roles in the process [42]. The interviewees were pseudonymised; their names will not be published in this article or anywhere else. However, it was emphasised to the respondents that due to their positions they might be recognised by the readers and thus their anonymity could not be guaranteed. Approval by the Research Ethics Committee of the University of Tartu was obtained.

Qualitative content analysis was used to analyse the interviews and documents in order to capture the main emergent themes. Three themes stood out in the material: (1) reasons for creating the code, (2) development and process (3) views of the final version.

57.5 Creation of the Code (Fig. 57.1)

The first meeting was held in March 2016 and 11 meetings were held all over the period of 20 months; additional work was conducted

[2] Estonian RDIs are regularly evaluated. At the time of ratification of the code (1 November 2017), there were 22 evaluated RDIs; in 2021, there were 20. See more [39].

[3] The organisations included public universities (1) University of Tartu, (2) Tallinn University, (3) Tallinn University of Technology, (4) Estonian University of Life Sciences, (5) Estonian Academy of Arts, (6) Estonian Academy of Music and Theatre (RDI), (7) National Institute of Health Development, (8) Estonian Academy of Sciences, (9) Estonian Research Council and (10) Republic of Estonia Ministry of Education and Research.

[4] The last interview was carried out during the emergency situation in Estonia due to COVID-19.

Table 57.1 Pseudonymised list of interviewed partners

Organisation/position	Role	Time of the interview	Reference in the article
Representative of Estonian Research Council (ETAg)	One of the coordinators of the process [1]	November 2019	Rep. ETAg
Professor, University of Tartu	Contributed significantly [2]	January 2020	Professor, Tartu
Lecturer, Tallinn University of Technology (TalTech)	Contributed significantly [2]	January 2020	Lecturer, TalTech
Representative of Estonian University of Life Sciences	Joined the agreement and has implemented the code [3, 4]	November 2019	Rep. EULS
Representative of University of Tartu	Partner in the process, joined the agreement and has implemented the code [1, 3, 4]	April 2020	Rep. Tartu
Representative of Tallinn University	Partner in the process, joined the agreement and has implemented the code [1, 3, 4]	December 2019	Rep. UTal
Representative of Tallinn University of Technology (TalTech)	Partner in the process, joined the agreement and has implemented the code [1, 3, 4]	January 2020	Rep. TalTech

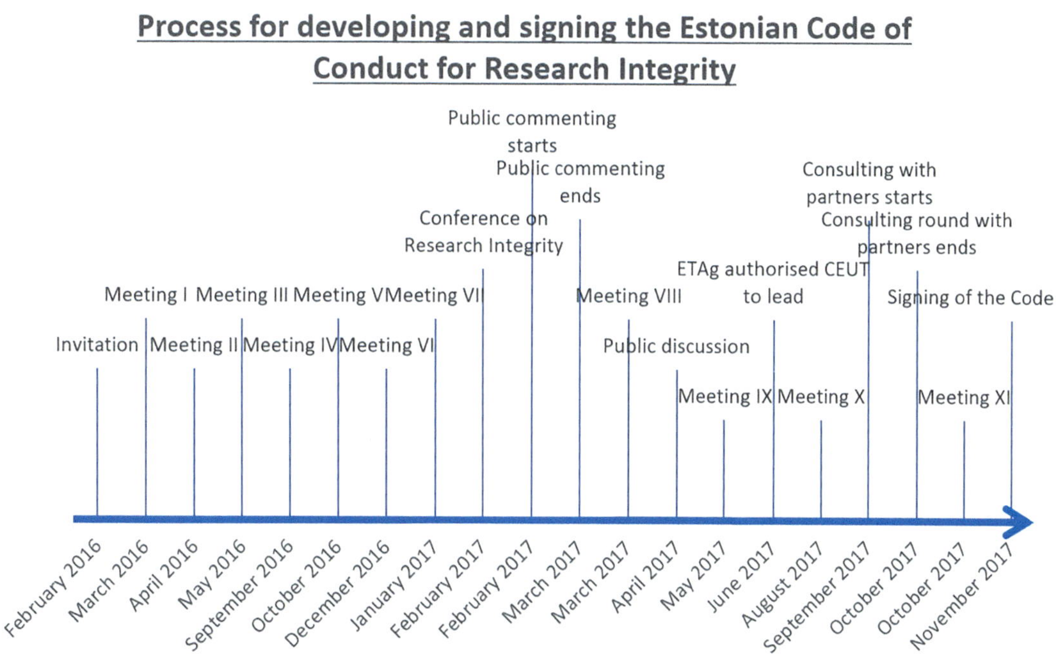

Fig. 57.1 Timeline of the process of drafting the Estonian Code of Conduct for Research Integrity

between the meetings on different documents as well as drafting the key points for the code (see more from Fig. 57.1). In February 2017, an international conference on RI was held, and the first draft version of the code was introduced to the wider public for comments in the space of 1 month. In June 2017, ETAg authorised CEUT to take over the leadership of the process. Before the adoption of the code, the second round of consultations was held (from September 2017 to October 2017), this time with partners (positively evaluated RDIs in Estonia) who commented on the next-to-final version of the code. The invitation to adopt the code and sign it formally was sent to all of Estonia's 22 nationally accredited research institutions, of which 21 signed it. In addition, ETAg and the ministry also joined the agreement.

57.5.1 Drafting the Integrity Code

Several issues were related to the actual writing process of the code: where to begin, which examples to use or how long the code should be. One of the first questions that arose was whether the code should be created from scratch or whether, perhaps, a translation of an already existing international document would suffice. The main candidate for translation was the European Code of Conduct for Research Integrity of the All European Academies (ALLEA) [43]. Two other codes—the Danish [44] and Dutch [45] national codes—were also considered as examples. The Danish code highlighted the obligations of both research institutions and funding agencies, a view shared by ETAg:

> For example, what primarily impressed me in the case of Denmark was that the obligations of research institutions and organisations for the financing of research had been clearly outlined. In our 2002 Code of Ethics, which was made by the Academy of Sciences, all the obligations had been placed on the researcher. This is why it caught my eye—it is not the individual responsibility of the researcher, actually it is a collective responsibility. (Rep. ETAg)

ETAg looked at the code from a wider perspective and tried to see its role among all the other national and international documents. From this perspective, both European and country level codes are necessary, because they have somewhat different aims. The European code is necessary for offering a similar framework for all European countries, whereas the value of the Estonian code resides in the fact that it is created by Estonian researchers:

> I think that the Estonian Code is also most necessary. The European one is, so to say, for advanced users. It is concentrated and, as far as I know, its aim is somewhat different. /.../ But the Estonian Code shows that each country must have its own. For our research community, our own Code of Conduct for Research Integrity is definitely necessary, as it has been compiled by our own researchers on such a broad basis. Both are necessary. (Rep. ETAg)

Similar ideas were shared by other interviewees—writing offers a chance to adapt the code to Estonian needs and the process of writing it would also lead to a greater sense of responsibility:

> This is the room for manoeuvre that makes our Code personalised for Estonia. Why not simply take it over from Europe and say we are part of Europe? The idea of this agreement is that we are not writing another document of principles that we can really take over from there in a somewhat more and less detailed way, but we signed it ourselves, and it is our responsibility to make it work. (Rep. UTal)

57.5.2 Lost in Translation

The term *research integrity* does not translate very well into Estonian, and therefore a suitable term was needed. The final translation was *hea teadus* which would directly translate as *good science*. According to the representative of ETAg, the other designations considered were *honest science* and *trustworthiness of science*. It was also necessary to distinguish *research ethics* from *research integrity* as all the discussions in Estonian had previously been only about *ethics*. According to the representative of ETAg, it took a few meetings to come to an agreement about the concept of *research integrity*, what it includes and how to translate it:

> Research ethics already had its specific content, but what we were going to deal with was research integrity. Initially, it was difficult to make the difference or understand what each included, perhaps they coincide or maybe not, to understand the limits of these concepts. What exactly are we talking about, if we talk about this in Estonian, what is it then? In the beginning, there was a polishing of concepts on which several meetings were spent. Finally, we reached an understanding of what we were speaking about. Maybe this doesn't exactly coincide with the English term, but at least in the Estonian language and the Estonian community we understand it in the same way when we speak about

research integrity and what it includes. (Rep. ETAg)

57.5.3 Mixed Approach

Both the developers of the code and the interviewees saw value in drafting the code from scratch, because this created a sense of responsibility and it was directed towards the specific cultural setting. However, it was also important to consider the best practices and experiences from other countries and their codes.

57.6 Motivations Behind the Code

From the organisational side, the interviews highlighted that the process of creating a code of conduct needed to be based on a broad constituency, including representatives from all universities and research institutions with an interest in RI. Therefore, the head of ETAg visited the Council of Rectors of Estonian Universities and invited them to assign a member to the working group:

> The first thing needed for forming our working group was a broad-based membership. We wanted to involve all the universities and research institutions who had a greater interest in the question or whom it concerned more. Therefore, [XXX] went to the Council of Rectors and asked all the rectors to appoint a representative to our working group. (Rep. ETAg)

RDIs mentioned the need to share common understanding and rules within and between universities in order to agree on the best practices; all Estonian universities should comply with the need for a common understanding and regulations so that each institution individually would not have to come up with regulations, explanations, arguments and solutions to potential problems:

> We do not want to invent these things ourselves; we do not want to cope with them on our own. We want to have a university-centred conception and regulation in the framework of which to interpret and discuss these situations. (Rep. UTal)

> Often questions crop up at an operational level about the border between black, white and grey. And often such boundaries cannot be drawn, and then the best practice, the exchange of information between the universities is the most important thing. (Rep. Tartu)

From the perspective of RDIs, the need to catch up to more developed countries was mentioned, making itself felt most in cases of funding applications and institutional accreditations, where the expectations of the research institutions have increased:

> Secondly, the times and expectations, the changes and catching up with the developed world have been so rapid. The expectations of us as an organisation when we're applying for research financing and we are seeking institutional accreditation are very topical today. (Rep. UTal)

The code also had a regulative motivation: ETAg felt the need for a national body responsible for promoting RI. Such a body could handle misconduct cases beyond just one research institution or offer help and guidance in dealing with institutions' cases. A national code of conduct for all research institutions was seen as a necessary part of national regulation. It was also hoped that all Estonian research institutions would adopt the new code:

> Actually, the overarching aim was that it would include whole Estonia. All the institutions already have their regulations; they can manage, but if a situation occurs when outside advice is needed, there should be a body or a person; we rather kept this body in mind who would solve such problems that the university cannot cope with alone. If there are conflicts of interest or a second opinion or experience is needed, a place where to discuss it is needed. For this a larger body with a larger, more broad-based membership is needed. (Rep. ETAg)

57.6.1 Well-Defined Responsibilities

From the RDI perspective, the need for institutionalised regulations of RI was stressed, with the note that ethics inevitably is and should be institutionalised to clarify responsibilities. If there is a shared responsibility, no one person would feel responsible. This way, the responsible persons should also cooperate with similar persons from other institutions and create common ethical requirements for research within this broader community:

> It seems to me that it can't be helped—ethics is institutional. /. . ./ If we say that everyone is responsible for it, then no one is responsible. When we say that for raising awareness about such things there should be a council of academic ethics /. . ./, then a very specific group of people is formed to deal with it. There will be a specific group of people who should cooperate among themselves. . . Let's say, if someone gets a bachelor's degree from Tallinn University, a master's from the University of Tartu and a doctoral degree from the Tallinn University of Technology, there should be a package of ethical requirements or even ethical preconditions for research that is uniform everywhere. (Lecturer, TalTech)

The preventive aim of the code was also mentioned, in hopes of avoiding large-scale scandals of research misconduct. Since many of the foreign codes were adopted after public scandals, the representatives of ETAg wished to adopt a code of conduct before such a scandal happened in Estonia:

> And we adopted the decision that this should be done before great blunders have happened. (Rep. ETAg)

57.6.2 Educational Role

> As I understand, it has not only a regulative but also an educational aim. Everyone who reads it should be aware that there are not only regulations about what can be done and what must not be, but there are also very general things that must be followed. (Rep. ETAg)

In line with the educational aim, the code also had an awareness-raising aim which was seen by the ETAg representative as a necessary step to prevent research misconduct:

> /. . ./ a large part of the Code of Conduct for Research Integrity and all these principles were raising people's awareness, namely because this prevents misconduct most effectively. (Rep. ETAg)

57.7 Adopting the Code

One interviewee mentioned that there had been some doubts about whether or not to sign the code. The main source of doubt was the code's unclear legal status and what kinds of obligations institutions would be assumed to fulfil by signing it:

> These hesitations did not reach me specifically, but I know that there may have been hesitations, namely because it was not understandable what the actual obligations were and how binding they were. The cause for hesitations could have been how much I'd be bound to it if I sign it, what my specific legal obligations resulting from it would be. (Rep. ETAg)

According to the interviews, the decision to sign the code was similar in all the larger universities: it was discussed in the senate and then decided by the rector. However, most of the larger universities were already committed to the code. The representative of UTal said that signing the code was not a question, as a member of the university's rectorate participated actively in the working group. It was already more or less decided that the university would sign the code. She also added that the signing of the code was probably not widely acknowledged in the university:

> Before signing it, we did not organise a discussion where the general public could have voted. As I had been an active member of the working group, questions would not have been raised about it by the Rector's Office. As a member of the Rector's Office, I was one of the essential leaders in the process. (Rep. UTal)

57.8 Views of the Final Version

The representatives of universities were mostly content with the final code, and the main concern was with the length of the document. Although a lengthier code does raise awareness and educate, the representatives had hoped to achieve an even greater degree of raising awareness, offering training programmes and creating an independent committee for handling issues of RI.

Those who found that ethical conduct in academic fields is an extension of basic moral conduct, taught early in life at home and in school, also found that a shorter, more concentrated code would suffice. Those who found that the code should make explicit all ethical issues and various topics and dimensions of research that could raise ethical concerns found that a lengthier document had its advantages. In addition, two of the universities hinted at the possibility that, because a longer, more detailed national code had already been adopted, they could draft a shorter, more concentrated version of these principles for their institution.

57.9 Final Remarks

So far academic attention has been directed towards the finalised codes and their content—which principles they contain and how they are framed—and not that much on the process of their creation. However, looking at the process can help understand a code's context, therefore explaining some of the choices and compromises that were made and giving guidance to others in a similar position of starting to develop a code.

The partners in the current process highlighted the same reasons brought up in the literature (e.g. [24]) for creating a national code: it regulates RI topics on a national level, raises awareness and helps prevent misconduct and keeps Estonia up to date with other countries which have national RI regulations. On the other hand, crafting a code from scratch valued the specific cultural setting and augmented the sense of responsibility for the code. The working group of the code also validated consideration of best practices and experiences from other countries. The interviews confirmed that the stakeholders had different expectations for the code which could be fulfilled only if the next steps were followed. For successful implementation of the code, one has to organise training, awareness raising, to work out procedures of reporting and investigating misconduct cases, as well as the system of counselling.

57.9.1 Trickle-Down Effect

By the time of the interviews, 2–5 years after the adoption of the code, most of the universities were creating or updating their institutional policies and procedures for ensuring RI. Since the effectiveness of the code is related to its implementation and communication (e.g. [7, 26]), CEUT also followed what happened after the adoption of the code. In media coverage, the first record of the code is dated 22 September 2015. From September 2015 to August 2020 (the last entry for 2020 before finishing the analysis for this article), there are 81 entries in the media monitoring. The coverage of the code includes explanations of why it is needed, the breadth of the signatories to the code (17 entries) and different ways in which the code has been used in explaining or justifying matters related to science. Therefore, it can be concluded that the code is quite well accepted by the Estonian science community and used in public discussions.

57.9.2 Code Offspring

A report on establishing a national system for monitoring and supporting RI in Estonia, ordered by ETAg, was published at the beginning of 2020. This report stated two things: first, that half of the respondents had read the code (which compared to Davies's [46] notes on Denmark which is an excellent result), and second, that Estonian scientists feel that research ethics training needs improvement [47]. The report indicated that researchers would like to see a publicly organised, harmonised and clear RI system in Estonia [47]. The results of the aforementioned report showed that RI is considered important but that there is a lack of awareness concerning it and a need to improve qualifications in the field of RI. The respondents found that the institutions do not offer sufficient training and training materials in RI. Therefore, the Estonian Ministry of Research and Education assembled a working group at the end of 2020 to initiate policy changes.

57.9.3 Downstream Changes

Since the code also focuses on the responsibilities of research organisations, several universities in Estonia have proceeded to develop their procedures for applying the code. The University of Tartu has finalised its instructions and appointed an RI advisor in each of the four faculties. Other universities are also setting up their instructions to fulfil the commitment they made when signing the code.

57.9.4 Necessary Tensions

Some interviewees argued that it was the right decision to write our new code and not translate a foreign one, although the revised edition of the ALLEA European Code of Conduct for Research Integrity was translated into Estonian in 2018. One interviewee shared the opinion that Estonia should not excessively imitate what is done in other countries. This tension could also be regarded from the perspective of international harmonisation of RI: such initiatives may meet resistance on a national level where local research communities show reluctance to imitate rules created abroad. On the other hand, the need to look at the best practices of other countries and to learn from their experience was considered an important part of creating the Estonian code.

Similar tensions exist between national and institutional levels. On the one hand, a national framework is needed to create similar rules for all the institutions. On the other hand, institutions regard themselves as autonomous and adopt their internal documents and procedures that may vary to a great degree. What, then, is the role of a national code? First, it helps bridge the national and international levels by adopting similar principles and, as a minimum, ensuring that the national principles do not deviate too much from the international ones. Second, the national code helps to harmonise institutional practices by offering them similar principles, concepts and procedures. And third, the code creates the basis for establishing the RI system in Estonia. All in all, it seems that the independence of institutions and academic autonomy need to be considered in the wider context of regulating research.

At the end, it is important to trace how the code will be implemented and used and whether "watchdogs" are fulfilling their role [27].

Acknowledgement We would like to thank all the people willing to be interviewed for this study. We also thank Ilmar Anvelt for his help with translation of the interview citations and Tiina Kirss for improving English expression.

Financial Support This study is funded by the Horizon 2020 project "PRO-RES—PROmoting integrity in the use of RESearch results in evidence-based policy: a focus on non-medical research" and by the Centre of Excellence in Estonian Studies (European Union, European Regional Development Fund). Five of the authors (Mari-Liisa Parder, Marten Juurik, Kristi Lõuk, Katrin Velbaum and Margit Sutrop) were part of the working group developing the Estonian Code of Conduct for Research Integrity.

References

1. Levine FJ, Skedsvold PR (2008) Behavioral and social science research. In: Emanuel EJ, Grady CC, Crouch RA, Lie RK, Miller FG, Wendler DD (eds) The Oxford textbook of clinical research ethics. Oxford University Press, Oxford, pp 336–355
2. Emanuel EJ, Wendler D, Grady C (2008) An ethical framework for biomedical research. In: Emanuel EJ, Grady C, Crouch RA, Lie RK, Miller FG, Wendler D (eds) The Oxford textbook of clinical research ethics. Oxford University Press, Oxford, pp 123–135
3. Israel M, Hay I (2006) Research ethics for social scientists. SAGE, London
4. Kitchener KS, Kitchener RF (2008) Social science research ethics. In: Donna MM, Pauline EG (eds) The handbook of social research ethics. Los Angeles, CA, SAGE, pp 5–23
5. Drenth PJDD (2010) Research integrity; protecting science, society and individuals. Eur Rev 18(3): 417–426
6. Schuurbiers D, Osseweijer P, Kinderlerer J (2009) Implementing the Netherlands code of conduct for scientific practice — a case study. Sci Eng Ethics 15: 213–231
7. Macaulay AC, Cross EJ, Delormier T, Potvin L, Paradis G, Mccomber A (1998) Developing a code of research ethics for research with a native community in Canada: a report from the Kahnawake Schools Diabetes Prevention Project. Int J Circumpolar Health 57(1): 38–40
8. Korkut Y (2010) Developing a national code of ethics in psychology in Turkey: balancing international ethical systems guides with a nation' s unique culture. Ethics Behav 20(3–4):288–296

9. Kelman I, Dodds R (2009) Developing a code of ethics for disaster tourism. Int J Mass Emerg Disasters 27(3): 272–296
10. Lozano JF (2006) Developing an ethical code for engineers: the discursive approach. Sci Eng Ethics 12(2):245–256
11. Sinclair C (2020) Developing and revising the Canadian Code of Ethics for Psychologists: key differences from the American Psychological Association code. Ethics Behav 30(4):249–263
12. Estonian Code of Conduct for Research Integrity (2017) Centre for Ethics, University of Tartu. etag.ee/wp-content/uploads/2017/11/HEA-TEADUSTAVA_eng.pdf. Accessed 8 June 2021
13. Campbell A, Glass KC (2001) The legal status of clinical and ethics policies, codes, and guidelines in medical practice and research. McGill Law J 46(2): 473–489
14. Sutrop M (2015) Can values be taught? The myth of value-free education. Trames J Humanit Soc Sci 19(2): 189–202
15. Resnik DB (1998) The ethics of science: an introduction. Routledge, London
16. Sutrop M, Parder M, Juurik M (2020) Research ethics codes and guidelines. In: Iphofen R (ed) Handbook of research ethics and scientific integrity. Springer Nature, Switzerland AG, pp 1–23
17. Lach HW (2019) Research integrity: "doing the right thing, even when no one is watching". Clin Nurs Res 28(6):655–657
18. DuBois JM, Anderson EE, Chibnall J, Carroll K, Gibb T, Ogbuka C et al (2013) Understanding research misconduct: a comparative analysis of 120 cases of professional wrongdoing. Account Res 20(5–6): 320–338
19. Shamoo AE, Resnik DB (2015) Responsible conduct of research, 3rd edn. Oxford University Press, New York, NY
20. Werhane P, Doering J (2007) Conflicts of interest and conflicts of commitment. In: Elliot D, Stern JD (eds) Researcher ethics: a reader. University Press of New England, London, pp 165–189
21. Braxton JM (2010) Norms and the work of colleges and universities: introduction to the special issue — norms in academia. J Furth High Educ 81(3):243–250
22. Koepsell D, Brinkman WP, Pont S (2015) Human participants in engineering research: notes from a fledgling ethics committee. Sci Eng Ethics 4:1033–1048
23. Spielthenner G (2015) Why comply with a code of ethics? Med Health Care Philos 18(2):195–202
24. Resnik DB (2009) International standards for research integrity: an idea whose time has come? Account Res 16(4):218–228
25. Sutrop M (2016) Kuidas panna eetikakoodeksid paremini toimima. [How to make ethics codes work better]. In: Sutrop M (ed) Eetikakoodeksid Väärtused, normid ja eetilised dilemmad. [Ethics codes: values, norms and ethical dilemmas]. Eesti Keele Sihtasutus, Tartu, Finland, pp 85–103
26. Stevens B (2008) Corporate ethical codes: effective instruments for influencing behavior. J Bus Ethics 78(4):601–609
27. Bramstedt KA (2020) Integrity watchdogs, lap dogs, and dead dogs. Account Res:1–5
28. Anderson MS (2014) Global research integrity in relation to the United States' research-integrity infrastructure. Account Res 21(1):1–8
29. Six F, Lawton A (2013) Towards a theory of integrity systems: a configurational approach. Int Rev Adm Sci 79(4):639–658
30. Resnik DB, Shamoo AE (2011) Singapore statement on research integrity. Account Res 18(2):71–75
31. Forsberg E-M, Anthun FO, Bailey S, Birchley G, Bout H, Casonato C et al (2018) Working with research integrity — guidance for research performing organisations: the Bonn PRINTEGER statement. Sci Eng Ethics 24:1023–1034
32. Zwart H, Ter Meulen R (2019) Addressing research integrity challenges: from penalising individual perpetrators to fostering research ecosystem quality care. Life Sci Soc Policy 15(1)
33. Boesz CC (2011) Developing research integrity structures: nationally and internationally. In: Mayer T, Steneck N (eds) Promoting research integrity in a global environment. World Scientific, Singapore, pp 7–16. worldscientific.com/doi/abs/10.1142/9789814340984_0001. Accessed 8 June 2021
34. (2010) Fostering research integrity in Europe: a report by the ESF Member Organisation Forum on Research Integrity. esf.org/fileadmin/user_upload/esf/ResearchIntegrity_Report2011.pdf. Accessed 6 June 2021
35. Bouter LM (2018) Fostering responsible research practices is a shared responsibility of multiple stakeholders. J Clin Epidemiol 96:143–146
36. Marušić A (2019) Mutual learning exercise (MLE) on research integrity. Processes and Structures. hceres.fr/sites/default/files/media/downloads/MLE%20RI_Final%20Report_0.pdf. Accessed 6 June 2021
37. Allik J (2015) Eesti teaduse lugu. [The story of Estonian science]. Akadeemia 6:963–985
38. Raudvere K (2016) Ülevaade uuringutest, mis käsitlevad Eesti elanike suhtumist teadusesse. [An overview or the studies on the attitudes of the Estonian population towards science]. etag.ee/wp-content/uploads/2014/02/%C3%9Clevaade-teaduse-mainet-Eestis-k%C3%A4sitlevatest-uuringutest.pdf. Accessed 6 June 2021
39. Research and development | Estonian Ministry of Education and Research. hm.ee/en/activities/research-and-development. Accessed 6 June 2021
40. Van Buggenhout M, Christiaens J, Gutwirth S (2018) PRINTEGER D3.1.2. Final report on the incidence of misconduct. pp 1–23. printeger.eu/wp-content/uploads/2019/01/D3_2-On-the-Incidence-of-Misconduct.pdf. Accessed 6 June 2021
41. Code of Ethics of Estonian Scientists (2002) akadeemia.ee/wp-content/uploads/2020/06/code_ethics2002-3.pdf. Accessed 6 June 2021

42. Alumäe T, Tilk O, Ullah A (2018) Advanced rich transcription system for Estonian speech. Front Artif Intell Appl 307:1–8
43. All European Academies (ALLEA) (2011) The European Code of Conduct for Research Integrity. pp 1–20. allea.org/code-of-conduct/. Accessed 6 June 2021
44. Ministry of Higher Education and Science (2014) Danish Code of Conduct for Research Integrity. ufm.dk/en/publications/2014/the-danish-code-of-conduct-for-research-integrity. Accessed 6 June 2021
45. VSNU (2014) The Netherlands Code of Conduct for Academic Practice. Assoc Univ Netherlands, p 11. vsnu.nl/files/documenten/Domeinen/Onderzoek/The_Netherlands_Code of_Conduct_for_Academic_-Practice_2004_(version2014).pdf. Accessed 6 June 2021
46. Davies SR (2018) An ethics of the system: talking to scientists about research integrity. Sci Eng Ethics 25(4):1235–1253
47. Espenberg S, Juurik M, Lõuk K, Parder M, Remmik M, Sutrop M (2020) Teaduseetika järelevalve ja toetamise riikliku süsteemi loomine Eestis. [Establishment of a national system for monitoring and supporting research ethics in Estonia] Tartu. etis.ee/Portal/Publications/Display/c8c14829-c0c4-4e47-9b25-36c9d609ac00. Accessed 6 June 2021

Part XIV

Tools for Integrity Assessment, General Guidelines and Supplementary Material

58 Scientific Electronic Library Online/SciELO: Good Practices Guide for the Enhancement of Ethics in Scientific Publication (Version 09/2018)

Scientific Electronic Library Online SciELO

Abstract

The guide promotes integrity and transparency in the process of assessment of manuscripts, reproducibility of research, possible manipulation or fabrication of data, noncited copy of text or data from another author, duplicate publication of the same text or investigation, and conflicts of interest (COI) or authorship.

This document serves as benchmark for all publications of the collections of SciELO Net. The SciELO program follows rules and recommendations about ethics and responsibility in scientific communication established by national and international institutions, notably the COPE (Committee on Publication Ethics) [1], CSE (Council of Science Editors) [2], Equator Network [3], ICMJE (International Council of Medical Journal Editors) [4], CNPq (National Research Council) and FAPESP (Research Funding Foundation of the State of Sao Paulo) [5] and Manual of Good Practices for the Enhancement of Ethics in Scientific Publication of SciELO.

The guide promotes integrity and transparency in the process of assessment of manuscripts, reproducibility of research, possible manipulation or fabrication of data, noncited copy of text or data from another author, duplicate publication of the same text or investigation, and conflicts of interest (COI) or authorship.

Everything that is published in the periodical as well as the corrective actions that could be necessary is the responsibility of the editor in chief. This guide includes concepts and actions that promote integrity in the publication process and further initiatives in case of suspicion or confirmation of misconduct.

58.1 Responsibilities of the Editor in Chief

They encompass implementation of the editorial policies, surveillance of the editorial process, and relationship of the periodical with authors, reviewers, indexing offices, research funding agencies, the scientific community, and the general public [1]. Transparency and quality control are essential features of such editorial process.

Translated from the original publication (free reproduction): Guia de boas práticas para o fortalecimento da ética na publicação científica [online]. SciELO, Versão setembro 2018 [Viewed on March 15th, 2021]. Available from: http://old.scielo.org/local/File/Guia_de_Boas_Praticas_para_o_Fortalecimento_da_Etica_na_Publicacao_Cientifica.pdf

Editorial note: SciELO is one of the biggest open access libraries in the world. It publishes over 150 journals from nearly all Latin American countries, as well as from Portugal, Spain, and South Africa. Financial support is provided by Brazilian as well as Latin American official funding agencies.

Scientific Electronic Library Online SciELO (✉)
São Paulo, Brazil
e-mail: scielo@scielo.org

J. Faintuch, S. Faintuch (eds.), *Integrity of Scientific Research*,
https://doi.org/10.1007/978-3-030-99680-2_58

Also with regard to the good practices for the enhancement of ethics in scientific publication, the editorial process should warrant, after all formal procedures are conducted, that authors revise the texts and accept responsibility for the content, with the contribution of each author indicated at the end of the manuscript [6].

Confirmation can be by signature or digital method including COI, which must be disclosed in the text [3]. If conflicts of authorship arise, the first author should be contacted and if necessary all authors. In case of persisting conflict or deadlock, the responsible institutions as well as funding agencies should be addressed.

If the study involves research participants, ethical approval, informed consent, trial registry, and other documents can be requested. Text similarity software can be adopted to establish the originality of the material and should be informed to article's authors. Analogous to authorship conflict, if duplicate publication is suspected, the first author, all authors, and also institutions and funding agencies should be contacted as required [1]. The same procedures should be adopted if doubts concerning reference citations occur, as the editor in chief should contact the first author and, however, other parties as well if deemed appropriate. The same care is required in circumstances of excessive self-citation or overcitation of specific journals, when clarifications should be requested.

Editors and reviewers should prioritize impartiality, integrity, and confidentiality in their reports, with emphasis on constructive criticism and respect to journal deadlines. If necessary, the editor in chief should intervene on other editors or reviewers.

Data fabrication or falsification are serious forms of misconduct, and investigation of suspicious cases should be thorough. Additional data to confirm methods and results can be requested from the authors. Should misconduct be demonstrated, responsible institutions and funding agencies should be reported.

58.2 Decision-Making Framework in Case of Misconduct

The periodical should inform in the author's instructions how misconduct questions arise and are handled. In questionable circumstances, the flux diagrams of COPE for identification and action against misconducts are followed [1]. If the misconduct decision is appealed, a committee comprising both members of the editorial staff and external participants can be organized.

58.3 Decision-Making Framework Concerning Retractions and Errata/Corrigenda

If misconduct is identified after publication, the article remains in the SciELO database with a retraction note. The retraction note should point out the reason for retraction in a documented fashion, as reported by the editor, author, or other authorized persons, and should be published in the same periodical. Partial retractions are possible, whenever misconduct applies to a portion of the article only, without compromising the whole research publication [7].

Article removal is not possible. Errors and shortcomings of any nature or origin not involving misconduct should be addressed by errata or corrigenda [8–10]. The periodical should publish as rapidly as possible all errata, corrigenda, and retractions.

References

1. COPE (2018) Committee on Publication Ethics. www.publicationethics.org
2. CSE (2018) Council of Science Editors. https://www.councilscienceeditors.org/resource-library/editorial-policies/whitepaper-on-publication-ethics/
3. Equator Network (2008) www.equatornetwork.org
4. ICMJE (2018) International Committee of Medical Journal Editors. www.icmje.org
5. FAPESP (2014) Código de boas práticas científicas (Research Funding Foundation of the State of Sao

Paulo/FAPESP Code of good scientific practives, 2014). www.fapesp.br/boaspraticas/
6. SciELO (2017) Critérios, políticas e procedimentos para admissão e a permanência de periódicos científicos na Coleção SciELO Brasil (Scientific Electronic Library Online/SciELO Criteria, policies and procedures for the admission and maintenance of scientific periodicals in the SciELO Collection, Brazil 2017). http://www.scielo.br/avaliacao/Criterios_SciELO_Brasil_versao_revisada_atualizada_outubro_20171206.pdf
7. SciELO (2017) Guia para o registro e publicação de retratação (SciELO – Guide for registration and publication of retractions, 2017). https://wp.scielo.org/wp-content/uploads/guia_retratacao.pdf
8. SciELO (2017) Guia para o registro e publicação de Errata (SciELO – Guide for registration and publication of errata, 2017). https://wp.scielo.org/wp-content/uploads/guia_errata.pdf
9. ABEC BRASIL (2012) Diretrizes do CSE para promover integridade em publicações de periódicos científicos (Directives of the Council of Science Editors to promote integrity in publication of scientific periodicals, 2012). https://www.abecbrasil.org.br/arquivos/whitepaper_CSE.pdf
10. World Economic Forum (2018) Code of ethics for researchers. https://widgets.weforum.org/coe/

What Research Institutions Can Do to Foster Research Integrity

59

Lex Bouter

Abstract

In many countries, attention for fostering research integrity started with a misconduct case that got a lot of media exposure. But there is an emerging consensus that questionable research practices (QRPs) are more harmful due to their high prevalence. QRPs have in common that they can help to make study results more exciting, more positive and more statistically significant. That makes them tempting to engage in. Research institutions have the duty to empower their research staff to steer away from QRPs and to explain how they realise that in a Research Integrity Promotion Plan. Avoiding perverse incentives in assessing researchers for career advancement is an important element in that plan. Research institutions, funding agencies and journals should make their research integrity policies as evidence based as possible. The dilemmas and distractions researchers face are real and universal. We owe it to society to collaborate and to do our utmost best to prevent QRPs and to foster research integrity.

59.1 Introduction

Traditionally, research integrity has focussed on the prevention, identification and handling of the three deadly sins of scientific and scholarly research: fabrication, falsification and plagiarism [1]. In many countries, the attention for research integrity was fuelled by a misconduct case that got a lot of media exposure. In the Netherlands, it was the Diederik Stapel case [2] that served as a call to arms. It shocked many within and outside academia but turned out to be a blessing in disguise as well. Stapel's successor as dean of the faculty of social sciences of Tilburg University acted according to the dictum 'never waste a good crisis' [3]. The other Dutch universities followed, and local and national measures were taken. This sequence of events seems typical for many countries.

In recent years, attention has shifted to the lesser breaches of research integrity that are commonly referred to as questionable research

This chapter is a reprint of the previously published Open Access (Creative Commons Attribution 4.0 International License) article Bouter, L. What Research Institutions Can Do to Foster Research Integrity. Sci Eng Ethics 26, 2363–2369 (2020). https://doi.org/10.1007/s11948-020-00178-5.

Based on the Dies Lecture on the occasion of the 92nd *Dies Natalis* of Tilburg University, Tilburg, The Netherlands, 21 November, 2019.

L. Bouter (✉)
Department of Epidemiology and Data Science, Amsterdam University Medical Centers, Amsterdam, The Netherlands

Department of Philosophy, Faculty of Humanities, Vrije Universiteit, Amsterdam, The Netherlands
e-mail: lm.bouter@vu.nl

J. Faintuch, S. Faintuch (eds.), *Integrity of Scientific Research*,
https://doi.org/10.1007/978-3-030-99680-2_59

practices or QRPs [4, 5]. The idea is that these are much more prevalent and thus collectively do more harm to the validity of and the trust in the results of research [1, 6, 7]. Examples are selective reporting, P-hacking and hypothesising after the results are known or HARKing. In an excellent paper from the Meta-Research Center of Tilburg University, 34 QRPs are identified as researcher degrees of freedom that should be avoided in hypothesis testing research [8]. QRPs have in common that they can help to make study results more exciting, more positive and more statistically significant, which in its turn increases the likelihood to be accepted by a high impact journal, to get many citations and to obtain the next grant or academic tenure.

Almost all researchers want to deliver good quality science, to avoid QRPs and to follow their moral compass to steer a course of research integrity. Like any compass, the functioning of a moral compass depends on its quality and on external factors. The quality is determined by the virtuousness of the individual at issue. Major external factors that can corrupt the moral compass concern the local research climate and the perverse incentives of the science system as a whole. Researchers need help from their research institution in optimising the functioning of their moral compass. That help involves adequate education and skills training, good facilities and expert help and clear codes and procedures. That being said, we should realise that research institutions experience perverse incentives that concern the way research is financed and evaluated by governments and research funders which ultimately trickle down the researchers themselves [9, 10]. Consequently, research institutions need help from the other stakeholders in the research system [11].

59.2 Duties of Care

The Netherlands Code of Conduct for Research Integrity specifies 61 standards for good research that mirror in fact as many QRPs to be avoided [12]. A unique feature of the code is that it also contains a chapter on the duties of care research institutions which have to empower their research staff to steer away from QRPs. This idea is not new and was already contained in the Singapore Statement [13] that says in responsibility 13: 'Research institutions should create and sustain environments that encourage integrity through education, clear policies, and reasonable standards for advancement, while fostering work environments that support research integrity'. In other words, research institutions need to have a Research Integrity Promotion Plan. The Horizon 2020 funded consortium Standard Operating Procedures for Research Integrity [14] will offer research institutions help to formulate this plan. Having implemented such a plan might become a contractual obligation for institutions accepting grants from the next EU framework programme Horizon Europe.

The idea is that the Research Integrity Promotion Plan explains what the research institution sets out to do—in the context of its mission, disciplinary focus and type of research it performs—to promote research integrity. The plan needs to cover a set of mandatory topics and will typically describe a mix of educational programmes, codes, manuals, policy measures, regulations, facilities, audit schemes and support systems. SOPs4RI will produce a toolbox filled with standard operating procedures (SOPs) and guidelines that can help research institutions to formulate their Research Integrity Promotion Plan (e.g. [15–18]). A preliminary version of the SOPS4RI toolbox will become available by the end of 2020, and the final version will be ready in 2022. The difference between a SOP and a guideline is gradual, with SOPs being more strict step-by-step recipes and guidelines offering some freedom of choice. It is important to make this not another box-ticking exercise but to ensure that researchers appreciate and use the guidance offered by their institution.

Initiatives of research institutions and other stakeholders to improve the quality of research and research integrity are by no means unique to the Netherlands. Some important initiatives are the US Centre for Open Science [19], the UK Reproducibility Network [20], the European Quality In Preclinical Data Innovative Medicine

Initiative [21] and the German Quality, Ethics, Open Science, Translation Center [22]. Taken together, currently there is a lack of solid guidance for research institutions that want to improve responsible research practices. The examples are scattered and not all evidence based and fit for application. The hope is that the results of the SOPs4RI consortium will improve this situation substantially.

59.3 Perverse Incentives

Arguably, one of the most important things research institutions should do is to avoid perverse incentives in assessing researchers for career advancement. The current dominant focus on bibliometric indicators derived from publication and citation counts sends a strong message that only these things really matter [23]. During recent years, the myopic use of quantitative indicators in research evaluations has been criticised. This led to initiatives like the Leiden Manifesto [24] and the San Francisco Declaration on Research Assessment [25]. In line with this, the Hong Kong Principles for assessing researchers [26] were formulated and endorsed at the 6th World Conference on Research Integrity [27]. These principles will help research institutions that adopt them to minimise perverse incentives that invite to engage in questionable research practices or worse.

The Hong Kong Principles are chosen with a view to explicitly recognise and reward researchers for behaviour that leads to trustworthy research by avoiding QRPs. The principles have been developed with the idea in mind that their implementation could assist in how researchers are assessed for career advancement with a focus on behaviours that strengthen research integrity. Five principles were formulated: assess responsible research practices, value complete reporting, reward the practice of open science, acknowledge a broad range of research activities and recognise essential other tasks like peer review and mentoring. For each principle, a rationale for its inclusion is provided, and examples of research institutions where these principles are already being adopted are given.

59.4 Meta-research

The little empirical evidence on interventions to improve responsible research practices we have is often of poor quality, negative or both. This is illustrated by the Cochrane review that summarises the evidence on interventions to prevent misconduct and promote integrity in research and publication [28]. Research institutions should make their research integrity policies as evidence based as possible. In hindsight, it is difficult to understand why it took us so long to establish a solid tradition in research on research integrity. That only started to happen recently and was fuelled by granting programmes like the Horizon 2020 Science with and for Society (SwafS) calls for research ethics and research integrity [29]. In the Netherlands, the programs on Fostering Responsible Research Practices [30] and Replication Studies [31] contributed to the emerging field of research on research. At the 5th World Conference on Research Integrity, the Amsterdam Agenda was adopted that strongly encourages research on research integrity especially focussing on solutions that really work and effect change [32, 33].

That being said, there is still a lot we do not know about research integrity in research institutions. To fill this gap, in May 2020, all researchers in Dutch universities and university medical centres will be invited to participate in the National Survey on Research Integrity. The survey is expected to provide valid and reliable knowledge on how often specific QRPs occur and what their underlying explanatory variables are. This will provide insights that help research institutions to improve their policies and to fulfil their duties of care in fostering research integrity better. Given the sensitivity of some of the questions, the survey will pay particular attention to ensuring the protection of the identity of the participants and their research institutions. The randomised response technique that will be used is expected to elicit a strong sense of trust in respondents because their answers can never be linked to them [34]. And to keep the time to complete the survey short, we make use of missingness by design.

But let me be clear: surveys, focus group interviews and Delphi studies can only guide us towards potentially effective measures research institutions can take to improve responsible research practices. How good, for instance, SOPs and guidelines [14] or the Hong Kong Principles [26] really are in comparison to alternative approaches needs to be sorted out in future studies designed to demonstrate effectiveness in terms of outcomes that matter.

59.5 Conclusion

Finally, it is important to note that there are many stakeholders with a responsibility to foster research integrity. First and foremost, the researchers themselves are responsible to behave well and to refrain from QRPs. Researchers should also be a good role model and help others to keep on track. Second, research institutions must empower researchers to act according to the standards of good research. But also funding agencies and scientific journals have important roles to play. There is no magic pill or a quick fix. The dilemmas and distractions researchers face are real and universal. We owe it to society to collaborate and to do our utmost best to prevent QRPs and to foster research integrity.

References

1. National Academies of Sciences, Engineering and Medicine (2017) Fostering integrity in research. National Academies, Washington, DC
2. Levelt, Noort and Drenth Committees (2012) Flawed science: the fraudulent research practices of social psychologist Diederik Stapel. Tilburg University. https://www.rug.nl/about-us/news-andevents/news/news2012/stapel-eindrapport-eng.pdf. Accessed 3 Jan 2020
3. Sijtsma K (2017) Never waste a good crisis: towards responsible data management. In: Keynote lecture on the 5th world conference on research integrity. https://www.youtube.com/watch?v=bddRx-LN8lo. Accessed 3 Jan 2020
4. Bouter LM, Tijdink J, Axelsen N, Martinson BC, ter Riet G (2016) Ranking major and minor research misbehaviors: results from a survey among participants of four World Conferences on Research Integrity. Res Integ Peer Rev 1:17
5. Haven T, Tijdink J, Pasman HJ, Widdershoven G, ter Riet G, Bouter L (2019) Do research misbehaviours differ between disciplinary fields? A mixed methods study among academic researchers in Amsterdam. Res Integr Peer Rev 4:25
6. Editorial (2019) Integrity for all: considering research integrity to be confined to misconduct stops scientists from improving. Nature 570:5
7. Macleod M, Mohan S (2019) Reproducibility and rigor in animal-based research. ILAR J. https://doi.org/10.1093/ilar/ilz015
8. Wicherts JM, Veldkamp CLS, Augusteijn HEM, Bakker M, van Aert RCM, van Assen MALM (2016) Degrees of freedom in planning, running, analyzing, and reporting psychological studies: a checklist to avoid p-hacking. Front Psychol 7:1832
9. Anderson MS (2019) Shifting perspectives on research integrity. J Empir Res Hum Res Ethics 13:459–460
10. Bagioli M, Kenney M, Martin BR, Walsh JP (2019) Academic misconduct, misrepresentation and gaming: a reassessment. Res Policy 48:401–413
11. Bouter LM (2018) Fostering responsible research practices is a shared responsibility of multiple stakeholders. J Clin Epidemiol 96:143–146
12. Netherlands code of conduct on research integrity (2018) https://www.vsnu.nl/files/documents/Netherlands%20Code%20of%20Conduct%20for%20Research%20Integrity%202018.pdf. Accessed 3 Jan 2020
13. Singapore Statement (2010). https://www.wcrif.org/documents/327-singapore-statement-a4size/file. Accessed 3 Jan 2020
14. Website of Standard Operating Procedures for Research Integrity. https://www.sops4ri.eu/. Accessed 3 Jan 2020
15. European Network of Research Integrity Offices (2019) Recommendations for the investigation for research misconduct. http://eneri.eu/wp-content/uploads/2019/03/INV-Handbook_ENRIO_web_final.pdf. Accessed 3 Jan 2020
16. Forsberg EM, Anthun FO, Bailey S, Birchley G, Bout H, Casonato C et al (2018) Working with research integrity—guidance for research performing organizations: the Bonn PRINTEGER statement. Sci Eng Ethics 24:1023–1034
17. Office of Research Integrity (1995) Guidelines for institutes and whistleblowers: responding to possible retaliation against whistle blowers in extramural research. https://ori.hhs.gov/ori-guidelines-institutions-and-whistleblowers-responding-possible-retaliation-against. Accessed 3 Jan 2020

18. Penders B, Shaw D, Lutz P, Townend D, Akrong L, Zvonareva O (2018) ENERI manual of research integrity and ethics. Maastricht University. http://eneri.eu/wp-content/uploads/2018/10/ENERI-e-Manual.pdf. Accessed 3 Jan 2020
19. Website of Center for Open Science. https://cos.io/. Accessed 3 Jan 2020
20. Website of UK Reproducibility Network. http://www.dcn.ed.ac.uk/camarades/ukrn/. Accessed 3 Jan 2020
21. Website of European Quality in Preclinical Data Innovative Medicine Initiative. https://www.imi.europa.eu/projects-results/project-factsheets/eqipd. Accessed 3 Jan 2020
22. Website of Quality, Ethics, Open Science, Translation Center. https://www.bihealth.org/en/research/quest-center/mission-approaches/. Accessed 3 Jan 2020
23. Moher D, Naudet F, Cristea IA, Miedema F, Ioannidis JPA, Goodman SN (2018) Assessing scientists for hiring, promotion, and tenure. PLoS Biol 16: e2004089
24. Hicks D, Wouters P, Waltman L, de Rijcke S, Rafols I (2015) The Leiden Manifesto for research metrics. Nature 520:429–431
25. Website of San Francisco Declaration on Research Assessment. https://sfdora.org/. Accessed 3 Jan 2020
26. Moher D, Bouter L, Kleinert S, Glasziou P, Sham MH, Barbour V, Coriat AM, Foeger N, Dirnagl U (2019) The Hong Kong principles for assessing researchers: fostering research integrity. OSF Preprints. https://osf.io/m9abx
27. Webpage of 6th WCRI on World Conference on Research Integrity Foundation website. https://www.wcrif.org/wcri2019. Accessed 3 Jan 2020
28. Marusic A, Wager E, Utrobicic A, Rothstein HR, Sambunjak D (2016) Interventions to prevent misconduct and promote integrity in research and publication. Cochrane Database Syst Rev. https://doi.org/10.1002/14651858.mr000038.pub2
29. Webpage of Horizon 2020: science with and for society (Swafs) on European Commission website. https://ec.europa.eu/research/swafs/index.cfm?pg=funding. Accessed 3 Jan 2020
30. Webpage of Fostering Responsible Research Practices on ZonMw website. https://www.zonmw.nl/en/research-and-results/fundamental-research/programmas/programme-detail/fostering-responsible-research-practices/. Accessed 3 Jan 2020
31. Webpage of Replication studies on NWO website. https://www.nwo.nl/en/funding/our-funding-instruments/sgw/replication-studies/replication-studies.html. Accessed 3 Jan 2020
32. Amsterdam Agenda (2015). https://www.wcrif.org/documents/42-amsterdam-agenda/file. Accessed 3 Jan 2020
33. Mayer T, Bouter L, Steneck N (2017) Addressing scientific integrity scientifically. Science 357:1248–1249
34. Lentsvelt-Mulders GJLM, Hox JJ, van der Heijden PGM, Maas CJM (2005) Metaanalysis of randomized response research: thirty-five years of validation. Sociol Methods Res 33:319–348

Useful Online Resources and Guideline 60

Joel Faintuch and Jacob J. Faintuch

Abstract

The Internet has become such an omnipresent service, that in the view of many, it has already upgraded from slave to master, and quite a tyrannical one. Whatever the feelings, scientific information would not flow worldwide without the tool, in such unlimited amounts. It is currently estimated that 4.7 billion people use its electronic resources (60% of the planet population). Just the four big names in the field (Google, Amazon, Microsoft and Facebook) store a total of 1200 petabytes of files, each petabyte corresponding to 10^{15} bytes. If it corresponded to text, only that would represent over 670 billion pages. The MEDLINE database is a more pertinent paradigm as it deals with indexed scientific journals. Other publications such as books, theses, congress proceedings, nonindexed journals, court minutes and press reports are with few exceptions uncovered. It still contains close to 30 million articles or over 300 million pages. Given that a human being rarely reads more than 700 pages/week (a reasonably thick book), that would signal over 8000 millennia to peruse such files, if all of them were available online. These ruminations could look rather aimless and out of focus, yet they point towards the incontrovertible need for careful curation of the key Internet addresses, so that the interested researcher will minimize wasted time and likely find helpful material. That is the objective of the chapter listing a choice of useful sites as well as general Internet resources, nearly all of them costless.

See Tables 60.1–60.8.

J. Faintuch (✉)
Department of Gastroenterology, Sao Paulo University Medical School, Sao Paulo, Brazil
e-mail: j.faintuch@hc.fm.usp.br

J. J. Faintuch
Department of Internal Medicine, Hospital das Clinicas, Sao Paulo University Medical School, Sao Paulo, Brazil
e-mail: jacob.faintuch@hc.fm.usp.br

J. Faintuch, S. Faintuch (eds.), *Integrity of Scientific Research*,
https://doi.org/10.1007/978-3-030-99680-2_60

Table 60.1 Sites on ethics and misconduct in social sciences

publicationethics.org	Publication ethics in the arts, humanities and social sciences
encyclopedia.com/science/encyclopedias-almanacs-transcripts-and-maps/misconduct-science-social-science-cases	Misconduct in science: social science cases
forskningsetikk.no/en/guidelines/social-sciences-humanities-law-and-theology/guidelines-for-research-ethics-in-the-social-sciences-humanities-law-and-theology/	Guidelines on research ethics in the social sciences and related fields
ed.ac.uk/arts-humanities-soc-sci/research-ke/support-for-staff/res-ethics-policies/ethics	Research integrity in the arts, humanities and social sciences
ignet.gov/content/national-endowment-humanities	Integrity and efficiency in the humanities
gao.gov/about/what-gao-does/fraudnet/	US Govt. Accountability Office financial fraud hotline
bitssblog.wordpress/prizes	Prizes for open social science

Table 60.2 Accessible guidelines and publications about fraud, ethics and research

nlm.nih.gov/bsd/bioethics.html	Links to clinical trials, cloning, end of life issues, genes and gene therapy, health fraud, organ transplantation, patient rights, personal health records and stem cells, among others
ori.hhs.gov	Office for Research Integrity of the US Department of Health and Human Services Guidelines including misconduct cases
globethics.net	Large non-profit organization. Links to over 200 journals and 2 million documents
Ethicshare.org	Non-profit organization. News, discussions and publications
thehastingscenter.org	Large non-profit organizations with their own journals on ethics and misconduct
ethics.harvard.edu	Research and courses on misconduct, corruption and ethics
nottingham.ac.uk/research/groups/centre-for-applied-bioethics/	Ethics of research with animals and other topics
tuskegee.edu/about-us/centers-of-excellence/	National Center for Bioethics in Research and Health Care—NIH, USA
Elsevier.com_data/assets/pdf_file/0008/653885/Ethics-in-research-and publication-brochure.pdf	Guidelines of the International Committee of *Medical* Journal Editors concerning research and publication

Table 60.3 Ethical databases and official statements [1]

bioethics.georgetown.edu/library-materials/bioethics-research-library-databases/	ETHXWEB platform with >300,000 documents on bioethics, GENETHX on human genetics and others
ama-assn.org/delivering-care/ethics	Ethics code and links of the American Medical Association
acponline.org/clinical-information/ethics-and-professionalism	Ethical publications and links of the American College of Physicians (USA)
Drze.de/belit-1/belit	German Reference Centre for Ethics in the Life Sciences and the BELIT database
en.unesco.org/themes/ethics-science-and-technology/ibc	Ethics and bioethics declarations of UNESCO (UN)
singaporestatement.org/statement.html	Singapore Statement on Research Integrity
ethics.elsevier.com	Guidelines and links about ethics in publication
Publicationethics.org	Committee on Publication Ethics/COPE including conflicts of interest, article retraction and allegation of misconduct; advises over 12,000 journals
icmje.org	International Committee of Medical Journal Editors. Role of authors and contributors
publicationethics.org/files/A-legal-perspective.pdf	Legal perspectives of publication ethics
interacademies.org/sites/default/files/publication/9780691170756_secured.pdf	Guidelines on doing global science
equator-network.org/EURODAT	Reporting guidelines (health research)
Consort-statement.org	Consolidated standards of reporting trials, adopted by most medical journals
wma.net/what-we-do/ medical-ethics/declaration-of-helsinki/	Helsinki Declaration concerning research with human beings
wma.net/what-we-do/medical-ethics/declaration-of-taipei	Declaration about research with biobanks and datasets
acmedsci.ac.uk/file-download/41514-572ca1ddd6cca.pdf	UK Academy of Medical Sciences—conflict of interest
sci-com.eu/main/docs/Brussels-Declaration.pdf?58b6e4b4	American Association for the Advancement of Science—Ethics and Principles for Science and Society Policy-Making
eudat.eu/data-management	Guidelines on data management
ema.europe.eu	European Agency for the Evaluation of Medicinal Products
wcrif.org/documents/354-montreal-statement-english/file	The FAIR Guiding Principles—cross-boundary research collaborations
fda.gov	Food and Drug Administration, USA, with guidelines concerning clinical trials
mhra.gov.uk	UK Drug Regulatory Agency
ichgcp.net	Good clinical practices in medical research

Table 60.4 Misconduct, fraud, retractions and related policies

Ithenticate.com	Plagiarism screening site
Cross-check.com	Double publication and plagiarism check
turnitin.com	Similarity and plagiarism check
Retractionwatch.com	Largest site for article retractions
retractiondatabase.org/retractionsearch.aspx?	Most complete searchable database with 18,000 retracted articles and abstracts
federalregister.gov/documents/2019/08/26/2019-18305/findings-of-research-misconduct	US official reports about research misconduct
obamawhitehouse.archives.gov/blog/2016/12/19/scientific-integrity-policies-update	Scientific integrity policies of the White House, USA
peer.org/legal-resource-center/scientific-integrity-policies/	Scientific integrity policies of Federal Agencies, USA
nsf.gov/oig/report-fraud/	National Science Foundation fraud policies
Cgsnet.org	Council of Graduate Schools, US policies on scholarly integrity
grants.nih.gov/policy/research_integrity/overview.htm	Research integrity and misconduct activities of the NIH, USA
fda.gov/medical-devices/reporting-allegations-regulatory-misconduct/allegations-regulatory-misconduct-form	FDA, US mechanism for misconduct related to medical devices
cdc.gov/maso/Policy/misconductprocedures.pdf	Misconduct procedures of the centres for disease control and prevention, USA
epa.gov/osa/allegations-loss-scientific-integrity-2016	Issues regarding scientific integrity, environmental protection agency, USA
usda.gov/our-agency/staff-offices/office-chief-scientist-ocs/scientific-integrity-and-research-misconduct	Research misconduct handling at the US Department of Agriculture
oig.usaid.gov/report-fraud	Fraud prevention at the US agency for international development
goodjobfirst.org/violation-tracker	Database of fraud by healthcare companies
academic.sexual.misconduct.database.org	>1000 cases documented in the USA
academicintegrity org	International non-profit (USA/Canada): publications and educational activities
allea.org/code-of-conduct/	The integrity code of the All European Academies of Science/ALLEA
scienceeurope.org/our-priorities/research-integrity-and-ethics	The integrity code of Science Europe (27 countries, 37 funding agencies)
oecd.org/sti/sci-tech/40188303.pdf	Scientific integrity and prevention of misconduct
sops4ri.eu/toolbox	Standard operating procedures and guidelines (to promote research integrity) All European Academies of Science/ALLEA
ucl.ac.uk/academic-manual/	University College London procedures for student and staff academic misconduct
ed.ac.uk/academic-services/students/conduct/academic-misconduct/academic-misconduct-procedure	University of Edinburgh rules for student academic misconduct

Table 60.5 Student and academic qualification databases

myequals.net/#/user/login	Australia/New Zealand academic credentials database
nscverifications.org	National student clearinghouse (USA) qualification platform
groningendeclaration.org	A group of >80 worldwide institutions aiming to improve academic and professional mobility
eaie.org	European Association for International Education
etico.iiep.unesco.org/	UNESCO group for prevention of fraud and corruption in education
hedd.ac.uk	Higher Education Degree Datacheck, UK
naces.org	National Association of Credential Evaluation Services, USA

Table 60.6 Educational databases of integrity for students and scientists

ethicsresearchcore.org/education/rcr	Research ethics course
ethicsresearchcore.org/education/case-studies	Research ethics case discussion
icts.wustl.edu/icts-researchers/education/	Research ethics course
radboudumc.nl/en/research/principles-of-research/scientific-integrity/courses	Research ethics course for PhD candidates
wgs.crs.wur.nl/courses/details/117	Scientific integrity course for students
vumc.nl/educatie/onze-opleidingen/opleidingsdetail/courses-on-research-integrity.htm	Scientific integrity courses
acta.nl/en/research/phd-traning-graduate-school/Courses-for-PhD-students/Scientific-Integrity/index.aspx	Scientific integrity course for dentists
tudelft.nl/ethics/ethics/teaching-activities/courses-for-phd-students/scientific-integrity/	Scientific integrity course for PhD candidates

Table 60.7 Ethics training sites and resources

training.cochrane.org/resource/committee-publication-ethics-cope-elearning-course	Committee on publication ethics/COPE e-learning
ori.hhs.gov/education/products/montana_round1/research_ethics.html	US official ethics course
un.org/en/ethics/ethics-training/index.shtml	United Nations ethics training
ethics.org/resources/free-toolkit	Ethics and compliance toolkit
pathlms.com/naccho	Public health ethics e-learning
aspph.org/teach-research/#curricula	Public health ethics curricula
brown.edu/research/research-ethics/northeast-ethics-education-partnership/training-materials/training-materials	Research ethics resources
umass.edu/sts/digitallibrary	Ethics in engineering site
ethics.iit.edu/projects/modules-applied-ethics	IIT/US applied ethics
ethicscodescollection.org	IIT/US codes of ethics
cdc.gov/os/integrity/phethics/resources.htm	CDC education/training tools

Table 60.8 Research protocol registration sites

clinicaltrials.gov	Largest dataset of clinical studies worldwide
alltrials.net/find-out-more/	Trials Registration and Reporting Platform
crd.york.ac.uk/prospero	Systematic reviews and meta-analyses
Osf.io	Qualitative studies, psychology and social sciences
Aspredicted.org	Social sciences and quantitative studies
crd.york.ac.uk/prospero/	Systematic review data repository (PROSPERO)
socialscienceregistry.org/	Social sciences and economy randomized trials
who.int/clinical-trials-registry-platform (ICTRP)	ICTRP works in collaboration with trial registries worldwide
isrctn.com	General recognized registry linked to Springer/Nature group
cos.io	General recognized registry linked to the Center for Open Science

Reference

1. Kretser A, Murphy D, Dwyer J (2017) Scientific integrity resource guide: efforts by federal agencies, foundations, nonprofit organizations, professional societies, and academia in the United States. Crit Rev Food Sci Nutr 57(1):163–180

Correction to: Preregistration in Animal Research

Céline Heinl, Gilbert Schönfelder, and Bettina Bert

Correction to:
Chapter 37 in: J. Faintuch, S. Faintuch (eds.),
***Integrity of Scientific Research*,**
https://doi.org/10.1007/978-3-030-99680-2_37

Chapter 37 was previously published non-open access. It has now been changed to open access under a CC BY 4.0 license and the copyright holder updated to 'The Author(s)'. The book has also been updated with this change.

The updated original version of this chapter can be found at https://doi.org/10.1007/978-3-030-99680-2_37

J. Faintuch, S. Faintuch (eds.), *Integrity of Scientific Research*,
https://doi.org/10.1007/978-3-030-99680-2_61

Glossary

Joel Faintuch and Jacob J. Faintuch

Abstract
Along with unfettered ethics and science expansion come new domains, terms, definitions, concepts, and classifications. Students and academics are knee-deep in university and professional routines, with little time left for clarifying and understanding the cascade of new expressions. This glossary aimed at selecting the most useful words and expressions, providing handy explanations for daily practice.

Disclaimer This Glossary should be understood as an informal introduction to the principles, domains, and procedures related to fraud and misconduct. No definitions are provided, just basic concepts, and the material is not intended for lawsuits or other legal use. The authors will accept no responsibility for divergences or conflicts between the Glossary and other published or unpublished terminologies or references.

J. Faintuch (✉)
Hospital das Clinicas, University of Sao Paulo, São Paulo, Brazil
e-mail: j.faintuch@hc.fm.usp.br

J. J. Faintuch (✉)
Beth Israel Deaconess Medical Center & Harvard Medical School, Boston, MA, USA
e-mail: jacob.faintuch@hc.fm.usp.br

J. Faintuch, S. Faintuch (eds.), *Integrity of Scientific Research*,
https://doi.org/10.1007/978-3-030-99680-2

Glossary

Accountability Research teams should keep all their procedures registered and transparent, as they are responsible (answerable) for them.

Anonymity In principle, all studies involving human beings should assure anonymity, including images, videos, or voice recordings, thus preventing identification of the participants. Data should be published in the forms of tables, graphs, or other collective representations. Exceptions should be approved by a Research Ethics Committee. See also Confidentiality.

Article retraction See Retraction.

Assent It is accepted that even research participants unable to fully consent (minors, cognitive impairment, indigenous populations) should assent, after a simplified explanation consistent with age, cultural/clinical conditions, and social values. See also Informed Consent and Social Values.

Authorship kidnapping Including non-involved and unauthorized respected names among the authors of a study, in order to gain prestige or pave the way for journal acceptance.

Bias An external and sometimes suspicious variable outside the study protocol, which could interfere with accuracy and representativity of the results. Biases can be recognized during any phase of the study, namely experimental design, recruitment, randomization, group allocation, statistical analysis, and others. Serious biases are common causes for research unreliability and article retraction.

Blinding For research purposes, indicates the concealment of group allocation, so that participants are not aware whether they belong to the control (placebo) or experimental (active) group. An important precaution during studies with interventions, as it prevents several forms of bias. The most common modality in clinical trials is double blinding, so that neither the participant nor the investigator knows to which group each participant belongs. See also Placebo.

Bullying Frequent offensive jokes or remarks against somebody with a disability, belonging to a minority, or somehow different from the group. See also Harassment.

Cherry picking See Selective reporting.

Cohort study A study design that follows the same group of individuals for a given period, in order to gain insight about the course of a disease, condition, or intervention.

Compensation Payment by the sponsor to the investigator and/or to the research participant, according to the budget and to the laws of the country. Compensation can be a source of fraud, misconduct, and protocol or selection bias. It should be disclosed to the Research Ethics Committee, as well as in the case of the participant, in the Informed Consent form.

Confidentiality The responsibility of using research information only for the stated purposes of the research, not exposing the participants. See also Anonymity.

Conflict of interest Any financial or non-financial interest which could interfere with the honesty, transparency, objectivity, and independence of a research project or article.

Crossover study An experimental design consecutively submitting the same population to two different interventions, so that each individual is a control of himself or herself. Theoretically, this design eliminates the bias of inadequately matched cases and controls.

Data It encompasses all information, particularly personal information, collected during research or other academic activities. Images and biological samples (biomaterials) are also included, when part of a databank or biobank. See also Databank; Public databank.

Databank A large repository of information, collected by a single individual or more frequently by an organization or agency. Databanks are highly valued for research activities.

Data protection The European Union, and many countries around the world, have specific and often highly detailed and complex legislation concerning data storage, utilization, exchange, and protection. Scientific research is exempted in a few circumstances, however that is not the general rule.

Data sharing In studies involving humans it is expected that at the end of the investigation, principal findings are shared not only with professionals, in the form of technical publications, but also with all participants, as a form of moral compensation for their collaboration. See also Scientific data sharing.

Diversity A variety of ethnic origins, cultures, and minority groups. Diversity should be fostered and defended in all academic and professional environments.

Double blind See Blinding.

Dual publication See Self-plagiarism.

Ego The central component of personality. Individuals with a big ego are unable to hear or accept other points of view. This can be a driver of harassment, bullying, conflict of interest, and other forms of misconduct.

Empirical study A study that directly observes or monitors the individuals or phenomena of interest.

Equity Equal access to education, health, employment opportunities, and other benefits. Minorities and vulnerable populations are sometimes overlooked or mistreated in this regard.

Evidence See Evidence based medicine.

Evidence based medicine Medical science based on well designed and conducted investigations and clinical trials. It is considered superior and supersedes empirical medicine, which had a scientific background (observations), however deemed weak in comparison with modern controlled experiments. The same reasoning can be applied to nearly all fields of science.

Evidence based science See Evidence based medicine.

Exclusion Lack of access to personal, social, and professional progress. Human rights and inclusion should be particularly enforced among marginalized populations.

Fabrication Reporting a non-existent experiment, or one in which participants, images, or data were made up.

Falsification Purposeful selection, misinterpretation, distortion, or adulteration of findings and conclusions of a study.

Fraud The use of deception, misrepresentation, or other dishonest methods in science. It is a cause of potential or actual damage to research truth and integrity, in all fields of knowledge.

Ghost authorship See Ghost writer, Gift authorship.

Ghost writer A commercial or sometimes unpayed third party who writes an article, thesis, or other text for somebody else, without actual participation or engagement in the research process.

Gift authorship Inclusion among the authors of a study, of people who had no direct involvement with it. Often such encompasses department heads, supervisors, and other eminent persons. Also known as guest authorship.

Good clinical practices Internationally approved set of procedures and conducts when performing clinical trials, in order to assure honesty, transparency, reliability, and replicability of the investigation, at the same time protecting the study participants.

Guest authorship See Gift authorship.

Harassment Sexual, professional, or psychological aggression, pression, or intimidation against research colleagues and associates. Typically performed by senior investigators against junior staff or students. Some prefer the expression Bullying.

Hate Crimes Personal or community verbal, physical, or psychological violence based on ethnic origin, religion, culture, sexual orientation, and other stigmatized conditions. See also Microaggressions.

Hype See Scientific sensationalism.

Hypothesis A testable prediction about the question to be addressed during research and the general expected result.

Informed consent A voluntary agreement explaining the nature, aims, risks, benefits, responsible persons, and other features of a research protocol. In principle, no research with humans can be conducted without informed consent. See also Assent.

Institutional review board See Research ethics committee.

Journal kidnapping Unauthorized use by predatory journals of the name and even the logo of a renowned journal, in order to lure customers to their business.

Longitudinal study See Cohort study.

Meta analysis An investigation that combines results from multiple previous studies, all addressing the same problem, in order to achieve a much larger sample size and more consistent results.

Microaggressions A lesser form of hate crimes encompassing insulting, hostile, or humiliating comments and attitudes against certain individuals or minorities. See also Harassment; Prejudice;

Nepotism Inclusion of non-participating relatives among the authors of a scientific contribution. Also known as clientelism or favoritism.

Open Access A publication arrangement employed by scientific journals allowing free electronic access to the published articles by any reader. The costs of publication are defrayed by the authors. Predatory journals are always Open Access, whereas many reputable publications also adopt the system, in order to increase readership.

Oversight Research should be independent however not free from ethical oversight. The ethics in research committee, sites for protocol registration, and scientific integrity boards whenever available are examples of potential or actual monitoring and oversight.

Participant Any person who participates in a study protocol, or whose data (information, images, voice) are utilized for research. Participants are always voluntaries (compulsory engagement is not ethically acceptable). They can be payed or not, depending on the nature of the study and country regulations. In principle, informed consent is required for all participants.

Payment See Compensation.

Peer research A model of investigation in the social sciences, using trained students or professionals to interview colleagues and friends in their natural environment. It is believed to avoid the location bias related to a strange, laboratory setting. On the other hand, if collected data and images are mishandled, this can be a form of misconduct.

Peer review Critical analysis of a scientific article by qualified and experienced professionals, in order to assure that the research project was correctly conducted and the conclusions are supported by the results. Adopted by major journals worldwide.

Placebo A non-active drug, supplement, or intervention, which looks similar to the original one. It is administered to the control group, in order to conduct the study in both active and control participants as analogously as possible, avoiding biases. When the placebo is an intervention it is sometimes designated as Sham.

Plagiarism Undisclosed copying of ideas, methods, results, or images from self or from others. See Self-plagiarism.

Population For research purposes, this indicates the entire group that is addressed by a study. As populations are typically very large, representative samples are analyzed. See Sample.

Predatory journals Publications which pretend to conduct peer review, however do it perfunctorily or not at all. All require payment, article acceptance tends to be 100%, and publication is assured in weeks, if not in days. See also Open Access.

Prejudice A false belief that certain social, ethnic, cultural, or religious groups are intrinsically inferior and undeserving of fair and dignified treatment. See also Stereotype.

Protocol registration Journals dealing with medicine and health care professions often request that clinical trials and some other modalities be previously registered, in order to be considered for publication. Several countries have their own registration sites.

Public databank Databanks mostly provided by the government or its agencies, which are not confidential and can be accessed by anyone. Usually research conducted in such databanks does not require Informed Consent.

Racism Personal, institutional, or community persecutory initiatives against people of defined ethnic origin or skin color. Often includes or overlaps with religion, culture, sexual orientation, and other stigmatized conditions.

Randomization A research precaution allocating participants to the various study or intervention groups by random techniques, such as the flip of a coin, or preferably computer generated random numbers. Randomization prevents allocation bias and is also advocated for animal studies.

Research ethics committee In principle all studies involving human beings (as well as animals) should be submitted to and approved by such a committee (or an Animal Ethics Committee, whenever appropriate). Also known as Institutional review board (IRB).

Research participant See Participant.

Research payment See Compensation.

Research registration See Protocol registration.

Research sponsor See Sponsor.

Retraction Elimination or removal of a previously published scientific article, by the corresponding journal. It usually suggests serious ethical problems with the authors and the submitted material. See also Self retraction.

Risks Dangers, inconveniences, discomforts, or losses for the participant resulting from research. They could be physical, psychological, social, financial, professional, or of other natures. The informed consent document should fully disclose all risks of the study. See Informed consent.

Salami slicing Publication of a single investigation in the form of multiple articles, typically with partial or near-total overlap of contents. See also Self-plagiarism.

Sample The number of individuals or observations included in a study. The sample should follow the protocol inclusion and exclusion criteria, as well as be statistically calculated in order to be representative of the larger population.

Scientific data sharing Exchange or transfer of data among researchers is common nowadays, conducting to larger collections of information and more relevant studies. Nevertheless, in most circumstances, such initiatives have to be approved by a Research Ethics Committee.

Scientific sensationalism Scientific articles or publications aiming to shock, excite, or unduly impress the reader, at the expense of precision, transparency, honesty, and accuracy. This is a form of misconduct.

Selective reporting Suppressing part of the data of a study or writing on the basis of incomplete evidence, so that the conclusions could be misleading. Also known as cherry picking.

Self-plagiarism Copying from previous publications by the same author, without due reference.

Self-retraction Removal or elimination of a published article by request of the authors themselves. See also Retraction.

Sensationalism See Scientific sensationalism.

Sham intervention See Placebo.

Social justice A deliberate and structured effort to address biased beliefs and unfair treatments, in order to increase social harmony and equity. See also Equity.

Social values The beliefs and standards of a person, group, or ethnic population defining what is correct, proper, and desirable for the persons. Any research should be careful to respect the personal and social values of the participants, particularly in the case of vulnerable populations, as a breach would represent a form of misconduct.

Spin A form of misconduct featuring unscientific or sensationalistic language, specifically aiming to disguise or attenuate the fact that analysis of the research data failed to elicit statistically significant results. See Scientific sensationalism.

Sponsor Any person or organization, public or private, financially, technically, or materially contributing to the investigators or to the research budget. This can be an important source of fraud and misconduct, if not adequately formatted. Sponsors should be disclosed to and approved by the Research Ethics Committee.

Stereotype A false belief that certain social, ethnic, cultural, or religious groups automatically share certain characteristics or behaviors, most often negative and derogatory. See also Prejudice; Stigma;

Stigma A quality or characteristic of certain people, considered shameful or dishonorable. Physical deformities, poverty, color, and ethnic origin can be mentioned. Research and publications should be careful never to endorse stigma or to use stigmatizing language.

Text recycling See Self-plagiarism.

Transparency Research activities do not require just collection and storage of all pertinent informations. Full disclosure of all steps and procedures should be available to oversight agencies and other legitimate stakeholders. Transparency is also needed in the Informed Consent form.

Unscientific language See Scientific sensationalism.

Values See Social Values.

Voluntary See Participant.

Vulnerability See Vulnerable populations.

Vulnerable populations All groups that could suffer increased risks or harms as a consequence of research, because of cultural, financial, physical, social, or intellectual limitations. They include the poor, the illiterate, the homeless, indigenous people, minors, elderly people, drug addicts, prisoners, and others. Approval by the Research Ethics Committee is mandatory for studies with such populations. See also Social Values.

Whistleblower Somebody who publicly reveals fraud or misconduct occurring in his or her own institution, or somewhere else. False or malicious whistleblowing is a form of misconduct. The same is true for retaliation against honest and bona fide whistleblowing, which should receive institutional protection.

Xenophobia Unfounded hate or fear of foreigners, understood as any different ethnic, political, religious, social, or cultural group, on account of racism and stereotypes. See also Racism; Stereotype

Bibliography

1. coe.int/en/web/bioethics/guide-for-research-ethics-committees-members. Accessed December 27, 2020
2. corruptionwatch.org.za/glossary-of-corruption-related-terms/. Accessed December 27, 2020
3. .eupati.eu/glossary/
4. Hall J, Martin BR (2019) Towards a taxonomy of research misconduct: the case of business school research. Res Policy 48(2):414–427
5. icmje.org. Accessed December 7, 2020
6. Jordan SR (2013) Conceptual clarification and the task of improving research on academic ethics. J Acad Ethics 11:243–256
7. Kumar MN (2008) A review of the types of scientific misconduct in biomedical research. J Acad Ethics 6: 211–228
8. McClung EL, Schneider JK (2015) A concept synthesis of academically dishonest behaviors. J Acad Ethics 13:1–11
9. Plagiarism.org. Accessed December 20, 2020
10. pre.ethics.gc.ca/eng/policy-politique/initiatives/tcps2-eptc2/glossary-glossaire/. Accessed December 27, 2020
11. publicationethics.org. Accessed December 7, 2020
12. retractionwatch.org. Accessed December, 2020
13. Tauginiene L, Gaizauskaite I, Razi S, Glendinning I, Sivasubramaniam S, Marino F, Cosentino M, Anohina-Naumeca A, Kraviar J (2019) Enhancing the taxonomies relating to academic integrity and misconduct. J Acad Ethics 17:345–361

GPSR Compliance

The European Union's (EU) General Product Safety Regulation (GPSR) is a set of rules that requires consumer products to be safe and our obligations to ensure this.

If you have any concerns about our products, you can contact us on ProductSafety@springernature.com

In case Publisher is established outside the EU, the EU authorized representative is:

Springer Nature Customer Service Center GmbH
Europaplatz 3
69115 Heidelberg, Germany

Batch number: 08430492

Printed by Printforce, the Netherlands